Speed of light	$c = 2.997925 \times 10^{8}$ m/s
Gravitational constant	$G = 6.67 \times 10^{-11}$ N·m²/kg²
Avogadro's number	$N_A = 6.022 \times 10^{26}$ particles/kg·atom
Boltzmann's constant	$k = 1.3807 \times 10^{-23}$ J/K
Gas constant	$R = 8314$ J/(kg)(mol)(K) $= 1.9872$ kcal/(kg)(mol)(K)
Planck's constant	$h = 6.6262 \times 10^{-34}$ J·s
Electron charge	$e = 1.60219 \times 10^{-19}$ C
Electron rest mass	$m_e = 9.1095 \times 10^{-31}$ kg $= 5.4859 \times 10^{-4}$ u
Proton rest mass	$m_p = 1.67265 \times 10^{-27}$ kg $= 1.0072766$ u
Neutron rest mass	$m_n = 1.67495 \times 10^{-27}$ kg $= 1.0086650$ u
Permittivity constant	$\epsilon_0 = 8.85419 \times 10^{-12}$ C²/N·m²
Permeability constant	$\mu_0 = 4\pi \times 10^{-7}$ N/A²
Standard gravitational acceleration	$g = 9.80665$ m/s² $= 32.17$ ft/s²
Mass of earth	5.98×10^{24} kg
Average radius of earth	6.37×10^{6} m
Average density of earth	5.57 g/cm³
Average earth-moon distance	3.84×10^{8} m
Average earth-sun distance	1.496×10^{11} m
Mass of sun	1.99×10^{30} kg
Radius of sun	7×10^{8} m
Sun's radiation intensity at the earth	0.032 cal/(cm²)(s) = 0.134 J/(cm²)(s)

PRINCIPLES OF PHYSICS

PRINCIPLES OF PHYSICS

Third Edition

F. Bueche
Professor of Physics, University of Dayton

New York St. Louis San Francisco
Auckland Bogotá Düsseldorf
Johannesburg London Madrid Mexico
Montreal New Delhi Panama Paris
São Paulo Singapore Sydney
Tokyo Toronto
McGraw-Hill Book Company

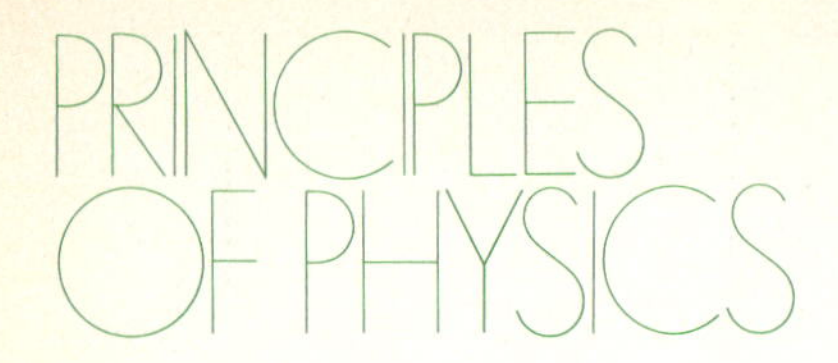

1 2 3 4 5 6 7 8 9 0 VHVH 7 8 3 2 1 0 9 8 7

This book was set in Times Roman by York Graphic Services, Inc.
The editors were C. Robert Zappa, Alice Macnow, and Michael Gardner;
the designer was Jo Jones;
the production supervisor was Charles Hess.
New drawings were done by J & R Services, Inc.
Von Hoffmann Press, Inc., was printer and binder.

Library of Congress Cataloging in Publication Data

Bueche, Frederick, date
Principles of physics.

Includes index.
1. Physics. I. Title.
QC23.B8496 1977 530 76-18706
ISBN 0-07-008848-9

Cover: Sketch of a ball bearing by Leonardo da Vinci. By kind permission of Biblioteca Nacional, Madrid. This drawing appears in THE MADRID CODICES OF LEONARDO DA VINCI, published by McGraw-Hill Book Company, 1974.

CONTENTS

PREFACE

The wide acceptance of the previous editions of "Principles of Physics" has provided a valuable resource, the critical comments of its many users. In preparing the third edition, comments have been solicited from both faculty and students and as a result many changes have been made although the basic features of the previous edition have been retained. The major changes are as follows.

All problems have been reexamined. Previous problems have been modified where necessary, and many new problems have been added. Many of the new problems deal with topics of interest to students in the biological sciences. The problems are classified into three groups—ordinary difficulty, medium difficulty, and difficult. In keeping with increasing student familiarity with the metric system, the British system of units is further deemphasized, being used only in the earliest sections of the text while the student is becoming accustomed to work in physics.

Many sections and chapters have been rewritten in the interest of clarity and increased appeal to the students. Although these changes have been so extensive that it is impossible to point them all out here, three of these should be mentioned: (1) Some teachers prefer to discuss statics early in the course. Others prefer to defer the treatment of rigid bodies until later. Both approaches may be used with this text. Statics of rigid objects is discussed in the first part of Chap. 9, but this topic and the end-of-chapter material which applies to it are designated so that they may be taken up as the last portion of chap. 1 if desired. (2) The material on thermodynamics has been expanded into a new chapter. (3) Although relativity is discussed briefly early in the text, the major portion of this discussion has been rewritten and moved to Chap. 26. You will notice many other beneficial changes as you compare the second and third editions.

The teaching effectiveness of the text has been increased in several ways. Each chapter has been provided with a list of Minimum Learning Goals, a chapter Summary, and a checklist of Important Terms and Phrases. The popular Questions and Guesstimates have been retained. A new feature is the technique used in the body of the text to point out important sentences and phrases. It, in effect, carries out the underlining chore used so successfully by many of the better students. Now even the student unskilled in underlining important concepts is provided with this valuable learning aid.

Many faculty and students have contributed to this new edition by their comments. I am indebted to them. Three persons contributed extensive detailed reviews, Professors Gerard P. Lietz, Marlo R. Martin, and George W. Parker. Their help is greatly appreciated. Suggestions that you may have for further improvement of the text will be welcomed. I am also grateful to my wife, Phyllis, for her aid in preparing the manuscript.

F. Bueche

PRINCIPLES OF PHYSICS

VECTORS AND BALANCED FORCES

1

The description and interpretation of physical phenomena constitute much of physics. In order to correlate, understand, and use experimental observations concerning the physical universe, we must deal with such concepts as forces, motions, and displacements. These concepts, as well as others which we shall soon encounter, belong to a class of quantities which are most conveniently described by means of special arrows, called **vectors.** In this chapter, we shall learn the nature of vector quantities and how they are used. It will be seen that the concept of a vector force is a convenient tool for describing situations in which all forces are balanced.

1.1 VECTORS

When you tell someone that you drove your car 30 mi (miles) east, you are speaking in vector language. The displacement of your car has both magnitude and direction. Its magnitude was 30 mi and its direction was east.

DEFINITION

Quantities which have both magnitude and direction are called **vector** *quantities. Typical vector quantities are displacements, forces, and velocities, as we soon shall see. Many quantities have no direction associated with them. They are called* **scalar** *quantities.* For example, the number of eggs in a carton is a scalar quantity since there is no direction associated with it.

One of the most important features of vector quantities is that it is possible to represent them by pictures. For example, suppose a car goes 30 mi east. This can be pictured as shown in Fig. 1.1. The 30-mi eastward displacement of the car is represented by an arrow. We call this arrow a vector. The direction of the arrow (or vector) shows that the displacement is eastward. The length of the arrow is made proportional to the magnitude of the displacement, 30 mi in this case. For example, we might represent a distance of 1 mi by a distance of 1 mm (millimeter)*. Then the arrow representing the 30-mi displacement would be taken to be (30)(1) mm, or 30 mm long. Now let us consider another example.

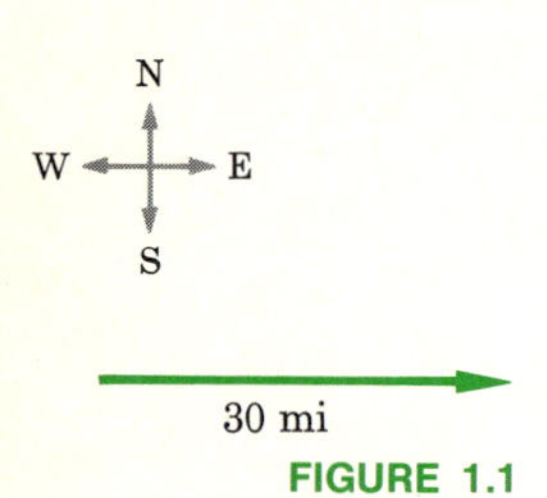

FIGURE 1.1

The vector indicates a displacement of 30 mi east.

Illustration 1.1 Suppose you want to illustrate the statement: I went 30 km (kilometers) east and then 10 km north.†

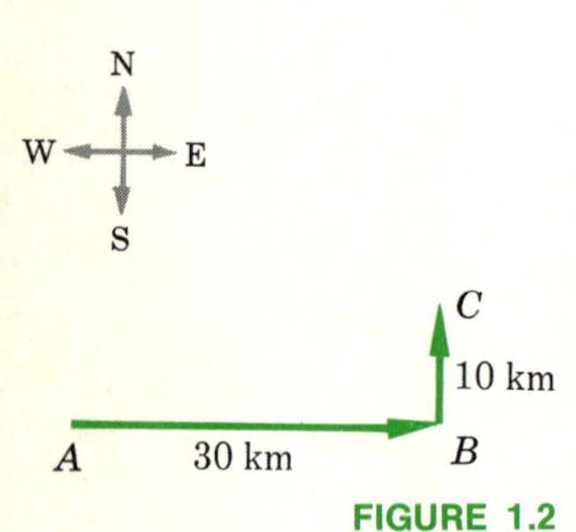

FIGURE 1.2

A vector diagram of a trip in which the traveler went 30 km east and then 10 km north.

Reasoning It is apparent that two vectors (arrows representing movements) are involved here. They are drawn in Fig. 1.2. You obviously started at point A, went first to point B, and ended up at point C. The diagram of the two vectors is a convenient way of picturing successive movements.

DEFINITION

In addition, Fig. 1.2 shows clearly that the end point of the trip was not 40 km from the starting point. The actual distance between A and C is the length of the outlined arrow R shown in Fig. 1.3. We call *this straight-line distance from the beginning point A to the end point C the* **displacement.** *It is a vector quantity,* of course, and is represented by the outlined arrow labeled R in Fig. 1.3.

From the pythagorean theorem for a right triangle, this distance can be calculated as

$$R = \sqrt{(10)^2 + (30)^2}\text{ km}$$
$$= \sqrt{1000}\text{ km} = 10\sqrt{10}\text{ km} \tag{1.1}$$

where *the straight-line distance R from starting point to* end *point is called the* **resultant displacement.**

FIGURE 1.3

A displacement of 30 km east and 10 km north is equivalent to a resultant displacement of $10\sqrt{10}$ km in the direction shown.

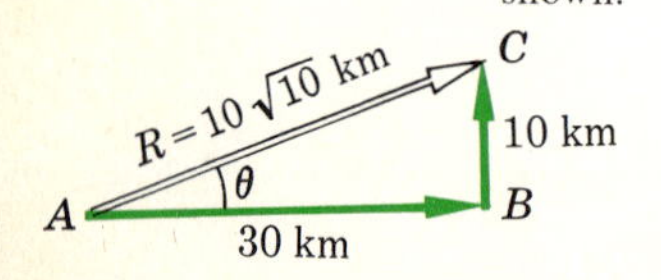

* 1 mm is roughly $\frac{1}{32}$ in. To be exact, 1 in = 25.4 mm, and 10 mm = 1.0 cm (centimeters).

† 1 km = 1000 m (meters) = 0.62 mi.

Not only is the resultant displacement from the starting point for the trip given by the outlined arrow of Fig. 1.3, but the direction is given as well. One would describe the direction by saying, "The resultant vector is at an angle θ (theta) north of east." If the diagram had been drawn accurately to scale, care being taken to make the angle between the two vector displacements exactly 90°, the length of the resultant could be measured directly on the diagram. The value so obtained should be the same as one would compute from Eq. (1.1) within the error of reading the ruler. A protractor could be used to find the value of θ.

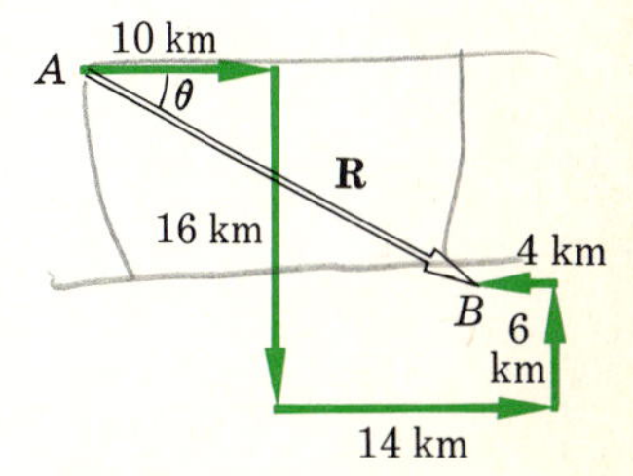

FIGURE 1.4
The path traveled by the car in Illustration 1.2. It started at point A and ended at point B.

Finding the resultant displacement (magnitude and direction) by use of ruler and protractor is referred to as the **graphical method** of solving the problem. It has general validity and can be used conveniently in many cases. It can be summarized as follows:

To add several vectors graphically, place them end to end with tail of the second on the tip of the first. The tail of the third is placed on the tip of the second, and so on. The resultant vector is an arrow with its tail at the tail of the first and its tip at the tip of the last vector. This method is applied in the following illustration.

Illustration 1.2 If a car travels 10.0 km east, 16.0 km south, 14.0 km east, 6.0 km north, and 4.0 km west, what is the resultant displacement from the starting point?

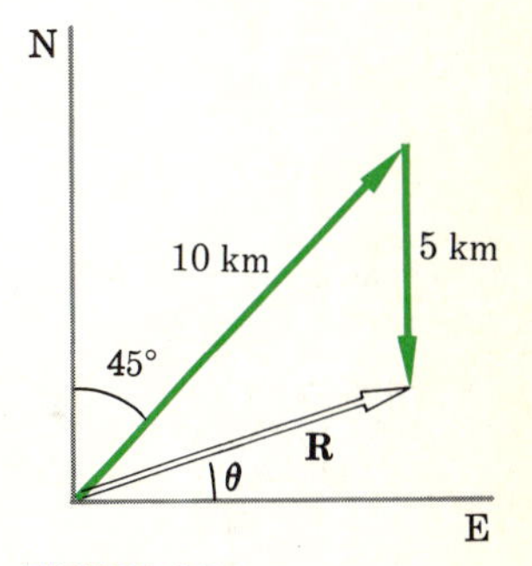

FIGURE 1.5
A vector diagram of a trip of 10 km northeast followed by one of 5 km south.

Reasoning The appropriate vector diagram is shown in Fig. 1.4. Once again, the resultant displacement is shown as the outlined arrow. From its length, one measures the resultant displacement to be 22.4 km. The angle θ can be measured with a protractor and is 26.5° south of east.

It is not necessary that all the vectors be at right angles for our method to work. Suppose you want to add a 5-km south displacement to a 10-km northeast displacement. This is done in Fig. 1.5 and is obviously no more difficult than for vector displacements at right angles. Extension to several vectors is done exactly as in Fig. 1.4.

1.2 VECTORS OTHER THAN DISPLACEMENTS

Vectors can be used to represent any quantity which has both magnitude and direction. For practical reasons, *we also require that the quantity obey the same mathematical laws which govern displacements*. All common quantities which have direction satisfy this requirement. For example, *forces are important vector quantities*. A typical situation which illustrates the vector nature of forces is shown in Fig. 1.6.

In Fig. 1.6*a* we see a string pulling on a ring fastened to a post. *The force with which a string pulls upon the object to which it is attached is called the* **tension** *in the string*. As indicated in the figure, we represent the tension in the string by F. This force exerted by the string on the ring has both magnitude F

and direction (to the right). It can therefore be represented by a vector (or arrow) as shown. We draw the length of the arrow proportional to the magnitude of the force F. Since the force pulls to the right on the ring, the arrow drawn to represent it also points toward the right.

A convenient schematic diagram applicable to the situation of Fig. 1.6*a* is shown in Fig. 1.6*b*. Here the essential feature of part *a*, the force on the circular ring, is at once evident without extraneous detail. We shall make frequent use of schematic diagrams like this. They are called **free-body diagrams.**

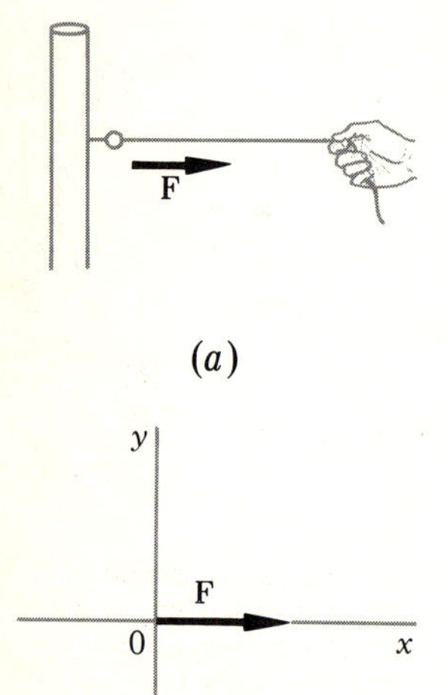

FIGURE 1.6
The pull of the string on the ring (the tension in the string) can be represented by the force vector **F**.

It is customary to represent force vectors by the symbol **F**, where boldface type is used to alert the reader to the fact that the quantity it represents has both direction and magnitude.* When we are not interested in the direction of the force, we commonly represent the magnitude by the symbol F, that is, without boldface type.

Another example of a vector quantity is the velocity of a moving object. We say that it is going 20 mi/h (miles per hour) east. Hence this quantity can be represented by an arrow pointing eastward.

Although other quantities can also be represented as vectors, the three mentioned above—displacements, forces, and velocities—are perhaps the most common. We shall see in the next section that all vectors can be treated in the same way. Hence, when one has learned to deal with displacement vectors, force and velocity vectors can be handled in a similar fashion and will present no difficulty.

1.3 RECTANGULAR COMPONENTS OF VECTORS

Suppose that a person travels 20 km northeast. The appropriate vector is shown in Fig. 1.7. But it is clear that the same final point would have been reached if the person had gone east to point B and then directly north to point C. From this we see that the two colored vectors, when added together, are equivalent to the single 20-km vector. Several vectors which when added end to end add up to give a single vector (as $\overline{AB}$ and $\overline{BC}$ do to give $\overline{AC}$) are said to be **components** of the vector. We now investigate how one can use the concept of components to simplify the addition of vectors.

DEFINITION

FIGURE 1.7
A displacement of 20 km northeast is resolved into component displacements AB east and BC north. AB and BC are components of vector AC.

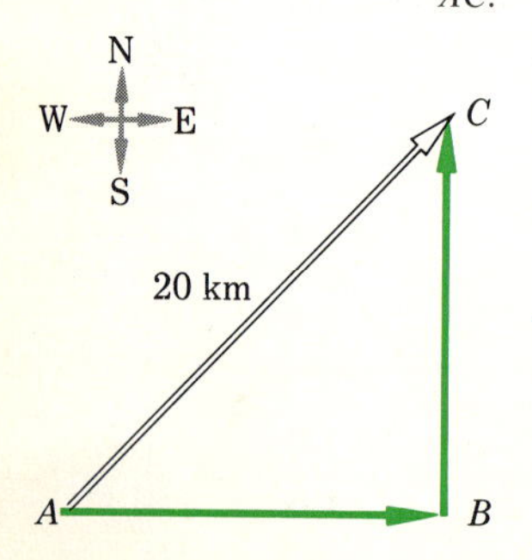

Illustration 1.3 The component method for adding vectors is based upon the fact that any vector can be thought of as being made up of components which are at right angles to each other. In order to see how the addition is performed, let us refer to the displacement vectors shown in Fig. 1.8. The first vector (*a*) is all in the eastward direction. But the second (*b*) consists of two parts as shown, 2 m toward the east and 2 m toward the north.† In the same

*In writing the symbol for a vector by hand on paper or the blackboard, two different methods are frequently used to distinguish vectors. A wavy line is placed below the symbol F, or an arrow is placed above it, $\vec{F}$.

†1 m is a little longer than 1 yd; precisely, 1 m = 3.28 ft.

manner the third vector (c) has a northward component of 3 m and a westward component of 1 m. But a displacement of 1 m westward is the same as subtracting 1 m from the eastward displacement. For this reason, a westward displacement can be represented as a negative eastward displacement. We shall make use of that fact by representing the 1 m westward-directed component by a -1 m eastward vector.

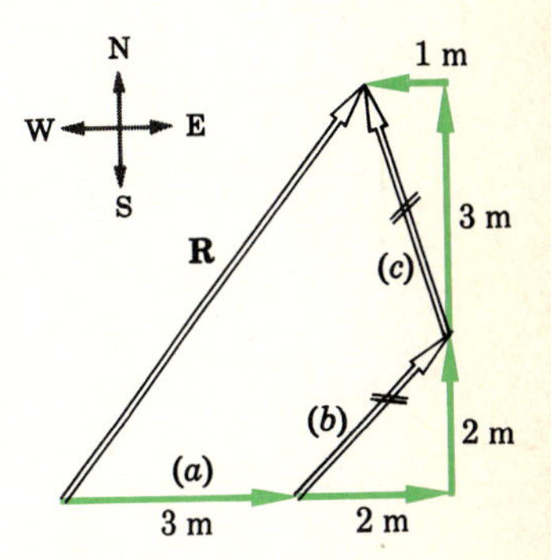

FIGURE 1.8
The sum of the several vectors is equivalent to the sum of the component vectors, as shown.

Reasoning If we wish to find the resultant of these three displacement vectors, we see that the total movement north has been $0 + 2 + 3 = 5$ m, where the zero represents the fact that vector (a) of Fig. 1.8 has no northward component. In the same way, the total eastward movement has been $3 + 2 - 1 = 4$ m. As a result, we have moved 4 m east and 5 m north. The resultant vector must have these components. It is shown in Fig. 1.9 and is seen to be the same as **R** in Fig. 1.8. Of course the magnitude of **R** is $R = \sqrt{4^2 + 5^2}$ m.

Since the only thing that matters in finding **R** is how far east and how far north one has moved altogether, it does not matter in what order the movements were made. Hence, *the order in which we add vectors is of no importance.* We simply compute *the total eastward* movement and *the total northward movement,* and these totals are the east and north components of the resultant movement. Therefore it is often convenient to replace a vector by its component vectors. To show in a diagram that this is what we do, the original vector is crossed off, leaving only its components. We shall illustrate this by placing slash marks on the original vector to be replaced by its components. That is why, in Fig. 1.8, vectors (b) and (c) are slashed out. During the computation of the resultant movement, we replaced these two vectors by their components.

Illustration 1.4 Add the displacement vectors shown in Fig. 1.10a. (Recall that 2.54 cm is equivalent to 1 in.)

Reasoning We shall use an x and y system of coordinates for our directions rather than east and north, but the method remains unchanged. For the present case we shall do the addition graphically even though the next section will show that using components is perhaps a better way of approaching the problem.

The actual graphical addition is carried out in Fig. 1.10b. Notice that the vectors are laid out tip to tail and that the resultant vector heads in the direction from the starting point to the end point, not vice versa. One might legitimately question whether the same resultant would be obtained if the vectors had been added in a different order. When this is tested in part c of the figure, we see that the resultant does not depend upon the order in which the vectors are laid down. We should not be surprised that this is true since **R** simply gives the net result of all the x and y component displacements. It will not depend upon the order in which the components are added.

FIGURE 1.9
Adding together the components of the vectors in Fig. 1.8 gives the same resultant as when the vectors themselves are added.

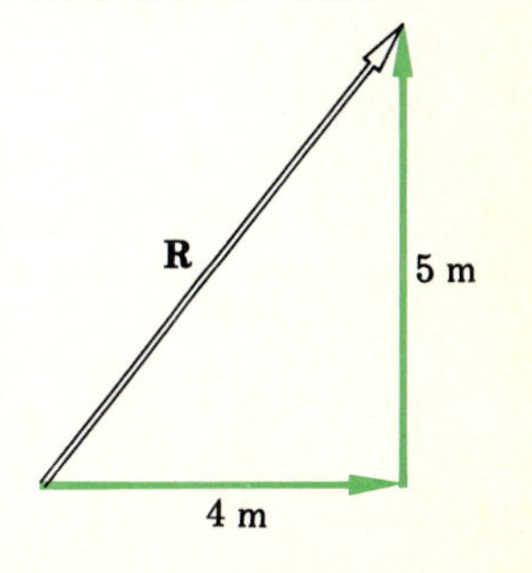

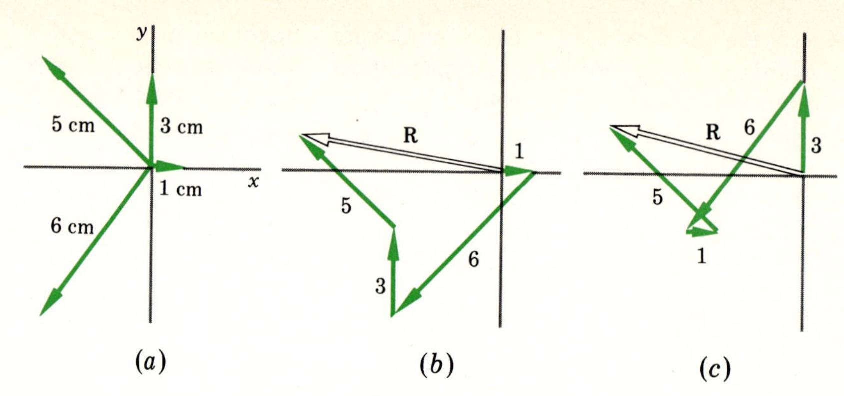

FIGURE 1.10

The four vectors shown in (*a*) can be added as in (*b*) or (*c*) to obtain their resultant. Notice that the order in which the vectors are taken is of no importance.

1.4 TRIGONOMETRIC METHODS

The graphical method for adding vectors is satisfactory if one is good at drafting, is patient, and has a ruler and protractor. For most of us, though, it is more convenient to have a quicker, less cumbersome method for adding vectors. A more convenient method is available to those who spend a few minutes learning a few basic rules of trigonometry, which we shall give here.

If one knows the east and north (or x and y) components of a group of vectors, the resultant is easily found. This is apparent from the fact that the x component of the resultant is the sum of all the x components of the vectors and the y component of the resultant is the sum of all the various individual y components. The virtue of trigonometry is that it provides an easy method for finding the components of vectors. Three quantities having to do with right triangles have been evaluated and are given in trigonometric tables, e.g., the table given in Appendix 5. If we refer to Fig. 1.11, we can see what these quantities are. In relation to the right triangle shown in Fig. 1.11 we define three trigonometric quantities:

DEFINITION

$$\sin\theta = \frac{a}{c} \qquad \cos\theta = \frac{b}{c} \qquad \tan\theta = \frac{a}{b}$$

Let us now refer to Fig. 1.12. From Fig. 1.12 and the definitions, it is apparent that the x component c_x of vector c is just $c\cos\theta$. Similarly, the y component c_y of vector c is $c\sin\theta$. If someone would tell us the value of $\cos\theta$ and $\sin\theta$, we could easily multiply the values by c and thereby obtain the x and y components of c. This is the purpose of the trigonometric tables. Let us now clarify this with some examples.

FIGURE 1.11

In terms of this right triangle, $\sin\theta = a/c$, $\cos\theta = b/c$, and $\tan\theta = a/b$.

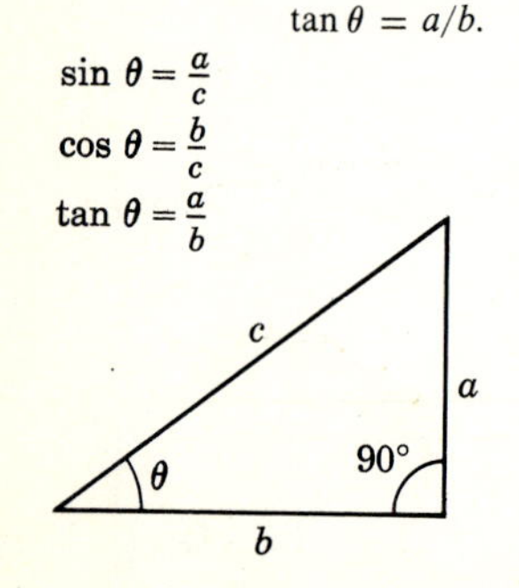

Illustration 1.5 Find the components of the vector shown in Fig. 1.13.

Reasoning The components are shown as c_x and c_y. From the definitions of the sine and cosine, we see from the figure that

$$\sin\theta = \frac{c_y}{20} \qquad \text{and} \qquad \cos\theta = \frac{c_x}{20}$$

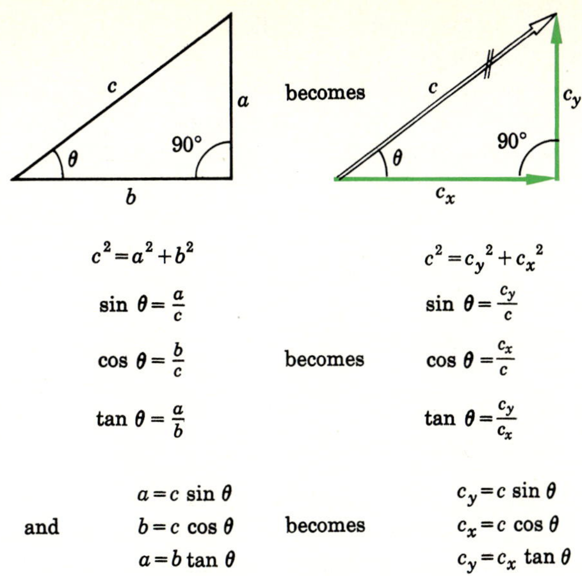

FIGURE 1.12
The y component of this vector is $c \sin\theta$, and its x component is $c\cos\theta$.

where $\theta = 37°$ in this case. The trigonometric tables (or a calculator) tell us that $\sin 37° = 0.60$ and $\cos 37° = 0.80$. Substituting these values, we solve and find $c_y = 12$ and $c_x = 16$.

FIGURE 1.13
The y component of the 20-unit vector is (20) (sin 37°), or 12, while the x component is (20) (cos 37°), or 16.
$c_x = 20\cos\theta = 20(0.80) = 16$
$c_y = 20\sin\theta = 20(0.60) = 12$

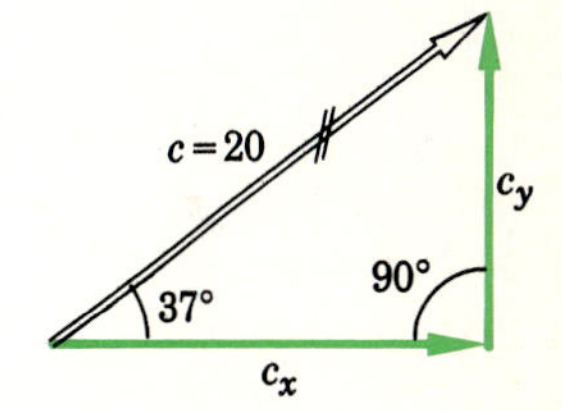

Add the vectors shown in Fig. 1.14*a* by the component method.

We have labeled the vectors as *a*, *b*, *c*, and *d*, as shown. Vectors *a* and *b* have only one nonzero component each. In part *b* of the figure we find the components of vector *c*. Since $\sin\theta$ = (opposite side)/hypotenuse, we have for the y component

$$y \text{ component} = (\text{hypotenuse})(\sin 37°) = (5)(0.60) = 3$$

Similarly, $\cos\theta$ = (adjacent side)/hypotenuse, which gives

$$\text{Adjacent side} = (5)(\cos 37°) = 4$$

FIGURE 1.14
With the component method, the vectors in (*a*) can be added to give the resultant shown in (*d*).

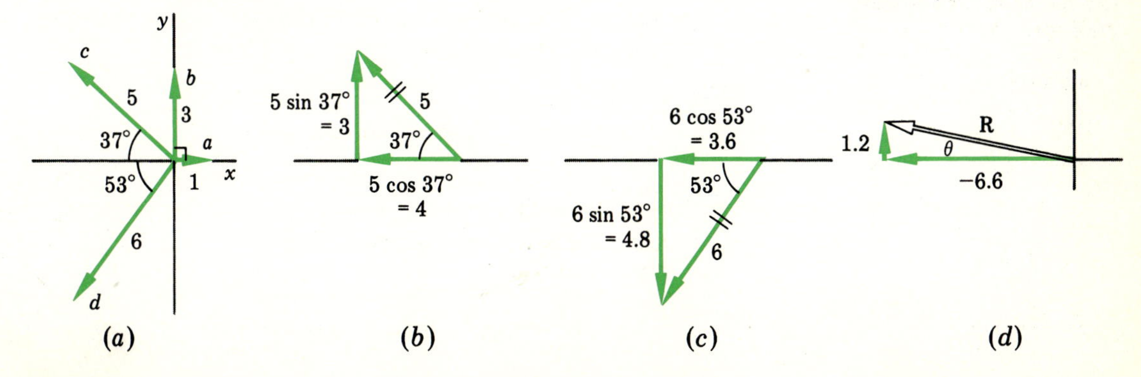

But this is the magnitude of the x component. However, the x component is negative since it points in the $-x$ direction. The x component of the 5-unit vector is therefore -4.

A similar procedure, shown in Fig. 1.14*c*, can be used to find the components of the 6-unit vector. Its x and y components are -3.6 and -4.8. Now the components of all the vectors are known. Until one becomes familiar with the method, it is helpful to arrange the components in a table as follows:

VECTOR	x COMPONENT	y COMPONENT
a	$a_x = +1.0$	$a_y = 0$
b	$b_x = 0$	$b_y = +3.0$
c	$c_x = -4.0$	$c_y = +3.0$
d	$d_x = -3.6$	$d_y = -4.8$
Resultant	$R_x = -6.6$	$R_y = +1.2$

In preparing this table, notice how components pointing in the negative x and y directions are taken as negative. Now that we know the components of the resultant, we can draw it, as shown in Fig. 1.14*d*. By use of the pythagorean theorem

$$R = \sqrt{R_x^2 + R_y^2} = \sqrt{(-6.6)^2 + (1.2)^2} = 6.7$$

We are not yet finished with this problem, since we still wish to find the value of the angle θ shown in Fig. 1.14*d*. This is easily done by referring to the definition of the tangent. Thus,

$$\tan\theta = \frac{R_y}{R_x} = \frac{1.2}{6.6} \approx 0.18$$

where $\approx$ is read "is approximately equal to." By examining the trigonometric table, we see that the angle whose tangent is 0.18 is approximately 10.5°. Therefore $\theta = 10.5°$ north of west. To obtain a more accurate value, we can use a process called **interpolation,** or we could consult a more complete table, such as those given in handbooks.* Many hand calculators also can be used to find trigonometric functions.

1.5 ADDITION OF FORCES

It is now a simple matter to add vectors of all kinds. We know that the order in which they are added is of no importance. We also now understand a

*"The Handbook of Chemistry and Physics" (Chemical Rubber Publishing Co.) contains a good set of mathematical tables. It also has a wealth of other information and is frequently consulted by persons working in the various sciences.

powerful method for adding vectors, the rectangular-component trigonometric method.

Illustration 1.7 Consider the problem illustrated in Fig. 1.15*a*. Several people are pulling on ropes attached to a post, and the figure shows a top view of the post and ropes. The force in pounds exerted on the post by each rope is indicated in part *b* of the figure. Our problem is to find the net result of the various forces exerted on the post by the ropes.

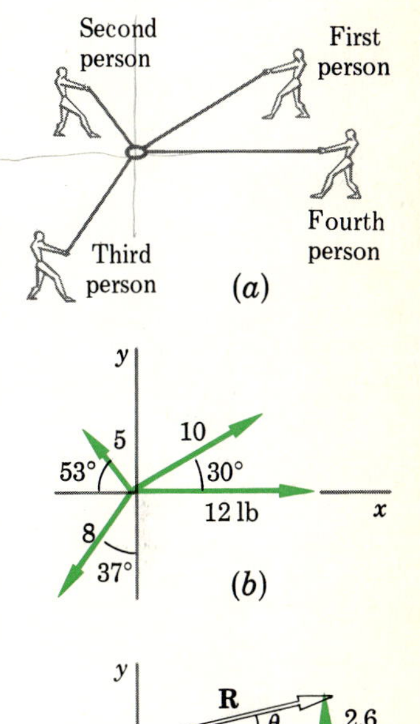

FIGURE 1.15
By adding the vectors shown in (*b*) the resultant illustrated in (*c*) is obtained.

Reasoning The appropriate component table is shown below, where the 12-lb vector is considered first and the others are taken in turn proceeding counterclockwise around the diagram (no particular reason for this, of course, since order doesn't matter):

VECTOR, lb	x COMPONENT, lb	y COMPONENT, lb
12	12.0	0.0
10	8.7	5.0
5	−3.0	4.0
8	−4.8	−6.4
Resultant	12.9	2.6

The pythagorean theorem then tells us that

$$R = \sqrt{(2.6)^2 + (12.9)^2}\ \text{lb} = 13.2\ \text{lb}$$

The resultant is shown in part *c* of Fig. 1.15. In addition, $\tan\theta = 2.6/12.9 \approx 0.20$, from which $\theta = 11.5°$.

One should always examine the diagram for the resultant, to make sure that no error has been made in magnitude or direction. In this case, since the 10- and 12-lb forces are pulling mostly in the $+x$ direction, and since the 5- and 8-lb forces tend to balance each other out, it seems reasonable that the resultant pull on the post should be in the direction found for **R**.

This completes our discussion of the methods for adding directed quantities, i.e., vectors. We are now ready to apply these methods to problems involving the effects of forces acting on objects. In the remainder of this chapter we shall be concerned only with the case where the forces on an object are balanced. Unbalanced forces cause an object to move. That situation will be discussed in later chapters.

1.6 OBJECTS AT REST

For an object to remain at rest, the effects of the forces pushing and pulling on it must exactly cancel. Even when all the forces acting on the object are balanced, the forces may still cause the object to rotate (this will be discussed fully in Chap. 8). Now, however, we concern ourselves only with the case where the object is not subjected to forces which can cause it to rotate. We consider bodies which are at rest and which remain at rest.

DEFINITION

An object or body which is at rest and remains at rest is in **equilibrium.** (A body *need not be at rest* to be in equilibrium, but this fact will be discussed further in Chap. 3.) In many cases, it is obvious why a body does not move. For example, the 100-lb object shown in Fig. 1.16 is at rest because the rope from the ceiling keeps it from falling.

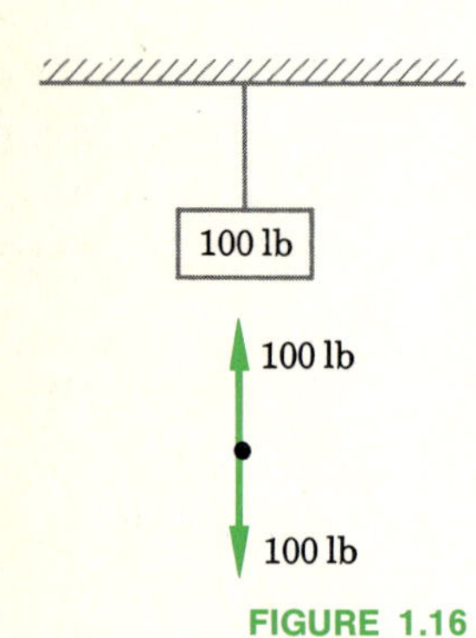

FIGURE 1.16
Two forces act upon the 100-lb object, the pull of gravity downward and the equal 100-lb pull of the rope upward.

If we examine this object in more detail, we see that *two* forces are acting upon it. We need only hold an object in our hand to know that something is pulling it straight down toward the earth. This downward pull, or force, on an object, is the result of the gravitational attraction of the earth for it. We shall not elaborate this point further at this time but discuss it more fully in a later chapter. Here we merely use the fact that *each body at rest on the earth is pulled toward the earth with a force which we call its* **weight.** This force is shown by the downward 100-lb vector in Fig. 1.16. *We shall often refer to it as the* **pull** *of gravity.*

Experimental observation shows us that unless something pushes or pulls upward on this body with a 100-lb force, it will fall to the earth. In this case, the rope pulls upward on the body with a force equal to its weight, 100 lb. This force is represented by the upward vector in Fig. 1.16. Hence we know from experiment that if the body is to move neither upward nor downward, the vertical forces acting on the body must just balance. In this case, the upward pull of the rope, 100 lb, exactly balances the 100-lb downward force resulting from the gravitational attraction of the earth.

Notice that we are speaking only about the forces acting on the body under consideration, namely, the object hanging at the end of the rope. There are, of course, other forces exerted on other objects in Fig. 1.16. For example, forces are exerted on the ceiling and on the rope. We are considering here, however, only the forces acting on the object hanging from the rope. *Experiment leads to the observation that if the resultant of all the forces pulling upward on the body exactly equals the resultant of all the forces pulling downward on the body, then the body will continue to hang at rest.* Since we are concerned only with the forces acting *on* the object, it is convenient to draw the free-body diagram as shown at the bottom in Fig. 1.16. It shows only the essentials, the forces acting *on* the body. We see at once that the upward force balances the downward force and hence the body will not move.

Consider next the cart shown in Fig. 1.17*a*. The cart weighs 50 lb. Someone is pulling to the left on the rope with a force of 30 lb, and the spring, attached to the wall, is pulling the cart to the right. As an approximation we shall consider the friction in the wheels to be negligible. What forces act *on* this body?

First, the pull of gravity downward on the cart is 50 lb. This downward pull must be balanced, or the cart will fall downward. Apparently the road on which the carts rests must be pushing up on the cart with a force of exactly 50 lb so as to balance the weight of the cart. Once again the upward forces must balance (or be equal to) the downward forces if the cart is to remain at rest.

Two other forces act *on* the cart. The rope pulls to the left with a force of 30 lb. Experience tells us that if the pull of the rope were not balanced by an equal force pulling the cart to the right, the cart would move to the left. Hence, if the cart is to be at rest, the spring must be pulling to the right *on the cart* with a force of 30 lb.

It is clear, then, from our common experience that *a body remaining at rest must have no unbalanced force acting on it*. All the forces pulling down on it must be balanced by upward forces. As much force must be pulling it to the right as to the left. In fact, there can be no net force in the third direction either, trying to pull it in a direction in or out of the page in Fig. 1.17. We can conveniently represent the forces acting on the cart of Fig. 1.17*a* by the free-body diagram of Fig. 1.17*c*. Notice that the resultant force on the cart is zero.

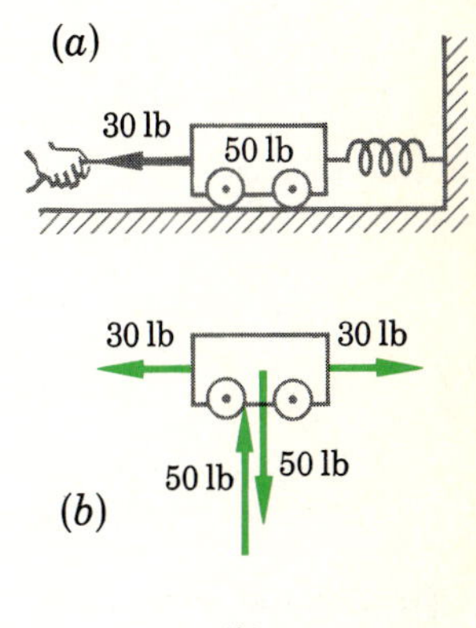

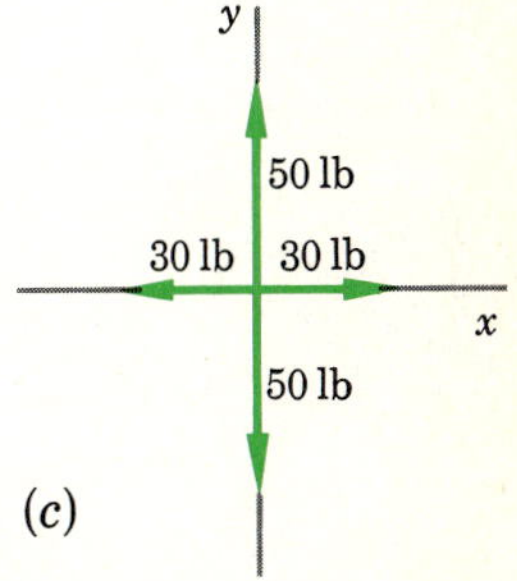

FIGURE 1.17
The 50-lb cart is held to the wall with a spring, and a rope pulls on it to the left with a force of 30 lb. All four forces acting on the body are illustrated in the free-body diagram (*c*). Since the cart is in equilibrium, all forces are balanced.

1.7 OBJECTS CONSTRAINED BY ROPES AND CABLES

Let us now examine a more complex example to see how our rule of balance of forces may be used.

Illustration 1.8 Consider the street lamp held by the cables shown in Fig. 1.18*a*. It weighs 500 N (newtons).*

Reasoning Notice carefully now that we are going to talk only about the forces on the lamp. Forces exist on many other objects in the figure, but we are concerned only with the lamp. Only two forces act on the lamp, the 500-N pull of gravity down and the balancing upward 500-N pull of the cable *AL*. Notice that the two forces acting on the lamp exactly balance, and so the lamp remains at rest.

Consider next the forces acting on the length of cable *AL*. Since the earth pulls down on the lamp with 500 N, the lamp in turn weighs 500 N and exerts a 500-N force downward on the rope *AL*. At the upper end, the ring at *A* must pull upward on the cable *AL* with a force of 500 N in order to keep the cable from dropping. Hence the cable *AL* remains at rest because a 500-N downward force on one end of the cable balances a 500-N upward force on the other end.

Let us now look at the small ring at *A*. There are three things pulling on it, the cable *AL*, the cable *AC*, and the cable *AB*. These forces *on the ring* are drawn in Fig. 1.18*b*. Notice that they are not so simple as the forces on the cable or on the lamp since they do not act along the same line. However, this is of very little consequence now that we know how to resolve force vectors into components.

*Like the pound, the newton is a unit of force. We shall learn more about it in Chap. 3. For now we simply state that 1 lb = 4.45 N. The object we are considering weighs 500 N, which is the same as 112 lb.

The pull of cable AC may be thought of as consisting of two pulls, an upward pull and a leftward pull. This is shown in Fig. 1.18*c*, where the original vector F_1 has been crossed out to show that we have replaced it by its components. Now it is clear how the forces must balance. The 500-N downward force is entirely balanced by the upward component of force F_1, namely, F_{1y}. We therefore have

$$F_{1y} = 500 \text{ N} \tag{1.2}$$

In addition, the rightward pull of F_2 must be balanced by the leftward pull of force F_{1x}. Therefore

$$F_{1x} = F_2 \tag{1.3}$$

Unfortunately, F_{1x} is not yet known, and so F_2 cannot be computed until F_{1x} is found. This is readily done, though, if we notice that the definition of the tangent of an angle says in this case

$$\tan 37° = \frac{F_{1y}}{F_{1x}}$$

or

$$F_{1x} = \frac{F_{1y}}{\tan 37°} = \frac{500 \text{ N}}{0.75} = 670 \text{ N} \tag{1.4}$$

where we have used the fact that the tangent of 37° is 0.75. From Eq. (1.3) we then have $F_2 = 670$ N also.

If we want the tension in cable AC and not just its components, we can write

$$F_1 = \sqrt{F_{1x}^2 + F_{1y}^2}$$

$$= \sqrt{(670 \text{ N})^2 + (500 \text{ N})^2} = 830 \text{ N}$$

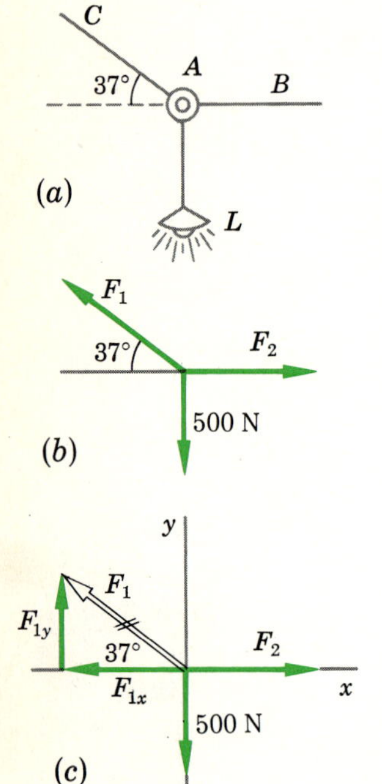

FIGURE 1.18
The forces on the junction of the three cables in (*a*) are drawn in (*b*). In (*c*) the components of the forces have been computed in preparation for writing the equilibrium relation for the junction point.

Or, more simply, we could notice from the definition of the sine

$$\sin 37° = \frac{F_{1y}}{F_1}$$

giving

$$F_1 = \frac{F_{1y}}{\sin 37°} = \frac{500 \text{ N}}{0.60} = 830 \text{ N} \tag{1.5}$$

We have therefore been able to find the tension in each of the ropes and the forces on several bodies by an application of the idea of balance of forces.

1.8 THE FIRST CONDITION FOR EQUILIBRIUM

Two basic conditions must be satisfied if an object is to remain at rest. In the previous sections we have been using the **first condition for equilibrium.** The second condition involves a statement concerning rotation of an object and will be discussed in Chap. 8. Assuming that the object is not made to rotate, the object will remain at rest provided all the forces are balanced. We can therefore state the first condition for equilibrium in the following rather crude way: *The upward forces must balance the downward forces, and the leftward forces must balance the rightward forces.* Although this is certainly true, and although it is clear to us what we mean, it is really not a very precise or concise scientific statement. We can improve it if we express it in terms of the three coordinate directions x, y, and z.

What we really mean is that if there is a force of 20 lb directed in the positive x direction, there must be a 20-lb force in the negative x direction if the body is to be at equilibrium. If we agree to call the positive-directed force $+20$ and the negative-directed force -20 lb, we can state our condition quite simply by saying "the x-directed forces must add up to zero." Or, using the mathematical symbol ΣF_x for the words "the sum of all the x-directed forces," *we have for the first condition necessary for equilibrium**

$$\Sigma F_x = 0 \qquad \Sigma F_y = 0 \qquad \text{at equilibrium} \tag{1.6a}$$

These two equations can be simplified even further by noticing that they merely state that the resultant of all the forces acting on the body must be zero. That is to say, the sum of all the vector forces acting on the body equals zero. Thus

$$\Sigma \mathbf{F} = 0 \tag{1.6b}$$

at equilibrium.

Equations (1.6) are the mathematical statement of the first condition for equilibrium. We should always remember they are to be applied to only one isolated body at a time. They refer to the forces on that body, and one must always be prepared to say by what means each force is being exerted on the body isolated for consideration. It must be the pull of a rope, the pull of gravity, the push of a table, the push of a beam, or some similar, real physical means of force exertion upon the body being considered.

1.9 SOLUTION OF PROBLEMS IN STATICS

With a little practice, the solution of many problems in statics (the study of bodies at rest) by using Eq. (1.6) becomes almost routine. But a few simple rules should be followed so that one does not become confused:

1 Isolate a body. What point or object are you going to talk about? The forces acting on this object are the only ones which will be needed in writing Eq. (1.6).

---

*This assumes all forces to have no component in the z direction. If z-directed forces exist, then we must add $\Sigma F_z = 0$ to the conditions listed in Eq. (1.6).

2 Draw the forces acting *on the body you have isolated,* and label them. (If their value is not known, give each one of them a symbol such as F_1, P, Q, etc.) We call this a **free-body diagram.**
3 Split each of the forces into its x, y, and z components, and label the components in terms of the symbols given in rule 2 and the proper sines and cosines.
4 Write down Eq. (1.6).
5 Solve the equations for the unknowns.

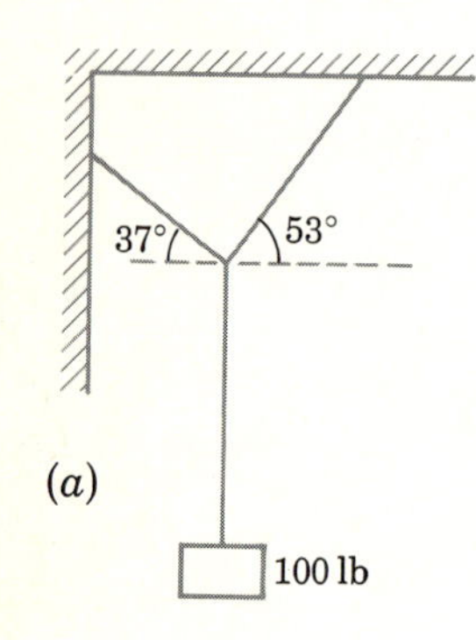

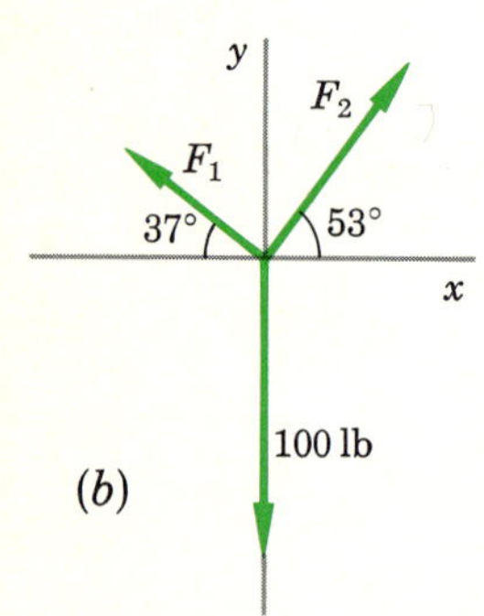

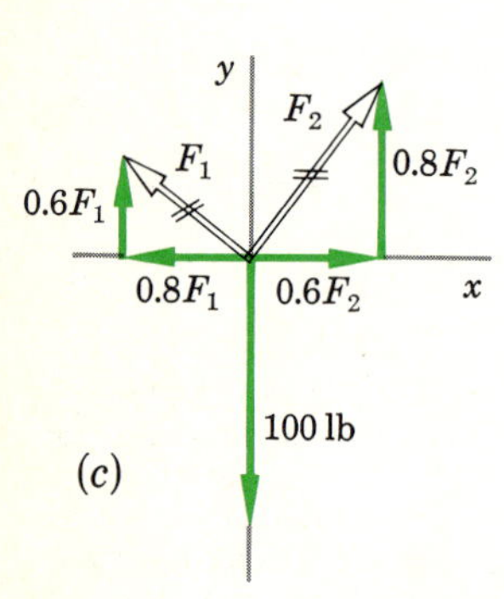

FIGURE 1.19
Since the junction point of the ropes in (*a*) is at equilibrium, the y forces must cancel each other in (*c*). The same holds true for the x forces.

Illustration 1.9 To illustrate this procedure, consider how we would find the tensions in the ropes shown in Fig. 1.19*a*. (Remember that the tension in a rope is the force with which it pulls upon the object to which it is attached.)

Reasoning We know at once that the tension in the lower rope is 100 lb, since it just holds the heavy object in place. Let us choose the junction of the three ropes as the body (or point) we shall isolate. It is reasonable to choose this point since one of the forces acting on it is already known and the other two are forces we actually seek.

In part *b* of Fig. 1.19 we have carried out step 2 of our solution procedure. Step 3 is carried out in part *c* of the figure. You would probably do this directly on the figure of part *b*, but we have separated the two here for clarity. In addition, we have written $F_1 \sin 37°$ as $0.60F_1$, since $\sin 37°$ is 0.60. The same notation is used for the other components.*

We are now ready to carry out step 4. Just by looking at Fig. 1.19*c*, we have

$$\Sigma F_x = 0: \qquad 0.60F_2 - 0.80F_1 = 0 \tag{1.7}$$
$$\Sigma F_y = 0: \qquad 0.80F_2 + 0.60F_1 - 100 = 0 \tag{1.8}$$

where we have omitted the units (lb) with the 100 in order to simplify the equation.

To carry out step 5, we now solve these equations for F_1 and F_2. Recall from your algebra that two general methods are often used.

Method 1: Multiply Eq. (1.7) by 0.6 and Eq. (1.8) by 0.8. Thus

$$\begin{aligned} 0.36F_2 - 0.48F_1 &= 0 \\ 0.64F_2 + 0.48F_1 - 80 &= 0 \end{aligned} \tag{1.9}$$

Adding the two equations gives

$$1.00F_2 - 80 = 0 \qquad \text{or} \qquad F_2 = 80 \text{ lb}$$

---

*As a convenience, we shall usually employ angles of 37 and 53° in illustrations. If we remember that $\sin 37° = \cos 53° = 0.60$ and $\cos 37° = \sin 53° = 0.80$, we shall not need to interrupt our train of thought to refer to the tables. A right triangle having 37 and 53° for its two acute angles is called a **3, 4, 5 right triangle** since its sides are proportional to these numbers.

Substitution in Eq. (1.9) gives

$$0.48F_1 = (0.36)(80) \qquad \text{or} \qquad F_1 = 60 \text{ lb}$$

Method 2: Solve Eq. (1.7) for F_1 in terms of F_2.

$$F_1 = 0.75F_2 \tag{1.10}$$

Substitute this in Eq. (1.8) to give

$$0.80F_2 + (0.60)(0.75)(F_2) - 100 = 0$$

or

$$F_2 = 80 \text{ lb}$$

and from Eq. (1.10)

$$F_1 = 60 \text{ lb}$$

The physics involved in finding the tensions in the ropes is mainly concerned with drawing the forces, finding their components, and applying the condition for equilibrium. The major portion of the work involved is algebra and occurs in solving the simultaneous equations for the unknowns. Do not allow the algebra to confuse you. Problems like this are quite straightforward and should present no difficulty. In the next chapter we shall learn how to describe motions of objects, and in Chap. 3 we shall see how forces acting on objects can cause motion.

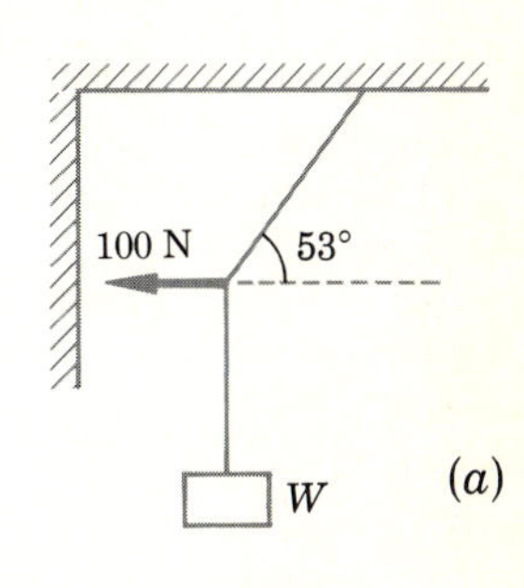

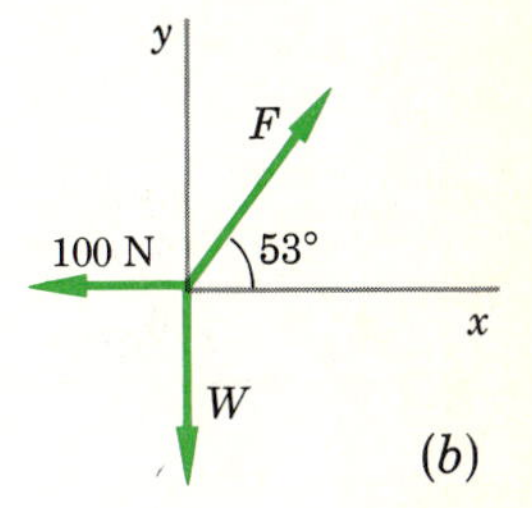

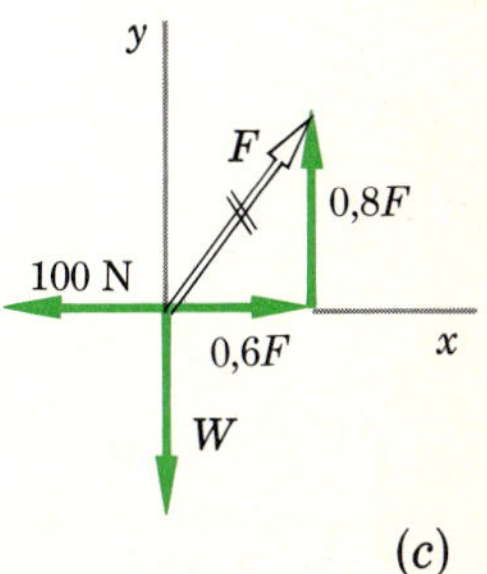

FIGURE 1.20

It being known that the tension in the horizontal rope is 100 N, the problem is to find the value of W. Appropriate force diagrams for the junction point are shown in (*b*) and (*c*).

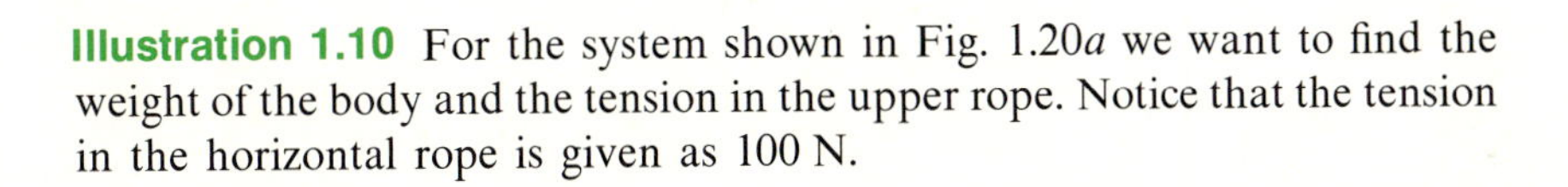

Illustration 1.10 For the system shown in Fig. 1.20*a* we want to find the weight of the body and the tension in the upper rope. Notice that the tension in the horizontal rope is given as 100 N.

Reasoning To find the tension in the rope, choose the junction of the three ropes as the body under consideration. We now isolate it and draw the appropriate free-body diagram as shown in Fig. 1.20*b*. The forces are resolved in part *c*.

Writing down the conditions for equilibrium gives

$$\Sigma F_x = 0: \qquad 0.60F - 100 = 0$$
$$\Sigma F_y = 0: \qquad 0.80F - W = 0$$

Solving the first of these equations gives $F = 167$ N. Substitution of this value in the second equation gives $W = 133$ N.

SUMMARY

Vector quantities have both magnitude and direction. Scalar quantities have no direction. We can represent a vector quantity by an arrow called a vector. The length of the vector is proportional to the magnitude of the vector quantity. The direction of the vector shows the direction of the vector quantity.

To add several vectors graphically, one places them end to end. The tail of the second is placed on the tip of the first. The tail of the third is placed on the tip of the second, and so on. The resultant vector is an arrow with its tail at the tail of the first and its tip at the tip of the last vector.

To add several vectors by the trigonometric component method, one first finds the rectangular components of each vector. This is done by the trigonometric method described in Sec. 1.4. The x component of the resultant is found by taking the algebraic sum of the x components of the vectors. Similarly, the y component of the resultant is found. Then

$$R = \sqrt{R_x^2 + R_y^2} \qquad \text{and} \qquad \tan\theta = \frac{R_y}{R_x}$$

where θ is the angle the resultant vector makes with the x axis.

In order for an object to be at equilibrium, the resultant force on the object must be zero. As an equation, this can be written as

$$\Sigma \mathbf{F} = 0$$

Or, in terms of the components of the forces acting on the object

$$\Sigma F_x = 0 \qquad \Sigma F_y = 0 \qquad \Sigma F_z = 0$$

This is called the first condition for equilibrium. There is a second necessary condition which concerns applied forces that can cause rotation.

To solve problems involving objects at equilibrium, one follows a five-step procedure, given in Sec. 1.9.

MINIMUM LEARNING GOALS*

Upon completion of this chapter, you should be able to do the following:

1. Explain the difference between scalar and vector quantities and give examples of each.
2. Draw a vector diagram showing a series of displacements described to you.
3. Find the resultant of several vectors by means of a scale drawing.
4. State what is meant by the tension in a string or rope.
5. Define sin θ, cos θ, and tan θ. Given an angle, use the trigonometric tables to find its sine, cosine, and tangent.
6. Use trigonometry to find the unknown sides or angles in a right triangle.
7. Find the rectangular components of a given vector; or, given the components of a vector, find the vector and its angle.
8. Add several given vectors by use of the trigonometric component method.
9. State the first condition for equilibrium in words and in equation form. Tell whether the given forces on an object satisfy this condition.
10. Use the first condition for equilibrium to solve simple problems such as those given in Sec. 1.9.
11. Give examples of distances which are about 1 mm, 1 cm, 1 km.
12. State your approximate weight in newtons.

*Notice the word minimum. All students should achieve these goals.

IMPORTANT TERMS AND PHRASES

You should be able to define or explain each of the following:

Scalar quantity
Vector quantity
Displacement vector
Pythagorean theorem
Resultant vector
Graphical method for vector addition
Sine, cosine, and tangent
Rectangular components of a vector
Trigonometric method for vector addition
Force vector

Newton (unit)
Tension in a string
Weight of an object
Free-body diagram
Isolating an object
First condition for equilibrium

QUESTIONS AND GUESSTIMATES

Questions marked with (E) require you to make an estimate based upon your own experience.

1. Which of the following can be represented by a vector: (*a*) the force due to gravity, (*b*) the number of people in a city, (*c*) the wind velocity, (*d*) the motion of a boat, (*e*) the books on a shelf, (*f*) the flow of water in a pipe?

2. Can a force directed east ever balance a force directed north?

3. Is it possible to obtain a force directed straight north by combining a northeast force and an equal northwest force? What if the forces are unequal?

4. Show how three equal-magnitude vectors would have to be oriented if they are to add up to zero. Can this be done in more than one way? Can this be done with three unequal-magnitude vectors? Two unequal-magnitude vectors?

5. Traffic lights hung from cables stretched across a street make the cable sag. Why don't the workers remove the sag when they adjust the cable?

6. A mother cat usually carries her kittens around by grabbing the skin at the back of the kitten's neck in her mouth and lifting. Describe the action of the forces involved and show why this is probably preferable to other methods, e.g., pulling the kitten's paw.

7. Represent each person in a city of 200,000 by a vector extending from toe to nose. Estimate the resultant of these vectors at (*a*) 12 noon and (*b*) 12 midnight. Repeat if the vector extended from left ear to right ear. (E) (You must estimate the average toe to nose distance of each person as well as how many are standing up.)

8. A common circus act is the so-called human pyramid. Typically three acrobats stand side by side on the floor, two more stand on their shoulders, and one acrobat is supported on the shoulders of these two. Estimate (*a*) the force exerted upon the shoulders of the bottom acrobat in the center and (*b*) the force each one of his feet exerts on the floor. Will your answer depend upon whether or not the bottom acrobats are standing close together? (E)

9. As a part of his act, a performer is lifted slowly by pulling upward on the hair of his head, but unfortunately he is rapidly going bald. Approximately how many hairs must remain on top of his head if the act can be performed safely? (E) (You must estimate the strength of a hair.)

PROBLEMS

Problems labeled (G) are to be done graphically; all others are to be done by using trigonometry.

1. (G) My home is six blocks east and eight blocks north from the building in which I work. What is the straight-line distance between these points? At what angle with east is this straight line?

2. (G) What is the actual straight-line distance between two points if one lies 20 km west and 16 km north of the other? What angle does this line make with east?

3. (G) If an airplane flies southwest for 100 km, how far west and how far south has it gone (see Fig. P1.1)?

4. (G) Detroit is approximately 400 mi northwest of Washington, D.C. How much farther south is Washington than Detroit (see Fig. P1.1)?

5. (G) In positioning a device on a lathe bench, it is given the following displacements: 5.0 cm at 0°, 12.0 cm at 80°, 7.0 cm at 110°, 9.0 cm at 210°. All angles are measured counterclockwise from the x axis. (See Fig.

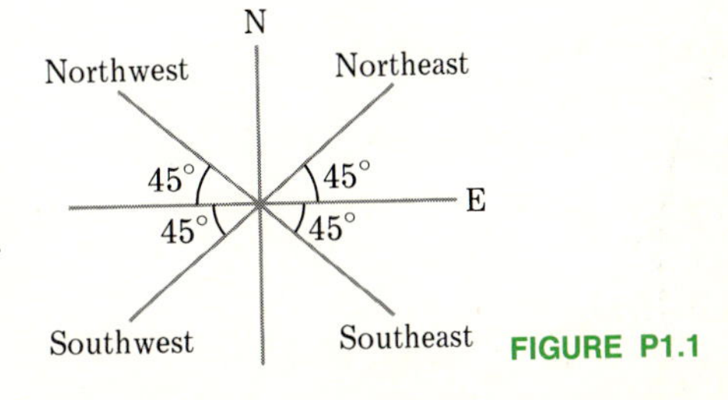

FIGURE P1.1

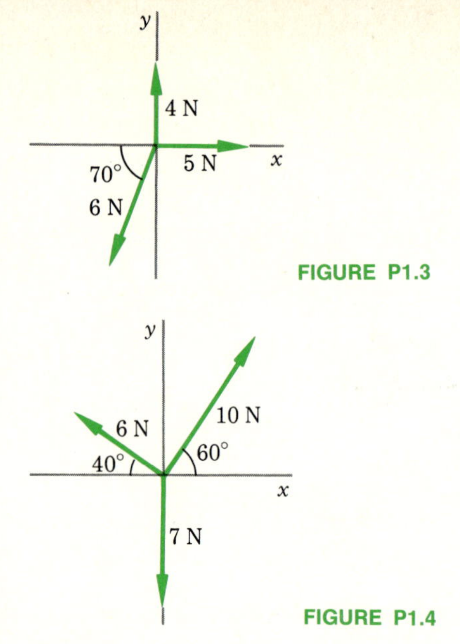

FIGURE P1.3

FIGURE P1.4

P1.2 for the 7.0-cm displacement at 110°.) Find the magnitude and angle of the resultant displacement.

6. (G) Point B is 300 km from point A and at an angle of 50° north of east. A road goes 40 km east from point A and then ends. Starting at the end of the road, how far and at what angle to the east must one travel in order to reach point B?

7. To get to Chicago from Miami, one must fly about 1100 mi at an angle of 30° west of north. How far north of Miami is Chicago? How far west?

8. A force of 50 N pulls at an angle of 110° measured counterclockwise to the $+x$ axis (see Fig. P1.2). What is the x component of this force? Its y component?

9. Find the resultant of the forces shown in Fig. P1.3. Give the counterclockwise angle relative to the x axis.

10. Find the resultant of the forces shown in Fig. P1.4. Give the counterclockwise angle relative to the x axis.

11. After breaking his leg in a freak accident, a student finds himself in a hospital with traction applied to his leg. He notes that his condition is as shown in Fig. P1.5. Assume that the pulleys are frictionless; then the tension in the cord is everywhere the same, namely, 7.0 lb. How large is the force which stretches his leg? How large an upward force does the device exert on his foot and leg together?

12. Find the resultant of the following three forces: 10.0 N at 30°, 20.0 N at 90°, and 30 N at 250°. All angles are measured counterclockwise from the $+x$ axis. What is the magnitude and direction of the force needed to balance these three?

13. A force of 20.0 lb at an angle of 270° to the $+x$ axis (angles being measured counterclockwise from the $+x$ axis) is needed to balance two forces, one of 15.0 lb at an angle of 90° and the other unknown. Find the magnitude and direction of the unknown force.

14. A bug sits on a board which is floating southward down a straight stream with a speed of 5.0 cm/s (centimeters per second). The bug now walks upstream (northward) along the board at a speed of 3.0 cm/s. What is the speed of the bug relative to the shore of the stream?

15.* Repeat Prob. 14 if the bug is walking with the same speed but going eastward on the board.

16. Two students are climbing the same rope in gym class. When they stop to rest, the one on top is 12 ft from the ceiling while the other is 18 ft from the ceiling.

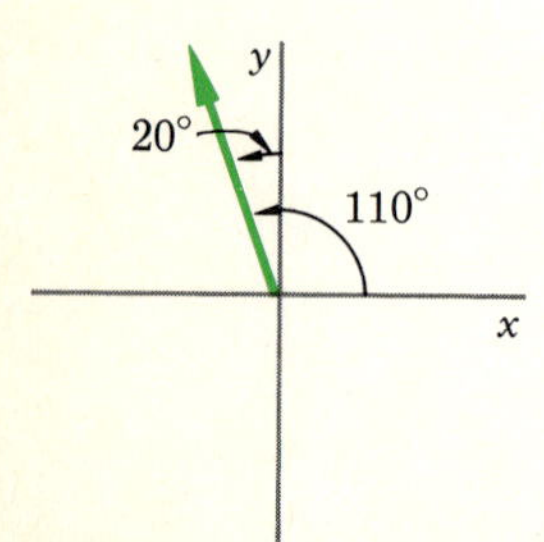

FIGURE P1.2

FIGURE P1.5

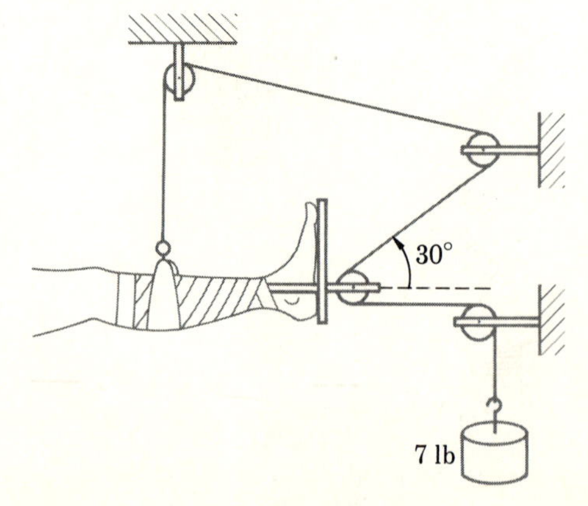

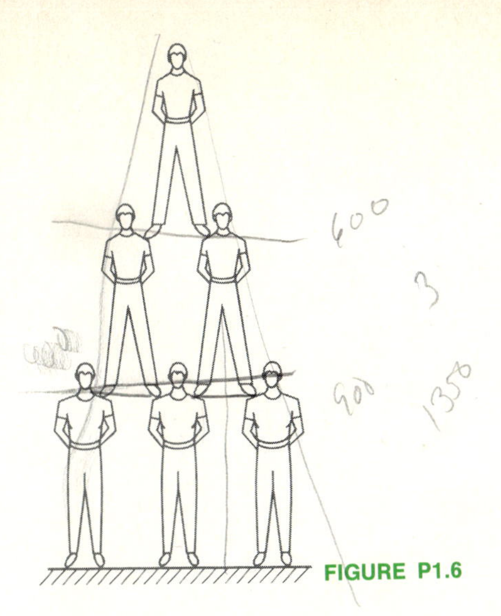
FIGURE P1.6

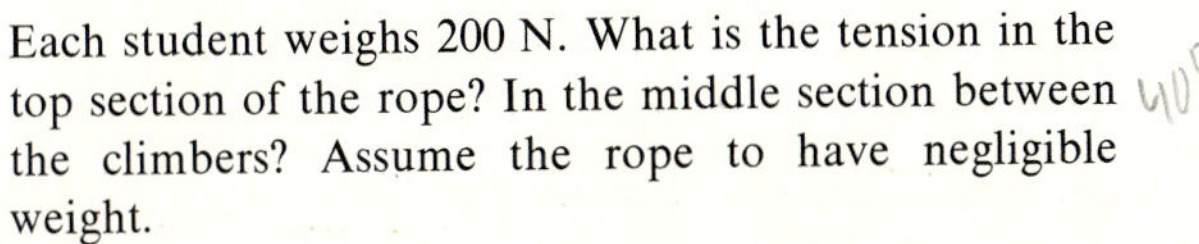
Each student weighs 200 N. What is the tension in the top section of the rope? In the middle section between the climbers? Assume the rope to have negligible weight.

17. A rope of negligible weight is thrown over a tree limb so that its two ends hang close to the ground. Two 50-lb girls climb part way up the two parts of the rope and hang there. What is the tension in either part of the rope? With how large a total force does the rope pull down on the limb?

18. Six performers stand as shown in Fig. P1.6. Each of the lower three weighs 900 N while each of the upper three weighs 600 N. Find the resultant force (magnitude and direction) exerted on the floor by them.

19. The tightrope walker shown in Fig. P1.7 weighs 150 lb. Find the tension in the rope.

20. The rope in Fig. P1.7 can stand a tension of only 2000 N. How heavy a tightrope walker can it support in the position shown?

21. In the equilibrium situation shown in Fig. P1.8, the pulleys are nearly frictionless, and so the tension is the same in all parts of the cord. Assume that the weights of the pulleys and cord are negligible and that the cords are essentially vertical. What is the tension in the cord in terms of W_2, the weight of the object on the right? What is the ratio of W_1 to W_2? This is a simple device used to lift large loads.

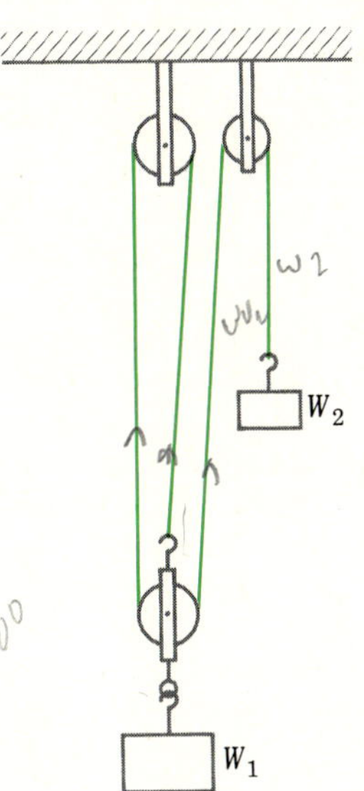

FIGURE P1.8

22. The weight W shown in Fig. P1.9 is 200 N. What is the tension in the horizontal rope? Assume the ropes to have negligible weight.

23. The tension in the left-hand rope in Fig. P1.10 is 60 N. Find W and the tension in the rope at the right.

24. For the equilibrium situation shown in Fig. P1.11, find W_2 and W_3. Assume $W_1 = 100$ lb and consider the pulleys to be frictionless.

FIGURE P1.7

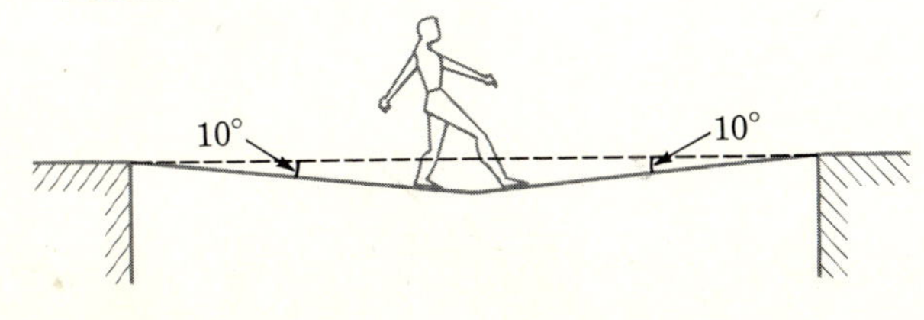

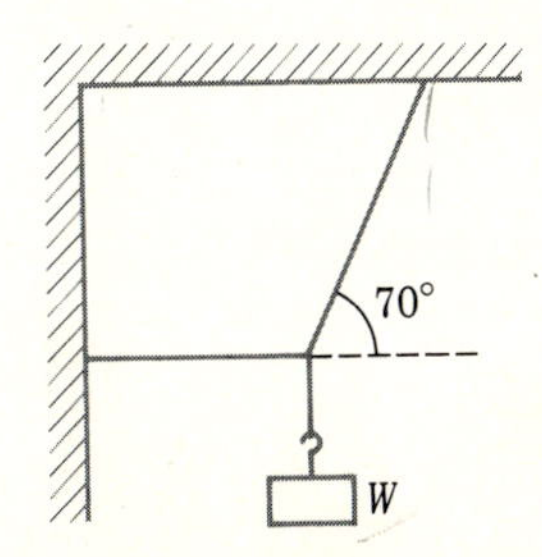

FIGURE P1.9

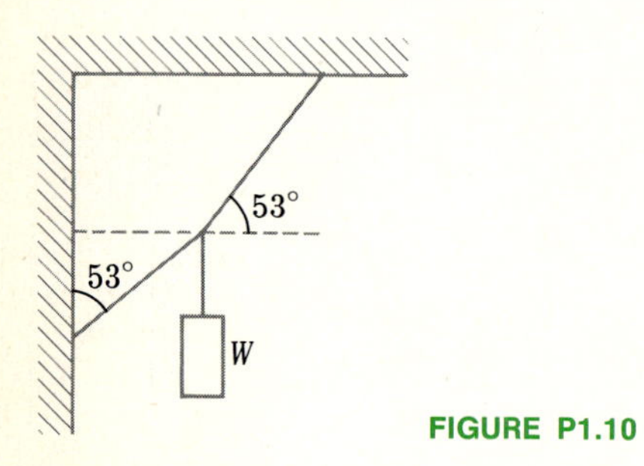

FIGURE P1.10

FIGURE P1.11

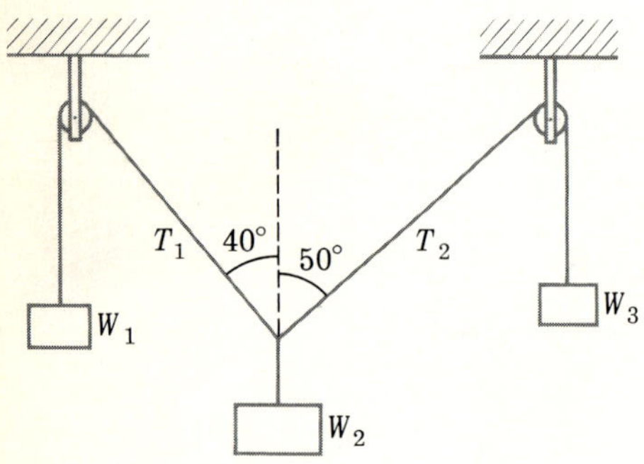

25.* For the equilibrium situation shown in Fig. P1.12, assume the pulley to be frictionless. Find W_1, W_2, T_1, and T_2.

* Problems so marked are slightly more complicated or require a little more ingenuity than average.

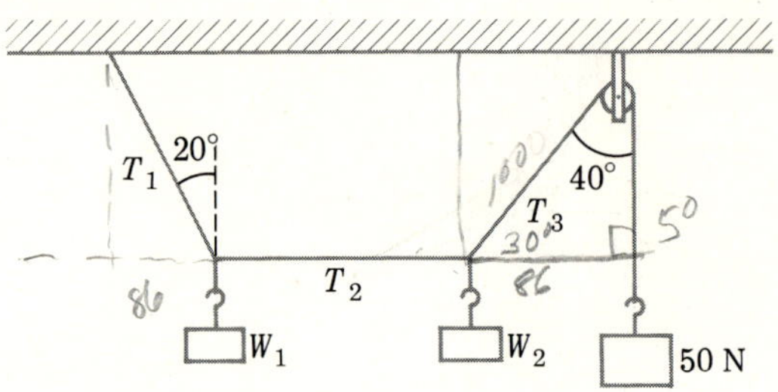

FIGURE P1.12

$\sin\theta = \frac{y}{r}$ $r\sin\theta = y$

$r = \frac{y}{\sin\theta}$

$T_2 = 86$

UNIFORMLY ACCELERATED MOTION

2

We learned in the preceding chapter how to deal with bodies at rest. It is now appropriate to consider the behavior of moving bodies. The present discussion will be directed toward methods for describing the motion of an object in terms of its velocity and acceleration. We shall find that these methods enable us to predict the motion of bodies in free fall. The role of forces in causing motion will be the subject of the following chapter.

2.1 VELOCITY AND SPEED

We all understand what is meant by the statement "the car was traveling at 50 mi/h." This means that the car would travel 50 mi in 1 h if it continued at this speed. However, it is very unlikely that the driver would travel exactly at this speed for a full hour. If he actually did complete a 50-mi trip in 1 h, he would surmise that his *average* speed had been 50 mi/h. But we have no way of knowing from the data given that he didn't actually break the 50-mi/h speed limit and take time off for coffee along the way. Before delving further into this complication, let us define precisely what we mean by average speed.

DEFINITION

Speed *is the distance traveled in unit time.* Its units are miles per hour, feet per second, meters per second, and so forth. If we travel 10 m in 2 s, our average speed is 5 m/s. We obtained this, perhaps unconsciously, by the equation

$$\text{Average speed} = \frac{\text{distance gone}}{\text{time taken}}$$

or, in symbols,

$$\bar{u} = \frac{d}{t} \tag{2.1}$$

where the bar above the u indicates that it is an average value.

Let us see how the units of speed are arrived at. Carrying out the computation for a woman moving 10 m in 2 s,

$$\bar{u} = \frac{10\ \text{m}}{2\ \text{s}} = 5\ \text{m/s}$$

which we read as "5 meters per second." Similarly, if a snail travels 2 cm in 0.4 day, its speed is

$$\bar{u} = \frac{2\ \text{cm}}{0.4\ \text{day}} = 5\ \text{cm/day}$$

The units of speed are always a length unit divided by a time unit.

When a car moves along a straight road, its motion is in a certain direction. If we wish to combine the two concepts, speed and direction, we are speaking of a vector. It is customary to *define the* **average velocity vector** $\bar{\mathbf{v}}$ *in the following way:*

DEFINITION

$$\bar{\mathbf{v}} = \frac{\text{displacement vector}}{\text{time taken}}$$

For example, suppose a car starts from a point A and at a time t later arrives at a point B. If the vector displacement from A to B is $\mathbf{s}$, we have for the definition of the average velocity

$$\bar{\mathbf{v}} = \frac{\mathbf{s}}{t} \tag{2.2}$$

For motion in a constant direction along a straight line, the length of the vector **s** would be simply the distance traveled d. As a result, the magnitude of the velocity *in this particular case* is equal to the speed since $\bar{u} = d/t = s/t = \bar{v}$. However, if the motion is not along a straight line, the total distance covered d may not be the same as the magnitude of the displacement **s**.

Illustration 2.1 A simple example of this is shown in Fig. 2.1*a*, which shows the path of a car traveling from A to B and then to C.

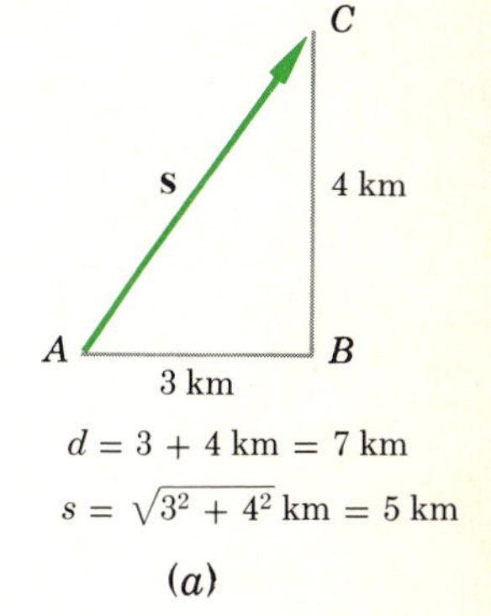

Reasoning We see that the total distance covered d is 7 km. However, the displacement **s** has a magnitude of only 5 km. We therefore find

$$\bar{u} = \frac{d}{t} = \frac{7\text{ km}}{t} \qquad \text{while} \qquad \bar{v} = \frac{s}{t} = \frac{5\text{ km}}{t}$$

Clearly, the magnitudes of the speed and velocity are not the same in this case.

Even for motion along a straight line these two quantities can differ. This is shown in Fig. 2.1*b* where the car goes from A to B and then back to C. In this case

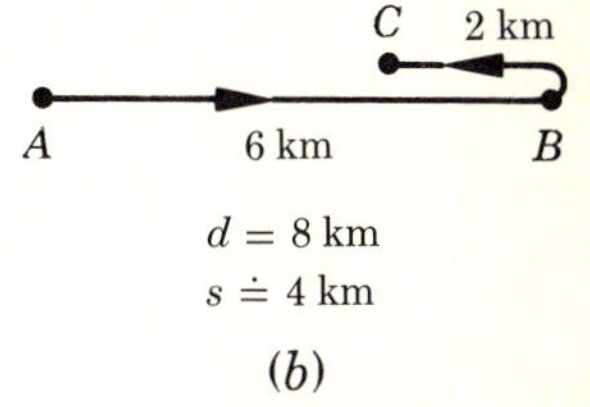

$$\bar{u} = \frac{8\text{ km}}{t} \qquad \text{while} \qquad \bar{v} = \frac{4\text{ km}}{t}$$

FIGURE 2.1
Since the magnitude of the average speed is $\bar{u} = d/t$ and for the velocity is $\bar{v} = s/t$, the two quantities are not always equal.

Even so, much of our attention in this chapter will be turned upon situations in which motion occurs in a single direction. In such cases $\bar{u}$ and $\bar{v}$ will be equal.

2.2 INSTANTANEOUS VELOCITY AND SPEED

Let us consider an object which is moving along a straight line, the car in the upper part of Fig. 2.2, for example. We shall take the x coordinate along the line on which the car is moving, as shown. The motion of the car is therefore in either the plus or minus x direction. (Its motion is negative if it travels in the direction opposite that shown.) Suppose the car is traveling with constant speed (and velocity) in the direction shown. Let us say that the magnitude of the speed (and velocity) is 20 m/s.

We can show the displacement of the car as a function of time on a graph. It is convenient to measure the displacement from the point $x = 0$. Then the magnitude s of the displacement vector is equal to x. Moreover, if the car is at a position $x = -10$ m, for example, the negative sign tells us the direction of the displacement. It is 10 m in the $-x$ direction in this case. We shall therefore represent the displacement along a straight line by x, the displacement of the object from the point $x = 0$.

The graph of x versus time for the car is shown in Fig. 2.2*b*. As we see,

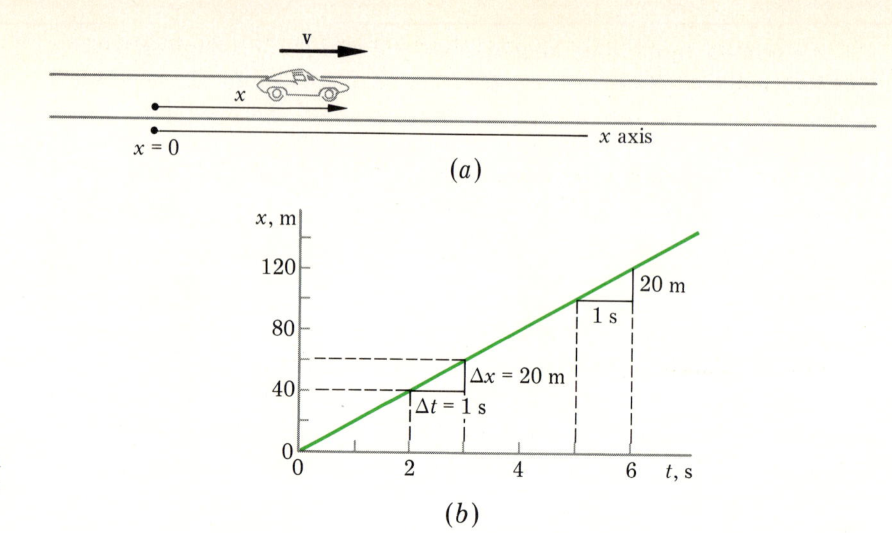

FIGURE 2.2

Motion along a straight line can be shown on a graph. In this case, the car's speed was constant at 20 m/s.

the time t is the independent variable and is measured along the horizontal axis of the graph. With the passage of time, the car's position changes. The straight line shown on the graph tells us where the car is at any particular time. For example, at $t = 0$ the car was at the origin, $x = 0$. But, at $t = 3$ s, $x = 60$ m; the car was 60 m from the origin. Notice that the little triangles shown on the graph tell us that the car goes a distance $\Delta x = 20$ m in a time of $\Delta t = 1$ s. (We frequently use Δx, read "delta x," and Δt, read "delta t," to represent very small distance and time intervals.) Our definition of average speed tells us that

$$\bar{u} = \frac{\text{distance gone}}{\text{time taken}} = \frac{\Delta x}{\Delta t} = \frac{20\text{ m}}{1\text{ s}} = 20\text{ m/s}$$

as it should.

The magnitude of the average velocity is given by

$$\bar{v} = \frac{\text{magnitude of displacement vector}}{\text{time taken}} = \frac{\Delta x}{\Delta t} = \frac{20\text{ m}}{1\text{ s}} = 20\text{ m/s}$$

As we see, the magnitude of the average velocity equals the average speed in this case. Moreover, the graph tells us that the car is moving with constant speed. Each little Δx, Δt triangle we draw similar to the ones shown will always give a speed of 20 m/s. *An object traveling with constant speed and velocity along a straight line will always give a straight-line x-versus-t graph.*

Before leaving the example of Fig. 2.2, we should notice that $\Delta x/\Delta t$ is the slope of the line shown.* In this case at least, the magnitudes of the speed

*To find the slope of a curve at a point on it, construct a tangent line at the point. The slope is the tangent of the angle this line makes with the horizontal axis. Or, colloquially, it is the rise divided by the run of the tangent line.

and velocity are equal to the slope of the x-versus-t graph. We shall soon see that this is always true.

Illustration 2.2 Let us now refer to Fig. 2.3, which shows the motion of the car on the same straight road but with nonconstant speed. Describe the car's motion.

Reasoning The graph tells us that the car is standing still at points A and C. Near these points on the graph, x is not changing with t. This must mean the car is not moving. However, near B the car is moving toward larger x's since x is increasing with time. Near B, the situation is much like that shown in Fig. 2.2. In fact, the average speed near B is

$$\bar{u} = \frac{\Delta x}{\Delta t} = \frac{50 \text{ m}}{20 \text{ s}} = 2.5 \text{ m/s}$$

The car's average *velocity* near B is 2.5 m/s in the $+x$ *direction*. (Remember, velocity is a vector.)

At point D, the situation is a little different. Near there, the car is moving towards smaller x values; the value of x is decreasing with time. In other words, the car is going in the reverse direction to that shown in Fig. 2.2*a*. Since x is becoming smaller in the region near D, the quantity Δx is negative since it is a *decrease* in x. The average speed of the car near D is

$$\bar{u} = \frac{\text{distance gone}}{\text{time taken}} = \frac{|\Delta x|}{\Delta t} = \frac{100 \text{ m}}{20 \text{ s}} = 5.0 \text{ m/s}$$

Notice that speed is not concerned with direction. Therefore we use only the magnitude of the distance Δx, which we represent by $|\Delta x|$. The negative sign shows direction and is dropped when computing speed.

But to compute the average velocity near D, we make use of the negative sign on Δx. We interpret the minus sign as a symbol which gives direction.

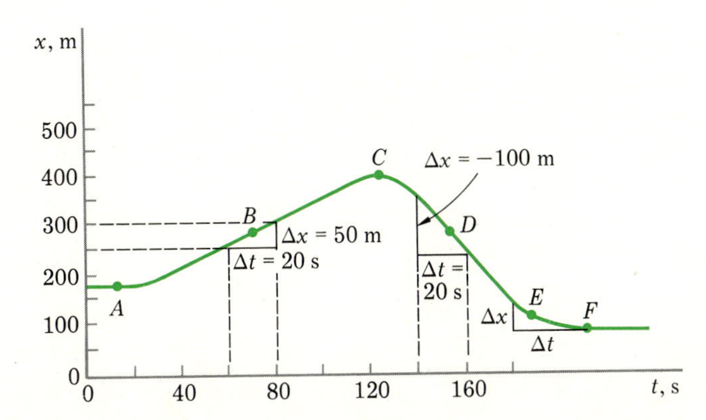

FIGURE 2.3

Can you show from the graph that the object is at rest at points A and C? That it is moving in opposite directions at points B and D?

Then

$$\text{Average velocity at } D = \frac{\Delta x}{\Delta t} = \frac{-100 \text{ m}}{20 \text{ s}} = -5.0 \text{ m/s}$$

The negative sign tells us the velocity is in the $-x$ direction.

A still different situation prevails at point E. There, the car is slowing down and is preparing to stop at F. If we now compute the average speed near E from the triangle shown, our result may not be the exact speed the car had at E. The computed value is only an average of the speeds during the interval for which this triangle is taken. However, if we take a very small time interval Δt centered on E, our result will be very close to the speed the car had at E.

The speed the car had as it passed point E in Fig. 2.3 was the car's speed for only an instant. Before the car reached E, it was moving faster than at E. Later it was moving slower than at E. We define the speed an object has at a certain instant to be the object's **instantaneous speed.** To find the car's instantaneous speed at point E, we must measure how long Δt it takes the car to pass through a very small distance Δx centered on point E. We can represent this statement in mathematical form as

DEFINITION

$$\text{Instantaneous speed} \equiv u = \lim_{\Delta t \to 0} \frac{\Delta x}{\Delta t}$$

The notation $\lim_{\Delta t \to 0}$ means to evaluate $\Delta x / \Delta t$ when Δt is taken small enough for its value to be close to zero. Clearly, the car cannot possibly change its speed much in this tiny time. Therefore, the speed we measure will be very close to the speed at point E. It is the instantaneous speed at point E.

Since an object cannot change its direction much during an extremely short time interval, it follows that the measured displacement s of the object during this interval will be very nearly the same as the distance moved d during the interval. Hence the magnitude of the velocity computed for this tiny interval will be very close to the instantaneous speed. We therefore define an instantaneous velocity for an object in the following way. *Suppose in a very short time interval* Δt *the object undergoes a vector displacement* $\Delta \mathbf{s}$. *Then the* **instantaneous velocity** *of the object is defined to be*

DEFINITION

$$\mathbf{v} = \lim_{\Delta t \to 0} \frac{\Delta \mathbf{s}}{\Delta t} \tag{2.3}$$

where the notation $\lim_{\Delta t \to 0}$ *means to take the limiting value of* $\Delta \mathbf{s}/\Delta t$ *as* Δt *approaches zero magnitude.* As pointed out above, *the magnitude of the instantaneous velocity* $\mathbf{v}$ *is equal to the instantaneous speed* u.

In our particular example, $\Delta s = \Delta x$, and so $v = \lim_{\Delta t \to 0} \Delta x / \Delta t$. As we see, instantaneous velocity is given by the ratio of Δx to Δt in a tiny triangle like those shown in Fig. 2.3. But if the triangle is taken small enough, the value of $\Delta x/\Delta t$ is the slope of the graph's curve even near a point such as E. We conclude from this that *the instantaneous velocity in the x direction is equal to the slope of the x-versus-t curve at the instant in question.* Moreover, *the sign of the slope (+ or −) tells us whether the velocity is in the plus or in the minus x direction.*

2.3 CONVERSION OF UNITS

Frequently we are given quantities in one set of units and we need the quantity in another unit. We might, for example, know a speed in miles per hour and wish to find its equivalent in meters per second. Or, as another example, old property deeds often express property dimensions in rods. One then needs to know how to change this length in rods to feet or meters. To do this type of conversion, we make use of conversion factors. For example, it is known that

$$1 \text{ rod} = 16.50 \text{ ft}$$

To make use of this fact we divide each side of the equation by 1 rod to find

$$1 = \frac{16.50 \text{ ft}}{1 \text{ rod}}$$

This quantity is the **conversion factor** between rods and feet.

Notice that the conversion factor is equal to unity. When we multiply or divide any number by it, the number remains unchanged. We see the utility of this in the following illustration.

Illustration 2.3 Change 30 rods to the equivalent length in feet.

Reasoning We have

$$30 \text{ rods} = (30 \text{ rods})\left(\frac{16.50 \text{ ft}}{\text{rod}}\right)$$

This is true because the multiplying factor, 16.50 ft/rod, is equal to unity. But the unit rods can be canceled to give

$$30 \text{ rods} = (30 \,\cancel{\text{rods}})\left(\frac{16.50 \text{ ft}}{\cancel{\text{rod}}}\right) = 495 \text{ ft}$$

By use of the conversion factor, we have succeeded in finding that 30 rods is 495 ft.

We give a table of useful conversion factors on the inside front cover of this book. Let us now see a few typical uses of them.

Illustration 2.4

1 Change 36 in (inches) to centimeters.
Conversion factor:

$$2.54\ \text{cm/in}$$

Method:

$$36\ \text{in} = (36\ \cancel{\text{in}})\left(\frac{2.54\ \text{cm}}{\cancel{\text{in}}}\right) = 91.44\ \text{cm}$$

2 Change 270 cm to inches.
Method:

$$270\ \text{cm} = (270\ \text{cm})\left(\frac{1}{2.54\ \text{cm/in}}\right)$$

Notice that we divide by the conversion factor in this case so that the unit cm will cancel. Inverting the denominator and multiplying by it gives

$$270\ \text{cm} = (270\ \cancel{\text{cm}})\left(\frac{\text{in}}{2.54\ \cancel{\text{cm}}}\right) = 106.3\ \text{in}$$

3 Change 60 acres to hectares.
Conversion factors:

$$10{,}000\ \text{m}^2/\text{hectare} \qquad 0.305\ \text{m/ft} \qquad 43{,}560\ \text{ft}^2/\text{acre}$$

Method:

$$\begin{aligned} 60\ \text{acres} &= (60\ \cancel{\text{acres}})\left(\frac{43{,}560\ \text{ft}^2}{\cancel{\text{acre}}}\right) = 2{,}613{,}600\ \text{ft}^2 \\ &= (2{,}613{,}600\ \text{ft}^2)\left(\frac{0.305\ \text{m}}{\text{ft}}\right)\left(\frac{0.305\ \text{m}}{\text{ft}}\right) \\ &= 243{,}100\ \text{m}^2 \\ &= (243{,}100\ \text{m}^2)\left(\frac{1\ \text{hectare}}{10{,}000\ \text{m}^2}\right) = 24.3\ \text{hectares} \end{aligned}$$

We shall encounter many other uses of conversion factors as we proceed with our studies.

2.4 ACCELERATION

DEFINITION

An object whose velocity is changing is said to be **accelerating.** *Acceleration is defined as the change in velocity* (not speed!) *per unit time.* As an equation,

$$\text{Average acceleration} = \frac{\text{change in velocity}}{\text{time taken}}$$

or

$$\bar{a} = \frac{\mathbf{v}_f - \mathbf{v}_0}{t} \tag{2.4}$$

where $\mathbf{v}_0$ is the starting velocity and $\mathbf{v}_f$ is the final velocity a time t later. Since $\bar{\mathbf{a}}$ is the sum of two vectors divided by a scalar, t,

$$\bar{\mathbf{a}} = \frac{\mathbf{v}_f + (-\mathbf{v}_0)}{t}$$

$\bar{\mathbf{a}}$ must also be a vector. In this chapter we shall discuss motion in a straight line only, and so only the magnitude of $\bar{\mathbf{a}}$, represented as $\bar{a}$, will usually be considered. However, a plus or minus sign will be appended to indicate whether it is in the positive or negative direction along the line.

Illustration 2.5 Suppose that a car starts from rest and accelerates (speeds up) to a velocity of 20 ft/s along a straight-line path in 4 s, as shown by the black line in Fig. 2.4. Find its average acceleration.

Reasoning From the graph, $v_0 = 0$ and at $t = 4$ s, $v_f = 20$ ft/s. Substituting in Eq. (2.4) (with vector notation omitted, since all motion is in the same direction)

$$\bar{a} = \frac{20\ \text{ft/s} - 0}{4\ \text{s}} = \frac{20\ \text{ft/s}}{4\ \text{s}} = \frac{5\ \text{ft/s}}{\text{s}}$$

which is read "5 feet per second each second" or "5 feet per second per second." In other words, the velocity of the car is increasing at a rate of 5 ft/s each second, and this rate of increase in velocity is the quantity which we have called the acceleration. The above result for $\bar{a}$ is often written as 5 ft/s^2 and is read "5 feet per second squared."

In general, the units of $\bar{a}$ will be a distance unit divided by the product of two time units. The units could be miles per second squared, feet per second per hour, centimeters per second per minute, miles per hour per second, and so forth.* In any event, *the physical meaning of acceleration should be clear. It is the change in velocity in a unit of time* as shown in Fig. 2.4.

FIGURE 2.4

The black line shows the time variation of the velocity of an auto undergoing a uniform acceleration of 5 ft/s each second.

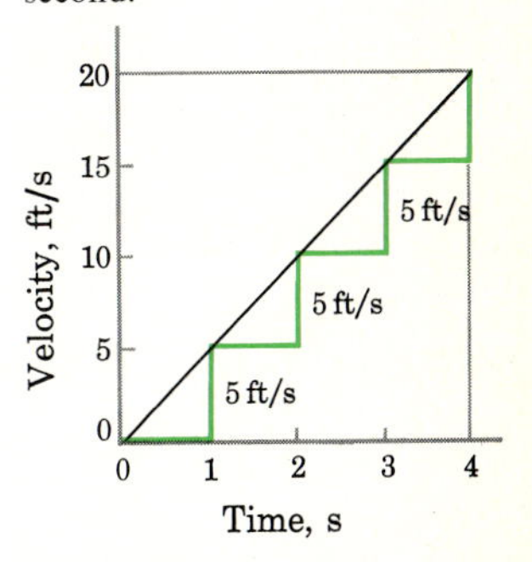

* When solving problems, one should usually not mix two time units in a problem.

2.5 UNIFORMLY ACCELERATED MOTION

Let us refer once again to Fig. 2.4. Notice that the velocity of the car represented by the graph increases 5 ft/s each second during the time covered by the graph. That is to say, the rate of increase of velocity, i.e., the acceleration, is uniform and is 5 ft/s^2. Although uniform or constant accelerations are achieved only in certain special cases, we shall see that these cases are of considerable importance. Since motion involving nonuniform acceleration usually becomes quite difficult to handle, we restrict the discussion in this chapter to instances of constant acceleration. Under these conditions, the average and instantaneous accelerations are the same, since the acceleration is constant. We shall therefore not distinguish between them in the remainder of this chapter.

DEFINITION

When the acceleration is constant, the speed and velocity of the moving body change uniformly with time. Because of this, *the* **average velocity** *is merely one-half the sum of the initial and final velocities; thus*

$$\bar{v} = \frac{v_f + v_0}{2} \tag{2.5}$$

where v_f is the final velocity and v_0 is the initial velocity. In the special case when the acceleration is constant and equal to zero, the velocity of the object is not changing, and so $v_f = v_0$. In this case Eq. (2.5) leads to the result that $\bar{v} = \frac{1}{2}(v_0 + v_0) = v_0$, as we would expect.

We now have three equations which apply to uniformly accelerated motion, Eqs. (2.2), (2.4), and (2.5). They are sufficient to solve any ordinary situation involving uniformly accelerated motion. Let us examine a typical case.

Illustration 2.6 Suppose that a car starts from rest and accelerates uniformly to a speed of 5.0 m/s in 10 s. Find its acceleration and the distance it traveled in this time.

Reasoning To begin the solution of this problem, let us first write down what is known and what is unknown.

$$v_0 = 0 \qquad v_f = 5.0 \text{ m/s} \qquad t = 10 \text{ s}$$
$$a = ? \qquad s = ?$$

From Eq. (2.4) we have at once

$$a = \frac{v_f - v_0}{t} = \frac{5.0 - 0}{10} \frac{\text{m/s}}{\text{s}} = 0.50 \text{ m/s}^2$$

From Eq. (2.5) we see that

$$\bar{v} = \frac{5.0 + 0}{2} = 2.5 \text{ m/s}$$

and finally, from Eq. (2.2),

$$s = \bar{v}t = (2.5 \text{ m/}\cancel{s})(10 \cancel{s}) = 25 \text{ m}$$

Illustration 2.7 As another example, suppose that a car traveling at 5.0 m/s is brought to rest in a distance of 20 m. Find its deceleration and the time taken to stop.

Reasoning Following the same procedure as before, we write down the knowns and unknowns.

$$v_0 = 5.0 \text{ m/s} \qquad v_f = 0 \qquad s = 20 \text{ m}$$
$$a = ? \qquad t = ?$$

We can first find the average velocity.

$$\bar{v} = \tfrac{1}{2}(v_0 + v_f) = 2.5 \text{ m/s}$$

Knowing this and the distance,

$$s = \bar{v}t \qquad \text{or} \qquad t = \frac{s}{\bar{v}}$$

we find

$$t = \frac{20 \text{ m}}{2.5 \text{ m/s}} = 8 \text{ s}$$

The acceleration is now obtained from Eq. (2.4).

$$a = \frac{0 - 5.0 \text{ m/s}}{8 \text{ s}} = -0.625 \text{ m/s}^2$$

The negative sign on the answer signifies that the car's velocity is decreasing rather than increasing. That is to say, the acceleration is in the negative direction along the line of motion.

These two illustrations show clearly that it is advantageous to follow a systemized approach to these problems. First, write down what is known and what is unknown. Then use the appropriate formula, Eq. (2.2), (2.4), or (2.5), to find the unknowns. There are some cases in which this is not as easily done as in the previous two illustrations. A case in point would be the following.

Illustration 2.8 A car starts from rest and accelerates at 4.0 m/s² through a distance of 20 m. How fast is it then going? How long did it take?

Reasoning Writing down the knowns and unknowns,

$$v_0 = 0 \qquad a = 4.0\ \text{m/s}^2 \qquad s = 20\ \text{m}$$
$$v_f = ? \qquad t = ?$$

Our three equations are

$$s = \bar{v}t \tag{2.2}$$
$$\bar{v} = \tfrac{1}{2}(v_0 + v_f) \tag{2.5}$$
$$a = \frac{v_f - v_0}{t} \tag{2.4}$$

None of these equations can be used directly to obtain one of the unknowns. Each equation contains at least two unknowns, and we would have to solve simultaneous equations to obtain any unknowns. In the next section, we will find how to avoid this computation.

2.6 TWO DERIVED EQUATIONS FOR UNIFORMLY ACCELERATED MOTION

The previous example can be solved easily if we first obtain two companion equations to the three given in (2.2), (2.4), and (2.5). These extra two equations are obtained by solving the three known equations simultaneously. Having once done this, we shall not have to repeat the process again but shall simply make use of the results found.

If we substitute Eq. (2.5) in Eq. (2.2), we find

$$s = \tfrac{1}{2}(v_0 + v_f)t \tag{2.6}$$

Now replace t by its value from Eq. (2.4) to give

$$s = \tfrac{1}{2}(v_f + v_0)\left(\frac{v_f - v_0}{a}\right) \tag{2.7}$$

After clearing fractions and rearranging, we have

$$v_f^2 - v_0^2 = 2as \tag{2.8}$$

(Notice, in passing, that this is just the equation we needed to solve the previous example.)

The other equation we shall find useful is found by making a different substitution in Eq. (2.6). Replace v_f in that equation by its value obtained from Eq. (2.4). Thus

$$s = \tfrac{1}{2}v_0 t + \tfrac{1}{2}(v_0 + at)t$$

which simplifies to

$$s = v_0 t + \tfrac{1}{2}at^2 \tag{2.9}$$

Equations of Uniform Acceleration

We now have five equations available to use in the solution of problems involving **uniformly accelerated motion.** *They are*

$$s = \bar{v}t \qquad (2.10a)$$
$$\bar{v} = \tfrac{1}{2}(v_f + v_0) \qquad (2.10b)$$
$$v_f = v_0 + at \qquad (2.10c)$$
$$v_f^2 = v_0^2 + 2as \qquad (2.10d)$$
$$s = v_0 t + \tfrac{1}{2}at^2 \qquad (2.10e)$$

Returning now to the problem posed at the end of the last section, we had there that

$$v_0 = 0 \qquad a = 4.0\ \text{m/s}^2 \qquad s = 20\ \text{m}$$
$$v_f = ? \qquad t = ?$$

Using Eq. (2.10*d*), we have at once

$$v_f^2 = 0 + (2)(4.0\ \text{m/s}^2)(20\ \text{m}) = 160\ \text{m}^2/\text{s}^2$$
$$v_f = \sqrt{160\ \text{m}^2/\text{s}^2} = \sqrt{160}\ \text{m/s} \approx 12.6\ \text{m/s}$$

Now we can use Eq. (2.10*c*) to find t.

$$(4.0\ \text{m/s}^2)t = 12.6\ \text{m/s} - 0$$
$$t = 3.15\ \text{s}$$

Illustration 2.9 Find the time taken for a car to travel 98 ft if it starts from rest and accelerates at 4.0 ft/s².

Reasoning Known:

$$v_0 = 0 \qquad a = 4.0\ \text{ft/s}^2 \qquad s = 98\ \text{ft}$$
$$t = ?$$

The appropriate equation is

$$s = v_0 t + \tfrac{1}{2}at^2$$
$$98 = 0 + 2t^2$$
$$t = \sqrt{49} = 7 \text{ s}$$

The units have been omitted from these equations. You should insert them and check to see that the units of t are as stated.

Illustration 2.10 A car is moving at 60 km/h when it begins to slow down with a deceleration of 1.50 m/s^2. How long does it take to travel 70 m as it slows down?

Reasoning Known:

$$v_0 = (60 \text{ km/h})\left(\frac{1 \text{ h}}{3600 \text{ s}}\right)\left(\frac{1000 \text{ m}}{1 \text{ km}}\right) = 16.7 \text{ m/s}$$
$$s = 70 \text{ m} \qquad a = -1.50 \text{ m/s}^2 \qquad t = ?$$

Notice that we have changed the units of velocity to meters per second so that we shall not be using two sets of distance and time units in the same problem. One must always make this kind of change.

If we now choose to solve this problem by using Eq. (2.10*e*), we run into trouble with algebra:

$$s = v_0 t + \tfrac{1}{2}at^2$$
$$70 = 16.7t + \tfrac{1}{2}(-1.50)t^2 = 16.7t - 0.75t^2$$

where once again the student should supply the units. The solution of this quadratic equation can be found by using the quadratic formula. However, it is usually simpler to find one of the other unknowns first so as to avoid the complicated algebra. For example, using Eq. (2.10*d*), we have

$$v_f^2 = v_0^2 + 2as = (16.7 \text{ m/s})^2 + (2)(-1.50 \text{ m/s}^2)(70 \text{ m})$$
$$= 279 \text{ m}^2/\text{s}^2 - 210 \text{ m}^2/\text{s}^2$$

from which

$$v_f = 8.3 \text{ m/s}$$

Now, using Eq. (2.10*c*),

$$v_f = v_0 + at$$
$$8.3 = 16.7 - 1.50t$$

Supply the proper units in this equation, and show that

$$t = 5.6 \text{ s}$$

2.7 A NOTE ABOUT EQUATIONS

Usually in physics we do not encourage "plugging" values into memorized equations. The reason for this is simple. It is much easier for you to memorize a very few equations and, knowing what they mean, use them to solve all sorts of problems. There is no need to memorize a different equation for each situation. So far in this book we have asked that you memorize only the defining equations for the sine, cosine, and tangent.

In the present chapter on Uniformly Accelerated Motion, we might well ask you to memorize Eqs. (2.10*b*) and (2.10*c*) since they define average velocity and acceleration. Equation (2.10*a*) is a fundamental equation which we already knew in a qualitative way. Whether or not we should memorize Eqs. (2.10*d*) and (2.10*e*) will depend on several factors.

They are, of course, nothing new, since they are merely Eqs. (2.10*a*), (2.10*b*), and (2.10*c*) combined in various ways. However, the algebra involved in formulating them is not negligible. We saw in some of the examples that we must obtain these equations to solve certain motion problems. The question therefore reduces to the following: Shall we be using these two equations enough for it to be advisable to learn them? I think you will use all five of Eqs. (2.10) frequently enough to find it to your advantage to memorize them.

It should be remembered, though, *that any equation worth memorizing is certainly worth understanding*. One must know the exact meaning of each of the symbols in these equations. In addition, Eqs. (2.10*a*), (2.10*b*), and (2.10*c*) summarize easily understood physical meanings and should not be looked upon as a mere collection of algebraic expressions. We also have pointed out that all except the first of Eqs. (2.10) apply only to situations where a is constant.

Sometimes it is worthwhile to examine the limiting forms of equations. Let us see what Eqs. (2.10) say when the acceleration is zero. First, look at Eq. (2.10*c*). It says that when $a = 0$, $v_f = v_0$. This is certainly true. If the acceleration is zero, the velocity is constant. The final and initial velocities are the same.

Under the same condition, Eq. (2.10*b*) says that

$$\bar{v} = \tfrac{1}{2}(v_f + v_0) = v_f = v_0$$

since v_f and v_0 are the same under these conditions. Of course this relation is true, since if the acceleration is zero, the velocity is constant and the average velocity is the same as the initial and final velocities. Similarly, Eqs. (2.10*d*) and (2.10*e*) reduce to obviously true statements when $a = 0$.

2.8 GALILEO'S DISCOVERY

Those who know the history of science are not surprised if a child gives the wrong answer to the question: Which will fall faster, a heavy ball or a light one? It does seem reasonable, doesn't it, that the heavier ball will fall faster than the light ball? Most children will give that answer.

In fact, even the great genius Aristotle (384–322 B.C.) gave this same answer. And he was not alone. It was commonly thought, until about A.D. 1600, that heavier bodies *do* fall more swiftly than light ones. After all, it is obviously true that a feather does not fall through the air nearly so fast as a stone. Aristotle himself showed by involved philosophical arguments that this should be true for all light bodies as compared with heavier ones. (In Aristotle's defense, however, we should state that he was considering the way the body falls in air, not vacuum.)

Natural scientists are usually not completely satisfied with philosophic proofs of statements which can be verified experimentally. It would be wonderful if our philosophic methods were so good that we need never doubt the conclusions to which they lead us. Then nearly all the time, effort, and money now devoted to expensive research could be diverted to other uses, and the problems of science could be solved by a few philosophers maintained in sumptuous solitude in velvet-carpeted ivory towers.

Galileo (1564–1642) had the temerity to question whether Aristotle had made a mistake in his philosophical arguments. It seems incongruous to us today that for nineteen centuries no one had troubled to carry out the simple experimental tests needed to answer this question. However, it remained for Galileo to measure the speed with which objects fall.* He recognized that situations such as the feather-stone problem were not a real test, because of the pronounced effect of the rush of air past the feather as it fell. Hence, he experimented with relatively heavy objects, all of the same size but differing in weight. His results are, of course, well known. He concluded that if one neglects air-friction effects, all bodies fall to earth in the same way, independent of their weight.

Galileo is often called the father of modern science, since he was instrumental in awakening the world to the fact that experimentation, when possible, is the most valid way of learning the facts of nature. Following in his footsteps, the great scientists Newton and Faraday, and many others, opened up a whole new area of knowledge with their experimental observations of nature's ways.

Even today, we sometimes find scientists following the path of Aristotle rather than Galileo. We still find, on occasion, that experiments disprove ideas which we supposed to be well founded. Sometimes experimental proofs are not undertaken because they involve methods beyond our capabilities. But in any event scientists must never forget that even the best minds make philosophic errors.

---

* Galileo actually increased the time of fall by rolling spheres down an incline. By making the incline steeper and observing the effect of this action, he was able to draw quantitative conclusions about the behavior of freely falling bodies.

There is another side to the coin, though. In many fields, experimental proofs are not possible for one reason or another. We have no alternative in such cases but to rely upon the conclusions of our most able philosophers. Even where experimental test is possible, it is sometimes necessary for scientists to postpone the test because of the pressure of other pursuits or prohibitive costs. We shall see in later chapters that some of the greatest scientific discoveries and theories were based upon not only unverified, but wrong, premises. The mere fact that the discoveries were made justifies the approach the scientist used. He would certainly never have been able to prove the starting premise; so if he had wasted many years on that proof, the discovery might well have been delayed a century. Of course, we tend to remember those who were lucky in such an approach. Those who failed because of wrong and unproved premises are soon forgotten, even though their number is certainly very large.

2.9 FREE FALL AND THE ACCELERATION DUE TO GRAVITY

Many experimenters, using highly refined techniques, have duplicated Galileo's experiments. It is not unlikely that in your laboratory work you will perform an experiment to check his results. We now recognize as undisputed fact that *a freely falling object will accelerate downward with essentially constant acceleration. Although the exact value of this acceleration* resulting from the pull of gravity *varies from place to place on the earth, it is close to 9.8* m/s^2, which is the same as 980 cm/s^2 or 32.2 ft/s^2. Table 2.1 shows some typical values. These results, of course, assume that proper care was taken to minimize the effects of the flow of air past the falling body.

Motion of a free body in an up and down direction will be uniformly accelerated motion; the acceleration is approximately* 9.8 m/s^2. Hence, in any such motion, Eqs. (2.10) apply, and a is known to be 9.8 m/s^2 or 32 ft/s^2

TABLE 2.1
ACCELERATION DUE TO GRAVITY g

PLACE	ELEVATION, m	g, m/s^2	g, ft/s^2
Beaufort, N.C.	1	9.7973	32.143
New Orleans	2	9.7932	32.130
Galveston	3	9.7927	32.128
Seattle	58	9.8073	32.176
San Francisco	114	9.7997	32.151
St. Louis	154	9.8000	32.152
Cleveland	210	9.8024	32.160
Denver	1638	9.7961	32.139
Pikes Peak	4293	9.7895	32.118

*For problem solutions in this text we shall use the approximate values 32 ft/s^2 and 9.8 m/s^2 for g.

or 980 cm/s^2. Only one extra precaution must be taken in the use of these equations. It has to do with the vector nature of acceleration. *The acceleration due to gravity is always down. If we choose to call down the positive direction, a is a positive number,* 9.8 m/s^2. In any case where we choose the upward direction as positive, however, the pull of gravity will actually decrease the upward velocity of the body, and so $a = -9.8\ m/s^2$. *For problems which involve both upward and downward motion,* such as a ball rising and then coming down again, *it is absolutely necessary at the outset to choose either up or down as the positive direction. This choice is arbitrary,* but once we have made it in a particular problem, we must retain that choice through the whole problem.

Illustration 2.11 A boy drops a stone from a bridge. If it takes 3.0 s for the stone to hit the water beneath the bridge, how high above the water is the bridge? Ignore air friction. (Notice here that our problem ends the instant *before* the stone hits the water. It is only during this interval that the stone is a freely falling body.)

Reasoning Known:

$$a = 9.8\ m/s^2 \qquad \text{down positive} \qquad s = ?$$
$$t = 3.0\ s \qquad v_0 = 0$$

Use

$$s = v_0 t + \tfrac{1}{2}at^2$$
$$= 0 + \tfrac{1}{2}(9.8\ m/s^2)(9\ s^2) = 44\ m$$

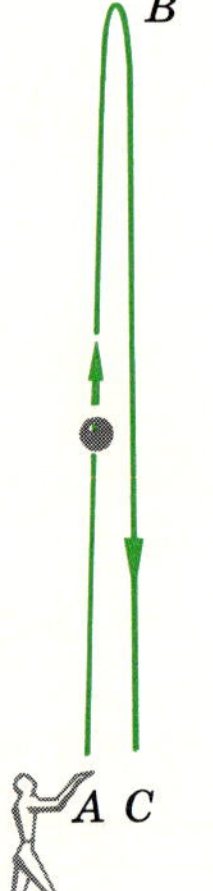

FIGURE 2.5
The ball is thrown upward at point *A* with a speed of 15 m/s. Since the ball stops at point *B*, its upward speed there is zero.

Illustration 2.12 A girl throws a ball upward with a speed of 15 m/s. How high does it go? What is its speed just before the girl catches it again? How long was it in the air? Neglect the effects of air friction.

Reasoning Method 1: It is important that we define the "trip" we are interested in. Consider the motion from *A* to *B* in Fig. 2.5. Known:

$$a = -9.8\ m/s^2 \qquad \text{up positive} \qquad s = ?$$
$$v_0 = 15\ m/s \qquad v_f = 0$$

Use

$$2as = v_f^2 - v_0^2$$
$$-19.6s = 0 - 225$$

Supply the units in this equation, and show that

$$s = 11.5\ m \qquad \text{from } A \text{ to } B$$

Notice at the top of the path, at *B*, that the upward speed of the ball has decreased to zero. Since it has not yet started to fall downward again, its speed is zero at point B.

For the second part of the problem, we shall be concerned with the trip from B to C. This is a new trip, and so we shall have new knowns. Known:

$$a = +9.8\ \text{m/s}^2 \qquad \text{down positive} \qquad v_f = ?$$
$$s = 11.5\ \text{m}$$
$$v_0 = 0$$

Use

$$2as = v_f^2 - v_0^2$$
$$(19.6\ \text{m/s}^2)(11.5\ \text{m}) = v_f^2 - 0$$
$$v_f = \sqrt{225} = 15\ \text{m/s} \qquad \text{downward}$$

Notice that the final speed is the same as the speed with which the ball was thrown. This is an example of a famous basic law of physics, called the **law of conservation of energy,** which we shall examine in considerable detail in a later chapter.

Method 2: There is yet another way in which we could have solved this latter problem. Consider the full trip from A to B to C. The starting point is A, and the end point is C. Take up as positive. Known:

$$a = -9.8\ \text{m/s}^2 \qquad v_0 = 15\ \text{m/s} \qquad s = 0$$
$$v_f = ? \qquad t = ?$$

We write that s, the vector displacement from the starting point A to the end point C, is zero, since the ball returned to the thrower. Use

$$2as = v_f^2 - v_0^2$$
$$0 = v_f^2 - (15\ \text{m/s})^2$$
$$v_f = \pm 15\ \text{m/s} \qquad \text{as before}$$

Notice that when the square root of a number is taken, the answer can be either plus or minus. For our trip, the minus sign is correct. Before investigating the reason for the choice of two answers, plus or minus, let us compute the time for the trip. Use

$$s = v_0 t + \tfrac{1}{2}at^2$$
$$0 = 15t - 4.9t^2$$

Supply the units to this equation, and show that

$$t = 0 \qquad t = 3.1\ \text{s}$$

Here again we have a choice of answers.

The reason for the choice of two answers in each case is not hard to find. Our knowns for this method of solution are actually true for two trips, the one from A to B to C and also the trip from A which ends just after its start. For this latter trip, all the known conditions are true, and the time taken for it is zero. In addition, the end velocity is still the same as the starting velocity.

Illustration 2.13 A ball is thrown upward with a speed of 40 ft/s, as shown in Fig. 2.6. How long will it take to reach a point 20 ft above the ground on its way down again? Neglect the effects of the surrounding air.

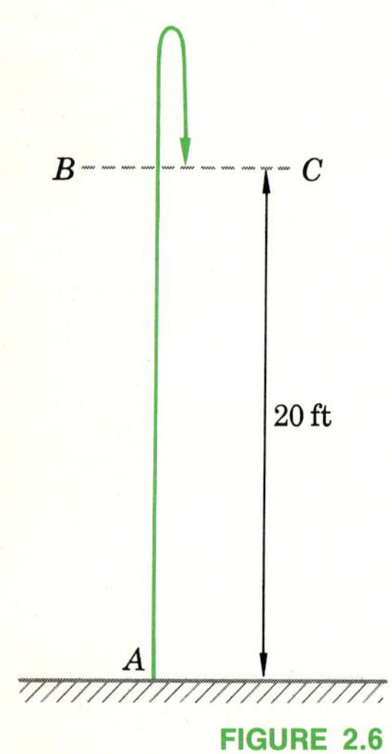

FIGURE 2.6
The ball is thrown upward from point A with a speed of 40 ft/s. How long does it take for it to reach point C?

Reasoning Known:

$$a = -32 \text{ ft/s}^2 \qquad \text{up positive} \qquad t = ?$$
$$s = 20 \text{ ft} \qquad v_0 = 40 \text{ ft/s}$$

If we used $s = v_0 t + \frac{1}{2}at^2$, we would need the quadratic formula to find t. Instead, first find v_f using

$$\begin{aligned} 2as &= v_f^2 - v_0^2 \\ (-64 \text{ ft/s}^2)(20 \text{ ft}) &= v_f^2 - 1600 \text{ ft}^2/\text{s}^2 \\ v_f^2 &= 320 \text{ ft}^2/\text{s}^2 \\ v_f &= \pm 17.9 \text{ ft/s} \end{aligned}$$

We choose the negative sign, since we are interested in the ball on its way down at C. The positive sign represents the ball on its way up at point B. Now use

$$\begin{aligned} v_f &= v_0 + at \\ -17.9 \text{ ft/s} &= 40 \text{ ft/s} - (32 \text{ ft/s}^2)t \\ 32t &= 57.9 \end{aligned}$$

Supply the units to this equation, and show that

$$t = 1.81 \text{ s}$$

Illustration 2.14 How fast must a ball be thrown straight upward if it is to return to the thrower in 3.0 s? Neglect the effects of the air on the ball.

Reasoning Let us use the metric system in this problem. We shall take up as positive, and we notice that the displacement vector from beginning to end point has zero length. Then

$$a = -9.8 \text{ m/s}^2$$
$$t = 3.0 \text{ s} \qquad s = 0 \qquad v_0 = ?$$

We can use

$$s = v_0 t + \tfrac{1}{2}at^2$$

to give

$$0 = (v_0)(3 \text{ s}) - (4.9 \text{ m/s}^2)(9 \text{ s}^2)$$

We then find

$$v_0 = 14.7 \text{ m/s}$$

SUMMARY

The average speed $\bar{u}$ of an object is the total distance traveled divided by the time taken for the trip. Average velocity $\bar{\mathbf{v}}$ is the displacement vector from beginning to end of a trip divided by the time taken. Speed is a scalar, and velocity is a vector. They both have the units of distance divided by time.

When an object moves along a straight line, its motion can be shown by an x-versus-t graph. The slope of the graph line at any point gives the velocity and speed at the point. Instantaneous velocity is given by

$$v = \lim_{\Delta t \to 0} \frac{\Delta x}{\Delta t}$$

In order to change the units of a quantity, use is made of conversion factors. Each conversion factor is unity, and so multiplication by it does not change the value of a quantity. Units are carried through computations just like algebraic symbols.

If an object changes its velocity from $\mathbf{v}_0$ to $\mathbf{v}_f$ in a time t, its average acceleration is given by

$$\bar{\mathbf{a}} = \frac{\text{change in velocity}}{\text{time taken}} = \frac{\mathbf{v}_f - \mathbf{v}_0}{t}$$

Acceleration is a vector.

If an object moves along a straight line and its acceleration is constant, the average velocity of the object is given by

$$\bar{v} = \tfrac{1}{2}(v_f + v_0)$$

By use of it and the definitions of average velocity

$$\bar{v} = \frac{s}{t}$$

and average acceleration

$$\bar{a} = a = \frac{v_f - v_0}{t}$$

we can derive two more useful equations.

$$v_f^2 = v_0^2 + 2as \qquad \text{and} \qquad s = v_0 t + \tfrac{1}{2}at^2$$

These are the five motion equations.

The earth pulls downward on each object with a force we call the object's weight. An object which is subject only to this one force is said to be freely falling. Freely falling objects experience an acceleration g toward the earth's center. The value of g, the acceleration due to gravity, is 9.8 m/s^2, which is the same as 32.2 ft/s^2.

MINIMUM LEARNING GOALS

Upon completion of this chapter you should be able to do the following:

1. Define average speed and average velocity. Give examples which show that sometimes the magnitudes of speed and velocity are not the same.

2. Use a graph of x versus t to describe the behavior of an object. Interpret the slope of the graph in terms of instantaneous velocity and speed. Distinguish between average and instantaneous velocity.

3. Change a quantity from one set of units to another when conversion relations between the units are given.

4. Define average acceleration in words and by equation. Give several possible sets of units for it.

5. Draw a graph relating v to t for an object which is undergoing a given constant (uniform) acceleration.

6. Write down the five uniform-motion equations. Define each symbol in them. Give the restrictions on the use of each equation.

7. Find one or more unknowns for an object undergoing uniform acceleration when sufficient data are given.

8. Define g, the acceleration due to gravity, and give its approximate value on earth.

9. Find one or more unknowns for an object undergoing free-fall vertical motion on the earth when sufficient data are given.

IMPORTANT TERMS AND PHRASES

You should be able to define or explain each of the following:

Average speed
Average velocity
Distance and displacement are different
Average acceleration
Instantaneous speed and velocity
Velocity in terms of slope
Free-fall acceleration g
Five uniform-motion equations
A positive direction must be chosen
Conversion factor

QUESTIONS AND GUESSTIMATES

1. Give an example of a case where the velocity of an object is zero but its acceleration is not zero.
2. Can the velocity of an object ever be in a direction other than the direction of acceleration of the object? Explain.
3. Sketch graphs of the velocity and the acceleration as a function of time for a car as it strikes a telephone pole. Repeat for a billiard ball in a head-on collision with the edge of the billiard table.
4. Consider a small rocket which is shot straight upward. Assume it to start from rest and accelerate uniformly until its fuel is exhausted. Sketch graphs showing the velocity and the acceleration as a function of time for the rocket as it rises and then falls to the earth.
5. A rabbit enters the end of a drainpipe of length L. Its motion from that instant on is shown in the graph of Fig. P2.1. Describe the rabbit's motion in words.
6. A stone is thrown straight up in the air. It rises to a height h and then returns to the thrower. For the time that the stone is in the air, sketch the following graphs: y versus t; v versus t; a versus t.
7. A high school girl who is an average runner completes a 100-yd dash by twice completing an indoor circular track which is 50 yd in circumference. Estimate her average speed. Repeat for her average velocity. (E)
8. Do all objects on the earth fall with the same acceleration? Explain.
9. A woman drops her car keys in an elevator in which she is standing. The elevator cable has just broken and the elevator is falling down the shaft in free fall. What happens to the keys?

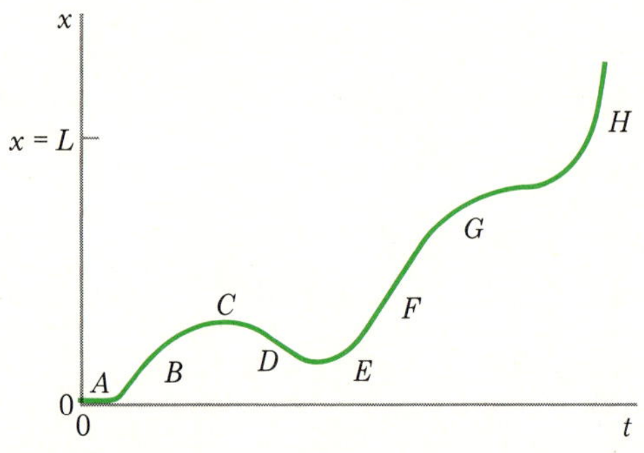

FIGURE P2.1

10. The driver of a car sees a child dart out into the road ahead. The car is going 20 m/s. About how far will the car go before the driver applies the brakes? (E) (We shall see in Chap. 4 how far the car will skid after the brakes are applied.)
11. The acceleration due to gravity on the moon is about 1.6 m/s^2. Estimate how high you could throw a baseball straight up on the moon. (Assume the space suit you would be wearing does not hinder you.) (E)
12. About what is the maximum average acceleration capability of a standard car in the range 0 to 40 mi/h? (E)
13. Suppose you are running at a speed of about 5 m/s on a flat lawn. Estimate the least distance in which you could stop without hitting something. Using your estimate, about how large would be your deceleration, i.e., negative acceleration? (E)

PROBLEMS

Unless otherwise stated, assume uniform acceleration. Use $g = 9.8\ \text{m/s}^2 = 32\ \text{ft/s}^2$. Neglect air friction.

1. In 1904, Henry Ford set the world speed record for the mile. He drove the mile in a Ford "999" and took a time of 39.4 s. What was his average speed in miles per hour and in kilometers per hour?

2. In the 1932 Olympic Games, the 3000-m steeplechase was won by Volmari Iso-Hollo of Finland. He ran 3450 m in 10 min 33.4 s. (They ran an extra lap since someone made a counting error.) What was his average speed in meters per second? His average velocity?

3. The earth's crust is still adjusting because of the loss of ice since the last Ice Age. The Lake Superior region is still rising at about 16 ft each 1000 yr. Find its speed due to this cause in centimeters per year. Find its average velocity in centimeters per century.

4. According to the theory of continental drift, South America was once attached to Africa. They have been drifting apart for about 200 million years. Their present separation is about 4000 mi. Assuming these numbers to be exact, what was the average rate of separation of the continents in centimeters per year?

5. Refer to Fig. P2.2. It shows the motion of an object along a straight line. Find the average velocity of the object during the following intervals: (*a*) *A* to *E*; (*b*) *B* to *E*; (*c*) *C* to *E*; (*d*) *D* to *E*; (*e*) *C* to *D*.

6. A can run at a top speed of 5.0 m/s while B can run at a speed of only 3.0 m/s. They are to race 200 m. To make the race more even, A is required to start *t* seconds later than B. How large should *t* be if they are to end in a tie?

7.* For the situation in Prob. 6, the handicap is to be made this way. B is to be given a head-start distance *s* while A must run the full 200 m. Both start at the same time. How large should *s* be if the race is to end in a tie?

8. Refer to Fig. P2.2 for the straight-line motion of an object. Find the instantaneous velocity of the object (*a*) at point *F*, (*b*) at point *B*, (*c*) at point *E*.

9. A car manufacturer claims that his car, starting from rest, can attain a speed of 28 m/s in a time of 20 s. Find the average acceleration of the car and the distance it travels in this time.

FIGURE P2.2

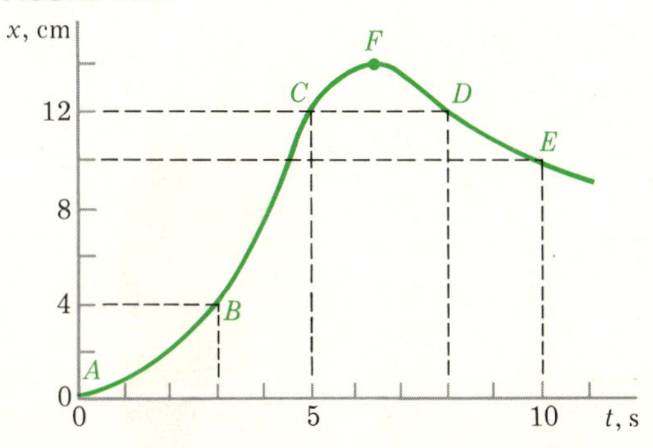

10. A rough value for the deceleration of a skidding auto is about 7.0 m/s^2. Using this value, how long does it take for a car going 30 m/s to stop after the skid starts? How far does it go in this time?

11. A bullet moving with speed 150 m/s strikes a tree and penetrates 3.5 cm before stopping. Find its deceleration and the time taken to stop.

12. In a TV tube, electrons shoot down the tube and strike the fluorescent material on the screen at the end of the tube. Their impact with the material causes light to be given off, thereby producing the picture we see. The electrons in the tube are accelerated from rest to speeds of about 200 million meters per second in a distance of about 1.50 cm. (We can write this speed as 2×10^8 m/s using the notation explained in Appendix 2.) What is the acceleration of an electron during the acceleration process? How long does the acceleration take?

13.* A heavily loaded train blocks a crossing. Having nothing better to do, a waiting motorist notices that it takes 20 s for a railway car to pass completely through the crossing just as the train starts from rest and begins to move. Find the acceleration of the train in terms of the length *L* of the railway car. Assuming the acceleration to remain the same, how long after the train starts will it be before the following 60 railway cars have passed?

14. A truck initially traveling at 50 ft/s decelerates at 4 ft/s^2. Find (*a*) how long it takes to stop, (*b*) how far it moves in that time, and (*c*) how far the truck moves in the third second after the brakes are applied.

15. A stone is thrown straight up with an initial speed of 80 ft/s. How high does the stone go, and how long does it take to reach its highest point?

16. A boy throws a ball straight down from the top of a 50-ft building with a speed of 20 ft/s. How long does it take for the ball to reach the ground, and how fast is it going just before it hits?

17. A bullet shot straight upward from a gun is found to rise to a height of 2 km. What is the least possible speed with which it could have left the gun? (In practice, air friction is important.)

18. A stone is thrown straight upward from the ground and goes as high as a nearby building. The stone returns to the ground 3.0 s after it was thrown. How high (in meters) is the building?

19. A girl is standing on the top edge of a 18-m-high building. She tosses a coin upward with a speed of 7.0 m/s. How long does it take for the coin to hit the ground 18 m below? How fast is the coin going just before it strikes the ground?

20.* A physics student who always uses a calculator to evaluate 2×2 comes up with the following scheme to measure the height of a building. A timing mechanism is set up. It measures the time taken for an object dropped from the top of the building to fall the last 2.0 m before it hits the ground. By experiment, the object dropped from the building top is found to take 0.150 s to move this last 2.0 m. How high is the building?

21. A truck traveling 15 m/s due west is on a collision course with a car going due east at 30 m/s. They are 400 m apart. How long does it take them to hit? How far are they from the original position of the truck when they finally collide? Assume their speeds to remain constant.

22.** A car is traveling at 60 mi/h along a road parallel to a railroad track. How long does it take the car to pass a $\frac{1}{2}$-mi-long train traveling at 40 mi/h in the same direction? In the opposite direction?

23.** The driver of a car notes the car's speed to be 30 m/s in a 20-m/s zone. Just then, the driver notices a police car sitting behind a tree. Hoping that the police are not using radar, the driver at once releases the accelerator. The police are actually timing cars between two check points 150 m apart. The first check point coincides with the point where the accelerator was released. What must be the deceleration of the car if the police are to time its speed as 20 m/s?

24.** A boy wants to throw a can straight up in the air and then hit it with a second can. He wants the collision to occur 5.0 m above the throwing point. In addition, he knows that the time he needs between throws is 3.0 s. Assuming he throws both cans with the same speed, what must be the initial speed of the cans?

25.** An elevator in which a woman is standing is moving upward at a speed of 4.0 m/s. If the woman drops a coin from a height of 1.2 m above the elevator floor, how long does it take for the coin to strike the floor? What is the speed of the coin relative to the floor just before it hits?

* Difficulty somewhat above average.

** The problems thus marked require considerably more ingenuity than the average problem.

FORCES AND LINEAR MOTION

3

In the last chapter we discussed the concepts of velocity and acceleration without specifically considering what caused the object to move. The role of forces in causing motion was deferred until this chapter. We shall now investigate how forces cause accelerations. In carrying through this study we shall state and discuss Newton's three laws of motion. These laws will be found to be of primary importance in physics.

3.1 THE DISCOVERY OF PHYSICAL LAWS

A physical law is a statement of the way in which matter behaves. These are laws over which we have no control; they have existed and will exist forever. The purpose of all basic research in the physical sciences is to discover these laws. Understanding in science is equivalent to knowing the laws of nature and their consequences.

Although people sometimes make mistakes in what they think to be physical laws, the incorrect statements which they believe to be laws are, of course, not laws of nature at all. For example, Aristotle believed that he had discovered a law of nature when he stated that heavier bodies accelerate toward the earth faster than lighter bodies. In fact, he had not discovered a law of physics. There is no such physical law. The actual law of nature which applies to this situation was discovered many centuries later by Galileo, as we have seen.

Even the law which Galileo discovered for the acceleration of falling bodies is now known to be far from complete and general. We all know that objects in a spaceship behave much differently from Galileo's falling weights. In an earth-satellite ship, objects do not seem to the occupant of the ship to fall at all. They appear to him to be completely weightless. Of course Galileo had no way of knowing this, and so it is only natural that his proposed law of nature was incomplete. Nor were his measurements accurate enough to show that the same body accelerates differently under gravity at various places on the earth.

It is a general principle in science that no law of nature is ever fully and completely known. Scientists do their best to state nature's laws as they know them. However, no scientist would ever be so presumptuous as to say that this or that law of nature, as now understood, will not have to be modified as we learn more and more about our universe. There is an excellent chance that nearly all the laws of nature which we shall propose and use in this text are correct in all essentials or that their limitations are known. We know this because the laws have been tested against experience in all manner of ways. However, there is always the chance that someday one of these laws as presently stated and understood will be found to disagree with some new and ingenious experiment. Our statement of the law will then be changed so as to conform with that new experience, as well as with the results which we already know today. So when we state a physical law, it can be considered correct only in the light of present-day knowledge.

You will notice as you progress in your study of physics that only rarely is an accepted idea concerning a physical law found to be wrong. It is, however, quite common to find that the accepted idea must be extended, amplified and modified as our knowledge of the universe becomes wider. We shall see that the great reputation of the scientist who proposes a law sometimes does not ensure the correctness of the law. Even such a great scientist as Newton had some badly mistaken ideas about the way in which natural objects behave, as we shall see in our study of light. For this reason, the wise scientist does not accept a physical law because of the greatness or reputation of its promulgator alone. A law discovered by a young "unknown" through careful experiments is likely to be more reliable than the philosophic opinion of a "great" scientist.

Finally, we should state that *there is a hierarchy of physical laws. It is the ultimate aim of scientists to reduce the number of physical laws needed to describe the universe to a minimum.* We could have a different physical law for each different physical situation. This would be an intolerable situation, since no one could remember such a large number of laws. However, many of these individual laws would be found to conform to a more general law which would encompass them all. For example, the individual laws

1 A 10-lb object accelerates in free fall on the earth at 32 ft/s^2.
2 A 12-lb object accelerates in free fall on the earth at 32 ft/s^2, etc.

were encompassed in a single law by Galileo:

All bodies accelerate in free fall on the earth at approximately 32 ft/s^2.

We shall see that Galileo's statement of a physical law is only a portion of a much more general law first stated by Newton some years later. *A statement of a law of nature which contains in it the substance of many lesser laws is preferable to any of the individual lesser laws.*

3.2 NEWTON'S FIRST LAW OF MOTION

Isaac Newton (1642–1727) was one of the foremost physicists of all time. While still in his twenties he invented the branch of mathematics known as **calculus,** discovered the law of gravitation, and found that white light is a combination of colors. We shall discuss some of these matters in later chapters. In this chapter, we discuss the three laws concerning motion which he discovered. He first published these laws in 1687 in a classic compendium of his scientific discoveries entitled "Principia mathematica philosophiae naturalis."

Newton's first law of motion, translated from the Latin, *can be stated as:*

Every body perseveres in its state of rest or of uniform motion in a straight line unless it is compelled to change that state by forces impressed thereon.

The first part of this law seems reasonable and in agreement with our everyday experience. We know full well that bodies at rest will remain at rest until some external force causes them to move. Everyone recognizes this fact. Great consternation arises when, as sometimes happens, a person claims to have seen a body move with no apparent force causing the motion. Physicists, together with most other rational persons, accept this part of the law as being self-evident. We state it as follows:

An object at rest will remain at rest unless a nonzero resultant force acts on it.

Newton's First Law

The second part of the law is somewhat more subtle. It is as follows:

A moving object will continue its motion along a straight-line path at constant velocity unless a nonzero resultant force acts on it.

Common experience seems to contradict this. We know that nothing continues to move forever without change. A ball rolled across the ground soon stops. A metal object sliding across a smooth table eventually slows and stops. There are many other similar cases.

However, each of the examples cited is not a valid test of Newton's law. A force is acting on each of these bodies trying to stop its horizontal motion. It is the force of friction. We all know that the more precautions we take to eliminate this force, the less rapidly the moving object is brought to rest. Newton mentally generalized this observation to the case where no friction forces exist and concluded that if this case were possible, the moving object would never stop. Although an example of perfect constant-velocity motion has never been attained, of course, all our experience leads us to believe that Newton's conjecture, stated as a law of nature, is valid.

It should be pointed out here that *a body in motion with no resultant force acting on it is frequently said to be in* **equilibrium.** To a neophyte, this may seem like a misuse of the word. However, if we define a body to be in equilibrium when the sum of all the vector forces acting upon it is zero, the above usage of the word is justified. Perhaps you would have it otherwise. It is basically a matter of definition. You would do well to accept the definition physicists have given it (at least when speaking with physicists), even though you may feel the definition not to be a "natural" one.

3.3 NEWTON'S THIRD LAW OF MOTION

Because of its greater complexity, Newton's second law will be discussed last. The **third law,** as Newton stated it, is as follows:*

Newton's Third Law

> To every action there is always opposed an equal reaction; or the mutual actions of two bodies upon each other are always equal and are directed oppositely.

We can restate the law in terms of two objects and the forces they exert on each other. In those terms we can say

> When one object exerts a force on a second object, the second object exerts an equal but oppositely directed force on the first object. One of these forces is called the **action force.** The other is called the **reaction force.**

Several examples of the third law are shown in Fig. 3.1. We list below the action and reaction forces:

EXAMPLE	FORCE 1	FORCE 2
Part (*a*)	Block pushes down on table	Table pushes up on block
Part (*b*)	Woman's head pushes down on man	Man's head pushes up on woman
Part (*c*)	Hand pulls rope to right	Rope pulls hand to left
Part (*d*)	Fist exerts force on jaw	Jaw exerts force on fist

*This passage, together with the writings of many other great scientists, can be found in "Source Book in Physics," by W. F. Magie, Harvard University Press, Cambridge, Mass., 1963. I think you might enjoy perusing this excellent book.

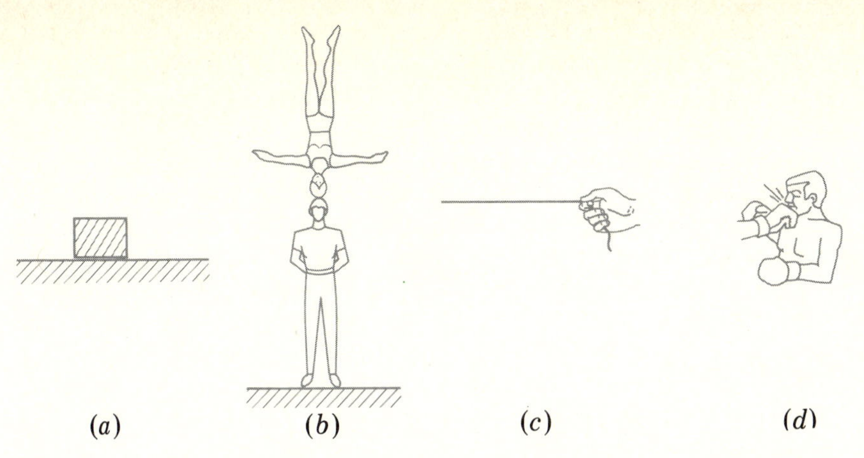

FIGURE 3.1
Notice that the action and reaction forces act on two different objects.

You should notice that *Newton's third law deals with two distinct bodies. An action force is exerted on one (body A) by the other (body B), while a reverse reaction force is exerted by A on B.* From time to time we shall make use of this law to tell us the magnitude and direction of a force on a body when the force on some other body is known.

3.4 NEWTON'S SECOND LAW OF MOTION

The first and third laws are rather simple statements of behavior which are familiar to all of us. They seem quite reasonable, even to the uninitiated. It will be seen, however, that the second law is not quite so obvious. This is true for two reasons: (1) It is a precise mathematical statement of relationships between quite different physical quantities. (2) Its precise statement involves the introduction of the concept of mass, which, though not really complex in itself, is sometimes not clearly recognized as being different from the weight of an object. We state the law here in the following way:

Newton's Second Law

> If a nonzero resultant force acts upon a body, that force will produce an acceleration of the body in the direction of the force. The magnitude of the acceleration is in direct proportion to the magnitude of the resultant force and in inverse proportion to the quantity of matter in the body.

It is simple to illustrate this law qualitatively. If one wishes to push a child in a wagon, it is obvious to everyone that the harder one pushes, the larger the acceleration of the wagon will be. The quantitative fact that for a given body this acceleration is exactly proportional to the force has been substantiated many times by experiments like the one shown in Fig. 3.2. Hence we can write that

$$\mathbf{a} \propto \mathbf{F}$$

Where $\propto$ is read "is proportional to."

But the force required to accelerate an object depends also upon the

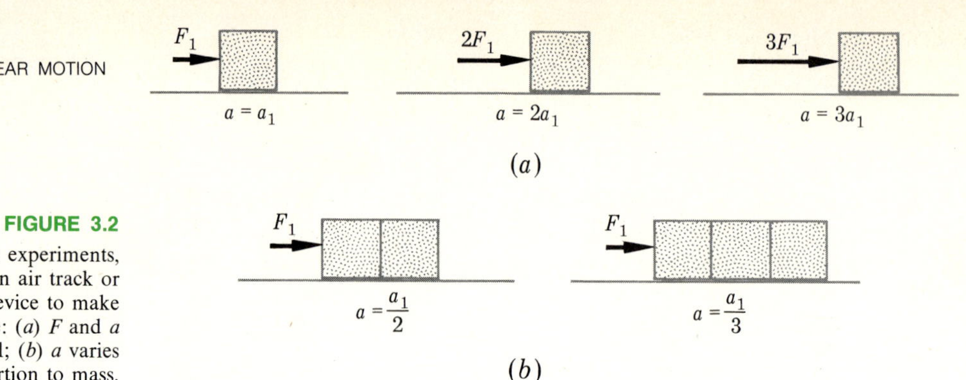

FIGURE 3.2
In doing these experiments, one would use an air track or some similar device to make friction negligible: (*a*) F and a are proportional; (*b*) a varies in inverse proportion to mass.

massiveness of the object. Each object possesses a quality we call **inertia.** All objects tend to remain at rest unless an unbalanced force acts on them. Moreover, an object in motion tends to remain in motion. We say each object possesses inertia. The inertia of an object is related to its massiveness. For example, a football coach chooses very massive players for the line since they are not easily knocked out of position. The coach knows that massive objects have more inertia than less massive objects. They are more difficult to set in motion if they are originally at rest. Moreover, they are more difficult to stop if they are already in motion. Somehow this property of massiveness or inertia must enter into determining the acceleration produced by a force. To see how this factor enters in, we can do an experiment like the one shown in Fig. 3.2*b*.

As we see there, the acceleration produced by a given force varies inversely with the massiveness of the object being accelerated. If the massiveness of the object is doubled, the acceleration is cut in half. If the massiveness is tripled, the acceleration is reduced to one-third its original value. We therefore write

$$a \propto \frac{1}{m}$$

DEFINITION

where *m is a measure of the massiveness of the object. We shall call m the* **mass** *of the object.* The mass of an object is an intrinsic property of the object and is a measure of the unbalanced force needed to accelerate the object. Mass is a measure of inertia.

We have now found that $\mathbf{a} \propto \mathbf{F}$ when m is constant but $a \propto 1/m$ when $\mathbf{F}$ is constant. These two proportionalities can be combined into one:

$$\mathbf{a} \propto \frac{\mathbf{F}}{m}$$

Notice that for m constant this says $\mathbf{a} \propto \mathbf{F}$. Similarly, when F is constant, this says $a \propto 1/m$. It therefore correctly represents the experimental results shown in Fig. 3.2.

A proportion can be made into an equation by using a proportionality constant. We then can write

$$\mathbf{a} = \frac{(\text{const})(\mathbf{F})}{m}$$

To simplify the equation, scientists long ago agreed to measure F and m in such a way that the constant will be unity. With that proviso (and we shall see its consequences later), we write

$$\mathbf{a} = \frac{\mathbf{F}}{m} \qquad \text{or} \qquad \mathbf{F} = m\mathbf{a} \tag{3.1}$$

Mathematical Statement of Newton's Second Law

This is a mathematical statement of Newton's second law. In writing this *we should note that* **F** *is the resultant force on the object whose mass is m.* Recall also that *m is a measure of the inertia of the object. It measures how hard it is to set the object into motion or to stop it if it is already moving.* Moreover, *the acceleration* **a** *is in the same direction as* **F**.

At this point you are likely to associate the mass of an object with its weight. There is a relation between the two, of course, since massive objects are heavy. But *weight and mass are not the same even though they are related.* It is of great value to know the relation between mass and weight. Let us now find out what this relation is.

Consider the experiment shown in Fig. 3.3. The object shown there is falling freely on the earth under the effect of the earth's gravitational pull upon it. We know that its acceleration is the free-fall acceleration due to gravity g. Moreover, we know what the unbalanced force is on the object. It is the pull of gravity on it and we have called this force the weight of the object. Therefore, in Fig. 3.3, the resultant force on the object is W, the objects' weight. This unbalanced force causes the object to have an acceleration g. Let us now place these values in Eq. (3.1). We then find

$$W = mg \qquad \text{or} \qquad m = \frac{W}{g} \tag{3.2}$$

Relation between Mass and Weight

This is a very important equation. It tells us the relation between the mass and the weight of an object.

FIGURE 3.3
The unbalanced force on the object is W. It causes an acceleration g.

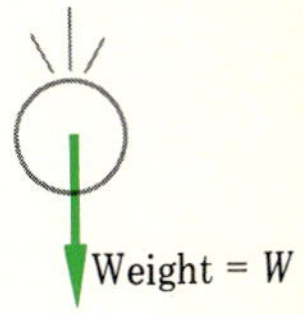

Weight = W

The mass of a body turns out to be a more characteristic constant of the body than is its weight. Careful experiments have shown that measurements of g at various places on the earth do not give exactly the same value. This was pointed out in the previous chapter. However, it is found that the weight of the body also changes at different points on the earth. The changes in W and g are always found to behave in such a way that the ratio W/g is always a constant. Hence, *though the weight of a body changes, depending upon where it is measured, Newton's law indicates that the mass of a body is everywhere constant.**

*That this statement is not precisely true was first pointed out by Einstein in 1905. We shall discuss this matter in more detail later.

This is indeed a comforting fact of nature in our modern age. When an object appears weightless to an astronaut in an earth satellite, we know that its mass m is still the same as when it was on earth. This being true, we see from Eq. (3.1) that the same force will be required to accelerate an object within a satellite as would be required on earth. To a high order of precision the mass of a body may be considered constant (except in certain cases to be mentioned later), and the acceleration of a body produced by a given force is independent of whether the body is on the earth or elsewhere in the universe.

Since the mass is proportional to the weight of the object on earth, it is common to attribute certain characteristics of bodies to their weight when properly we should be speaking of their mass. For example, if it were possible to put the locomotive of a railroad train in a satellite orbiting the earth, the locomotive would be as difficult to accelerate as when it was on earth, since Eq. (3.1) would still apply. However, the locomotive would appear weightless to a person in the satellite, and so it is not strictly proper to say that a light object is easy to accelerate. It all depends upon where the object is weighed and who does the weighing.*

The property of the locomotive which is constant on earth and also in orbit about the earth is its mass m. It is clear, then, that the mass of an object is a much better measure for the forces needed to accelerate it than its weight is. *We call this property of a body, which requires that a force be exerted upon it to accelerate it,* the **inertia** *of the body. The mass of a body is a direct measure of its inertia.*

DEFINITION

3.5 STANDARD UNITS OF MEASUREMENT

The units of measurement which we have used in the previous sections of this book are familiar to us all. Even so, it is now necessary to define precisely the basic quantities we have used in previous sections.

Units of Length By international agreement, for many years the standard unit of length was the distance between two engraved marks on a platinum-iridium-alloy bar kept at the International Bureau of Weights and Measures near Paris. When the temperature of the bar was 0°C, this length was defined as one meter. One centimeter is one-hundredth of the standard meter. As a precaution, the length of this bar was compared with the wavelength of a particular color of light so that if the bar should ever be destroyed, the wavelength of light could be used as a standard in its stead. Subsequently the standard meter was redefined in terms of this wavelength of light, and so the bar kept in Paris is now only a secondary standard. We shall discuss this in more detail when we take up our study of light.

The British units (foot, inch, and yard) are defined in terms of the standard meter. One inch is by definition exactly 0.0254 m. One yard is 0.9144 m.

* We have not attempted to explain *why* the locomotive appears weightless to someone in the satellite. The reasons for this will become apparent after we complete a study of circular motion in Chap. 8.

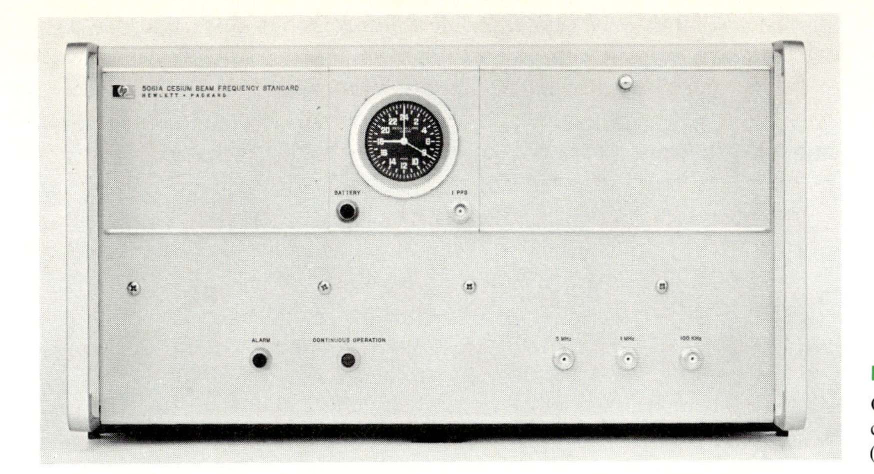

FIGURE 3.4
Commercially produced cesium standard clock. (*Hewlett-Packard.*)

Units of Time For many years the basic time unit was defined in terms of the length of an average day. But over the centuries the motion of the earth changes slightly, and so the average was specified for the year 1900. In order to have a more easily accessible time standard, the time unit was redefined in 1967. To define a basic unit of time, a certain comparatively easily measured frequency of vibration associated with cesium atoms was utilized. The second is now defined to be exactly 9,192,631,770 times as long as the time taken for one of these vibrations. A commercial version of this standard atomic clock is shown in Fig. 3.4. Of course the minute is 60 s, and the hour, day, year, etc., are defined as usual.

Units of Mass A particular platinum-iridium cylinder kept near Paris is defined to have a mass of one kilogram. A copy of this standard mass is shown in Fig. 3.5. Since weight and mass are proportional, we can compare other objects with the standard mass by weighing. An object which weighs twice as much as the standard kilogram has a mass of 2 kg. (Of course, the weighings must be done under identical conditions.) We define the gram (g) to be $\frac{1}{1000}$ kg.

FIGURE 3.5
The platinum-iridium cylinder shown here (prototype kilogram number 20) is a copy of the standard kilogram mass. This photograph was furnished by the U.S. National Bureau of Standards, whose responsibility it is to preserve this secondary standard of mass.

3.6 DERIVED UNITS

The system of units based on the *m*eter, *k*ilogram and *s*econd is called the **mks system.** Its official name is the **SI** (for Système International). It has been adopted as the units system recommended for all scientific publications. Moreover, the metric system, adopted as the units system for commerce and industry by most countries, corresponds closely to the SI system. The United States is one of the last major countries in the world to adopt the metric system. It is likely that the units of the British system such as the foot, pound, etc., will soon be obsolete.

One of the great advantages of the metric system and the SI is that the units are related by powers of 10. In naming these units, the following prefixes are used (see Appendix 2 if you are not familiar with the notation 10^n and 10^{-n}):

PREFIX	SYMBOL	FACTOR
exa	E	10^{18}
peta	P	10^{15}
terra	T	10^{12}
giga	G	10^9
mega	M	10^6
kilo	k	10^3
deci	d	10^{-1}
centi	c	10^{-2}
milli	m	10^{-3}
micro	μ	10^{-6}
nano	n	10^{-9}
pico	p	10^{-12}
femto	f	10^{-15}
atto	a	10^{-18}

For example, 1 kg is 10^3 g while 1 pm (picometer) is 10^{-12} m. As you see, the prefix can be appended to any unit to make a new unit.

SI Unit of Force

In addition to the units of length, mass, and time, we need other units. One of the most important is the unit of force. The fundamental units of force are defined in terms of Newton's second law, $F = ma$. This definition is easily made in the SI. We call the unit of force in this system the newton (N).

DEFINITION

A **newton** is that unbalanced force which will give a 1-kg mass an acceleration of 1 m/s².

Writing this in terms of $F = ma$, we have

$$1\ \text{N} = (1\ \text{kg})(1\ \text{m/s}^2)$$

so the newton unit is equivalent to $1\ \text{kg}\cdot\text{m/s}^2$.

CGS Unit of Force

Some people still use the units *c*entimeter, *g*ram, and *s*econd as fundamental units. In that, the **cgs system,** the unit of force is called the dyne (dyn). It, too, is defined in terms of $F = ma$.

A **dyne** is that unbalanced force which will give a 1-g mass an acceleration of 1 cm/s².

As we see from $F = ma$,

$$1\ \text{dyn} = 1\ \text{g}\cdot\text{cm/s}^2$$

Further, as you can show by converting $1\ \text{kg}\cdot\text{m/s}^2$ to gram-centimeters per second squared,

$$1\ \text{N} = 10^5\ \text{dyn}$$

British Unit of Force

The unit of force in the British system is the pound (lb). It is defined in terms of the newton. By definition,

$$1 \text{ lb} = 4.45 \text{ N}$$

At present, the pound is fast becoming obsolete except in the United States.

British Unit of Mass

We need yet to discuss the British unit for mass (corresponding to the kilogram in the SI). It is defined in such a way that $F = ma$ is fundamental. The unit is called the **slug** and is chosen such that

> A 1-lb unbalanced force on an object of mass 1 slug produces an acceleration of 1 ft/s^2.

Or you may wish to use the equivalent definition that

$$1 \text{ slug} = 14.6 \text{ kg}$$

Of course, since $W = mg$, we have that 1 slug weighs 32.2 lb at a place where $g = 32.2 \text{ ft/s}^2$.

You may well question why you have not previously been much concerned about the British unit of mass, the slug. There is a very simple reason for this. Those who use the British system of units usually write Newton's law in the form

$$F = \frac{W}{g}a$$

where m has been replaced by its value from Eq. (3.2), W/g. Since in most engineering practice the small variation in weight of an object from place to place on the earth is of little concern, the engineer uses the weight of the object as measured, even though this will ordinarily not be the exact weight, i.e., the weight measured at the same position on earth where g is measured. By this device, one can avoid the use of the term mass and its unit, the slug.

Units Systems

To summarize this discussion of derived units, we see that *there are three basic sets of units.* (There are others, but they are seldom used and will not be discussed in this text.) *They can be systematized through the use of Newton's second law as follows:*

	F =	m ×	a
UNITS SYSTEM	FORCE UNIT	MASS UNIT	ACCELERATION UNIT
SI	newton	kilogram	m/s^2
cgs	dyne	gram	cm/s^2
British	pound	slug	ft/s^2

It should be noticed that in $F = ma$ a single system of units is always used at one time. For example, if the force is expressed in pounds, the mass must be in slugs and the acceleration must be in feet per second per second, not miles per second per second. *Mixed units should never be used in this relation.*

Before leaving this section, it is well to point out the following facts, the indicated approximate values for g being used:

1. A 1-kg mass weighs 9.8 N on earth.
2. A 1-g mass weighs 980 dyn on earth.
3. A 1-slug mass weighs 32.2 lb on earth.
4. A 1-kg mass weighs 2.21 lb on earth.
5. A letter cut out of the text on this page weighs about 1 dyn.
6. 1 N is exactly 10^5 dyn.

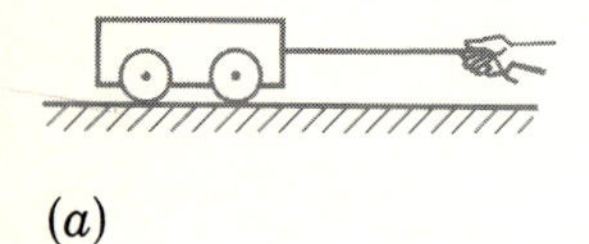

(a)

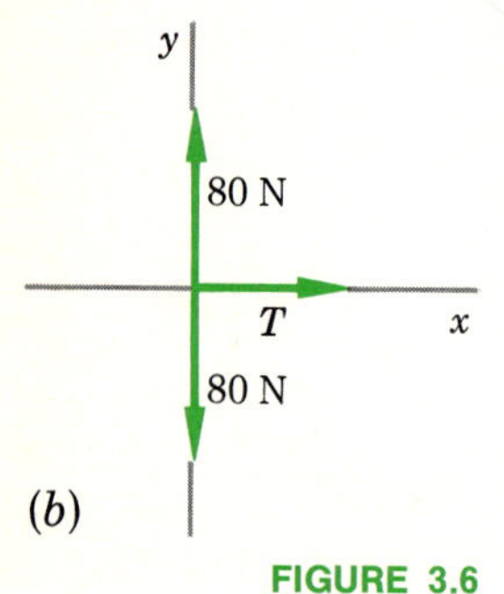

(b)

FIGURE 3.6
The free-body diagram for the forces acting on the wagon is in (b).

3.7 USE OF NEWTON'S SECOND LAW

Illustration 3.1 In this section we shall see how Newton's second law is used in practical situations. Let us consider the situation shown in Fig. 3.6. How large must the tension be in the rope pulling the wagon in Fig. 3.6 if the wagon is to accelerate at $0.50\ \text{m/s}^2$? The wagon weighs 80 N.

Reasoning To solve this problem, we first draw a free-body diagram showing the forces acting on the wagon. This is done in Fig. 3.6. Notice that three forces act on the wagon, the pull of gravity down, the push of the floor up, and the pull of the rope to the right. Since the wagon will neither fall through the floor nor float up into the air, the vertical forces must add up to zero; i.e., they must cancel. Therefore, the unbalanced (or resultant) force on the wagon is T. Further, the acceleration will be in the direction of T.

To write Newton's second law, $F = ma$, we need the mass of the wagon. We are told that the wagon weighs 80 N. Then, since $m = W/g$, we have

$$m = \frac{80\ \text{N}}{9.8\ \text{m/s}^2} = 8.2\ \text{kg}$$

Using this value, we then find from

$$F = ma$$

that

$$T = (8.2\ \text{kg})(0.50\ \text{m/s}^2) = 4.1\ \text{N}$$

We shall now examine several other examples to illustrate various points concerning the use of Newton's second law.

Illustration 3.2 A child pulls with a 10.0-lb force on a wagon weighing 64 lb, as shown in Fig. 3.7. (1) At what rate will the wagon accelerate? (2) With how large a force is the ground pushing up on the wagon?

Reasoning As in the previous example, we draw the free-body diagram as in part *b* of the figure. P is the push of the ground up on the wagon. Notice that the pull of the rope now has both a vertical and a horizontal component. After splitting this force into x and y components, the force diagram is as shown in part *c*. Since the wagon is to remain on the ground, the sum of the vertical forces must be zero. Hence

$$P + 6.0 - 64 = 0$$

or

$$P = 58 \text{ lb}$$

which is the answer required for part 2 of the problem.

The acceleration of the wagon will occur in the direction of F, the unbalanced force, and this will be in the horizontal direction. One has

$$F = ma$$

$$8.0 \text{ lb} = \left(\frac{64 \text{ lb}}{32 \text{ ft/s}^2}\right)(a)$$

or

$$a \approx 4.0 \text{ ft/s}^2$$

Notice that we replaced m by W/g.

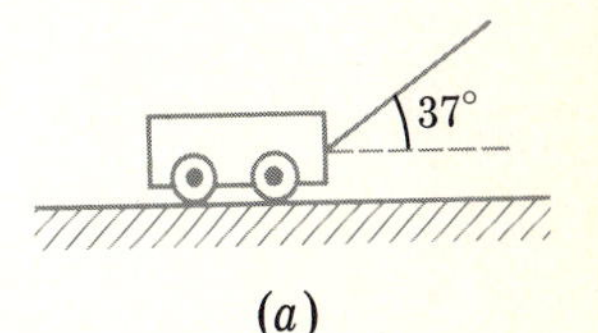

(*a*)

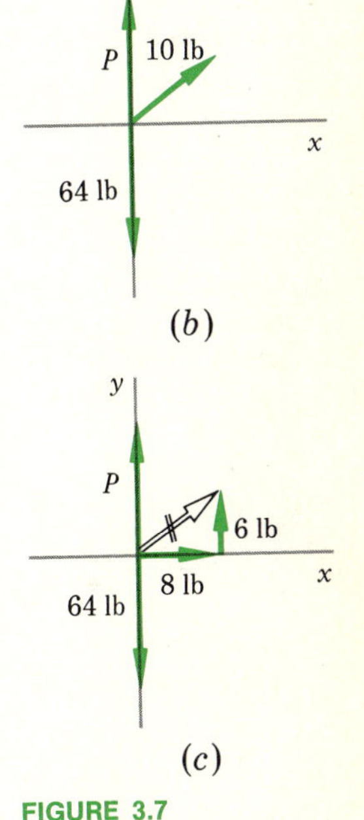

FIGURE 3.7
The rope pulling on the wagon helps P balance the weight of the wagon as well as causing the wagon to accelerate to the right.

Illustration 3.3 An object weighting 64 lb hangs by a rope from the ceiling of an elevator (Fig. 3.8*a*). What will be the tension in the rope if the elevator is accelerating (*a*) upward at 12 ft/s^2; (*b*) downward at 12 ft/s^2; (*c*) downward at 32 ft/s^2?

Reasoning The free-body diagram is shown in Fig. 3.8*b*, where T is the tension in the rope, i.e., the pull of the rope upward on the object. In the previous examples, very little ambiguity in direction was involved. This problem is a little more complicated, and we should now adopt a more methodical plan of attack. It is wise to choose a given direction as positive and take forces and accelerations in that direction to be positive. Forces and accelerations in the opposite direction are then negative.

a. Taking up as positive, we have

$$F = ma$$

$$T - 64 = (\tfrac{64}{32})(12)$$

You should insert appropriate units in the equation and show that

$$T = 88 \text{ lb}$$

b. Taking up as positive, we have

$$T - 64 \text{ lb} = \left(\frac{64 \text{ lb}}{32 \text{ ft/s}^2}\right)(-12 \text{ ft/s}^2)$$

$$T = 40 \text{ lb}$$

Notice the negative sign on the acceleration. It is needed because the acceleration is directed downward in this case and we are taking up as the positive direction.

c. Taking down as positive, we have

$$64 - T = (\tfrac{64}{32})(32)$$
$$T = 0$$

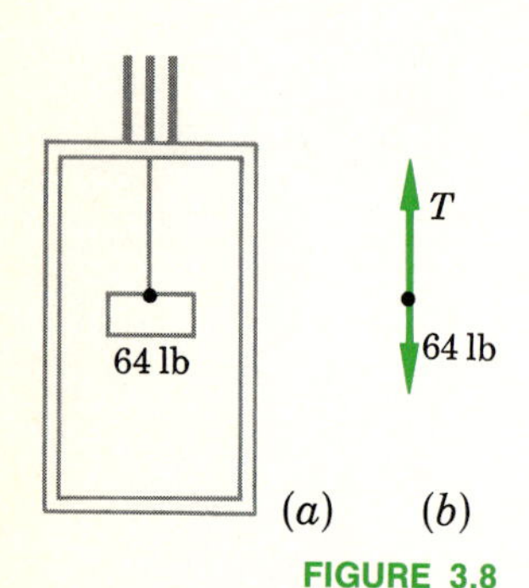

FIGURE 3.8
When the elevator is accelerating upward, T is larger than 64 lb. Under what condition will the object appear to be weightless?

Notice that it is a matter of personal choice whether up or down is called positive. The same result is obtained either way. Let us now check to see whether our answers are reasonable.

In part *a* the weight was to be accelerated upward. Hence, the unbalanced force must be upward. Gravity was pulling down with a force of 64 lb, and the rope was pulling up with a force of 88 lb. The unbalanced force was 24 lb *upward,* as it should be.

In part *b*, the unbalanced force should be down, since the acceleration is down. We see from our answer that the pull of gravity down is 24 lb stronger than the pull of the rope up.

The answer to part *c* may seem a little strange until one recognizes that the acceleration required is just g. This acceleration is attained only in free fall, and hence the tension in the rope must be zero.

It is worthwhile noticing at this point that a person who was supporting an object in the falling elevator in part *c* would notice that the object in question appeared to be weightless.* As we have seen, the object is in free fall, and so the tension in the cord is zero. No force is required to support it. If the object had been hanging from a spring scale or sitting on a scale, the scale would read zero. Clearly, the object would appear to be weightless. The apparent weightlessness of objects in earth satellites and in spaceships coasting through space is very closely allied to this example. As we shall see in Chap. 8, all these cases are equivalent to the elevator in free fall, and so the freely falling objects (the ship and everything in it) must appear to be weightless to a person who is also falling freely.

Illustration 3.4 A 1500-kg car is traveling at 60 km/h (about 37 mi/h) when its brakes are applied and it skids to rest. The skidding tires experience a friction force about seven-tenths the weight of the car. How far does the car go before stopping?

Reasoning This is a typical combination of an $F = ma$ problem and a motion problem. We can use $F = ma$ to find a. Then the five motion equations can be used to find the stopping distance.

We are told that the car has a mass of 1500 kg. Its weight is

$$\text{Weight} = mg = (1500\ \text{kg})(9.8\ \text{m/s}^2) = 14{,}700\ \text{N}$$

---

*However, the weight of the object, i.e., the pull of the earth on it, is not zero. Hence $W = mg$ *does not* imply a change in mass.

We are further told that the stopping force is

$$(0.70)(\text{weight}) = 10{,}290 \text{ N}$$

This is the unbalanced force acting on the car. Placing it in $F = ma$ and taking it to be negative since it is opposite to the direction of motion gives

$$-10{,}290 \text{ N} = (1500 \text{ kg})a$$

From this

$$a = -6.9 \text{ m/s}^2$$

It is a deceleration and is therefore negative.

Now we are able to solve the motion problem for the skidding auto. As knowns we have

$$v_0 = 60 \text{ km/h} = 16.7 \text{ m/s} \qquad v_f = 0$$
$$a = -6.9 \text{ m/s}^2 \qquad s = ?$$

We can use $v_f^2 - v_0^2 = 2as$ to find $s = 20.2$ m.

It is instructive to notice that the stopping force we have used is close to the maximum value possible. We therefore conclude that a car moving at about 60 km/h (37 mi/h) requires a distance of about 20 m to stop. Can you show that the stopping distance increases as the *square* of the initial speed of the car?

Illustration 3.5 The masses shown in Fig. 3.9 are tied to opposite ends of a massless rope, and the rope is hung over a massless and frictionless pulley. Find the acceleration of the masses. (This device is called **Atwood's machine.**)

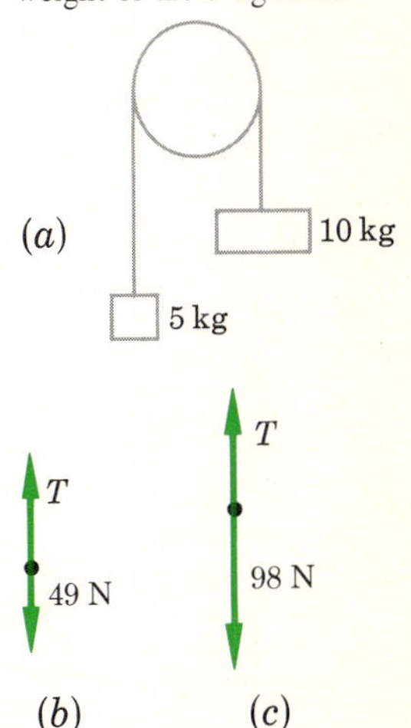

FIGURE 3.9 Since the 10-kg mass will fall as the pulley turns clockwise, the tension in the rope must be less than the weight of the 10-kg mass but more than the weight of the 5-kg mass.

Reasoning Obviously the 10-kg mass will fall and the 5-kg mass will rise. The frictionless pulley does nothing but provide a support for the rope. The tension in the rope is the same throughout its whole length. Call it T.

To solve a problem such as this, involving two or more bodies, we write down $F = ma$ for each of the bodies separately. First, we must choose a positive direction for the motion. We know in this case that the pulley will turn clockwise, and so it is reasonable to take motions of the objects as positive when moving in that sense.

The free-body diagrams for the 10- and 5-kg masses are shown in parts *b* and *c* of the figure. Notice that the 10-kg body weighs (10)(9.8) N* and that its mass is 10 kg. Applying $F = ma$ to each of these bodies in turn gives

$$98 - T = 10a$$

and

$$T - 49 = 5a$$

* Since $W = mg = (10 \text{ kg})(9.8 \text{ m/s}^2) = 98 \text{ kg}\cdot\text{m/s}^2 = 98 \text{ N}$

Notice that the forces in the positive direction of motion are taken as positive.
Adding the lower equation to the upper yields

$$49\ \text{N} = (15\ \text{kg})(a)$$

or

$$a = 3.3\ \text{m/s}^2$$

where use has been made of the fact that a newton is $1\ \text{kg} \cdot \text{m/s}^2$. By substituting this value for a in either of the two equations one finds

$$T = 65\ \text{N}$$

To check the answer, we see that the tension in the rope is larger than the weight of the lighter object, which thus will rise. In addition, the tension is smaller than the weight of the larger object, which thus will fall.

Illustration 3.6 The two objects shown in Fig. 3.10 are connected by a string, with one object hanging from a frictionless pulley and the other sitting on a table. We are told that the friction force retarding the motion of the object on the table top is 0.098 N. Find the acceleration of the bodies.

Reasoning Sometimes when students are presented with this problem, they state that the objects will not move since the 200-g object is much lighter than the 400-g object. This is a fallacy, however. The weight of the 400-g object is downward. This force is balanced by the upward push of the table. The friction force is the only force pulling backward on the object, as is easily seen from the free-body diagram in part *b* of the figure.

Isolating each body in turn and writing $F = ma$ for each, after changing all quantities to SI units we have (see Fig. 3.10*b* and *c*)

$$T - 0.098 = 0.40a$$

and

$$(0.20)(9.8) - T = 0.20a$$

where the forces are in newtons and the masses in kilograms. In writing these we have taken the direction of motion which occurs as the 200 g falls downward to be positive. Notice also that since $W = mg$, the weight of a 0.20-kg object is (0.20)(9.8) N, while its mass is 0.20 kg.

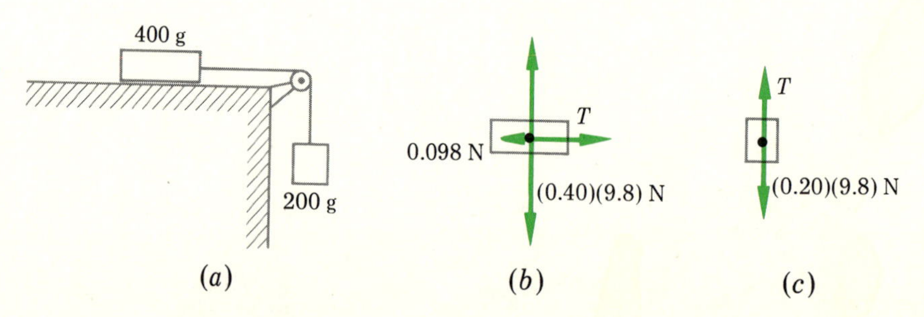

FIGURE 3.10
Although the 0.098-N friction force retards the motion of the 400-g object, the weight of the 200-g object is large enough to cause the objects to move. The weight of the 400-g object is balanced by the push of the table.

Adding the two equations and solving for a gives

$$a = 3.1 \text{ m/s}^2$$

Using the fact that a newton is 1 kg · m/s², you should show that a really is in the units stated.

Substitution in either of the equations yields

$$T = 1.34 \text{ N}$$

3.8 FRICTION FORCES

Let us now investigate the nature of the friction force which retards an object as it slides along a surface. You can learn quite a bit concerning friction forces by pushing your textbook across a table as shown in Fig. 3.11. Suppose you slowly increase your pushing force as shown in the graph in Fig. 3.11*b*. At first, the book does not move. However, when the pushing force reaches a certain critical value, the book suddenly begins to move. We see from this that there is a maximum friction force which resists the onset of sliding. It is equal in magnitude to the critical pushing force which causes the object to slide. We represent this friction force which resists the start of sliding by f_s. (The subscript s stands for "static," or stationary, since the object is at rest at the instant sliding begins.)

Once an object has begun to slide, a force much less than f_s will keep it moving. This tells us the friction force resisting the motion of the moving object is less than f_s. We represent it by f_k. (The subscript k stands for "kinetic," or moving.) You can easily verify what we are saying by pushing a book along a table. Try it and see.

The major reasons for this behavior can be seen from Fig. 3.12. As you see, the surfaces in contact are far from smooth. Their jagged points penetrate each other and cause the surfaces to resist sliding. Once sliding has begun, however, the surfaces do not have time to settle down onto each other completely. As a result, less force is required to keep them moving than to start their motion.

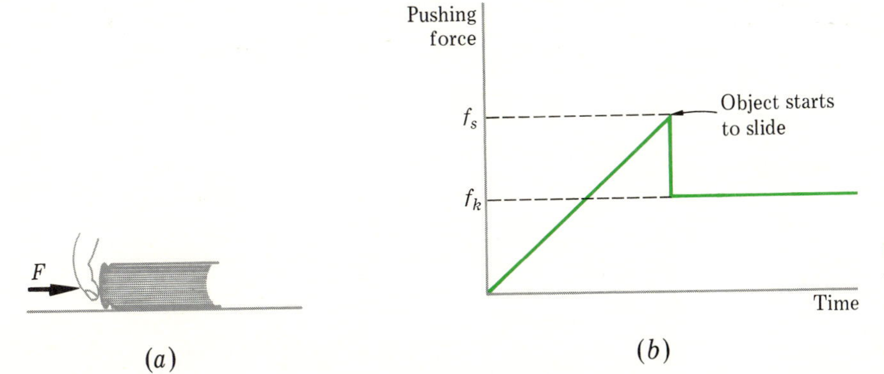

FIGURE 3.11
The friction force which opposes the motion of the book drops suddenly as the book begins to slide.

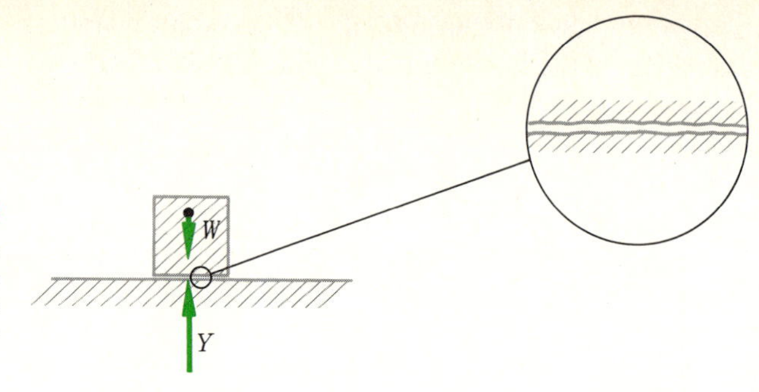

FIGURE 3.12
When highly magnified, the surfaces show considerable roughness; this makes it difficult for them to slide over each other.

It is helpful in various circumstances to be able to estimate the magnitude of the friction force which impedes motion. For this purpose one frequently defines a coefficient of friction μ (Greek mu) which, experiment shows, relates the friction force f to what is called the **normal force** Y. The normal force is simply the perpendicular force with which the supporting surface pushes on the sliding object. Typical situations and the corresponding values of Y are shown in Fig. 3.13. You should examine each to make sure why Y is as indicated.

Coefficient of Friction

If one measures the friction force f impeding the motion of an object which is sliding along a surface, one finds the approximate relation

$$f = \mu Y \tag{3.3}$$

This equation is used to define the **coefficient of friction** μ. In words, *the friction force increases in proportion to the normal force.* It is found that *the coefficient of friction μ varies widely from surface to surface.* A few typical values are given in Table 3.1.

The values given in Table 3.1 are designated kinetic (or dynamic) coefficients of friction since they apply only to an object which is already sliding across the surface. As we discussed in relation to Fig. 3.11, a somewhat larger friction force exists just before the object begins to slide, and the coefficient which applies to that case is usually called the static coefficient of friction. All these coefficients depend upon the exact condition of the surface, and so one should use them only as a means for obtaining an approximate friction force.

FIGURE 3.13
Notice that Y, the normal force, need not be equal to the weight of the object.

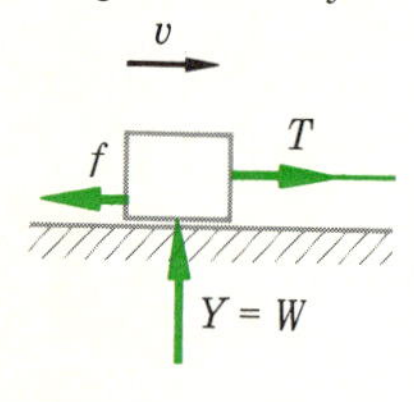

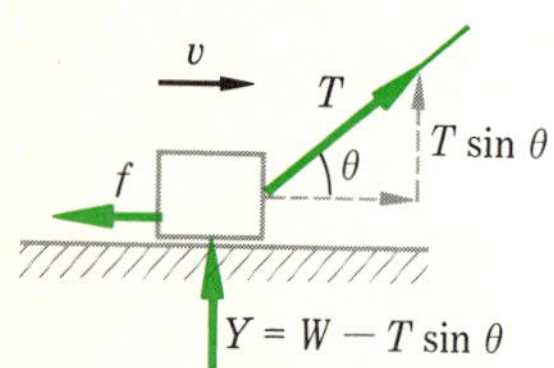

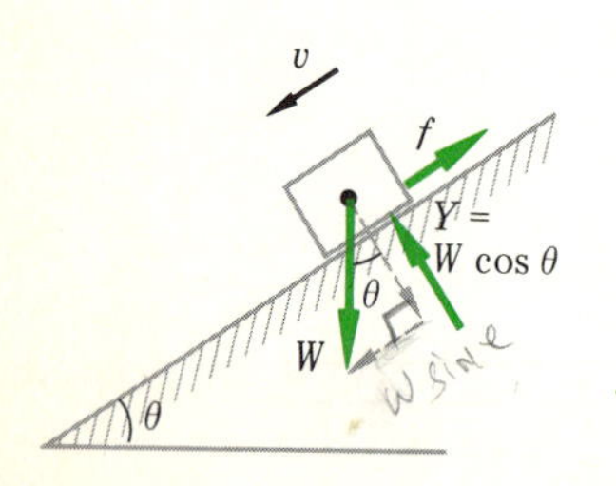

TABLE 3.1
KINETIC COEFFICIENTS OF FRICTION*

SURFACE 1	SURFACE 2	μ
Wood	Snow	~0.06
Brass	Ice	0.02–0.1
Metal	Metal (lubricated)	~0.07
Oak	Oak	0.25
Rubber	Concrete (wet)	0.5–0.9
Rubber	Concrete (dry)	0.7–1.0

*The sign ~ is read "approximately."

Illustration 3.7 In an investigation of an accident, a police officer notices that one car shows skid marks on the dry, level pavement. The skid marks are measured to be 7.0 m long. Estimate a lower limit for the speed of the car before it started the skid.

Reasoning Let us assume that the car decelerated uniformly to rest. To find its near largest possible deceleration, let us assume (from Table 3.1) that the coefficient of friction between car and road is 0.90.

If the car has a mass m, its weight is mg. The normal force Y must balance this, so $Y = mg$. From

$$f = \mu Y$$

we then know that the friction force stopping the car is

$$f = 0.90\ mg$$

This is the unbalanced force on the car and it causes the car to decelerate.

To find the deceleration, we write $F = ma$ for the car with $F = \mu Y = f$. Then

$$F = ma$$

becomes

$$-0.9\ mg = ma$$

Notice that the mass of the car cancels and we obtain $a = -0.90\ g$, where g is the acceleration due to gravity.

We can now solve the motion problem involving the skidding car. The knowns (or assumed knowns) are

$$v_f = 0 \qquad s = 7.0\text{ m} \qquad a = -0.90\ g = -8.8\text{ m/s}^2$$

We wish to find v_0, the initial velocity. To do so we can use $v_f^2 = v_0^2 + 2as$. Then

$$0 = v_0^2 - (17.6)(7.0)\text{ m}^2/\text{s}^2$$

which gives

$$v_0 = 11.1\text{ m/s}$$

In all probability, the car was traveling at an initial speed larger than this. Why can we conclude this?

3.9 TERMINAL VELOCITY

Until now we have assumed that freely falling bodies accelerate at 9.8 m/s^2. Actually, of course, a falling feather or slip of paper does not accelerate at this rate since we easily see that they both fall less rapidly than a coin or a stone. We have already pointed out that the friction force due to the rush of air past a falling body causes this effect. The object can be treated as falling freely only if the friction force is negligible.

Very soon after a feather is dropped, it reaches a terminal velocity. In fact, *any object allowed to fall for a long distance through the air will reach a constant velocity called the* **terminal velocity.** Since the object is no longer accelerating (its velocity is constant), there can be no unbalanced force acting on it. The pull of gravity on the object must be balanced by some upward force. This upward force is simply the friction force of the air rushing past the falling object.

Since a feather is very light and extended in space, only a small speed and air friction force are needed to balance the pull of gravity on it. Hence the terminal downward average velocity of a feather is quite small, perhaps only a fraction of a meter per second. It reaches this velocity in less than $\frac{1}{10}$ s after being dropped. After that time it no longer accelerates.

A similar situation exists for all falling bodies. The more compact and dense the object, the higher its terminal speed will be. Typical (very approximate) examples are the following: raindrop, 25 ft/s; smoke particle, 0.1 cm/s; human being, 250 ft/s. Of course these values depend upon the shape and orientation of the falling object. An interesting example of terminal velocity is shown in Fig. 3.14.

FIGURE 3.14
Example of terminal velocity. (*London Daily Express—Pictorial Parade.*)

SUMMARY

Experimental results lead us to state laws of nature. These laws summarize the way in which the world about us behaves. We must be ready to modify the laws as presently stated if future experiments show that the statement is not completely correct.

Newton's first law of motion has two parts. The first part tells us that an object at rest will remain at rest unless there is a nonzero resultant force acting on it. The second part states that the vector velocity of an object will not change unless there is a nonzero resultant force acting on the object.

Newton's third law of motion is the action-reaction law. It states that any force exerted by body A upon body B is accompanied by an equal but opposite force exerted by B upon A.

Newton's second law is summarized by $\mathbf{F} = m\mathbf{a}$. It applies to an object of mass m upon which a resultant force $\mathbf{F}$ acts. The acceleration $\mathbf{a}$ of the object is in the same direction as $\mathbf{F}$.

The mass m of an object measures the inertia of the object. If m is large, the object has large inertia: it is difficult to set the object into motion; it is difficult to stop or deflect the object if it is in motion. The units of mass are kilograms (in the SI), grams (in the cgs system), and slugs (in the British system).

When using $F = ma$, the units must be consistent. In the SI the units which must be used are newtons, kilograms, and meters per second squared. In the cgs system, they are dynes, grams, and centimeters per second squared. The British system uses pounds, slugs, and feet per second squared. No other units should be used.

If the weight W of an object whose mass is m is measured where the free-fall acceleration is g, then $W = mg$. This is simply $F = ma$ rewritten for the case of a free-fall experiment.

When an object is sliding, it experiences a retarding force f called the friction force. The supporting surface pushes upon the sliding object with a force Y, where Y is directed perpendicular to the surface. It is then found that $f = \mu Y$, where μ is called the coefficient of friction for the two surfaces involved.

An object falling through air or any other gas or fluid experiences a retarding friction force. Since this force increases with speed of fall, it may eventually become equal to the pull of gravity. Then no unbalanced force exists on the object. It then falls with constant velocity called its terminal velocity (or speed).

MINIMUM LEARNING GOALS

Upon completion of this chapter, you should be able to do the following:

1. State Newton's first law and give several examples to illustrate each part of it.

2. State Newton's third law. Point out the action-reaction pair of forces in any given simple situation.

3. List several given objects so that they are listed in sequence from smallest to largest inertia values. Select from the following list those quantities which, when changed, always cause comparable changes in inertia: size, shape, weight, speed, velocity, mass, motion.

4. State Newton's second law both in words and in equation form. Explain clearly what is meant by F in the law.

5. List the basic units of time, length and mass in each of our three units systems. Also list the unit of force used in each of these three systems.

6. Give the equation which relates mass and weight. Explain how this equation is related to $F = ma$. Point out how W, g, and m vary as one goes from the earth to the moon.

7. Compute F, m, or a for an object providing sufficient data from the following list are given: forces acting on the object, weight, mass, enough motion data to compute a.

8. Write a quantity given in prefix form, for example, 1.2 ng, in power of 10 form, for example, 1.2×10^{-9} g. Also be able to reverse the process. (See Appendix 2 as well as this chapter.)

9. Determine the normal force acting on a sliding object provided sufficient data are given. Use Y to determine f or μ provided one or the other is given.

10. Compute how far an object of known mass will slide when its initial speed and the forces acting on it are given.

11. Analyze situations involving connected objects similar to the situations discussed in Illustrations 3.4 and 3.5.

12. Explain under what conditions an object falling through the air may attain a terminal velocity.

IMPORTANT TERMS AND PHRASES

You should be able to define or explain each of the following:

Physical law
Inertia
Newton's three laws of motion
The SI and cgs and British units systems
Mass units (kilogram, gram, and slug)
Force units (newton, dyne, pound)
Normal force
Coefficient of friction μ

QUESTIONS AND GUESSTIMATES

1. A car at rest is struck from the rear by a second car. The injuries (if any) incurred by the drivers of the two cars will be of distinctly different character. Explain what will happen to each driver.
2. What happens to a spring in the seat of a car upon which a woman is sitting as the car goes over a large bump? Explain.
3. Clearly identify the action and reaction forces in the following situations: (*a*) a football player kicks the ball; (*b*) the sun holds the earth in orbit; (*c*) a raindrop hits a window; (*d*) a bullet shoots from a gun; (*e*) a rocket accelerates upward; (*f*) a bullet enters a tree trunk and stops.
4. Is it possible for an object to accelerate downward on the earth at a rate greater than g?
5. It is generally believed that, on the average, a drunk will be less injured after a fall from a window than a sober man would be. Can you explain why the belief might be valid?
6. Why is it more dangerous for a diver to hit the concrete walk around the pool than to hit the water? Explain.
7. Suppose a brick is dropped from a height of several inches onto your open hand. Why may severe injury to your hand ensue if your hand is lying flat on a tabletop at the time, even though you can easily catch a falling brick without injury under other circumstances?
8. If there are living beings on some distant body in the universe, are they likely to use the same time unit as we do? The same length unit? The same weight unit? The same mass unit? Which unit would probably be the easiest for them to understand and duplicate?
9. When a Ping-Pong ball is dropped onto a table top, it bounces back up into the air. What knocked it upward?
10. People frequently have a peculiar inner sensation (usually a disturbing feeling in the pit of the stomach) in a moving elevator. Discuss the cause of this effect.
11. A man standing in an elevator with a pipe in his hand is so surprised when the elevator cable breaks that he drops his pipe. What will the pipe do? (Do not neglect friction effects on the elevator system.)
12. Estimate the minimum distance in which a car can be accelerated from rest to 10 m/s if its motor is extremely powerful. (E)
13. When a high jumper leaves the ground, where does the force come from which accelerates the jumper upward? Estimate the force which must be applied to the jumper in a 2-m-high jump. (E)
14. A flower pot weighing 2 lb falls from a window and has a speed of 30 ft/s just before it hits a man on the head. Estimate how large a force the pot exerts on the man's head. Does it matter whether the pot breaks? What if the pot was simply a plastic bag filled with dirt? (E)
15. Estimate the terminal speed of a mercury droplet. Mercury is 13.6 times denser than water. (E)
16. Estimate the force your ankles must exert as you strike the floor after jumping from the top of a 2.0-m-high ladder. Why should you let your legs flex in such a situation? (E)

PROBLEMS

Change all problems stated in the cgs system to the SI before working them.

1. A water skier is being pulled at a constant speed of 12 m/s by a speedboat. The tension in the cable pulling the skier is 140 N. How large a friction force is opposing the skier's motion?

2. A 150-lb parachutist is gliding to the earth with a constant speed of 25 ft/s. The parachute itself weighs 20 lb. How large a force upward does the air rushing past the parachutist and chute exert upon them?

3. A 3200-lb car is to be given an acceleration of 0.50 ft/s^2 on a level road. If friction can be neglected, how large must the unbalanced force on the car be to produce this acceleration?

4. A 1300-kg car moving at 20 m/s is to be stopped in a distance of 80 m. If deceleration is assumed to be uniform, how large is it? How large an unbalanced force is needed?

5. A 1500-kg car is to be towed by another car. If the towed car is to be accelerated uniformly from rest to a speed of 2.0 m/s in a distance of 3.0 m, how large a force must the tow rope hold? Give your answer in both newtons and pounds.

6. The friction force retarding the motion of a particular 400-lb box across the level ground is 100 lb. How large an acceleration can be given to the box if it is pulled by a force of 200 lb by means of a rope inclined at an angle of 53° to the horizontal?

7. A 3200-lb automobile traveling at a speed of 60 ft/s collides with a tree and stops within a distance of 5 ft. How large was the average retarding force exerted by the tree on the automobile?

8. If one pulls straight upward on a 5-kg mass with a string just capable of holding a 20-kg mass at rest, what is the maximum acceleration one can impart to the 5-kg mass?

9. A 160-lb man stands on an accurate spring scale inside an elevator. What does the scale read when the elevator is accelerating (*a*) upward at 10 ft/s^2 and (*b*) downward at 10 ft/s^2?

10. A 100-g mass is hung from a thread, and from the bottom of the 100-g mass a 200-g mass is hung by a second thread. Find the tensions in the two threads if the masses are (*a*) standing still; (*b*) accelerating upward at 20 cm/s^2; (*c*) falling freely under the action of gravity.

11. The coefficient of friction between a box and the floor is 0.40. If the box is given a shove so that it is sliding at 5 ft/s, how far will it slide before it stops?

12. Find the acceleration of the 4-kg block shown in Fig. P3.1 if the coefficient of friction between block and surface is 0.60.

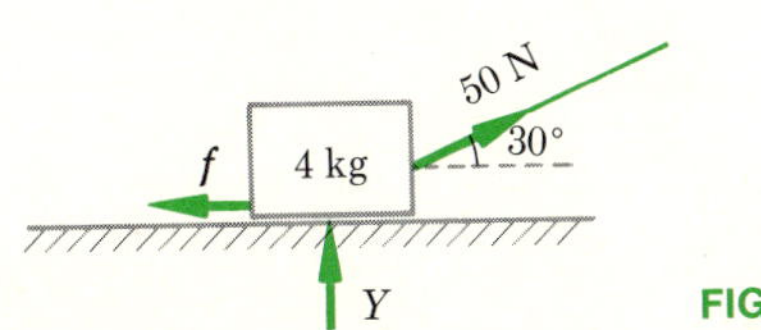

FIGURE P3.1

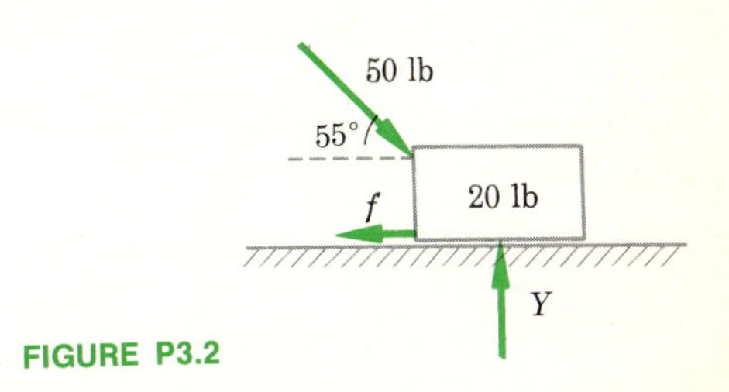

FIGURE P3.2

13. If the coefficient of friction between block and surface shown in Fig. P3.2 is 0.40, find the acceleration of the block.

14. A truck is moving along a flat road when it decelerates to stop. In the rear of the truck is a large box. If the coefficient of friction between the box and truck bed is 0.55 when sliding just starts, what is the maximum deceleration the truck can have if the box is not to slide?

15. In Fig. P3.3, block 1 has a mass of 2.50 kg while block 2 has a mass of 1.60 kg. Ignoring friction forces, what would be the acceleration of the blocks and the tension in the connecting cord? Repeat if a friction force of 12.0 N retards the motion of block 1.

16. In Fig. P3.3, object 1 is a 3000-g mass while object 2 is 2000 g. Upon release of the system, object 2 falls 50 cm in 1.50 s. How large a friction force opposes the motion of object 1? Assume no friction forces exist in the rest of the system.

17. Two 4-kg masses are held at the ends of a cord hung over a frictionless pulley in an Atwood's-machine arrangement. If 2000 g is now added to one of these masses, (*a*) what will be the tension in the cord, and (*b*) how long will it take for the heavier mass to fall 2 m?

18. For the situation shown in Fig. P3.4, find the tension in the cord and the time needed for the masses to

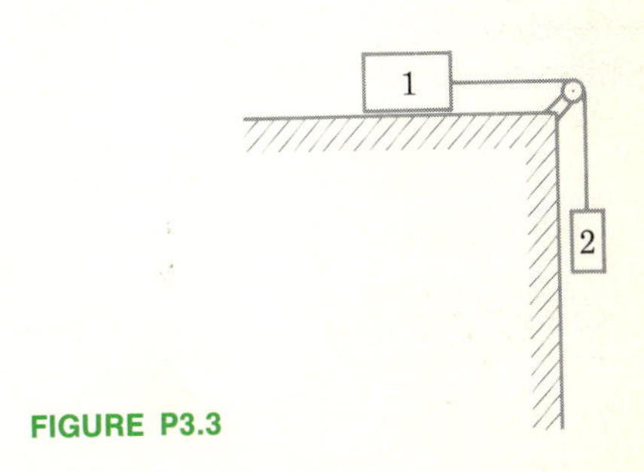

FIGURE P3.3

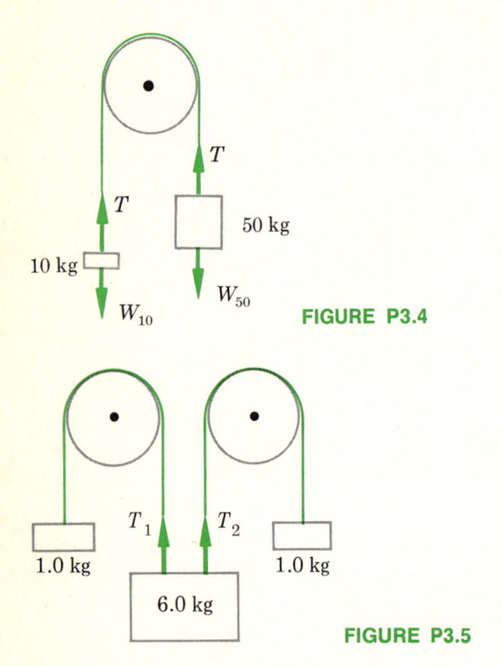

FIGURE P3.4

FIGURE P3.5

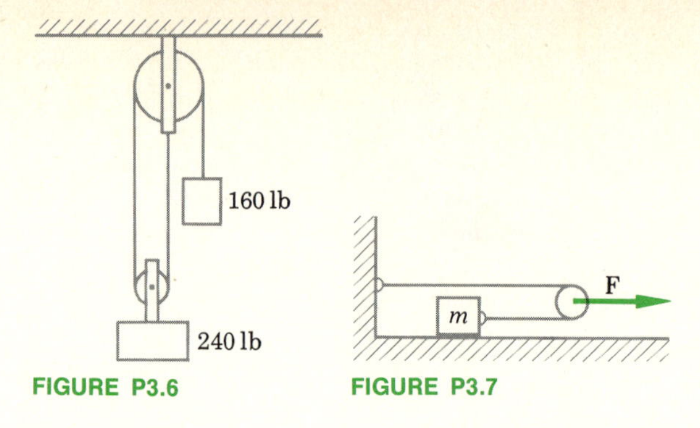

FIGURE P3.6

FIGURE P3.7

move 50 cm starting from rest. Assume the pulley to be frictionless and massless.

19.* Referring to Fig. P3.5, find T_1 and the acceleration of the 6-kg block.

20.* A 500-g mass is suspended from the end of a light spring. The spring stretches 10 cm more when 200 g is added to the 500-g mass. If the 200-g mass is suddenly removed, what will be the acceleration of the 500-g mass at the instant just after removal?

21.* A physicist is sealed in a closed boxcar on a train. She hangs an 8-lb object as a pendulum from the ceiling. As the train begins to accelerate she notices that the pendulum hangs steadily at an angle of 10° with the vertical. How large is the acceleration of the train? Does the physicist really need to know the weight of the object to work this problem?

22.** In Fig. P3.6 the large weight is 240 lb, and the small weight is 160 lb. Ignoring the friction and weight of the pulleys, find the tension in the string and the acceleration of each of the weights. *Hint:* Notice that the large weight moves only half as fast as the smaller weight. Since the pulleys are frictionless and massless, the tension in the string is everywhere the same.

23.** In Fig. P3.7 the pulley is assumed massless and frictionless. Find the acceleration of the mass m in terms of F if there is no friction between the surface and m. Repeat if the friction force on m is f.

24.** In Fig. P3.8 assume there is negligible friction between the blocks and table. Compute the tension in the cord and the acceleration of m_2 if $m_1 = 300$ g, $m_2 = 200$ g, and $F = 0.40$ N. *Hint:* Note that $a_2 = 2a_1$.

25.** Find the tensions in the two cords and the accelerations of the blocks shown in Fig. P3.9 if friction is negligible. Assume the pulleys to be massless and frictionless. ($m_1 = 200$ g; $m_2 = 500$ g; $m_3 = 300$ g.)

26.** If the friction force on each of the blocks sliding on the table shown in Fig. P3.9 is 2 lb, find the acceleration of each block and the tensions in the strings. (The weights of masses 1, 2, and 3 are 8, 32, and 16 lb, respectively.)

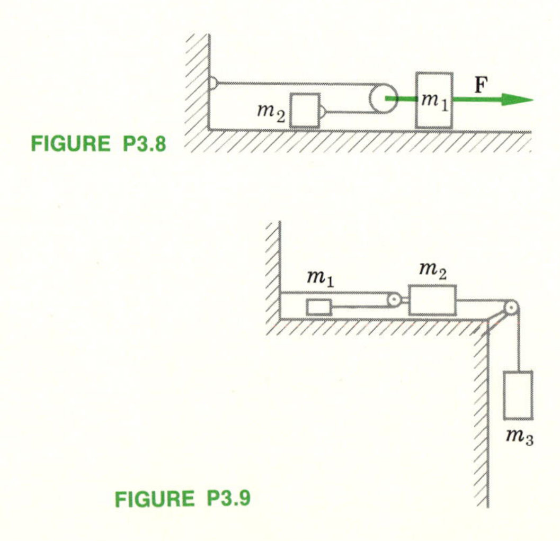

FIGURE P3.8

FIGURE P3.9

4 MOTION UNDER THE ACTION OF GRAVITY

In previous chapters we have made use of the fact that the earth exerts a gravitational attractive force upon all objects close to it. In this chapter we shall find that a gravitational attraction occurs between any two pieces of matter. The mathematical form of this force is presented and discussed. Armed with this knowledge, we shall be able to understand the motion of bodies traveling at an angle to the surface of the earth. This type of motion, projectile motion, for example, will also be investigated.

4.1 NEWTON'S LAW OF GRAVITATION

Long before the time of Newton, astronomers had discovered how the planets of our solar system revolve about the sun. Their motions were known with high precision, and very accurate predictions could be made for their future motions. All this was accomplished without any real understanding of the reasons for these motions.

It was Newton who first showed the nature of the forces involved in planetary motion. Since he knew that "a moving body will continue to move in a straight line forever unless acted upon by some external force," it was clear to him that an external force acted upon each of the planets. For the planets to move in nearly circular orbits, there must be some external force deflecting them from a straight-line path. Experience tells us that a ball revolving in a circle at the end of a string is held in that circular path by the pull of the string. The pull of the string on the ball is toward the center of the circle. This pull on the object toward the center of the circle is called the **centripetal force,** and we shall discuss it more fully in Chap. 8.

Newton concluded that *the planets experience a force directed toward the sun and that this force holds them in their nearly circular paths, or orbits, about the sun.* He pictured the sun as exerting some type of attractive force on the planets. Moreover, since the moon circles about the earth, there must be an attraction of the moon by the earth. This attractive force upon the moon holds it in its orbit about the earth. There are moons circling others of the planets, and so *this attraction of one body for another appears to be a general phenomenon.*

FIGURE 4.1
The law of action and reaction tells us that the gravitational attraction force on one sphere is equal and opposite to that on the other sphere.

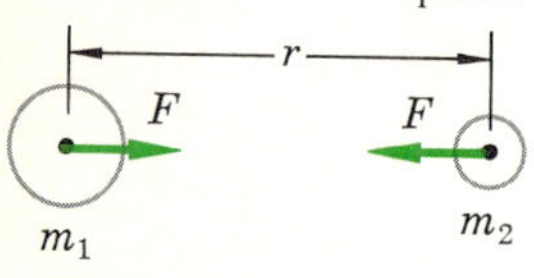

The exact mathematical form of the attractive force between two bodies was first stated by Newton. He compared the way in which the planets should move under various assumed attractive forces with their known motion. Only one type of force could duplicate the observed motion. This force is proportional to the product of the masses of the two bodies and inversely proportional to the square of the distance between them. If, as shown in Fig. 4.1, the distance between the centers of two spherical bodies is r and the masses of the bodies are m_1 and m_2, then the attractive force exerted by one of the spheres on the other is given by the equation*

$$F = G\frac{m_1 m_2}{r^2} \tag{4.1}$$

where G is a proportionality constant, the value of which will be given presently. *This is* **Newton's law of universal gravitation.**

Newton's Law of Universal Gravitation

Not only does body m_1 attract m_2 with a force F, but Newton's law of action and reaction tells us that body m_1 will feel an equal and opposite reaction force. This is why we show the equal and opposite forces acting on the two spheres in Fig. 4.1.

*If the bodies are not spherical, the value to be used for r is more complicated. However, if the bodies are small compared with their distance of separation, r can be taken equal to the distance of separation between any two points on the bodies, to a fair approximation.

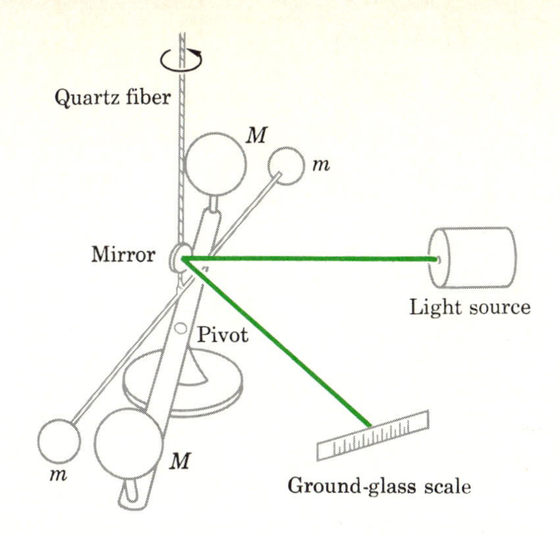

FIGURE 4.2
Schematic diagram of the Cavendish balance. Observe how the beam of light is used to detect the twist of the fiber.

The value of the gravitational constant G in Eq. (4.1) is not predicted by theory and can be determined only by experiment. Its value was first found by Henry Cavendish in 1798 with a device called the **Cavendish balance.** One form of this device is shown schematically in Fig. 4.2. Two identical masses m are suspended from an extremely thin and delicate quartz fiber. The two large masses M can be moved close to the small masses m, and the attraction between M and m will cause the fiber to twist. After calibrating the system so that the amount of force needed to produce a given twist is known, the attraction force between m and M can be computed directly from the observed twist of the fiber. Then, since m, M, r, and F are all known, one can substitute in Eq. (4.1) to obtain

$$F = G\frac{mM}{r^2}$$

and solve for the single unknown G.

In practice, this is an extremely delicate experiment since the attraction force is so very small. To measure the very small twist of the fiber, a beam of light is reflected from a mirror attached to the fiber system. By this means the slight twist of the fiber results in a more easily measured deflection of the light beam. (Using a light beam in this way is referred to as an **optical lever.**) Because of the delicacy of the fiber, even the slightest movement of the air close to it will disrupt the measurements. Therefore great care is needed to eliminate air movement and vibration if reliable results are to be obtained. The most accurate measurements of G presently available give the following value:

$$G = 6.672 \times 10^{-11}\ \mathrm{N \cdot m^2/kg^2}$$

Illustration 4.1 Once the value of G is known, it becomes possible to determine the mass of the earth. Let us now do so.

Reasoning To do this, one simply observes the force with which the earth attracts an object on its surface. For example a mass m weighs an amount $W = mg$. But this is the force with which the earth of mass M_{earth} attracts the object. Since the distance between centers of the two masses, M_{earth} and m, is simply the radius of the earth R_{earth}, Newton's law of gravitation can be written

$$W = G\frac{M_{\text{earth}}m}{R^2_{\text{earth}}} \tag{4.2}$$

Since $W = mg$, this becomes

$$mg = G\frac{M_{\text{earth}}m}{R^2_{\text{earth}}}$$

or, after canceling m from each side,

$$g = G\frac{M_{\text{earth}}}{R^2_{\text{earth}}} \tag{4.3}$$

The acceleration due to gravity $g = 9.8\ \text{m/s}^2$ while $R_{\text{earth}} = 6.4 \times 10^6$ m. Using these together with the value of G, one finds $M_{\text{earth}} = 6 \times 10^{24}$ kg. Can you devise a method which could be used to determine the masses of the moon, the sun, and the planets?

It is interesting to note that Eq. (4.3) tells us how the acceleration of gravity should depend upon the distance R from the earth's center provided we replace R_{earth} by R. Clearly, the larger the R, the smaller g will be. This fact (that g depends upon distance from the center of the earth) can be and has been checked by measuring the difference between g at sea level and on mountaintops. Now that space travel is an accomplished fact, it has been verified that to high precision *g varies inversely as the square of the distance from the center of the earth.* Notice, however, that *this is true only outside the surface of the earth.* Beneath the surface of the earth, Eq. (4.3) is no longer correct.

DEFINITION

Often we hear the term **gravitational field.** In colloquial use this term refers to a region in which a gravitational force exists. But it has a more precise meaning to the scientist. *The* **gravitational field strength** *at a point is defined to be the gravitational force exerted on a unit mass at the point in question.* Throughout a region, the gravitational field strength may vary. But *it has a definite value at each point in the region. These values for the gravitational field strength constitute the gravitational field for the region in question.*

Gravitational Field

Before leaving this section we should point out that Newton's law of gravitation encompasses Galileo's law of freely falling bodies. As we see from Eq. (4.3), the mass of the falling object has canceled out, and so the acceleration due to gravity will be the same for all bodies in free fall. Of course Newton's law is much more fundamental than Galileo's since it includes not only Galileo's results but many others as well.

Illustration 4.2 Find the ratio of an object's weight on the moon to its weight on the earth. Also find the acceleration in free fall on the moon.

Reasoning When we speak about the weight of an object on a planet or the moon we mean the force of attraction the planet or moon has for the object. In these cases, the earth is far enough away for its force of attraction to be very small. For example, an object on the moon is approximately 60 earth radii from the center of the earth. Since the gravitational attraction decreases as $1/R^2$, the earth's pull on the object is only about $\frac{1}{3600}$ as large as its weight on the surface of the earth. This very small force of the earth on the moon object can usually be ignored.

We know that the attraction of the moon for an object on its surface will be

$$F = G\frac{mM}{R^2} = W_{\text{moon}}$$

where m is the object's mass, M is the mass of the moon ($M \approx 0.0123M_{\text{earth}}$, where M_{earth} is the earth's mass), and R is the radius of the moon ($R \approx 0.27R_{\text{earth}}$, where R_{earth} is the earth's radius). Therefore the object's weight on the moon is

$$W_{\text{moon}} = 0.168G\frac{mM_{\text{earth}}}{R_{\text{earth}}^2}$$

But the weight of the object on the earth, the earth's attraction for it, will be

$$W_{\text{earth}} = G\frac{mM_{\text{earth}}}{R_{\text{earth}}^2}$$

Comparing these two relations, we see that

$$\frac{W_{\text{moon}}}{W_{\text{earth}}} \approx 0.168 \approx \tfrac{1}{6}$$

In other words, *an object weighs only about one-sixth as much on the moon as it does on the earth.*

To find the free-fall acceleration on the moon, we simply make use of $F = ma$, where F is now the weight of the object and a is the acceleration due to gravity. Then

$$W_{\text{moon}} = mg_{\text{moon}}$$

while

$$W_{\text{earth}} = mg_{\text{earth}}$$

Dividing one equation by the other yields

$$\frac{g_{\text{moon}}}{g_{\text{earth}}} = \frac{W_{\text{moon}}}{W_{\text{earth}}} \approx \tfrac{1}{6}$$

and so
$$g_{\text{moon}} \approx \frac{g_{\text{earth}}}{6} = 1.6 \text{ m/s}^2$$

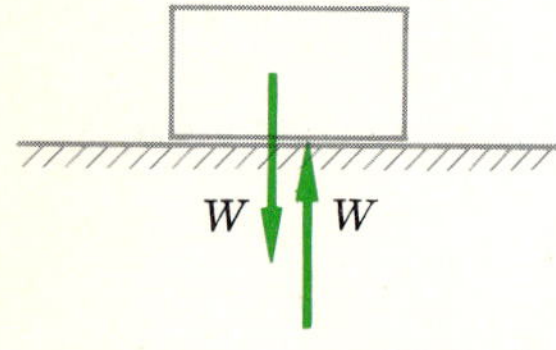

FIGURE 4.3 Since the block is at equilibrium, the pull of gravity down on it must be balanced by the push of the floor upward.

4.2 MOTION ON AN INCLINE

If a block is at rest on level ground, it will remain at rest unless someone pushes it. The fact that it does not move does not mean that no forces act upon it. We know that gravity pulls down on it. There is no unbalanced force upon it, though, because the push upward of the ground just balances the pull of gravity. This is illustrated in Fig. 4.3. Application of $F = ma$ to this case is trivial, since F, the unbalanced force, is zero and hence a is zero.

Illustration 4.3 Consider now the case when the block is on an incline. This is shown in Fig. 4.4*a*. What forces now act upon the block?

Reasoning As before, gravity pulls straight down, with force W. The forces at the inclined surface are now somewhat more complicated. It is customary and convenient to consider the force exerted by the incline upon the block to consist of two parts. One part, a push directed perpendicular to the surface of the incline, is shown as the vector **Y** in the free-body diagram of Fig. 4.4*b*. This force we called the **normal,** i.e., perpendicular, force in Sec. 3.8. In addition, there will be a friction force at the surface of the incline pushing on the block so as to keep it from sliding down the incline. This force is shown as the vector **f** in the diagram.

It is clear that if the block moves at all, i.e., if the friction force is too small to keep it from slipping, the block will move down the incline parallel to its surface. Since the motion is going to be along the incline, *it is convenient to take our x and y axes parallel and perpendicular to the incline,* as shown in Fig. 4.4*c*. We then split the various forces into x and y components. (At first, until you are accustomed to inclined axes, it might be well to rotate the page

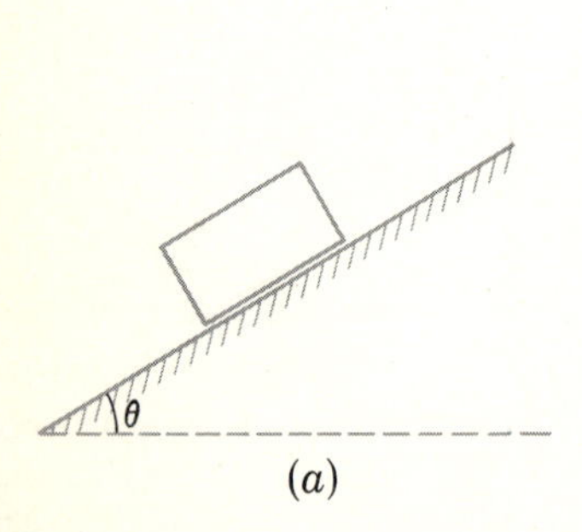

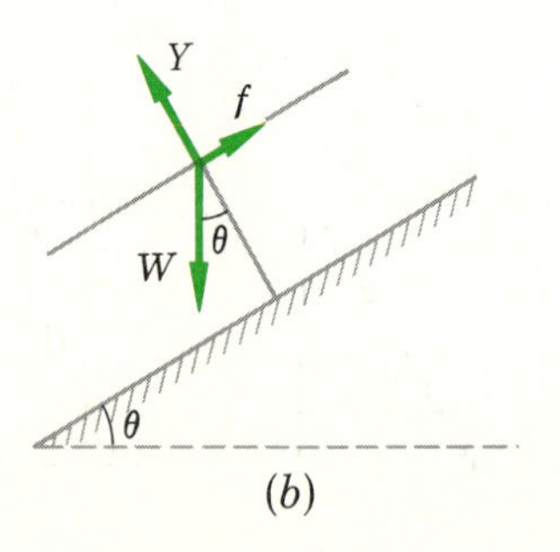

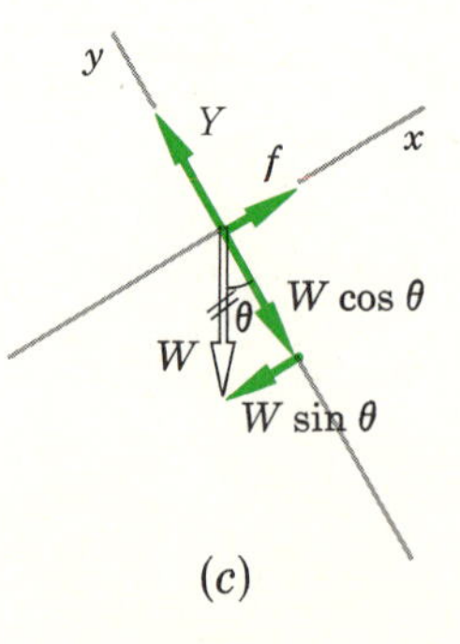

FIGURE 4.4 In dealing with a body on an incline, it is convenient to take x and y axes parallel and perpendicular to the incline, as shown. The forces are then split into components along these axes.

so that the x axis lies in a horizontal direction.) Both **Y** and **f** are already in the proper directions, and only **W** need be split into its components. They are shown in the figure.

Now, since we know that the block will not crash through the incline or float above it, no motion occurs in our y direction. Hence, the sum of the y forces must be zero. From Fig. 4.4c this means

$$Y - W\cos\theta = 0 \tag{4.4}$$

If there is enough friction to hold the block in place, the sum of the x forces must also be zero. One would then have

$$W\sin\theta - f = 0 \tag{4.5}$$

But if the friction force were too small to hold the block from sliding, there would be an unbalanced force on the block. It would be

$$F = W\sin\theta - f \tag{4.6}$$

This unbalanced force would cause the block to accelerate. From Newton's law we have

$$F = ma$$

which becomes

$$W\sin\theta - f = ma \tag{4.7}$$

Hence the acceleration of the block can be computed.

It is interesting to notice that since $W = mg$, Eq. (4.7) for the acceleration becomes

$$mg\sin\theta - f = ma$$

and if there is no friction so that $f = 0$,

$$a = g\sin\theta \qquad \text{no friction} \tag{4.8}$$

Hence, it appears that the acceleration of a body on an incline does not depend upon the nature of the body, provided that friction is absent. This means that a child's wagon will move down a hill with the same acceleration as an automobile if friction forces are negligible.

In the limiting case, when $\theta = 0$ and the ground is flat, Eq. (4.8) says that the acceleration is zero, since $\sin\theta = 0$. Of course this is true. On the other hand, if $\theta = 90°$, that is, if the incline is straight up and down, the block will fall straight down. If there is no friction, the body should fall with its free-fall acceleration g. Equation (4.8) is also true in that limit, since $\sin 90° = 1$, and Eq. (4.8) becomes $a = g$.

You should be cautioned at this point that it usually is not worthwhile to memorize most of the above equations. Equation (4.8) is perhaps sufficiently important to be memorized, but the others are easy to work out in any given case. Moreover, in many situations there will be other forces acting on the body, and so the problem will have to be worked through from the beginning. As we shall see, this is not at all difficult to do.

Illustration 4.4 Let us consider the following situation. A child in a wagon starts from rest at the top of a hill which is 20 ft high and 100 ft long, as shown in Fig. 4.5. (1) What will be the wagon's acceleration? (2) How fast will it be going at the bottom of the hill? (Assume that the friction forces are negligible.)

Reasoning To answer these questions, let us examine the free-body diagram shown in part *b* of the figure. Y must just balance $W \cos \theta$. Hence, the only unbalanced force is in the x direction and is $W \sin \theta$. Using Newton's law, $F = ma$, yields

$$W \sin \theta = ma$$
$$mg \sin \theta = ma$$

or

$$a = g \sin \theta$$

To find $\sin \theta$, we notice from Fig. 4.5*a* that the definition of the sine as the opposite side over the hypotenuse gives, in the triangle formed by the incline,

$$\sin \theta = \frac{20 \text{ ft}}{100 \text{ ft}} = 0.20$$

Since $g = 32 \text{ ft/s}^2$, the acceleration becomes

$$a = (32 \text{ ft/s}^2)(0.20) = 6.4 \text{ ft/s}^2$$

The child's motion is uniformly accelerated, and hence the motion

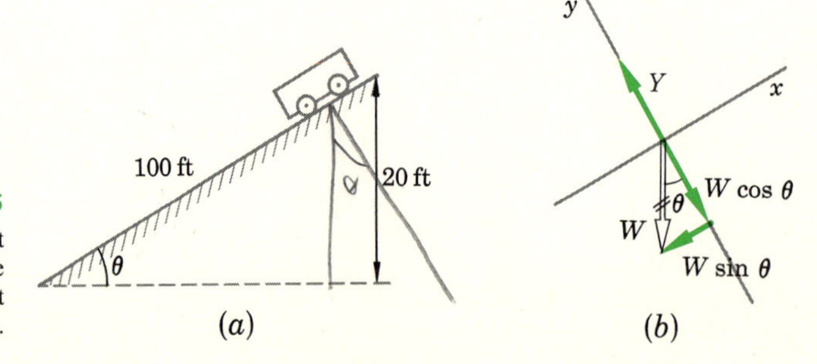

FIGURE 4.5
The component of the weight along the incline causes the wagon to accelerate (not drawn to scale).

equations of Chap. 2 apply to this problem. We have

$$v_0 = 0 \qquad a = 6.4\ \text{ft/s}^2$$
$$s = 100\ \text{ft} \qquad v_f = ?$$

The appropriate equation is

$$v_f^2 - v_0^2 = 2as$$
$$v_f^2 = (2)(6.4\ \text{ft/s}^2)(100\ \text{ft})$$
$$v_f = 10\sqrt{12.8} = 36\ \text{ft/s}$$

Notice that the weights of the child and the wagon were not needed to solve this problem. If we had considered a friction force to be present, the mass would not have canceled out.

Illustration 4.5 A particular 1200-kg automobile is capable of accelerating at 0.50 m/s² up an incline which rises 4.0 m in each 40 m. With friction ignored, how large a force must be exerted on the car, pushing it up the incline?

Reasoning Refer to Fig. 4.6*a* and *b*. The force P pushing the car up the incline is the force desired. Obviously the unbalanced force in the x direction is $P - W\sin\theta$. From Newton's law we have, since $m = 1200$ kg and $W = mg$,

$$F = ma$$
$$P - W\sin\theta = ma$$

or

$$P - (1200 \times 9.8)\left(\frac{4.0}{40}\right) = (1200\ \text{kg})(0.50\ \text{m/s}^2)$$

from which

$$P = 1780\ N$$

Can you describe how the wheels furnish this force to the car?

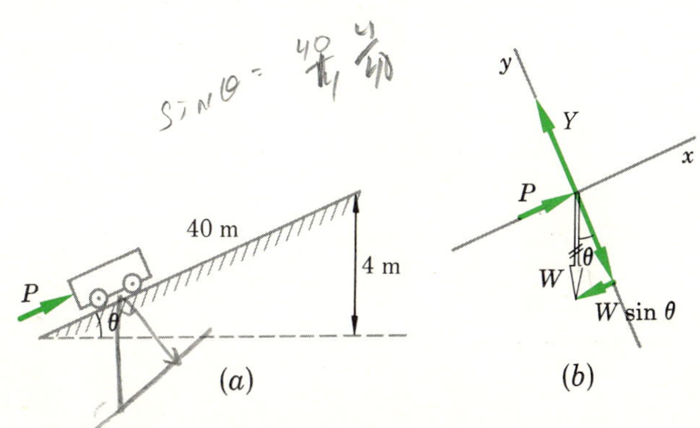

FIGURE 4.6
The force P is partly balanced by the component of the weight acting down the incline. Acceleration up the incline results from the unbalanced portion of P.

Illustration 4.6 A motor is to pull a 50-kg block up the incline shown in Fig. 4.7. The coefficient of friction between the block and incline is 0.70. What is the tension in the rope if the cart is moving at constant speed?

Reasoning Notice here that the acceleration is zero. Hence there is no unbalanced force on the cart, and the cart is in equilibrium (traveling at constant speed). Since the sum of the x forces is zero, one has from Fig. 4.7b

$$T - f - W \sin \theta = 0$$

or, since $W = mg$,

$$T = f + (50 \times 9.8 \text{ N})(\tfrac{6}{10})$$

But the friction force $f = \mu Y$, and since the y forces must balance, $Y = W \cos \theta$. Hence

$$f = (0.70)(50 \times 9.8 \text{ N})(\tfrac{8}{10})$$

Substitution of this value gives $T \approx 570$ N.

Illustration 4.7 Find the acceleration for the system shown in Fig. 4.8a. The friction force on the 5-kg block is 20 N.

Reasoning As in the similar problems in the last chapter, we isolate each body and write down $F = ma$ twice, once for each body. The free-body diagrams are shown in parts b and c of Fig. 4.8. When friction forces are present, it is necessary to know the direction of motion, since f will oppose the motion. In this case, we surmise that the 7-kg mass will fall. Taking the motion direction as the positive direction, we have the following $F = ma$ equations:

$$(7 \times 9.8) - T = 7a$$
$$T - 29.4 - 20 = 5a$$

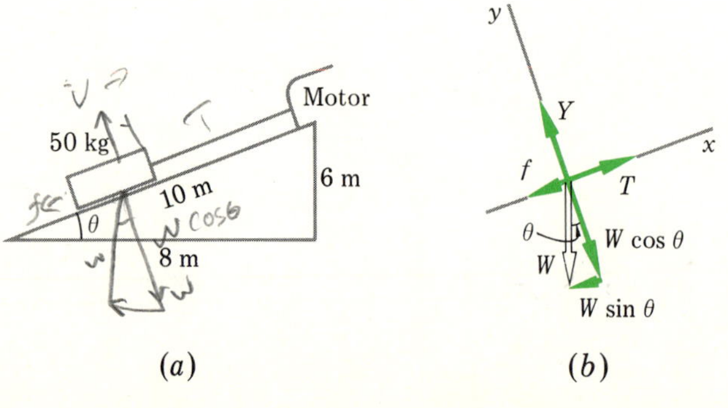

FIGURE 4.7
Since the block is to move up the incline at constant speed, the pull resulting from the motor must exactly balance the sum of the friction force and the component of the weight acting down the incline.

(a) (b)

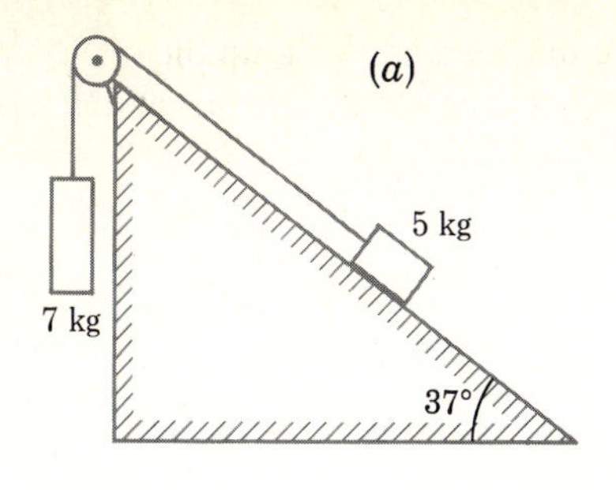

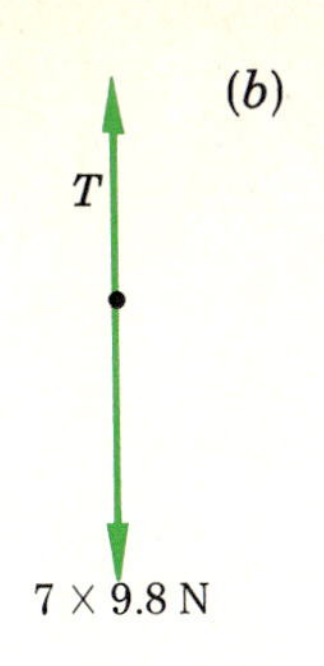

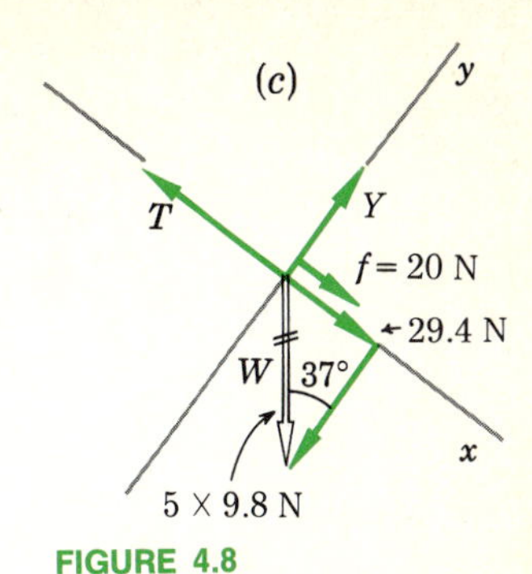

FIGURE 4.8

The 7-kg weight falls downward, thereby pulling the other weight up the plane. (*b*) and (*c*) give the free-body force diagrams for the two weights.

Adding the equations gives

$$19.2 = 12a$$
$$a = 1.6 \text{ m/s}^2$$

You should supply the units for these equations and show that the stated units for a are correct.

The tension is found by substitution in the second equation, giving

$$T = 49.4 + 8.0 = 57.4 \text{ N}$$

4.3 PROJECTILE MOTION

In this section we shall consider the motion of a freely falling body which has a velocity component parallel to the surface of the earth. The freely falling bodies considered in Chap. 2 are special cases of the motion we now wish to discuss. In that chapter we treated the case of a body which moved only in a vertical direction. This is not the most common case.

For example, when a baseball player throws the ball to a teammate, the ball does two things at once after leaving the player's hand. Since it is attracted downward by the earth, it will accelerate downward under this unbalanced force. At the same time, however, it will travel horizontally with a certain speed. This may be understood by reference to Fig. 4.9.

Suppose that the ball leaves the pitcher's hand at point A and that its velocity at that instant is completely horizontal. Call its velocity at that instant v_H. According to Newton's law, there will be no acceleration in the

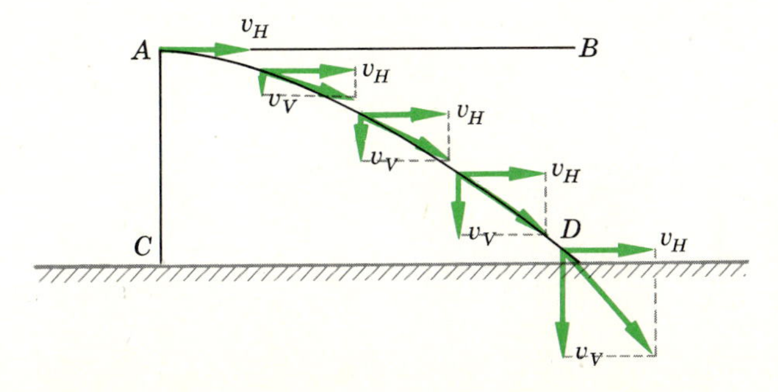

FIGURE 4.9

The initial horizontal speed of the ball v_H does not change. As the ball moves with this speed to the right, it also falls under the action of gravity, as shown by the vectors representing the vertical component of the velocity. Note that $v = \sqrt{v_H^2 + v_V^2}$ and is tangent to the trajectory.

THE DISCOVERY OF X-RAYS

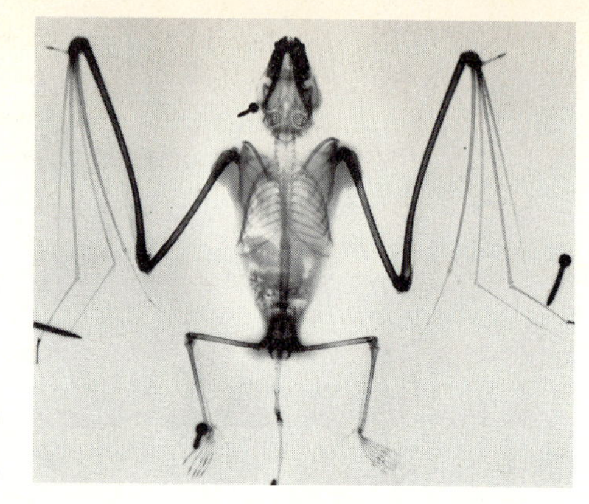

An x-ray of a bat taken with the apparatus of Fig. 27.15*b*. (*Hewlett-Packard.*)

Many of the laws of physics have been discovered by interpreting the results of experiments carefully designed to determine these laws. This was the case in Galileo's discovery of the law of falling bodies. Newton's laws are also of this same general type. However, sometimes in physics a curious, unexpected laboratory circumstance leads to the discovery of an important phenomenon. This was how Wilhelm Konrad Röntgen (1845–1923) discovered x-rays.

Carrying out experiments with high-voltage discharges or sparks, he applied a potential difference of several thousand volts to electrodes in the two ends of a partly evacuated tube. Under such conditions, a discharge or glow much like that observed in neon signs occurs. However, if the pressure of the gas in the tube is reduced low enough, the glow nearly ceases. While carrying out experiments with a highly evacuated discharge tube in his darkened laboratory in 1895, Röntgen observed that a nearby fluorescent screen (much like that on the end of present-day TV tubes) was also glowing in the darkness. By moving the screen about the room, he was able to show that the light given off by the fluorescent screen resulted from something taking place in the discharge tube. Since a light-tight cover could be placed over the tube without greatly affecting the glow on the screen, it was clear that the fluorescent glow was caused by something other than the light given off by the tube. Röntgen named this unknown radiation striking the screen and coming from the tube ***x-rays.***

He carried out many experiments with the rays from the tube and found that they were highly penetrating, even being able to pass through a book. Heavy metals and bone (among other materials) were not nearly so transparent to the rays as such materials as wood and paper. He was even able to cast a shadow with these rays. Shadows could be produced showing the bones of his wife's hand and the ring on her finger. Because of this ability to produce shadows, Röntgen held open the possibility that these rays were short-wavelength light. However, he was reluctant to state this as fact because, unlike light, the rays were not deflected, i.e., refracted, appreciably in passing from air to water.

As we shall see in later chapters, the x-rays discovered by Röntgen are indeed similar to light waves but have much shorter wavelengths. Unfortunately, as we know today, x-rays damage human tissue and cause severe burns through overexposure. Even when visible burns do not occur, other damage may be present. Many early workers in the field of x-rays suffered deteriorating health and even death because of their ignorance of the harmful properties of these little-understood rays.

horizontal direction unless a horizontally directed force acts on the ball. But if we ignore friction with the air, the only force acting on the ball once it is free from the hand of the pitcher is the force of gravity. There is no horizontal force acting on the ball. Hence, its horizontal velocity will remain unchanged and will be v_H until it hits something.

It should be clear from this that the horizontal motion of the ball is extremely simple. The ball moves with constant horizontal velocity component v_H until it strikes something. For the horizontal motion we therefore have (since the initial velocity v_0 is equal to the final velocity v_f and average velocity $\bar{v}$)

Horizontal:
$$v_0 = v_f = \bar{v} = v_H$$
$$a = 0 \qquad s = \bar{v}t = v_H t$$

The vertical motion of the ball is not much more complicated. It will accelerate downward under the force of gravity, and hence $a = g$. This vertical motion is exactly the same as we discussed for a freely falling body in Chap. 2. The solution to the problem of the vertical motion of the ball is not new to us, therefore, and we should be able to carry it through with no difficulty.

Our procedure, then, will be to recognize that the free motion of a ball, bullet, or any other projectile contains within it two separate problems. The horizontal problem is motion with constant velocity and is therefore quite simple. The vertical motion is identical to the motion of a free body in a vertical line. We shall compute each portion of the projectile-motion problem separately. The solutions to the individual problems will then be combined to obtain the answer which is requested. We shall, in effect, have found the horizontal v_H and vertical v_V components of the velocity of the projectile. The magnitude of the resultant velocity is found by using the usual method for vectors at right angles, $v = \sqrt{v_H^2 + v_V^2}$. This procedure is best seen by means of an example.

Illustration 4.8 Consider the situation in Fig. 4.9. Suppose that the ball leaves the hand of the pitcher, traveling horizontally with a velocity of 15 m/s, and suppose that it is 2.0 m above the ground at that instant. Where will it hit the ground? (That is, how far is point D from point C in Fig. 4.9?)

Reasoning We start the solution by splitting the problem into two parts:

HORIZONTAL	VERTICAL (DOWN POSITIVE)
$v_0 = v = \bar{v} = 15$ m/s $s_H = \bar{v} = 15t$	$v_0 = 0$ $a = 9.8$ m/s^2 $s_V = 2.0$ m
To find t, we solve the vertical problem	To find t, we use $s_V = v_0 t + \frac{1}{2}at^2$ $2.0 = 4.9t^2$ $t = 0.64$ s

Now that the time of flight, i.e., the time to drop to the ground, has been found from the vertical problem, the result can be used in the horizontal problem. One has

$$s_H = (15\ \text{m/s})(0.64\ \text{s}) = 9.6\ \text{m}$$

In other words, the ball travels only 9.6 m in a horizontal direction before the pull of gravity causes it to fall to the ground. Actually, one usually throws a ball somewhat upward* if one wants it to travel any great distance. If the ball has an initial upward component to its velocity, it will take longer for it to fall to the ground and so it will have more time to travel in a horizontal direction.

Illustration 4.9 A ball is thrown with a velocity of 30 ft/s at an angle of 37° above the horizontal. It leaves the pitcher's hand 4.0 ft above the ground and 15 ft from a wall, as shown in Fig. 4.10. (1) At what height above the ground will it hit the wall? (2) Will it still be going up just before it hits, or will it already be on its way down?

Reasoning Consider the trip from the starting point to the wall. For part 1:

HORIZONTAL	VERTICAL (UP POSITIVE)
$v_0 = v = \bar{v} = 24$ ft/s	$v_0 = 18$ ft/s
$s = 15$ ft	$t = \frac{5}{8}$ s
$= \bar{v}t$	$a = -32$ ft/s^2
$15 = 24t$	$s = ?$
$t = \frac{5}{8}$ s	$s = v_0 t + \frac{1}{2}at^2$
We now use this value	$= (18)(\frac{5}{8}) - (16)(\frac{5}{8})^2$
in the vertical problem	$= 11.2 - 6.2$
	$= 5.0$ ft above the starting point

Hence, the ball strikes the wall 5 ft above the starting point, or 9 ft above the ground.

*But not at an angle to the horizontal greater than 45°. Why not?

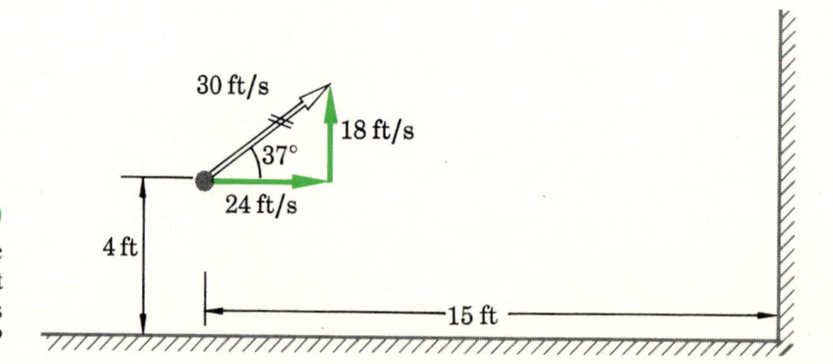

FIGURE 4.10
Where will the ball hit the wall? Is it going up when it hits, or is it already on its way down?

For part 2 let us continue the vertical part of the problem and find the vertical component of the velocity of the ball just before it hits the wall. We have

$$v = v_0 + at = 18 - (32)(\tfrac{5}{8}) = 18 - 20 = -2 \text{ ft/s}$$

Since the upward direction was taken as positive, a negative velocity means a downward velocity. Hence, the ball was on its way down when it hit.

The total velocity of the ball just before it strikes has a downward component of 2 ft/s and a horizontal component of 24 ft/s. These components can be added in the usual way to find the magnitude of the velocity at that point. It is $\sqrt{(24)^2 + 2^2}$ ft/s.

SUMMARY

Every piece of mass in the universe attracts every other piece of mass. If two separated spheres have masses m_1 and m_2, the attractive force either one exerts on the other is given by Newton's law of gravitation:

$$F = G\left(\frac{m_1 m_2}{r^2}\right)$$

In this relation, G is a constant of nature $6.67 \times 10^{-11}\ \text{N} \cdot \text{m}^2/\text{kg}^2$. The distance between the centers of the spheres is r. This relation also applies to nonspherical masses which are tiny in comparison to their separation r.

The acceleration due to gravity g can be found from $F = ma$ and Newton's law of gravitation. Assuming the earth to be a sphere, a mass outside the earth at a distance r from the earth's center will fall with an acceleration $g = GM_{\text{earth}}/r^2$, where M_{earth} is the earth's mass. A similar expression applies to free fall on the moon with M_{earth} replaced by the moon's mass. It is found that g_{moon} is about one-sixth as large as g_{earth}.

When considering motion on an incline, the situation is analyzed in terms of component motions. The x axis is taken up along the surface of the incline while the y axis is taken perpendicular to the inclined surface.

Projectile motion, without friction, is analyzed in terms of its component motions. For the horizontal motion along the earth, the acceleration is zero. As a result, $v_0 = v_f = v_H$ and the five motion equations reduce to $s = v_H t$. The vertical component of the motion is treated independently and is exactly the same as that which would occur in the absence of the horizontal motion.

MINIMUM LEARNING GOALS

Upon completion of this chapter, you should be able to do the following:

1. Write the equation which states Newton's law of gravitation. By means of a diagram, explain each symbol in the equation. State what restrictions apply to the equation.

2. Find the gravitational force exerted by one sphere upon another provided m_1, m_2, r, and G are given. Repeat in the case of two tiny masses (point masses) of any shape provided r is much larger than the dimensions of the masses.

3. Compute the acceleration due to gravity on any large planet provided the planet's mass and radius are given. Assume G is also given.

4. Resolve the forces acting on an object supported by an incline into components along and perpendicular to the incline. Point out which component force(s) cause the object to slide down the incline.

5. Find the friction force retarding the motion of an object along an incline provided θ, μ, and m are given. Repeat for the case where a known force on the object assists or retards the motion.

6. Compute the acceleration of an object up or down an incline when the forces acting on the object are given.

Similarly, if the acceleration and all but one of the forces are given, find the unknown force.

7. Analyze the motion of systems similar to the one described in Illustration 4.7.
8. Find the distance traveled over level ground by a projectile shot at a known angle and speed from a given height above the ground.
9. Calculate where a projectile will strike a distant wall when the initial velocity and position of the projectile are given.

IMPORTANT TERMS AND PHRASES

You should be able to define or explain each of the following:

Newton's law of universal gravitation

The gravitation law can be used to derive Galileo's results concerning g.

Take the components of an object's weight along and perpendicular to the incline.

Projectile motion consists of two separate motions occurring simultaneously.

In the horizontal motion of a projectile,

$$\bar{v} = v_0 = v_f \qquad \text{and} \qquad a = 0$$

QUESTIONS AND GUESSTIMATES

1. It has been proposed that a space colony be established on one of the asteroids in our solar system. These tiny planetoids have masses only a small fraction of that of the moon. How would the lives of the inhabitants of the colony be changed because of gravitational effects?
2. Suppose the earth had its present geometrical size but had a mass 100 times larger. For reasons we shall see in later chapters, its atmosphere would differ from ours. But, for discussion, suppose that the atmosphere was the same as ours and that humanoid beings existed on it. How would their bodies have to differ from our own?
3. If you kick your shoe into a mound of dry, powdery dirt, dust flies in all directions. Explain why this effect is much more dramatic on the moon than on earth.
4. Airplane enthusiasts sometimes hold meets where they try to show their skills. One event is to drop a sack of sand exactly in the center of a circle on the ground while flying at a predetermined height and speed. What is so difficult about that? Don't they just drop the sack when they are directly above the circle?
5. Parents should be prepared for nearly anything. Suppose your child wants to find out how fast a slingshot can shoot a stone. Devise a method for finding out. Assume the only tool you have is a meterstick.
6. If you want to hit a distant stationary object with a rifle, you do not aim the gun with the object in line with the hole in the barrel of the gun. How is the gun aimed?
7. About how high could you throw a baseball on the moon? About how high could you jump? About how far could you throw a quarter?
8. Even though the diameter of the planet Neptune is about four times as large as that of earth, the acceleration due to gravity is the same on Neptune as it is on the earth. What can you conclude about Neptune from these data?
9. The faster you throw a stone parallel to the earth, the farther it will go. Explain what would happen to a stone as you throw it at high speeds parallel to the surface of the moon. Assume you can throw it at very high speed.
10. Suppose you have a very heavy box you wish to lift onto the back of a station wagon, but it is too heavy for you to lift. Explain why you can get it into the station wagon by sliding it up a board acting as an incline even though the box is too heavy to lift.
11. Two 100-g balls hang side by side at the ends of two strings. If the strings are 2 m long and the balls are spaced 10 cm from center to center, give an order-of-magnitude estimate of the angle at which the strings will hang to the vertical if all except gravitational effects are ignored.
12. In Fig. P4.1 is shown a sequence of photos of a breakaway-type aluminum lamp pole. Why does a pole of this type decrease damage to the auto? Estimate the speed of the auto during the collision. Each photo is separated by the same time interval. (E)

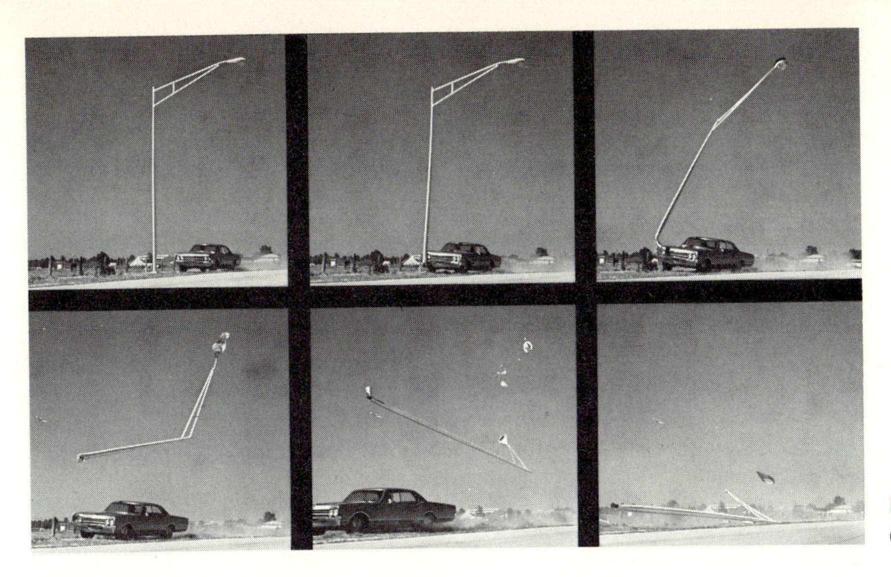

FIGURE P4.1
(*Photo courtesy Alcoa.*)

PROBLEMS

1. A neutron is an uncharged particle with a mass of 1.67×10^{-27} kg and a radius of the order of 10^{-15} m. Find the gravitational attraction between two neutrons whose centers are separated by 10^{-12} m. Compare this with the weight of a neutron on the earth.

2. Two coins, each of mass m, lie 50 cm apart on a table top. (*a*) Find the approximate gravitational pull (in terms of m) of one upon the other. (*b*) One coin is given a push toward the other. How small must the coefficient of friction be between coin and table if the coin is to continue its motion? Assume $m = 10$ g for part (*b*).

3. Compare the pull of gravity on a spaceship at the surface of the earth with the gravitational pull when the ship is orbiting 1000 km above the surface of the earth. (The radius of the earth is 3960 mi, or 6370 km.)

4. The planet Jupiter has a mass 314 times larger than that of the earth. Its radius is 11.3 times larger than the earth's radius. Find the acceleration due to gravity on Jupiter.

5. A 3600-lb car is to be pushed up a hill which is inclined 8° to the horizontal. If friction forces are negligible, what is the minimum pushing force needed?

6. The brakes on a car are released by a child while the car is parked on an 11.5° hill (an inclined plane with angle 11.5° to the horizontal). What is the acceleration of the car, and how fast is it going after 10 s? (Ignore friction forces; use SI units.)

7. A car is moving up a 15° incline at a speed of 20 m/s when the driver disengages the motor. How far up along the incline will the car coast if friction forces are negligible?

8. If, in Prob. 7, the 1400-kg car coasts 30 m before stopping, how large is the average (assumed constant) friction force which stopped it?

9. In Prob. 6, the car weighs 2400 lb and is actually moving at 9.0 ft/s after 10 s have elapsed. How large is the friction force which retards the car's motion?

10. With how large a force must the wheels of a 3200-lb car push backward upon the roadway if the car is to climb 200 ft along an 11.5° incline in 40 s? Assume that the car starts from rest and that the average friction force is 140 lb.

11. How large a force must the trailer hitch on a car be able to supply if the car is to pull a 2000-kg trailer up a 11.5° hill as the car is decelerating at a rate of 0.5 m/s²? Ignore friction effects.

12. A marble rolls off the edge of a horizontal table 1.2 m high, as shown in Fig. P4.2. Find the distance labeled x if the marble is rolling with a speed of 2 m/s. What are the vertical and horizontal components of its velocity just before it hits the floor?

13. A fire hose shoots water horizontally from the top of a tall building toward the wall of a building 20 m away. If the speed with which the water leaves the hose is 5.0 m/s, how far below the hose level does the water strike the wall? *Hint:* Consider the water to consist of a series of particles shooting along the stream.

14. An electron ($m = 9.1 \times 10^{-31}$ kg) traveling hori-

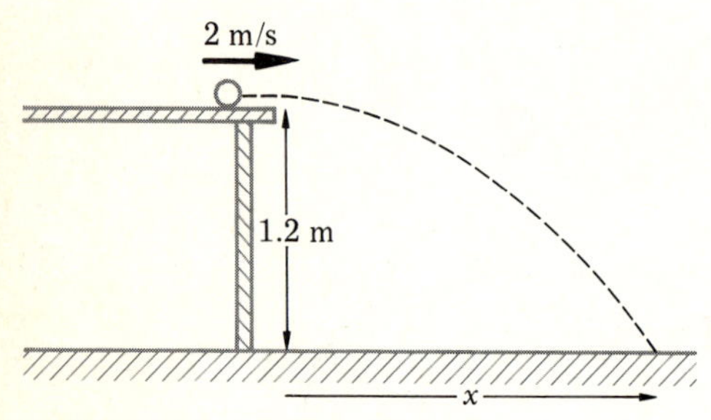

FIGURE P4.2

zontally leaves an electron gun at the end of a TV tube with a speed of 10^8 cm/s. If the fluorescent screen on the opposite viewing end of the tube is 40 cm away, how far below its original level has the electron fallen by the time it hits the screen?

15. At a circus, the "human cannonball" is shot out of the cannon with a speed of 18 m/s. The cannon barrel is pointed 40° above the horizontal. How far from the end of the cannon should the net (used to catch the person) be placed? Assume it is at the same level as the end of the barrel. If the person is accelerated uniformly from rest to 18 m/s in a distance of 3.0 m in the barrel, how large a force must push on the person? Give your answer in terms of W, the weight of the person.

16. A ball is thrown with a speed of 20 m/s at an angle of 30° below the horizontal from a bridge which is 30 m above the water. (*a*) Where, relative to a point on the water directly below the throwing point, does the ball hit the water? (*b*) How long is the ball in the air?

17. Repeat Prob. 16 for a ball thrown at an angle of 30° above the horizontal.

FIGURE P4.3

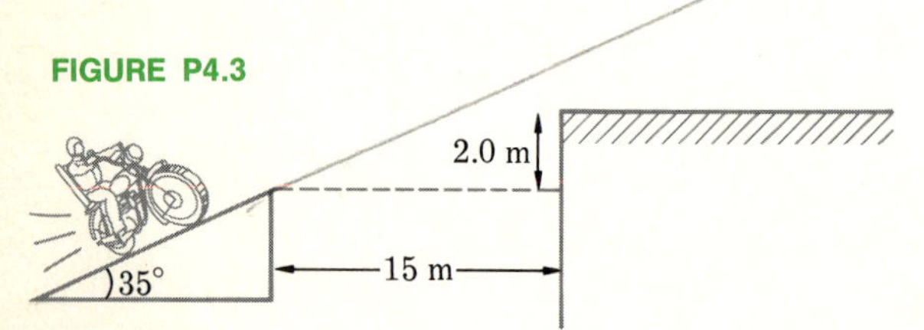

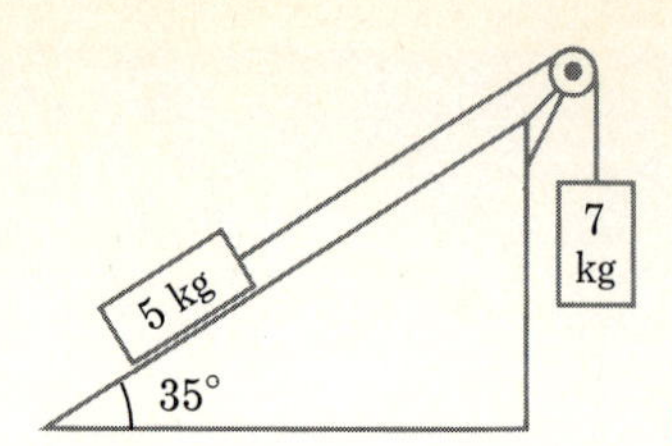

FIGURE P4.4

18.* As shown in Fig. P4.3, a stunt driver wishes to shoot off the incline and land on the platform. How fast must the motorcycle be moving if the stunt is to succeed?

19.** A projectile is shot from the ground with a velocity v at an angle θ above the horizontal level ground. It returns to the ground at a distance R from the shooting point. Show that the range R of the projectile is given by

$$R = \frac{2v_0^2 \sin\theta \cos\theta}{g}$$

provided friction forces are negligible. Using the trigonometric formula $2 \sin\theta \cos\theta = \sin 2\theta$, show that the range is maximum when $\theta = 45°$.

20. Consider the two blocks shown in Fig. P4.4. Find the acceleration of the blocks and the tension in the cord if there is negligible friction.

21.* Repeat Prob. 20 for $\mu = 0.20$.

22.* Each of the two blocks in Fig. P4.5 experiences a friction force of 5 lb. How long will it take, starting from rest, to pull the 32-lb block through a distance of 19 ft?

23.** Find the tension in the various cords shown in Fig. P4.6 and the acceleration of the blocks if μ for each block is 0.20.

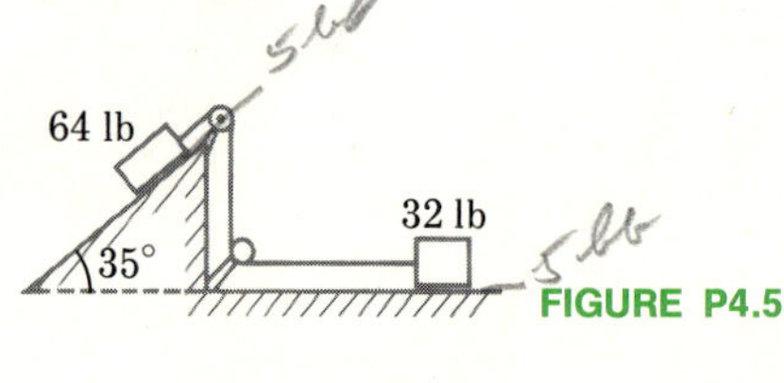

FIGURE P4.5

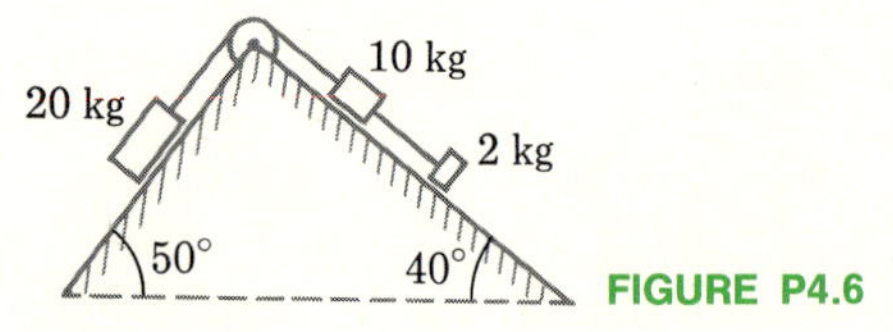

FIGURE P4.6

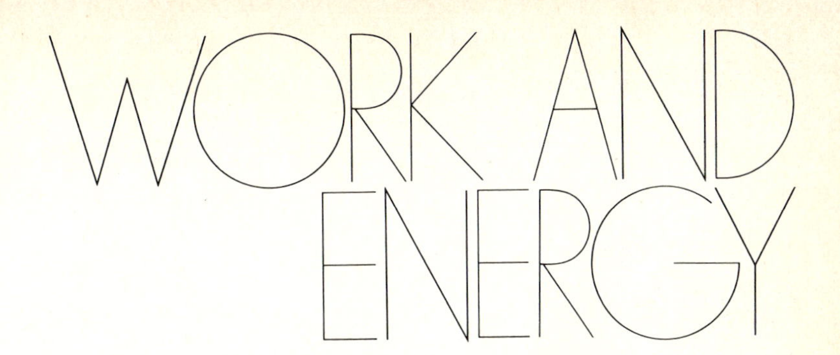

5

One of the goals of scientists is to discover ways of unifying and simplifying the various facts and concepts in their field of study. In previous chapters we have been discussing forces and their effect in causing motion. We can in principle describe all motions in terms of the forces which cause them. However, a concept to be introduced in this chapter, the conservation of energy, greatly unifies and simplifies the description of motion in many instances. The principle of conservation of energy will be found to be a unifying concept important not only in mechanics but in other branches of physics as well. In order to understand the principle, we shall first discuss the concept of work and how it leads to the concept of energy.

5.1 DEFINITION OF WORK

When you sit at your desk studying this book, you are not doing work. This statement does not mean that you are lazy or that learning physics is an effortless process. It is simply stating a fact which arises from the definition of work as used by the scientist. There are so many colloquial ways in which the word *work* is used that it becomes particularly important to give it a precise meaning.

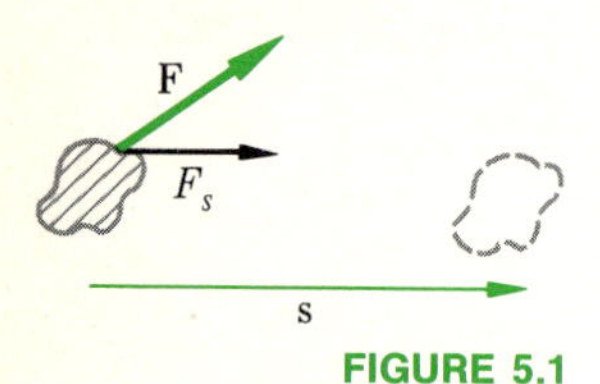

FIGURE 5.1
The work done by **F** during the displacement **s** is $F_s s$.

Does a baseball player work when he is playing baseball? Many people would say that since he is playing a game he is not working. But what if he were being paid to play baseball? Is the ground underneath a house doing work? It is holding the house. Is it, therefore, basically different in its function from a pillar holding the roof over the porch of the house? Yet some would insist that the pillar was doing work. Clearly, if we are to use the term *work* in physics, we need to define it in a precise way.

Physicists and all other scientists define the work done by a force in the following way. Suppose a force **F** acts on an object and the object is given a displacement **s**. Suppose further that F_s is the component of **F** in the direction of **s** as shown in Fig. 5.1. *We then define the* **work done by the force F** *during the displacement* **s** *to be*

DEFINITION

Work

$$\text{Work done by } \mathbf{F} = F_s s \tag{5.1}$$

Upon accepting this definition, it follows that the supporting force exerted by a post holding a load is not doing work; the load is not moved by the force, and so **s** is zero. Similarly, if you hold a pail of water stationary at your side for an hour, the supporting force of your hand has done no work on the pail. The pail has not been moved. Hence **s** is zero, and so is the work done by the supporting force.

Suppose that you carry the pail of water over level ground and pass two points a $\frac{1}{2}$ mi apart. The work done on the pail by the supporting force is still zero according to our definition. As seen in Fig. 5.2, your hand pulls upward on the pail with the supporting force **F**. But the displacement of the pail is perpendicular to **F**. Hence F_s is zero, and so the work done during this displacement by the supporting force is zero.*

Illustration 5.1 As another illustration, suppose that you were to pull a sled along the level ice, as illustrated in Fig. 5.3*a*. You exert a force **F** upon the sled by means of the rope on which you are pulling. After you have pulled the sled through a distance *s*, how much work have you done?

FIGURE 5.2
No work is done on the pail since there is no component of the force in the direction of the displacement.

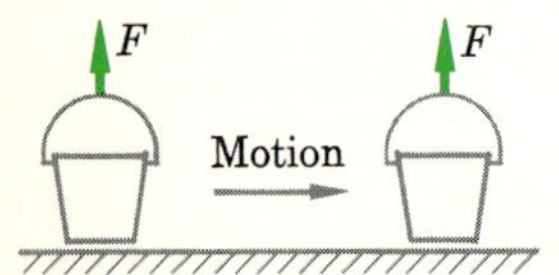

*If the pail moved with constant velocity between the two points, it is easy to see that no work was done by horizontal forces on the pail. Recall that, for constant-velocity motion, the resultant force on the object is zero. Hence, the net horizontal force on the pail was zero, and so no work was done by horizontal forces. If the pail started and stopped along the way, some work was done in starting the motion. But that work is recovered when the motion stops, as we shall see in following sections. Therefore, the net work done by horizontal forces acting on the pail is also zero.

Reasoning To answer this question, notice that the force F can be thought of as two forces, its components, as shown in Fig. 5.3*b*. Since the vertical component, $F \sin \theta$, is perpendicular to the direction of motion, it does no work on the sled. The horizontal component, $F \cos \theta$, is parallel to the direction of motion, and so it does do work on the sled. The work done is

Equation for Work

$$\text{Work} = (F \cos \theta)s = Fs \cos \theta \qquad (5.2)$$

Equation (5.2) can be used as a defining equation for work done, since it is Eq. (5.1) written in a different way.* It says merely that *the work done is equal to the following product:* (*the applied force*) · (*the displacement*) · (*the cosine of the angle between the force and the displacement*). Sometimes Eq. (5.2) will be used instead of its equivalent, Eq. (5.1).

The units in which we measure work are obtained from its defining equation, Eq. (5.1) or (5.2). There will be three different units of work, one each for the British system, SI, and cgs system. They are defined as follows:

	WORK	= F_s	× s
British	foot-pound	pound	foot
SI	joule	newton	meter
cgs	erg	dyne	centimeter

We should never mix the units in this equation. All quantities must be in a single system of units.

Units for Work

Sometimes, in analogy to the way the British unit is named, people call a joule (J) of work a **newton-meter** of work. Similarly, an erg of work is sometimes called a **dyne-centimeter** of work. The joule and erg are the preferred forms, however. Notice that an erg of work is very small, since both

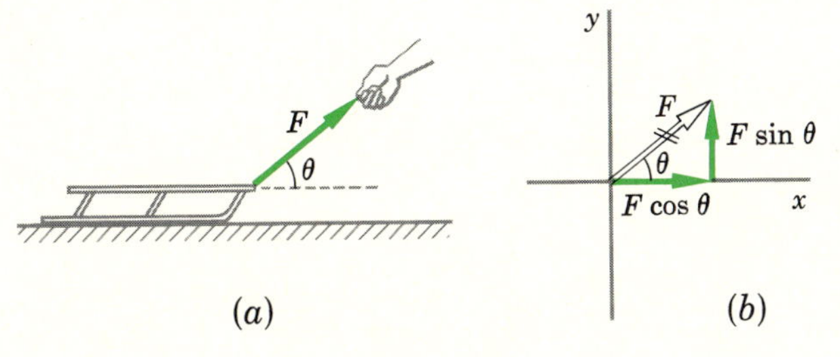

FIGURE 5.3
The vertical component of **F** does no work, since the sled does not move up or down. All the work is done by $F \cos \theta$, the component parallel to the direction of motion.

* In physics the product $Fs \cos \theta$ is often written $\mathbf{F} \cdot \mathbf{s}$. This symbolism means, in words, "Take the product of the magnitude of the force vector with the magnitude of the displacement vector, and multiply this product by the cosine of the angle between the two vectors." With this symbolism, the defining equation for work would be

$$\text{Work} = \mathbf{F} \cdot \mathbf{s}$$

Note that $\mathbf{F} \cdot \mathbf{s} = F(s \cos \theta) = (F \cos \theta)s = Fs \cos \theta$. The product of two vectors calculated in this manner is called the **dot** (or **scalar**) **product.**

the dyne and the centimeter are rather small units of force and distance. Since a newton is 10^5 times larger than a dyne and since a meter is 10^2 times larger than a centimeter, the joule is 10^7 times larger than an erg. Also, one finds that a joule of work is somewhat smaller than a foot-pound of work, the approximate relation being

$$1\ \text{J} = 0.738\ \text{ft} \cdot \text{lb}$$

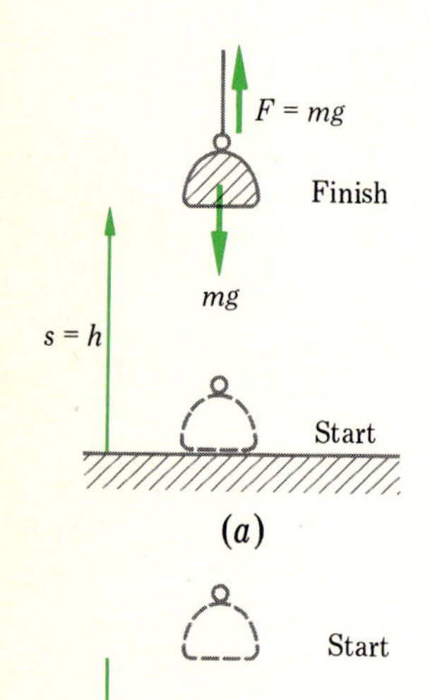

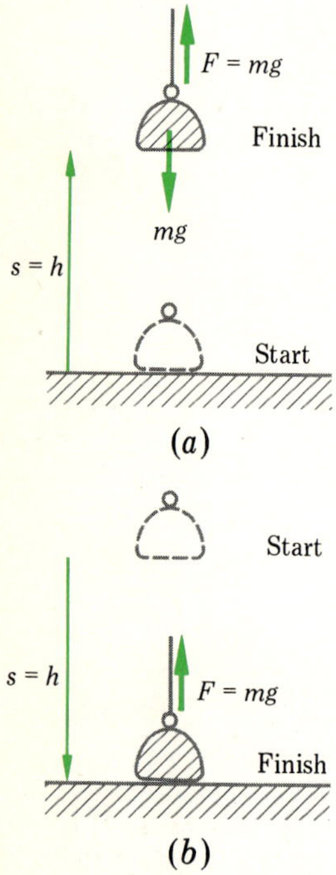

FIGURE 5.4
The work done by the lifting force in (*a*) is *mgh*, and in (*b*) it is $-mgh$.

Let us now summarize what we have learned about work.

> The work done by a force **F** acting on an object during a displacement **s** is
>
> $$F_s s \qquad \text{or} \qquad Fs\cos\theta$$
>
> In these equivalent expressions, F_s is the component of **F** in the direction of the displacement **s**. The angle θ is the angle between **F** and **s**.

The units of work are joules, ergs, or foot-pounds, as indicated in the table given on the previous page.

Illustration 5.2 How much work do you do on an object of weight mg as you slowly lift it a distance h straight up? Repeat for the case where it is slowly lowered through this same distance.

Reasoning The situations are shown in Fig. 5.4. To slowly lift the object, a force equal to its weight mg must pull up on it. (A force slightly larger than this is needed to give it an initial acceleration upward. But once the object is moving, no unbalanced force is needed to continue the motion. As a result, a pulling force equal to the weight of the object is used except at the very first instant.) As we see in Fig. 5.4*a*, the lifting force is mg, equal to the object's weight. The displacement is h upward and the lifting force is in the same direction. Therefore, from either Eq. (5.1) or (5.2) one has

$$\text{Work} = F_s s = Fs\cos 0° = (mg)h$$

This is the work you must do to lift the mass a distance h.

In Fig. 5.4*b* we show what happens when you lower the mass. Now **F** and **s** are in opposite directions. Therefore $F = mg$ and $\theta = 180°$. We then find from Eq. (5.2) that

$$\text{Work} = Fs\cos\theta = (mg)(h)\cos 180° = -mgh$$

Negative Work

Notice that negative work is done by a force when the displacement is opposite in direction to the force.

Illustration 5.3 As shown in Fig. 5.5, a box is pulled along the floor at constant speed by a force F. Let us say that the friction force opposing the motion is 20 N and the box has a mass of 30 kg. Find the work done by the pulling force as the object is moved 5.0 m.

Reasoning Since the object is moving with constant speed, there can be no unbalanced force acting on it. As a result, $\Sigma F_x = 0$. The appropriate free-body diagram is shown in Fig. 5.5*b*. From it we see that $\Sigma F_x = 0$ becomes

$$0.8F - 20 \text{ N} = 0$$

This gives

$$F = 25 \text{ N}$$

To find the work done by this force, we can use

$$\text{Work} = F_s s$$

to give

$$\text{Work} = (0.80F)(5.0 \text{ m}) = 100 \text{ N} \cdot \text{m} = 100 \text{ J}$$

Or we could use

$$\text{Work} = Fs \cos \theta$$

to give

$$\text{Work} = (25 \text{ N})(5 \text{ m})(0.80) = 100 \text{ J}$$

Notice that the y component of **F** does no work. No displacement occurs in the direction of it.

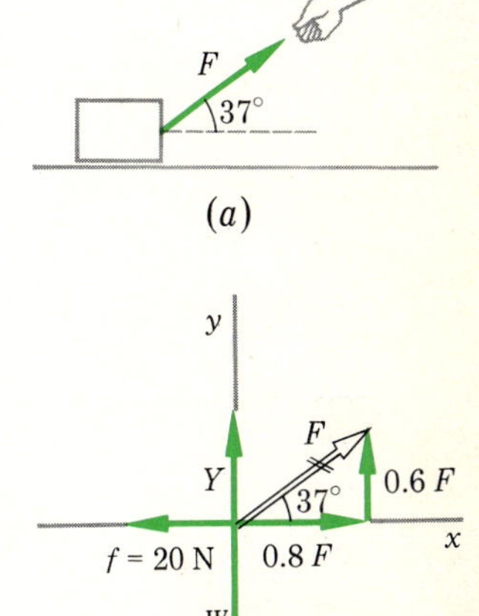

FIGURE 5.5
The horizontal component of **F** does work on the object, while the work done by the vertical component is zero.

5.2 POWER

DEFINITION

The rate of doing work is called **power,** *and it is defined as the work done in unit time.* As a formula, this would be

Power

$$\text{Power} = \frac{\text{work done}}{\text{time taken to do the work}}$$

$$P = \frac{\text{work}}{t} \tag{5.3}$$

The units can be any unit of work divided by any unit of time. *The more common units are*

$$P \rightarrow \frac{\text{J}}{\text{s}} = \text{watts (W)}$$

$$P \rightarrow \frac{\text{ft} \cdot \text{lb}}{\text{s}}$$

Other units are also used at times. For example, a kilowatt is 1000 times larger than a watt. *The unit* **horsepower** (*hp*) *is frequently used in the British* *system. This quantity is defined to be 550 ft · lb/s.* Since 1 ft · lb/s is equivalent to 1.36 W, it follows that *1 hp is the same as 746 W of power.* Since the definition of power is

$$P = \frac{\text{work}}{\text{time}}$$

one has

$$\text{Work} = (\text{power})(\text{time})$$

This equation for work forms the basis for a work or energy unit we shall encounter in electricity, the kilowatthour (kWh). That is, if the power output of a machine is measured in kilowatts and its time of operation is stated in hours, the product of power and time would give the work in kilowatthours.

Illustration 5.4 Suppose that the motor shown in Fig. 5.6 lifts the 480-lb object at a constant speed of 2 in/s. What is the horsepower being developed by the motor? What is the power in watts?

Reasoning The motor does the following amount of work in 1 s:

$$\begin{aligned}\text{Work in 1 s} &= (\text{force})(\text{distance lifted in 1 s})\\ &= (480\ \text{lb})(\tfrac{2}{12}\ \text{ft}) = 80\ \text{ft}\cdot\text{lb}\end{aligned}$$

But from its definition

$$\begin{aligned}\text{Power} &= \frac{\text{work}}{t} = \frac{80\ \text{ft}\cdot\text{lb}}{1\ \text{s}} = 80\ \text{ft}\cdot\text{lb/s}\\ &= (80\ \text{ft}\cdot\text{lb/s})\left(\frac{1\ \text{hp}}{550\ \text{ft}\cdot\text{lb/s}}\right) = 0.145\ \text{hp}\end{aligned}$$

Since 1 hp = 746 W, we have

$$\text{Power} = (0.145\ \text{hp})\left(\frac{746\ \text{W}}{1\ \text{hp}}\right) = 108\ \text{W}$$

FIGURE 5.6
We wish to find the power output of the motor as it lifts the object with a speed of 2.0 in/s.

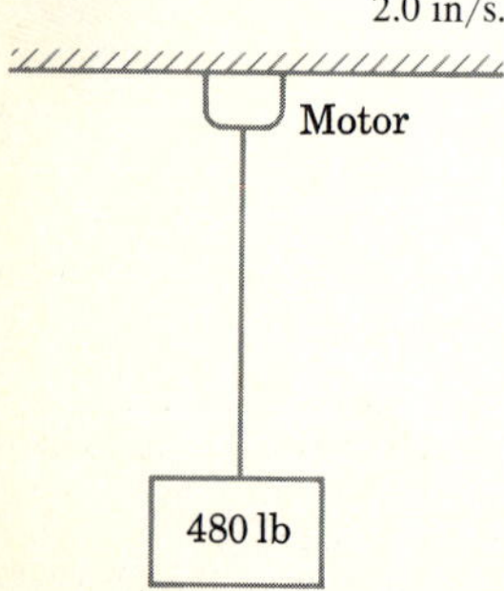

5.3 KINETIC ENERGY

An object which is moving can do work. For example, a hammer does work on a nail as the hammer drives it into a piece of wood. Or a baseball can break a window as it crashes through it. There are many other examples we could cite where a moving object does work as it slows down. As we shall see, this is a very important fact, and we shall make much use of it. Let us now examine how work and motion are related.

To begin our discussion, let us see how an object's motion depends upon the unbalanced force which acts on it. Suppose an object of mass m is subjected to a constant resultant force $\mathbf{F}$ in the direction of a displacement $\mathbf{s}$. During this displacement, the force does work given by

$$\text{Work} = Fs$$

We wish now to relate this work to the change in motion of the object caused by this unbalanced force.

We know that the unbalanced force F is related to the acceleration of the object by $F = ma$. Therefore, the above equation for work becomes

$$\text{Work} = mas$$

Suppose the object had an original velocity v_0 and a final velocity v_f after the work was done. Then, since

$$v_f^2 - v_0^2 = 2as$$

we can substitute to obtain

$$\text{Work} = \tfrac{1}{2}mv_f^2 - \tfrac{1}{2}mv_0^2$$

This is a very important relation. It tells us that

> The work done by the resultant force on an object leads to a change in motion of the object. One has
>
> $$\text{Work done by resultant force} = \tfrac{1}{2}mv_f^2 - \tfrac{1}{2}mv_0^2 \qquad (5.4)$$

Although this relation was derived for a constant resultant force, it proves to have general validity. Whenever an object is subjected to a resultant force, the work done by the force is related to the object's change in velocity by Eq. (5.4).

In obtaining Eq. (5.4) we tacitly assumed that the force speeded up the object. However, nothing in the derivation requires this. Equation (5.4) also applies if the object is being slowed down. We can easily see this by reference to Fig. 5.7. There we see a mass being slowed as it slides across a table. Notice that the stopping force, the force of friction f, is opposite in direction to the displacement. As a result, $Fs\cos\theta$ becomes negative in this case since $\cos\theta = -1$. If we now look at Eq. (5.4) we see that, since negative work is done in this case, v_f must be smaller than v_0. The equation tells us the object was slowed by the friction work done on it.

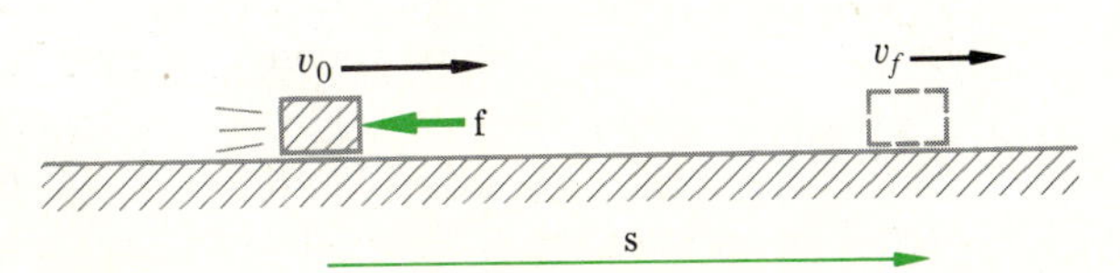

FIGURE 5.7
The friction force does negative work on the block. It slows it according to the relation $fs\cos\theta = \tfrac{1}{2}mv_f^2 - \tfrac{1}{2}mv_0^2$.

DEFINITION

Kinetic Energy

As we see from this discussion, work is related to the quantity $\frac{1}{2}mv^2$. This quantity is given the name **kinetic energy** (KE). The name comes from the Greek words *kinetikos* (which means "for putting in motion") and *energeia* (which means "activity"). We define the following: *Kinetic energy (KE) is $\frac{1}{2}mv^2$, and it represents the maximum work an object of mass m can do as a result of its motion with speed v.*

To clarify this, recall that we just saw in connection with Fig. 5.7 that an object loses kinetic energy as it does work against a friction force. Indeed, the KE lost was equal to the work done against the friction force. The stopping force need not have been a friction force. For example, it might have been the result of a spring fastened to it, as shown in Fig. 5.8, where the sliding block stretches the spring as it slows and comes to rest. The work done against the stopping force in this case is used to stretch the spring. There are many other examples we could give where a moving object does work as it is brought to rest or slowed down. In each and every such case the following is found to hold true:

> When an object loses kinetic energy, it does an amount of work equal to the KE lost.

Let us now apply this result to a few examples.

Illustration 5.5 A 2000-kg car is traveling at 20 m/s and coasts to rest on level ground in a distance of 100 m. How large was the average friction force tending to stop it?

Reasoning The original KE of the car was lost in doing work against the average friction force $\bar{f}$. Therefore

$$\begin{aligned}\text{Loss in KE} &= \text{work done against friction force}\\ \tfrac{1}{2}mv^2 &= \bar{f}s\\ \tfrac{1}{2}(2000\text{ kg})(20\text{ m/s})^2 &= (\bar{f})(100\text{ m})\\ \bar{f} &= 4000\text{ kg}\cdot\text{m/s}^2 = 4000\text{ N}\end{aligned}$$

The interrelation of units in this last step is obtained by writing the units in the equation $F = ma$.

Illustration 5.6 If the friction force on the car of Illustration 5.5 was constant at 4000 N, how fast was the car going after it had gone 50 m?

Reasoning The loss in KE of the car by the time it reached the 50-m mark was used up doing work against friction. Therefore

$$\begin{aligned}\text{Loss in KE} &= \text{friction work done}\\ \tfrac{1}{2}mv_0^2 - \tfrac{1}{2}mv_f^2 &= \bar{f}s\\ \tfrac{1}{2}(2000\text{ kg})(20\text{ m/s})^2 - \tfrac{1}{2}(2000\text{ kg})v^2 &= (4000\text{ N})(50\text{ m})\end{aligned}$$

FIGURE 5.8

By the time it comes to rest, the mass will have done work equal to $\frac{1}{2}mv^2$ in stretching the spring.

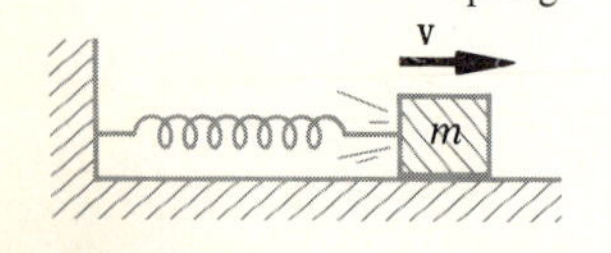

or, after rearranging

$$v^2 = 400 - 200 \text{ m}^2/\text{s}^2$$
$$v = \sqrt{200} \text{ m/s} = 14.1 \text{ m/s}$$

Once again use was made of the fact that 1 N is 1 kg · m/s².

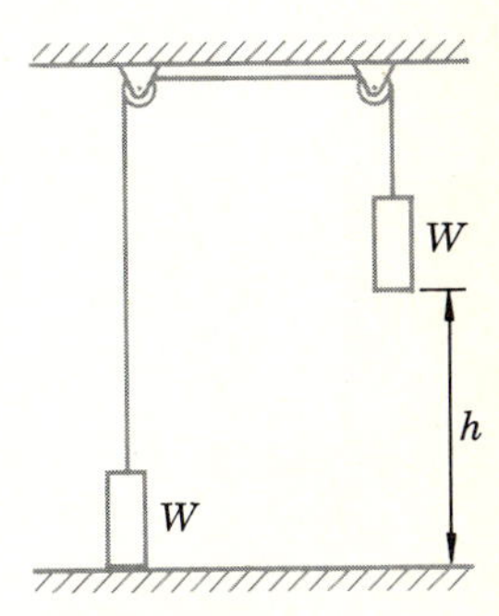

FIGURE 5.9
As the object on the right falls, it does work by lifting the object on the left. The loss of PE by the object on the right is compensated by a gain in PE by the other object.

5.4 POTENTIAL ENERGY

Consider the system shown in Fig. 5.9. The pulleys are taken to be frictionless. Since the two objects are exactly the same, each having weight $W = mg$, if the top object is given a very slight downward push, the object on the right will drop to the floor at constant speed. The one on the left will rise to a height h above the floor. Its height is increased by an amount h.

We now ask: How much work was done on the object at the left by the rope as the object was raised above the floor? Clearly, the work is equal to the tension mg in the rope times the height through which the object was lifted.

$$\text{Work done} = mgh$$

Who or what external agent did this work? It was the weight of the object on the right which pulled the other object upward, and so it did the work. We must therefore conclude that the right-hand object possessed the ability to do work when it hung in its original position above the floor. It is said to have possessed gravitational **potential energy** (PE) when it was in this position. *The potential energy possessed by an object is the amount of work the body is capable of doing because of its position. For a body in the earth's gravitational field, one has*

DEFINITION

$$\text{Gravitational potential energy} = mgh \tag{5.5}$$

Gravitational Potential Energy

where m is the mass of the body and g is the acceleration due to gravity.* *The units of potential energy are the same as those for work,* i.e., joules, ergs, or foot-pounds, as you may easily verify from Eq. (5.5). Notice in Eq. (5.5) that h is a difference between two heights. It measures how far the object has been lifted above a certain reference level.

In connection with this fact, we should mention one other important fact about potential energy. It may be stated this way:

The zero level for computing potential energy is arbitrary.

For example, it is a matter of personal choice where you choose to measure the position of an object from. One person might say an object was 50 cm above a table and so its height is 50 cm. Another might say that since the table top is 90 cm from the floor, its height is 140 cm. Still a third might measure the object's position as being 60 cm below the ceiling of the room.

*For positions far from the earth, g is not equal to 9.8 m/s². In that case, we can no longer use mgh for PE.

For that person, its height is -60 cm. All these choices are possible and correct. As we shall see, *computations involve only the changes in potential energy of objects. This fact allows us to take the zero level of potential energy wherever it is convenient.*

5.5 THE GRAVITATIONAL FORCE IS A CONSERVATIVE FORCE

We have already seen several situations in which work was done in lifting an object. To lift an object straight up at constant speed a force equal to the weight of the object mg is required. As a result, the work done in lifting an object straight up through a distance h is $(mg)h$. We shall now show that even if the object is not lifted straight up, this same result is true.

Suppose we wish to lift the mass shown in Fig. 5.10*a* from the floor to the table top. How much work must be done? Let us lift it along the path shown by the black line from A to B.

To compute the work done in lifting the mass from A to B, we approximate the actual path by the jagged path shown in *b*. By making the jag lengths very small, the two paths can be made identical for all practical purposes. Notice that the lifting force is vertical. Therefore, it does no work in the tiny horizontal movements of the jagged path.* Work is done by the lifting force only on the vertical movements. When the object is raised, positive work is done. But when it is lowered (as near point C) negative work is done. The effect of this is that the downward movements cancel the work done on equivalent upward movements.

We therefore conclude that the work done is dependent only upon the net effect of all the vertical movements. In going from A to B, the object was lifted a net distance h. As a result, the work done is mgh.

But this is the same as the work done in lifting the object straight up from A through a distance h and then moving it sideways to point B. In fact,

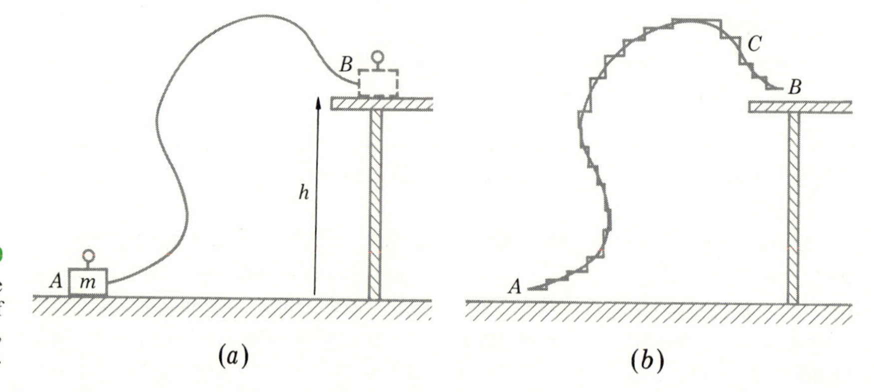

FIGURE 5.10
The path in (*a*) can be approximated by a series of horizontal and vertical steps, as shown in (*b*).

*Recall that only a negligibly small force is required to cause the horizontal motion.

since the path shown from A to B is perfectly arbitrary, we conclude the following:

> If point A is a distance h below point B, the work done against gravity in lifting a mass m from A to B is mgh. This result holds for any path taken between A and B.

Of course, if the object is lowered from point B to point A, the work done against the gravitational force is $-mgh$.

DEFINITION

Conservative Force

The gravitational force is an example of a **conservative force.** *A force is conservative if the work done in moving an object from point A to point B against the force is not dependent upon the path taken for the movement.* We shall see later that the electrostatic and nuclear forces are also conservative. *Friction forces,* on the other hand, *are not conservative.* You can easily verify this by sliding your textbook from one point to another across a table. You obviously have to do more work when you slide it by way of a complicated, long path than when you follow a straight-line path. This would not be the case if the friction force were conservative.

5.6 INTERCONVERSION OF KE AND PE

Each time you toss an object into the air or drop an object you see an example of the interchange of KE and PE. For example, when you toss a coin upward, as shown in Fig. 5.11, its original KE changes to PE. This is quantitatively correct as we shall now show.

At the instant the coin leaves your hand it has an upward velocity v_0. It then rises to a height h, where its velocity is zero. To find h we can use

$$v_f^2 - v_0^2 = 2as$$

to give

$$0 - v_0^2 = -2gh$$

from which $h = v_0^2/2g$. But its PE at this height h is mgh. Substituting the above value for h gives

$$mgh = mg\frac{v_0^2}{2g} = \tfrac{1}{2}mv_0^2$$

As we see, the coin's original KE is changed to PE as the coin rises. Notice that the lost KE was used to do an equivalent amount of work against the gravitational force.

In this same way we can show that a frictionless falling object gains KE equal to its lost PE. This is a general result. Before stating it, however, we shall extend it further to include the effects of other forces. For example, in addition to the force of gravity, friction forces might be acting on an object. Or perhaps some other external forces provide a resultant force to the object. What is true in those cases?

We have already seen that KE and PE can be lost in doing work. An object moving at high speed can do work as it is slowed. An object like that in

FIGURE 5.11

The KE of the coin is changed to PE as the coin rises. When it falls, the PE is changed back to KE.

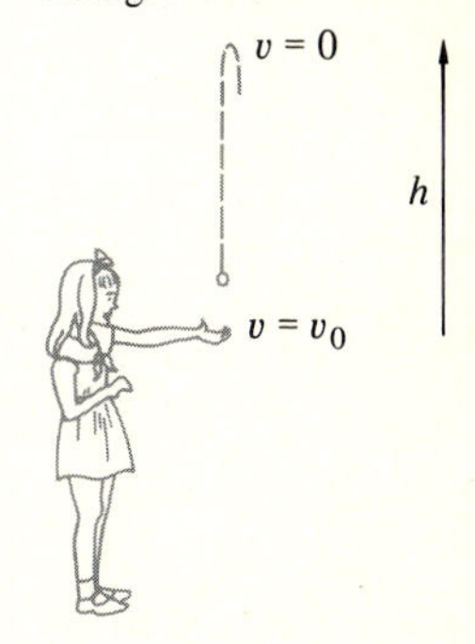

Fig. 5.9 can do work as it falls. In each of these cases, the energy lost was equal to the work done. A similar situation exists when an external force accelerates or lifts an object. The increase in KE and PE equals the work done by the external force.

All these results can be summarized as follows:

$$\text{Loss in KE} + \text{loss in PE} = \text{work}_{\text{dec}} - \text{work}_{\text{inc}}$$

Work-Energy Theorem

where inc stands for increase and dec for decrease. *This statement is one form of the work-energy theorem.* The quantity work_{dec} is the sum of work of all kinds done on the object which decreases the KE and PE of the object. Most commonly, this will be work done by friction forces and other forces which tend to slow the object. The quantity work_{inc} is the sum of work of all kinds which increase the KE and PE of the object. Work done by forces which lift the object or speed it up is included here. This term is preceded by a negative sign since it adds energy to the object instead of causing an energy loss. In writing each of these terms, we do not include the work done by the gravitational force. That work is already accounted for in the PE term on the left side of the equation.

The work-energy theorem is a statement of the fact that work and energy can be accounted for. When an object's KE or PE changes, this change must be the result of an equivalent amount of work done. The work-energy theorem is essentially a balance-sheet equation showing that these quantities must balance each other.

The theorem can be stated in symbols in the following way provided we recall that the loss in a quantity is given by initial value − final value:

Work-Energy Equation

$$(\tfrac{1}{2}mv_0^2 - \tfrac{1}{2}mv_f^2) + (mgh_0 - mgh_f) = \text{work}_{\text{dec}} - \text{work}_{\text{inc}} \tag{5.6}$$

Notice once again that the work terms are not to include the work done by the gravitational force since that work is accounted for in the PE term. Let us illustrate the use of the work-energy theorem by means of a few examples.

Illustration 5.7 A 3.0-kg object falls 4.0 m. How fast is it going just before it hits the ground?

Reasoning Method 1: We could solve this by using the motion equations. We have

$$a = 9.8 \text{ m/s}^2 \qquad \text{(down positive)}$$
$$v_0 = 0 \qquad s = 4.0 \text{ m} \qquad v_f = ?$$

Since

$$v_f^2 - v_0^2 = 2as$$

we have

$$v_f^2 = (2)(9.8 \text{ m/s}^2)(4 \text{ m})$$

from which

$$v_f = 8.85 \text{ m/s}$$

Method 2: From the work-energy theorem [Eq. (5.6)] (since only gravity does work),

$$\text{Loss in PE} + \text{loss in KE} = \text{work}_{\text{dec}} - \text{work}_{\text{inc}}$$

becomes

$$mg(h_1 - h_2) + \tfrac{1}{2}m(v_0^2 - v_f^2) = 0$$

Substituting 4.0 m for $h_1 - h_2$ and $v_0 = 0$ gives

$$(9.8\ \text{m/s})(4.0\ \text{m}) + \tfrac{1}{2}(0 - v_f^2) = 0$$

Solving for v_f gives $v_f = 8.85$ m/s.

Notice that the mass of the object was not needed.

Illustration 5.8 Suppose that the object in the last example was moving at only 6.0 m/s just before it hit. How large was the average friction force acting on it?

Reasoning Notice that we must now take account of the work done by the friction forces. In this case, the object moved 4.0 m against an average friction force f. Therefore, the work done by friction was $4.0f$ joules.

From the work-energy theorem we have (since friction decreases the object's energy),

$$\text{Loss in PE} + \text{loss in KE} = \text{friction work}$$

Substitution gives

$$mg(h_1 - h_2) + \tfrac{1}{2}m(v_0^2 - v_f^2) = \text{friction work}$$

or

$$(3.0\ \text{kg})(9.8\ \text{m/s}^2)(4.0\ \text{m}) + \tfrac{1}{2}(3\ \text{kg})(36\ \text{m}^2/\text{s}^2) = (4.0\ \text{m})f$$

Notice that the object's mass does not cancel in this case. Solving for f gives $f = 43$ N.

Illustration 5.9 A roller coaster which weighs 640 lb starts from rest at point A in Fig. 5.12 and begins to coast down the track. If the friction force is 5.0 lb, how fast will the coaster be going at point B? At point C?

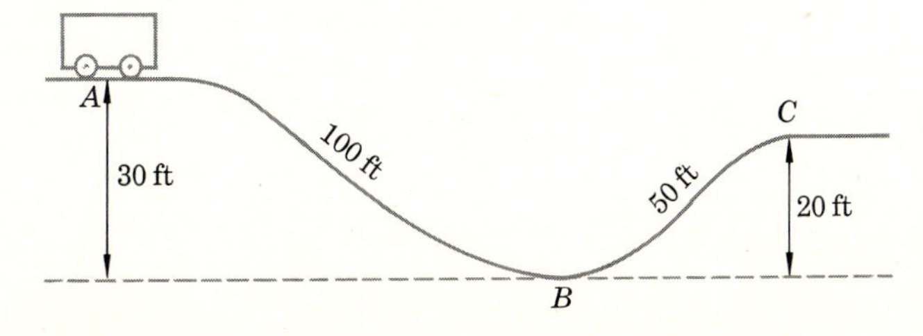

FIGURE 5.12 The PE which the coaster has at A is at least partly changed to KE and work against friction forces as it moves to points B and C.

Reasoning The work-energy theorem in this case is

$$\text{Loss in PE} + \text{loss in KE} = \text{friction work}$$

Or, for A to B,

$$mg(h_A - h_B) + (0 - \tfrac{1}{2}mv_B^2) = (f)(100 \text{ ft})$$

This gives

$$(640 \text{ lb})(30 \text{ ft}) - \left(\frac{1}{2}\right)\left(\frac{640 \text{ lb}}{32 \text{ ft/s}^2}\right)v_B^2 = (5 \text{ lb})(100 \text{ ft})$$

and so $v_B = 43$ ft/s. In going from A to C, the equation becomes

$$mg(h_A - h_c) + (0 - \tfrac{1}{2}mv_c^2) = (f)(150 \text{ ft})$$

Since $h_A - h_c = 10$ ft, we solve and find $v_c = 24$ ft/s.

Illustration 5.10 The 2000-kg car shown in Fig. 5.13 is at point A and moving at 20 m/s when it begins to coast. As it passes point B, its speed is 5.0 m/s. How large is the average friction force which retards its motion; assuming the same friction force, how far beyond B will it go before stopping?

Reasoning From the work-energy theorem we have

$$(\text{PE at } A - \text{PE at } B) + (\text{KE at } A - \text{KE at } B) = \text{friction work}$$

or

$$mg(h_A - h_B) + \tfrac{1}{2}m(v_A^2 - v_B^2) = fs$$

Putting in the values and taking $h_A - h_B = -8$ m gives

$$(2000 \text{ kg})(9.8 \text{ m/s}^2)(-8 \text{ m}) + \tfrac{1}{2}(2000 \text{ kg})(375 \text{ m}^2/\text{s}^2) = (f)(100 \text{ m})$$

Solving yields $f = 2180$ N.

Now let us apply the same method from point B to the stopping point. Since both points are at the same level, there is no change in PE and we have

$$0 + \tfrac{1}{2}m(v_B^2 - 0) = fd$$

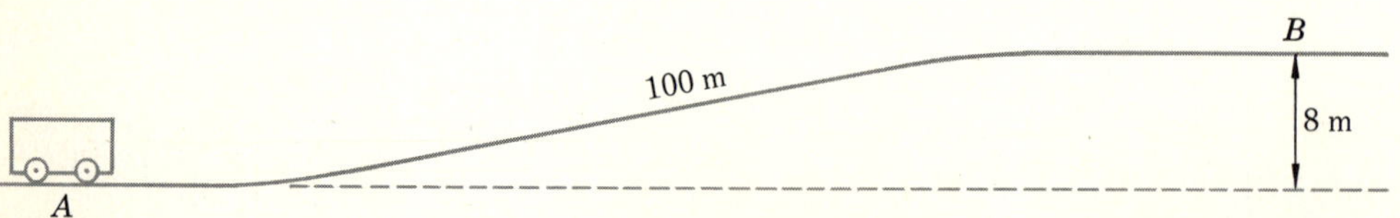

FIGURE 5.13 The KE of the car when it is at A is partly lost to PE and work against friction forces as it moves to B.

where d is the distance from B to the stopping point. Putting in the known values gives $d = 11.5$ m.

Illustration 5.11 A small 2-kg object falls from a height of 10 m into a box of sand (see Fig. 5.14). It comes to rest 3.0 cm beneath the surface of the sand. How large was the average force exerted upon it by the sand?

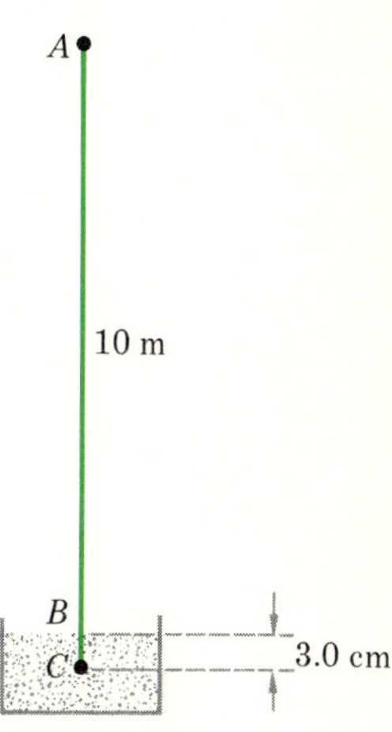

FIGURE 5.14
The PE of the ball is all lost to friction work by the time the ball comes to rest at point C.

Reasoning We notice that the PE at A is all changed to KE at B. This KE is then lost to friction work as the object penetrates to C. Let us apply the work-energy theorem to the complete process from A to C. At both these points KE = 0. Therefore

$$\text{Loss in PE} + \text{loss in KE} = \text{work}_{\text{dec}} - \text{work}_{\text{inc}}$$

becomes

$$mg(h_A - h_C) + 0 = fs - 0$$

where $s = 0.030$ m and $h_A - h_C = 10$ m. Using these values, we find $f = 6500$ N.

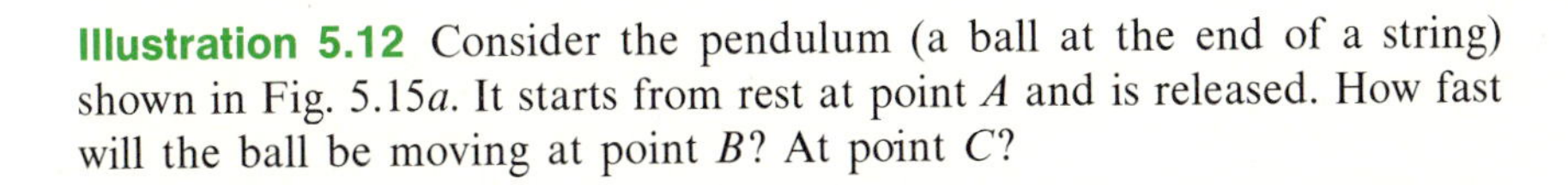

Illustration 5.12 Consider the pendulum (a ball at the end of a string) shown in Fig. 5.15a. It starts from rest at point A and is released. How fast will the ball be moving at point B? At point C?

Reasoning As the ball falls from A to B, it loses PE. In the process, the ball speeds up. We see from this that at least some of the lost PE was changed to KE. Since the string's tension pulls perpendicular to the arc in which the ball moves, no component of the force exists in the direction of motion. There-

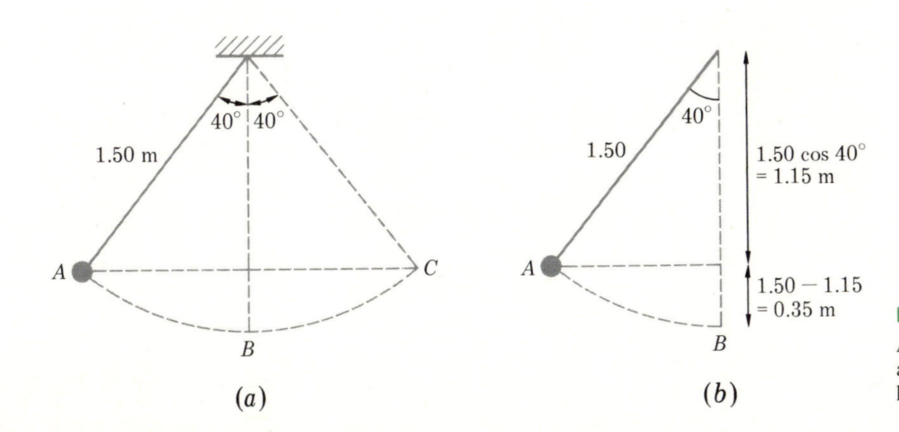

FIGURE 5.15
As the pendulum swings back and forth, its KE and PE keep interchanging.

fore, the string does no work on the ball. If we ignore friction, no work is done on the ball except by the gravitational force.

The work-energy theorem tells us that

$$\text{Loss in PE} + \text{loss in KE} = \text{work}_{\text{dec}} - \text{work}_{\text{inc}}$$

As we have just seen, the work term is zero. Moreover, the loss in PE of the ball is mgh, where m is the ball's mass and h is the height difference between the points A and B. From Fig. 5.15b we see that $h = 0.35$ m. Therefore the above equation becomes

$$m(9.8\ \text{m/s}^2)(0.35\ \text{m}) + (\tfrac{1}{2}mv_0^2 - \tfrac{1}{2}mv_f^2) = 0$$

But the initial speed of the ball v_0 was zero. The equation therefore becomes

$$m(3.43\ \text{m}^2/\text{s}^2) - \tfrac{1}{2}mv_f^2 = 0$$

After dividing through the equation by the mass of the ball, we can solve for v_f and find $v_f = 2.6$ m/s. This is the speed of the ball as the pendulum swings through point B.

The situation at point C is very simple. By the time the ball has reached C, it has regained all of its original PE. Therefore no energy can remain as KE. We can show this by use of the work-energy theorem applied to points A and C. The equation

$$\text{Loss in PE} + \text{loss in KE} = \text{work}_{\text{dec}} - \text{work}_{\text{inc}}$$

becomes (since only the gravitational force does work on the ball)

$$0 + \text{loss in KE} = 0$$

In other words, the KE is the same at A and C. Since the ball was at rest at A, it is also at rest at C.

As we see, the pendulum is an interesting example of interconversion of PE and KE. Over and over again the PE and KE interchange as the pendulum swings back and forth. At any point in its motion, the PE lost in the fall from point A has been changed to KE. Of course, all this assumes friction forces to be negligible.

Illustration 5.13 How large a force is needed to accelerate a 2000-kg car from rest to a speed of 15.0 m/s in a distance of 80 m? Assume that an average friction force of 500 N opposes the motion.

Reasoning We assume the car to move along level ground so that its change in PE is zero. The work-energy theorem in this case is

$$0 + (\tfrac{1}{2}mv_0^2 - \tfrac{1}{2}mv_f^2) = \text{friction work} - \text{work done by accelerating force}$$

Notice that since the accelerating force (call it F) increases the KE of the car, it belongs in the work_{inc} term. As usual, the friction work belongs in the work_{dec} term since it decreases the KE of the object. With $v_0 = 0$, $v_f = 15$ m/s, $m = 2000$ kg, and $s = 80$ m, this equation becomes

$$-\tfrac{1}{2}(2000 \text{ kg})(15 \text{ m/s})^2 = (f)(80 \text{ m}) - (F)(80 \text{ m})$$

where f is the friction force, 500 N. Putting the value for f in the equation and solving for F, we find the accelerating force to be about 3300 N.

5.7 CONSERVATION OF ENERGY

In this chapter we have introduced the concept of energy, the ability to do work. As we have seen, *KE represents the work an object can do because of its motion. Gravitational PE tells us how much work an object can do because of its position.*

There are other types of energy in addition to these two. A stretched spring is able to do work, and so it, too, has energy. The battery on your car can turn the starter, and so it contains energy, chemical energy. A hot gas in the combustion chamber of a car's cylinder can force out the piston and cause the car to move. The gas therefore has energy, heat energy. Other forms of energy are electrical or nuclear in nature. As you see, *energy exists in many forms.* We shall see later that mass itself is a form of energy.

An extremely important law describes the behavior of energy. Called the **law of conservation of energy,** it states that:

Law of Conservation of Energy

> Energy can neither be created nor destroyed. When a loss occurs in one form of energy, an equal increase must occur in other forms of energy.

Notice that the law does *not* state that KE is conserved, that PE is conserved, or that even the sum of KE and PE is conserved. Losses of KE and PE may result in gains in many other forms of energy.

One of the most important cases where KE is lost is in doing friction work. Where does the KE go in such a situation? It must appear as some other form of energy. It appears as heat energy. Indeed, you certainly know of many examples where friction work leads to heating effects. A hot bearing results when the bearing is not greased. The American Indian sometimes started a fire by rubbing pieces of wood together. Your skin suffers a floor burn when you accidentally slide across the gym floor. These are but a few of the many ways in which friction work leads to heating. The important fact is that:

> When friction work is done, an equivalent amount of heat energy is generated.

As a result, when other forms of energy are lost in doing friction work, an equivalent amount of another form of energy is gained. *Energy is never created, nor is it ever destroyed.*

An interesting aspect of the law of conservation of energy has to do with perpetual-motion machines. For example, from time to time an inventor will propose a design for a perpetual-motion machine. This may be an ingenious mechanism which he suggests will run a car forever without fuel. Or, more subtly, it may be a machine which uses fuel to start but which generates power within itself. The inventor may claim that it generates enough energy to keep itself running and to do work in addition.

If we know about the law of conservation of energy, we can say without further thought that these machines will not work. No matter how complicated they are, they cannot create energy. Therefore, if they lose energy in any way, by friction or by doing work, that energy must be supplied to the machine from some inside or outside source. *Since any machine will have some friction, it will not run forever unless energy is furnished to it from another source.*

There are many other interesting applications of the energy-conservation law. It is a very important law and has widespread application throughout all branches of science. We shall use it many times during our study of physics.

5.8 THE EARTH'S ENERGY SUPPLY

As we are well aware, we people of the earth are concerned about the earth's energy supply. Since energy is conserved, one might assume that the problems arise from our increasing use of energy. Such is not the case, however. A possible future shortage of energy would exist even if we continue to use energy at the present rate. The trouble has to do with the usefulness of the energy which exists on earth, not only upon the quantity of energy which exists here.

The usefulness of energy will be discussed in more detail in Chap. 12, but for now, a few examples will suffice to point out the major problem. To take a simple case, consider the burning of coal, oil, or gas to generate steam, which in turn is used to generate electricity. As the fuel is burned, only a portion of the chemical energy in the coal is changed to electrical energy. A great deal of the heat energy produced in the burning goes up the chimney with the combustion products. Other losses of the heat energy also occur. In fact, we shall see in Chap. 12 that it is impossible, even in theory, to change all of the combustion heat to useful work. As a result, not all the energy of the fuel is usable.

Another example showing this effect is as follows. Suppose you give KE to a book by pushing it across the table. (In passing, you might consider the fact that this energy came from you and you obtained it from the food you ate.) The book soon loses this energy because of friction work. It appears as heat energy, a slight heating of the two sliding surfaces. The original KE of the book has been transformed to another form of energy. But this new energy is less usable than the book's original KE. *This type of phenomenon is widespread in nature. Energy is transformed from a useful form to a less usable form.* For that reason, *the earth's energy becomes less usable as time goes on—even though the total amount of energy does not decrease significantly.*

Indeed, the quantity of energy on the earth is probably being increased constantly because of the energy from the sun; certainly it is not decreasing.

There are two basic energy sources available to the earth. One of these existed in the early years when the earth was first formed. At that time the elements found on the earth were formed. Locked in their atoms and nuclei is the energy we refer to as **nuclear energy.** We make use of it in nuclear reactors. This supply of energy is only now being tapped to any great extent. Depending upon future technological advances, it may provide us with a large portion of our needed energy for centuries to come. (It is also the source of the heat deep within the earth, which is basic to the operation of geothermal energy sources.)

The other major source of the earth's energy is the sun. Nearly all our present energy comes from this source. Oil, coal, and gas are the result of the decay of vegetable matter, which grew in the light of the sun. It was from the sun that it obtained its useful energy. Presently we are beginning to use sunlight directly in solar heating and solar cells for electricity.

Even the energy of our hydroelectric plants comes from the sun indirectly. Evaporation occurs from the lakes and oceans of the earth because of the sun's heat. This water vapor then falls as rain or snow, often upon regions above sea level. As the water loses gravitational PE while falling to the seas, it can be made to do work, turning turbines and similar equipment.

There are other minor sources of energy available to the earth. The tides, for example, are the result of gravitational effects exerted on the earth by the moon. The wind, ocean currents, temperature differences in the ocean, and so on are also possible energy sources, but they are uncertain at this time.

We see, then, that *the earth is receiving energy from two major sources, nuclear energy sources and the sun.* (Actually, the sun is also a nuclear energy source; this is discussed in more detail in Chap. 28.) The energy presently locked in fossil fuels (oil, gas, coal) will become less usable as it is changed from its present chemical form. We hope that this source can be partly replenished as we grow fuel materials such as plants, algae, etc., in the light of the sun. Our nuclear sources of energy are extremely vast, and it appears that we shall eventually be able to make wide use of them. But on a scale of tens of thousands of years, these too will be depleted. And on a scale of billions of years, even the sun will become a poorer source of energy. These and similar topics are discussed in the final chapter of this book.

5.9 MACHINES

Nearly any mechanical device which helps us do work can be called a machine. Most of these devices fall into one of three general categories. A very important class of machines makes it possible to lift a very heavy load by the application of a comparatively small force. Examples of this type of machine are the car jack, the claw hammer, and the compound pulley, to name only a few.

A second type of machine is used to move an object very swiftly even though the driving agent is moving comparatively slowly. An example of this would be a gear or pulley system which changes the rotation of an input shaft into the much faster rotation of an output shaft.

A third type of machine merely makes it more convenient for the work to be done. One such machine is the single pulley shown in Fig. 5.16. It is often more convenient for a person to pull down on a rope at A in order to lift the weight W than it is to pull straight up on the weight. Another example of a machine of this general type would be a device which enables someone to move an object which is some distance away.

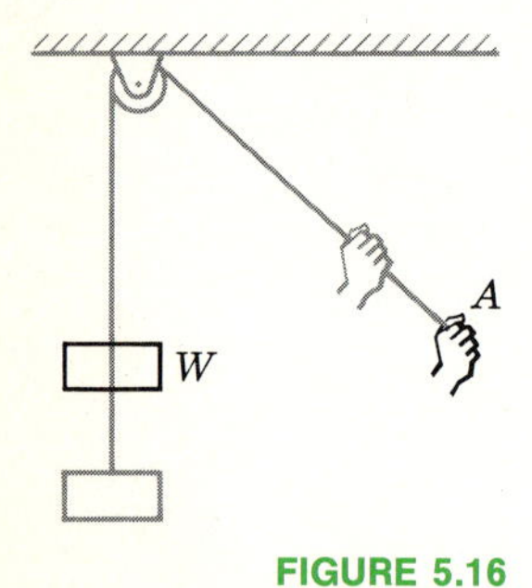

FIGURE 5.16
The type of machine shown here merely alters the direction of the applied force. By so doing, it makes it more convenient to lift the object.

None of these machines is capable of doing more work than the driving agent does on the machine. We know this to be true without examining the machine in detail, since energy must be conserved in these systems. If more work could be done by the machine than was done upon the machine to make it run, the machine would be creating the ability to do work within itself. This, of course, it cannot do. Moreover, since any practical machine will have some friction, energy will be lost doing friction work within the machine. Hence more work has to be done on the machine than is required to do the work without the machine. The purposes of machines are usually the three outlined above. Never does a machine do more work than is done on the machine to make it run.

5.10 THE BLACK-BOX MACHINE

There are certain characteristics common to all machines. In order to describe the operation of machines in general, it is convenient to imagine that the machine is enclosed in a black box so that its mechanisms will not distract us. This situation is shown in Fig. 5.17.

This machine has an input end, the rope at the left. When someone pulls on the rope with a driving force F, the machine runs. The output end of the machine is at the right; in this case, it does work by lifting the object having weight W.

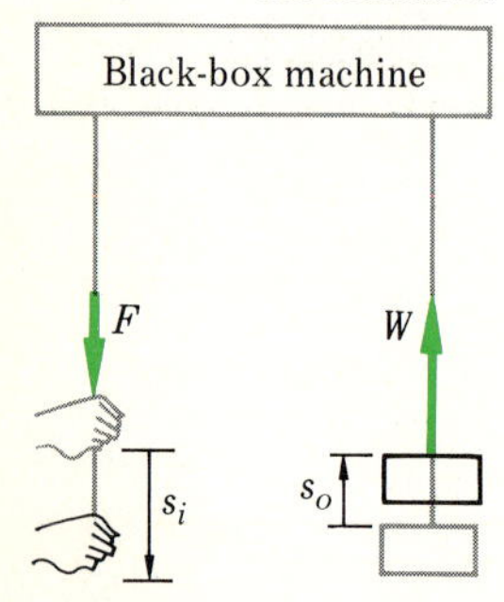

FIGURE 5.17
We can determine the IMA and AMA for this machine, as well as its efficiency, provided that F, W, s_i, and s_o are measured.

To describe the characteristics of this machine completely, only two types of experiments need to be done with it. (1) We must measure how far the point of application of the input force moves, s_i, when the weight is lifted a distance s_o by the output end. That is to say, the rope on the left comes out of the machine a distance s_i when a length s_o of the rope on the right is taken up into the box. (2) We must measure how large a force we need to apply to the input end in order to lift a known load. The first of these experiments tells us how great a load the machine can lift if the applied force is F and if there is no friction in the machine. In particular, since the energy put into the machine Fs_i is all used by the machine to do useful work lifting the load, we have

$$Fs_i = Ws_o \qquad \text{ideal, no friction}$$

From this it follows that

$$\frac{W}{F} = \frac{s_i}{s_o} \qquad \text{no friction}$$

Since the ratio of the load lifted W to the force required F is a measure of the weight-lifting advantage of the machine, we say that its **ideal mechanical advantage** (IMA) is given by the value this ratio would have if there were no friction, that is, s_i/s_o. Hence, for any machine

Ideal Mechanical Advantage

$$\text{IMA} = \frac{s_i}{s_o} \tag{5.7}$$

Equation (5.7) has the following meaning. If a given machine is examined and it is found that the input force moves through a distance of 10 ft when the load is lifted 2 ft, then

$$\text{IMA} = \frac{10 \text{ ft}}{2 \text{ ft}} = 5$$

This would mean that, if the machine had no internal friction, it would be capable of lifting an object weighing 500 lb if an input force of 100 lb were applied. Notice that the machine cannot do more work than the driving agent furnishes, even in this ideal case. Although the output force is five times as large as the input force, the output force moves the object only one-fifth as far as the input force moves. Force is multiplied by the machine, but energy is not.

A real machine is not entirely friction-free, of course. We need yet another experiment to tell us about this aspect of the machine. It is the second experiment mentioned above, in which we measure the force F needed to lift the weight W. The ratio W/F in this case is the **true,** or **actual, mechanical advantage** (AMA) of the machine. We have, therefore,

Actual Mechanical Advantage

$$\text{AMA} = \frac{W}{F} \tag{5.8}$$

Of course, *the AMA will always be smaller than the IMA,* since, for the same applied force F, friction would reduce the load W which could be lifted.

The **efficiency** *of a machine or a process is defined to be the ratio of the output work to the input work.* Since these two quantities are Ws_o and Fs_i, respectively, we have

DEFINITION

$$\text{Efficiency} = \frac{Ws_o}{Fs_i} = \frac{W/F}{s_i/s_o}$$

Hence

$$\text{Efficiency} = \frac{\text{AMA}}{\text{IMA}} \tag{5.9}$$

Efficiency of a Machine

where use has been made of Eqs. (5.7) and (5.8). If there is no friction, the AMA and IMA are identical and the efficiency would be 1.00 or, as more often stated, 100 percent. On the other hand, if the AMA = 4 and the IMA = 5, the efficiency would be $\frac{4}{5}$, which is 0.80, or 80 percent.

Notice that the basic qualities of a machine, AMA, IMA, and efficiency, are defined in the same way for all machines. To investigate these quantities in particular situations, we must now examine the inner workings of various machines.

5.11 THE SIMPLE PULLEY

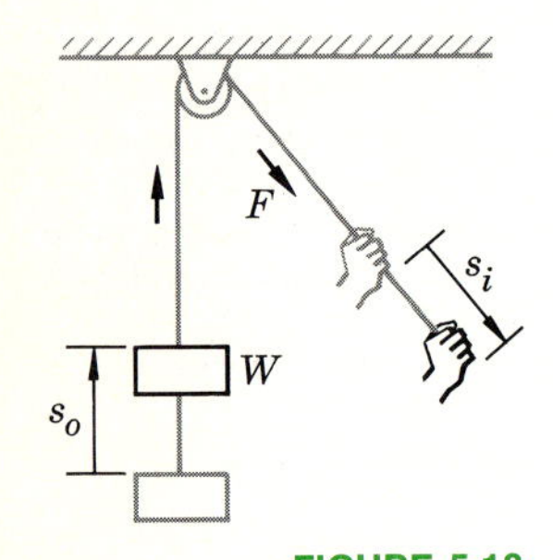

FIGURE 5.18
When used as shown, the simple pulley has an IMA of unity.

This is the machine shown in Fig. 5.18. It consists of a rope looped once over a wheel mounted on an axle. Since the rope is assumed not to stretch, if the force F pulls the rope out through a distance s_i, the object having weight W must rise a distance $s_o = s_i$. Therefore, for this machine, we have from Eq. (5.7)

$$\text{IMA} = \frac{s_i}{s_o} = 1.00$$

The actual mechanical advantage of the machine would depend upon the friction in the pulley. For a simple pulley such as this, the friction could be made quite small, and so the AMA would be nearly as large as the IMA. It could never be as large as 1.00, though. The efficiency of the machine would be somewhat less than 100 percent.

Illustration 5.14 For the system of Fig. 5.18, a force of 50 lb is needed to lift a 40-lb object. What is the efficiency of the machine?

Reasoning From the design of the machine, $s_i = s_o$, so its IMA = 1.00. The data given show AMA = 40 lb/50 lb = 0.80. Hence

$$\text{Efficiency} = \frac{0.80}{1.00} = 0.80, \text{ or } 80\%$$

FIGURE 5.19
This pulley system has an IMA of 2, since the applied force moves twice as far as the load.

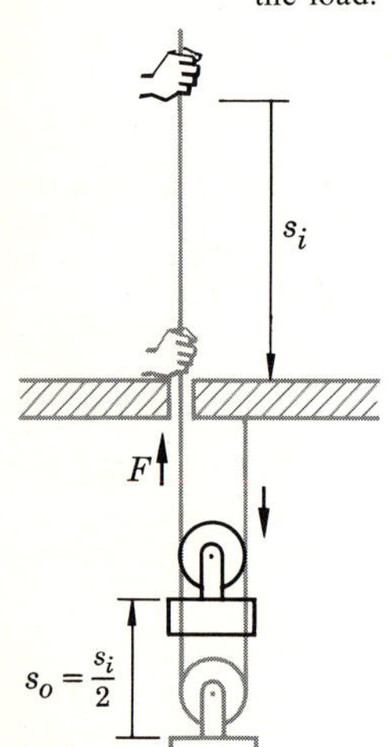

5.12 OTHER PULLEY SYSTEMS

In Fig. 5.19 we consider the case of a single movable pulley. Notice that for each foot that the load moves up, the force F must pull the rope up 2 ft. This is the result of the fact that when the pulley moves 1 ft closer to the ceiling, both the right and left ropes to the ceiling will be 1 ft shorter. This means that 2 ft of rope must be pulled up through the hole to shorten the support by 1 ft.

For this system, then, $s_i = 2s_o$, and so

$$\text{IMA} = \frac{s_i}{s_o} = 2.00$$

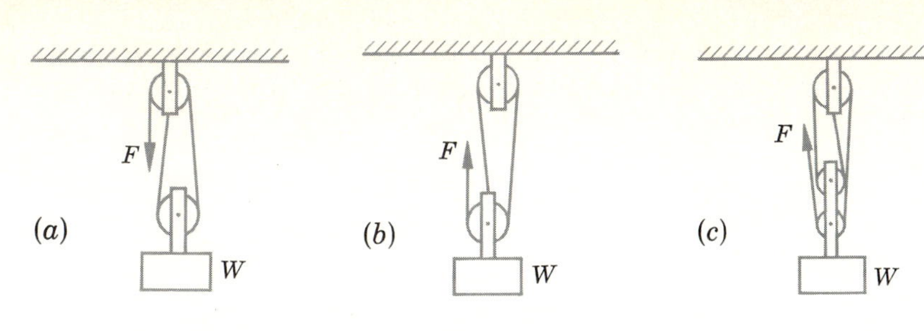

FIGURE 5.20

The IMAs for the systems shown are 2, 3, and 4, respectively. Note that these are also the number of ropes pulling up on the load-bearing pulley.

It is clear that if the system were not moving, both ropes would be pulling up on the pulley and weight. Hence, if the pulley were weightless, the tension in each rope would be $W/2$, so that their combined upward pull on the pulley would be W. In that case, F would be equal to $W/2$. Actually, of course, F must be large enough so that the ropes also support the weight of the pulley and overcome the action of friction as well. Hence, it would be impossible for this machine to have an AMA = 2.00. If the pulley weighed nearly as much as W, the machine would have a very low efficiency, even though the friction in the pulley itself were very small. Nevertheless, it would be an improvement over a single rope in some respects.

More complicated pulley systems are shown in Fig. 5.20. Using the same reasoning as for the previous case, we find that the IMA for systems (*a*), (*b*), and (*c*) are 2, 3, and 4, respectively. Notice that, in this case as well as in Fig. 5.19, the IMA is numerically equal to the number of ropes pulling up on the free pulley. This fact provides one with a simple rule of thumb for determining the IMA of block-and-tackle systems such as these.

5.13 THE WHEEL AND AXLE

The device shown in Fig. 5.21 consists of an input rope wound around a big wheel of radius b. The load is held by the axle of the wheel, and the axle has a radius a, much smaller than b.

To find the IMA of this simple machine, consider what happens as the wheel and axle turn counterclockwise through one revolution. The input rope will unwind a length equal to the circumference of the wheel. Therefore

$$s_i = 2\pi b$$

The rope holding the load will wind up one turn on the axle, and this length is just the circumference of the axle. Thus

$$s_o = 2\pi a$$

We therefore have

$$\text{IMA} = \frac{s_i}{s_o} = \frac{b}{a}$$

FIGURE 5.21

The wheel and axle have an IMA given by the ratio of the radius of the wheel to the radius of the axle.

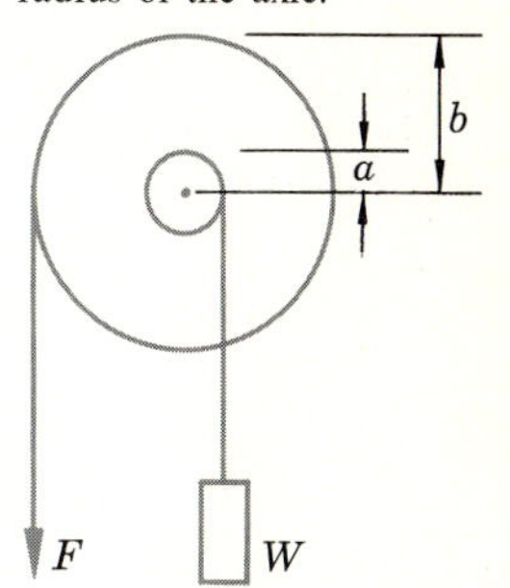

This is a very useful machine, since it does not have too much inherent friction and yet b can be made much larger than a. It would not be uncommon for a to be $\frac{1}{2}$ in and b to be 12 in. For this combination, the IMA is 24, a very large mechanical advantage for such a simple device.

SUMMARY

The work done by the force **F** on an object as the object moves through a straight-line displacement **s** is given by work $= Fs \cos \theta$. θ is the angle between **F** and **s**. Alternatively, in terms of the component of **F** along **s**, work $= F_s s$. The units of work are joules, ergs, and foot-pounds. These units are related through $1\ \text{J} = 10^7$ ergs $= 0.738$ ft · lb.

Power measures the rate of doing work. It is defined as the work done per unit time. Typical units are the watt (which is joules per second), the foot-pound per second, and the horsepower. Relations between these units are 746 W = 1 hp and 1.36 W = 1 ft · lb/s.

When an object has energy, it possesses the ability to do work. Kinetic energy (KE) is the ability to do work because of an object's motion. An object with mass m and velocity v has KE $= \frac{1}{2}mv^2$. It can do this much work against external forces as the object is brought to rest. The units of KE are the same as those of work. If an object is subjected to a resultant force which does work on the object, the work done on the object equals the change in KE of the object.

An object which is capable of doing work because it is pulled downward by a gravitational force is said to possess gravitational potential energy (PE). Its potential energy is mgh; the object can do this much work while descending a distance h. The reference level for measuring the object's height is arbitrary.

If the work done in moving an object from point A to point B against a force is not dependent upon the particular path taken for the movement, then the force is said to be conservative. The gravitational force is conservative. As a result, the work done against gravity in lifting an object from A to B is mgh independent of the path followed. Friction forces are not conservative.

The work-energy theorem can be stated as follows for any object:

$$\text{Loss in KE} + \text{loss in PE} = \text{work}_{\text{dec}} - \text{work}_{\text{inc}}$$

The law of conservation of energy states that energy can neither be created nor destroyed. When a loss occurs in one form of energy, an equal increase must occur in other forms of energy. Work done against friction forces produces heat energy equal in magnitude to the friction work done. Many forms of energy exist. Typical are chemical, electrical, and nuclear energy.

Machines are devices which help us do work. They never do more work than the energy given to them. To characterize a simple machine one makes use of its ideal and actual mechanical advantages. If the input force F moves through a distance s_i when the output force W moves through a distance s_o, then IMA $= s_i/s_o$ while AMA $= W/F$. The AMA is always less than the IMA. A machine's efficiency is defined by efficiency = (output work)/(input work). This is equivalent to eff = AMA/IMA.

MINIMUM LEARNING GOALS

Upon completion of this chapter, you should be able to do the following:

1. Define work and its units. Give several examples of a force which does no work because (*a*) $s = 0$ or (*b*) $F_s = 0$. Give examples where a force does negative work.

2. Compute the work done on an object by a specified constant force when the object is moved through a given straight-line distance.

3. Define power and its units. Compute the power from data for the rate at which work is being done.

4. Convert between the units watts, foot-pounds per second, and horsepower.

5. Define KE. Compute its value provided m and v are given.

6. Compute the change in KE of an object which is subjected to a given resultant force acting through a known distance.

7. Define gravitational PE. Compute its value for an object provided the necessary data are given.
8. Explain what is meant by a conservative force. Point out the importance of the fact that the gravitational force is conservative.
9. Give several examples of situations in which KE and PE are interchanged.
10. State the work-energy theorem in words and in equation form. Apply it to examples similar to those given in Sec. 5.6.
11. State the law of conservation of energy in your own words. Explain why the loss of KE in doing friction work does not contradict the law.
12. Compute the IMA, AMA, and efficiency of a black-box machine provided sufficient data are given.
13. Give the IMA of a simple machine such as a pulley system provided the construction of the machine is made clear.

IMPORTANT TERMS AND PHRASES

You should be able to define or explain each of the following:
Work $= F_s s = Fs \cos \theta$
Joule, erg, foot-pound
Power = work/time
Watt, horsepower
Kilowatthour
KE $= \frac{1}{2}mv^2$
Gravitational PE $= mgh$
Conservative forces
Work-energy theorem
Law of conservation of energy
IMA $= s_i/s_o$
AMA $= W/F$
Efficiency = (output work)/(input work) = AMA/IMA

QUESTIONS AND GUESSTIMATES

1. The moon and most man-made earth satellites circle the earth in essentially circular paths with center at the earth's center. How much work does the earth's gravitational force do on them?
2. Reasoning from the standpoint of KE, why is a loaded truck likely to be much more damaging than a Volkswagen in a collision with a massive stationary object? Assume equal initial speeds.
3. A ball hangs at the end of a thread and the system swings as a pendulum. Describe what happens to the KE of the ball as the pendulum swings back and forth. Repeat for the PE. How are the two related? What happens to the energy of the system as the pendulum loses its energy? Does the pull of the thread on the ball do any work?
4. A person holds a bag of groceries while standing still talking to a friend. A car sits stationary with its motor running. From the standpoint of work and energy, how are these two situations similar?
5. A conscientious hobo in a boxcar traveling from Chicago to Peoria pushes on the front wall of the car all the way. Having at one time been a physics student, he thinks his pushing did a great deal of work since both F_s and s were large. Where was the flaw in his reasoning?
6. Since the earth is moving with respect to the sun, everything on it has KE, at least in the opinion of an observer who thinks the sun is at rest. Why did we not need to consider this when working the examples in this chapter?
7. Is KE a vector or a scalar quantity?
8. A ball of mass m is held at a height h_1 above a table. The table top is a height h_2 above the floor. One person says that the ball has a PE mgh_1, but another person says that its PE is $mg(h_1 + h_2)$. Who is correct?
9. As a rocket reenters the atmosphere, its nose cone becomes very hot. Where did this heat energy come from?
10. If a piece of chalk falls off a table, its speed just before it hits the floor can be found by equating the original PE to its total energy just before striking the floor. Show that the same result is obtained if the zero of PE is taken as (*a*) the floor, (*b*) the table top, (*c*) the ceiling.
11. Automobiles, tractors, etc., have gear systems which

can be changed by shifting. Considering these to be ideal machines, discuss why the shifting process is used.

12. It has been suggested that tides flowing in and out of harbors could be used as sources of energy. Another suggestion is to use the ocean waves for this purpose. Discuss the pros and cons of either proposal from a practical standpoint. Where does the energy come from in each case?

13. About what horsepower is a human being capable of producing for a short period, as in climbing a flight of stairs? (E)

14. Estimate the amount of useful work (as defined in physics) which an average human being might perform in 1 day. For comparison purposes, a typical diet might furnish the person with 2000 kcal ($\approx 8.4 \times 10^6$ J) of energy each day. Where does the rest of the energy go? (E)

15. Estimate the force a driver experiences when the car being driven hits another car head on. Assume both cars to be similar and traveling at 60 mi/h. Discuss the effect of seat belts, position of the person in the car, and similar factors. (E)

PROBLEMS

Many of the problems in this set are solved most easily by energy methods.

1. A horizontal string is to be used to pull a 3.0-kg object along the floor at constant speed. If the coefficient of sliding friction between the floor and object is 0.60, how much work is done by the string in pulling the object 5.0 m?

2. In order to pull a certain child in a wagon, a force of 30 lb at an angle of 20° above the horizontal sidewalk is required. How much work is done in pulling the wagon a distance of 50 ft?

3. A 6000-lb cart filled with coal is being pulled up an incline within a coal mine. The pulling cable is parallel to the incline and exerts a force of 200 lb on the cart. How much work is done on the cart by the cable as the cart moves through a distance of 100 ft?

4. An 80-kg hiker climbs a 600-m-high hill. How much work does the hiker do against gravity? Does this amount of work depend upon the path the hiker takes? If the hike takes 90 min, what average horsepower was expended in this way by the hiker?

5. A 30,000-kg truck takes 40 min to climb from an elevation of 1800 to 2900 m along a mountain roadway. (*a*) How much work did the truck do against gravity? (*b*) What average horsepower did the truck expend against gravity in the process?

6. A 2-kg object dropped from a height of 12 m is moving with a speed of 7 m/s just before it hits the ground. How large was the average friction force retarding its motion?

7. What constant force is required to accelerate a car which weighs 3200 lb from rest to a speed of 60 ft/s in a distance of 80 ft?

8. If a 2000-kg car is traveling at a speed of 20 m/s, hits a cement wall, and stops within 3 m, how large was the average retarding force?

9. A motor is to lift a 2000-lb elevator from rest at ground level in such a way that it has a speed of 8 ft/s at a height of 40 ft. (*a*) How much work will the motor have done? (*b*) What fraction of the total work went into kinetic energy?

10. Starting from rest, a locomotive pulls a series of boxcars up a 5° incline. After the train has moved 6000 ft, its speed is 30 ft/s. Assume the entire train weighs 1.50×10^6 lb. How much work must the locomotive do? What fraction of this work is done against gravity? Assuming the acceleration to be uniform, how long did the process take? What average horsepower did the locomotive expend?

11. An electric motor is to drive a small laboratory pump which lifts 500 cm^3 of water through 2 m each minute. Find the minimum wattage for a motor capable of doing this (1 cm^3 of water has a mass very close to 1 g). *Hint:* How much work must the motor do each second?

12. Suppose that 10^{16} electrons strike the screen of a TV tube each second. If each electron is accelerated through a voltage large enough to give it a speed of 10^9 cm/s, starting from rest, how many watts of power are expended in maintaining this beam of electrons? ($m_e = 9.1 \times 10^{-31}$ kg.)

13. An atom-smashing machine known as a Van de Graaff generator (discussed in later chapters) is capable of accelerating a beam of protons ($m_{proton} = 1.67 \times 10^{-27}$ kg) from rest to a speed of 10^9 cm/s. If the machine accelerates 10^{16} protons per second, how many watts of power is it producing? (A proton is a hydrogen atom from which an electron has been removed.)

14. The brakes on a 3200-lb car are accidentally released when the car is on a hill. How fast will the car be moving when it reaches a point 30 ft lower than its starting point, assuming an average friction force of 200 lb retards the motion. Assume the road distance to be 90 ft.

15. A 2000-kg car is traveling at 20 m/s up a hill when the motor stops. (*a*) If the car is a vertical distance of 8 m from the top of the hill at that instant, will it be able to reach the top? (*b*) How far below the top of the hill could the car be and still reach the top? Ignore friction.

16. A 1.8-g bullet with velocity 360 m/s strikes a block of wood and comes to rest at a depth of 6 cm. (*a*) How large is the average decelerating force, and (*b*) how long does it take to stop the bullet?

17. A 0.70-g bead slides along a wire, as shown in Fig. P5.1. If it starts from rest at point A and the friction forces are negligible, how fast will it be going at points B, C, and D?

18. In Fig. P5.1 the distance along the wire from A to D is 400 cm. If the bead starts from rest at A and finally stops just as it reaches D, how large is the average friction force retarding its motion?

19.* A body of weight W is pulled up a frictionless incline of angle θ with a speed v. Show that the work done on the body in a time t is given by

$$\text{Work} = Wvt \sin\theta$$

20.* A car of mass m rests on a hill of height h and length L. Show that the speed of the car when it reaches the bottom of the hill is

$$v = \sqrt{2gh - \frac{2LF}{m}}$$

where f is the average friction force retarding the motion.

21. A ball of mass M is suspended as a pendulum bob from a cord 4.0 m long. The cord is pulled to one side by a force on the mass until it makes an angle of 70° to the vertical, and the system is then released. With what speed is the mass moving as it passes directly underneath the point of suspension? (Ignore air friction.)

22.* If the ball described in Prob. 21 broke the cord just as it passed below the point of suspension, where would it hit the floor, which is 2.45 m below it? Give your answer in terms of the distance from the point on the floor directly below the suspension point.

23.* For the pendulum described in Prob. 21, how fast is the ball moving when the cord makes an angle of 20° to the vertical?

24.** A man performs the following experiment. Using his 3200-lb car on level ground, he finds that the car will accelerate from rest to 40 ft/s in 10 s. The car will coast to rest from 40 ft/s in a distance of 1600 ft. Compute the average horsepower delivered by the car. (Notice that the friction work done in stopping is not the same as the friction work in starting. Why? Assume instead that the average friction force is the same in the two cases.)

25. A 600-lb object is to be lifted by a pulley system using a force of 40 lb. The machine found suitable for this lifts the load $\frac{1}{2}$ ft when the applied force moves 10 ft. Find the (*a*) IMA, (*b*) AMA, and (*c*) efficiency of the machine.

26. Estimating an efficiency of 75 percent, design the simplest pulley system capable of lifting 200 N by the application of a 40-N force.

27. A force of 60 N is applied to a wheel-and-axle device in order to lift a 500-N load. If the radius of the wheel is 1.0 ft and that of the axle is 1.0 in, and if the diameter of the rope wound on the axle is 0.40 in, find the (*a*) AMA, (*b*) IMA, and (*c*) efficiency of the system.

28. For a particular type of car jack, the operator moves his hand, i.e., the input force, through a distance of 2.0 ft

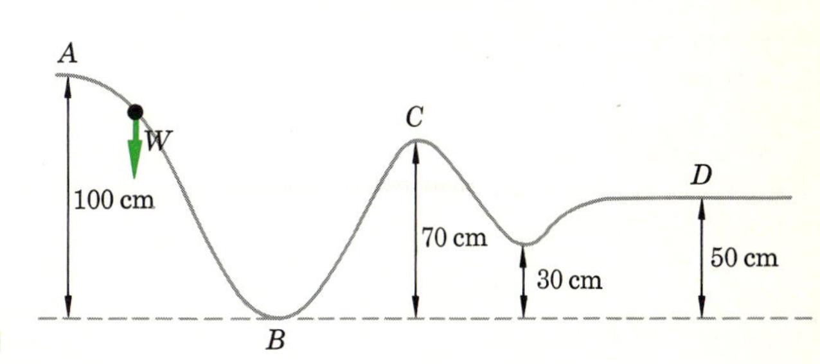

FIGURE P5.1

for every $\frac{1}{2}$ in the load is lifted. (*a*) What is the IMA of the jack? (*b*) On the assumption of a 20 percent efficiency, how large an applied force is needed to lift 2000 lb?

29. An electric motor is labeled as a 0.55-kW motor. On the assumption that it is 80 percent efficient, how many horsepower can it deliver?

30.* A $\frac{1}{4}$-hp motor has a 8.0-cm-diameter pulley attached to its shaft. If the shaft rotates at 1800 rev/min (revolutions per minute), how large a load is a belt running on the pulley capable of pulling? Assume that the motor is 80 percent efficient.

31.* A certain 50-W motor runs with a shaft speed of 1800 rev/min. By use of reducing gears the final, or output, shaft rotates at 18 rev/min. (*a*) If the machine is 30 percent efficient, with what force can the machine pull a belt on a 3.0-cm-radius pulley at the output shaft? (*b*) If the gear system on the motor were reversed so that the output shaft rotated at 180,000 rev/min, what force would be available to pull the belt on the same pulley?

MOMENTUM AND THE PRESSURE OF GASES

6

The law of conservation of energy, with which we have been dealing in the previous chapter, is not the only conservation law known to physics. A second example, the law of conservation of momentum, will be introduced in this chapter. It has far-reaching consequences. We shall make use of it by deriving a relation for the pressure of an ideal gas. Both the gas-pressure relation and the momentum-conservation law will be of importance to us in the following chapters.

6.1 THE CONCEPT OF MOMENTUM

An experience common to us all is that a moving object possesses a quality which causes it to exert a force upon anyone who tries to stop it. The faster the object is traveling, the harder it is to stop. In addition, the more massive the object is, the more difficulty we have in stopping it. For example, an automobile moving at 2 m/s is much less easily stopped than a kiddie car traveling at the same speed.

DEFINITION

Newton called this quality of a moving body the *motion* of the body. We today term it the **momentum** of the body and define it by the relation

Momentum

$$\text{Momentum} = m\mathbf{v} = \mathbf{p} \tag{6.1}$$

In this expression, $\mathbf{v}$ is the vector velocity of the mass m. The *momentum is* therefore *a vector quantity* and has the direction of the motion, i.e., of the velocity. It is customary to use $\mathbf{p}$ to represent momentum.

Obviously, the momentum of a body is the result of forces which accelerated the body from rest to velocity v. Similarly, if we are to slow down the body, i.e., cause it to lose momentum, we must apply a retarding force to it. A large retarding force will cause the body to lose more momentum in a given length of time than a small retarding force will. We now wish to find the exact mathematical relation between force and change of momentum.

6.2 NEWTON'S SECOND LAW RESTATED

Consider the acceleration of a body by a constant force $\mathbf{F}$ as shown in Fig. 6.1. The force is the net unbalanced force on the body and will cause a uniform acceleration in the x direction. Newton's second law gives

$$\mathbf{a} = \frac{\mathbf{F}}{m}$$

If the body had an original velocity $\mathbf{v}_0$ and the force acts for a time t, our motion equations yield the result

$$\mathbf{v} = \mathbf{v}_0 + \mathbf{a}t$$

Replacing $\mathbf{a}$ by $\mathbf{F}/m$ and rearranging leads to

$$\mathbf{F}t = m\mathbf{v} - m\mathbf{v}_0 \tag{6.2}$$

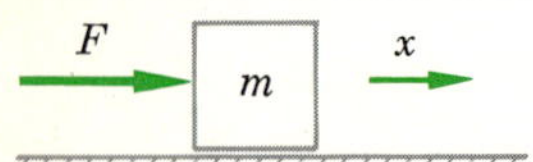

FIGURE 6.1
The net applied force F increases the momentum of the mass. Momentum has direction, and the increase will be in the direction of F.

We see from this that the change in momentum of the object, $m\mathbf{v} - m\mathbf{v}_0$, is equal to the product of the applied force and the time for which the force was applied. This is actually closer to the way Newton wrote his second law than the form $F = ma$ which we previously gave. In some respects it is more general than the $F = ma$ form, since it can easily be extended to account for changes in mass as the body accelerates. For example, as a rocket accelerates, it loses mass because its fuel is being used up and ejected to provide thrust. We shall also see in Chap. 28 that the mass of an object increases at high speeds. If the mass changes with velocity, Eq. (6.2) should be written as

$$\mathbf{F}t = m\mathbf{v} - m_0\mathbf{v}_0 = \Delta\mathbf{p} \tag{6.3}$$

where $\Delta\mathbf{p}$ is the change in momentum. Notice that *the direction of the change in momentum is in the direction of the force. The units of momentum are the product of a force and a time unit* or, fully equivalent, a mass and a velocity unit. Examples are the newton-second or kilogram-meter per second.

Very often we wish to apply the above concept of momentum to cases where the applied force is not constant. For example, when a bat hits a baseball, the force certainly varies from instant to instant during the collision. In such cases, we are usually not much interested in anything but an average force. We therefore define an average force by Eq. (6.2) or (6.3) as the steady force required to cause the observed change in momentum. Generally, the force obtained by application of Eq. (6.2) or (6.3) to collision processes will be an average force defined in the above sense.

Since the time of collision as well as the force are usually not well known, *it is common to lump the product* **F***t into a single quantity called the* **impulse.** In this terminology, *the change in momentum of a body is equal to the impulse applied to it.*

Illustration 6.1 A 1500-kg car traveling at a speed of 20 m/s reduces its speed to 15 m/s in a time of 3.0 s. How large was the *average* retarding force?

Reasoning From Eq. (6.2) or (6.3) we have

$$Ft = mv - mv_0$$
$$F(3\text{ s}) = (1500\text{ kg})(15\text{ m/s}) - (1500\text{ kg})(20\text{ m/s})$$

From this we find $F = -2500$ N. The negative sign indicates that the force is a retarding one.

Illustration 6.2 An electron traveling at a speed of 2.6×10^8 m/s has a mass of 18.0×10^{-31} kg, which is about twice the rest mass of the electron, 9.1×10^{-31} kg. How large an average force F is needed to accelerate the electron to this speed from an initial speed of 1.0×10^8 m/s in a time of 1.0×10^{-7} s? The mass of the electron at this slower speed is 9.5×10^{-31} kg. (We shall discuss the change in mass of particles moving with high speeds in Chap. 26.)

Reasoning From Eq. (6.3) we have

$$Ft = mv - m_0 v_0$$
$$F(10^{-7}\text{ s}) = (18 \times 10^{-31}\text{ kg})(2.6 \times 10^{8}\text{ m/s}) - (9.5 \times 10^{-31}\text{ kg})(1 \times 10^{8}\text{ m/s})$$

or $$F \approx 47 \times 10^{-16} - 9.5 \times 10^{-16}\text{ N} \approx 37 \times 10^{-16}\text{ N}$$

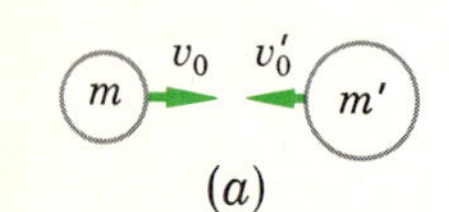

(a)

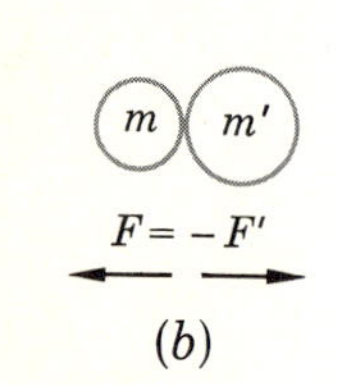

(b)

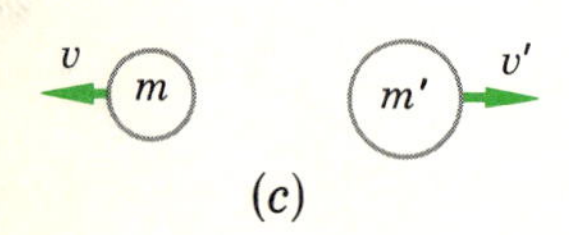

(c)

FIGURE 6.2
When two bodies collide, the forces on each are equal but oppositely directed. Momentum is conserved in a collision.

6.3 CONSERVATION OF MOMENTUM

Consider the collision of two particles as shown in Fig. 6.2. The particles might be balls, spherical molecules, or something similar. For our purposes, we can consider the two particles to be a universe of their own, since they are isolated from all other bodies. Hence, according to Newton's third law, if there is a force on one of the particles, there must be an equal and opposite reaction force on some other body in the universe. In this particular case, the reaction force would act on the other particle. This fact leads to an interesting and valuable conservation law.

Let us compute the change in momentum of the particle on the left in Fig. 6.2 as a result of the collision. From Eq. (6.2) or (6.3) we have for the *average* force

$$\mathbf{F}t = m\mathbf{v} - m\mathbf{v}_0$$

Similarly, for the particle on the right

$$\mathbf{F}'t = m'\mathbf{v}' - m'\mathbf{v}_0'$$

If we add these two expressions, we obtain

$$(\mathbf{F} + \mathbf{F}')t = (m\mathbf{v} - m\mathbf{v}_0) + (m'\mathbf{v}' - m'\mathbf{v}_0')$$

But since the vector force $\mathbf{F}$, the action force, is equal and opposite to the reaction force $\mathbf{F}'$, we have $\mathbf{F} = -\mathbf{F}'$ and the left side of this equation is zero. Hence

$$0 = (m\mathbf{v} - m\mathbf{v}_0) + (m'\mathbf{v}' - m'\mathbf{v}_0')$$

In words this says

0 = change in momentum of 1st ball + change in momentum of 2d ball

Consequently, the total change in momentum of the isolated universe consisting of the two particles was zero.

We can extend this line of reasoning to much more complicated systems. To do so, we define what is called an isolated system. *An* **isolated system** *is a group of objects upon which the net resultant force from outside is zero.* For

DEFINITION

example, the universe as a whole is an isolated system. In it, for every force upon one object there is an equal but opposite force on some other object. As a result, the change in momentum of the objects taken as a whole is always zero.

These considerations apply to any isolated system and are summarized in the **law of conservation of linear momentum,** stated as follows:

Law of Conservation of Linear Momentum

The total momentum of an isolated system of objects is a constant.

Even if the objects are not completely isolated, the law is often useful. For example, in a collision of two cars, the skidding of the wheels along the pavement causes external forces to act on the two-car system. Even so, the forces of one car on the other at the instant of collision are often extremely large, much larger than the skidding forces on the road. Therefore, the large changes in momentum which occur at the instant of collision are almost all the result of the force of one car on the other. As a result, the law of conservation of momentum can still be applied to the two-car system at the instant of collision even though the system is not isolated.

In applying the conservation law we must notice that the momentum of a body is a vector. To illustrate the importance of this, refer to Fig. 6.3. Taking the x direction to be positive, the total momentum of the system before collision (Fig. 6.3*a*) is

$$\begin{aligned}\text{Momentum before} &= mv_0 + m'v_0' \\ &= (2\text{ kg})(6\text{ m/s}) + (3\text{ kg})(-4\text{ m/s}) \\ &= 12 - 12 = 0\end{aligned}$$

Notice that even though each of the individual bodies had momentum prior to the collision, the system as a whole had zero momentum. This is, of course, a very special case, chosen because it emphasizes so dramatically the fact that momentum is a vector. However, this particular case of zero total momentum is interesting in several other respects.

What must be true after the collision? The law of conservation of momentum tells us that the momentum of this isolated system will not be changed by the collision. Hence, in this case, the momentum after collision must still be zero. One possible way in which this could be achieved is shown in part *b* of Fig. 6.3. Notice that the momentum of each body has a magnitude of 9 kg · m/s, one being positive, the other negative. This is definitely a possible solution for the problem, since momentum is conserved. We have the right, though, to ask whether this is the only possible solution to the problem.

It is a simple matter to show that the solution of Fig. 6.3*b* is not the correct solution in a certain case. Suppose that one of the bodies had a wad of chewing gum stuck on the side where the collision occurred. If the gum were sticky enough, the two bodies would remain stuck together after the collision. (If you don't like this way of fastening the bodies together, we could use magnets on the two bodies to hold them together after collision. You can perhaps think of other ways this could be accomplished.) What can these bodies do if they stick together?

FIGURE 6.3

The situations shown in (*b*) and (*c*) are physically possible results of the collision of the bodies shown in (*a*). In both instances the momentum is the same as before collision, namely, zero. Hence, momentum is conserved, although KE is not.

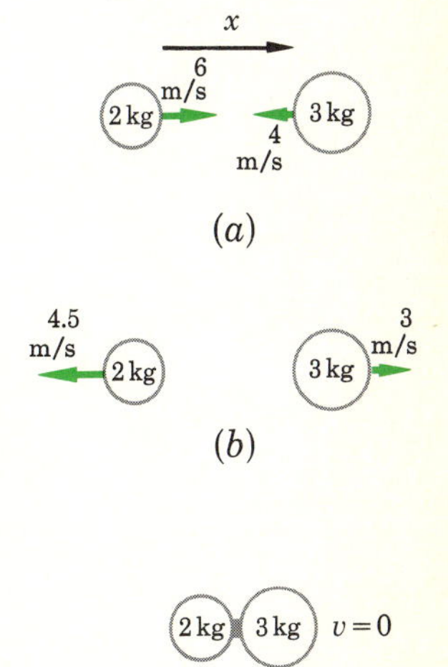

The law of conservation of momentum allows only one answer in this case. Since before collision the momentum of the system was zero, it must still be zero after collision. But now, since the bodies are stuck together, they must move as a unit, and their velocities will be in the same direction. Hence, there is no possibility at all that their momenta can cancel each other after the collision. We therefore conclude that the only way the total momentum can be zero after collision, if the bodies stick together, is for the bodies to be at rest. Therefore, in this case, the moving bodies will collide, stick together, and stop dead.

Of course, the above is a very special case. It is one of the simplest cases we can imagine, however. But even here, we are unable to give a completely definitive answer to the question: What happens when two objects with equal and opposite momenta collide? The situation is even more intriguing if we notice that in neither solution given in Fig. 6.3 is the KE of the system the same after collision as before collision. It appears that *KE is usually not conserved in collisions.* We shall need to know how much of the KE was lost in the collision before we can find the correct answer to the question posed above.

6.4 ELASTIC AND INELASTIC COLLISIONS

What happened to the KE of the two objects discussed in the last section when they collided and stuck together? Probably most of the energy was lost doing work on the gum between the bodies, causing it to flow under the applied force of the collision. This energy eventually appears as heat energy. In addition, probably some energy was given off in the form of sound waves. In any event, the KE was *entirely* transformed into other types of energy. A collision in which this occurs is said to be a **perfectly inelastic collision.**

If the colliding objects had been tennis balls, only a portion of the KE would have been lost. Hence, the balls would have rebounded from each other. Even so, the energy lost to friction as the molecules in the ball move past each other as the balls distort, together with the energy lost in sound and by other means, will cause the KE to be less after impact than before impact. We term *a collision in which KE is lost an* **inelastic collision.**

Under certain special conditions, scarcely any energy is lost in the collision. *In the ideal case, when no KE is lost, the collision is said to be* **perfectly elastic.** An example is a hard ball dropped onto a hard, massive object, such as a marble floor, and rebounding to very nearly the same height as its starting point, with a negligible amount of energy lost in the collision with the floor.

FIGURE 6.4
In a perfectly elastic collision, KE is conserved. When the collision is perfectly inelastic, the KE is completely lost.

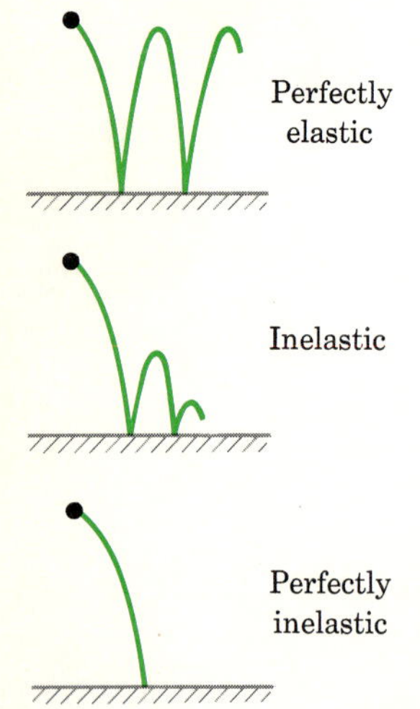

The three types of behavior in collision with a massive object are shown in Fig. 6.4. In fact, the decrease in height on rebound is a common method for testing the resiliency of various materials. Obviously, if a ball dropped from height h rises to height h_2 on rebound, the difference in PE of the ball at these heights is the energy lost by the ball upon collision. Even for a highly elastic object such as a solid rubber ball, the final height on rebound is only about 75 percent as large as the original height. (A Superball does better than this.)

Considerable heat was generated in the ball by the rubbing of the molecules against each other when the ball distorted upon collision. Since this effect is essentially missing in a very hard material, it is not surprising that a small metal ball often loses less energy upon collision than a rubber ball. It therefore rebounds *better* than a rubber ball.

(*a*) Before

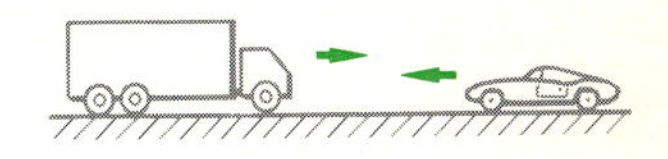

(*b*) After

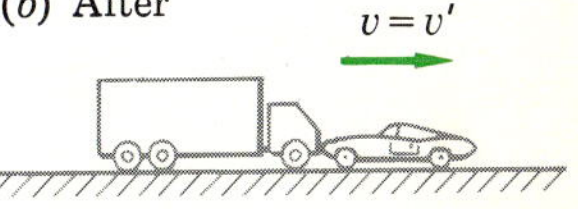

FIGURE 6.5

The collision of the truck and the car is perfectly inelastic in this case. Momentum must still be conserved, however.

Illustration 6.3 A 64,000-lb truck traveling at 30 ft/s collides with a 3200-lb car traveling at 90 ft/s in the opposite direction. If they stick together after the collision, how fast and in what direction will they be moving? (See Fig. 6.5.)

Reasoning Calling the x direction positive, we apply the law of conservation of momentum (KE is obviously not conserved in this collision). Let v be their combined speed after collision.

$$\text{Momentum before} = \text{momentum after}$$
$$(m_1 v_{01})_{\text{truck}} + (m_2 v_{02})_{\text{car}} = (m_1 + m_2)v$$
$$\left(\frac{64{,}000}{32}\right)(30) + \left(\frac{3200}{32}\right)(-90) = \left(\frac{64{,}000 + 3200}{32}\right)v$$

Solving for v gives $v = 24.3$ ft/s. The positive sign for v indicates that the final motion is in the positive x direction, i.e., in the direction in which the truck was moving.

Notice that the truck was slowed down only slightly, while the direction of movement of the car was actually reversed. It is instructive to estimate the average force on the driver of the car during the collision, but this is left as an exercise. (If you wish to make this estimate, note that the impulse exerted on the driver equals the driver's momentum change. You will also need to estimate the time taken for the impact. One way to do this is to find the time it takes for the truck to move a distance about equal to the hood length of the car. Why?)

Illustration 6.4 Figure 6.6*a* shows an x-ray photo of a pistol just after the bullet has been fired. The hot combustion gases from the exploded gunpowder are accelerating the projectile part of the bullet down the barrel. If the masses of the gun and projectile are M and m, respectively, and the exit velocity of the projectile is v', find the recoil velocity of the gun.

Reasoning If you look carefully at the figure, you can see a hand holding the gun. But the external force it exerts on the gun system is small in comparison to the internal forces exerted by the exploding powder. Therefore, for the instant of the explosion, we can assume that the gun is isolated and that momentum is conserved. The situation is shown in Fig. 6.6*b*. We have

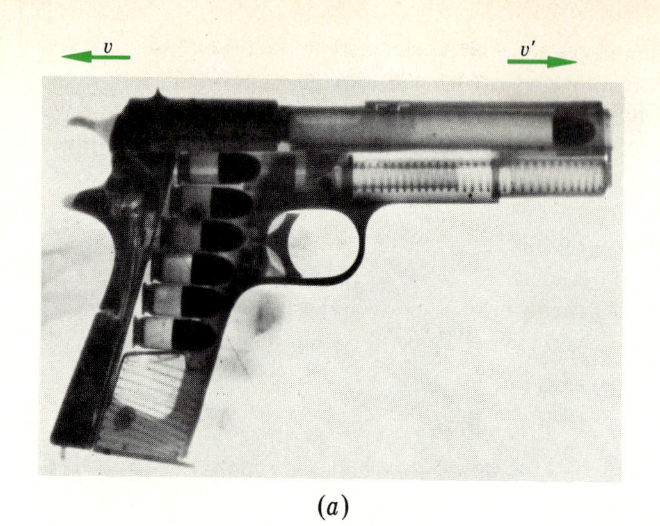

(*a*)

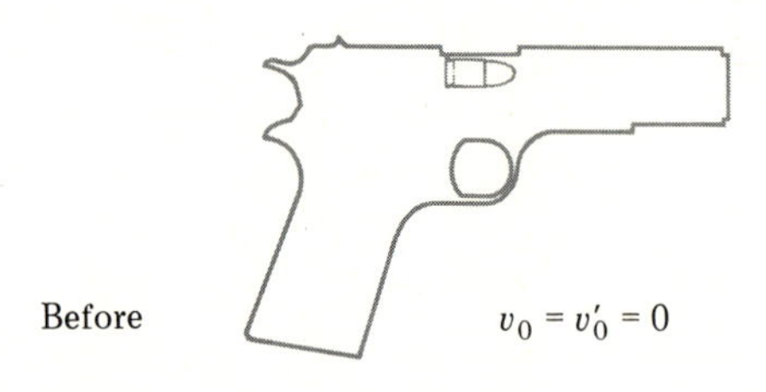

v v'

After

(*b*)

FIGURE 6.6
Before the gun was fired, its momentum was zero. Hence the sum of the momenta must still be zero after it is fired. [(*a*), *Hewlett-Packard Co., was taken using an x-ray flash system.*]

$$\text{Momentum just before} = \text{momentum just after}$$
$$mv'_0 + Mv_0 = mv' + Mv$$

Since $v'_0 = v_0 = 0$, this becomes

$$0 = mv' + Mv$$

Therefore the recoil velocity of the gun is

$$v = -\frac{m}{M}v'$$

Notice that the more massive the gun, the less the recoil will be.

Illustration 6.5 In the previous illustration suppose the ratio $m/M = 1/150$ and that $v' = 200$ m/s. Suppose further that the gun recoils 2.5 cm as the hand holding it brings it to rest. What average force does the gun exert on the hand? Take the mass of the gun to be 1.25 kg.

Reasoning After the discharge, the gun has KE given by

$$\text{Gun's KE} = \tfrac{1}{2}Mv^2 = \tfrac{1}{2}M\left(\frac{m}{M}\right)^2 v'^2 = 0.889M \qquad \text{joules}$$

$$= 1.11 \text{ J}$$

This energy is lost doing work against the hand which holds it. Taking the average restraining force exerted by the hand to be F and the recoil distance to be 0.025 m, we have

$$\text{Work done} = \text{gun's KE}$$

$$(F)(0.025) = 1.11$$

from which

$$F = 44 \text{ N}$$

Can you list some of the factors upon which the kick of a rifle or shotgun depends?

Illustration 6.6 A 40-g ball traveling to the right at 30 cm/s collides head on with an 80-g ball which is at rest. If the collision is perfectly elastic, find the velocity of each ball after collision. (By head on we mean that all motion takes place on a straight line.)

Reasoning During the collision, momentum is conserved. Hence, letting the velocities of the 40- and 80-g balls after collision be v and v', respectively,

$$\text{Momentum before} = \text{momentum after}$$

$$(40 \text{ g})(30 \text{ cm/s}) + 0 = (40 \text{ g})v + (80 \text{ g})v'$$

or

$$2v' + v = 30 \text{ cm/s}$$

We need yet another equation since we have two unknowns, v and v'. Since the collision was perfectly elastic, KE was also conserved. Therefore

$$\text{KE before} = \text{KE after}$$

$$\tfrac{1}{2}(40 \text{ g})(30 \text{ cm/s})^2 + 0 = \tfrac{1}{2}(40 \text{ g})v^2 + \tfrac{1}{2}(80 \text{ g})v'^2$$

or

$$2v'^2 + v^2 = 900 \text{ cm}^2/\text{s}^2$$

Solving for v in the first equation and substituting in this equation yields

$$6v'^2 - (120 \text{ cm/s})v' = 0$$

RELATIVITY

In 1905, Albert Einstein presented his special theory of relativity. We shall discuss this theory in detail in Chap. 26. This note summarizes some of the important features of this very important advance in our understanding of nature.

Experiments of many types have convinced us of the following facts:

1 A pulse of light moving through vacuum always has a speed $c = 2.9979 \times 10^8$ m/s. This speed is not changed even if the light source is moving.

2 Consider two laboratories moving at different, but constant, velocities. All accurate experiments in either laboratory will lead to the same physical laws. Because of this, it is impossible to determine the absolute velocity of a laboratory from measurements confined to that laboratory.

These two facts were shown by Einstein to lead to the following conclusions. They have been tested by making predictions based upon them. All such predictions, when tested, have been found to be correct. We call the theory which derives them the **special theory of relativity.**

1 No object which carries energy can be accelerated to speeds beyond the speed of light in vacuum c.

2 Suppose an object moves past you with speed v. Then all processes occurring within the object will appear to you to be slowed down. For example, if the object flying past is a clock, then to you it will appear to be ticking out time too slowly. If your clock ticks out a time t_0, then the moving clock will appear to you to tick out a time

$$t_0\sqrt{1 - \left(\frac{v}{c}\right)^2}$$

This effect is called **time dilation.**

3 Suppose an object is moving with speed v past you along the x-axis. Then you will measure the x dimensions of the object to have shrunken. A length L_0 will be measured to have become

$$L_0\sqrt{1 - \left(\frac{v}{c}\right)^2}$$

This contraction occurs only along the direction of motion. No change occurs in the y and z dimensions.

4 Suppose an object has a mass m_0 when at rest beside you. This is called its **rest mass.** If the same object moves past you with speed v, its mass will appear to be greater. Its mass m is then given by

$$m = \frac{m_0}{\sqrt{1 - (v/c)^2}}$$

Notice, at $v = c$, its mass would be infinitely large. For this reason, infinite forces would be needed to accelerate an object to a speed c. In part, this explains result 1 given above.

5 Mass and energy are interchangeable. If the energy of an object is increased by an amount ΔE in any way, the object's mass will increase by an amount Δm. Or, as in a nuclear reactor, mass is destroyed to produce energy. The two are related by

$$\Delta E = \Delta m\, c^2$$

In particular, if an object has KE, its mass will be m, where

$$\text{KE} = (m - m_0)c^2$$

If you substitute $m = m_0/\sqrt{1 - (v/c)^2}$ for m and approximate for $v/c \ll 1$, then the usual expression $\text{KE} = \frac{1}{2}m_0v^2$ is found.

In most of these results, the relativistic factor $\sqrt{1 - (v/c)^2}$ is important. If this factor is very close to unity, then $m \approx m_0$, $L \approx L_0$, $t \approx t_0$, and $\text{KE} \approx \frac{1}{2}m_0v^2$. Only when $v/c \to 1$ do the effects mentioned here become important. As a rough rule of thumb, we can take $v = 3 \times 10^7$ m/s as the dividing line. For speeds less than this (one-tenth the speed of light), relativistic effects are usually not important. We shall examine this topic in more detail in Chap. 26.

Hence there are two possible answers for v', namely, 0 and 20 cm/s. Substitution of these values in the first equation gives 30 cm/s and -10 cm/s for v.

The first of these answers implies that the first ball went right through the second. Since this is a physical impossibility, we discard this answer. We therefore find that after the collision the first ball is going to the left at 10 cm/s while the other ball is going to the right at 20 cm/s.

Illustration 6.7 A 10-g bullet of unknown speed is shot into a 2.000-kg block of wood suspended from the ceiling by a cord. The bullet hits the block and becomes lodged in it. After the collision, the block and bullet swing to a height 30 cm above the original position (see Fig. 6.7*a*, *b*, and *c*). What was the speed of the bullet? (This device is called a **ballistic pendulum**.)

Reasoning *KE is not conserved upon impact*. However, *after impact*, the KE of the bullet plus that of the block is transformed into PE. Hence, in going from Fig. 6.7*b* to *c* we can write (remember, this is all *after* the collision)

$$\text{KE at bottom} = \text{PE at top}$$
$$\tfrac{1}{2}(2.000 + 0.010)v^2 = (2.000 + 0.010)(9.8)(0.30)$$
$$v \approx 2.4 \text{ m/s}$$

This is the speed of the block and bullet just after collision. You should

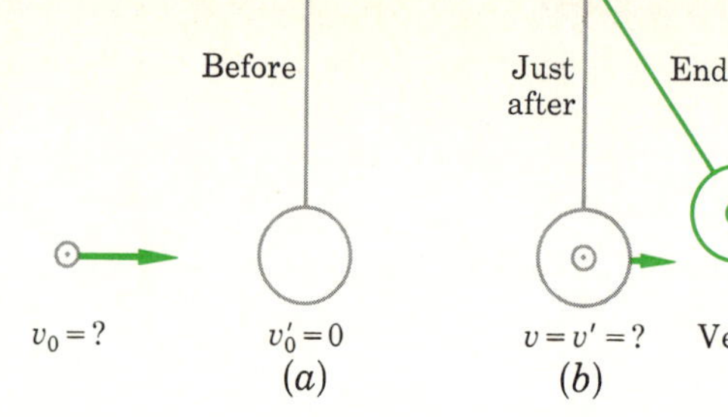

FIGURE 6.7
The momentum is the same in (a) and (b) but not in (c). KE is changed to PE in going from (b) to (c).

supply the units in the above equation and verify the units given for v.

Now in going from part a to part b in Fig. 6.7, a collision occurred. Momentum was conserved, but KE was not. We have, writing v_0 for the original velocity of the bullet,

$$\text{Momentum before} = \text{momentum after}$$
$$(0.010\ \text{kg})v_0 + 0 = (2.000 + 0.010)(2.4)\ \text{kg}\cdot\text{m/s}$$
$$v_0 \approx 480\ \text{m/s}$$

Hence, the original speed of the bullet was 480 m/s.

6.5 ROCKET AND JET PROPULSION

One of the most spectacular examples of the use of the law of momentum conservation occurs in the jet propulsion of rockets, spaceships, and jet airplanes. Operation of all these devices depends upon the fact that recoil occurs when an object shoots part of its mass out from itself. All these devices resemble a gun or a cannon which recoils when it shoots a projectile.

In jet engines and rockets, fuel is burned, and in the process very hot gases are formed. The swiftly moving gas molecules are shot out the rear of the jet or rocket engine much as a stream of bullets would be shot from a fantastically fast repeating gun. Like the gun recoiling, the rocket and aircraft are made to recoil in the direction opposite to the motion of the gas. Since the gas molecules were given momentum in the direction toward the rear, the rocket must acquire an equal momentum in the opposite direction (forward) because momentum is conserved.

A close examination of this kind of propulsion system shows that the interior of the engine pushes on the hot gas molecules in such a way that they are shot preferentially in a rearward direction. But in the process, according to Newton's law of action and reaction, the molecules exert a force forward on the engine, thereby throwing the rocket forward. Both these forces occur *within the engine itself*. No force is exerted on the craft from outside. The craft is not propelled by interaction of the expelled hot gases with the atmosphere, and in fact a rocket operates best in outer space, where there is no air. Air exerts a friction force which retards the motion of the rocket and is therefore undesirable.

Illustration 6.8 A Centaur rocket shoots hot gas from its engine at a rate of 1300 kg/s. The speed of the molecules is 50,000 m/s relative to the rocket. How large a forward push (or **thrust**) is given to the rocket by the exiting gases?

Reasoning The impulse exerted on the gas thrown out each second will be

$$Ft = mv - mv_0$$

If we take $t = 1$ s, m is the mass of gas expelled in a second (1300 kg), $v = 50{,}000$ m/s, and $v_0 = 0$. Putting in the values gives $F = 65{,}000{,}000$ N, or about 15 million pounds. This is the action force needed to expel the hot gases. An equal and opposite reaction force exerts a forward thrust on the rocket.

6.6 PRESSURE OF AN IDEAL GAS

In this section we shall compute the pressure on the walls of a container resulting from the bombardment of the walls by the gas molecules. Qualitatively, the picture is very simple. In Fig. 6.8, the gas molecule shown will be bouncing around inside the container. Following the dotted path shown, it will bounce off the wall. Of course it exerts a force on the wall during the time of collision. The equal and opposite reaction force of the wall on it is responsible for its momentum change.

Since there are such a large number of molecules in a gas at most reasonable pressures, there will be billions of collisions with the wall in a time of a second or less. Hence, the force due to these collisions will appear about constant. Moreover, the average force will be perpendicular to the wall. We define the perpendicular force on a unit area, i.e., 1 square meter, 1 square inch, etc., of the wall to be the **pressure** of the gas. In symbols, the pressure P is

$$P = \frac{F}{A} \tag{6.4}$$

FIGURE 6.8

Huge numbers of gas molecules colliding with the wall in the manner shown give rise to the observed pressure of the gas.

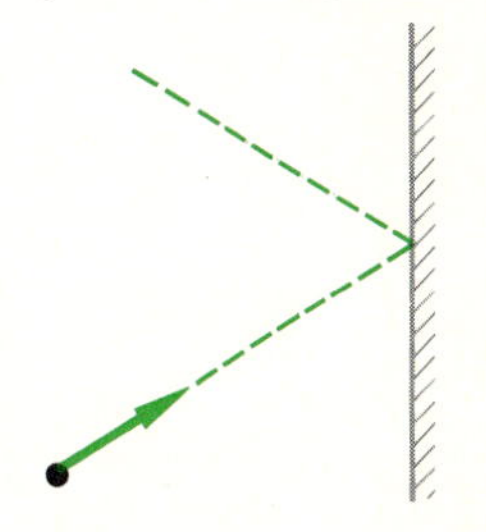

where F is the force on and perpendicular to the area A. The units of pressure are newtons per square meter, pounds per square inch, etc.

The exact computation of the pressure of a gas on the container wall is an extremely complicated procedure for any real gas. However, it turns out that even very simplified versions of the actual physical situation give nearly the correct answer for most real gases. It is common procedure in physics to carry through a grossly simplified computation for a physical situation first. As scientists grow in their understanding of the physical processes, they refine their computation, using more and more realistic pictures or models of the true physical situation. Sometimes, many years may go by before the original, unrealistic model is successfully modified to conform more exactly to the actual process as evidenced by experiment. These delays are usually the result of the scientist's inability to devise a more realistic model which would be susceptible to mathematical analysis. In addition, the true physical situation itself is often obscure, and the scientists must first conceive a model for the physical process which could give rise to the experimentally observed facts.

In the present case, the pressure of the common gases, even quite unrealistic models of the physical situation yield essentially correct results, i.e., results which agree with experiment. We shall find the result for the pressure of an ideal gas by using a very simple (and unrealistic) model. However, our result will be correct in all essentials.

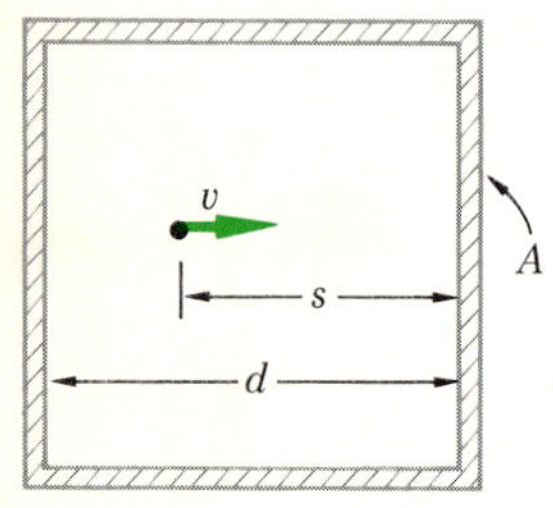

FIGURE 6.9
One-sixth of the molecules in the box are assumed to be traveling in the direction shown. They will all strike the wall in a time d/v. We assume the number of molecules in unit volume to be ν_0, read "nu sub zero."

Suppose first that we have a gas in a container, there being ν_0 molecules per unit volume. Suppose further that one-third of the molecules are traveling in each of the three directions x, y, and z. We shall take their speeds to be identical, v. Of the third of the molecules traveling in the x direction, half will be headed toward the $+x$ direction and half will be going in the $-x$ direction. Hence, one-sixth of the molecules will be traveling in the $+x$ direction, and each will have a speed v. We assume further that the molecules do not collide with each other.

Consider the situation shown in Fig. 6.9, where a particle is traveling in the $+x$ direction toward the wall of area A. The time taken for this particle to reach the wall, a distance s away, and hit it will be s/v. As shown in the figure, the width of the box is d. Therefore in a time $t = d/v$, all the molecules in the box which were traveling in the $+x$ direction will hit the area A. Since the volume of the box, and therefore of the gas, is Ad, and since there are $\frac{1}{6}\nu_0$ molecules in unit volume traveling in the $+x$ direction, there will be a total of $(\frac{1}{6})(\nu_0)(Ad)$ molecules traveling in the $+x$ direction. All these will hit the right-hand wall of the box in time d/v.

If we assume the collisions with the walls to be perfectly elastic, when any single molecule hits the wall, its velocity will completely reverse. Hence, for each molecule hitting the wall

$$\text{Change in momentum} = 2mv$$

where m is the mass of the molecule. That is to say, the momentum of the particle was changed by an amount mv when it was stopped by the wall, and then it was given an additional reverse momentum of mv, making a total change of $2mv$.

Since there will be $(\frac{1}{6})(\nu_0)(Ad)$ molecules hitting the wall in time $t = d/v$, the total momentum change caused by the wall in this time is

$$(2mv)(\tfrac{1}{6})(\nu_0)(Ad)$$

We know that the impulse Ft on the wall is equal to this change in momentum, where F is the average force exerted on the wall by the molecules.*

*Notice that t as used here is *not* the time of impact for a single molecule. It is the time taken for a large number of molecules to strike the wall and, as used here, is given by d/v.

Using the fact that $t = d/v$, we have

$$F\frac{d}{v} = (2mv)(\tfrac{1}{6})(\nu_0)(Ad)$$

After simplification of this expression we have the following equation for the average pressure on the wall:

$$P = \frac{F}{A} = (\tfrac{1}{3})(\nu_0)(mv^2) \tag{6.5}$$

In terms of the kinetic energy of the molecule, this is

$$P = (\tfrac{2}{3})(\nu_0)(\tfrac{1}{2}mv^2) \tag{6.6}$$

Pressure Is Proportional to KE of Molecule

Notice that the pressure of a gas is equal to two-thirds of the kinetic energy per unit volume, since ν_0 is the number of molecules in unit volume. We shall see in a later chapter that gas pressure can also be used as a measure of the temperature of a gas. Hence it appears that we shall find temperature and kinetic energy to be intimately related. In fact, we shall learn that a hot object is one whose constituent molecules possess a great deal of kinetic energy.

Although Eq. (6.6) was obtained by using a greatly oversimplified model, the result is general. *As long as a gas is far removed from conditions under which it will liquefy, Eq. (6.6) is found to be applicable, to good approximation. A gas which obeys this equation exactly would be an* **ideal gas,** *and Eq. (6.6) is one form of the* **ideal-gas law.** It is fortuitous that the crude method of computation we have used (and which was used by the early workers in physics) leads to a result which agrees so well with experiment. In addition to the artificiality of the assumptions of equal speeds, perpendicular motions, and elastic collision, the assumption that the molecules do not collide with each other is far from true. *In air at ordinary pressures, a molecule collides with another after traveling a distance of about* 10^{-5} *cm.*

Illustration 6.9 *Standard atmospheric pressure is* 1.01×10^5 *N/m*^{2}. At standard pressure and temperature (0°C), there are 2.7×10^{25} molecules in 1 m^3 of volume. Find the average speed of the nitrogen molecules in air under standard conditions. One molecule of nitrogen has a mass of 4.7×10^{-26} kg.

Reasoning We make use of Eq. (6.6). From the data we have $\nu_0 = 2.7 \times 10^{25}$ m^{-3},* $m = 4.7 \times 10^{-26}$ kg, and $P = 1 \times 10^5$ N/m^2, all in the proper SI units. Hence, from Eq. (6.6),

* We use a common method of abbreviating "per cubic meter" and write this 2.7×10^{25} m^{-3}.

$$1 \times 10^5\ \mathrm{N/m^2} = \tfrac{1}{3}(2.7 \times 10^{25}\ \mathrm{m^{-3}})(4.7 \times 10^{-26}\ \mathrm{kg})v^2$$

from which, remembering that a newton is $1\ \mathrm{kg \cdot m/s^2}$,

$$v \approx 490\ \mathrm{m/s}$$

Illustration 6.10 Commercial jet aircraft often fly at altitudes in excess of 7000 m. At that altitude, air pressure is only about half as large as it is at sea level. Nearly all aircraft are pressurized to compensate for this. What effect would it have on the occupants of an aircraft at this altitude if the cabin was at the same pressure as the atmosphere?

Reasoning From Eq. (6.6) we know that $P \sim \nu_0$, where ν_0 is the number of molecules per unit volume. At an altitude of 7000 m, the pressure (and ν_0) is only half as large as at sea level. Consequently, each time a person breathes, only about half as many oxygen molecules enter the lungs as compared to sea level. The oxygen supplied to the body is cut in half. To compensate for this, a person needs to breathe much faster than ordinary. Often, this in itself is very tiring. But if it is not done, the person will be extremely limited in body activity. This effect is often noticed by visitors to mountainous regions. The natives of regions at high altitude have undergone bodily adaptations, such as increased red-blood-cell count, which allow them to function better with this smaller supply of oxygen.

SUMMARY

The linear momentum of an object of mass m and velocity $\mathbf{v}$ is $m\mathbf{v}$. It is a vector quantity and is often represented by $\mathbf{p}$.

When an object is subjected to a force $\mathbf{F}$ for a time t, an impulse $\mathbf{F}t$ has been applied to it. The impulse causes a change in momentum of the object. We have $\mathbf{F}t = m\mathbf{v} - m_0\mathbf{v}_0$. This is an alternate way for stating Newton's second law of motion.

A group of objects upon which no resultant external force exists can be considered isolated as far as its translational motion is concerned. For an isolated system, the linear momentum of its constituent objects is constant. This is the law of conservation of linear momentum. It is of particular value in analyzing the collisions of objects.

In a perfectly elastic collision, KE is conserved. More often, much of the KE is lost to heat and other forms of energy during the collision. Such a collision is said to be inelastic.

The pressure P on a surface of area A due to a force F perpendicular to the surface has magnitude $P = F/A$. A gas in a container exerts a pressure on the walls by virtue of the collisions of the gas molecules with the walls. The pressure due to an ideal gas containing ν_0 molecules per unit volume is $P = (\frac{2}{3})(\nu_0)(\frac{1}{2}mv^2)$. In this expression v is an average speed for a molecule of mass m.

MINIMUM LEARNING GOALS

Upon completion of this chapter you should be able to do the following:

1. Define linear momentum and compute it for an object provided sufficient data are given. Give examples showing how momentum depends upon velocity and/or mass.

2. Find the change in momentum for an object subjected to a given average force for a stated time. Define impulse and show how it is related to linear momentum. Compute the impulse required to cause a given change in momentum.
3. State the law of conservation of linear momentum being careful to point out the importance of the words "isolated system." Explain why, under certain conditions, the law is useful even though the system is not completely isolated.
4. Analyze the collision of two objects which stick together upon impact; i.e., situations similar to Illustration 6.3.
5. Analyze situations in which an object originally at rest explodes into two pieces or recoils; i.e., similar to Illustration 6.4.
6. Analyze situations where two objects move along a straight line, undergo a perfectly elastic collision, and then continue to move along the same straight line.
7. Explain the difference between an elastic and an inelastic collision. In your explanation, describe how a bounding ball would behave in each case. Give plausible reasons for the fact that KE is not conserved in most collisions.
8. Analyze situations in which an object strikes a stationary object and becomes embedded in it and the remaining KE is changed to PE or lost doing friction work. The ballistic pendulum would be a typical situation.
9. Explain the principle of operation of rockets, jet engines, and similar devices which are propelled by means of a jet.
10. Define pressure and explain why a gas in a container exerts a pressure on the container walls. In your explanation, make it clear why the pressure should depend upon v, m, and ν_0.
11. Given the relation between P, ν_0, v, and m, be able to use it to solve simple problems such as that typified by Illustration 6.9.

IMPORTANT TERMS AND PHRASES

You should be able to define or explain each of the following:

Linear momentum: $\mathbf{p} = m\mathbf{v}$
Impulse: $\mathbf{F}t = m\mathbf{v} - m_0\mathbf{v}_0$
Isolated system
Conservation of linear momentum
Elastic, inelastic collisions
Recoil
Ballistic pendulum
Jet propulsion
Pressure: $P = F/A$
Ideal gas
Ideal-gas law: $P = (\frac{2}{3})(\nu_0)(\frac{1}{2}mv^2)$

QUESTIONS AND GUESSTIMATES

1. Reasoning from the impulse equation, explain why it is unwise to hold your legs rigidly straight when you jump to the ground from a wall or table. How is this related to the commonly held belief that a drunken person has less chance of being injured in a fall than one who is sober?
2. Explain, in terms of the impulse equation, the principle of operation of impact-absorbing car bumpers and similar impact-absorbing devices.
3. When a large cannon is fired, it recoils for some distance against a cushioning device. Why is it necessary to make the support so it "gives" in this way?
4. A ball dropped onto a hard floor has a downward momentum, and after it rebounds, its momentum is upward. Clearly the momentum of the ball is not conserved in the collision, even though the ball may rebound to the height from which it was dropped. Does this contradict the law of momentum conservation?
5. Contrive a device which, momentarily at least, can have KE but no momentum. Is it possible to design a device having momentum but no KE? Explain.
6. When a toy balloon filled with air is released, allowing the air to escape, the balloon shoots off through the air. Explain why it moves in this way. In what way is a rocket like the balloon?
7. As your heart beats, it causes blood to surge through

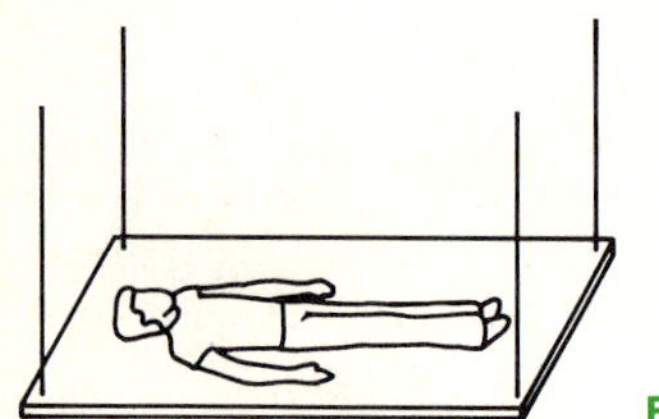

FIGURE P6.1

your body. This motion can be monitored by placing a person on a platform suspended by wires as shown in Fig. P6.1. The table undergoes slight movements which reflect the action of the heart. Sensitive instruments record the table movements and give information concerning heart action. (This device is called a **ballistocardiograph**.) Explain why the motion of the table gives information concerning heart action.

8. A baseball player has the following nightmare. He is accidentally locked in a railroad boxcar. Fortunately, he has his ball and bat along. To start the car moving, he stands at one end and bats the ball toward the other. The impulse exerted by the ball as it hits the end gives the car a forward motion. Since the ball always rebounds and rolls along the floor to him, the player repeats this process over and over again. Eventually the car attains a very high speed, and the player is killed as the boxcar collides with another car sitting at rest on the track. Analyze this dream from a standpoint of the physics involved.

9. Explain how a Mexican jumping bean can cause itself to jump.

10. A fly is held in a lightweight container which is completely closed. Can we tell when the fly is resting or flying by weighing the container? Explain.

11. A 200-lb man jumps from a roof 30 ft above the ground. About how large a force must his legs withstand when he lands on the ground? (E)

12. A 10-kg child falls from a window and is caught by a woman 15 m below. Estimate the force experienced by the child as it is caught. (E)

13. How large a force does a moderate-size raindrop exert on the head of a bald man? (E)

14. A hose squirts a horizontal beam of water against a window at a speed of 5 m/s and a rate of 30 g/s. About how large a force is exerted by it on the window? (E)

15. The wind is blowing against a sign with a speed of 30 m/s. If the mass in 1 m^3 of air (the density) is 1.29 kg/m^3, estimate how large a force is exerted upon a unit area of the sign by the wind. (E)

16. Estimate the force exerted by a woman's head upon her neck if her stationary car is hit from the rear by a loaded truck going 20 mi/h (9 m/s). Why does this type of accident often lead to the so-called whiplash injury? You will need to make an estimate of the mass of the woman's head and the time taken for her head to be accelerated. (E)

17. Suppose you lay your hand flat on a table top and then drop a 1.0-kg laboratory mass squarely on it from a height of 0.50 m. Estimate the average force exerted on your hand by the mass. (E) Why is injury very likely in this case even though you can catch the mass easily when dropped from this height?

PROBLEMS

1. A mass m is dropped from a height h. What is its momentum just before striking the ground? Neglect friction effects.

2. Show that the linear momentum p and KE of a mass m are related through $\text{KE} = p^2/2m$.

3. By use of the impulse equation, (*a*) determine how large an average force is required to stop a 3200-lb car in 5.0 s if the car's initial speed is 80 ft/s. (*b*) Assuming uniform deceleration, how far would the car go in this time?

4. A 1500-kg car going 20 m/s strikes a wall and stops in a distance of 3.0 m. Assuming uniform deceleration, how long did it take to stop? Use the impulse equation to find the stopping force which acted on it.

5. A 120-g ball moving at 18 m/s strikes a wall perpendicularly and rebounds straight back with this same speed. The center of the ball moves 0.27 cm farther toward the wall after touching it. Assuming that the deceleration produced by the wall is uniform, show that the time of contact with the wall is 2×0.00030 s. How large is the average force which the ball exerts on the wall?

6. Two identical balls collide. Ball 1 is traveling to the right at 10 m/s, and ball 2 is standing still. Find the

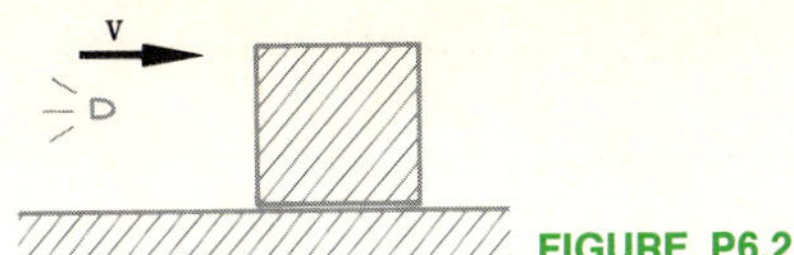

FIGURE P6.2

direction and magnitude of their velocity if they stick together after collision.

7. As shown in Fig. P6.2, a 20-g bullet with speed 5000 cm/s strikes a 7000-g block resting on a table. The bullet embeds in the block after collision. Find (*a*) the speed of the block after collision and (*b*) the friction force between the table and block if the block moves 1.5 m before stopping.

8.* In Fig. P6.2 a horizontal force of 0.70 N is required to pull the 5000-g block across the table at constant speed. Find the speed of the 20-g bullet shown if the bullet embeds in the block and causes the block to slide 1.50 m before coming to rest.

9. A 2.0-kg block rests over a small hole on a table. A man beneath the table shoots a 15.0-g bullet through the hole into the block, where it lodges. How fast was the bullet going if the block rises 1.30 m above the table?

10. A 60-kg astronaut becomes separated in space from her spaceship. She is 15.0 m away from it and at rest relative to it. In an effort to get back, she throws a 500-g wrench with a speed of 8.0 m/s in a direction away from the ship. How long does it take her to get back to the ship?

11. A 2.0-kg melon is balanced on a bald man's head. His wife shoots a 50-g arrow at it with speed 30 m/s. The arrow passes through the melon and emerges with a speed of 18 m/s. Find the speed of the melon as it flies off the man's head.

12. Refer to Fig. P6.3. The pendulum on the left is pulled aside to the position shown. It is then released and allowed to collide with the other pendulum, which is at rest. (*a*) What is the speed of the ball on the left just before collision? After collision, the two stick together. (*b*) How high in terms of h does the combination swing? Assume the two balls have equal masses.

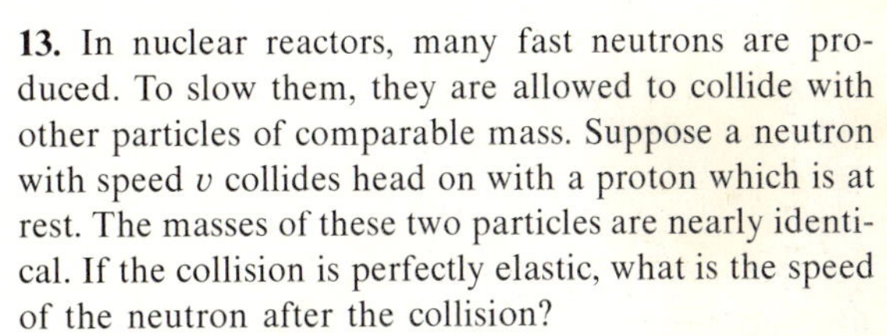

13. In nuclear reactors, many fast neutrons are produced. To slow them, they are allowed to collide with other particles of comparable mass. Suppose a neutron with speed v collides head on with a proton which is at rest. The masses of these two particles are nearly identical. If the collision is perfectly elastic, what is the speed of the neutron after the collision?

14. In Fig. P6.3 the two masses are identical. The mass on the left is displaced as shown and released. It collides perfectly elastically with the other mass. (*a*) How high (in terms of h) does the mass on the right swing after collision? (*b*) The mass on the left?

15. In Fig. P6.3, both masses are displaced to a height h, one to the left as shown and the other to the right. They are released simultaneously and undergo a perfectly elastic collision at the bottom. How high does each swing after collision? Both masses are identical.

16.* As shown in Fig. P6.3, the mass on the left is pulled aside and released. Its speed at the bottom is v_0. It then collides with the mass on the right in a perfectly elastic collision. Find the velocities of the two masses just after collision if the mass on the left is three times as large as the mass on the right.

17.** Repeat Prob. 16 if the mass on the left is one-third the mass on the right.

18.** An electron ($m = 9 \times 10^{-31}$ kg) traveling at 2.0×10^7 m/s undergoes a head-on collision with a hydrogen atom ($m = 1.67 \times 10^{-27}$ kg) which is initially at rest. Assuming the collision to be perfectly elastic and the motion to be along a straight line, find the final velocity of the hydrogen atom.

19. The device shown in Fig. P6.4 is sold as a novelty. All the masses are identical. When one of the masses is pulled back as shown and released, the mass at the opposite end flies out while all the others retain their

FIGURE P6.3

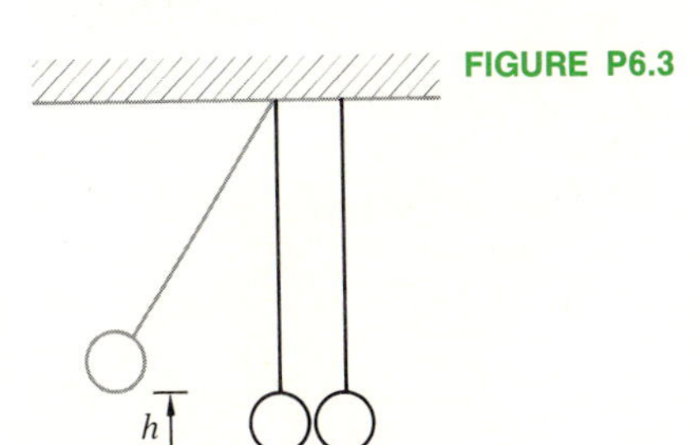

FIGURE P6.4

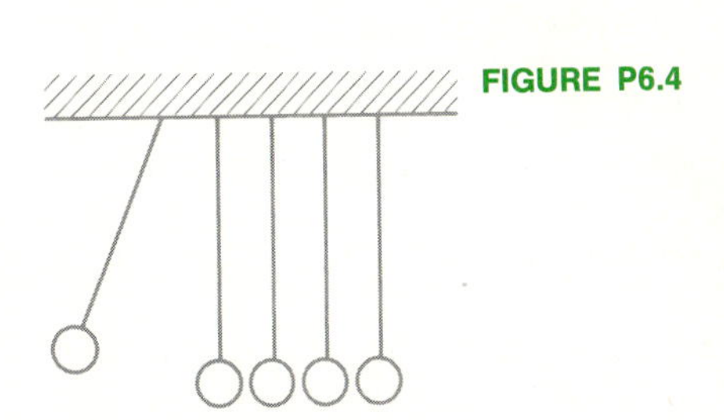

position. (*a*) Show that this result is to be expected if the collisions are perfectly elastic. (*b*) What will happen if two masses are pulled back and released instead of one?

20. Suppose that half the molecules of a gas are removed from a container without changing the average KE per molecule within the container. By what factor does the gas pressure change.

21. A number of molecules are contained at atmospheric pressure in a chamber closed by a piston device, as shown in Fig. P6.5. If the piston is pushed down in such a way that the KE of the molecules is not changed appreciably, what is the pressure in the container when the volume is only one-fourth its original size?

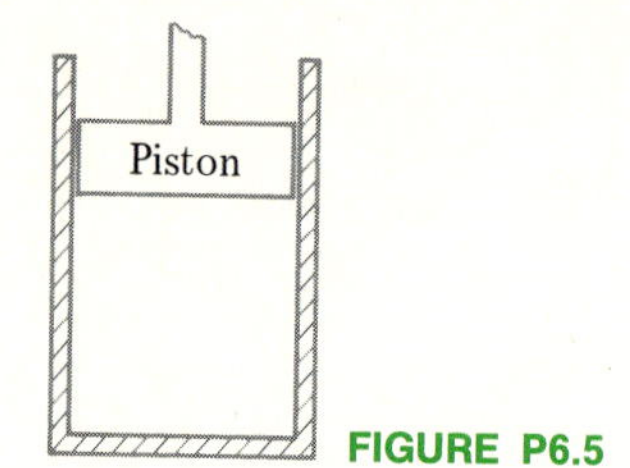

FIGURE P6.5

ANGULAR MOTION AND CENTRIPETAL FORCE

In this chapter we shall extend our discussion of motion to objects which travel in circular paths. This type of motion is frequently encountered in nature, and a knowledge of it is necessary for an understanding of such diverse topics as the motion of bodies in the solar system and the structure of atoms. We shall apply the principles of circular motion to various physical situations and include a discussion of weightlessness.

7.1 ANGULAR DISTANCE θ

Until this chapter, we have usually discussed motion in a straight line only. However, motion in a circle is extremely common and is of great importance. It is this type of motion which we shall discuss in this chapter. Suppose that a wheel is mounted on an axle through its center, as shown in Fig. 7.1. In going from Fig. 7.1*a* to *b* the wheel has been rotated through an angle θ. There are three different ways in which this rotation is measured.

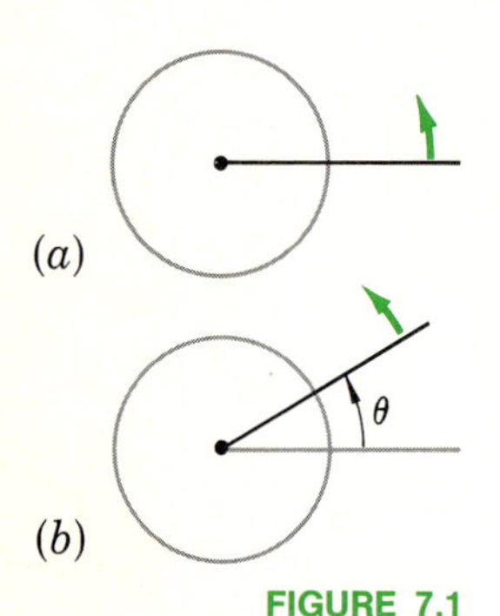

FIGURE 7.1
The angle θ describes the angular distance through which the wheel has turned.

One way is to measure θ in degrees. If the wheel turns through a full circle, θ is 360°. This way of measuring how far the wheel has rotated is well known to you. The second way for measuring θ is also common in everyday life. When a wheel turns through a full circle, we say that $\theta = 1$ rev (revolution). Clearly,

$$1 \text{ rev} = 360°$$

The third method for measuring θ is of particular value in science. This way for measuring angles, called **radian measure,** is defined in terms of a ratio of two lengths. If we refer to Fig. 7.2, we see that the point A on the wheel has moved through an arc length s as the wheel turned through the angle θ. Calling the radius of the wheel r, we define

Angles in Radians

$$\theta \text{ in radian measure} = \frac{\text{arc length}}{\text{radius}}$$

$$\theta = \frac{s}{r} \qquad \text{radians} \tag{7.1}$$

Notice that θ is simply a ratio, so, strictly speaking, it has no units. Even so, we shall say, for example, "the angle is π rad (radians) or 180° or $\frac{1}{2}$ rev" to make it clear how we are measuring angles. Since $s = 2\pi r$ for one full turn of the wheel, we see that

$$1 \text{ rev} = 360° = 2\pi \text{ rad}$$

Illustration 7.1 A certain angle is 70°. Find its equivalent in radians and in revolutions.

FIGURE 7.2
In radian measure, $\theta = s/r$.

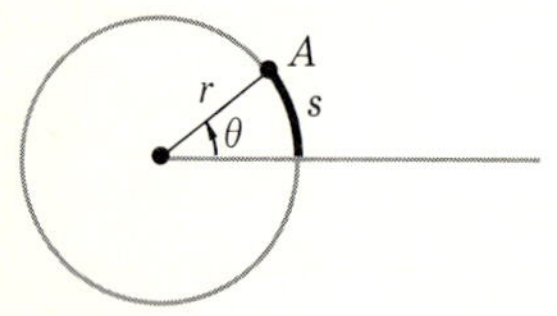

Reasoning The relations between the various angular measures tell us that

$$70 \text{ deg} = (70 \cancel{\text{deg}})\left(\frac{2\pi \text{ rad}}{360 \cancel{\text{deg}}}\right) = 1.22 \text{ rad}$$

and

$$70 \text{ deg} = (70 \cancel{\text{deg}})\left(\frac{1 \text{ rev}}{360 \cancel{\text{deg}}}\right) = 0.194 \text{ rev}$$

Illustration 7.2 As shown in Fig. 7.3, a string is wound around a wheel which has a radius of 20 cm. Through how large an angle does the wheel turn in order to unwind 30 cm of string?

Reasoning If you examine Fig. 7.3, you can convince yourself that the following is true: as a point on the rim of the wheel turns through an arc length s, the length of string unwound is also s. Further, Eq. (7.1) tells us that

$$\theta \text{ in radians} = \frac{s}{r}$$

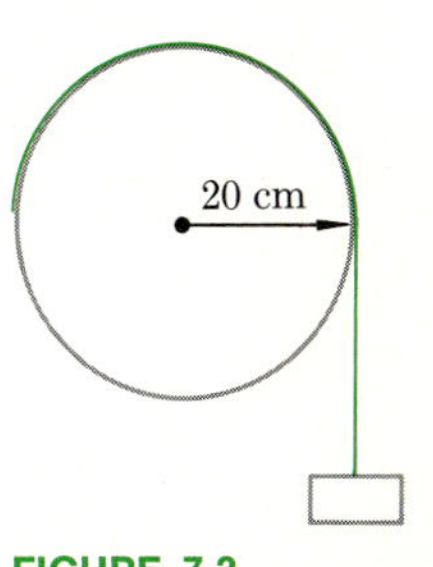

FIGURE 7.3
How far must the wheel turn in order to unwind 30 cm of string?

In the present case, $s = 30$ cm and $r = 20$ cm. Therefore

$$\theta \text{ in radians} = \tfrac{30}{20} = 1.50 \text{ rad}$$

or

$$\theta = (1.50 \text{ rad})\left(\frac{360 \text{ deg}}{2\pi \text{ rad}}\right) = 86 \text{ deg}$$

7.2 ANGULAR VELOCITY

When we state that a wheel is rotating at 800 rev/min, we are giving its angular speed. We are telling how far it rotates in a given length of time. The **average angular velocity** of a rotating wheel is defined to be the angle turned divided by the time taken to turn through the angle. The defining equation for average angular velocity is*

DEFINITION

$$\bar{\omega} = \frac{\theta}{t} \tag{7.2}$$

Average Angular Velocity

where θ is the angle through which the wheel rotates in time t. (ω is Greek omega.) As we see, *the units for ω are those of an angle divided by a time.* For example, the units might be degrees per second, revolutions per minute, radians per second, etc.

The definition for average angular velocity is very similar to our definition of velocity for linear motion. We have $\bar{v} = s/t$, where s is the linear distance moved in time t. In that which follows, we shall see that each of our five linear-motion equations has an analog in circular motion.

In linear motion we made a distinction between average and instantaneous velocity. This is also an important distinction in angular motion. The average velocity, in the linear case, is defined by the analog of Eq. (7.2). The instantaneous velocity is obtained by measuring the distance moved in such a small time that the velocity could not change appreciably during that time.

*In more advanced work, angular velocity is given direction so it is a vector quantity. However, for rotation in a plane, angular speed and angular velocity have the same magnitude, and the direction of the angular velocity does not change.

This is written mathematically in the following way, where ω is the **instantaneous angular velocity:**

Instantaneous Angular Velocity

$$\omega = \lim_{\Delta t \to 0} \frac{\Delta\theta}{\Delta t} \tag{7.3}$$

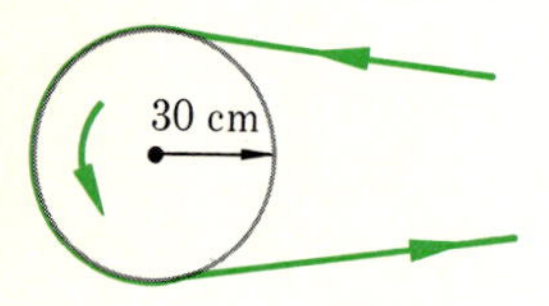

FIGURE 7.4
As the wheel rotates, it winds up one end of the belt and unwinds the other.

In this expression, $\Delta\theta$ is the small angular distance moved by the wheel in the small time Δt, and $\lim_{\Delta t \to 0}$ means to take the value of the ratio as the time interval Δt approaches zero, as discussed in Chap. 2.

Illustration 7.3 The wheel in Fig. 7.4 is turning at 240 rev/min. (*a*) Through how large an angle will it turn in 10 s? (*b*) If the wheel has a radius of 30 cm, how far will a belt which runs on this wheel be pulled in this time?

Reasoning *a*. Changing all time units to minutes, we have from Eq. (7.2), since $\bar{\omega} = 240$ rev/min and $t = 10$ s.

$$\theta = \bar{\omega}t = (240 \text{ rev/min})(\tfrac{10}{60} \text{ min}) = 40 \text{ rev} = 80\pi \text{ rad}$$

b. We recall that Eq. (7.1) can be used to tell us how much string or belt a wheel winds. As in Illustration 7.2, the amount wound in time t is

$$s = \theta r$$

provided θ is measured in radians. Since we found $\theta = 80\pi$ rad, and we were told that $r = 30$ cm, we have

$$s = (80\pi \text{ rad})(0.30 \text{ m}) = 75.4 \text{ m}$$

Notice that the term radian does not carry through the equation. As we saw from the definition, it is not a unit in the usual sense of the word.

7.3 ANGULAR ACCELERATION α

The average linear acceleration was defined in Chap. 2 by the relation

$$a = \frac{v_f - v_0}{t}$$

It measured the rate at which the moving object was speeding up or slowing down. The quantity $v_f - v_0$ was the change in velocity during the time t. You will recall that typical units for acceleration are meters per second squared and feet per second squared.

In the case of rotating objects we are often interested in how they speed up or slow down. Hence we are concerned with angular acceleration, i.e., rate of change of angular velocity. We *define the* **average angular acceleration** α (alpha) of a rotating wheel or any other object by the relation

Average Angular Acceleration

$$\alpha = \frac{\omega_f - \omega_0}{t} \tag{7.4}$$

Its units will be those of angular velocity divided by time. For example, if t is measured in seconds and ω is measured in radians per second, the angular acceleration would be expressed in radians per second per second. Although it is not wrong to measure ω in radians per second while t is in minutes, α therefore having units of radians per second per minute, it is generally preferable to use the same unit for t.

If the angular acceleration is uniform, we have, as in the case of linear motion, that the average angular velocity is given by

$$\bar{\omega} = \tfrac{1}{2}(\omega_f + \omega_0)$$

Illustration 7.4 A wheel starts from rest and attains a rotational velocity of 240 rev/s in a time of 2.0 min. What was its average acceleration?

Reasoning Known:

$$\omega_0 = 0 \qquad \omega_f = 240 \text{ rev/s} \qquad t = 2.0 \text{ min} = 120 \text{ s}$$

Using Eq. (7.4), we have

$$\alpha = \frac{\omega_f - \omega_0}{t} = \frac{240 - 0}{120} = 2 \text{ rev/s}^2$$

7.4 THE ANGULAR-MOTION EQUATIONS

It is clear from the foregoing that *the angular-motion equations* thus far written *are exactly the same as the analogous linear-motion equations, except that θ, ω, and α replace s, v, and a, respectively.* Since the remaining two linear-motion equations were obtained by algebraic manipulation of the first three, it follows that analogous equations will also apply to angular motion. We list all five equations here, together with their linear counterparts.

Five Angular-Motion Equations

LINEAR	ANGULAR	
$s = \bar{v}t$	$\theta = \bar{\omega}t$	(7.5)
$v_f = v_0 + at$	$\omega_f = \omega_0 + \alpha t$	(7.6)
$\bar{v} = \frac{1}{2}(v_f + v_0)$	$\bar{\omega} = \frac{1}{2}(\omega_f + \omega_0)$	(7.7)
$2as = v_f^2 - v_0^2$	$2\alpha\theta = \omega_f^2 - \omega_0^2$	(7.8)
$s = v_0 t + \frac{1}{2}at^2$	$\theta = \omega_0 t + \frac{1}{2}\alpha t^2$	(7.9)

The last four equations (7.6) to (7.9) are not always applicable. It will be recalled that Eqs. (7.6) and (7.7) were true only for constant acceleration. Since Eqs. (7.8) and (7.9) make use of these equations in their derivation, Eqs. (7.7) to (7.9) are restricted to cases of uniform acceleration.

Illustration 7.5 An electric fan is turning at 3.0 rev/s when it is turned off. It coasts to rest in 18 s. Assuming its deceleration to be uniform, what was its deceleration? How many revolutions did it turn through while coming to rest? (In practice, the deceleration would not be constant.)

Reasoning This is a typical angular-motion problem. We know

$$\omega_0 = 3.0 \text{ rev/s} \qquad \omega_f = 0 \qquad t = 18 \text{ s}$$
$$\alpha = ? \qquad \theta = ?$$

From the defining equation for α [Eq. (7.6)]

$$\alpha = \frac{\omega_f - \omega_0}{t} = \frac{0 - 3.0 \text{ rev}}{18 \text{ s}^2} = -0.167 \text{ rev/s}^2$$

To find θ let us use Eq. (7.9),

$$\theta = \omega_0 t + \tfrac{1}{2}\alpha t^2$$
$$= (3.0 \text{ rev/s})(18 \text{ s}) + \tfrac{1}{2}(-0.167 \text{ rev/s})(18 \text{ s})^2 = 27 \text{ rev}$$

7.5 TANGENTIAL QUANTITIES

When a wheel unwinds a string or rolls along the ground, both rotation and linear motions occur. We wish now to find how these two types of motion are related. The relation between linear and angular distances, s and θ, was inherent in Eq. (7.1), the definition of angular measure. To see this, let us look at Fig. 7.5.

As we see in Fig. 7.5*a*, a point on the rim of the wheel traces out an arc length s as the wheel turns through the angle θ. From Eq. (7.1), the definition of radian measure, we have that

$$s = r\theta$$

provided θ is measured in radians. We call the distance s a **tangential distance** *since it is measured tangential to the wheel's rim.*

Relation between Tangential and Angular Distances

If we look at Fig. 7.5*b*, we see that the wheel rolls a linear distance $s = r\theta$. This fact allows us to relate linear motion to angular motion for a rolling wheel. We have already seen in Figs. 7.3 and 7.4 that a tangential

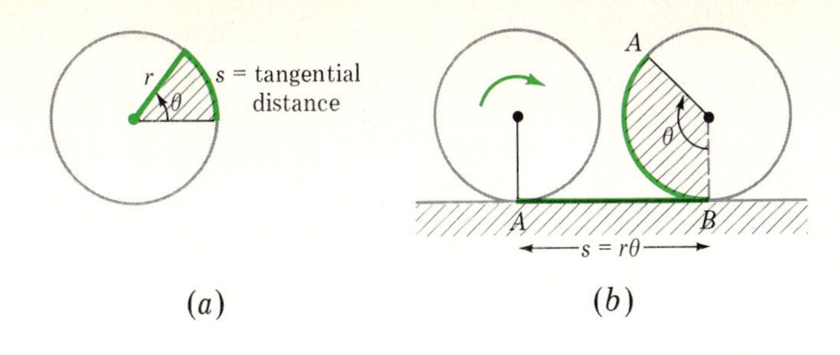

FIGURE 7.5

As the wheel turns through an angle θ, it lays out a tangential distance $s = r\theta$.

length $s = r\theta$ of belt or string is wound by a wheel as it rotates through an angle θ. In all cases like this

$$\text{Tangential distance} = s = r\theta \tag{7.10}$$

Once again *we emphasize that θ must be measured in radians.*

Suppose a wheel is rotating with a constant angular speed ω. As it does so, the angle θ keeps changing as shown in Fig. 7.6*a*. From the definition of ω we have

$$\omega = \frac{\theta}{t}$$

where θ is the angle turned in time t. But θ is related to the tangential distance through $s = r\theta$, and so, upon substitution, we obtain

$$\omega = \frac{s/r}{t} = \left(\frac{s}{t}\right)\left(\frac{1}{r}\right)$$

Relation between Tangential and Angular Velocities

But s/t is simply the speed of point A on the rim of the wheel due to the turning motion of the wheel. Or, alternatively, it is the linear speed with which the wheel's center point is moving in Fig. 7.6*b*. We call this the tangential speed v_T. Substituting $v_T = s/t$ and rearranging, we find

$$\text{Tangential speed} = v_T = \omega r$$

This is also the tangential velocity of point A labeled in Fig. 7.6*c*.

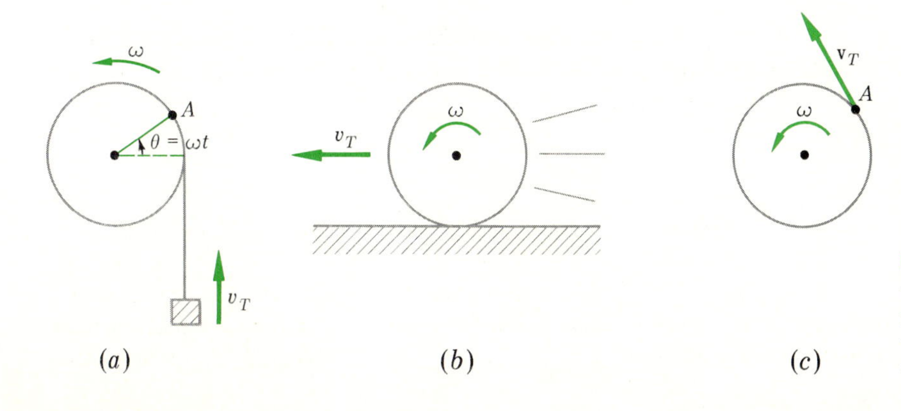

FIGURE 7.6

The angular velocity ω is related to the tangential velocity v_T by $v_T = \omega r$. In this relation, ω must be in radians.

If ω is increasing for a rotating wheel, then v_T must also be increasing. The angular acceleration α is given by

$$\alpha = \frac{\omega_f - \omega_0}{t}$$

where $\omega_f - \omega_0$ is the change in ω during the time t. But since $\omega = v_T/r$, we can write this as

$$\alpha = \frac{v_{Tf} - v_{T0}}{rt} \qquad \text{or} \qquad \frac{v_{Tf} - v_{T0}}{t} = \alpha r$$

Relation between Tangential and Angular Accelerations

But this is simply the rate of change of tangential velocity. We call it the **tangential acceleration** a_T. Therefore

$$\text{Tangential acceleration} = a_T = \alpha r \tag{7.12}$$

This is also the **linear acceleration** *of the center of a rolling wheel or of an unwinding string.*

Illustration 7.6 A car with 80-cm-diameter wheels starts from rest and accelerates uniformly to a velocity of 20 m/s in 9 s. Find the angular acceleration and final angular velocity of one of its wheels.

Reasoning We know that the linear velocity and acceleration of the center of the wheel is given by the tangential relations (7.11) and (7.12) to be

$$v_T = \omega r \qquad \text{and} \qquad a_T = \alpha r$$

In our case, the final value for v_T was 20 m/s. Therefore

$$\omega_f = \frac{v_{Tf}}{r} = \frac{20\ \text{m/s}}{0.40\ \text{m}} = 50\ \text{rad/s}$$

Notice how we must insert the proper angular measure since angular units do not carry through the equations. Why is it radians per second and not revolutions per second?

To find the acceleration, we shall first solve the linear-motion problem. We know that

$$v_0 = 0 \qquad v_f = 20\ \text{m/s} \qquad t = 9.0\ \text{s} \qquad a = ?$$

We have

$$a = \frac{v_f - v_0}{t} = \frac{20 - 0}{9.0}\ \text{m/s}^2 = 2.22\ \text{m/s}^2$$

Then, since this value of a is really a_T, we can write

$$\alpha = \frac{a_T}{r} = \frac{2.22\ \text{m/s}^2}{0.40\ \text{m}} = 5.55\ \text{rad/s}^2$$

Notice that, once again, we must furnish the proper angular units (radians) for our result.

7.6 RADIAL QUANTITIES AND CENTRIPETAL FORCE

Until now we have been concerned simply with motion on a circle. We have said nothing about the forces which cause an object to follow a circular path. We shall see in this section that a radial force, i.e., a force directed along a radius, is needed if an object is to move on a circle. This radial force causes the object to undergo a radial acceleration.

It is not natural for an isolated object to travel in a circle. Newton knew this when he formulated his laws of motion. He stated, in effect, that a body will travel *in a straight line* unless forced to do otherwise. We all know that a ball twirled in a horizontal circle at the end of a string would not continue in a circular path if the string should break. Careful observation will show at once that if the string breaks when the ball is at point A in Fig. 7.7b, the ball will follow the straight-line path AB, just as Newton said it would.

The fact is that unless a string or some other mechanism pulls the ball toward the center of the circle with a force F as shown in Fig. 7.7, the ball will not continue along the circular path. *This force needed to bend the normally straight path of the particle into a circular path is called the* **centripetal force.** Notice that *it is a pull on the body and that it is directed toward the center of the circle. It is a radial force.*

Newton's laws of motion tell us another fact which it is sometimes difficult to accept. Since the only thing pulling on the ball in Fig. 7.7 is the string (neglect gravity for now), the centripetal force F exerted by the string on the ball is an **unbalanced force.** By Newton's second law,

$$\mathbf{F} = m\mathbf{a}$$

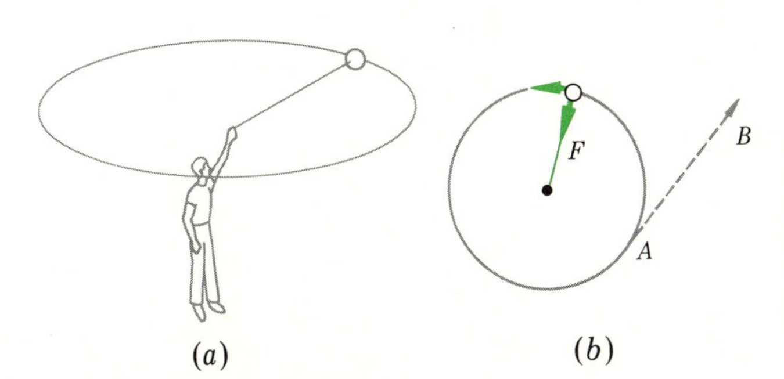

FIGURE 7.7

The tension F in the string holds the ball in a circular path. If the string breaks when the ball is at A, the ball will continue in a straight line toward B provided gravity is neglected.

and since the unbalanced force F is not zero, the ball must be accelerating. This is true even though the ball is going around the circle with constant angular speed!

Since $\mathbf{F}$ and $\mathbf{a}$ are vectors and m is a scalar, the acceleration predicted above must be in the direction of the force and hence must be directed toward the center of the circle. Clearly, the acceleration predicted from Newton's law is not in the correct direction either to speed up or to slow down the ball in its circular orbit. To learn the nature of the acceleration produced by the centripetal force, we must return to the basic concepts involved. Let us consider the definitions of velocity and acceleration.

First, we defined velocity to be a vector, and, second, we defined acceleration to be the rate of change of that vector velocity. Notice that the velocity of an object, defined to be a vector, can change even though the object's speed, a mere number, does not change. For an example of this, consider a car which is going at a speed of 50 km/h toward the west and then turns and continues with a speed of 50 km/h, but now traveling south. The appropriate velocity vectors are shown in Fig. 7.8*a* and *b*.

Clearly the vector velocity has been changed. To find the change in the velocity vector, we must find the velocity vector $\Delta\mathbf{v}$ which must be added to the original velocity vector $\mathbf{v}_0$, to obtain the final velocity vector $\mathbf{v}$. It is shown in Fig. 7.8*c*.

Notice that the speed of the moving object in Fig. 7.8 has not changed but that the velocity has changed. The direction of the motion has changed. By the definition of the average acceleration,

$$\mathbf{a} = \frac{\Delta\mathbf{v}}{t}$$

where t is the time taken for the change $\Delta\mathbf{v}$ in velocity. This is the type of acceleration present when an object travels in a circle. Even though the speed of the object is not changing, the direction of motion of the object is changing under the action of the centripetal force. Let us now examine this case more closely in order to obtain a mathematical expression for the force needed to hold an object in a circular path.

Consider what happens when the ball travels from A to B in Fig. 7.9*a*. Since the velocity of the ball has changed its direction but not its magnitude, the change in velocity is as shown in Fig. 7.9*b*. Hence the acceleration of the ball will be

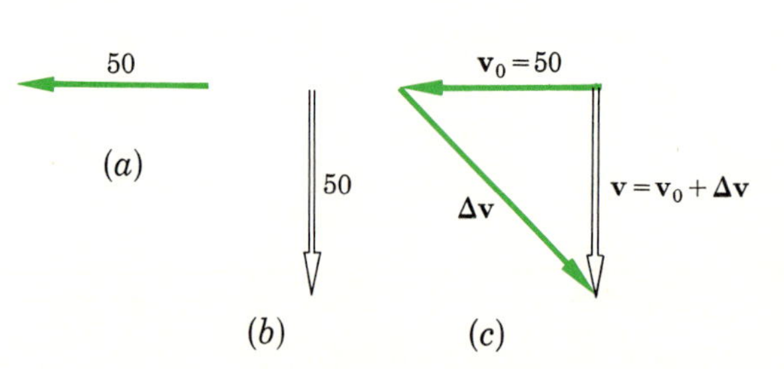

FIGURE 7.8

To change a velocity $\mathbf{v}_0$ of 50 km/h west, as shown in (*a*), to a velocity $\mathbf{v}$ of 50 km/h south, as shown in (*b*), one must add to $\mathbf{v}_0$ a velocity change $\Delta\mathbf{v}$, as shown in (*c*).

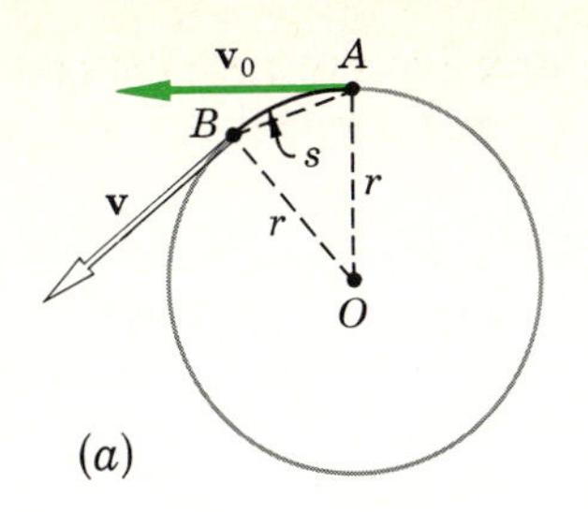

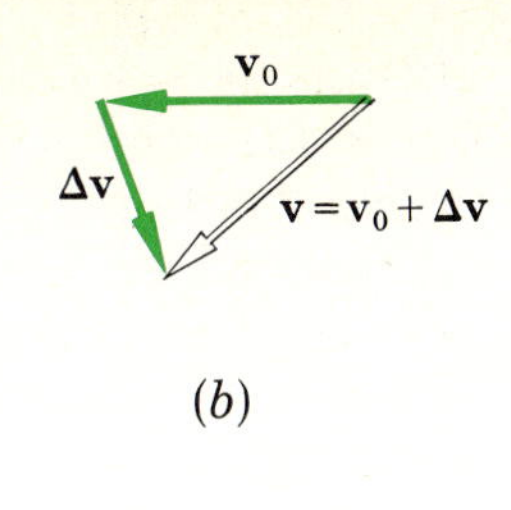

FIGURE 7.9
The vector triangle in (*b*) is similar to triangle *OAB* in (*a*), since the sides of the two triangles are mutually perpendicular.

$$a = \frac{\Delta v}{\Delta t}$$

where Δt is the time taken for the ball to travel from A to B. Eventually we shall take the limit as Δt approaches zero, in which case a will be the instantaneous acceleration.

Since the speed of the ball is v and the distance traveled by it is s, as shown in Fig. 7.9*a*, the time taken is just

$$\Delta t = \frac{s}{v}$$

from which we have, after substitution in the equation for a,

$$a = \frac{v \Delta v}{s}$$

Let us now compare the triangle OAB with the triangle in part *b* of Fig. 7.9. Clearly OA is perpendicular to v_0, OB is perpendicular to v, and BA is perpendicular to Δv. Hence these two triangles are similar. We can therefore write

$$\frac{\Delta v}{v} = \frac{\overline{BA}}{r}$$

But if point A is close to point B on the circle, as it will be if $\Delta t \to 0$, the arc from B to A is nearly the same length as the line $\overline{BA}$. To that approximation we can write $\overline{BA} = s$, and then the above expression becomes

$$\Delta v = \frac{sv}{r}$$

in the limit as $\Delta t \to 0$.

Placing this value for Δv in the equation for the acceleration a yields

Centripetal Acceleration

$$a = \frac{v^2}{r} \tag{7.13}$$

where a is now the instantaneous acceleration, since it is evaluated in the limit as $\Delta t \to 0$. *This acceleration, which is caused by the centripetal force, is called the* **centripetal acceleration.** Finally, *the centripetal force is given by* $F = ma$, and we have

Centripetal Force

$$F_{\text{centripetal}} = \frac{mv^2}{r} \tag{7.14}$$

Moreover, *the direction of* **a**, *and therefore the direction of* $\mathbf{F}_{\text{centripetal}}$, *is the same as* $\Delta\mathbf{v}$. From Fig. 7.9*b* we see that $\Delta\mathbf{v}$ is perpendicular to the perpendicular bisector of the angle between $\mathbf{v}$ and $\mathbf{v}_0$ and is therefore directed along a radius of the circle at the midpoint of the arc between A and B in Fig. 7.9*a*. When $\Delta t \to 0$, the arc between A and B also approaches zero. We therefore conclude that *the instantaneous acceleration of an object traveling with uniform speed in a circle is directed inward toward the center of the circle. That is why it is sometimes called the* **radial acceleration.** *The centripetal force has this same direction.*

Equation (7.14) tells us that the force needed to hold an object in a circular path is proportional to the square of the velocity of the object times its mass. Hence, the larger the mass of the object and the faster it is circling, the greater must be the tension in the rope which constrains it to move in a circle. This is a common experience for most of us. Even the fact that the centripetal force increases as the radius of the circle decreases is easily demonstrated by whirling a piece of chalk tied to a string around a circle. Remember while doing this experiment, though, that, if v is held constant in magnitude, it will take the piece of chalk less time to travel around a small circle than around a large one. Hence the angular speed of the particle will be greater for a small circle even if the linear speed of the object is held constant.

Illustration 7.7 A 1200-kg car is turning a corner at a speed of 8.0 m/s, and it travels along an arc of a circle in the process. If the radius of the circle of which the arc is a portion is 9.0 m, how large a horizontal force must be exerted by the pavement on the tires to hold it in the circular path? (See Fig. 7.10.)

Reasoning The force required is the centripetal force.

$$F = m\frac{v^2}{r} = (1200\text{ kg})\left(\frac{64\text{ m}^2/\text{s}^2}{9.0\text{ m}}\right) = 8530\text{ N}$$

This force must be supplied by the friction force of the pavement on the wheels. Can you show that the coefficient of friction must be at least 0.73 in order to provide such a large force? If the pavement is wet so that there is little friction between the tires and the road, the friction force on the tires will perhaps not be this large. In that event the car will skid out of a circular path (into more nearly a straight line) and may not be able to make the curve.

FIGURE 7.10
For the car to turn the corner as shown, the friction force F between the tires and pavement must furnish the centripetal force needed to hold the car in a circular path.

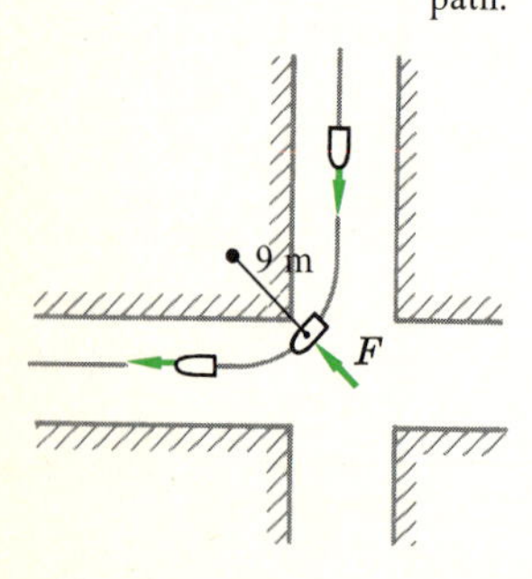

Illustration 7.8 A ball tied to the end of a string is swung in a vertical circle of radius r under the action of gravity, as shown in Fig. 7.11. What will be the tension in the string when the ball is at point A of the path, as shown in Fig. 7.11, if the ball's speed is v at that point?

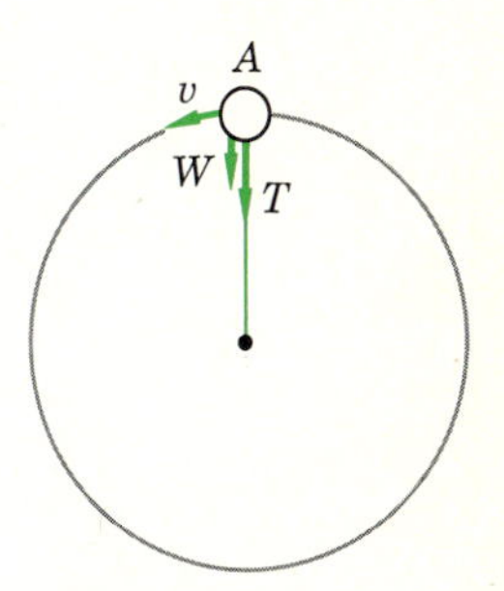

FIGURE 7.11
When the ball is in the position shown, its weight provides part of the necessary centripetal force.

Reasoning For the ball to travel in a circle, the forces acting on the ball must have, as their resultant, a force directed toward the circle's center. In the present case, at point A, two forces act on the ball, the pull of the string T and the pull of the earth (or the weight of the body). The resultant of these two forces must be directed toward the center of the circle. It is this resultant which furnishes the centripetal force required to keep the ball moving on a circular path. We can therefore write, for point A, that the resultant force on the ball $(T + W)$ equals the required centripetal force (mv^2/r):

$$T + W = \frac{mv^2}{r}$$

Therefore the tension in the string is

$$T = \frac{mv^2}{r} - W = m\left(\frac{v^2}{r} - g\right)$$

Notice that if $v^2/r = g$, the tension in the string will be zero. Then the centripetal force is just equal to the weight. This is the case where the ball just makes it around the circle. If the speed v of the ball is less than the value given by this relation, that is, $v < \sqrt{rg}$, the required centripetal force is less than the ball's weight. In that case the ball will fall down out of the circular path.

7.7 CENTRIFUGAL FORCE

Newton's law of action and reaction tells us that for every force on a body there is an equal and opposite force acting *on some other body or bodies.* Therefore, if a centripetal force is exerted on a body in order to hold it in a circular path, there must be some other body which experiences an equal and opposite force. This second force (acting on a *second* body and *not* on the body traveling in the circle) is called the **centrifugal force.**

In Fig. 7.12, the centripetal force acts on the ball to hold it in the circular path. This force must act in the direction toward the center of the circle. However, the person holding the string at the center of the circle experiences a pull outward toward the ball. This force is equal to the centripetal force but is oppositely directed. Moreover, it acts, not on the ball, but on the person or object holding the ball. This latter force is the one we have called the centrifugal force.

FIGURE 7.12
The centripetal force acts on the ball and pulls it toward the center of the circle. The equal but opposite centrifugal force acts on the hand of the person holding the string at the center of the circle and pulls away from the center, as shown.

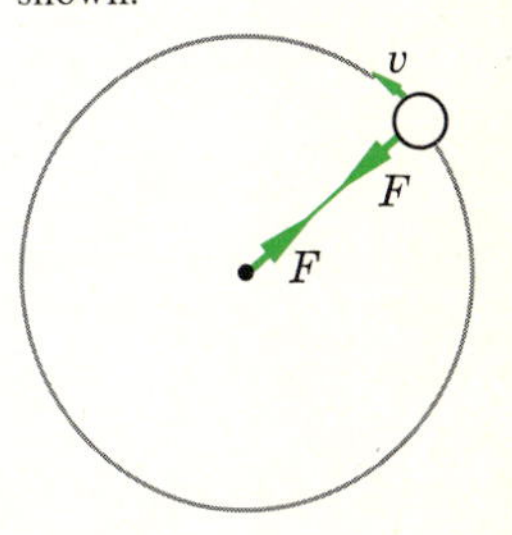

7.8 A COMMON MISCONCEPTION

People are likely to jump to completely erroneous conclusions when interpreting their experiences. For example, a person seated in the center of a car seat sometimes thinks that he has been pushed to the side of the car as it rounds a corner. He might even assert that the force pushing him sideways was so great that it threw him against the side hard enough to injure him. This is nonsense, of course. There was no mysterious ghost pushing him toward the side of the car. Certainly no material object was pushing him in that direction. He must therefore be mistaken.

The same person would not claim that a mysterious force suddenly acted on him to throw him violently against the dashboard of the car as it suddenly stopped. He knows that his forward momentum could be lost only if some force retarded his motion. Hence, when the car suddenly stopped, he continued going forward until the dashboard of the car began to exert a force on him to stop him from moving forward. This is merely an example of Newton's idea that things continue in motion until a force acts on them to stop them.

Similarly with the car turning the corner; the friction between the pavement and the tires pushed horizontally on the car and altered its straight-line motion. It is too bad about the man sitting in the middle of the nearly frictionless seat. The friction force between the seat of his pants and the seat was too small to alter his straight-line motion. Hence he slid along in a straight line until he hit the side of the car, which then exerted a force on him so that he could travel in the same curved path as that followed by the automobile.

7.9 WEIGHTLESSNESS

We often hear that objects appear to be weightless in a spaceship circling the earth or on its way to a distant point in space. Let us examine this effect in some detail. First, we should restate our definition of weight. It was defined to be the pull of gravity on the object in question. *On the earth, the weight of an object is the gravitational pull of the earth upon the object.* Similarly, *an object's weight on the moon is taken to be the gravitational pull of the moon upon the object.*

Ordinarily one measures the weight of an object by placing it on a scale. Or if only a rough measure is needed, one simply *notices the force the object exerts on one's hand when it is held fixed.* Usually the force read by the scale, i.e., the force exerted by the object on the scale, and the force on one's hand are equal to the pull of gravity on the object, i.e., to the weight of the object. This is not always true, however, as we shall now see, and so *we reserve the phrase* **apparent weight** *for the reading of the scale and the force on one's hand together with other common nonbasic ways for judging the weight of an object.*

To illustrate this point, let us consider the apparent weight of an object of mass m in an elevator. This question was discussed in Chap. 3, but let us consider it again here. In Fig. 7.13*a*, if the elevator is at rest, Newton's second

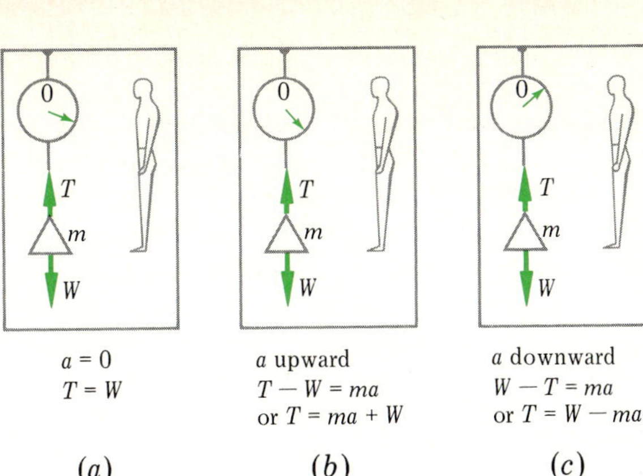

FIGURE 7.13 The weight of an object in an elevator seems to a person in the elevator to vary, depending upon the motion of the elevator.

law tells us that since the acceleration is zero, the resultant force on the object is zero. Calling the gravitational force on the body (its weight) W and the tension in the string holding the object T, we have

$$T - W = 0 \qquad \text{or} \qquad T = W$$

when $a = 0$. In this instance, the tension in the string is W, and the apparent weight of the object, the reading of the scale, is equal to its actual weight W.

This same situation will prevail *as long as* $a = 0$. Under that condition $T = W$, *and the apparent weight is equal to the actual weight*. Even if the elevator is moving up or down at constant speed, the acceleration is still zero and the apparent weight will equal the actual weight.

Let us now examine the situation shown in Fig. 7.13*c*. The elevator is accelerating downward in this case. If we apply Newton's second law as before, we find

$$W - T = ma$$

which gives

$$T = W - ma$$

Notice that the tension in the string (and therefore the scale reading) is less than W by the amount ma. To the person in the accelerating elevator, the object appears to weigh less than W. Its apparent weight is $W - ma$.

The most spectacular observation occurs when the elevator is freely falling, so that $a = g$, the acceleration in free fall. Then, since $W = mg$ and since $a = g$ for a freely falling body, the tension in the string

$$T = W - ma$$

becomes

$$T = mg - mg = 0$$

The object appears weightless in a freely falling elevator! If we think about it a little, this is not strange at all. Since the elevator (and everything in it) are supposedly accelerating with the acceleration of free fall, by the definition of what we mean by free fall, there can be no force supporting the objects (elevator and everything in it) or in any way retarding their free fall. Hence all support forces on the elevator and everything in it must be zero. The tension in the cord supporting the object must be zero. All objects within the elevator appear to be weightless.

We see from these considerations that *apparent weights of objects are not necessarily equal to the true weights of the objects in accelerating systems. In particular, if the system is freely falling,* all support forces must be zero and all objects appear to be weightless.* This means that *whenever a spaceship is falling freely in space,* i.e., when its rocket engines are not being operated, *everything within this freely falling system will appear to be weightless.* It does not matter where the object is, whether it is falling under the force of attraction of the earth, the sun, or some distant star; as long as it is freely falling, everything in it will appear weightless.

An earth or moon satellite ship is simply an example of a freely falling ship. At first this statement may surprise you, but it is easily seen to be correct. Consider the behavior of a projectile shot parallel to the horizontal surface of the earth in the absence of air friction. (At satellite altitudes, the air is so thin as to be almost negligible.) The situation is shown in Fig. 7.14. In part *a*, the projectile is thrown at successively larger speeds, and one sees that during its free fall to the earth the curvature of the path decreases with

* Recall that a freely falling object is one which is subject to only one type of unbalanced external force, the gravitational pull of the earth and similar bodies.

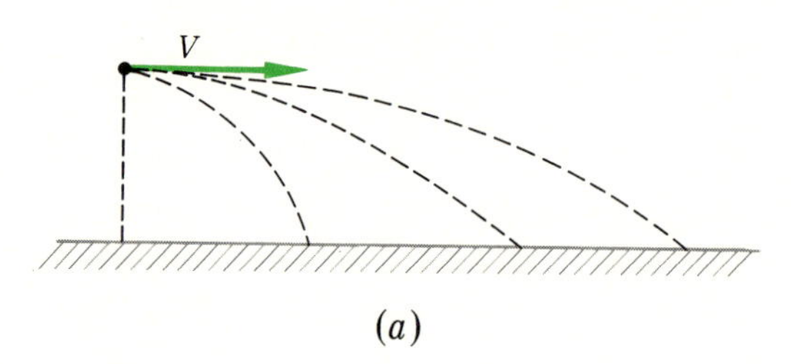

(*a*)

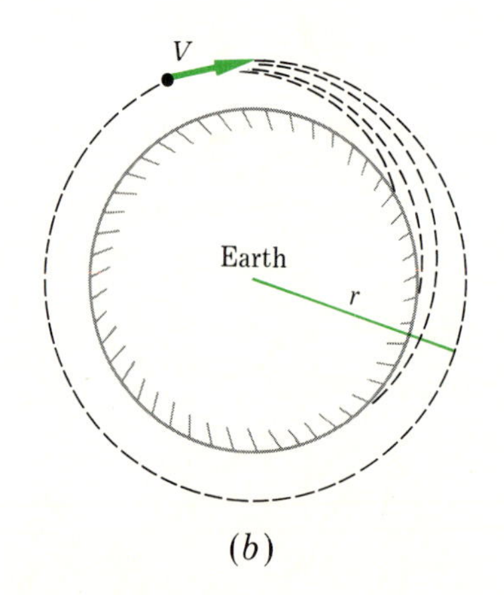

(*b*)

FIGURE 7.14
If an object is shot fast enough parallel to the earth, it will circle the earth. (Newton was probably the first to recognize this fact.)

increasing horizontal speed. If the object is thrown fast enough parallel to the earth, the curvature of its path will match the curvature of the earth, and this is shown in part *b*. In this case, the object (a spaceship perhaps) will simply circle the earth. Near the surface of the earth, i.e., a few hundred miles above the surface, this critical speed is about 29,000 km/h. Earth satellite ships circle the earth with this speed. The satellite is accelerating toward the center of the earth at all times since it circles the earth. This radial acceleration is simply g, the free-fall acceleration. In effect, the satellite is falling toward the center of the earth at all times, but the curvature of the earth prevents it from hitting. Since it is in free fall, all objects within it appear weightless.

The equations governing this motion are really quite simple. In order for the ship to follow a circular path, a centripetal force must pull it toward the center of the circle. This force is furnished by the earth's gravitational pull on the ship, namely,

$$F_G = G\frac{mM_{\text{earth}}}{r^2}$$

where m is the spaceship's mass, M_{earth} is the mass of the earth, and r is the radius of the orbit. Equating this to the formula for centripetal force, mv^2/r, gives

$$G\frac{mM_{\text{earth}}}{r^2} = \frac{mv^2}{r}$$

from which

$$v = \sqrt{\frac{GM_{\text{earth}}}{r}} \tag{7.15}$$

We see that the mass of the ship cancels out and so, as pointed out above, the speed with which it must be thrown, i.e., injected, into orbit, is independent of what the projectile happens to be. The required speed does, however, decrease with increasing radius of the orbit as the above equation shows.

SUMMARY

Angles are measured in degrees, revolutions, and radians; 1 rev = 360° = 2π rad. We represent angles by θ. The time rate of change of θ is called the angular velocity ω. The angular acceleration α is the time rate of change of ω.

As in linear motion, there are five angular-motion equations which apply if α is constant. They are

$$\theta = \bar{\omega}t \qquad \alpha = \frac{\omega_f - \omega_0}{t} \qquad \bar{\omega} = \tfrac{1}{2}(\omega_0 + \omega_f)$$

$$\omega_f^2 - \omega_0^2 = 2\alpha\theta \qquad \theta = \omega_0 t + \tfrac{1}{2}\alpha t^2$$

The first equation defines average angular velocity, and the second defines average angular acceleration.

When a wheel turns on its axle, a point on its circumference can be described in terms of tangential quantities. The tangential distance moved by the point s when the wheel turns through an angle θ is given by $s = r\theta$, where r is the radius of the wheel. Similarly, the tangential velocity and acceleration of the point are given by

$$v_T = r\omega \qquad a_T = r\alpha$$

In all these equations relating angular to tangential quantities, radians must be used for the angular measure.

When a wheel on an axle winds a string on its rim, the displacement, speed, and acceleration of a point on the string are given by the tangential quantities s, v_T, and a_T. Similarly, when a wheel rolls without slipping along a level surface, the displacement, speed, and acceleration of the center of the wheel are given by s, v_T, and a_T.

A centripetal force is required to cause an object to move in a circular path. This force is needed to pull the object from its normal straight-line motion. When following a circle, the vector velocity of an object is constantly changing. It is undergoing a radial acceleration toward the center of the circle. This acceleration is given by v^2/r, and the centripetal force required to cause this acceleration is mv^2/r.

The force needed to support an object is not always equal to the weight of the object. We call the required supporting force the apparent weight of the object. If the object and its support are accelerating, the weight and apparent weight usually differ. In the special case of free fall, the apparent weight is zero. As a result, objects appear weightless when coasting in space. Similarly, since an earth satellite is freely falling, objects within it appear weightless.

MINIMUM LEARNING GOALS

Upon completion of this chapter you should be able to do the following:

1. Convert an angle in degrees, radians, or revolutions into each of the other units.

2. Make use of the equation $\theta = \bar{\omega}t$ in simple situations in which two of the three quantities (θ, $\bar{\omega}$, t) are given.

3. Compute the angular acceleration of a wheel when ω_0, ω_f, and t are given.

4. Write down the five angular-motion equations and define each quantity in them. State the major restriction upon their use.

5. Make use of the five angular-motion equations to solve problems.

6. Explain how θ for a rotating wheel is related to the distance the wheel rolls or the amount of string which winds onto the wheel. When θ or s is given for a wheel of known radius, be able to find the other.

7. Explain the difference between tangential velocity and angular velocity. When one is given for a wheel of known radius, find the other. Explain why tangential velocity is important for a rolling wheel and for a wheel which is winding something on its rim.

8. Explain what is meant by a centripetal force and why it must be furnished to an object if the object is to follow a circular path. State the relation for centripetal force and be able to use it in simple situations where an object is moving in a horizontal or vertical circle.

9. Using a diagram, show why an object traveling on a circular path is accelerating even though its angular speed is constant. State the direction and value for the radial acceleration.

10. Calculate the supporting force needed on an object of known mass if the object is moving with constant speed; accelerating upward; accelerating downward. Explain what is meant by apparent weight in such circumstances and show why it differs from the object's weight.

11. Explain why an object orbiting the earth (or in some similar situation) is said to be freely falling. Use your explanation to point out why objects appear weightless under certain circumstances.

IMPORTANT TERMS AND PHRASES

You should be able to define or explain each of the following:

Radian measure
Angular analogs to s, v, a
The five angular-motion equations
Tangential quantities: $s = r\theta$, $v_T = r\omega$, $a_T = r\alpha$
Centripetal acceleration
Centripetal force
Even though an object is going around a circle at constant speed, it is accelerating
Apparent weight and weightlessness
An earth satellite is continuously falling toward the earth

1. The earth follows a path which is nearly circular with the sun at the center of the circle. Assuming this to be exactly true, explain why the sun's gravitational pull on the earth does no work. What does the sun's gravitational force do to the earth?

2. When mud flies off the tire of a bicycle, in what direction does it fly? Explain.

3. Figure P7.1 shows a simplified version of a cyclone-type dust remover. It is widely used to purify industrial waste gases before venting them to the atmosphere. The gas is whirled at high speed around a curved path, and the dust particles collect at the outer edge and are removed by a water spray or by other means. Explain the principle behind this method for removing particulate matter from dirty air.

4. Discuss the principle of the spin-dry cycle in an automatic washing machine.

5. An airplane pilot pulling out of a steep dive experiences a force of several g's; that is, the pilot's support in the plane must push on the pilot with a force several times the pilot's weight. Why must this force exist? If the pullout is too quick, the pilot blacks out. Why?

6. An insect is sitting on a smooth flat wheel which can be rotated about a vertical axis perpendicular to the plane of the wheel and through its center. Describe qualitatively the motion of the insect as the wheel begins to rotate. Assume that the insect is quite close to the axis and that there is some, but not much, friction between it and the wheel. (The wheel might be a phonograph turntable, for example.)

7. A person in an earth satellite is said to be weightless. Explain what is meant by this. Why do objects not fall closer to the earth in the ship since they experience the pull of gravity?

FIGURE P7.1

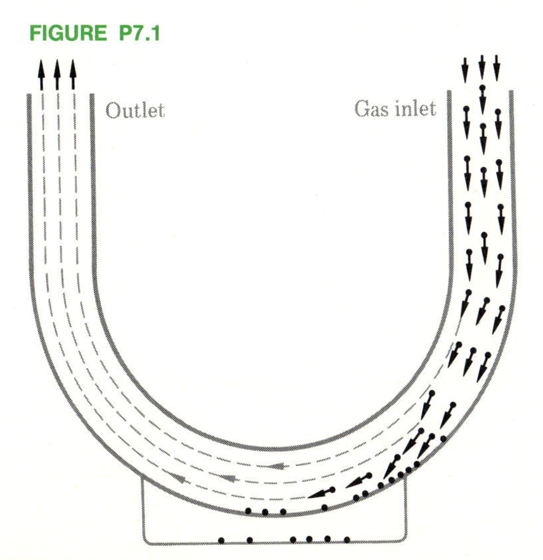

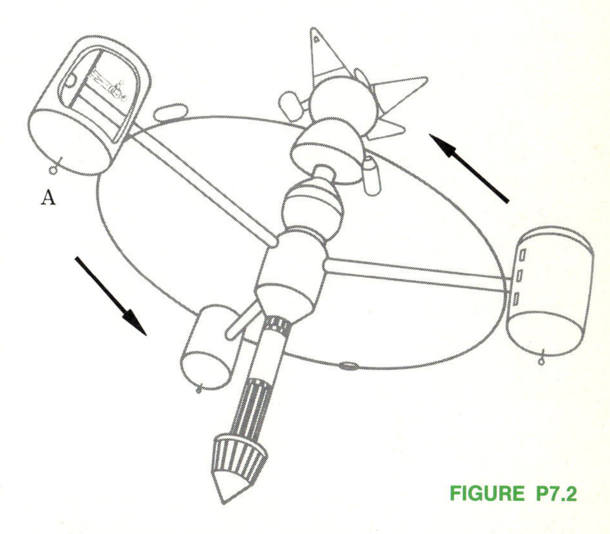

FIGURE P7.2

8. Figure P7.2 is an artist's conception of a rotating space station. Assuming the rotation to take place as shown, did the artist choose the proper direction for "up" in the space capsule at A? Explain.

9. A woman weighs herself daily on a spring-operated bathroom scale. Suppose that the earth stopped rotating about its axis. Would she weigh more, less, or the same? Would it matter where she lived on the earth?

10. A bug sits on the very top of a freshly waxed bowling ball. It loses its footing and slides freely down the surface of the ball. Explain why the bug will leave the surface *before* it falls halfway down the ball.

11. From the fact that the moon circles the earth at a radius of about 3.8×10^8 m, estimate the mass of the earth.

12. About how fast can a car be going if it must negotiate a turn from one street into a perpendicular street? Assume both streets to be made of concrete and to be of average size, carrying one lane of traffic each way. (E)

13. During the 1970 flight of Apollo 13 to the moon, serious trouble developed when the ship was about halfway, and it returned to earth without executing its moon mission. However, the ship continued toward the moon, passed on the other side of it, and only then returned to earth. Why didn't they simply turn around at the time the decision was taken to return instead of continuing toward the moon?

PROBLEMS

1. Express each of the following angles in degrees, revolutions and radians: (*a*) 25°; (*b*) 8.1 rad; (*c*) 0.73 rev.

2. A pulley on a motor is rotating at 1400 rev/min. What is its angular speed in (*a*) radians per second and (*b*) degrees per second?

3. In a certain large motor, it takes 20 s for the motor shaft to accelerate uniformly from rest up to its operating speed of 16 rev/s. Find (*a*) the angular acceleration of the motor shaft (revolutions per second squared) and (*b*) the number of revolutions it turns in this time.

4. How large an angular acceleration (radians per second) must be given a wheel if it is to be accelerated from rest to a rotation speed of 520 rad/s after completing 7.5 rev?

5. The time taken for a certain roulette wheel to coast to rest is 15 s. If the wheel turns through 8.5 rev in that time, how fast was it originally turning? (Assume uniform deceleration.)

6. There is a speck of dust 5.0 cm from the center of a $33\frac{1}{3}$-rev/min phonograph record. (*a*) When the phonograph is running, what is the speed in radians per second of the record? (*b*) How fast, in centimeters per second, is the dust speck moving?

7. The radius of the earth is 6.37×10^6 m. (*a*) How fast, in meters per second, is a tree at the equator moving because of the earth's rotation? (*b*) A polar bear at the North Pole?

8. Two gear wheels which are meshed together have radii of 0.50 cm and 0.15 cm. Through how many revolutions does the smaller turn when the larger turns through 3 rev?

9. A wheel turning at a speed of 1800 rev/min has a diameter of 2 in. If a string is wound on the wheel, how much string does the wheel wind up in 3 s?

10. A vehicle is traveling along the road at 20 m/s. If the diameter of its wheels is 80 cm, how fast are the wheels rotating in revolutions per second, radians per second, and degrees per second?

11. A car with 80-cm-diameter wheels starts from rest and accelerates uniformly to a speed of 15 m/s in 30 s. Through how many revolutions did each wheel turn in this time?

12. A motor turning at 1800 rev/min coasts uniformly to rest in 15 s. (*a*) Find its angular deceleration and the number of revolutions it turned before stopping. (*b*) If the motor had a 5-cm-radius wheel attached to its shaft, what length of belt did the wheel wind in the time taken for it to stop?

13. How large a force is needed to hold a 3600-lb car in an arc of 20 ft radius when the car is turning a corner at 40 ft/s?

14. In a thrilling ride at an amusement park, carts on a horizontal surface are whipped about at the end of a long rod. Suppose the cart and occupants weigh 2000 N and the rod is 3.0 m long. How large a stretching force (tension) must the rod withstand as it whips the cart around a 3.0-m-radius circle at a rate of 1 rev in 2.0 s?

15. A 20-mg bug sits on the smooth edge of a 25-cm-radius phonograph record as the record is brought up to its normal rotation speed of 45 rev/min. How large must the coefficient of friction between bug and record be if the bug is not to slip off? (It is a very compact bug and so air friction can be ignored.)

16. In a certain research device a person is subjected to an acceleration of 5 *g*'s, that is, five times the acceleration due to gravity. This is done by rotating the person in a horizontal circle at very high speed. The seat in which the subject is strapped is 20 ft from the rotation axis. How fast is the person rotating (revolutions per second) if the centripetal force on the person is five times the person's weight?

17. An old trick is to hold a pail of water with your hand and swing it in a vertical circle. If the rotation rate is large enough, water will not fall out of the pail when the pail is upside down at the top of its path. What is the minimum speed your hand must have at the top of the circle if the trick is to succeed? Assume your arm to be 0.60 m long.

18. A Ferris wheel is a large vertical circular device which carries riders in horizontal seats around the vertical circle. One such wheel has a radius of 20 m. (*a*) How fast (in revolutions per second) must the wheel be turning if the rider is to push down on the seat with a force $1\frac{1}{2}$ times as large as the rider's weight when the rider is at the bottom of the circle? (*b*) How fast must the wheel be turning if the rider is to exert no force on the seat at the top of the circle?

19. In one model of the hydrogen atom (the Bohr model) an electron is pictured rotating in a circle (0.5×10^{-10} m radius) about the positive nucleus of the atom. The centripetal force is furnished by the electrical attraction of the positive nucleus for the negative electron. How large is this force if the electron is moving with a speed of 2.3×10^6 m/s? (The mass of an electron is 9×10^{-31} kg.)

20. Use the law of gravitation to compute the speed of the earth rotating about the sun by assuming the earth to travel in a circular path with the sun at its center. ($M_{earth} = 6.0 \times 10^{24}$ kg; $M_{sun} = 3.3 \times 10^5 M_{earth}$; and earth-to-sun distance $= 1.5 \times 10^{11}$ m.)

21. A certain disk oriented horizontally starts from rest and begins to rotate about its axis (vertical) with an acceleration of 0.50 rev/s^2. After 20 s a 3.0-g mass cemented to the rim of the wheel breaks loose from the wheel. How large was the force holding it in place? The wheel's radius is 40 cm.

22.* A 2.0-kg ball hangs as a pendulum from a cord 10.0 m long. If the pendulum is pulled to one side so that it makes an angle of 37° with the vertical and is then released, find the tension in the cord (*a*) just after its release and (*b*) when it passes through the bottom of its swing.

23. A device used to test the strength of adhesives makes use of a mass m cemented to the rim of a horizontal wheel of radius b. Show that the force withstood by the cement just before the mass breaks loose is given by $F = 4\pi^2 n^2 bm$, where n is the speed of the wheel in revolutions per second when the mass breaks loose.

24. When a satellite is in circular orbit, the force of gravity exactly equals the required centripetal force. With what speed must a satellite travel (meters per second) if the radius of its orbit is 7.00×10^6 m; that is, it is 0.63×10^6 m, or 400 mi, above the earth?

25.* As shown in Fig. P7.3, a boy on a rotating platform holds a pendulum in his hand. The pendulum is at a radius of 6.0 m from the center of the platform. The rotation speed of the platform is 0.020 rev/s. It is found that the pendulum hangs at an angle θ to the vertical as shown. Find θ.

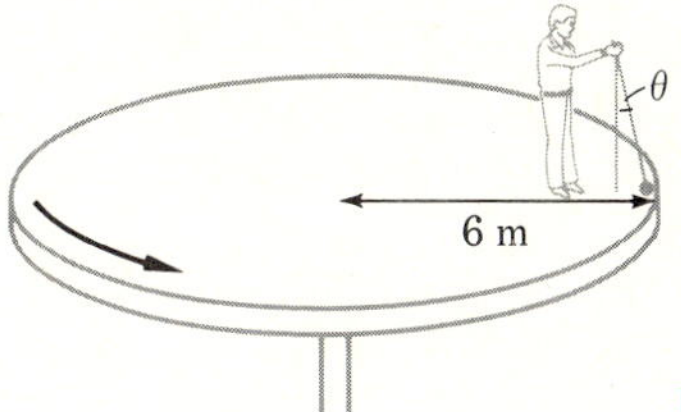

FIGURE P7.3

26.** The bug shown in Fig. P7.4 has just lost its footing while near the top of the bowling ball. It slides down the ball without appreciable friction. Show that it will leave the surface of the ball at the angle θ shown, where θ is given by $\cos\theta = \frac{2}{3}$.

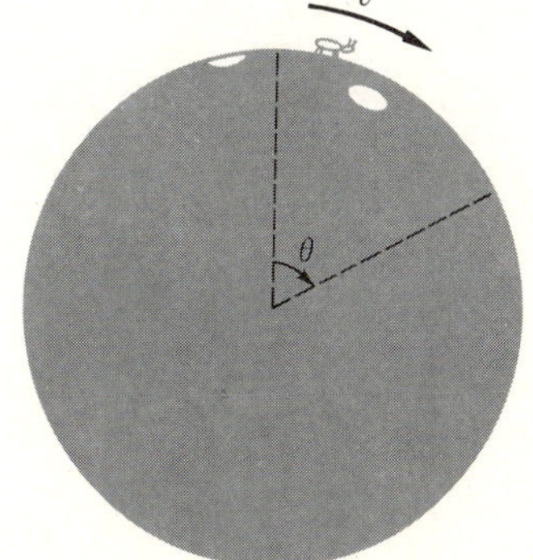

FIGURE P7.4

ROTATION OF RIGID BODIES

8

The past several chapters have been devoted to the motion of objects which are not rotating. Although the last chapter considered the motion of objects on circular paths, the objects were small enough for us to ignore the fact that the objects themselves might be spinning. In this chapter we consider the motion of large rigid objects which are spinning. We shall see that an analog of Newton's second law applies to the rotation of objects. The concepts of rotational kinetic energy and angular momentum will also be discussed.*

---

*Sections 8.1 to 8.5 can be combined with Chap. 1 if the instructor chooses. Items at the end of the chapter marked with a dagger † apply to these sections.

8.1 CENTER OF GRAVITY AND CENTER OF MASS

The translational and rotational motion of rigid, i.e., nondeformable, bodies can be analyzed most easily with the help of the concepts of center of mass and center of gravity. **DEFINITION** *We define the* **center of gravity** *of an object to be the single point at which the pull of gravity (the weight) of an object may be considered to act when discussing the behavior of the object.* Putting this in another way, for many purposes a large body may be replaced by a very small body (or point mass) which has the same weight. As an example, if a uniform sphere is hung from a string, as in Fig. 8.1, the center point of the sphere is always found to hang directly below the string. It follows from this that to describe the pull of the ball on the string in such an experiment, the whole weight of the ball can be considered to act at *C*, its center. We call this point the center of gravity of the sphere. *When an object is suspended by a single cord, the center of gravity lies below the suspension point and on the line of the cord. The center of gravity is the point at which the weight of the object may be considered to act.*

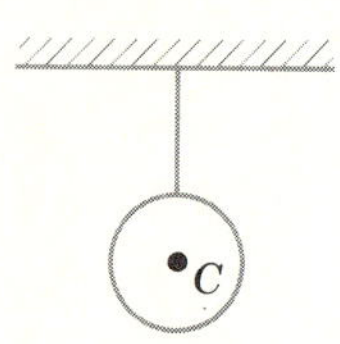

FIGURE 8.1
The center of gravity *C* of the sphere hangs directly below the suspension point.

It is quite a simple matter to find the center of gravity for symmetrical objects which have an easily recognized center point. For example, the center of gravity of a flat rectangle of cardboard is at its geometrical center. The same holds true for a cube. It is not quite so easy to see where the center of gravity of a jagged rock would be—or of a person.

But from the definition of center of gravity and Fig. 8.1, we know that the center of gravity of a rigid body will always hang directly below its point of suspension. This fact furnishes an easy method for finding centers of gravity. As an example, consider the flat, irregular-shaped object shown in Fig. 8.2*a*. If it is supported on a frictionless pivot at *A*, the center of gravity will hang directly below the pivot point, somewhere on the line *AB*. Now if the mass is hung from a new point such as *C*, it will hang as in Fig. 8.2*b*. The center of gravity must now be somewhere on the line *CD*.

FIGURE 8.2
After determining the plumb lines *AB* and *CD* by suspending the object from points *A* and *C*, we know that the center of gravity is at the intersection *O* of the two lines.

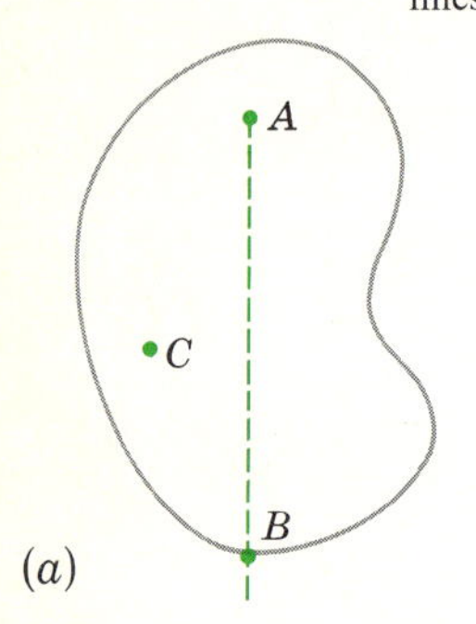

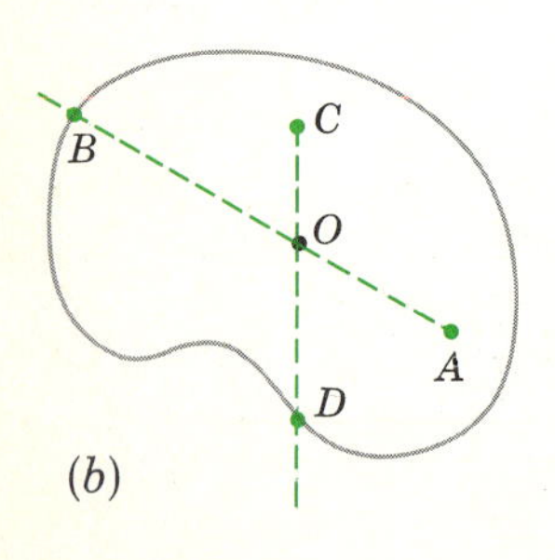

Since the center of gravity must be on both lines *AB* and *CD*, it must be at the intersection of these two lines, namely, at *O*. If the object is hung from a still different pivot point, the point *O* will be found to be directly below that pivot point as well. This, then, provides a simple experimental method for locating the center of gravity of a rigid object. *As long as the body exists in a uniform gravitational field, the center of gravity coincides with another important point, the* **center of mass** *of the body.*

We make use of the center of mass because it has a very important property.

> All the mass of an extended object may be assumed to be at the center of mass of the object when considering the translational behavior of that object under the action of external forces.

This means that the rigid body is equivalent to an equal point mass placed at the mass center insofar as its translational motion under a given external force is concerned. For example, consider the rotating projectile shown in Fig. 8.3. Its center of mass follows a smooth curve, the same curve a tiny object would follow if subjected to the same external forces. As we shall see, the importance of the center of mass is that it allows us to separate the

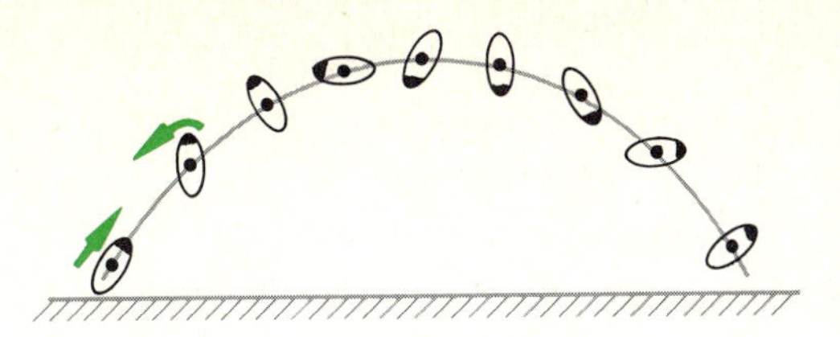

FIGURE 8.3

The center of mass follows the same path a point-mass projectile would follow when subjected to the same external forces.

translational and rotational motion of an object. The object will be translated as though its mass were all concentrated at its mass center.

8.2 TORQUE AND THE SECOND CONDITION FOR EQUILIBRIUM

Let us now investigate what condition is necessary if a rigid object is to be prevented from rotating. You will recall that in Chap. 1 we discussed the condition under which an object would be at equilibrium provided we ignored all rotation effects. We concluded that the vector sum of all the forces acting on the object must be zero if the object is to be at equilibrium. This was summarized in the first condition for equilibrium, namely, $\Sigma \mathbf{F} = 0$, or

$$\Sigma F_x = \Sigma F_y = \Sigma F_z = 0$$

We shall now find the condition which must be satisfied if an object is to be in rotational equilibrium.

Consider a very simple experiment. Drill two small holes in a meterstick or yardstick, one of which is through the center of the stick. Now place a small nail through the off-center hole and use it as a pivot point, as shown in Fig. 8.4*a*.

It is common experience that the stick will not hang at rest at an angle, as indicated in Fig. 8.4*a*, but will fall to rest with the center of gravity below the pivot point, as shown. On the other hand, if the pivot point is at the exact center, as in Fig. 8.4*b*, the stick will remain at rest at any angle we place it, for example, the position shown in part *b*. This is to be expected, since the center of gravity is coincident with the pivot point. Only two forces act on the stick, the pull of gravity down and the push of the pivot up. When the stick is at rest, these forces must be equal in magnitude since at equilibrium $\Sigma F_y = 0$.

Consider now what will happen if two equal and opposite forces F are applied to the stick, as shown in Fig. 8.4*c*. Notice that the first condition for equilibrium is still satisfied, since $\Sigma F_y = 0$ and there are no x- or z-directed forces. But experience tells us that the stick will not remain at rest now. It will begin to rotate counterclockwise, as indicated in the diagram. From this it is apparent that even though the first condition for equilibrium is satisfied, the body is not at equilibrium. There must be a second condition, which we shall now discuss.

The example just given shows that not only must $\Sigma F_x = 0, \Sigma F_y = 0$, and $\Sigma F_z = 0$ but there must be no net turning effect of forces trying to turn the

FIGURE 8.4

The uniform stick will hang at rest in the positions indicated by black lines in (*a*) and (*b*). In (*c*), however, the stick will rotate counterclockwise.

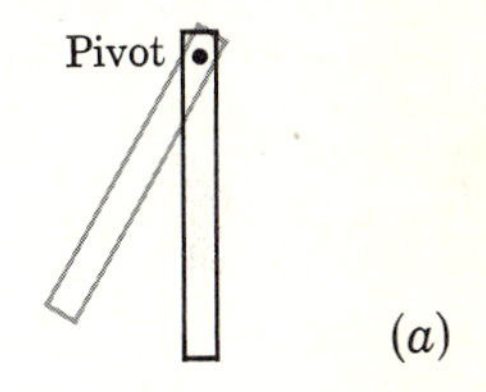

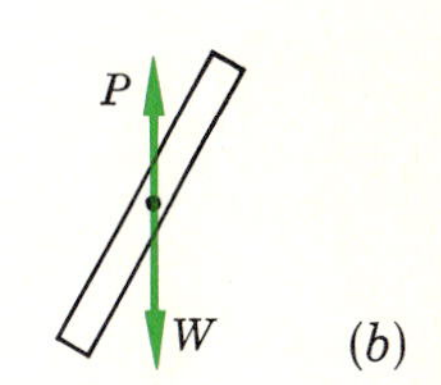

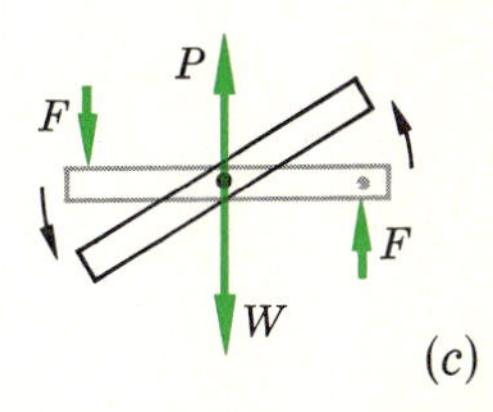

object about a pivot point if the object is to be at equilibrium. Let us examine turning actions in more detail so that we can formulate this second condition more precisely. To do this, consider the situations shown in Fig. 8.5*a*, *b*, and *c*, where the stick is pivoted at its center in each case.

If one pulls in line with the stick as in part *a*, experiment shows that the stick will remain in equilibrium and so F has no turning action in this case. On the other hand, in a seesaw type of arrangement such as that shown in part *b*, the force F would cause the stick to rotate counterclockwise. To prevent it from doing so, a force can be applied to the opposite side of the stick. Careful experiments show that a force F at the end of a seesaw can be balanced by a force twice as big placed as shown on the other side of the seesaw. The force $2F$, trying to turn the stick clockwise about the pivot point, exactly balances the force F, trying to turn the stick counterclockwise. Notice that the force F times its distance from the pivot, $L/2$, just equals the product $(2F)(L/4)$, the similar product for the other force.

As a third example, Fig. 8.5*c* shows a counterclockwise turning force also. It is found from experiment that a force acting in the direction shown causes more of a turning effect than the one in part *a* but less than the effect of F in part *b*. In fact, quantitative experiments show that the turning effect of a force about a pivot point is equal to the product of the force and a distance called the lever arm. *The* **lever arm** *is the length of a perpendicular dropped from the axis, or pivot point, to the line of the force.* It is length OB in Fig. 8.5*c*. The *turning effect is given the name* **torque**. The precise definition of torque is

DEFINITION

Torque

$$\text{Torque} = (\text{force})(\text{lever arm}) \tag{8.1}$$

or in symbols

$$\tau = Fl$$

where the symbol τ (tau) has been used for torque.*

One has to be careful in finding the lever arm. It is well to go back to its definition each time: the length of a perpendicular dropped from the axis, or pivot point, to the line of the force. *The* **line of the force** *is the line of which the force vector is a part.*

Illustration 8.1 Find the lever arm in Fig. 8.5*a*, *b*, *c*.

Reasoning The lever arm is zero in Fig. 8.5*a*, since the line of the force passes right through the axis. The turning effect, or torque, is zero in this case, since

$$\tau = (F)(0) = 0$$

* Notice that torque has the units of force times length. These are also the units of work. The concept, of work and torque are quite different, however.

In part b of the figure, the lever arm for F is $L/2$. Hence its torque, or turning effect, is

$$\tau = F\frac{L}{2} = \tfrac{1}{2}FL \qquad \text{counterclockwise}$$

Notice that the torque due to the force $2F$ is

$$\tau = (2F)(\tfrac{1}{4}L) = \tfrac{1}{2}FL \qquad \text{clockwise}$$

In this case at least, equilibrium is maintained if the clockwise torque equals the counterclockwise torque.

A more complicated case in which to find the lever arm is shown in Fig. 8.5c. In this case, the line of the force is the line of which $\overline{AB}$ is a part. The perpendicular to this line from the axis O is the line $\overline{OB}$. Hence, by definition, $\overline{OB}$ is the lever arm. Since

$$\cos 53° = \frac{\overline{OB}}{L/2}$$

it is apparent that the lever arm in this case is

$$\overline{OB} = (0.60)\left(\frac{L}{2}\right) = 0.30L$$

The torque is therefore

$$\tau = (F)(0.30L) = 0.30FL$$

Notice, as we said before, that the torque, or turning action, is in this case intermediate in size when compared with the cases shown in the other two diagrams.

Let us review what we have learned thus far. *The torque, or turning action, of a force in respect to an axis, or pivot point, is equal to the product of the force and its lever arm. The lever arm is the length of a perpendicular dropped from the axis to the line of the force.*

We have now acquired enough insight so that we can make a precise statement of the second (and last) condition for equilibrium. *Careful experiments show that the counterclockwise torques must equal the clockwise torques if no turning is to occur.* This may be written in a simple equation if we *call counterclockwise torques positive and clockwise torques negative* (or vice versa, since it makes no difference for our purposes). Then, with the notation $\Sigma\tau$ used to mean the "sum of all the torques," *the second condition for equilibrium becomes*

Second Condition for Equilibrium

$$\Sigma\tau = 0$$

We have tacitly assumed in this discussion of torques that motion of the

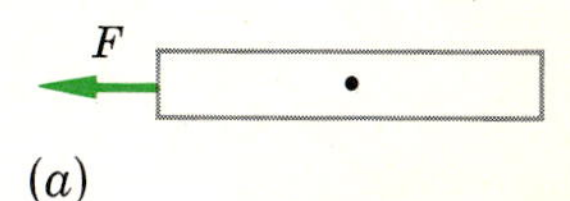

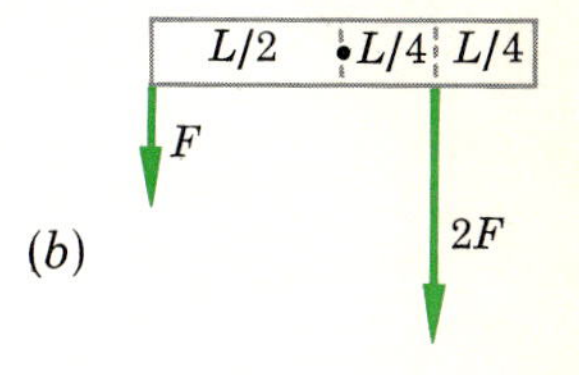

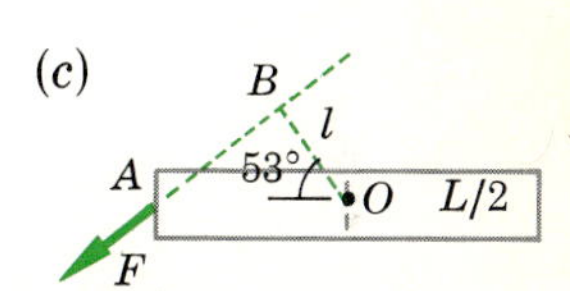

FIGURE 8.5
There is no net turning action upon a bar pivoted at its center, even when supporting the forces shown in (a) and (b). The torque, or turning action, in (c) is equal to F multiplied by the length OB.

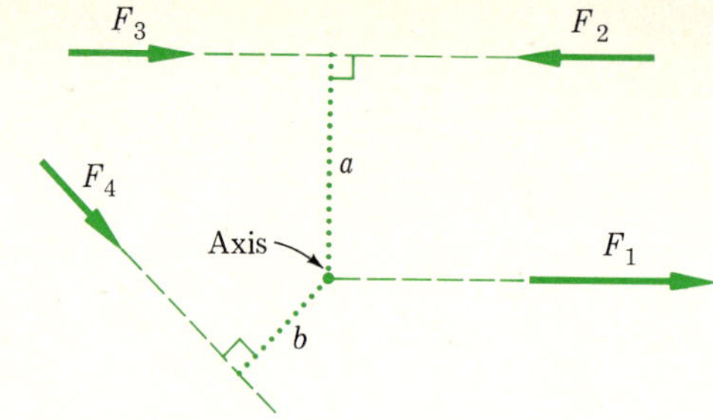

FIGURE 8.6
Find the lever arms for the forces with respect to the axis shown.

body under consideration is restricted to a plane. A large share of the cases of interest are of this type.

Conditions for Equilibrium

The total requirements for a body to be in equilibrium are now known. To summarize, in two dimensions, they are

$$\Sigma F_x = 0 \qquad \Sigma F_y = 0 \qquad \Sigma \tau = 0 \tag{8.2}$$

Before leaving this section, we should mention that the words **moment** or **moment of force** are sometimes used instead of torque. In that case, the lever arm is frequently referred to as the **moment arm.** The concepts are still the same, of course.

Illustration 8.2 Find the lever arms for the various forces shown in Fig. 8.6 about the axis point indicated.

Reasoning The lever arm in each case is shown by the dotted line. Notice that it is the length of the perpendicular dropped from the axis to the line of the force. We have:

FORCE	LEVER ARM	TORQUE
F_1	0	0
F_2	a	$+aF_2$
F_3	a	$-aF_3$
F_4	b	$+bF_4$

Notice also that the torques due to F_2 and F_4 are counterclockwise while that due to F_3 is clockwise. In a typical case the units of the torque would be newton-meters or pound-feet.

8.3 RIGID OBJECTS IN EQUILIBRIUM

Now that we know the second condition for equilibrium, we are able to treat all ordinary situations in which a rigid body is at equilibrium. To analyze such situations we follow the same procedure as outlined in Chap. 1. You will recall that there were five basic steps:

1 Isolate a body for discussion.
2 Draw the forces acting *on* the body.
3 Split the forces into components.
4 Write the equations $\Sigma F_x = 0$, $\Sigma F_y = 0$, and $\Sigma\tau = 0$.
5 Solve the equations for the unknowns.

Let us illustrate the procedure by means of the following example.

Illustration 8.3 Consider the 150-lb sign painter shown in Fig. 8.7. He is standing on a uniform 50-lb plank supported by two ropes. Find the tensions in the two ropes.

(*a*)

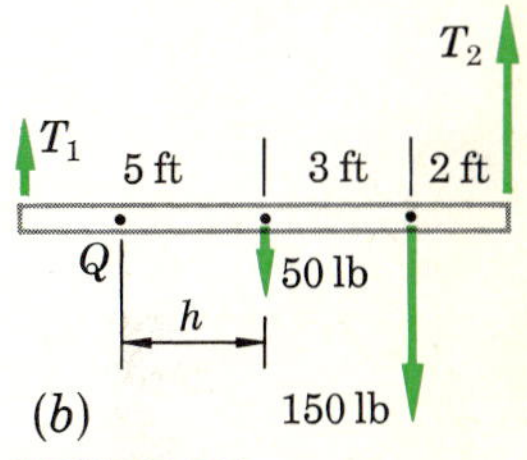

(*b*)

FIGURE 8.7

We wish to find the tension in the ropes holding the 50-lb plank upon which the 150-lb man stands. The forces acting on the plank are shown in (*b*).

Reasoning We isolate the plank as the body in equilibrium. The forces acting on it are shown in part *b* of Fig. 8.7. Since all the forces are in the *y* direction, $\Sigma F_y = 0$ becomes

$$T_1 + T_2 - 50 - 150 = 0$$

We have omitted the units from this equation in order to avoid too cumbersome an expression. You should check through equations where we do this and make sure you know which units are being used. In order to write the torque equation, a pivot point or axis must be chosen. The matter of its choice will be considered in detail in the next section. For now, we will simply choose it at the center of gravity of the board. Then $\Sigma\tau = 0$ becomes

$$(T_2)(5.0) - (150)(3.0) - (T_1)(5.0) = 0 \tag{8.3}$$

Solving these two equations simultaneously yields $T_1 = 55$ lb and $T_2 = 145$ lb.

8.4 THE POSITION OF THE AXIS IS ARBITRARY

In the last example, we tacitly assumed that the axis for writing the torque equation could be taken at the center of the board. Although this assumption can be proved correct by subjecting it to experimental test, it is inconvenient to carry out an experiment to justify a choice of axis in every situation which may confront us. Hence we seek to prove in a general way where it is allowable to choose a pivot point, or axis, for writing the torque equation.

Referring to Fig. 8.7*b*, let us determine whether the general point Q on the board is an allowable point for choosing the axis. Careful experiments show that the center of the board is an allowable point for the axis, and we may make use of that fact. Let us tentatively write the torque equation for Fig. 8.7*b*, using Q as axis. We have

$$(T_2)(5 + h) - (50)(h) - (150)(3 + h) - (T_1)(5 - h) \stackrel{?}{=} 0$$

where the question mark over the equality sign indicates that, at this stage, we are uncertain whether it is allowable to use point Q as axis.

Upon grouping terms we find

$$-(T_1)(5) - (150)(3) + (T_2)(5) + (h)(+T_1 + T_2 - 200) \stackrel{?}{=} 0 \qquad (8.4)$$

But the first condition for equilibrium ensures that the coefficient of h in this equation is zero, since the sum of the vertical forces must be zero, i.e.,

$$-T_1 - T_2 + 200 = 0$$

Thus we find that the term containing h in Eq. (8.4) is zero, no matter what value we used for h. Hence the value of h could have been anything. Moreover, since the term in Eq. (8.4) containing h is zero, the tentative equation becomes

$$-(T_1)(5) - (150)(3) + (T_2)(5) \stackrel{?}{=} 0$$

But this is exactly Eq. (8.3), which we already know to be correct. We can therefore dispense with the question mark on these equations and conclude that it is proper to take as axis any point Q we wish along the board. In fact, a more general proof can be given which shows that *the axis for writing the torque equation may be taken anywhere and is completely arbitrary. We are therefore justified in taking the axis anywhere we think convenient.*

Illustration 8.4 Now that we know that the choice of axis is arbitrary, let us again find the tensions in Fig. 8.7.

Reasoning We can solve for the tensions in the ropes more easily than we did in the last section if we choose our axis at the end of the board to which the rope with tension T_1 is fastened. Since the line of the T_1 force passes directly through this new axis, the lever arm for T_1 is zero and therefore it exerts no torque about this axis. As a result, the unknown T_1 will not appear in the torque equation.

Since the torque due to T_1 is zero, there are only two clockwise torques about this new axis, those due to the 50- and 150-lb forces. T_2 causes a counterclockwise torque. Writing down the torque equation for this new axis gives

$$\Sigma\tau = 0: \qquad (T_2)(10) - (50)(5) - (150)(8) = 0$$

This equation gives T_2 at once to be

$$T_2 = 145 \text{ lb}$$

Notice that the second equation, $\Sigma F_y = 0$, can now be used to find T_1.

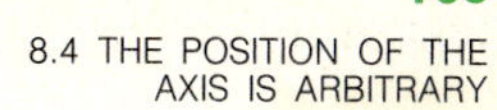

By choosing the axis in this particular position, we have considerably simplified the algebra. At the same time, we have illustrated the fact that the axis can be chosen wherever convenient.

We shall generally choose the axis so that one or more of the unknown forces has its line acting through the axis. This simplifies the computation.

> The torque equation can be simplified greatly by choosing the axis point on the line of an unknown force. That force, then, does not appear in the torque equation.

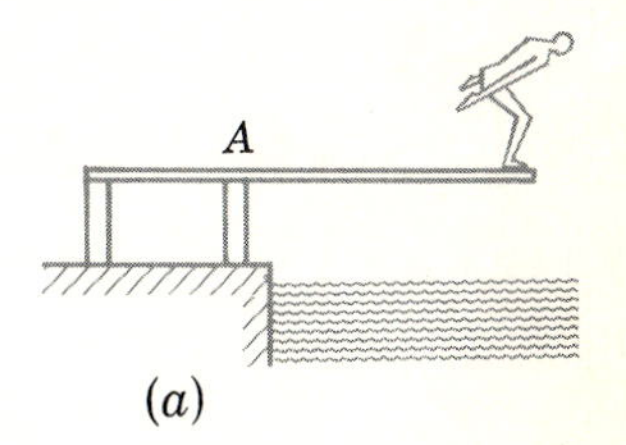

(a)

Illustration 8.5 As shown in Fig. 8.8, a 900-N man is about to dive from a diving board. Find the forces exerted by the pedestals upon the board. Assume the board has negligible weight. (900 N is about 200 lb.)

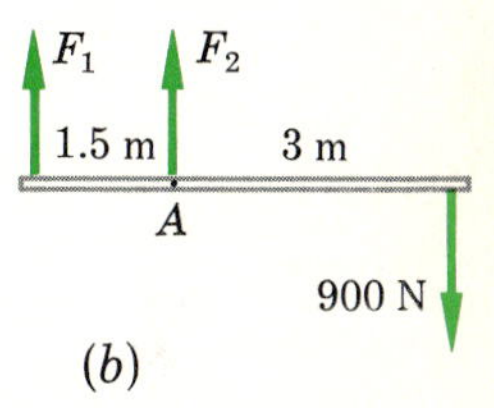

(b)

FIGURE 8.8

A 900-N man stands at the end of a light diving board. We guess, incorrectly, that the forces exerted by the two pedestals upon the board are as shown.

Reasoning We isolate the board and draw the forces on it as shown in part *b* of the figure. A little reflection will convince you that we have drawn the force F_1 in the wrong direction. This is a deliberate error, and we shall see how it affects the result. Choosing point A as pivot, we have

$$\Sigma\tau = 0: \qquad -(900)(3)-(F_1)(1.5) = 0$$
$$\Sigma F_y = 0: \qquad F_1 + F_2 - 900 = 0$$

Solving these gives $F_1 = -1800$ N and $F_2 = 2700$ N. Notice that the negative sign found for F_1 tells us our force was drawn in the wrong direction. Its magnitude is 1800 N.

Illustration 8.6 For the uniform 50-lb beam shown in Fig. 8.9, how large is the tension in the supporting cable, and what are the components of the force exerted by the wall upon the beam?

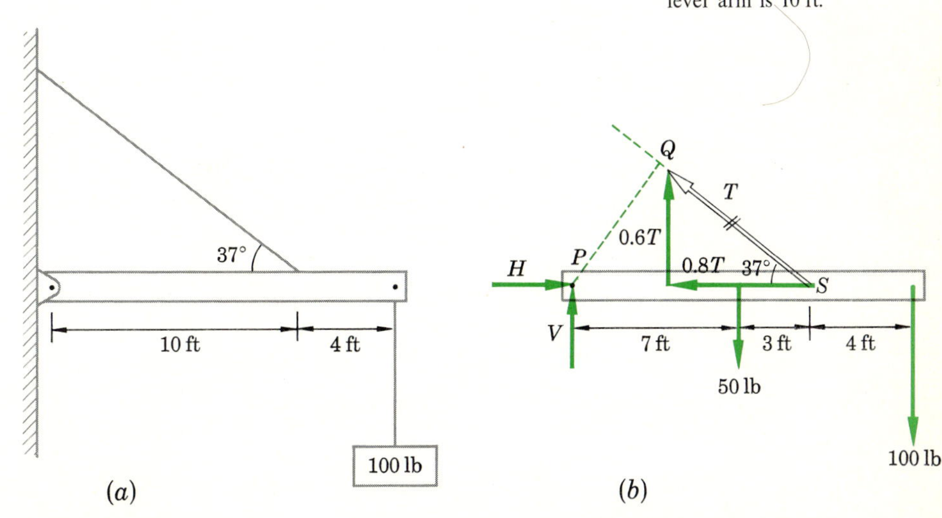

FIGURE 8.9

The forces acting on the beam in (*a*) are shown in detail in (*b*). Notice that the force component $0.6T$ pulls on the board at point S, and so its lever arm is 10 ft.

Reasoning Isolating the beam for consideration, the forces acting on it are shown in Fig. 8.9*b*. Notice that the weight of the beam, 50 lb, is taken to act at the board's center of gravity. Further, notice how the tension in the cable has been replaced by its components. We can eliminate the force components at the wall, H and V, by taking point P as axis. Then

$$\Sigma\tau = 0: \qquad (0.6T)(10) - (50)(7) - (100)(14) = 0$$
$$\Sigma F_x = 0: \qquad H - 0.8T = 0$$
$$\Sigma F_y = 0: \qquad V + 0.6T - 50 - 100 = 0$$

Solving first for T in the torque equation and then substituting in the other two equations gives

$$T = 292 \text{ lb} \qquad H = 234 \text{ lb} \qquad V = -25 \text{ lb}$$

What does the minus sign on V tell us?

Illustration 8.7 A person holds a 20-N weight as shown in Fig. 8.10*a*. Find the tension in the supporting muscle and the component forces at the elbow.

Reasoning The system can be replaced by the simplified model shown in part *b*. We assume the lower arm to weigh 65 N. The dimensions given are

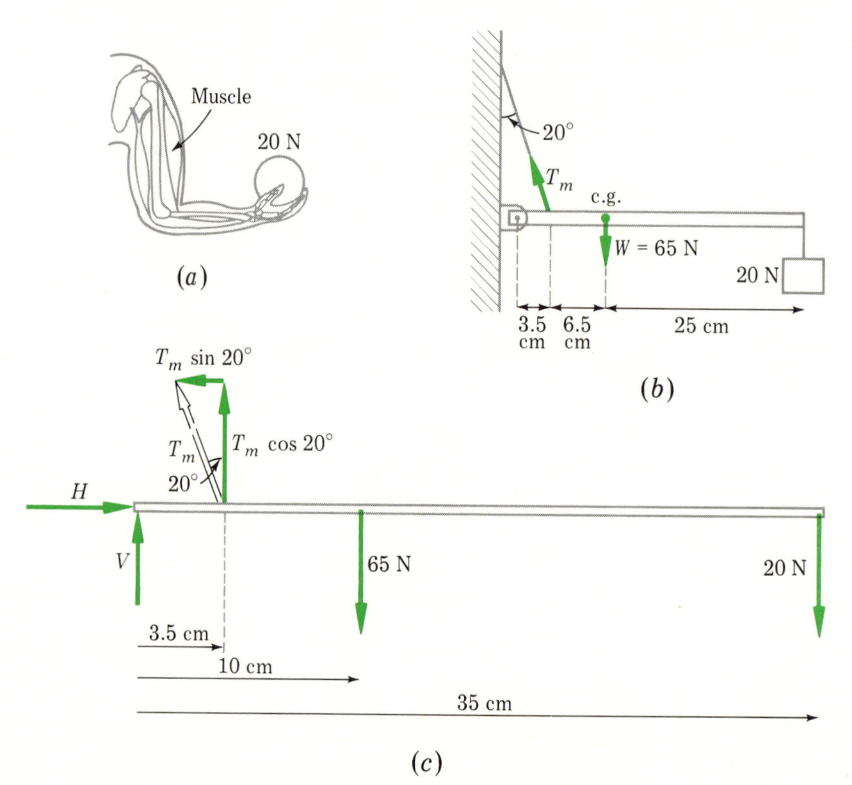

FIGURE 8.10

We can analyze the forces in the human arm by use of the models in (*b*) and (*c*).

typical. Notice that the situation is very similar to that of the previous illustration, the case of a beam supported by a cable. The appropriate free-body diagram is shown in part *c*. We use it to write the equilibrium conditions:

$$\Sigma F_x = 0: \qquad H - T_m \sin 20° = 0$$
$$\Sigma F_y = 0: \qquad V + T_m \cos 20° - 65 - 20 = 0$$
$$\Sigma \tau = 0: \qquad (T_m \cos 20°)(0.035) - (65)(0.10) - (20)(0.35) = 0$$

where the left end of the arm (the elbow) is taken as axis. Notice that the force $T_m \sin 20°$ actually acts through the axis and so its torque is zero.

Solving these equations, we find

$$T_m = 410 \text{ N} \qquad H = 140 \text{ N} \qquad V = -300 \text{ N}$$

All of these forces are much larger than the weight being held. Can you show that T_m becomes very large as the arm is outstretched? Why is it very tiring to hold a weight in your outstretched hand?

Illustration 8.8 A uniform 50-lb ladder leans against a smooth wall as in Fig. 8.11. (By the term **smooth,** we mean the force at the wall is perpendicular to the wall surface. No friction exists.) If a 100-lb boy stands on the ladder as shown, how large are the forces at the wall and at the ground?

Reasoning Isolating the ladder, the forces acting on it are as shown in part *b* of the figure. We then have (taking point *A* as axis)

$$\Sigma F_x = 0: H - P = 0$$
$$\Sigma F_y = 0: V - 50 - 100 = 0 \qquad \text{or} \qquad V = 150 \text{ lb}$$
$$\Sigma \tau = 0: \quad (P)(0.8 \times 20) - (50)(0.6 \times 10) - (100)(0.6 \times 15) = 0$$

In this problem, the lever arms should be noticed particularly. By definition, the lever arm is the perpendicular dropped from the axis *A* to the line of the force. The lever arms are *AB*, *AC*, and *AD* for the forces 50, 100, and *P*,

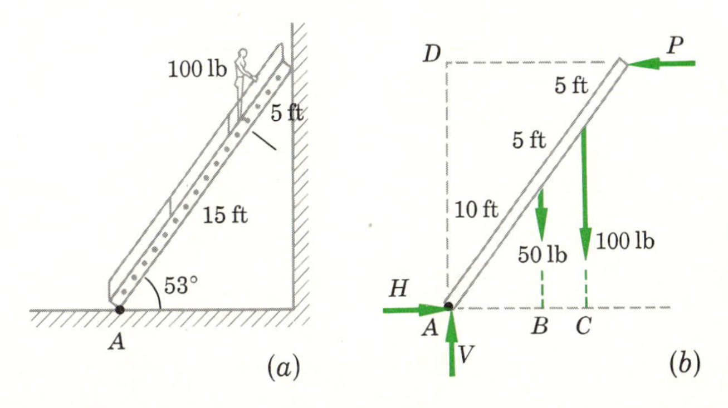

FIGURE 8.11
A 100-lb boy stands on the 50-lb ladder as shown. On the assumption that the wall is smooth, the forces acting on the ladder are as shown in (*b*).

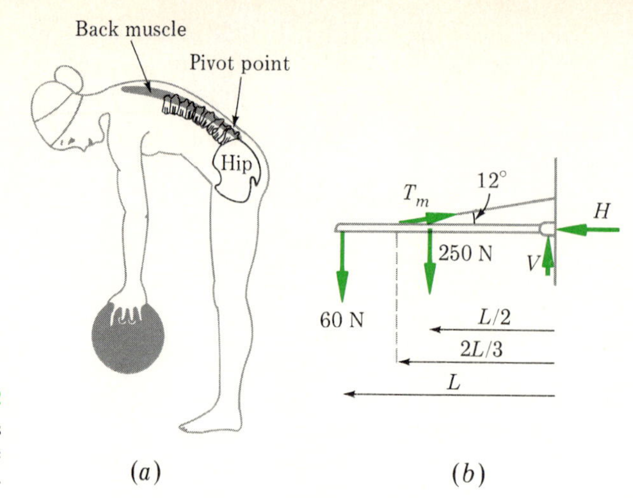

FIGURE 8.12 The forces in the woman's back can be found using the model shown in (*b*).

respectively. Solving the equations simultaneously gives

$$V = 150 \text{ lb} \qquad P = H = 75 \text{ lb}$$

Illustration 8.9 Consider the woman lifting a 60-N bowling ball as shown in Fig. 8.12*a*. Find the tension in her back muscle and the compressional force in her spine. Assume the upper part of her body to weigh 250 N.

Reasoning The horizontal upper portion of her body can be represented as a beam as shown in Fig. 8.12*b*. To a rough approximation, the dimensions can be taken as shown. When the pivot point is taken as shown, the equilibrium equations become

$$\Sigma F_x = 0: \qquad H - T_m \cos 12^\circ = 0$$
$$\Sigma F_y = 0: \qquad V - 60 - 250 = 0$$
$$\Sigma \tau = 0: \qquad (T_m \sin 12^\circ)\left(\frac{2L}{3}\right) - (250)\left(\frac{L}{2}\right) - (60)(L) = 0$$

How do the factors containing T_m arise? Solving these equations yields

$$T_m = 1335 \text{ N} \qquad V = 310 \text{ N} \qquad H = 1305 \text{ N}$$

The compression in the lower portion of the spine will be H. Notice that $H = 1305 \text{ N} \approx 290 \text{ lb}$. The tension in the back muscle is even larger than this. Although she is only lifting 60 N (about 15 lb), the forces are already large. Why should one not lift a very heavy object in the way shown? How should one lift a heavy object?

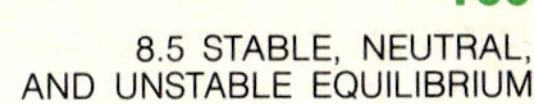

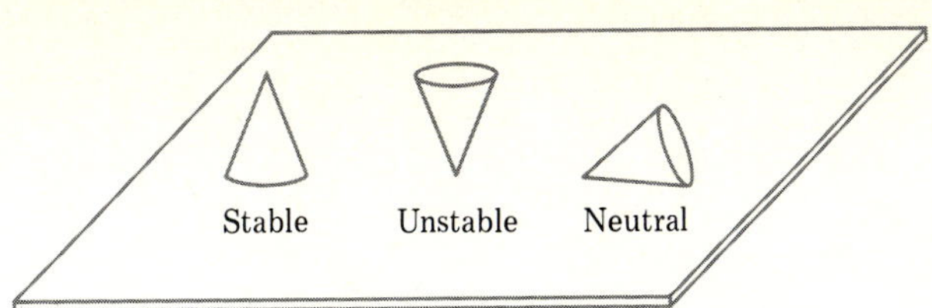

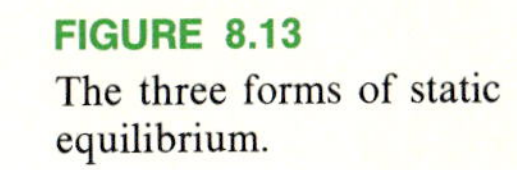
FIGURE 8.13
The three forms of static equilibrium.

8.5 STABLE, NEUTRAL, AND UNSTABLE EQUILIBRIUM

As we have seen, a rigid object is in equilibrium if $\Sigma \mathbf{F} = 0$ and $\Sigma \tau = 0$. However, there are three distinct types of static equilibrium. They are illustrated in Fig. 8.13. It is reasonably obvious how the stable and unstable forms obtain their names. If the stable cone is slightly tilted away from the position shown, it will return to position when released. But the unstable cone will topple over if even slightly tilted. The characteristic which distinguishes the third case, the case of neutral equilibrium, is that the object has no special preference for any one position.

DEFINITION

An object is in **stable equilibrium** *with respect to a slight disturbance if after the cause of the disturbance is removed the object returns to its nondisturbed condition.* For example, in Fig. 8.14*a*, the cone has been slightly disturbed. When released, the cone will fall back to its original position. This return is caused by the torque, due to the weight mg, about the point of contact A. In passing we should also note that the center of gravity of the object was raised during the disturbance. As a result, the stable position has a lower PE than the disturbed position.

DEFINITION

An object is in **unstable equilibrium** *with respect to a slight disturbance if after the cause of the disturbance is removed the object by itself increases the disturbance.* For example, in Fig. 8.14*b*, the cone will continue to fall over if released. It is pulled over by the torque, due to the weight mg, about the point of contact A. Moreover, we should notice that the disturbance lowers the center of gravity and PE for the object in part *b*.

DEFINITION

An object is in **neutral equilibrium** *if it is indifferent to a disturbance.* For example, in part *c*, the cone will remain in any position to which it is rolled. In this case the torque due to mg is zero about the points of contact as axis. The line of the force goes through the contact line. In addition, this type of disturbance neither raises nor lowers the object's center of gravity. The PE of the object is not changed by the disturbance.

This particular physical example is not unique. All rigid bodies behave

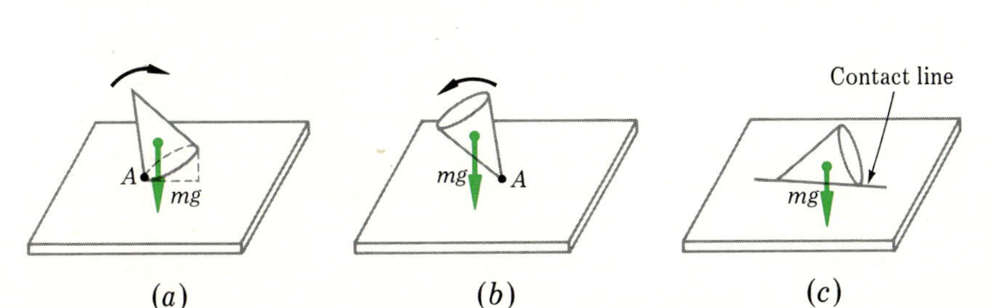

FIGURE 8.14
The weight of the object causes a torque about the point of contact as axis. This torque determines the type of equilibrium that exists.

in a similar way. We can therefore make the following statement in regard to static equilibrium:

Static Equilibrium

An object's equilibrium can be characterized as (1) stable, (2) unstable, or (3) neutral depending upon whether a slight disturbance (1) raises, (2) lowers, or (3) leaves unchanged the PE of the object.

Indeed, we shall see later that this statement applies to other forms of PE in addition to gravitational PE.

8.6 MORE ANALOGIES BETWEEN LINEAR AND ROTATIONAL MOTION

Anyone who has ever spun a wheel on an axle knows that the wheel possesses a quality closely akin to inertia. The fact that a body possesses inertia makes it necessary for us to exert a force upon it if it is to be set in motion. Similarly, if the body is already in motion, a force is required to bring it to rest. These evidences of inertia in linear motion are also easily seen in the rotation of a body about an axis.

A spinning fan blade continues to spin for some time after the fan motor has been shut off. The person who tries to stop the blade with a finger is likely to find that the blade has enough rotational inertia to injure the finger. It is also apparent that the fan blade resists being put into rotation, since it takes several seconds for the motor to get the fan up to speed. These typical evidences of inertia are present to a greater or lesser extent in all rotating objects.

It was found in earlier chapters that the force needed to overcome inertia effects and to set a body in linear motion is expressed by Newton's law

$$F = ma$$

In the last chapter, we saw that the equations of linear motion have analogous rational counterparts. Clearly, for rotational motion, the linear acceleration a in Newton's law should be replaced by the angular acceleration α. We might even guess, since torques cause bodies to rotate, that the unbalanced force F should be replaced by the unbalanced torque τ. This will be seen to be true in a later section.

However, it is not quite so obvious what should be done with m to obtain the rotational form of Newton's law. We shall soon see that it must be replaced by a quantity called the **moment of inertia.** This quantity must be dependent upon mass, since we all know that objects with large mass are more difficult to set into rotation than small masses. However, we shall see that the geometry of the object is also of considerable importance.

Finally, the center of mass of a rotating wheel mounted on a stationary axle has zero velocity, and so the wheel has no kinetic energy of translation. Nor does it possess any translational momentum. Yet we know that a rotating flywheel can do work because of its rotation. For example, it could wind up a rope with a weight hanging from it. Therefore, it does possess kinetic energy, i.e., energy of motion. In the following portions of this chapter, we shall find an expression for this KE of rotation, as well as for the rotational momentum of the rotating object.

8.7 RELATION BETWEEN TORQUE AND ANGULAR ACCELERATION

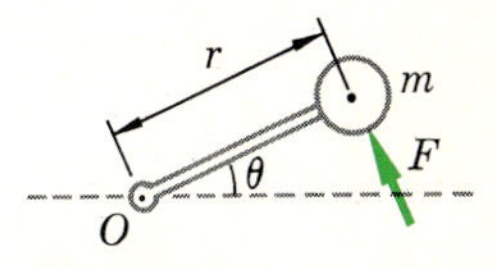

FIGURE 8.15
The force F causes a torque about the axis O and gives the mass m an angular acceleration about the pivot point.

Consider the situation shown in Fig. 8.15. A relatively large mass m is held to an axle at O by a very light rod. Assume that the bearing at the pivot point O is frictionless, and further assume that the mass of the rod is negligible. Let us take the system to be in a horizontal plane so that gravity will cause no torque about the pivot point.

The force F acting as shown on the mass m will not be opposed by any other force, since there is no friction in the pivot. It then is an unbalanced force acting on m, and it will accelerate the mass according to

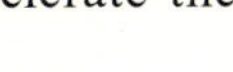

$$F = ma$$

However, in so doing, it will cause the mass to rotate about the pivot point.

Since we saw in the last chapter that the linear acceleration a was related to the angular acceleration α by the equation

$$\alpha r = a$$

we have at once

$$F = mr\alpha$$

Notice that the relation between α and a was true only in radian measure, and so we must measure angles in radians when using this equation.

If the force F in Fig. 8.15 had not been perpendicular to the bar, only the component of F acting on the mass in a direction perpendicular to the bar would cause rotation of the object. This is merely another way of saying that rotations are caused by torques, since a torque is defined to be the product of a force by its lever arm. Hence, it would be better to write our equations for rotations in terms of torques. We may do this by multiplying both sides of the last equation by r, since the torque is just Fr. Thus

$$Fr = \tau = \text{torque} = mr^2\alpha \qquad \alpha \text{ in rad/s}^2 \tag{8.5}$$

Equation (8.5) is one form for the analog to Newton's equation $F = ma$. As we expected, F is replaced by the torque on the system τ, and a is replaced by α, the angular acceleration. In place of the mass m, we have what is called the moment of inertia for this system, mr^2. Notice that the rotational inertia of the system depends not only on m but on r as well. More will be said about this later.

In practice, most rigid rotating bodies have the mass distributed at many different values of r from the pivot point. This is shown in Fig. 8.16a. We can easily modify Eq. (8.5) to apply in this case as well. The only change will be

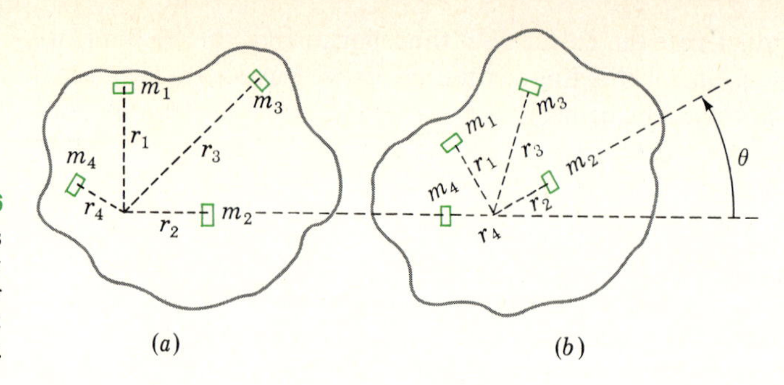

FIGURE 8.16
Each small piece of mass within a large rigid body undergoes the same angular acceleration about the pivot point.

to split the rigid object up into many little pieces at various radii and write Eq. (8.5) for each mass individually.

In Fig. 8.16*a*, several of the small pieces of mass from which the object is constituted are illustrated. The torque about the pivot, acting on the first mass, is the result of all forces acting on that mass, whether due to forces external to the object or to forces exerted by the masses adjacent to the first mass. Let us call this torque τ_1. Applying Eq. (8.5) to the mass, we have

$$\tau_1 = m_1 r_1^2 \alpha_1$$

where r_1 is shown in the figure. A similar equation can be written for any mass in the object. Let us write the sum of these equations, to give

$$\tau_1 + \tau_2 + \cdots + \tau_N = m_1 r_1^2 \alpha_1 + m_2 r_2^2 \alpha_2 + \cdots + m_N r_N^2 \alpha_N \tag{8.6a}$$

where it is assumed there are N masses composing the object. This equation can be written more concisely by abbreviating Eq. (8.6*a*) as follows:

$$\sum_{i=1}^{N} \tau_i = \sum_{i=1}^{N} m_i r_i^2 \alpha_i \tag{8.6b}$$

the symbol $\sum_{i=1}^{N}$ means to take a sum of terms. The first term has i replaced by 1. The second has i replaced by 2, etc. Finally, the last term has i replaced by N.

Since the body under discussion is rigid, each mass within it will rotate about the axis in the same time. As indicated in Fig. 8.16*b*, each mass rotates through the same angle as the body rotates. Consequently, the angular velocity and angular acceleration will be the same for each. Therefore each of the α values in Eq. (8.6*b*) is the same and can be factored out of the sum to give

$$\sum_{i=1}^{N} \tau_i = \alpha \sum_{i=1}^{N} m_i r_i^2 \tag{8.6c}$$

Let us now investigate the meaning of the left-hand side of Eq. (8.6*c*). It is the sum of all the torques acting on the individual masses. In order to express it in terms of the torque exerted by external forces on the body as a whole, we consider the following fact: If we did not wish the object to rotate at all under the action of external torques, we could stop the body from rotating in either one of two ways. We could apply an equal and opposite external torque to the body in order to cancel the original external torque. Or we could apply small torques to each of the masses so as exactly to cancel each of the individual torques acting on them. Therefore we see that an external torque on the body as a whole is equivalent to N small torques applied to the individual masses of which the body is made. As a result, we see that the left-hand side of Eq. (8.6*c*) is equivalent to the external torque τ applied to the object. We can therefore write Eq. (8.6*c*) as

$$\tau = \alpha \sum_{i=1}^{N} m_i r_i^2 \tag{8.6d}$$

DEFINITION

It is customary to define a quantity I, the **moment of inertia** *of the body, as*

Moment of Inertia

$$I \equiv \sum_{i=1}^{N} m_i r_i^2 \tag{8.7}$$

When this is placed in Eq. (8.6*d*), we find

$$\tau = I\alpha \tag{8.8}$$

This equation is the rotational analog of the relation $F = ma$ in the case of extended rigid bodies. We see that torque replaces force and angular acceleration replaces linear acceleration. *The inertial effect corresponding to mass is contained in I, the moment of inertia.* To see the meaning of the moment of inertia more clearly, let us rewrite Eq. (8.7) for a special case. Suppose that the object were split into pieces such that each of the m_i were the same. Then Eq. (8.7) would become

$$I = mr_1^2 + mr_2^2 + mr_3^2 + \cdots + mr_N^2$$

or

$$I = m(r_1^2 + r_2^2 + r_3^2 + \cdots + r_N^2)$$

This may be written

$$I = Nm\frac{r_1^2 + r_2^2 + r_3^2 + \cdots + r_N^2}{N}$$

But Nm is just the total mass of the object M, and the second term is just the average value of the various r^2 values for the masses composing the body. We represent this average square radius by K^2, and *we call K the* **radius of gyration** *of the body. K^2 is an average measure of the square of the distance*

Radius of Gyration

from the pivot point to the various pieces of matter composing the object. In symbols we have

$$I = MK^2$$

From the definition of I we see that it is a measure of the rotational inertia of the object. It is proportional to the mass of the object. The rotational inertia is also proportional to the average square of the distance from the pivot to the pieces of mass comprising the body.

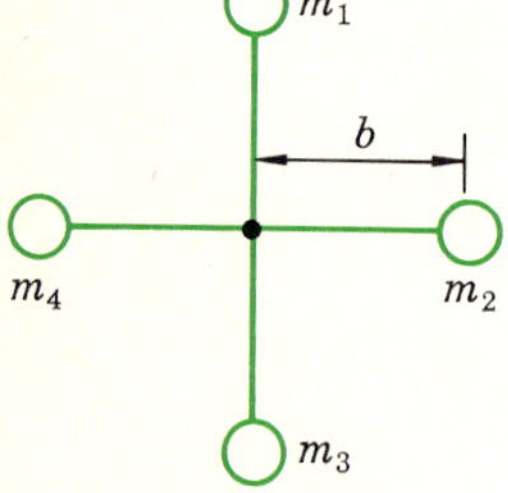

FIGURE 8.17
The radius of gyration of this body for a pivot at its center is just b. The size of the spheres is exaggerated.

8.8 MOMENTS OF INERTIA FOR VARIOUS BODIES

As defined in the last section, *the moment of inertia of a body may be considered to be the product of the mass of the body and the square of an average distance of the mass from the pivot point, or axle.*

In certain cases, this distance, the radius of gyration for a body, is very easily found. For example, in Fig. 8.17 the four masses shown mounted on an essentially massless rigid frame present a rather simple situation. Since $r_1 = r_2 = r_3 = r_4 = b$, the average distance from the pivot point at the center to the various pieces of mass is just b. Hence, for this device, I would be just Mb^2, where M would be the sum of m_1, m_2, m_3, and m_4.

Another interesting case is that of a hoop, or rim. This is shown in Fig. 8.18. Notice once again that all the mass is at a distance b from the pivot point. If m is the mass of the hoop, I would be just mb^2.

TABLE 8.1
MOMENTS OF INERTIA

OBJECT (AXIS AS SHOWN)		I	RADIUS OF GYRATION K
Hoop	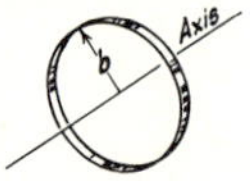	mb^2	b
Solid disk	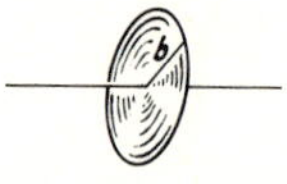	$\frac{1}{2}mb^2$	$b/\sqrt{2}$
Solid sphere	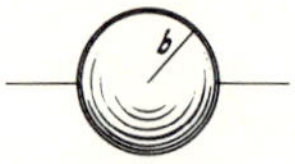	$\frac{2}{5}mb^2$	$b\sqrt{\frac{2}{5}}$
Solid cylinder (radius b)	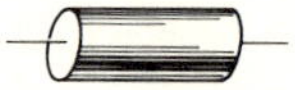	$\frac{1}{2}mb^2$	$b/\sqrt{2}$
Solid cylinder (thin) (length L)	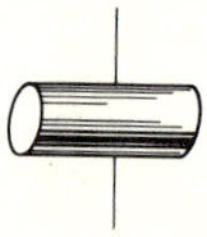	$\frac{1}{12}mL^2$	$L/\sqrt{12}$

There are many other fairly simple cases in which I can be found without too much trouble. Several of these are shown in Table 8.1. Notice in this respect that the gyration radius K is indeed a measure of the average distance of the mass from the axle, or pivot point. For the thin hoop, $K = b$, since the mass is all at a distance b from the axis. However, for a disk, the value of K will be less than b, since only the outermost mass is a distance b from the axis. In this case $K = b/\sqrt{2}$.

The units of I are mass units times the square of distance units. Hence, they will be slug-feet squared, kilogram-meters squared, etc. Also, it should be pointed out again that the use of the relation $a = r\alpha$ in deriving Eq. (8.8) restricts the units used in that equation. The angular acceleration α should be measured in radians per second per second.

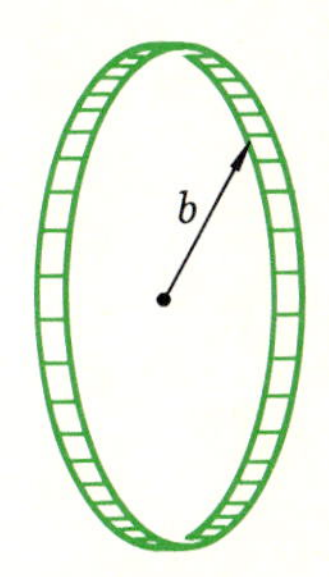

FIGURE 8.18
The radius of gyration of a hoop about its own axis is equal to the radius of the hoop.

Illustration 8.10 The large 80-kg wheel shown in Fig. 8.19 has an actual radius of 25 cm and a radius of gyration of 20 cm. It is driven by the belt shown running to a small pulley on a motor. Find the tension in the belt necessary to uniformly accelerate the wheel to a speed of 2.0 rev/s in 20 s.

Reasoning Before we can apply $\tau = I\alpha$, we must find α from the motion problem. Known:

$$\omega_0 = 0 \qquad t = 20 \text{ s}$$
$$\omega_f = 2.0 \text{ rev/s} = 4\pi \text{ rad/s}$$

We use

$$\omega_f = \omega_0 + \alpha t$$

$$4\pi \text{ rad/s} = (20 \text{ s})(\alpha) \qquad \alpha = \frac{\pi}{5} \text{ rad/s}^2$$

(Notice that we changed to radian measure so that α would be in proper units for the torque equation.)

Now we need I, and this is simply mK^2, where K is the radius of gyration. Hence, $\tau = I\alpha$ becomes

$$\tau = (80 \text{ kg})(0.20 \text{ m})^2(0.20\pi \text{ rad/s}^2) = 0.64\pi \text{ N}\cdot\text{m}$$

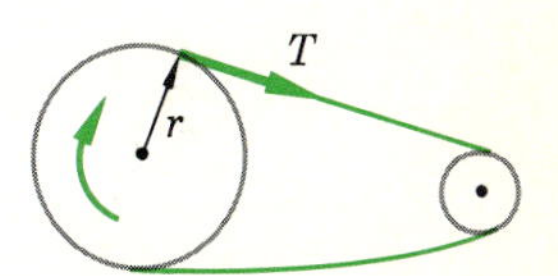

FIGURE 8.19
Angular acceleration is imparted to the large wheel by the torque resulting from the tension T in the upper belt.

However, the torque is force times lever arm. From the figure, the force is the tension T in the belt, and the lever arm is the radius of the wheel. Therefore

$$\tau = (T)(0.25 \text{ m})$$

giving

$$T = 2.56\pi \text{ N} \approx 8.0 \text{ N}$$

Illustration 8.11 The 3-kg object shown in Fig. 8.20 hangs from a rope wound on a 40-kg wheel. The wheel has an actual radius of 0.75 m and a radius of gyration of 0.60 m. Find (*a*) the angular acceleration of the wheel;

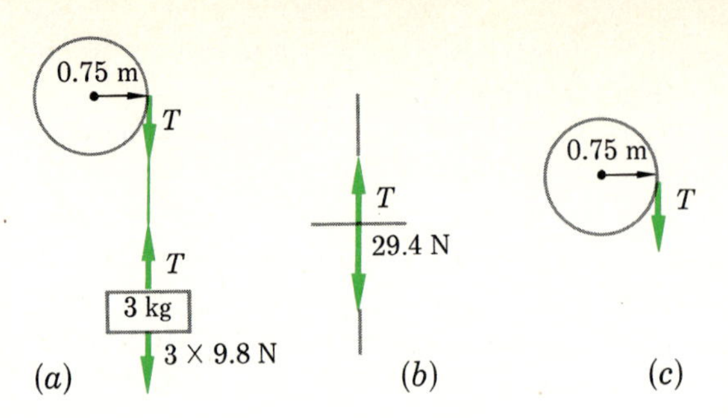

FIGURE 8.20 As the 3-kg object accelerates under the pull of gravity, the tension in the rope imparts an angular acceleration to the wheel.

(*b*) the distance the weight will fall in the first 10 s after it is released. Notice that the 3-kg object weighs 3×9.8 N.

Reasoning Problems involving two bodies must be solved by isolating each in turn.

a. For the body hanging from the rope we have that the unbalanced force acting on it is $29.4 - T$, as shown in Fig. 8.20*b*. Using Newton's law, $F = ma$, we have

$$29.4 - T = 3a \tag{8.9}$$

Isolating the wheel as in part *c* of the figure, we have

$$\tau = I\alpha$$
$$T(0.75) = (40)(0.60)^2\alpha$$

or

$$T = 19.2\alpha \tag{8.10}$$

The general idea is to solve these two equations, (8.10) and (8.9), simultaneously. To do this, we make use of the fact that $a = r\alpha$, which in this case gives $a = 0.75\alpha$. Equation (8.9) then becomes

$$29.4 - T = 2.25\alpha$$

Substitution from Eq. (8.10) gives

$$29.4 = 21.45\alpha$$

from which

$$\alpha = 1.37 \text{ rad/s}^2$$

where the student should carry through the units of the problem and verify the units of the answer.

b. We make use of the usual linear-motion equations together with the fact that

$$a = r\alpha \approx 1.03 \text{ m/s}^2$$

Known:

$$a = 1.03 \text{ m/s}^2 \qquad v_0 = 0 \qquad t = 10 \text{ s}$$

Using

$$s = v_0 t + \tfrac{1}{2}at^2$$

gives

$$s \approx 51.5 \text{ m}$$

If the distance had been measured and were known in a situation such as this, we could have reversed the procedure and computed the moment of inertia of the wheel. This type of experiment is sometimes used to determine moments of inertia and radii of gyration.

8.9 KINETIC ENERGY OF ROTATION

Consider an experiment in which a rope wrapped around a wheel is used to accelerate the wheel from rest. This situation is shown in Fig. 8.21, where the tension in the rope is designated F. We now ask: How much work does the person pulling on the rope do on the wheel as the wheel is set into rotation?

If the rope is pulled out a distance s, the work done on the wheel is clearly

$$\text{Work done} = Fs \tag{8.11}$$

To find s, we must first determine how far the wheel turned during the time of application of the force F. We must therefore compute the angular acceleration α from

$$\tau = I\alpha$$

Substituting Fr for the torque gives

$$Fr = I\alpha \qquad \text{or} \qquad \alpha = \frac{Fr}{I}$$

Now, working an angular-motion problem to find θ, the angle through which the wheel turned, we have

$$2\alpha\theta = \omega_f^2 - \omega_0^2$$

and after substituting for α we have

$$2\frac{Fr}{I}\theta = \omega_f^2 - 0$$

or

$$\theta = \frac{\omega_f^2 I}{2Fr}$$

But since $r\theta = s$, we have in place of Eq. (8.11)

$$\text{Work done} = \tfrac{1}{2}I\omega_f^2$$

FIGURE 8.21

The person pulling on the rope with force F does work as the rope is unwound. This work results in increased rotational KE of the wheel.

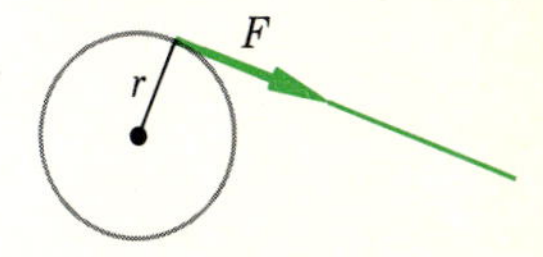

This is a very important relation. It states that the amount of work which was done on the wheel ended up in rotational energy of the wheel and is $\frac{1}{2}I\omega^2$. We

DEFINITION

define the **kinetic energy of rotation** *of the wheel to be* $\frac{1}{2}I\omega^2$.

Total Kinetic Energy

The total KE of an object which is both translating and rotating is $\frac{1}{2}mv^2 + \frac{1}{2}I\omega^2$. The moment of inertia must be about the mass center, and v is the velocity of the mass center.

We therefore have the following two relations for KE, one for translational KE, the other for rotational KE:

$$\begin{aligned} \text{Translational KE} &= \tfrac{1}{2}mv^2 \\ \text{Rotational KE} &= \tfrac{1}{2}I\omega^2 \end{aligned} \tag{8.12}$$

Once again we see that the linear velocity v has been replaced by the angular velocity ω. In addition, the rotational-KE equation is obtained by replacing m by I.

For all objects which can rotate, we see that a rotational KE $= \frac{1}{2}I\omega^2$ *exists due to the rotational motion of the object. As was true for translational KE, rotational KE is interconvertible with work and PE.* This fact is made use of in the following illustration.

Illustration 8.12 The spherical ball of radius r and mass m shown in Fig. 8.22 starts from rest at the top of the incline of height h and rolls down it. How fast is it moving when it reaches the bottom (assume that it rolls smoothly and that friction losses are negligible)?

Reasoning Problems of this sort are most easily solved by use of the work-energy theorem. We have

$$\text{Loss in KE} + \text{loss in PE} = \text{work}_{\text{dec}} - \text{work}_{\text{inc}}$$

In our case this becomes, since only gravitational work is done,

$$\text{Loss in KE of translation} + \text{loss in KE of rotation} + \text{loss in PE} = 0$$

which is

$$(0 - \tfrac{1}{2}mv^2) + (0 - \tfrac{1}{2}I\omega^2) + mgh = 0$$

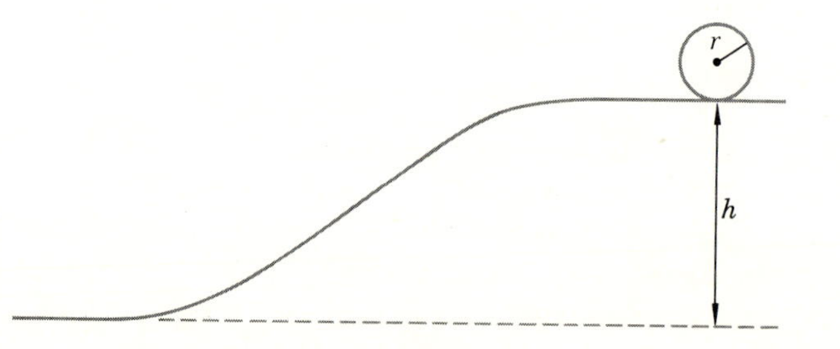

FIGURE 8.22

As the ball rolls to the bottom of the hill, its PE is changed to KE of rotation and translation.

where v and ω are the linear and angular speeds at the bottom. Now, for a sphere, Table 8.1 gives $I = \frac{2}{5}mr^2$, and so

$$mgh = \tfrac{1}{2}mv^2 + \tfrac{1}{5}mr^2\omega^2$$

Notice that the mass cancels and is of no importance in this instance.

If we recall that $v = r\omega$, we have at once that

$$gh = \tfrac{1}{2}v^2 + \tfrac{1}{5}v^2$$

or, after solving for v,

$$v = \sqrt{\frac{10\,gh}{7}}$$

It is interesting to notice that the radius of the sphere cancels out also. Moreover, most of the energy ended up as translational energy, as one sees from the first relation given above,

$$mgh = \tfrac{1}{2}mv^2 + (\tfrac{1}{2})(\tfrac{2}{5})(mv^2)$$

If the rolling object had been a hoop, the factor $\frac{2}{5}$ would be replaced by unity, in which case half the final energy would have been rotational KE.

8.10 CONSERVATION OF ANGULAR MOMENTUM

In view of the many analogies found thus far between linear and rotational phenomena, it should come as no surprise that *linear momentum has a rotational counterpart. Rotational, or angular, momentum is associated with the fact that a rotating object persists in rotating.* As one might expect from the fact that linear momentum is given by mv, *the defining equation for angular momentum is*

DEFINITION

$$\text{Angular momentum} = I\omega \tag{8.13}$$

Angular Momentum

The angular momentum of an object or a system obeys a conservation law much like the one obeyed by linear momentum. It may be stated as follows:

> Unless external torques act on the body or system, its angular momentum will remain constant; the system will continue to rotate about the same axis without change.

Law of Conservation of Angular Momentum

It is called the **law of conservation of angular momentum.** *Notice that the law not only states that ω will be constant for an isolated body; it also states that the orientation of the axis about which it rotates will remain unchanged.* This law governs the operation of stabilized gyroscopes and similar devices. For example, if a large wheel is set rotating about a north-south axis, the wheel will not change its orientation readily unless very large forces are applied to it. When a torque is applied to a rotating system such as this, the resulting motion of the system is interesting since it appears to contradict what one would expect to happen. Although the analysis of these effects is too complicated for us to pursue in this course, the effects themselves are easily demonstrated and your instructor may show you some.

The axis of rotation of an object will not change its orientation unless an external torque causes it to do so. This fact is of great importance for the earth as it circles the sun. No sizable torque is experienced by the earth since the major force on it, the pull of the sun, is a radial force. The earth's axis of rotation therefore remains fixed in direction with reference to the universe around us, i.e., with respect to the distant stars. We can see this behavior in Fig. 8.23.

Notice that the earth's path around the sun is nearly circular. But the rotation axis of the earth is not perpendicular to the plane defined by this orbit. Instead, the axis makes a fixed angle to the plane and, because of the conservation of angular momentum, maintains this orientation as the earth circles the sun. As you can see from the figure, the North Pole of the earth is in continuous daylight during summer (July). In midwinter (January), the North Pole experiences continual darkness. Of course, the reverse is true for the South Pole. The seasons we observe are a less striking example of this same effect.

Many other examples of the conservation of angular momentum can be seen in the universe about us. By use of it, the rotation of planets in their

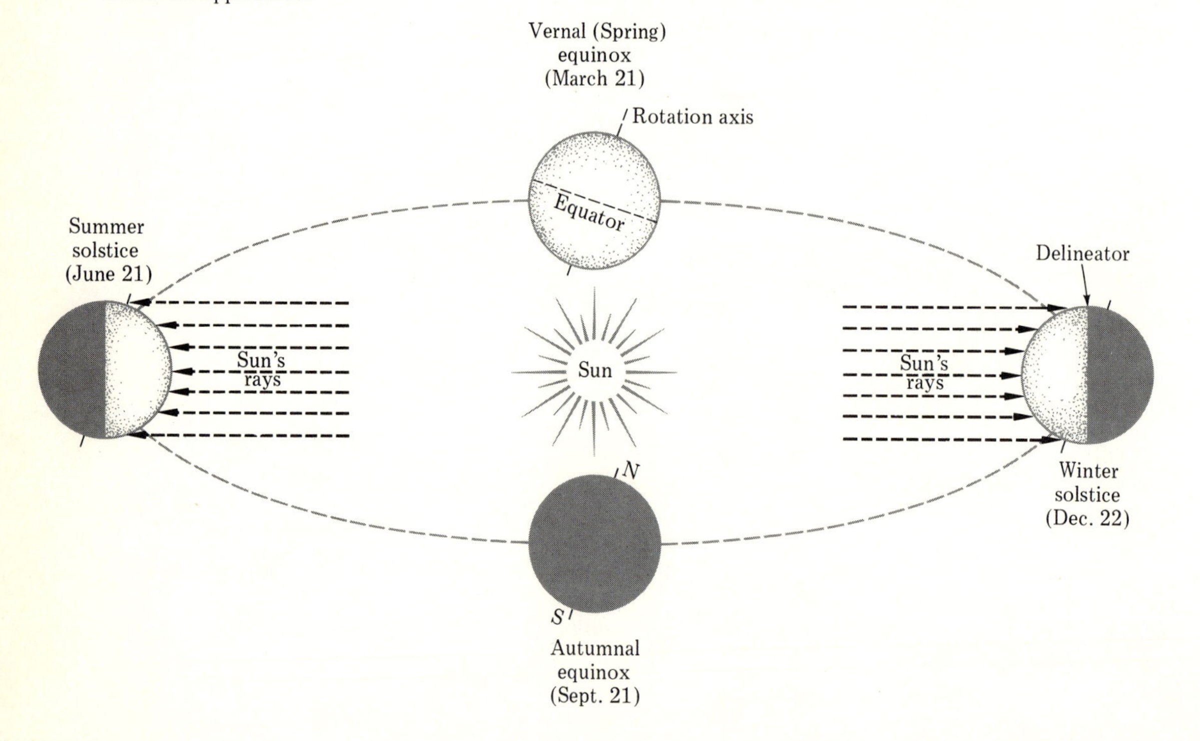

FIGURE 8.23

As an example of the conservation of angular momentum, the earth's rotation axis retains its same orientation relative to the rest of the universe. The dates shown are approximate.

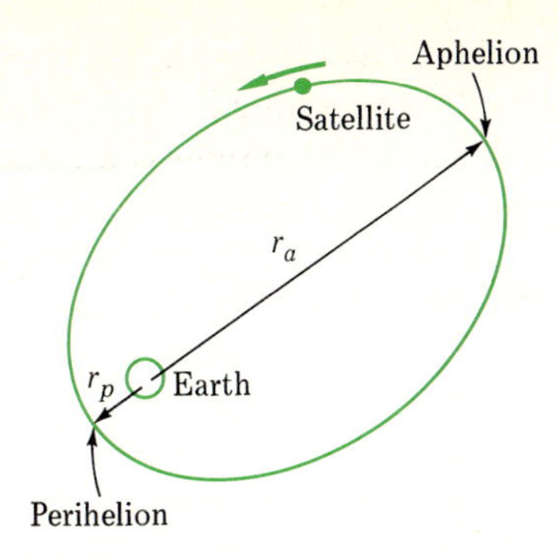

FIGURE 8.24

Find the ratio of the satellite's speed at perihelion to that at aphelion.

orbits and the motions of the stars in the heavens can be predicted. This same law is influential in determining the behavior of atoms in molecules and of electrons in atoms. Its scope is unlimited. It applies to both the smallest and largest objects in the universe.

Illustration 8.13 Consider the earth satellite circling the earth as shown in Fig. 8.24. Find the ratio of its speed at perihelion to that at aphelion.

Reasoning The satellite circles the earth in an ellipse with the center of the earth at one focus of the ellipse. Since the earth's force on the satellite is radial, the angular momentum of the satellite about the earth's center as axis must be conserved. When we denote perihelion and aphelion by subscripts p and a respectively, the conservation of angular momentum tells us that

$$I_p\omega_p = I_a\omega_a$$

But the moment of inertia of a point mass m (the satellite) at a rotation axis r is simply mr^2. Therefore this relation becomes

$$mr_p^2\omega_p = mr_a^2\omega_a$$

This gives

$$\frac{\omega_p}{\omega_a} = \left(\frac{r_a}{r_p}\right)^2$$

We know that $v_T = \omega r$. Since the velocity is tangential at both perihelion and aphelion, we therefore have

$$\frac{v_p/r_p}{v_a/r_a} = \left(\frac{r_a}{r_p}\right)^2$$

which simplifies to

$$\frac{v_p}{v_a} = \frac{r_a}{r_p}$$

Illustration 8.14 The earth rotates on its axis once each day. Suppose that by some process the earth contracts so that its radius is only half as large as at present. How fast would it be rotating then?

Reasoning Let us assume as an approximation that the earth is a uniform-density sphere in each case. The moment of inertia of the earth will change from $I_0 = \frac{2}{5}MR_0^2$ to $I_f = \frac{2}{5}MR_f^2$, where M is the mass of the earth and $R_0 = 2R_f$.

Since no outside torque is applied to the earth during the shrinking process, its angular momentum must be conserved. Hence,

$$I_0\omega_0 = I_f\omega_f$$

where $\omega_0 = 1$ rev/day and ω_f is the final speed of rotation. Putting in the values of I_0 and I_f, we find

$$\frac{2}{5}MR_0^2\omega_0 = \frac{2}{5}M\frac{1}{4}R_0^2\omega_f$$

From this we find

$$\omega_f = 4\omega_0 = 4 \text{ rev/day}$$

In other words, the length of a day would be reduced to 6 h.

SUMMARY

† The center of gravity of an object is the point at which the weight of an object may be considered to act. For nearly all cases, the center of gravity and center of mass may be considered coincident. All the mass of an object may be assumed to be at the center of mass when considering the translation of the object caused by external forces.

† The turning action of a force is measured by the torque τ due to the force. Torque about a given axis point is the force multiplied by its lever arm. The lever arm is the length of a perpendicular dropped from the axis to the line of the force.

† To be in equilibrium, the following conditions must apply to an object:

$$\Sigma F_x = \Sigma F_y = \Sigma F_z = 0 \qquad \text{and} \qquad \Sigma\tau = 0$$

In taking the sum of the torques, counterclockwise torques are positive and clockwise torques are negative. A five-step procedure is used in solving equilibrium problems. The five steps are given in Sec. 8.3.

† When writing the torque equation for an equilibrium situation, the axis may be taken wherever convenient. Usually the axis is taken so that the line of an unknown force passes through it. In this way, the unknown force is made absent from the torque equation.

† Static equilibrium can be characterized as (1) stable, (2) unstable, or (3) neutral depending upon whether a slight disturbance (1) raises, (2) lowers, or (3) leaves the PE of the object unchanged.

Unbalanced external torques give rise to angular accelerations of objects in accordance with the equation $\tau = I\alpha$. The quantity I measures the rotational inertia of the object about the same axis as that taken for the torque. It is called the moment of inertia of the object and is given by $I = \Sigma m_i r_i^2$. Or, in terms of the total mass M of the object, $I = MK^2$, where K is the radius of gyration. Values of I for simple objects are given in Table 8.1.

A rotating object has rotational KE equal to $\frac{1}{2}I\omega^2$. The total KE of an object which is simultaneously rotating and translating is $\frac{1}{2}mv^2 + \frac{1}{2}I\omega^2$. In this expression, both v and I must refer to the mass center.

The angular momentum of a rotating object is $I\omega$. In the absence of external unbalanced torques on the system, the law of conservation of angular momentum states that the angular momentum of the system is constant. Not only will the magnitude of the angular momentum remain unchanged but so will the direction of the rotation axis.

MINIMUM LEARNING GOALS

Upon completion of this chapter, you should be able to do the following:

1.† Point out the center of gravity and center of mass for any uniform, simple object such as a sphere, hoop, rod, cube, etc.
2.† Perform an experiment to locate the center of gravity of any simple, rigid, irregular object.
3.† Locate the lever arm for a given force with reference to a given axis.
4.† Compute the torque about a given axis due to a given force.
5.† State the necessary conditions for equilibrium in both words and by equation.
6.† Solve simple problems involving equilibrium, typical examples being those in Illustrations 8.4 to 8.9.
7.† Point out whether a given object is in stable, unstable, or neutral equilibrium.
8. Give the moment of inertia of a point mass m at a given distance from an axis. State in words how this result is related to the definition of moment of inertia for a complex object.
9. Find one of the following quantities if all the others are given: M, I, K.
10. Write the rotational analog to $F = ma$ and define each quantity in it. Use the relation to solve simple problems in rotational acceleration such as those typified in Illustrations 8.10 and 8.11.
11. Give the formula for rotational KE.
12. Find the total KE of a rolling object provided its radius, speed, and I are given.
13. Solve simple situations involving the work-energy theorem in the case of rotating objects. Typical of these are Illustration 8.11 and Probs. 19 to 21.
14. State the law of conservation of angular momentum. Give the formula for angular momentum. Use the law in simple problems such as those typified by Illustrations 8.13 and 8.14.

IMPORTANT TERMS AND PHRASES

You should be able to define or explain each of the following:

† Center of gravity; center of mass
† Lever arm
† Torque
† Second condition for equilibrium: $\Sigma\tau = 0$
† The position of the axis is arbitrary
† Three types of static equilibrium
Moment of inertia: $I = \Sigma m_i r_i^2$
Radius of gyration: $I = MK^2$
$\tau = I\alpha$
Kinetic energy of rotation: $\frac{1}{2}I\omega^2$
Total KE = $(\text{KE})_{\text{trans}} + (\text{KE})_{\text{rot}}$
Angular momentum: $I\omega$
Law of conservation of angular momentum

QUESTIONS AND GUESSTIMATES

1.† Suppose you are given some string, a meterstick, a 1-kg mass, and an unknown mass of the order of a few kilograms. How could you use the principles of this chapter to evaluate the unknown mass?
2.† There is a right and a wrong way to lift a heavy weight. Refer to Fig. 8.12 and use it to explain why the method shown there is the wrong way. What is the right way? Explain.
3.† Using the diagram and data in Fig. 8.10, explain why it is easier to hold a pail of water from your arm hanging at your side than when your arm is outstretched horizontally.
4.† Hold your body rigid with your feet together and try to lean forward at an angle θ to the vertical. Notice how small θ must be if you are to retain your balance. Explain what limits the size that the angle can be. Why are people with big feet able to lean farthest?
5.† We are told that slender people are less apt to have back trouble than obese people. Explain how this factor influences posture, muscle strain, and similar factors.
6.† As shown in Fig. P8.1, a large number of books are piled on top of each other at uniform offsets. Discuss the conditions which influence whether or not the pile will topple over.

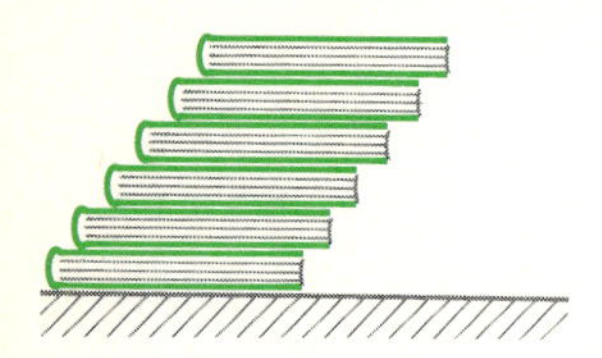

FIGURE P8.1

7.† You wish to mow a steep hill using a riding lawn mower or tractor. Sketch the situation for the mower going up the hill. Under what conditions will this be safe? Repeat for going down the hill. Repeat for riding sideways along the hill. In answering this, you will have to estimate the position of the center of gravity of mower and rider. Why?
8.† A common trick is to prop open a door by placing a wedge of wood in the crack next to the hinge. Why does this usually ruin the hinge?
9. Which bicycle wheel would be easier to stop rotating about its axle, one with a tire filled with air or one filled with water? Explain.
10. Which has the larger speed after rolling down the same incline, a sphere or a hoop of equal mass and radius? Explain. Discuss the case of a hoop and a disk of identical radius and mass.
11. Suppose that the sun's attraction for the earth suddenly doubled. What can you say about the rate of rotation and orbit of the earth about the sun?
12. Suppose that an internal explosion suddenly opened a huge cavity in the earth by pushing its surface outward. How would this affect the rotation of the earth about its axis and about the sun?
13. We maintain that angular momentum is conserved. Isn't this contradicted by the fact that almost all rotating objects eventually slow and stop?

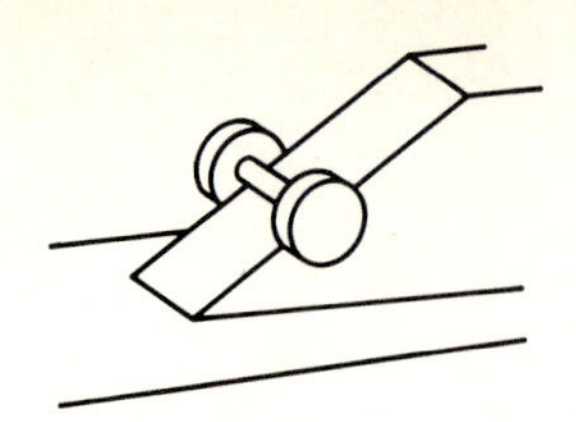

FIGURE P8.2

14. The spool shown rolling down the incline in Fig. P8.2 takes off with greatly increased translational motion as soon as the large-diameter disks on the sides of the spool touch the floor. Explain why, paying particular attention to the way the KE is apportioned.
15. The spool shown in Fig. P8.3 moves to the left when the cord is in position 1 and to the right when the cord is in position 2. Why? What happens when the cord is in position 3?
16. When a car rolls down a hill, about what fraction of its total KE is the rotational energy of the wheels? (E)
17. An automobile is facing east on an east-west highway. It accelerates from rest to its top speed. Is the rotation of the earth slowed or speeded up by this? Estimate the fractional change in the earth's rotation speed. (E)

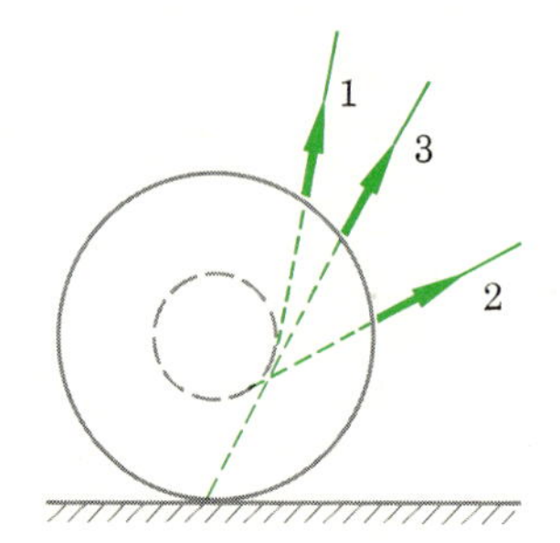

FIGURE P8.3

PROBLEMS

1.† Refer to Fig. P8.4. (*a*) Find the lever arm for the force F in each case if point A is taken as axis. (*b*) What is the torque caused by F about A in each case?
2.† The two vertical ropes shown in Fig. P8.5 support a uniform 50-lb plank and the weights, as shown. If T_1 is 80 lb and W_2 is 100 lb, find W_1 and T_2.
3.† The uniform 50-lb plank of Fig. P8.5 is supported by the two ropes shown. If the ropes supporting the plank can each withstand a tension of only 200 lb, and if W_2 is to be twice as heavy as W_1, what is the largest value W_1 can have? (Assume the ropes holding W_1 and W_2 are very strong. *Caution:* T_1 is not equal to T_2. Which will be 200 lb?)
4.† In order to determine the position of a person's center of gravity, the person is placed on two scales as shown in Fig. P8.6. Suppose the scale on the left reads 260 N while the one on the right reads 200 N. Find the distance x indicated to the center of gravity in terms of

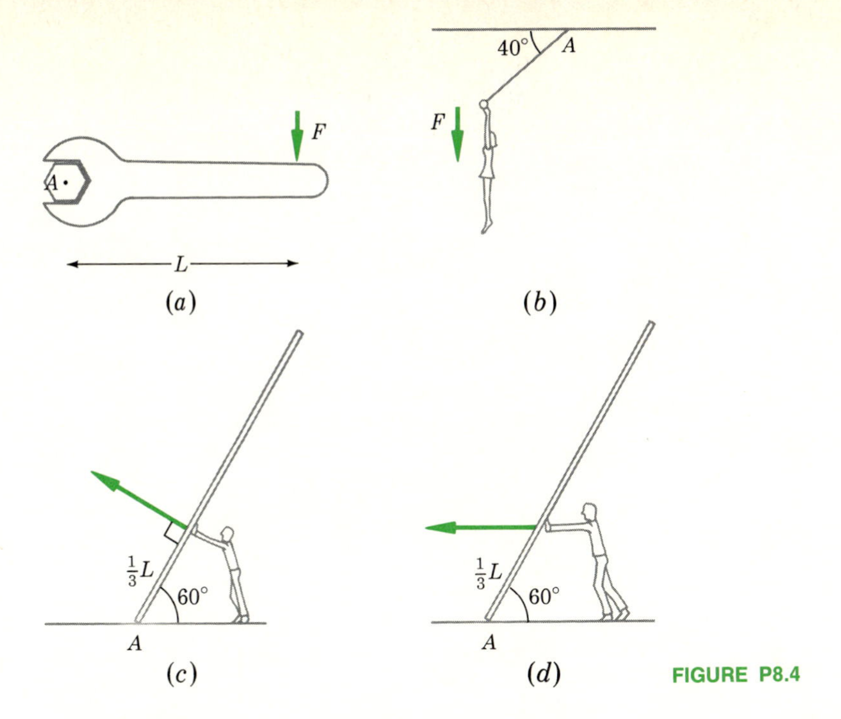

FIGURE P8.4

L. Assume all scale readings have been corrected by subtracting off the scale readings when the woman was not in place.

5.† In Fig. P8.7 the beam is uniform and weighs 200 N. Find (*a*) the tension in the rope and (*b*) the H and V component forces exerted by the pin if $W = 800$ N.

6.† The uniform 500-N beam shown in Fig. P8.8 supports a load, as shown. (*a*) How large can the load be if the horizontal rope is able to hold 2000 N? (*b*) What are the components of the force at the base of the beam?

7.†* When one stands on tiptoe, the situation is much like that shown in Fig. P8.9. We can replace the actual situation by the model shown in part *b*. The force F, the push of the floor, will be the person's weight if the person is standing on one foot. Find (*a*) the tension in the Achilles tendon and (*b*) the forces H and V at the ankle in terms of F for the situation shown.

8.† The uniform board in Fig. P8.10 weighs 120 N and supports a weight $W = 700$ N. Find the tensions in the three supporting ropes.

9.†* In Fig. P8.11, the beam weighs 750 N, and the value of T_3 is 870 N. Find T_1, T_2, W, and the force with which the beam pushes down on the frictionless pin at its base.

10. What is the moment of inertia of the earth with respect to the sun as axis? ($M_{earth} = 6 \times 10^{24}$ kg, $M_{sun} = 2 \times 10^{30}$ kg, and earth-sun distance $= 1.5 \times 10^{11}$ m.)

11. The four point masses shown in Fig. P8.12 are attached rigidly to a bar of negligible mass. (*a*) Find the moment of inertia of the system with respect to the axis shown. (*b*) Repeat for a similar axis through point A, midway between m_1 and m_2. (*c*) What is the radius of

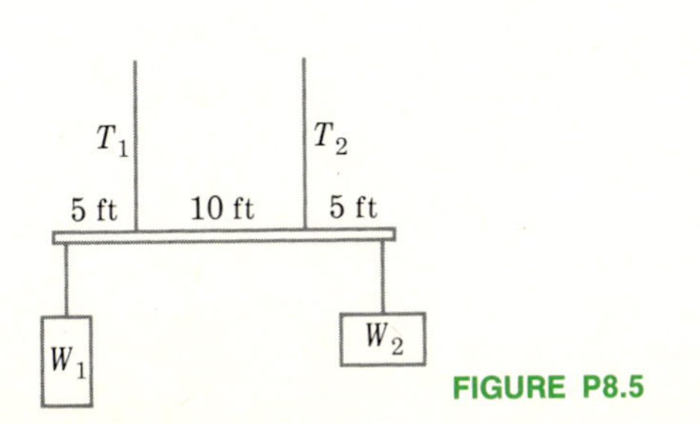

FIGURE P8.5

FIGURE P8.6

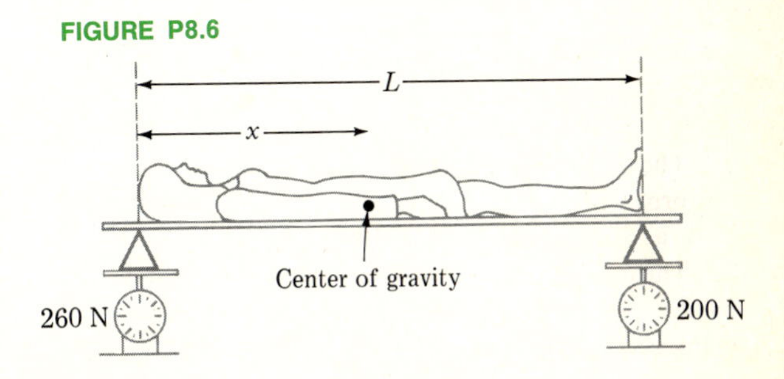

gyration in each case? Use $m_1 = 0.50$ kg, $m_2 = 2.0$ kg, $b = 0.50$ m.

12.* Two hoops are mounted on a frame of negligible mass as indicated in Fig. P8.13. The inner hoop has a mass M_1 and a radius a, and the outer hoop has a mass M_2 and a radius b. Find the moment of inertia and radius of gyration for rotation about an axis through the center perpendicular to the page if $M_2 = \frac{1}{2}M_1$ and $b = 2a$.

13. Find the moment of inertia of a solid disk having a radius of 10 cm and a mass of 200 g.

14. A hooplike wheel weighs 240 lb and has a radius of 2.0 ft. Find its approximate moment of inertia.

15. How large a torque is required to give an angular acceleration of 4.0 rad/s² to a wheel having a moment of inertia of 0.20 kg · m²?

16. A wheel having $I = 0.20$ kg · m² is spinning with a speed of $120/\pi$ rev/min when the power source is shut off. It coasts uniformly to rest in 100 s. How large was the average torque which stopped it?

17. (*a*) How large a torque is required to accelerate a 160-lb wheel having a $\frac{1}{2}$-ft radius of gyration from rest to a speed of $1/\pi$ rev/s in 20 s? (*b*) How far will the wheel turn in this time?

18. (*a*) How long will it take to accelerate a 50-kg 30-cm-radius solid disk (rotating about its usual axis) from rest to 4.0 rad/s, provided a force of 3.0 N acts

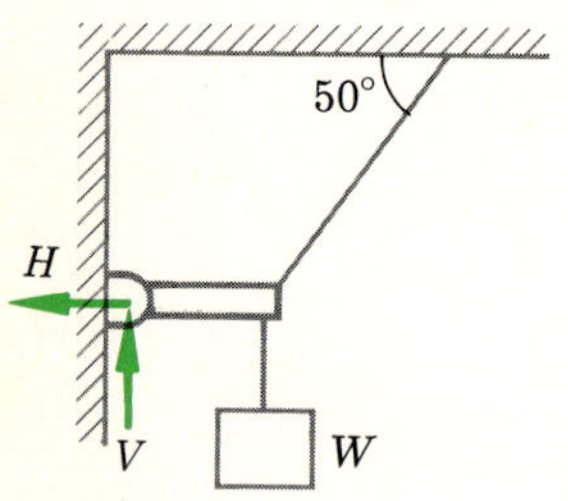

FIGURE P8.7

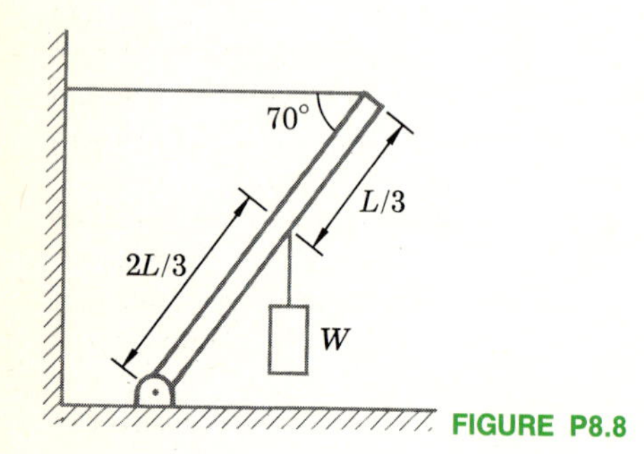

FIGURE P8.8

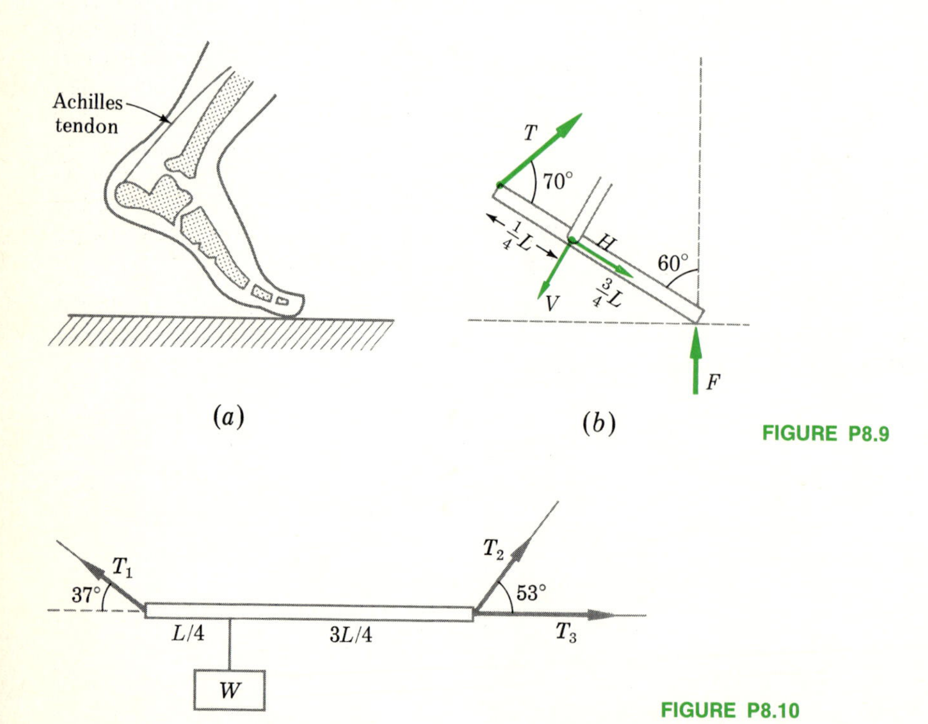

FIGURE P8.9

FIGURE P8.10

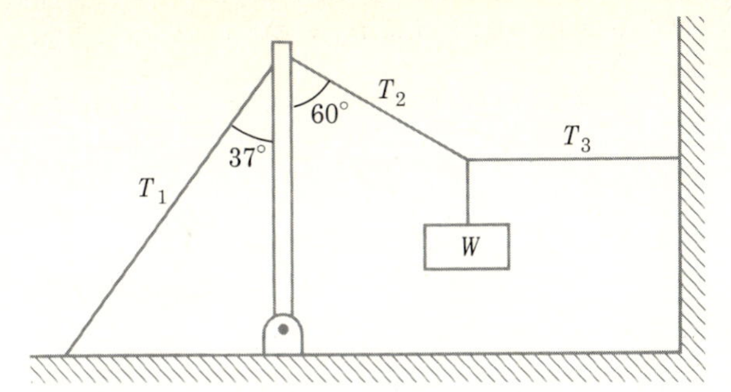

FIGURE P8.11

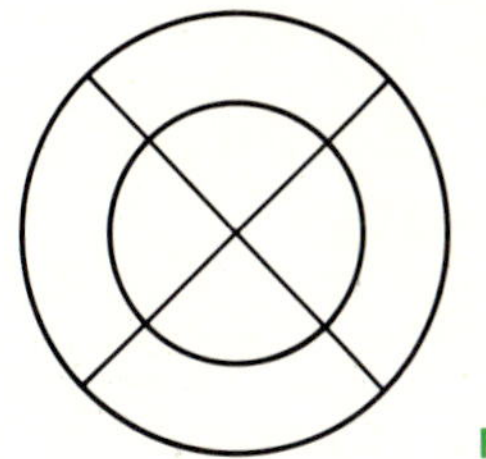

FIGURE P8.13

tangential to its rim? (*b*) Through how many revolutions will it rotate in this time?

19. Refer to Fig. 8.20. Suppose in a similar situation the falling weight is 16 lb and the outer radius of the wheel is 3.0 ft. It is found that the weight drops 4.0 ft in the first 20 s after being released. Find (*a*) the moment of inertia of the wheel and (*b*) the tension in the cord.

20. A hoop has a radius of 5.0 cm. It starts from rest and rolls down a slope. (*a*) What is its linear speed when it reaches a point which is 30 cm lower than its starting point? (*b*) How fast is it rotating in revolutions per second at that time?

21. Repeat Prob. 20 if the object is a wheel which has a radius of gyration of 4.0 cm and an actual radius of 5.0 cm.

22. As shown in Fig. P8.14, two identical small balls, each of mass 3.0 kg, are fastened to the ends of a light metal rod 1.00 m long. The rod is pivoted at its center point and is rotating with a speed of 7.0 rev/s. An internal mechanism is capable of moving the balls in toward the pivot. (*a*) Find the moment of inertia of the original device. (*b*) If the balls are suddenly moved in until they are 25 cm from the pivot, what is the new speed of rotation?

23.* Refer back to Fig. P8.12. Prove that the moment of inertia for rotation about an axis parallel to the one shown but a distance d away from it is given by $I = Md^2 + I_0$, where M is the total mass and I_0 is the moment of inertia about an axis through the center of mass. Neglect the mass of the bar. (This relation is a general one and is called the **parallel-axis theorem.**)

24.* A solid wheel rotating about an axis through its center has a moment of inertia I_0. Two identical masses are placed at opposite points on the rim of the wheel at opposite ends of a diameter. Calling the wheel radius a, and each mass M, show that the wheel now has a moment of inertia given by $I = I_0 + 2Ma^2$.

25.** A children's merry-go-round in a park consists of an essentially uniform 150-kg solid disk rotating about a vertical axis. The radius of the disk is 6.0 m, and a 90-kg teacher is standing on it at its outer edge when it is rotating at a speed of 0.20 rev/s. How fast will the disk be rotating if the teacher walks 4.0 m toward the center along a radius? *Hint:* Momentum must be conserved.

26.** Suppose that the merry-go-round described in Prob. 25 has no one on it but is moving at a speed of 0.20 rev/s. If a 90-kg teacher quickly sits down on the edge of it, what will be its new speed?

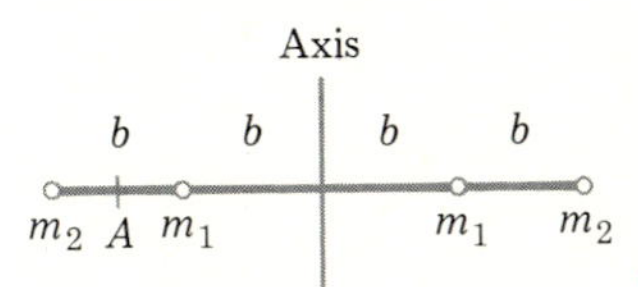

FIGURE P8.12

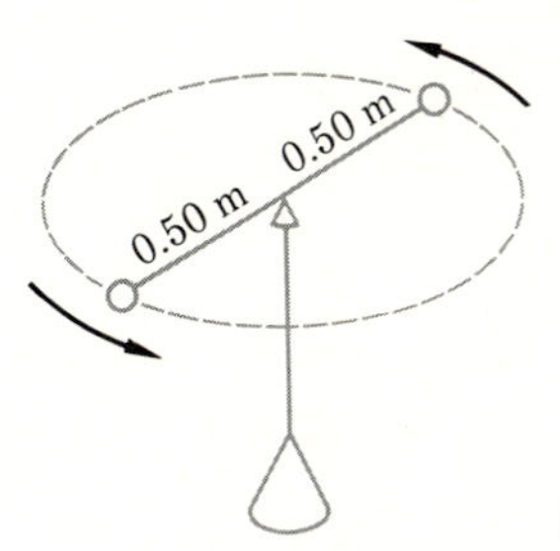

FIGURE P8.14

MECHANICAL PROPERTIES OF MATTER

The action of forces and torques on rigid bodies has been discussed in the past several chapters. We now wish to investigate some of the properties of bodies and materials which deform or flow when subjected to forces. This we do in the present chapter. In addition, an introduction to the static and dynamic properties of liquids is presented.

9.1 THE THREE STATES OF MATTER

We are all familiar with the three general classifications of matter: solids, liquids, and gases. In gases the molecules are essentially free from each other and travel around through space, colliding with each other but not sticking together. Their individual collisions with the walls of a container in which they are placed give rise to a pressure on the walls. The pressure of a gas was computed in Chap. 6 and was found to depend directly upon the KE of the molecules. The KE of the extremely light gas molecules is large enough for the difference in PE of a molecule at the top and bottom of any reasonable container to be negligible in comparison with the KE. As a result the *gas molecules fill the entire container in which they are placed.*

The situation is quite different in a liquid, however. Although we know the molecules in a liquid are in continuous motion, they do not possess enough KE to overcome the attractive forces of the neighbor molecules. They are unable to break loose completely from each other, and they therefore exist as a fluid aggregate of molecules. From time to time, a molecule may obtain enough KE to tear itself loose from the surface of the liquid. This is what happens when molecules evaporate from the surface of a substance. More will be said about this in a later chapter. *Liquids and gases both undergo flow quite easily. They are therefore grouped together in a classification called* **fluids.** *Using that classification system, there are two groups of substances, fluids and solids.*

Solids are much like liquids in many ways. The chief difference is largely one of degree rather than kind. *While the molecules in a liquid are still able to slip by their neighbors, the molecules in a solid are essentially locked in place.* The forces between molecules in a solid are so strong that, to a good approximation, each molecule is held tightly in place by its neighbors. Hence, the substance is hard and rigid, since the molecules are unable to slide past each other. This may be seen diagrammatically in Fig. 9.1.

If the molecules such as A and B in the figure are tightly bound together, they will not easily be torn apart. The forces shown in the figure will be unable to separate the molecules, and the material will not flow under their action. However, if the attractive forces between molecules A and B are relatively small, the applied forces will cause the molecular layers to slide past each other. In this case the material will be fluid.

In certain cases it is difficult to say whether a substance should be called a solid or a liquid. For example, if molasses is cooled, it becomes very "thick," or viscous (hence the phrase "slower than molasses in January"). At still lower temperatures it becomes quite hard. Yet is it a solid, or is it merely a very viscous liquid?

Another example is window glass. Modern window glass is harder than that used many years ago. It is not uncommon to find a very old window pane that is thicker at the bottom than at the top. It has obviously flowed ever so slowly during the passage of the years. Many plastics behave in this way also. It is clear from these examples that *the border line between solids and liquids is not a sharp one.* We shall see in the next section that there are actually two types of solids, one being merely a very viscous liquid.

FIGURE 9.1
The ease with which the molecular layers will flow over each other under the action of the forces shown depends upon the attractive force between molecules such as A and B.

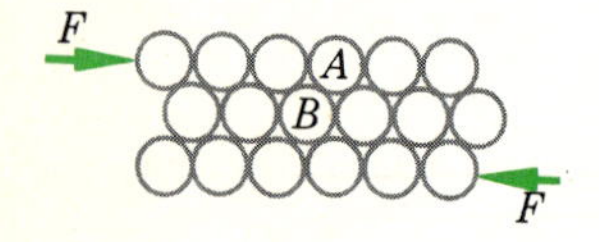

9.2 CRYSTALLINE AND GLASSY SOLIDS

DEFINITION

A **crystal** *is a collection of atoms or molecules in which each molecule is precisely placed in a definite pattern with respect to its neighbors. This pattern is repeated over and over again throughout the crystal.* An example of this would be the sodium chloride crystal (table salt), a small portion of which is pictured in Fig. 9.2*a*.

Notice that the sodium and chloride atoms are placed in perfect geometric order on the corners of a cube. The whole crystal consists of a multitude of these cubes packed in essentially perfect order. A sodium atom anywhere in the crystal has its neighbor atoms arranged exactly the same as every other sodium atom does.

Many crystals do not have as simple a structure as the cubic-lattice form of the sodium chloride crystal. However, in any crystal there is a definite, precise arrangement pattern of the atoms. This pattern is repeated over and over, much in the way that the pattern in many types of wallpaper repeats itself. In the crystal, though, the pattern is three-dimensional rather than existing only in a plane. Other typical examples are shown in Fig. 9.2.

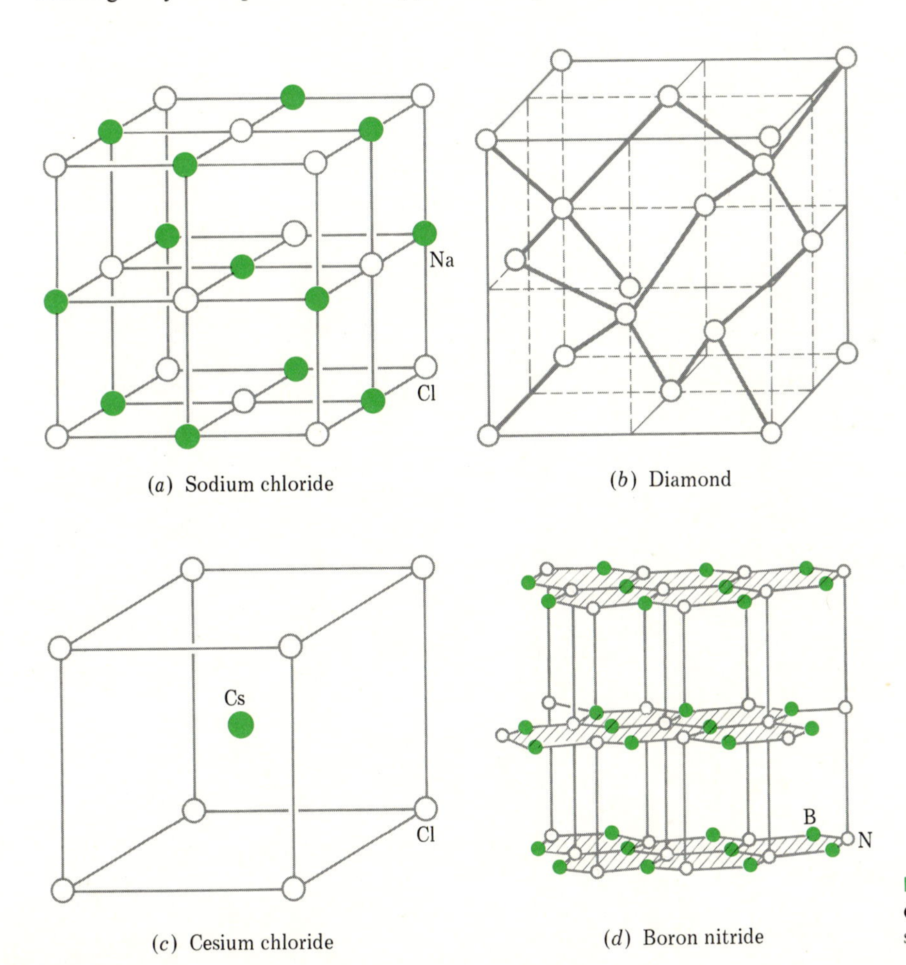

FIGURE 9.2 Crystal lattices of several solids.

In highly viscous liquids, so viscous that they are hard and brittle, the situation is much different. The molecules are no more ordered than they are in the molten liquid. Some slight amount of order may exist in the neighborhood of a given molecule, but this order does not persist throughout the liquid. In them, the molecules are locked in place but are not arranged in any precise pattern. *Supercooled liquids are called* **glasses.**

DEFINITION

Although most window-type glasses are noncrystalline, supercooled liquids, there are many other substances which can exist as a glass. Many plastics are amorphous, i.e., noncrystalline, glasses. However, some plastics, such as polyethylene, are partly crystalline and partly amorphous. It is often difficult to determine whether a given solid material is crystalline or amorphous. Usually this is best decided using x-rays in a manner which will be briefly touched upon later in this book.

9.3 DENSITY AND SPECIFIC GRAVITY

There are many ways in which a substance can be characterized. One of the most useful tells what quantity of substance occupies a unit of volume. This method of characterization is in terms of the **density** of the substance.

Mass Density

The mass density of a material is defined to be the mass of a unit volume of the material. Hence mass density = mass/volume, or in symbols

$$d = \frac{m}{V} \tag{9.1}$$

Clearly, the units of mass density are grams per cubic centimeter, kilograms per cubic meter, etc. The concept of mass density is seldom employed in the British system of units.

Weight Density

It is common in the British system to make use of the weight density rather than the mass density. The weight density is defined to be the weight of a unit volume of material, or, in equation form,

$$D = \frac{W}{V} \tag{9.2}$$

It is commonly expressed in units of pounds per cubic foot. Other units are possible, of course, but they are not ordinarily used. Since $W = mg$, we see that

$$D = dg$$

Many times the same symbol is used for weight and mass density. Which quantity is meant must be determined from the context. Instead of D or d, another symbol frequently used to represent mass density is the Greek letter rho, ρ.

Many materials expand as they are heated. This is a result of the fact that the molecules are vibrating through larger distances at the higher

TABLE 9.1
DENSITIES OF SUBSTANCES

SUBSTANCE	TEMPERATURE, °C	*d*, kg/m³	*D*, lb/ft³
Air (normal pressure)	0.0	1.29	0.0805
Benzene	20.0	879	56.1
Water	20.0	998	62.3
Water	3.98	1,000	62.4
Bone	20.0	~1,800	106–125
Aluminum	20.0	2,700	168
Iron	20.0	7,860	491
Copper	20.0	8,920	557
Lead	20.0	11,340	708
Mercury	0.0	13,600	849

temperature and hence are holding each other farther apart. On the average, since the mass in a unit volume will change if the molecules move farther apart, the density of a substance will change with temperature. Although the densities of most substances decrease with increasing temperature, there are several common exceptions in which the density actually increases as the temperature is raised through a certain temperature range. Water in the range 0 to 4.0°C is such a substance. A representative list of densities is given in Table 9.1. In using these data it is convenient to know that

$$1\ \text{kg/m}^3 = 10^{-3}\ \text{g/cm}^3$$

DEFINITION

The **specific gravity,** *or* **relative density,** *of a substance is defined to be the ratio of the density of the substance to the density of water.* A temperature at which this ratio is to be taken must be specified. One usually takes the ratio at 3.98°C, where the density of water is 1000 kg/m³. In this case, the specific gravity is equal numerically to the mass density.

9.4 HOOKE'S LAW

Another way of characterizing a material is in terms of its deformability. Two major types of deformation occur. In one, the substance flows under the action of a force. This behavior is characteristic of fluids. The other type of deformation, which is only temporary, is elastic in nature, like the stretching of a spring. When the deforming force is removed, the deformation returns to zero. Let us now examine this latter type of deformation.

FIGURE 9.3
The amount ΔL the bar stretches under the load W is proportional to the load as well as to L_0, provided that Hooke's law is satisfied.

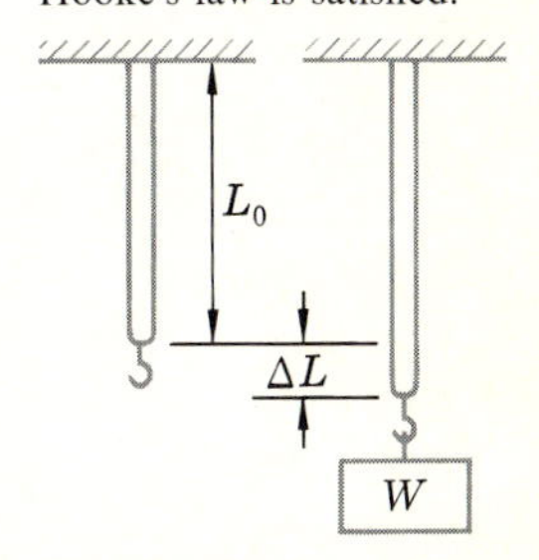

If a rigid bar such as that shown in Fig. 9.3 is subjected to a stretching force by hanging a weight on it, the bar will stretch a distance ΔL. For most solid materials, a graph relating applied load F to elongation ΔL will appear much like Fig. 9.4. In other words, if the applied load is doubled, the amount of stretch ΔL will double. Expressed as an equation, this becomes

$$F = (\text{const})(\Delta L) \tag{9.3}$$

where F is the applied force. In addition, if the weight is not too large, the bar will return to its original length when the load is removed. The bar is said to be **elastic** in this range of loads. Equation (9.3) is a statement of **Hooke's law:**

Hooke's Law

The distortion is proportional to the distorting force.

We further emphasize that *in the elastic range, the distortion returns to zero when the distorting force is removed.*

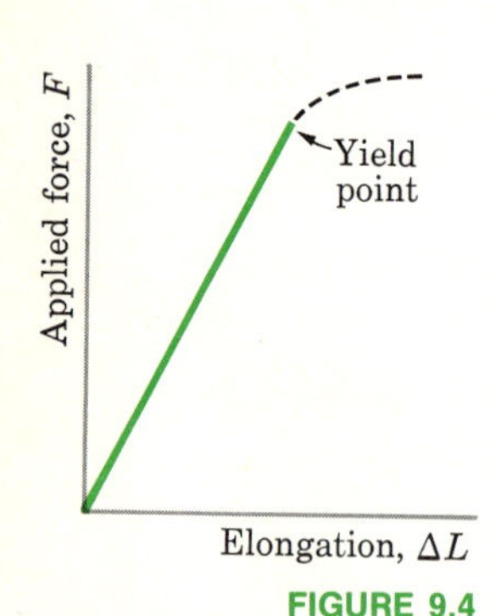

FIGURE 9.4
A typical stress-strain curve. Hooke's law applies in the linear region.

However, if too large a load is applied to the bar, the graph shown in Fig. 9.4 will begin to deviate from a straight line—perhaps as shown by the dashed curve. Moreover, the bar will not usually return completely to its original length if the load is now removed. We say that the bar has been stretched beyond its **elastic limit** in this case. The bar may even break under the tension if the load is increased much beyond the elastic limit.

In terms of the molecules or atoms in the bar, the applied stretching, or tensile, force is merely pulling them apart. In Fig. 9.5, the tensile force has separated the atoms slightly in the direction of the force. These very small separations between the atoms, when added up along the length of the bar, give rise to the observed elongation of the bar. As long as these separations between atoms are kept small, the separation distance is proportional to the applied stretching force and Hooke's law, Eq. (9.3), applies. However, the range of validity of Hooke's law varies widely from substance to substance and is usually difficult to predict without actually plotting a graph such as shown in Fig. 9.4.

FIGURE 9.5
A tensile force on a bar tends to separate the molecules in the way shown.

9.5 STRESS AND STRAIN

Hooke's law can be stated in a more useful form. The statement can be made applicable to many different situations if we make the statement in terms of stress and strain.

The terms stress and strain have precise meaning in physics. Although we often employ such phrases as the "stresses of everyday life" and the "strain of taking a test," such uses are colloquial. They do, however, bear some relation to the exact definition of the words as used by scientists, as we shall now see.

DEFINITION

Stress *is defined to be the applied force per unit of area to which the force is applied.* As an example, consider the rod shown in Fig. 9.6. Here a tensile (or stretching) force F is applied to the end surface of the bar. The cross-sectional area of the bar is A, as indicated. We define the stress on the bar to be

Stress

$$\text{Stress} = \frac{\text{force}}{\text{area}} = \frac{F}{A} \tag{9.4}$$

Notice in particular that the area involved is not the area of the outside of the bar but only of the cross section.

The units of stress are typically pounds per square foot, pounds per square inch, newtons per square centimeter, etc. You may recall that pressure also has these same units. It is seen that stress and pressure are both defined as force per unit area. Hence, the pressure on the wall of a container is merely a compressive force per unit area on the wall, rather than a stretching stress as shown in Fig. 9.6. We shall see examples of other types of stresses with different geometries in later sections of this chapter.

DEFINITION

We define the **strain** of an object to be *the distortion of the object divided by the original dimension before distortion.* For example, the *strain* in the bar of Fig. 9.6 is a measure of how much the bar has been stretched. The elongation ΔL of the bar under the applied force is not a good measure for the strain in the bar. This follows from the fact that a bar twice as long as the one shown would elongate twice as much, since the bar would have twice as many atom separations adding to give the observed elongation. Hence, ΔL varies in proportion to the length of the bar for a given stress. To remove this dependence on length, the strain is actually defined to be the elongation of a bar having original length of 1 unit. In equation form

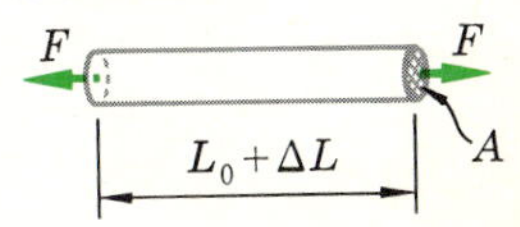

FIGURE 9.6
The stress is F/A, and the strain is $\Delta L/L_0$.

Strain

$$\text{Strain} = \frac{\text{elongation}}{\text{original length}} = \frac{\Delta L}{L_0} \tag{9.5}$$

There are no units for the strain; since it is simply the ratio of two lengths, the units cancel.

We shall see in later sections that there are many types of strain, depending upon the geometry of the system. In the present case we are speaking about tensile strain. If the bar were being compressed, the strain would be the ratio of the decrease in length of the bar to its original length. We are now able to restate Hooke's law. Since the law states that $F \propto$ distortion, we can replace F by the stress since the two are proportional. Similarly, we can replace the distortion by the strain. Then, Hooke's law becomes

Hooke's Law

$$\text{Stress} \propto \text{strain}$$

In this form, the law may be applied to many situations other than the stretching of a bar. For example, Hooke originally proved its applicability to the stretching, bending, and twisting of numerous springs and other objects.

9.6 THE CONCEPT OF MODULUS

If a large stress is needed to give a small strain to an object, the object is hard and rigid. *We measure this quality of hardness, or rigidity, by the* **modulus** *of the material.* There are several different kinds of moduli, depending upon the exact way the material is being stretched, bent, or otherwise distorted. Several of these moduli will be discussed in this section. They are all defined by the following relation for special geometries:

Modulus

$$\text{Modulus} = \frac{\text{stress}}{\text{strain}}$$

Since strain has no units, the modulus has the units of the stress.

TABLE 9.2
YOUNG'S MODULUS Y AND BULK COMPRESSIBILITY k

	Young's Modulus*		Compressibility	
MATERIAL	10^{10} **N/m²**	10^{6} **lb/in²**	10^{-11} **m²/N**	10^{-7} **in²/lb**
Tungsten	35	51	0.5	0.3
Steel	19–20	27–29	0.6	0.4
Iron (wrought)	18–20	26–29	0.7	0.5
Femur bone	~14	~20	—	—
Copper	10–13	14–19	0.8	0.5
Brass (cold rolled)	9	13	1.6	1.1
Iron (cast)	8–10	12–14	1.0	0.7
Aluminum	5.6–7.7	8–11	1.5	1.0
Polystyrene	~0.14	~0.20	20	15
Water	—	—	50	34
Benzene	—	—	100	69

*The label 10^{10} N/m² means that the numbers listed are in units 10^{10} times larger than N/m². Therefore, Y for tungsten is 35×10^{10} N/m².

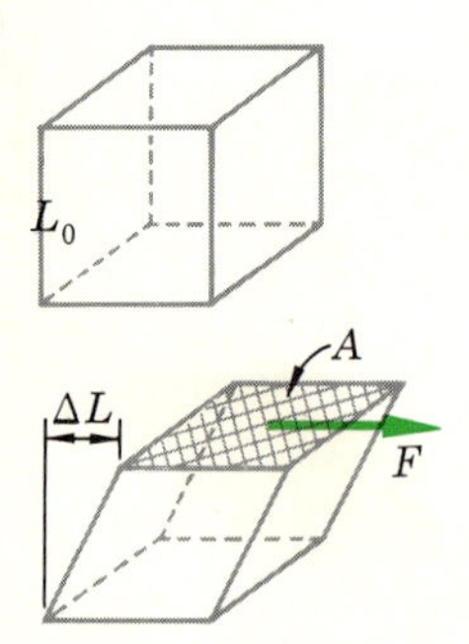

FIGURE 9.7
ΔL is exaggerated here so it can be seen. The shear modulus is given as $\dfrac{F/A}{\Delta L/L_0}$.

Young's Modulus This modulus is used to describe situations such as the stretching of the bar shown in Fig. 9.6. This modulus would be of interest if one wished to compute how much a wire or a rod would stretch under a tensile force. By definition

Young's Modulus

$$\text{Young's modulus} = Y = \frac{F/A}{\Delta L/L_0} \tag{9.6}$$

Typical values of Y for various materials are given in Table 9.2.

The Shear Modulus Suppose that we try to distort a cube of material in the manner shown in Fig. 9.7. A force F is applied parallel to the top face of the cube, the face having an area A. In this case, the stress is still F/A, and the strain is $\Delta L/L_0$, but observe how these symbols are defined in the figure. We have

Shear Modulus

$$\text{Shear modulus} = \frac{F/A}{\Delta L/L_0} \tag{9.7}$$

Although this equation is identical to Eq. (9.6) for Young's modulus, notice that the symbols are defined differently in the two cases.

For many substances it turns out that the shear modulus is approximately one-third as large as Young's modulus. Exceptions to this rule occur chiefly in cases where the volume of the material changes upon deformation or the material shows different properties in different directions.

One frequently makes use of the reciprocals of the shear and Young's (tensile) moduli. They measure how *easily* the material deforms rather than how hard the material is to deform. They are given the descriptive names **shear compliance** and **tensile compliance.**

The Bulk Modulus Suppose that a block having volume V_0 is subjected to a pressure increase ΔP on all sides, as shown in Fig. 9.8. The cube will shrink in volume by an amount ΔV. In this case, the strain is defined to be $\Delta V/V_0$, and the stress is just the applied pressure. As usual, the bulk modulus E is defined to be the ratio of stress to strain. Therefore

Bulk Modulus

$$E = \text{bulk modulus} = \frac{\Delta P}{\Delta V/V_0} \tag{9.8}$$

The reciprocal of this quantity, the compressibility, is discussed below.

The Bulk Compressibility *k* *While the bulk modulus is a measure of how hard it is to compress a substance, the compressibility k is a measure of how easy it is to compress a material. The* **compressibility** *k is just the inverse, or reciprocal, of the bulk modulus.* Usually, the equation by which it is defined is written as

DEFINITION

$$\frac{\Delta V}{V_0} = k\Delta P \tag{9.9}$$

It has the units of reciprocal pressure. Some typical values are given in Table 9.2. Notice that liquids have a much higher compressibility than the crystalline solids. This merely reflects the fact that the molecules in a liquid are fairly widely separated. The compression simply pushes them closer together.

9.7 PRESSURE IN A FLUID

You and I walk and live at the bottom of a vast sea of air. Our bodies are constantly under pressure from the great weight of the tremendous height of air above us. As we shall see in the next chapter, each square inch of our bodies experiences a force of about 14.7 lb (1×10^5 N/m^2). And yet we are not even aware in most cases that the force exists. Why is this?

The body, in some respects, is like a paper bag filled only with air. There are cavities within us, the lungs for example, and these body cavities are, in a rough sense, equivalent to the inside of the paper bag. When we close the top of the bag, it does not collapse under the pressure of the air upon it. The air within the bag exerts the same force outward on the inside of the bag as the outside air exerts trying to collapse the bag (see Fig. 9.9). These forces balance, and hence the bag appears as though there were no air pressure being exerted upon it. This is also true of our bodies.

However, if the air within the sack were removed, there would be no outward-directed forces against the inside of the bag and it would collapse. The great pressure of the air is dramatically shown when the air is pumped

FIGURE 9.8

The cube of original volume V_0 will contract by an amount ΔV under the action of the increased external pressure ΔP.

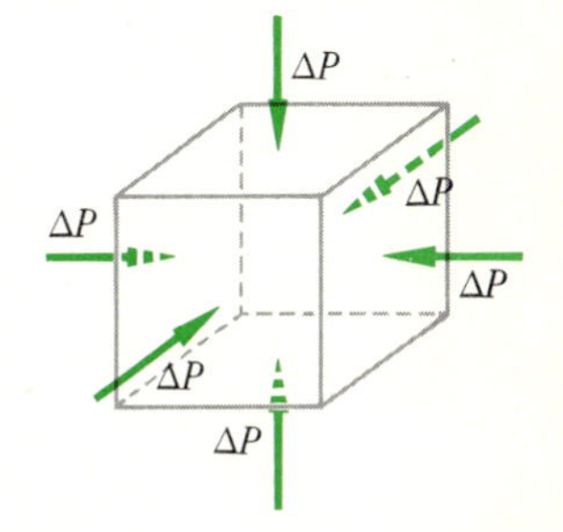

out of a metal can. Unless the can is extremely rigid, it will collapse under the unbalanced pressure of the air in which we live. This is shown in Fig. 9.10.

In this age of space flight, and even with jet airplanes at high altitudes, the reverse of the phenomenon just discussed is of life-and-death importance. Suppose that the bag of Fig. 9.9 were in a spaceship which by some accident suffered a pressure failure. The air pressure within the ship would drop to zero, and the outside forces on the bag would no longer exist. As a result, the pressure of the air on the inside of the bag would no longer be balanced. The bag would explode. It is clear that human beings are also in dire peril in such a situation.

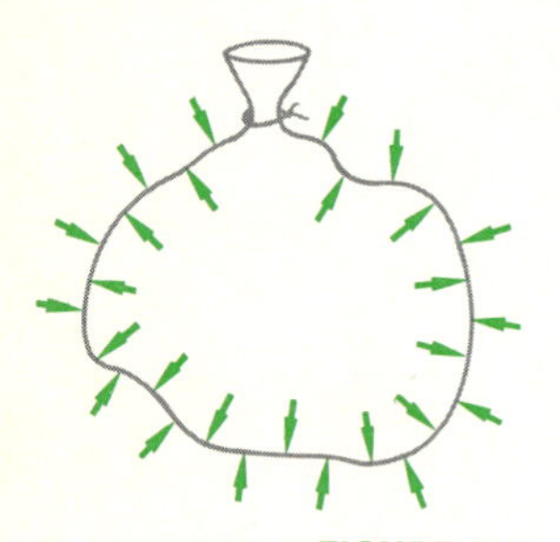

FIGURE 9.9

The sack does not collapse because the air pressure on the inside balances the pressure of the air on the outside.

Now let us turn to the pressures resulting from columns of fluids. A simple case in point is shown in Fig. 9.11. The liquid of height h is maintained in a tube of cross-sectional area A, as shown. Our aim is to compute the pressure on the bottom area A *due to the fluid above it.*

The definition of pressure was given in Chap. 6. It is the force on a unit area, or, in symbols,

$$P = \frac{F}{A}$$

where F must be perpendicular to A. The force due to the fluid on the bottom of the container is just equal to the weight of the fluid above it. But the weight of a unit volume of fluid is defined to be the weight density D of the fluid, as previously discussed. Hence, the force on the bottom of the container due to the fluid is the volume hA of the fluid above the bottom multiplied by the weight per unit volume D. As an equation

$$F = (hA)(D)$$

After dividing through this equation by the cross-sectional area A, we

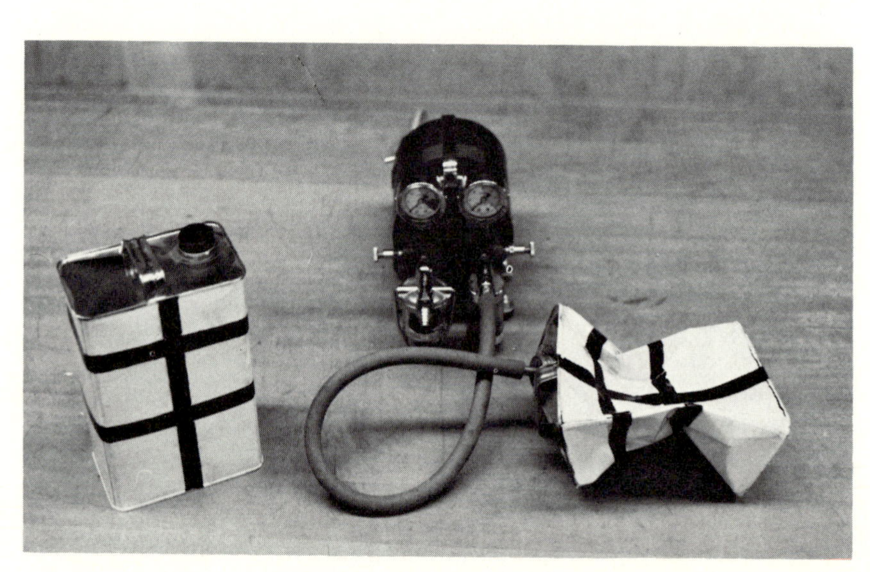

FIGURE 9.10

As the pump removes the air from inside the metal can, the can collapses under the unbalanced forces due to the atmospheric pressure outside it.

have

$$\Delta P = \frac{F}{A} = hD$$

for the pressure due to the height h of fluid. We use ΔP to represent this quantity rather than P because the quantity we have computed is only a portion of the pressure. Additional fluid above that being considered and the pressure of the atmosphere contribute to make the total pressure P larger than the quantity ΔP we have calculated.

In the expression for ΔP we have used the *weight* density D. If we wished to express the relation in terms of the mass density d, we would need to find the weight of a given mass of fluid. The mass of the fluid column is the volume hA multiplied by the mass per unit volume d. But the weight of a given mass is obtained by multiplying the mass by g, the acceleration due to gravity. Hence we have for *the pressure due to a height h of fluid*

$$\Delta P = hD$$

or

$$\Delta P = hdg \tag{9.10}$$

Pressure of a Fluid

Illustration 9.1 Find the pressure due to a column of water 34 ft high.

Reasoning The weight density of water was given in Table 9.1 as 62.4 lb/ft³. Hence, from Eq. (9.10),

$$\Delta P = (34\ \text{ft})(62.4\ \text{lb/ft}^3) = 2120\ \text{lb/ft}^2$$

Or if we prefer the pressure in pounds per square inch, we make use of the fact that $1\ \text{ft}^2 = 144\ \text{in}^2$; so

$$\Delta P = \left(2120\frac{\text{lb}}{\text{ft}^2}\right)\left(\frac{1\ \text{ft}^2}{144\ \text{in}^2}\right) = 14.7\ \text{lb/in}^2$$

Notice that this is just equal to the pressure at the surface of the earth, resulting from the weight of air above the earth, which was given earlier in this section. *The pressure of the air on the surface of the earth is equivalent to the pressure due to a 34-ft (10.4-m) column of water.*

FIGURE 9.11

The force on the bottom of the cylinder due to the fluid is equal to the weight of the fluid.

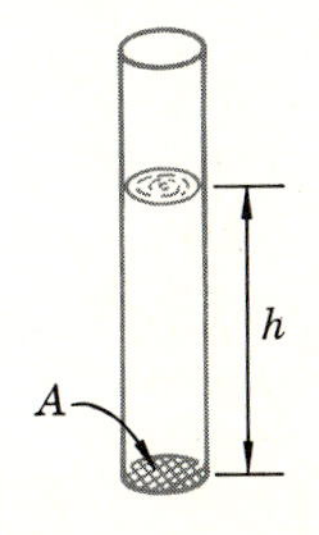

Illustration 9.2 Find the pressure due to a column of mercury 76 cm high.

Reasoning From Table 9.1, the mass density of mercury is 13,600 kg/m³. Since $\Delta P = hdg$,

$$\Delta P = (0.76\ \text{m})(13{,}600\ \text{kg/m}^3)(9.80\ \text{N/kg}) = 1.01 \times 10^5\ \text{N/m}^2$$

(You should show by use of the equation $F = ma$ that the units of g can be expressed as newtons per kilogram as well as the more usual units of meters per second squared.) Although we shall not pursue the matter further at this time, this pressure is equivalent to 14.7 $\mathrm{lb/in^2}$ and so *a column of mercury 76 cm high gives a pressure equivalent to atmospheric pressure.*

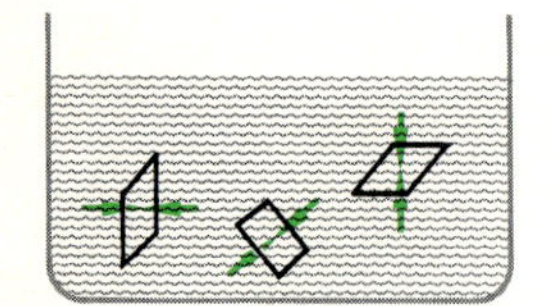

FIGURE 9.12
Since the sheet of paper can exist at equilibrium in the positions shown, we conclude that the pressure at a point within a fluid is the same in all directions.

9.8 PROPERTIES OF PRESSURE IN FLUIDS

The pressure in a fluid acts in all directions. This is easily seen if one considers the experiment illustrated in Fig. 9.12. If one takes a small piece of tissue paper and places it near the bottom of a tank of water as shown in the figure, the tissue paper will not be bent or moved appreciably by the pressure of the still water. We must therefore conclude that the force resulting from the pressure of the water on one side of the paper is balanced by the equal and opposite force on the other side. Since this is true no matter what the orientation of the paper, the pressure in a liquid at a given point must be the same in one direction as in the reverse direction.

Since the pressure due to the liquid at a point beneath the surface of the liquid in Fig. 9.12 is hD from Eq. (9.10), all points at a given depth will have the same pressure and this pressure will be exerted on any surface at this depth, no matter what its orientation. This is all quite obvious for the case of Fig. 9.12. But what about the case shown in Fig. 9.13? Is the pressure the same at point A as it is at B and C?

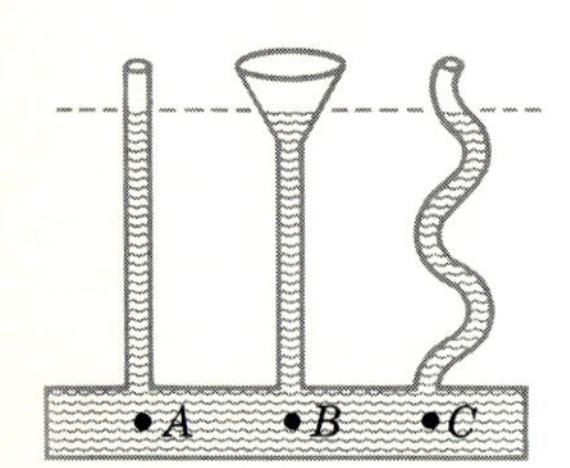

FIGURE 9.13
Why does the liquid stand at the same height in all three tubes?

The pressure must be the same at each of these points, or the liquid would flow in one direction or the other in the lower pipe. Clearly, then, the shape of the vessel is of no importance. Equation (9.10) for the pressure due to a height h of fluid is correct no matter what type of container encloses the fluid. Try to answer for yourself why the small total force on the bottom of tube B is able to support a large amount of fluid while the same force at A supports less fluid. *Hint:* Does the funnel support any of the liquid?

Suppose now that we have a liquid in a closed container like that shown in Fig. 9.14. This device is actually one modification of a hydraulic press. The two pistons shown have cross-sectional areas A_1 and A_2. If there are no forces on the pistons, and if the weight of the pistons is negligible, the liquid will stand at the same height in each of the tubes.

When an external force F_1 is applied to piston 1, the other piston would be pushed up unless a force F_2 were applied to it. If F_2 were made just large enough for no movement to occur, the liquid in the container would remain at rest. Clearly, then, the extra pressure at piston 1 must be balanced by equal pressures everywhere within the fluid, for if this were not true, the unbalanced pressure would cause the liquid to flow. This is an example of **Pascal's principle.** The principle may be stated as follows:

Pascal's Principle

If a pressure is applied to a confined liquid, the pressure is transmitted to every point within the liquid.

Of course, the liquid must remain at rest for this to be true.

It is instructive to compute the force F_2 needed to balance the force F_1.

The additional pressure in the liquid resulting from F_1 is

$$\Delta P = \frac{F_1}{A_1}$$

This increase in pressure, according to Pascal's principle, is also exerted on piston 2. This causes the liquid to exert a force F_2 on the piston, where

$$F_2 = \Delta P\, A_2$$

Substituting for ΔP, we find

$$F_2 = \frac{A_2}{A_1} F_1$$

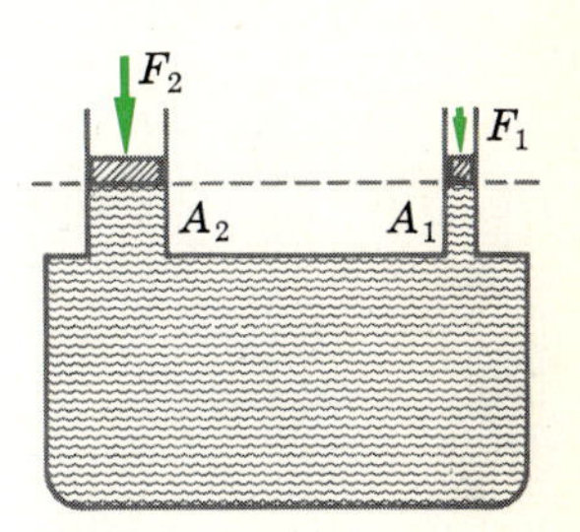

FIGURE 9.14
A small force on the small piston can balance a large force on the large piston.

If the area of the second piston is much larger than the area of the first, F_2 will be much greater than F_1. Hence, a device such as this is capable of lifting a large weight by the exertion of a small force. You should convince yourself, though, that the small force F_1 must still do an amount of work $F_2 s$ to lift the second piston a distance s. If this were not the case, the device could be made into a perpetual-motion machine.

9.9 ARCHIMEDES' PRINCIPLE

As you know, objects often float on fluids. Even if they sink, they appear to weigh less than when they are not submerged. These effects reflect the fact that *an upward force helps to support a submerged object. We call this a* **buoyant force.**

The **buoyancy principle,** *first discovered by Archimedes, is as follows:*

Archimedes' Principle

A body partially or wholly immersed in a fluid is buoyed up by a force equal to the weight of the fluid which it displaces.

For example, if a particular object which has a volume of 3 ft^3 is submerged in water, the buoyant force (BF) on it will be equal to the weight of the displaced water, namely, the weight of 3 ft^3 of water. Since the weight density of water is 62.4 lb/ft^3, the buoyant force will be (3)(62.4), or 187 lb.

In certain cases it is a fairly simple matter to show that Archimedes' principle should be true. Consider a cylindrical piece of material immersed in a liquid, as shown in Fig. 9.15. The buoyant force on the cylinder will be equal to the difference between the force on the bottom, P_2A, and the force on the top, P_1A:

FIGURE 9.15
The buoyant force on the cylinder is equal to the weight of the displaced fluid.

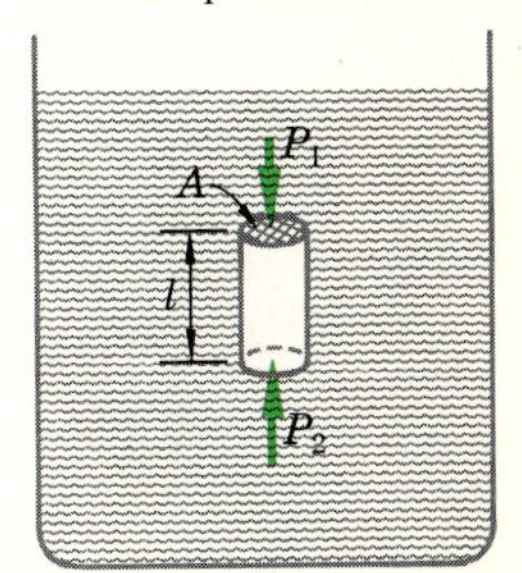

$$\text{BF} = P_2A - P_1A = A(P_2 - P_1)$$

However, $P_2 - P_1$ is simply ΔP, the pressure difference due to the height $h = l$ of liquid shown. Since $\Delta P = lD$, where D is the weight density of the liquid, we find

$$BF = (Al)D$$

But Al is simply the volume of the submerged cylinder. Therefore we find that

$$\text{Buoyant force} = (D)(\text{volume of cylinder})$$

Since the volume of the cylinder is the same as the volume of the displaced liquid, the right-hand side of the above relation for the buoyant force is just the weight of the displaced liquid, or

$$BF = \text{weight of liquid displaced}$$

and *this is a statement of* **Archimedes' principle.**

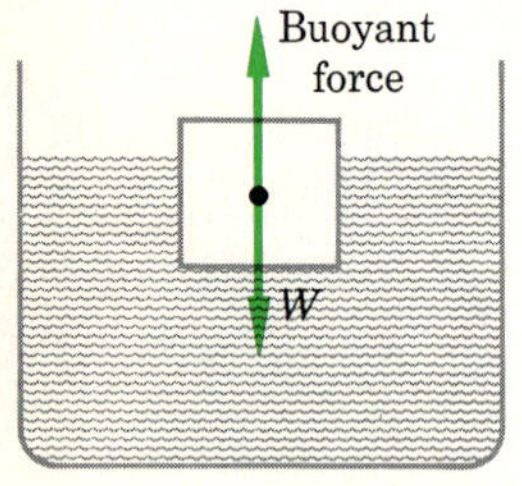

FIGURE 9.16

What must be true if the cube is not to sink?

Illustration 9.3 A cube of wood 2.0 ft on each edge floats in water with three-fourths of the wood submerged. How much does the cube weigh?

Reasoning The situation is shown in Fig. 9.16. Since the cube is floating, it is at equilibrium. Clearly, its weight is balanced by the buoyant force. Therefore

$$W = BF$$

Since three-fourths of the 8-ft^3 block is submerged, the volume of water displaced is $(\frac{3}{4})(8\ \text{ft}^3) = 6\ \text{ft}^3$. But $D = 62.4\ \text{lb/ft}^3$ for water, and so the weight of the displaced water is $(62.4\ \text{lb/ft}^3)(6\ \text{ft}^3) = 374\ \text{lb}$. According to Archimedes' principle, this is also equal to the BF. Then, since $W = BF$ from above,

$$W = BF = 374\ \text{lb}$$

FIGURE 9.17

The weight of the object balances the sum of the buoyant force and the tension in the cord.

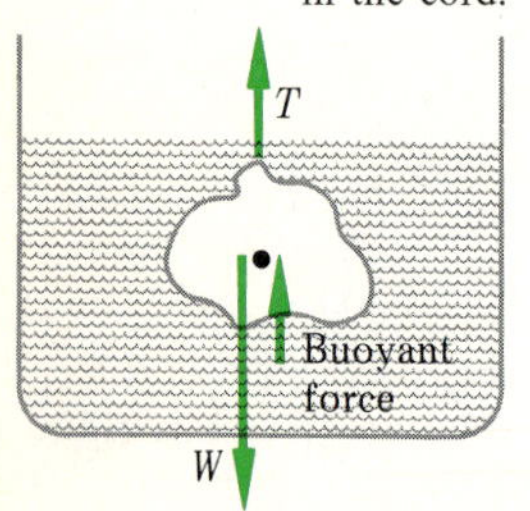

Illustration 9.4 A 20-g piece of metal has a density of 4.0 g/cm^3. It is hung in a jar of oil ($d = 1.50\ \text{g/cm}^3$) by a thread, as shown in Fig. 9.17. What is the tension in the thread?

Reasoning As we see in Fig. 9.17, three forces act on the object. Since the object is at equilibrium, we have $\Sigma F_y = 0$. This gives

$$T + BF - W_0 = 0$$

where W_0 is the weight of the object. We wish to find T,

$$T = W_0 - \text{BF}$$

Since $W = mg$, we have that $W_0 = m_0 g$, where m_0 is 0.020 kg.

To find the BF, we note that it is equal to the weight of fluid displaced by the object. We first need the volume of the displaced fluid. It is equal to the volume of the object. From the definition of density, we can write for the object that

$$d_0 = \frac{m_0}{V_0} \qquad \text{or} \qquad V_0 = \frac{m_0}{d_0}$$

This is also the volume of the displaced fluid V_f. Therefore

$$V_f = \frac{m_0}{d_0}$$

From the definition of density, the mass of the displaced fluid is $V_f d_f$. Then, since $W = mg$, the weight of the displaced fluid is

$$W_f = m_f g = V_f d_f g$$

But this is also the buoyant force. So

$$\text{BF} = V_f d_f g$$

Replacing V_f by its value found above, m_0/d_0, gives

$$\text{BF} = m_0 g \frac{d_f}{d_0}$$

Now we can substitute in our original expression for T, namely,

$$T = W_0 - \text{BF}$$

to give

$$T = m_0 g - m_0 g \frac{d_f}{d_0}$$

Using the SI $m_0 = 0.020$ kg, $g = 9.8\ \text{m/s}^2$, and $d_f/d_0 = 1.50/4.0$. Substituting these values gives

$$T = (0.0125\ \text{kg})(9.8\ \text{m/s}^2) = 0.1225\ \text{N}$$

Notice that the tension is equivalent to the weight of a 0.0125-kg object.

9.10 DENSITY DETERMINATIONS

Let us now spend a short time discussing how densities of materials are determined. You will recall that

$$d = \frac{m}{V} \qquad \text{and} \qquad D = \frac{W}{V}$$

Therefore, *either m or W must be known besides the volume of the object if d or D is to be found.* The values of m and W for an object are easily found by use of a balance. In certain special cases, the volume can be measured using calipers or calibrated volumetric flasks. When this is not possible, volumes are often found by use of Archimedes' principle.

It is possible to find the volume of an object if the BF on it is known when it is submerged in a fluid of known density. Since Archimedes' principle tells us that

$$\text{BF} = V_f D_f$$

(where the subscript f stands for fluid), and since $V_f = V_0$ when the object is submerged, we have

$$\text{BF} = V_0 D_f \qquad \text{or} \qquad V_0 = \frac{\text{BF}}{\text{D}_f}$$

The BF is easily measured by weighing the object twice, once in air and once when submerged in the fluid. (In very exact work, a correction is needed for the buoyancy of the air.) Then, since this difference in weights is due to the BF, we have

$$\text{BF} = W \text{ in air} - W \text{ in fluid}$$

Knowing the BF from these two measurements, we can then substitute it and the known density of the fluid into

$$V_0 = \frac{\text{BF}}{D_f}$$

to obtain the volume of the object.

It is then a simple matter to find the density of the object. If the object's weight has been measured, then

$$D_0 = \frac{W_0}{V_0}$$

Or, if its mass is known, then

$$d_0 = \frac{m_0}{V_0}$$

Since $W_0 = m_0 g$, it is clear that

$$D = dg$$

Therefore, once D or d is found, the other is easily obtained.

Illustration 9.5 A piece of rock is "weighed" and a value of 9.173 g is found. It is then again "weighed" when submerged in a fluid of density 873 kg/m³ and found to have a value of 7.261 g. Find the density of the material of the rock. *Note:* We place quotation marks around the term "weigh" since this is colloquial, not scientific usage.

Reasoning We know that the mass of the object is $m_0 = 9.173 \times 10^{-3}$ kg. To find its volume, we use the fact that

$$\mathrm{BF} = V_0 D_f = V_0(873\ \mathrm{kg/m^3})(9.8\ \mathrm{m/s^2})$$

But
$$\mathrm{BF} = (9.173 \times 10^{-3}\ \mathrm{kg} - 7.261 \times 10^{-3}\ \mathrm{kg})(9.8\ \mathrm{m/s^2})$$

Substituting gives

$$V_0 = \frac{9.173 - 7.261}{0.873} \times 10^{-6}\ \mathrm{m^3} = 2.190 \times 10^{-6}\ \mathrm{m^3}$$

Then, since $d = m/V$, we have

$$d_0 = \frac{9.173 \times 10^{-3}\ \mathrm{kg}}{2.19 \times 10^{-6}\ \mathrm{m^3}} = 4190\ \mathrm{kg/m^3}$$

Illustration 9.6 An object which "weighs" 24 g in air and 16 g when submerged in water "weighs" only 12 g when submerged in a particular type of fluid. What is the density of the fluid?

Reasoning Since $(\mathrm{BF})_w = V_0 D_w$, we have

$$V_0 = \frac{(0.024\ \mathrm{kg} - 0.016\ \mathrm{kg})(9.8\ \mathrm{m/s^2})}{(1000\ \mathrm{kg/m^3})(9.8\ \mathrm{m/s^2})} = 8.0 \times 10^{-6}\ \mathrm{m^3} = 8.0\ \mathrm{cm^3}$$

where the subscript w refers to water. This is also the volume of the displaced fluid V_f.

The weight of the displaced fluid W_f is equal to the BF when the object is submerged in the fluid:

$$W_f = (0.024\ \mathrm{kg} - 0.012\ \mathrm{kg})(9.8\ \mathrm{m/s^2})$$

Then, since $D_f = W_f/V_f$, we have

$$D_f = \frac{(0.024\ \mathrm{kg} - 0.012\ \mathrm{kg})(9.8\ \mathrm{m/s^2})}{8.0 \times 10^{-6}\ \mathrm{m^3}} = 14{,}700\ \mathrm{N/m^2}$$

Or, since $d_f = D_f/g$, we have

$$d_f = \frac{14{,}700\ \mathrm{N/m^2}}{9.8\ \mathrm{m/s^2}} = 1500\ \mathrm{kg/m^3}$$

9.11 THE BAROMETER

Let us now return to the discussion of fluid pressure and its effects. The air surrounding us, the atmosphere, is a very important fluid. The pressure within it is of considerable importance to us. As we shall see, atmospheric pressure is an influential variable in many processes. In this section we shall discuss ways in which atmospheric pressure is measured.

We are all familiar with the fact that increasing barometric pressures usually precede good weather, while decreasing and low barometric pressures are characteristic of bad weather. The daily barometric pressure is given in many weather reports and is one of the primary pieces of data in predicting weather. Not only is it important in this application, but it also is used for many other purposes. For example, the boiling points of liquids change depending upon the pressure, and so the barometric pressure is often needed in the laboratory. Let us now examine the operation of the barometer.

Consider the situation shown in Fig. 9.18*a*. The beaker is filled with mercury, and an open glass tube is immersed in it. Since both the mercury surfaces, inside and outside the tube, are open to the air, the air pressures on them are equal. The air pressure on the surfaces is the result of the weight of the air above the earth.

If an ideal vacuum pump is connected to the end of the tube as shown in Fig. 9.18*b*, the pump will remove the air from the tube and so the pressure on the mercury surface within the tube will be reduced to zero. (Recall that pressure exerted by a gas on a surface is the result of the collisions of the gas molecules with the surface. If no molecules are present, the pressure is obviously zero. This is what we mean by a perfect vacuum.) There will now be an unbalanced force on the mercury which tends to push it up the tube. The mercury will rise to such a height that the pressure at level A due to the mercury column is equal to the pressure P of the atmosphere. This follows from the fact that the pressure at a given level in a liquid must be everywhere the same or the liquid would flow to equalize the pressure.

Finally, the top of the evacuated tube is sealed off as shown in Fig. 9.18*c* so that the pressure on the surface of the mercury inside the tube will always be zero. Since the pressure at the bottom of the column of mercury just equals atmospheric pressure, we have

$$P = hD = hdg$$

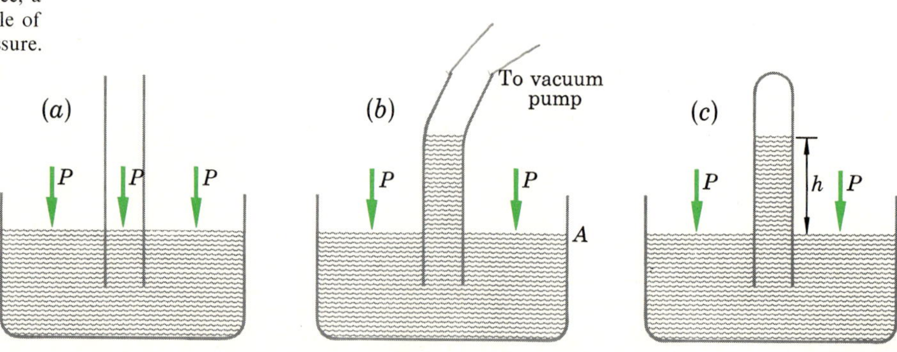

FIGURE 9.18
When the tube is evacuated, the mercury rises until $hD = P$. Hence the device, a barometer, is capable of measuring pressure.

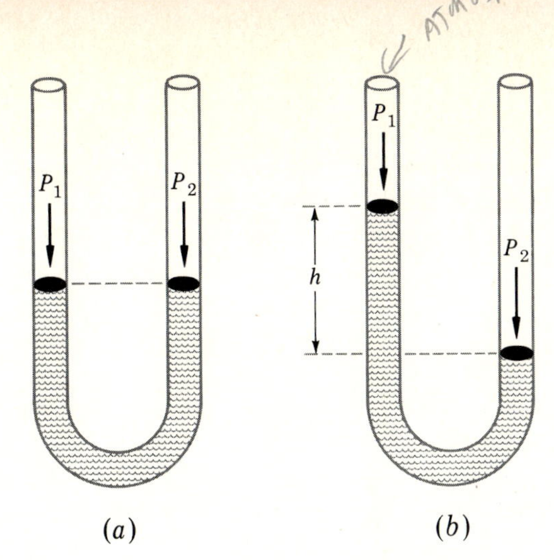

FIGURE 9.19

In a manometer, the pressure difference, $P_2 - P_1$, is measured by the height h.

Since mercury has a known density D, the atmospheric pressure P is easily found by measuring the height of the mercury column needed to balance the atmospheric pressure. In fact, more often than not, the pressure is quoted as so many inches or centimeters of mercury. *Standard atmospheric pressure is taken to be 760 mm Hg.* Of course, *this pressure is* really $hdg = (0.76)(13{,}600)(9.8)\ \text{N/m}^2 = 1.01 \times 10^5\ \text{N/m}^2$. Why? Since 1 in is 2.54 cm, 760 mm Hg is equivalent to about 30 in Hg.

Commercial mercury barometers are more refined than the simple device shown in Fig. 9.18*c*. They have an accurate scale beside the mercury column and special devices to adjust the level of the mercury in the cup. Basically, though, they are as shown. There are other types of barometers based upon different principles, but for accurate work the mercury barometer is preferred. However, it must be at least 76 cm long, and so there is often good reason to replace it by a smaller, but less accurate, device.

Another device often used to measure gas pressures precisely is called a **manometer** (Fig. 9.19). Although it has many variations, a manometer is basically a U-shaped tube partly filled with a liquid such as mercury. If the mercury stands at equal levels in the two tubes as shown in part *a*, we know the two gas pressures P_1 and P_2 above the columns must be equal. However, if P_2 is larger than P_1, the columns will adjust as in part *b*. The difference in heights h of the two columns when measured in centimeters, gives the pressure difference $P_2 - P_1$ in centimeters of mercury. Usually one column is open to the atmosphere; let us say P_1 is atmospheric pressure. Then to find P_2, one must add the barometric pressure to h. For small pressure differentials, it is often convenient to use a liquid which is less dense than mercury. If a liquid of density d is used to replace mercury, the difference in levels of the liquid will be increased by a factor $13{,}600/d$, where d is in kilograms per cubic meter. Can you explain why?

Illustration 9.7 A simple lung test is to have the patient blow with full force into one end of a manometer, as shown in Fig. 9.20. Suppose in a certain case

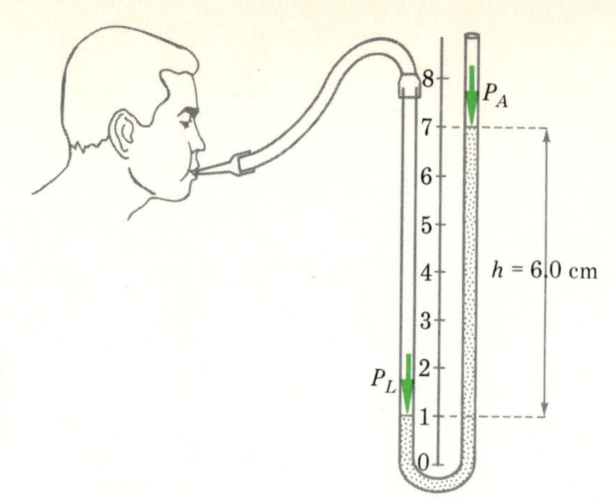

FIGURE 9.20
By blowing into the manometer, the person is able to support a column of fluid 6.0 cm high. How large is P_L?

a mercury manometer is used and the fluid level stands as shown. What is the pressure within the patient's lungs? (Ordinarily, a mercury manometer would not be used because mercury vapor is dangerous for repeated exposure.)

Reasoning Let us call the air pressure in the lungs P_L. The pressure exerted on the left side of the manometer is very nearly P_L. This pressure balances the pressure of the atmosphere plus the pressure due to the height $h = 6.0$ cm Hg. Therefore

$$P_L = 6.0 \text{ cm Hg} + P_A$$

Normally, P_A is about 76 cm Hg and so $P_L = 82$ cm Hg. (Of course, in accurate work P_A would be read by means of a barometer.) Since the pressure due to a height h of fluid is hdg, we have that

$$\begin{aligned} P_L = 82 \text{ cm Hg} &= (0.82 \text{ m})(13{,}600 \text{ kg/m}^3)(9.8 \text{ m/s}^2) \\ &= 1.093 \times 10^5 \text{ N/m}^2 = 1.08 \text{ atm} \end{aligned}$$

The conversions between these and other units of pressure are given in the table of conversion factors inside the front cover.

9.12 FLUIDS IN MOTION

Until now we have been discussing fluids which were at rest. Moving fluids are also of great importance. We can learn a great deal by examining the flow of fluids through pipes. Typical behavior is shown in Fig. 9.21.

As we see in part *a* of the figure, fluid does not simply move down a pipe in a pluglike form. Instead, the fluid close to the walls of the pipe moves scarcely at all. As shown by the velocity profile, the fluid near the center of the pipe moves most swiftly. This variation in speed across the pipe's cross section causes the fluid near the center of the pipe to rub past the outer

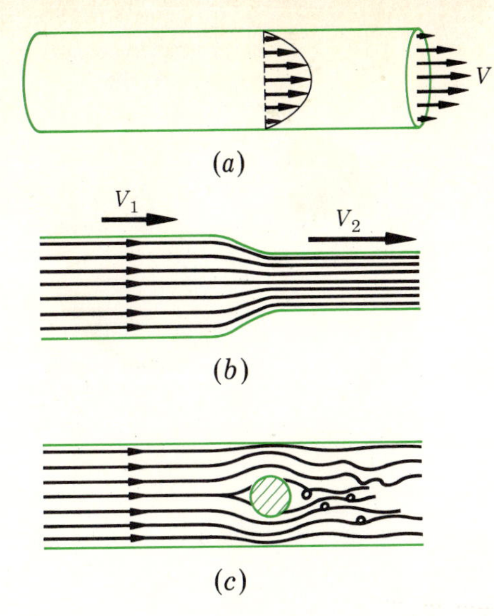

FIGURE 9.21
Examples of various features of flow in a tube: (*a*) velocity profile; (*b*) streamlines; (*c*) turbulent flow.

portion of the fluid. As a result, friction energy loss occurs in the flow process. More will be said about this later.

Part b of the figure shows the streamlines of simple flow. These lines show the path a tiny particle in the fluid follows as it moves along the pipe. Flow such as this is termed **laminar flow.** Notice also in this figure that the flow velocity changes as the cross-sectional area changes. The fluid velocity is lower in the large cross-sectional region.

Part *c* of the figure shows what happens if the flow becomes too swift past an obstruction. The smooth flow lines no longer exist. As the fluid rushes past the obstacle, it starts to swirl in erratic motion. No longer can one predict the exact path a particle will follow. *This region of constantly changing flow lines is said to consist of* **turbulent flow.** As you might expect, considerably more energy is lost to friction effects in turbulent flow than in laminar flow. Turbulent flow also occurs in a uniform tube even when an obstacle is not present provided the flow rate is large enough.

9.13 VISCOSITY

We saw in the last section that *friction effects occur in a flowing fluid. This effect is described in terms of the* **viscosity** *of the fluid. Viscosity measures how much force is required to slide one layer of the fluid over another layer. Substances which do not flow easily,* such as thick tar or syrup, *have large viscosity. Substances* (like water) *which flow easily have small viscosity.*

To give quantitative meaning to viscosity, we refer to the hypothetical shear-type experiment shown in Fig. 9.22. There we see two parallel plates, each of area A, separated by a distance l. The region between the plates is filled with a fluid whose viscosity we shall denote by η (Greek eta). In order to move the top plate with speed v relative to the bottom one, a force is

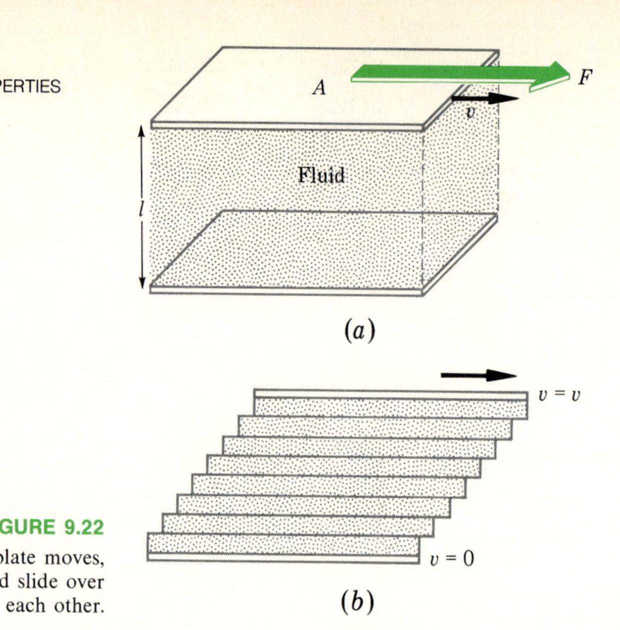

FIGURE 9.22
As the upper plate moves, layers of the fluid slide over each other.

required. The force will be large if the fluid has a large viscosity. *We define* η *by*

Viscosity

$$\eta = \left(\frac{F}{A}\right)\left(\frac{l}{v}\right) \tag{9.11}$$

In the SI, the units of viscosity [from Eq. (9.11)] are newton-seconds per square meter. Most frequently, η is tabulated in terms of a unit called the poise (P). These units are related to the SI unit by

$$1 \text{ SI viscosity unit} = 1 \text{ N} \cdot \text{s/m}^2 = 10 \text{ P} = 1000 \text{ cP}$$

The poise is obtained from Eq. (9.11) when the cgs system is used. Typical viscosities are given in Table 9.3.

TABLE 9.3
VISCOSITIES OF LIQUIDS AND GASES AT 30°C

MATERIAL	VISCOSITY,* cP
Air	0.019
Acetone	0.295
Methanol	0.510
Benzene	0.564
Water	0.801
Ethanol	1.00
Blood plasma	~1.6
SAE No. 10 oil	200
Glycerin	629
Glucose	6.6×10^{13}

*1000 cP = 1 N·s/m², the SI unit.

We can gain further insight into the meaning of viscosity by examining Fig. 9.22*b*. Notice that the fluid layers next to the two plates remain attached to the plates. We can think of the fluid between the plates as consisting of many thin layers, many more than shown. As the upper plate moves, these layers must slide over each other. In a high-viscosity fluid, the layers do not slide easily. A large amount of friction work is done as the layers are made to slide past each other. It is for this reason that *work done against viscous forces is equivalent to friction work.*

9.14 POISEUILLE'S LAW

It is often of value to know what volume of fluid will flow through a pipe in unit time. Let us call this quantity the **flow rate** and represent it by the symbol Q. From its definition,

$$Q = \text{volume flowing out pipe per second}$$

Let us refer to Fig. 9.23 and see if we cannot guess how Q should depend upon the variables shown there.

It is reasonable to think that the fluid will flow faster if the pressure differential $(P_1 - P_2)$ is large. We might guess that Q would be proportional to the driving pressure $(P_1 - P_2)$. The longer the pipe, the greater the resistance to flow should be. Therefore, Q should vary inversely with L. But the larger the pipe's cross section, the greater the flow rate should be. As a result, Q should increase as R increases. Finally, the larger the viscosity η, the smaller the rate of flow should be.

An exact mathematical relation for Q in terms of $P_1 - P_2$, L, R, and η was first found by Poiseuille (after whom the viscosity unit was named). He showed that *for laminar flow through a pipe of length L and radius R*

$$Q = \frac{\pi R^4 (P_1 - P_2)}{8\,\eta L} \tag{9.12}$$

This is called **Poiseuille's law.** Notice that Q increases as the fourth power of R, the pipe's radius.

Illustration 9.8 At 30°C, water has a viscosity of 0.801 cP (centipoise). How much water will flow each second through a 20-cm-long capillary which has a radius of 0.15 cm if the pressure differential across the tube is 3.0 cm Hg?

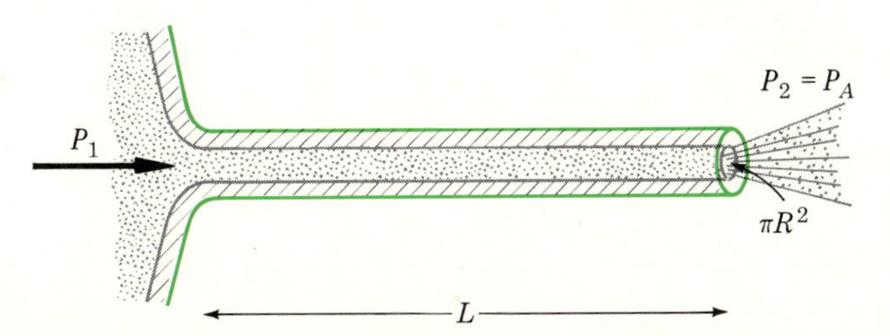

FIGURE 9.23
The flow rate is proportional to $(P_1 - P_2)R^4/\eta L$, according to Poiseuille's law.

Reasoning We have the following quantities known:

$$L = 0.20 \text{ m} \qquad R = 0.15 \times 10^{-2} \text{ m}$$
$$\eta = 0.801 \text{ cP} = 0.801 \times 10^{-3} \text{ N} \cdot \text{s/m}^2$$
$$P_1 - P_2 = 3.0 \text{ cm Hg} = (\tfrac{3.0}{76})(1.01 \times 10^5 \text{ N/m}^2) = 0.40 \times 10^4 \text{ N/m}^2$$

The conversion in pressure units is made here by noting that 1 atm = 76 cm Hg = 1.01×10^5 N/m^2. Substituting these values in Eq. (9.12) yields

$$Q = 5.0 \times 10^{-5} \text{ m}^3\text{/s} = 50 \text{ cm}^3\text{/s}$$

Illustration 9.9 A capillary viscometer consists of a vertical capillary tube through which a repeatable quantity of liquid flows in a measured time. The driving pressure is furnished by the weight of the liquid. In a certain viscometer at 30°C, the benzene flow time is 206.3 s, while a very dilute solution in benzene has a flow time of 309.7 s. Find the viscosity of the solution.

Reasoning The dilute solution will have very nearly the same density as benzene. Hence, the driving pressure ($P_2 - P_1$) will be the same for both solution and solvent. According to Poiseuille's law, the only factor influencing Q which varied in the two experiments was η. Therefore, we have

$$\frac{\eta \text{ for solvent}}{\eta \text{ for solution}} = \frac{\text{flow time for solvent}}{\text{flow time for solution}}$$

From Table 9.3, the viscosity of benzene at 30°C is 0.564 cP. Therefore, upon substitution of the known values,

$$\frac{0.564 \text{ cP}}{\eta \text{ for solution}} = \frac{206.3}{309.7}$$

giving

$$\eta \text{ for solution} = 0.847 \text{ cP}$$

Illustration 9.10 Older people often develop blood-circulation problems because of deposits building up in their arteries. By what factor is the blood flow reduced in an artery if the artery radius is cut in half? Assume the same pressure differential in the two cases.

Reasoning Poiseuille's law tells us that the volume of blood Q flowing through the artery each second is related to R by

$$Q \propto R^4$$

In the original artery, $Q_0 = (\text{const})(R_0^4)$, but in the constricted artery, $Q = (\text{const})(R_0/2)^4$. Taking the ratio Q/Q_0, we find $Q/Q_0 = \frac{1}{16}$. The flow rate is reduced by a factor of 16. It is clear from this strong dependence of Q on R why blood-circulation difficulties result from arterial deposits.

9.15 BERNOULLI'S EQUATION FOR LIQUIDS IN MOTION

As we have seen, all liquids have a characteristic viscosity. If the viscosity is large, a great deal of work is needed to push the liquid through a pipe. This energy is lost as the molecules rub against each other in the liquid. This lost energy appears eventually as heat energy.

Many liquids have such a small viscosity that their energy loss as a result of friction effects can be neglected, at least for certain purposes. When this is the case, an important, simple relation can be found for the pressure in the fluid. It is called **Bernoulli's equation** and was published by Daniel Bernoulli in 1738.

Consider the pipe system shown in Fig. 9.24. It is completely filled with liquid between the two frictionless pistons. We shall say that the lower piston is being pushed to the right with speed v_1 and that the upper piston is moving to the right with speed v_2. The force on the lower piston F_1 is balanced by the force resulting from the pressure of the liquid P_1A_1, where A_1 is the area of the lower piston. (The forces on the piston must balance, or it would be accelerating, and we have already specified that the piston is moving with constant speed.) Similarly, at the top piston, $F_2 = P_2A_2$. Now in a time t the lower piston will move a distance v_1t, thereby displacing a volume of liquid $(v_1t)(A_1)$. However, if the liquid is incompressible, the upper piston must make way for an equal volume of liquid. Hence

$$(v_1t)(A_1) = (v_2t)(A_2)$$

Bernoulli asked what happens to the work done by piston 1. The work done by piston 1 is just $F_1(v_1t)$, or

$$\text{Input work} = P_1A_1v_1t$$

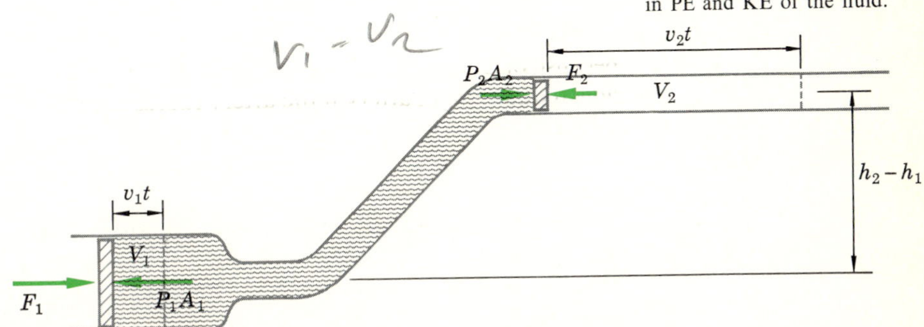

FIGURE 9.24
The work done by F_1 must equal the work done against F_2 plus the sum of the change in PE and KE of the fluid.

since $P_1A_1 = F_1$. Since piston 2 does an amount of work $F_2(v_2t)$, some of the input work is transformed there.

In addition, the liquid pressed to the right by piston 1 is essentially transferred to the upper tube. By so doing, that liquid (with volume A_1v_1t) is given some PE. Moreover, since it will now be traveling with a larger speed v_2, its KE will also be changed. Of course, some energy is lost in viscous-friction work, but we are assuming this to be small. We therefore have

$$\text{Input work} = \text{output work} + \text{change in PE} + \text{change in KE}$$

or, using the symbols of Fig. 9.24,

$$P_1A_1v_1t = P_2A_2v_2t + Mg(h_2 - h_1) + (\tfrac{1}{2}Mv_2^2 - \tfrac{1}{2}Mv_1^2)$$

But the volume of liquid involved is A_1v_1t, and its mass is found from the definition of density to be

$$M = dA_1v_1t = dA_2v_2t$$

Substitution of this in the above equation gives, after rearrangement,

Bernoulli's Equation

$$P_1 + \tfrac{1}{2}v_1^2d + gh_1d = P_2 + \tfrac{1}{2}v_2^2d + gh_2d$$

This equation is **Bernoulli's equation.** Clearly, the pistons need not be present. Points 1 and 2 could be any two points in the liquid. All that is needed are surfaces in the liquid; these surfaces can be imaginary and the computation will still be the same. *Notice,* however, *that the equation is only applicable if friction losses can be neglected.*

9.16 TORRICELLI'S THEOREM

Illustration 9.11 A simple application of Bernoulli's equation is shown in Fig. 9.25. Suppose that a large tank of fluid has a small spigot on it, as shown. Find the speed with which the water flows from the spigot at the right.

FIGURE 9.25
Torricelli's theorem tells us how fast the liquid is moving as it flows out the spigot.

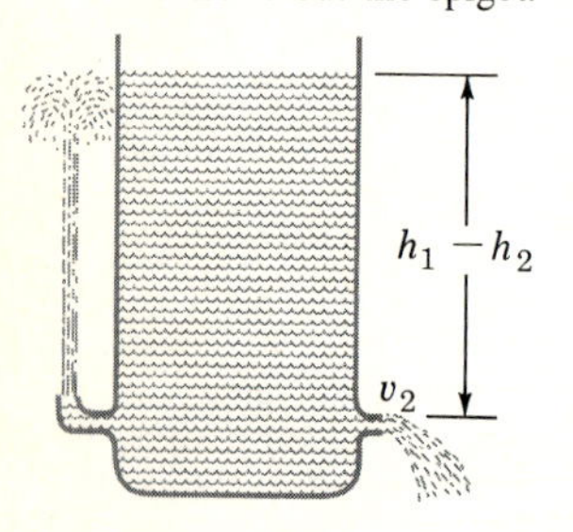

Reasoning Since the spigot is so small, the efflux speed v_2 will be much larger than the speed v_1 of the top surface of the water. We can therefore approximate v_1 as zero. Bernoulli's equation can then be written as

$$P_0 + gh_1d = P_0 + \tfrac{1}{2}v_2^2d + gh_2d$$

since $P_1 \approx P_2$ = atmospheric pressure = P_0.

Rearrangement of this equation gives

$$v_2 = \sqrt{2g(h_1 - h_2)}$$

Torricelli's Theorem

This is **Torricelli's theorem.** *Notice that the speed of the efflux liquid is the same as the speed of a ball which falls through a height $h_1 - h_2$.* This points out the fact that when a little liquid flows from the spigot, it is as though the same amount of liquid had been taken from the top of the tank and dropped to the spigot level. The top level of the tank has decreased somewhat, and the PE lost has gone into KE of the efflux liquid. If the spigot had been pointed upward, as at the left in Fig. 9.25, this KE would allow the liquid to rise to the level shown before stopping. In practice, viscous-energy losses would alter the result somewhat.

9.17 OTHER APPLICATIONS OF BERNOULLI'S EQUATION

Illustration 9.12 Suppose water to flow through a pipe system like the one shown in Fig. 9.26. Clearly, the water must flow faster at B than it does at A or C. Assuming the flow speed at A to be 0.20 m/s and at B to be 2.0 m/s, compare the pressure at B with that at A.

Reasoning Applying Bernoulli's equation and noting that the average PE is the same at both places, we have

$$P_A + \tfrac{1}{2}v_A^2 d = P_B + \tfrac{1}{2}v_B^2 d$$

Substituting $v_A = 0.20$ m/s, $v_B = 2.0$ m/s, and $d = 1000$ kg/m^3 gives $P_A - P_B = 1980$ N/m^2. Hence, *the fluid pressure within the constriction is much less than in the large pipes on either side.* This is probably opposite to what one would guess at first. However, it is true and has wide application. Aspirators, for example, obtain a partial vacuum by forcing water through a constriction where the pressure is greatly reduced.

It is easy to see in Fig. 9.26 that the pressure at A must be larger than at B. Since each little volume of fluid is accelerated as it moves from A to B, an unbalanced force towards the right must exist on it. Therefore, the pressure must decrease as one goes from A to B. Can you reverse this line of reasoning to show that the pressure at C is larger than at B?

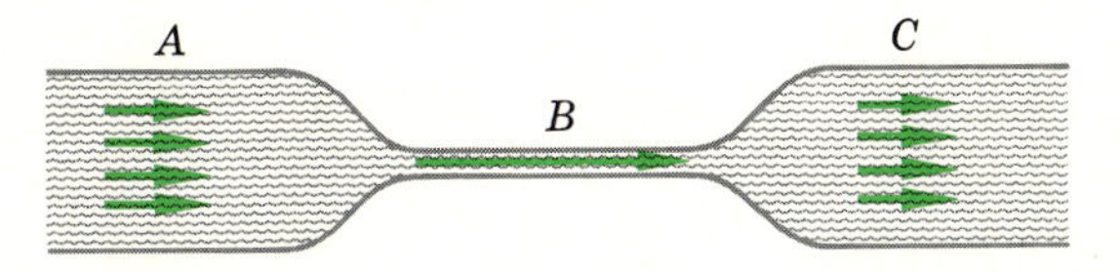

FIGURE 9.26
Since the fluid velocity is greatest at B, the pressure is smallest at that point.

This result—that *where the speed is high the pressure will be low*—affords an interpretation of such diverse facts as the lift on an airplane wing and the curve ball pitched by a good ballplayer. The flow around an airplane wing is illustrated in Fig. 9.27. In this case you will notice that the air is traveling faster on the upper side of the wing than on the lower. The pressure will be lower at the top of the wing, and the wing will be forced upward.

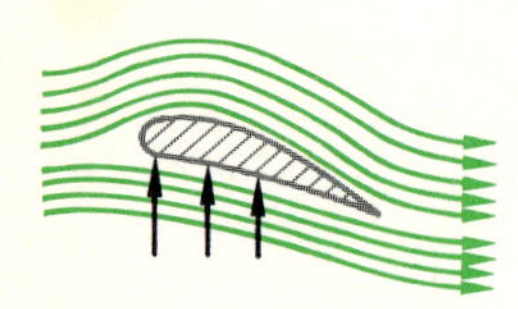

FIGURE 9.27
The airplane wing experiences a force from the low-velocity (high-pressure) region below the wing to the high-velocity (low-pressure) region above the wing.

9.18 BLOOD-PRESSURE MEASUREMENT

There are many other examples in which fluid flow is of importance and interest. We have space here for only one more, the measurement of blood pressure. The pressure of the blood in your arteries and veins varies widely both as a function of time and of position in your body. As the heart throbs, the blood pressure on the exit side of it alternately rises and falls. These pressure fluctuations persist throughout the artery system. But as the blood flows into smaller and smaller channels, friction effects and the elasticity of the channels themselves tend to even out the flow pattern. Finally, as the blood flows into the veins for the return trip to the heart, the flow is almost uniform.

The wide pressure fluctuations in the arteries is of importance in several respects although only two of these will be mentioned here: (1) Extremely high maximum pressures can lead to rupture of the channel walls through which the blood flows. Strokes are one evidence of such a rupture. (2) The magnitudes of the pressure peaks and valleys during the heart-beat cycle provide information concerning constrictions in the channels as well as other body factors which influence blood circulation.

Ordinary blood-pressure data give two numbers, the systolic pressure and the diastolic pressure. These are the pressure readings (usually expressed in millimeters of mercury) at the peak (systolic) *and low point (diastolic) of the blood-flow cycle.* To measure these values, use is made of the arrangement shown in Fig. 9.28.

An inflatable cuff is placed around the upper arm near the same level as the heart. The pressure exerted by the cuff when inflated is monitored using a mercury manometer. By pumping air into the cuff, the pressure can be

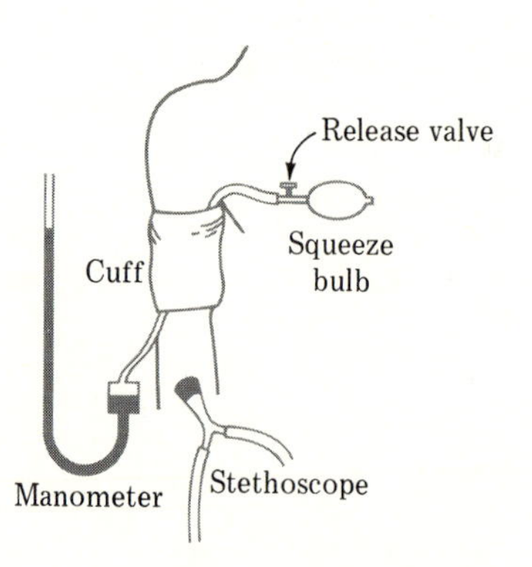

FIGURE 9.28
Apparatus used to measure blood pressure.

made to exceed the peak pressure in the arm artery. As a result, flow of blood into the lower arm is cut off. If one places a stethoscope (a listening device) on the artery below the cuff, no sound is heard since blood is not flowing through the artery.

The pressure in the cuff is slowly reduced by allowing it to deflate. Suddenly, at a well-defined pressure called the systolic pressure, one begins to hear the pulse beat in the stethoscope. At that point, the pressure is low enough in the cuff for the blood to surge past the cuff, through the artery, at the pressure peaks. This pressure reading for the cuff therefore gives the blood pressure at the peak of the heart-throb cycle. Actually, the sound one hears is the result of turbulent flow past the constricted artery at the cuff.

As the pressure is reduced further in the cuff, the blood flow becomes less turbulent. The sound in the stethoscope picks up this fact by changing to a less sharp sound. Eventually, the sound ceases. At that pressure, the diastolic pressure, the blood is able to flow past the cuff during all portions of the cycle. This reading therefore gives the lowest pressure during the pulse-beat cycle. In a normal young person, the systolic and diastolic pressures are about 120 and 80 mm Hg, respectively (usually reported as 120/80). As one ages, these pressures often change, although very high systolic pressures (200 mm Hg for example) are almost always reason for serious concern. Other pressure variations usually require careful consideration before their implications can be ascertained.

SUMMARY

Substances can be divided into two groups, fluids and solids. There are two types of fluids, namely, liquids and gases. The solids can also be subdivided into those which are crystalline and those which are not. In all these classifications, borderline examples exist. Glasses, for example, can be classed as solids or as very viscous liquids.

The mass density d of a substance is the mass of substance per unit volume: $d = m/V$. The weight density D is the weight per unit volume: $D = W/V$. Increases in temperature cause the density of most substances to decrease. However, water, in the range 0 to 4°C is an exception to this rule. Specific gravity is defined as the ratio of the density of the substance to the density of water. This ratio varies with temperature.

Hooke's law applies to many situations in which an elastic deformation occurs. It states that the distortion is proportional to the distorting force. When distortion exceeds the limit of elasticity, the system no longer returns to its undistorted form when the distorting force is removed. In terms of the stress and strain, Hooke's law can be stated as "stress is proportional to strain."

By definition, the ratio of stress to strain is a modulus of elasticity. For a tensile deformation, the modulus is called Young's modulus. In a shear-type distortion, the shear modulus applies. When distortion is caused by equal pressure on all sides, the ratio of pressure to $\Delta V/V$ is called the bulk modulus. The reciprocal of the bulk modulus is called the bulk compressibility.

The pressure increase due to a height h of fluid whose mass density is d is given by $\Delta P = hdg$. Or, since $D = dg$, one has $\Delta P = hD$. In a fluid, the pressure at a point acts equally in all directions. Moreover, if a confined liquid is subjected to a pressure, that pressure is transmitted to every point within the liquid. This is called Pascal's principle. It forms the basis for operation of the hydraulic press.

Archimedes' principle states that "a body partially or wholly immersed in a fluid is buoyed up by a force equal to the weight of the fluid which it displaces." The volume of an object can be found by use of this principle if the object is weighed when immersed in a fluid of known density.

The pressure of the atmosphere can be measured by use of a barometer. Differences in pressure can be measured by use of a manometer. Standard atmospheric pressure is 1.013×10^5 N/m^2, which is the same as 14.7 lb/in^2 and 76 cm Hg.

In laminar flow, the fluid follows a set pattern along definite flow lines called streamlines. When turbulent flow occurs, regions exist where stable flow lines are not followed.

Viscosity is a measure of the force required to cause a shearing type flow in a fluid. It is represented by η, and its SI units are newton-seconds per square meter. The cgs unit is the poise (P), where $10\ \text{P} = 1\ \text{N} \cdot \text{s/m}^2$.

The flow rate of a viscous fluid through a cylindrical tube is given by Poiseuille's law. Viscosities can be compared by measuring the flow rates through the same tube. With proper precautions, the flow rates vary in inverse proportion to the viscosities.

Bernoulli's equation describes the flow of fluids in which viscous effects are negligible. It can be used to obtain Torricelli's theorem and to explain the operation of aspirators, atomizers, and similar devices.

In measuring blood pressure, two pressure values are obtained. The highest pressure in the heart-beat cycle is the systolic pressure. The lowest pressure is the diastolic value. Normal values for a young person are about 120 and 80 mm Hg for the systolic and diastolic values, respectively.

MINIMUM LEARNING GOALS

Upon completion of this chapter you should be able to do the following:

1. Place each of a series of given substances in the category(s) appropriate for it: gas, liquid, solid, fluid, crystalline, amorphous, glass. Give the distinguishing characteristics of each category.

2. Define weight density and mass density. Compute each for a substance when the appropriate data are given. Use the density to compute the volume of a given mass or weight of substance. Give the relation between d and D.

3. Sketch an elongation-versus-force graph for a substance which obeys Hooke's law. Give the meanings (definitions) of stress and strain in the cases of tensile, shear, and bulk deformations. Define the modulus in each case. State Hooke's law in terms of stress and strain.

4. Relate bulk compressibility to bulk modulus and give appropriate units for each.

5. Explain why the can in Fig. 9.10 collapses as the pump removes air from it.

6. Find the pressure due to a column of fluid whose density is known. Explain how a manometer can be used to measure pressure differences. Explain the principle of the mercury barometer. Find the force due to a fluid on a given area at a known depth in the fluid.

7. State Pascal's theorem and explain how use is made of it in the hydraulic press.

8. State Archimedes' principle and use it to find the buoyant force on an object of known volume submerged in a fluid of known density.

9. Use Archimedes' principle to compute the density of a substance whose volume is found by immersion in a fluid of known density.

10. State the value of standard atmospheric pressure in newtons per square meter, pounds per square inch, and centimeters of mercury.

11. Explain what is meant by laminar flow, streamlines, and turbulent flow.

12. List several given substances in order of increasing viscosity. Explain what is meant by viscosity and how it is related to friction energy losses. Give the common units in which viscosity is measured.

13. Given Poiseuille's law, identify each quantity in it, and be able to use it for simple calculations.

14. Given Bernoulli's equation, identify each quantity in it, and be able to use it for simple calculations. Derive Torricelli's theorem from it. Show how it predicts that pressure decreases as flow rate increases.

15. Explain what a doctor is doing when he or she measures blood pressure.

IMPORTANT TERMS AND PHRASES

You should be able to define or explain each of the following:

Gases, liquids, solids
Fluids
Crystalline versus amorphous solid
Weight and mass densities
Hooke's law
Elastic, elastic limit

Stress, strain, and modulus
Tensile, shear, and bulk modulus
Compliance; bulk compressibility
$\Delta P = hdg$
Pascal's principle
Archimedes' principle; buoyant force
Barometer and manometer
Standard atmospheric pressure
Streamlines; laminar flow
Turbulent flow
Viscosity; poise unit
Poiseuille's law
Bernoulli's equation
High speed implies low pressure
Systolic and diastolic pressure

QUESTIONS AND GUESSTIMATES

1. Suppose you are given a solid cube of metal. How could you determine the density of it?
2. How could you determine the density of the glass from which tiny glass spheres are made? Assume the spheres are too small to be weighed and measured with the equipment available. However, a considerable quantity of the material is available.
3. How could you determine the density of a liquid? Of a gas?
4. Compare the pressure at a depth of 5.0 m in a small lake with the pressure 5.0 m below the surface of a large lake. How does the force which a dam must withstand depend upon the size of the lake which it dams?
5. Explain why "water seeks its own level."
6. How could you measure the shear modulus of a gelatin dessert? Why would the modulus depend upon whether or not the gelatin has pieces of fruit in it? Reinforced plastics such as fiber glass have glass fiber embedded in the plastic. What effect does the glass fiber have on the mechanical properties of the plastic? (Bone is reinforced by collagen fibers in this same way. As a result, the tensile modulus and strength of bone is higher than it otherwise would be.)
7. The viscosity of nearly all liquids decreases with temperature. As we shall learn in the next chapter, the KE of the molecules in the liquid increases with increasing temperature. How can this latter fact be used to justify the viscosity change with temperature?
8. Blood plasma consists of blood from which the platelets and other particulate matter have been removed. Whole blood has a viscosity of about 4 cP, while blood plasma has a viscosity of about 1.5 cP. Explain the reason for the difference.
9. A glass filled to the brim with water sits on a scale. A block of wood is gently placed in the water so it floats in the glass. Some of the water overflows and is wiped away but, at the end, the glass is still filled to the brim. Compare the initial and final readings of the scale.
10. Explain how you can determine the density of an irregular object by use of Archimedes' principle. Consider two cases: the material is more dense and less dense than the flotation fluid.
11. From your own experience in floating, estimate the density of your body. (E)
12. Explain the principle behind the operation of a siphon.
13. To make a baseball follow a curved path, the pitcher puts "spin" on the ball. Explain why a spinning ball should follow a curved path.
14. A Ping-Pong ball can be suspended in the air by blowing a jet of air just above it, as shown in Fig. P9.1. Explain.
15. Discuss the meaning of Bernoulli's equation when the liquid is not flowing.

PROBLEMS

1. A steel sphere has a radius of 0.50 cm and has a mass of 4.15 g. Find the mass density of the steel.
2. A certain can weighs 3.1 lb when empty and 46.2 lb when filled with water. What is the volume of the can?

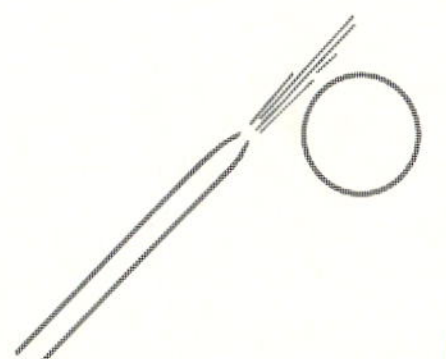

FIGURE P9.1

3. The human body has a density close to that of water. Find the volume occupied by a 70-kg person.
4. A pycnometer is a small flask used for the density determination of liquids. In a certain case, the pycnometer "weighs" 20.000 g when empty and 22.000 g when filled with water at 20°C. After being dried and refilled with a benzene solution, it "weighs" 21.760 g. From these measurements and the density of water, find the volume of the flask and the density of benzene solution. For very accurate measurements of density, the weight of air in the empty flask must be taken into account. How much does the air "weigh" in the pycnometer described in this problem? (Density of air = 0.00129 g/cm^3.)
5. A 5.0-kg mass is hung from the end of a 0.70-mm-diameter steel wire which is 1.40 m long. How far will the wire stretch under this load?
6. A certain broken rubber band is clamped at one end and held taut by a small load at its other. Its length is 9.0 in, and its cross-sectional area is 7.8×10^{-3} in^2. When a 2.0-lb load is added to its end, the rubber band stretches 0.27 in. Find the average Young's modulus in this range in units of pounds per square inch.
7.* An 80-lb stoplight hanging over the center of a street is held by two equal steel cables fastened to poles on either side of the street. If the cables make an angle of 20° with the horizontal, by what fraction are the cables stretched because of the weight of the light? Take the cross-sectional area of the cables to be 0.030 in^2.
8. A copper wire with cross-sectional area 0.0030 cm^2 is to be used in lifting a 100-g object. How fast can the object be accelerated if the wire is to stretch not more than 0.10 percent?
9.* Show that the fractional increase in length under load of a wire consisting of length L_1 of wire 1 attached to L_2 of wire 2 is given by

$$\left(\frac{F}{\pi a^2 Y_1 Y_2}\right)\left(\frac{L_1 Y_2 + L_2 Y_1}{L_1 + L_2}\right)$$

provided that they both have the same radius a.
10. A cube of gelatin has top dimensions of 4.0 by 4.0 cm and a height of 3.0 cm. A shearing force of 0.50 N applied to its upper surface causes the upper surface to displace 2.5 mm in the direction of the force. Find the shear modulus for the gelatin in newtons per square meter.
11. (*a*) By about what fraction will the volume of a bar of steel change as the air around it is removed in a vacuum chamber? (*b*) How large an increase in pressure is needed to decrease a volume of benzene by 1 percent?
12. What is the average pressure exerted by the floor on the foot of a 70-kg woman as she stands tiptoe on one foot? Assume the contact area between foot and floor to be 12 cm^2.
13. Standard atmospheric pressure is 1.013×10^5 N/m^2. Find the force exerted by the atmosphere on the side of a can if the area of the side is 600 cm^2. Express your answer in both newtons and pounds.
14. What is the pressure due to the water at a depth of 15.0 m in a freshwater lake? What is the total pressure at that depth, assuming atmospheric pressure is 1.00×10^5 N/m^2?
15. Standard atmospheric pressure is 1.013×10^5 N/m^2 (14.7 lb/in^2). (*a*) How high a column of water can be supported by this pressure? (*b*) To what maximum height can a vacuum-type water pump lift water in a well when the atmospheric pressure has the standard value?
16. By sucking on one side of a mercury manometer, a person is able to make one side 74 mm lower than the other. (This practice is not recommended since mercury vapor is poisonous.) Assuming atmospheric pressure to be 1.013×10^5 N/m^2, what is the pressure of air in the person's mouth when doing this experiment? Express your answer in both centimeters of mercury and newtons per square meter.
17. A U tube (much like the manometer shown in Fig. 9.19) has water in one side and oil in the other. The oil column on one side is 62 cm high and balances a column of water which is 54 cm high on the other side. What is the density of the oil?
18. Hydraulic stamping machines exert tremendous forces upon a sheet of metal to form it into the desired shape. Suppose the input pressure is 150 lb/in^2 on a piston which has a diameter of 0.50 in. The output force of the hydraulic press is exerted on a piston which has a diameter of 1.3 ft. How large a force does the press exert on the sheet being formed?
19. If a man floats with 95 percent of his volume beneath water, what is his mass density?
20. An irregular piece of metal "weighs" 10.00 g in air and 8.00 g when submerged in water. Find the volume of the metal and its density.
21. The piece of metal described in the previous problem "weighs" 8.50 g when immersed in a particular oil. Find the density of the oil.

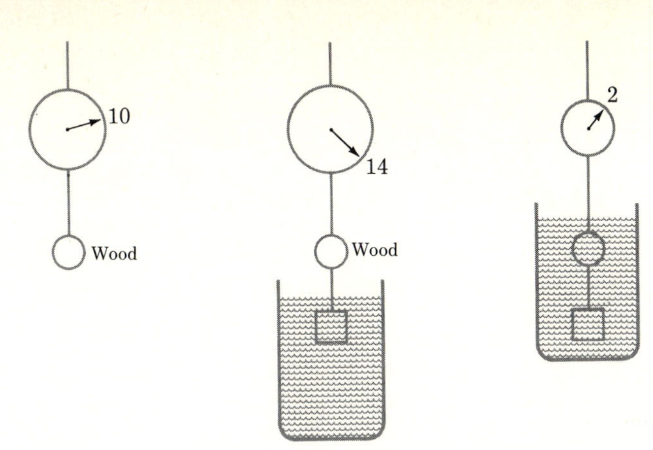

FIGURE P9.2

22. A piece of wood "weighs" 10.00 g in air. When a heavy piece of metal is suspended below it, the metal being submerged in water, the "weight" of wood in air plus metal in water is 14.00 g. The "weight" when both wood and metal are submerged in water is 2.00 g. Find the volume and the density of the wood. (See Fig. P9.2)

23. A beaker partly filled with water "weighs" 20.00 g. If a piece of wood having a density of 0.800 g/cm^3 and a volume of 2.0 cm^3 is floated on the water in the beaker, how much will the beaker "weigh?"

24.* A beaker partly filled with water "weighs" 20.00 g. If a piece of metal with density 3.00 g/cm^3 and volume 1.50 cm^3 is suspended by a thin string so that it is submerged in the water but does not rest on the bottom of the beaker, how much does the beaker "weigh?"

25. The density of ice is 0.917 g/cm^3, and the approximate density of the seawater in which it floats is 1.025 g/cm^3. What fraction of an iceberg is beneath the water surface?

26. What is the minimum volume of a block of material ($D = 50$ lb/ft^3) if it is to hold a 160-lb man entirely above the surface of the water when he stands on it?

27.* A block of material has a density d_1 and floats three-fourths submerged in a liquid of unknown density. Show that the density d_2 of the unknown liquid is given by $d_2 = 1.33d_1$.

28. According to the theory first developed by Stokes, the following equation (known as Stokes' law) gives the force needed to pull a sphere of radius a through a fluid of viscosity η with a speed v,

$$F = 6\pi\eta a v$$

How large a force is required to pull a 1.50-cm-diameter marble through water at a speed of 20 cm/s?

29.* A tiny glass sphere has a radius of 0.50 mm and a density of 2600 kg/m^3. It is let fall through a vat of oil ($d = 950$ kg/m^3, $\eta = 2.1$ P). Find (*a*) the buoyant force on the sphere and (*b*) the gravitational force on the sphere. Then (*c*), use Stoke's law (see Prob. 28) to find the terminal speed of the sphere as it falls through the oil.

30.** To determine the size of tiny spherical particles, their sedimentation speed is often measured. Show that a particle of density d which reaches a terminal speed v while falling through a fluid of density d_f and viscosity η has a radius b given by

$$b = \sqrt{\frac{9\eta v}{2g(d - d_f)}}$$

Make use of Stokes' law (see Prob. 28).

31. The pipe near the lower end of a large water-storage tank springs a small leak, and a stream of water shoots from it. If the top of the water in the tank is 50 ft above the point of the leak, (*a*) with what speed does the water gush from the hole? (*b*) If the hole has an area of 0.010 in^2, how much water flows out in 1 s?

32. Water is flowing smoothly through a closed pipe system. At one point the speed of the water is 3.0 ft/s, while at another point 10 ft higher the speed is 4.0 ft/s. If the pressure is 12 lb/in^2 at the lower point, (*a*) what is the pressure at the upper point? (*b*) What would the pressure at the upper point be if the water were to stop flowing and the pressure at the lower point were 10 lb/in^2?

33. A certain hypodermic needle has an inside diameter of 0.40 mm and a length of 5.0 cm. It is operated by a plunger which has an end area of 3.0 cm^2. How large a force must one apply to the plunger so that water flows from the needle at a rate of 0.20 cm^3/s? Assume the needle is squirting into the air.

34. The boundary between laminar and turbulent flow can be estimated by use of a number called the **Reynolds number,** given by

$$\text{Reynolds number} = \frac{Qd}{\pi r \eta}$$

for flow rate Q through a cylindrical pipe of radius r. The density of the fluid is d, and its viscosity is η. Turbulent flow sets in if the Reynolds number exceeds about 1000. Find the approximate maximum laminar flow rate for benzene through a 0.50-cm-diameter tube. Express your answer in cubic centimeters per second and use 1000 as the critical Reynolds number.

TEMPERATURE, MOTION, AND THE GAS' LAW

10

With this chapter we begin the study of heat energy and its effects. The definition of temperature will be given, and the nature of heat will be investigated. We shall be able to conclude, after a discussion of the gas law, that temperature is a measure of molecular KE and that heat energy is KE of random molecular motion of the gas. These concepts will be needed as a foundation for the chapter which follows, the study of thermal properties of materials and the equivalence of heat and other forms of energy.

10.1 EARLY IDEAS ABOUT HEAT

Ever since the discovery of fire by early man, the nature of heat has been a source of interest. We know that the wonderful properties of fire caused ancient man to deify it. Although the spiritual maturity of most of the human race soon reached a level such that fire was no longer considered a god, the feeling that heat and fire had mystical qualities persisted until relatively recent times.

In the eighteenth century, people believed that the heat which warmed them as they stood near a fire was actually a material quantity which flowed to them through space. This fluid was called **caloric.** When a hot piece of metal was cooled by water, the heat fluid, caloric, was believed to flow out of the metal into the water. A burning piece of wood was torn apart by the flames, and caloric was allowed to escape and flow to other bodies, or so the people of the 1780s believed. Each piece of material contained more or less of the caloric fluid, depending upon its temperature.

It was not until the 1790s that this prevalent concept of a heat fluid was effectively challenged. The most conclusive blow to the caloric theory was dealt by Benjamin Thompson (who was awarded the title Count Rumford, by which he is widely known). Rumford was an American who joined the British forces during the American Revolution. He became an armament expert and served for many years as an official in the government of Bavaria. During that time he performed a number of scientific experiments. The experiments in which we are interested here were conducted in a cannon workshop of the military arsenal at Munich.

He noticed that as cannon barrels were being bored, a tremendous amount of heat was given off. Since the metal chips cut off by the drill had lost caloric, i.e., the heat given off in the drilling process, the chips should not be the same as the original metal, which had not lost caloric. In spite of this reasoning, Rumford was unable to find any difference between the chips and the original metal in their ability to hold or give off heat. Seeking to investigate this effect further, he tried a very dull drill, which was unable to cut the metal. In spite of the fact that no metal was being cut, this borer was also found to generate heat as it was rotated while rubbing against the metal. In fact, the drill generated enough heat to boil water in a cavity of the metal. No matter how long the borer had previously been used on a piece of metal, heat was readily generated each time the borer was started once again. From this, Rumford concluded that the supply of heat was inexhaustible. He further concluded that the heat came, not from the metal, but from the rotation of the drill. As a result, in 1798 Rumford was led to discard the caloric concept of heat. Instead, he postulated that the motion of the drill transmitted motion to the particles within the metal and heat was actually this motion. As long as the drill kept transmitting motion to the metal by frictional forces, the motion and heat within the metal would keep on increasing. This is essentially the picture of heat which is accepted by physicists today. We shall return to this discussion of the nature of heat after we first examine the concept of temperature.

10.2 THERMOMETERS

Long before Rumford's discovery, practical means had been found for measuring how hot a body was. Qualitatively, we all know what is meant by hot and cold. We also know that a hot object when placed in contact with a cold object will cool, while the cold object warms. To measure the hotness of an object, some method must be found to place a number on this property of the body. We call this number the **temperature** of the object. There are a large number of ways in which a number may be associated with the property of hotness, but only four have been widely accepted.

FIGURE 10.1
As the mercury in the glass container is heated, it expands and rises in the capillary. The height at which it stands is a measure of the temperature.

Nearly all thermometers are based upon the principle that most liquids and solids expand as they are heated, a subject discussed more fully later in this chapter. If one fills a bulb with a liquid, such as mercury, and attaches that bulb to a uniform capillary tube, as shown in Fig. 10.1, the device can be used as a thermometer. As the device is made hotter, the mercury expands and rises higher in the capillary tube. Hence, the height of the mercury is a measure of the hotness, or temperature, of the device.

In order to place a numerical value on each level of hotness, one must define a scale of temperature. This has been done in several ways throughout the years. For example, a temperature scale may be conveniently expressed in terms of the boiling and freezing points of pure water.*

If the device of Fig. 10.1 is placed in a vessel containing water and melting ice, the mercury will reach a certain level in the capillary. A mark is made at that level. Now the device is placed in boiling water (atmospheric pressure being 760 mm Hg), and the capillary is marked at the new higher level of the mercury. There are now three common ways of placing numbers along the capillary tube so as to read temperature. These give rise to three different temperature scales. They are illustrated in Fig. 10.2, where the two scale marks are shown, together with three separate scales which can be superposed on them.

It should be noticed that both *the Celsius* (formerly called centigrade) *and Kelvin scales split the region from freezing to boiling of water into 100 parts called* **degrees.** *The Fahrenheit scale splits this region into 180 degrees. A Celsius degree is larger than a Fahrenheit degree by a factor of* $\frac{180}{100}$, *or* $\frac{9}{5}$.

The Celsius scale is used throughout the whole world. The United States will officially accept it when it converts to the metric system. Although once widely used in the English-speaking countries, the Fahrenheit scale is seldom used outside the United States today. The Kelvin scale makes use of Celsius-size degrees but its zero point is taken differently. It is often used by scientists.

We should point out that the mercury thermometer mentioned above will not function below $-39\,°\mathrm{C}$, because mercury freezes at this temperature. This difficulty is surmounted by using alcohol or pentane or some similar

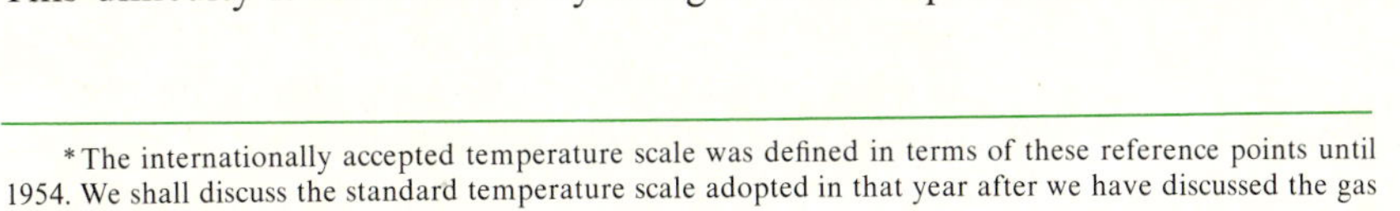

*The internationally accepted temperature scale was defined in terms of these reference points until 1954. We shall discuss the standard temperature scale adopted in that year after we have discussed the gas law. See Sec. 10.7.

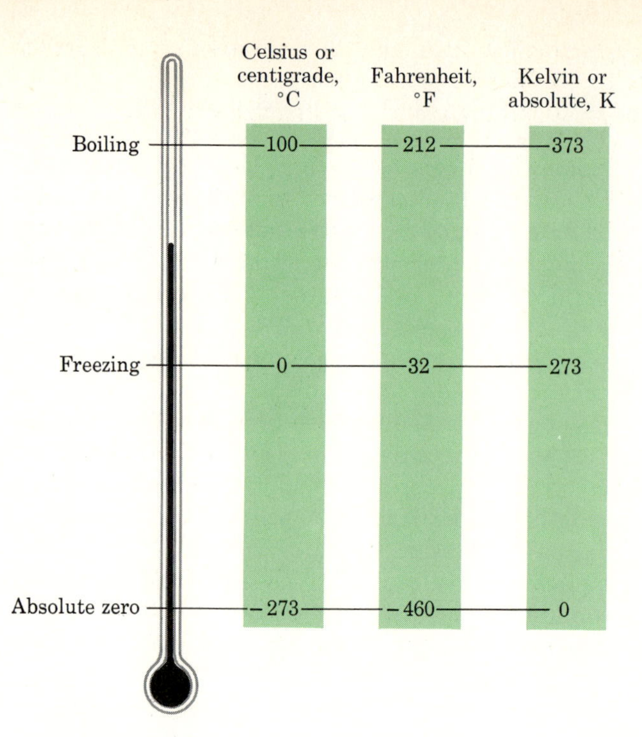

FIGURE 10.2
The boiling and freezing points of water can be used to illustrate the interrelations between the three usual temperature scales.

low-freezing-point liquid at low temperatures. In addition, no two liquids expand in exactly the same way over the whole temperature range. A more precise standard way of subdividing the scales is needed, and this is based upon a thermometer using a gas rather than a fluid.

Although the thermometers we have been discussing are most common, others also exist. *Any physical quantity which varies in a repeatable way with temperature can, in principle, be used as a temperature-measuring device.* Practical considerations determine which type of device should be used in a given circumstance. Perhaps one of the most useful of these alternate devices is the thermistor. Its electrical resistance changes with temperature, and this resistance change is converted electrically to a temperature reading on a meter scale. The thermistor itself can be very tiny and can be connected to the readout system by thin wires. Its response time is of the order of a second, and so readings are almost instantaneous. This device is in widespread use both in medicine and industry.

Illustration 10.1 The temperature of a room is 77°F. What is the Celsius temperature in the room?

Reasoning We always solve such problems in the following way.

1 Find out how many degrees the temperature is above or below the freezing point of water (32°F or 0°C).

2 Use the fact that a Fahrenheit degree is equivalent to $\frac{5}{9}$ Celsius degree to convert from one type of degree to the other.
3 Find the temperature on the new scale by adding to (or subtracting from) 32°F or 0°C.

In the present case we had 77°F. This temperature is $77 - 32 = 45$ Fahrenheit degrees above freezing. But 45 Fahrenheit degrees are equivalent to $(\frac{5}{9})(45) = 25$ Celsius degrees. Therefore the temperature is 25 Celsius degrees above freezing. Since freezing is 0°C, we see that the temperature in question is 25°C.

Illustration 10.2 What is the Fahrenheit temperature on a day when the temperature is −10°C?

Reasoning The temperature is 10 Celsius degrees below freezing. Since a Fahrenheit degree is $\frac{5}{9}$ as large as a Celsius degree, the 10 Celsius degrees are equivalent to $(\frac{9}{5})(10) = 18$ Fahrenheit degrees. The required temperature is 18 Fahrenheit degrees below freezing. Since freezing is 32°F, the temperature in question is $32 - 18 = 14$°F.

10.3 THE GAS LAW

In their laboratory work in science, many students perform simple experiments which illustrate the behavior of gases. The law which governs the behavior of air and many other gases is simple and easily found by experiment. There are three variables which enter into the law: pressure, temperature, and number of molecules per unit volume. Let us see what facts experiment tells us about the relation between these variables.

When a gas which is confined to a fixed volume is heated, its pressure increases. A graph showing the measured pressure versus temperature looks much like that shown in Fig. 10.3. Different gases and different initial pressures yield similar graphs. But, provided the gases are not near the conditions at which they liquefy, the graph *is that of a straight line.* Moreover, *the intercept of the line is always −273.15°C.*

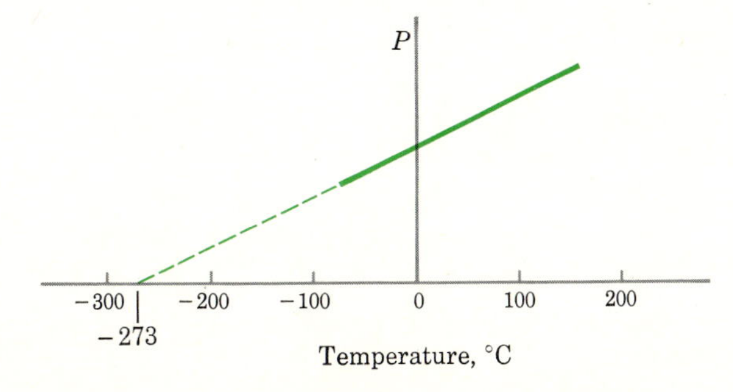

FIGURE 10.3
When a gas in a closed container is heated, the pressure of the gas varies linearly with the temperature. On the absolute temperature scale, the equation of the line is $P = (\text{const})(T)$.

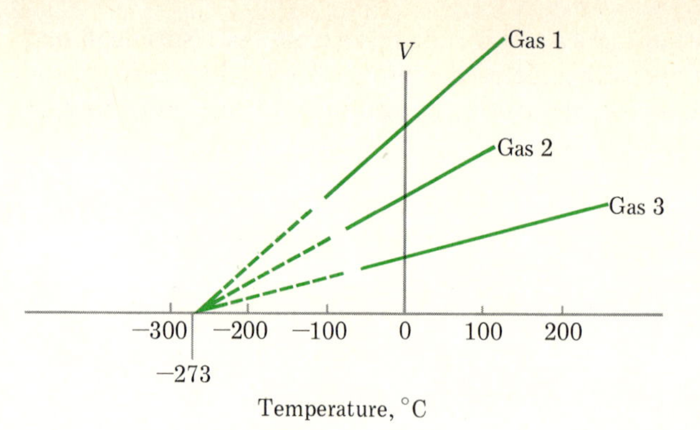

FIGURE 10.4

A gas maintained at constant pressure contracts as it is cooled. On the absolute temperature scale, the equation of the line for an ideal gas is $V = (\text{const})(T)$.

Another simple experiment is to measure the volume of a gas as a function of temperature but subject to a constant pressure. *The measured volume-versus-temperature graph is* typified by Fig. 10.4. Again, *a straight-line* relation is found *with an intercept at* $-273.15\,^\circ C$. Although different gases and different volumes give different straight lines (provided the gas is not near liquefaction), this intercept is always the same.

Notice in both Figs. 10.3 and 10.4 that *the temperature-axis intercept is* $-273.15\,^\circ C$. *This is taken as the zero point on the absolute* (*or Kelvin*) *temperature scale.* Let us represent absolute temperatures by the symbol T. Then the data of these two graphs can be represented by

$$PV = (\text{const})(T)$$

To check this against Fig. 10.3, recall that V is constant in that case. Then

$$P = \left(\frac{\text{const}}{V}\right)(T) = (\text{const})(T)$$

which is the equation of a straight line on the P-versus-T graph. When $T = 0$, that is, at $t = -273.15\,^\circ\text{C}$, $P = 0$. As we see, the equation and Fig. 10.3 agree. Similarly, for $P = \text{const}$, the equation becomes

$$V = \left(\frac{\text{const}}{P}\right)(T) = (\text{const})(T)$$

DEFINITION

This is the equation of a straight line with intercept at $T = 0$ (or $t = -273.15\,^\circ\text{C}$). It agrees with the data shown in Fig. 10.4. *We conclude that a gas which is far removed from conditions under which it liquefies obeys the equation*

$$PV = (\text{const})(T) \tag{10.1}$$

It is important to recognize that T is the absolute (*Kelvin*) *temperature of the gas.*

A gas which obeys Eq. (10.1) is defined to be an **ideal gas.** As we shall see, the constant in Eq. (10.1) has a definite, simply stated value. Before proceeding to discuss this value, we first introduce two numbers often used to express it.

10.4 THE MOLE AND AVOGADRO'S NUMBER

When dealing with atoms and molecules, it is often convenient to recall a number which is of fundamental importance. It is called **Avogadro's number** and is represented by N_A. Even before much was known about atoms, the chemists had already learned of this number. From the way that atoms combine into molecules, Avogadro inferred that an atomic mass* of substance always contains a certain number of atoms. This number is N_A.

For example, you will recall from chemistry that the atomic mass of hydrogen is 1.0. Therefore, 1 kg of hydrogen contains N_A hydrogen atoms. Or, since the atomic mass of chlorine is 35.5, we know that 35.5 kg of chlorine contains N_A chlorine atoms.

This same number carries over to molecules. Since the atomic mass of oxygen is 16, the molecular mass* of water, H_2O, is 18. Therefore 18 kg of water contains N_A oxygen molecules. As we see, N_A is a very meaningful number.

To determine the value of N_A, one must resort to experiment. The best experimental value we have at the present time for Avogadro's number is

Avogadro's Number

$$N_A = 6.022 \times 10^{26} \text{ particles/kg mol}$$

Notice the "units" we have appended to N_A. They are used for descriptive purposes only and should not be considered units in the usual sense. Their meaning is provided by the following definition: *A* **kilogram-mole** (kg mol) *of substance is that quantity of the substance in kilograms which contains N_A particles of the substance. Numerically, it is equal to the atomic or molecular mass for atoms or molecules.*

DEFINITION

Many people express N_A in a different set of units. They use the gram mole (g mol) rather than the kilogram mole. Since a gram is $\frac{1}{1000}$ kg, we see that

$$N_A = 6.022 \times 10^{23} \text{ particles/g mol}$$

Very frequently, in this set of units, the word "gram" is omitted. When a quantity is given in moles, the author or speaker usually means gram moles; however, the kilogram mole is appropriate for SI units, and so care should be taken to use kilogram moles in our equations.

---

*Many people use the terminology atomic weight and molecular weight instead of atomic mass and molecular mass.

Illustration 10.3 The atomic mass of carbon is 12.0. Find the mass of a carbon atom in kilograms. (Atomic masses are usually given as unitless. However, we shall append the "unit" kg/kg mol for descriptive purposes.)

Reasoning We know that 12.0 kg of carbon contains N_A atoms. Therefore, the mass per atom is

$$\text{Mass per carbon atom} = \frac{12.0\ \text{kg/kg mol}}{N_A} = \frac{12.0\ \text{kg/kg mol}}{6.02 \times 10^{26}\ \text{atoms/kg mol}}$$
$$= 1.99 \times 10^{-26}\ \text{kg/atom}$$

This same method can be used to find the mass of any atom or molecule whose atomic or molecular mass is known.

10.5 THE GAS-LAW CONSTANT

Let us now return to the law obeyed by ideal gases. It is

$$PV = (\text{const})(T)$$

where P is the pressure of the ideal gas in volume V at absolute temperature T. By means of the experiments described in Sec. 10.3, the constant can be evaluated. It is found to have the value nR, where n is the number of kilogram moles of gas in the volume V and

$$R = 8314\ \text{J/(kg mol)(K)} = 8.314\ \text{J/(g mol)(K)}$$

We can therefore write the **ideal-gas law** *as*

Ideal-Gas Law

$$PV = nRT \tag{10.2}$$

It is very important in using the law that the proper units for the various quantities be used. The temperature must always be expressed in the Kelvin scale. If n is the number of kilogram moles in V, then R must be 8314 J/(kg mol)(K). If for some reason you insist on using n in gram moles, then R must be taken as 8.314 J/(g mol)(K).

Illustration 10.4 Standard atmospheric pressure and temperature are $1.01 \times 10^5\ \text{N/m}^2$ and 0°C. Find the volume which 1 kg mol of N_2 occupies under these conditions. The molecular mass of N_2 is 28.

Reasoning To use the gas law, we note in the present case that

$$P = 1.01 \times 10^5\ \text{N/m}^2$$
$$T = 0 + 273.15\ \text{K} = 273.15\ \text{K}$$

where we read this latter value as "273.15 kelvins." Since we are concerned with the volume occupied by 1 kg mol, $n = 1$ kg mol. Then

$$PV = nRT$$

becomes

$$(1.01 \times 10^5 \text{ N/m}^2)V = (1 \text{ kg mol})\left[8314 \frac{\text{J}}{(\text{kg mol})(\text{K})}\right](273 \text{ K})$$

From this we find

$$V = 22.4 \text{ m}^3/\text{kg mol}$$

You should be able to generalize this example to show the following:

> One kilogram mole of ideal gas occupies a volume of 22.4 m^3 under standard conditions.

This is sometimes a convenient fact to remember.

10.6 MOLECULAR BASIS FOR THE GAS LAW

We have now considered the pressure of gases from two separate viewpoints. In Chap. 6 we discussed how gas molecules exert pressure on the walls of a container by collisions with the wall. We found that the pressure was related to the KE of the gas molecules. Our result was, in slightly different symbols,

$$P = (\tfrac{2}{3})(\nu_0)(\tfrac{1}{2}m_0 v^2) \tag{6.6}$$

In this equation, we assume there are ν_0 molecules per cubic meter, each molecule has a mass m_0, and its average translational KE is $\frac{1}{2}m_0 v^2$. Notice that the temperature of the gas does not appear in this equation.

On the other hand, the ideal-gas law, obtained by experiment, states that

$$PV = nRT \qquad \text{or} \qquad P = \frac{nRT}{V}$$

By comparing this equation with Eq. (6.6), we see that

$$(\tfrac{1}{3})\nu_0 m_0 v^2 = \frac{nRT}{V}$$

or, after rearranging,

$$T = \frac{(\nu_0 V)m_0 v^2}{3nR}$$

But $\nu_0 V$ is simply the total number of molecules in the volume V while n

is the number of kilogram moles of molecules in the volume. Since the total number of molecules in V is also given by nN_A, we can replace $\nu_0 V$ by nN_A to obtain

$$T = \frac{2N_A}{3R}\tfrac{1}{2}m_0v^2$$

Of course, N_A/R is simply a constant. Its value is

$$\frac{N_A}{R} = \frac{6.02 \times 10^{26}\ \text{kg mol}^{-1}}{8314\ \text{J/(kg mol)(K)}} = 7.24 \times 10^{22}\ \text{K/J}$$

Its reciprocal, R/N_A, is often called **Boltzmann's constant** and is represented by k. Then

$$k = 1.38 \times 10^{-23}\ \text{J/K}$$

In any case we see that *the* **absolute temperature** *of an ideal gas is given by*

Absolute Temperature

$$T = \tfrac{1}{2}m_0v^2\frac{2N_A}{3R} \qquad (10.3)$$

Absolute temperature is a measure of the translational KE of the gas molecules. This fact has great importance, as we shall soon learn.

Illustration 10.5 Estimate the average speed of the nitrogen molecules in the air under standard conditions.

Reasoning Under standard conditions, the temperature is 0°C, or 273.15 K. From Eq. (10.3) we have that

$$v^2 = \frac{3RT}{N_A m_0}$$

To find m_0 for nitrogen molecules, we note that the molecular mass of N_2 is 28. Therefore,

$$m_0 = \frac{28\ \text{kg/kg mol}}{N_A}$$

Substitution yields

$$v^2 = \frac{3RT}{28\ \text{kg/kg mol}}$$

Since $R = 8314$ J/(kg mol)(K) and $T = 273$ K, we find

$$v^2 = \frac{(3)(8314)(273)}{28} \frac{\text{m}^2}{\text{s}^2}$$

or

$$v = 493 \text{ m/s}$$

The molecules of the air have average speed of about 500 m/s.

10.7 USE OF THE GAS LAW

The ideal-gas law is found to apply to all gases provided the following two conditions are satisfied:

1 The volume occupied by the molecules themselves, i.e., the volume of a molecule multiplied by the number of molecules, must be a negligible fraction of the volume in which the gas is enclosed.

2 The translational KE of the molecules must be large compared with the energy needed to separate two gas molecules from each other.

This latter restriction has to do with the fact that uncharged atoms and molecules usually attract each other. As a result, two molecules tend to stick together. The second restriction tells us that the energy needed to tear them away from each other must be small compared with their average translational KE.

But we know that the average translational KE of a molecule is related to the absolute temperature through Eq. (10.3). We can therefore restate the second condition in the following way:

2′ The temperature must be high enough to ensure that only a negligible number of molecules stick together.

Of course, when the molecules stick together, the gas must condense to a liquid. This then allows us to state the second condition in still another way:

2″ The gas must be far removed from conditions under which it liquefies.

Condition 2 is normally satisfied for gases such as air, oxygen, nitrogen, helium, hydrogen, and all of the gases difficult to liquefy. Condition 1 is satisfied for all gases whose density is comparable with, or smaller than, that of air.

An interesting feature of Eq. (10.3) (and Figs. 10.3 and 10.4) can be seen if you notice what happens at absolute zero ($T = 0$ K). Since T is proportional to the translational KE of the molecules, when $T = 0$, the KE should also be zero. The molecules would no longer be moving, and so they would exert no pressure. This is why, in Fig. 10.3, we see that $P = 0$ when $T = 0$. Can you explain why, as shown in Fig. 10.4, $V = 0$ when $T = 0$?

But, of course, no gas can be cooled to absolute zero without condensing. Conditions 1 and 2 become invalid before we come close to $T = 0$, and so our equation (the ideal-gas law) no longer applies. In fact, queer behavior is observed near absolute zero. Classical physics no longer applies. Instead, as we shall see later, quantum effects become important and lead to such effects as superfluidity (zero-viscosity fluids) and superconductivity (zero-resistance electrical conductors).

In spite of these restrictions on the ideal-gas law, the law has a wide range of application. One of its most fundamental uses is made in defining the Kelvin temperature scale. To do this, we recognize that the gas law can be written as

$$P = (\text{const})(T)$$

if V is held constant. If we assign a value to the constant, then T is known from a measurement of P.

In order to evaluate this constant, it was agreed in 1954 that the triple point of water should be assigned the temperature 273.16 K. (The **triple point of water** is that unique temperature at which water, ice, and water vapor can coexist with each other.) Suppose now that the pressure of a fixed-volume container of ideal gas is measured at the triple point of water. P is then known, and $T = 273.16$ K by definition. Therefore the constant in the relation $P = (\text{const})(T)$ can be evaluated.

Now that the constant is known, the gas in the container can be used to measure temperature. One simply measures $P = (\text{const})(T)$ to find T. Of course, a gas thermometer is seldom used since it is inconvenient. But, when needed, it can be used to calibrate any other type of thermometer one wishes to use.

For most of us, the ideal-gas law is more useful in two other respects. We can use it to compute P, V, n, or T provided all but one of these quantities are known. Or, having P, V, n, and T under one set of conditions, we can compute the parameters under other conditions. These uses are illustrated in the following examples.

Illustration 10.6 An empty oil drum is closed at a temperature of 20°C. It is then set out in the sun, where it heats up to 200°C. If the original pressure was 1.0 atm, what is the final pressure in the drum?

Reasoning Write Eq. (10.2) twice,

$$P_1 V = nRT_1 \qquad P_2 V = nRT_2$$

where V is the constant volume of the drum and n is the number of moles of gas in the drum. Dividing one equation by the other yields

$$\frac{P_1}{P_2} = \frac{T_1}{T_2}$$

where T_1 and T_2 are absolute temperatures. We had $P_1 = 1$ atm,

$T_1 = 20 + 273 = 293$ K, and $T_2 = 473$ K. Therefore

$$P_2 = (1.0 \text{ atm})\left(\frac{473 \text{ K}}{293 \text{ K}}\right) = 1.61 \text{ atm}$$

Notice that absolute temperatures must be used. The pressure units are not important as long as they are the same.

Illustration 10.7 The gas in the piston of a diesel engine is originally at a temperature of 27°C and a pressure of 74 cm Hg when it is suddenly compressed. If the final pressure of the gas is 3700 cm Hg and the temperature has been increased to 547°C, what is the final volume of the gas in terms of the original volume?

Reasoning Write the gas law twice,

$$P_1V_1 = nRT_1 \qquad P_2V_2 = nRT_2$$

Dividing yields

$$\frac{P_1V_1}{P_2V_2} = \frac{T_1}{T_2}$$

Substituting gives

$$\left(\frac{V_1}{V_2}\right)\left(\frac{74}{3700}\right) = \frac{273 + 27}{273 + 547}$$

from which

$$V_2 = 0.058\,V_1$$

Illustration 10.8 A car tire is filled to a pressure of 32 lb/in^2 on a cold day when the temperature is −3°C. What is the pressure in the tire when the temperature rises to +47°C? (Assume the volume of the tire to remain constant.)

Reasoning Once again

$$\frac{P_1V_1}{P_2V_2} = \frac{T_1}{T_2}$$

But $V_1 = V_2$, $T_1 = 270$ K, $T_2 = 320$ K. *Notice that the pressure read by a tire gauge is the excess pressure within the tire.* (The gauge reads zero when the pressure is atmospheric pressure.) Assuming atmospheric pressure to be

14.7 lb/in², we have $P_1 = 46.7$ lb/in². Hence

$$P_2 = \frac{P_1 T_2}{T_1} = (46.7\ \text{lb/in}^2)\left(\frac{320\ \text{K}}{270\ \text{K}}\right) = 55.4\ \text{lb/in}^2$$

and the gauge would read 40.7 lb/in².

Illustration 10.9 A tank of oxygen gas at 0°C has a volume of 4 liters,* and the pressure is 50 atm. What mass of oxygen is in the tank?

Reasoning We shall solve this in two different ways.

Method 1: Let us first find what volume this gas would occupy at standard temperature and pressure:

$$\frac{P_1 V_1}{P_2 V_2} = \frac{T_1}{T_2} = 1$$

$$(50\ \text{atm})(4\ \text{liters}) = (1\ \text{atm})(V_2)$$

$$V_2 = 200\ \text{liters} = 0.20\ \text{m}^3$$

But 22.4 m³ will hold 1 kg mol of O_2 under these conditions. Since the atomic mass of oxygen is 16 and the molecular mass is 32, the mass of oxygen in the tank is therefore

$$\text{Mass} = \left(\frac{0.2\ \text{m}^3}{22.4\ \text{m}^3/\text{kg mol}}\right)(32\ \text{kg/kg mol})$$

$$= 0.285\ \text{kg} = 285\ \text{g}$$

Method 2: Working directly from the gas law, Eq. (10.2),

$$PV = nRT$$

we substitute $(50)(1.01 \times 10^5\ \text{N/m}^2)$ for P and $4 \times 10^{-3}\ \text{m}^3$ for V. Using $R = 8314$ J/(kg mol)(K) and $T = 273$ K, one finds

$$n = 8.9 \times 10^{-3}\ \text{kg mol}$$

But 1 kg mol of oxygen is 32 kg, and so the mass in the container is simply

$$\text{Mass} = (8.9 \times 10^{-3}\ \text{kg mol})(32\ \text{kg/kg mol}) = 0.285\ \text{kg}$$

Illustration 10.10 What is the mass of a molecule of carbon dioxide, CO_2?

---

*The liter, a unit familiar to many of you from chemistry, is defined to be a volume of 1000 cm³.

Reasoning The atomic mass of carbon is 12 and of oxygen is 16. Hence the molecular mass of CO_2 is 44. Since 44 kg of CO_2 will contain Avogadro's number of molecules, 6.02×10^{26}, the mass of a molecule will be

$$m_0 = \frac{44 \text{ kg}}{6.02 \times 10^{26}} = 7.33 \times 10^{-26} \text{ kg}$$

10.8 VARIATION OF MOLECULAR SPEEDS IN GASES

As we saw in the previous sections, temperature is a measure of the average molecular KE. In particular, from Eq. (10.3) we find that the average translational KE of a gas molecule is simply

$$\text{KE of translation} = \tfrac{1}{2}m_0v^2 = \left(\frac{3}{2}\right)\left(\frac{R}{N_A}\right)T = \tfrac{3}{2}kT \tag{10.4}$$

where k is Boltzmann's constant, R/N_A. But we know of course that not all molecules in a gas have the same KE and speed. Moreover, as they collide with each other, the speed of any given molecule is subject to change.

Long before the experimental skills of scientists had progressed far enough to enable them to measure the speeds of gas molecules directly, these speeds were predicted by James Clerk Maxwell. He showed that, in theory at least, all speeds between zero and infinity should be attainable by a gas molecule. We now know that no particle can exceed the speed of light c, and so the prediction concerning infinite speed is clearly in error. However, at all ordinary temperatures, his prediction showed that the chance (or probability) that a molecule should have speeds in excess of c was entirely negligible, although still finite.

Maxwell's prediction for molecular speeds is most easily seen from a graph. Suppose one selects a molecule in a gas and measures its speed at the instant of selection. The probability (or chance) that it will have any given speed is shown in Fig. 10.5. For the purposes of this example, the gas is assumed to be nitrogen at the indicated temperatures. We see that the molecules in nitrogen gas at 460 K (187°C) are most likely to have a speed near 500 m/s. However, a vast range of speeds is possible, and one should not be surprised to find the molecule selected to have any speed between about 50 and 1500 m/s. As stated earlier, the curves do not actually reach zero even when infinite speeds are reached; but, as we see, the chance of finding a molecule with such a high speed is extremely small—in fact, entirely negligible in the case shown.

It was not until 1926 that Maxwell's predictions (made in 1860) were susceptible to test by experiment. When the test was made, excellent agreement was found with the results of Maxwell's theory. Not only does experiment confirm that the average translational kinetic energy of a gas molecule is $\frac{3}{2}kT$, but the predicted distribution of speeds is also found.

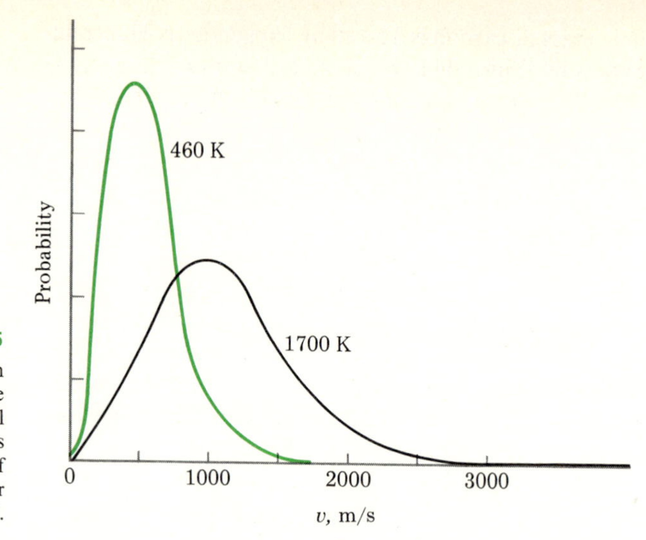

FIGURE 10.5
The chance that a nitrogen molecule in a gas at the temperature indicated will have a certain speed is proportional to the height of the curve at that particular speed.

10.9 BROWNIAN MOTION

The molecular concepts we have used in discussing the ideal-gas law have very wide applicability. As we have just seen, Maxwell was able to predict the variation of speeds within a gas by use of these concepts. Many other facets of gas behavior have been successfully treated using this model of an ideal gas. Indeed, these results of the so-called **kinetic theory of gases** have been amply confirmed by many different experiments.

In this and the two following sections, we discuss extensions of this model to systems we would not ordinarily think of as gases. Consider tiny smoke or dust particles in the air or, perhaps, colloidal particles such as those which make impure water slightly hazy or milk white. Each of these tiny particles acts like a "gas molecule" and has a translational KE given by Eq. (10.4), namely, $\frac{3}{2}kT$. If this KE is much larger than the gravitational PE of the particle, the particles will not settle to the bottom of the container. They will float through air (in the case of dust) or the liquid (for colloidal particles) just as though they were molecules of an ordinary gas.

If you observe a smoke particle or a colloidal particle with a microscope, you will see it follow a jagged path like the one shown in Fig. 10.6. *The motion is random. This type of motion* was first reported by a botanist, Robert Brown, and *is called* **Brownian motion.** The detailed theory of the effect was given by Albert Einstein. (He developed this theory in 1905, the same year he proposed two other famous theories, the theory of relativity and the theory of the photoelectric effect. He was 26 years old at the time.) Although the particles are much larger than gas molecules, the general ideas are applicable to both. Such particles behave like an ideal gas in many ways.

FIGURE 10.6
Tiny particles suspended in a fluid undergo a zigzag motion called Brownian motion.

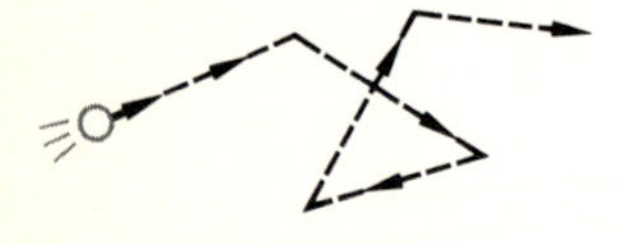

It is of interest to note the cause of the zigzag motion. Since the particles move past many surrounding molecules before changing course, the direction change points do not indicate points of collision. Instead, they represent points at which, by chance, the forces exerted on the particle by the surrounding molecules have a large resultant. Although, most often, the resultant force on the particle exerted by the surrounding moleclues is near zero, this is not always the case. Now and then, by chance, the resultant of these forces becomes large. At that time, the particle deflects under the large resultant force on it.

We can better understand this effect by an analogy. Suppose you throw 100 coins and count those which come up heads as $+1$ and tails as -1. Usually the number of heads and tails will be about the same. So the usual resultant is usually close to zero. But occasionally, by chance, many more heads will come up than tails. In that case the resultant will be large. As we see, random effects which usually balance out occasionally do not do so.

The Brownian motion effect is important in several respects. It occurs in all colloidal systems. But, from a fundamental standpoint, it provides a visible means for observing the molecular properties of an ideal gas. By means of the Brownian motion of colloidal particles, it was possible to obtain the first direct confirmation of the kinetic theory. In addition, one of the earliest measurements of Avogadro's number was made by means of an experiment involving the motion of colloidal particles.

10.10 OSMOTIC PRESSURE

Using the ideas of the kinetic theory of gases, we can draw an important conclusion concerning a mixture of two or more gases. The conclusion was first stated by Dalton and is known as **Dalton's law of partial pressures.**

Dalton's Law of Partial Pressures

> The total pressure due to a mixture of ideal gases is equal to the sum of the pressures each gas would exert if alone.

In other words, suppose certain amounts of N_2, H_2, and O_2 gas are placed in a container of volume V. Suppose further that this same amount of N_2 would exert a pressure P_N if placed in the same container by itself. The similarly determined pressures for H_2 and O_2 are P_H and P_O. Then, Dalton's law tells us that the total pressure due to the mixture of these gases will be

$$P = P_N + P_H + P_O$$

As we see, each gas is unaffected by the presence of the other gas as far as pressure is concerned. This is not surprising since the ideal-gas law applies at any gas pressure for ideal gases. It tells us that $P \propto n$, the number of moles of gas in the container. The law contains nothing which depends upon the particular type of molecule involved: 1 mol of N_2 plus 1 mol of H_2 in the container should give the same pressure as 2 mol of either N_2 or H_2. Indeed, the ideal-gas law predicts Dalton's law of partial pressures.

But we have also seen that the kinetic theory applies to colloidal particle motion. We expect, then, that the ideal-gas law should apply to both molecular and colloidal solutions. One might therefore guess that molecules dissolved to form a dilute solution should act like an ideal gas. This turns out to be true and gives rise to what is known as **osmotic pressure.** Let us now examine the very useful concept of osmotic pressure.

When you were in grade school you may have demonstrated the effects of osmotic pressure by means of a hollowed out carrot and colored sugar water. The colored sugar water is sealed inside the hollowed out carrot but is free to rise up in a capillary tube sealed in the top of the carrot. When the carrot is placed partly submerged in a beaker of pure water, a curious behavior is noted. Some of the water slowly passes through the wall of the carrot and enters the sugar water. This causes the sugar water to rise up in the capillary tube. When the flow stops, the sugar water in the capillary tube stands much higher than the water in the beaker. We conclude that the final pressure inside the carrot is much larger than the water pressure in the beaker. The water in the beaker has flowed from a region of low pressure to a region of higher pressure. We shall see that the difference in the pressures inside and outside the carrot is what is called the osmotic pressure. We shall describe here a more precise version of the experiment. It is used to measure molecular masses of proteins, polymer molecules, and of other rather large molecules. The experiment makes use of the device which will now be described.

An osmometer* consists of two chambers separated by a very thin, semipermeable membrane, as shown in Fig. 10.7. In the grade-school experiment, one chamber is the hollow space inside the carrot and the other is the beaker in which the carrot is immersed. The carrot wall acts as the semipermeable membrane. Modern osmometers make use of very thin sheets of plastic which swell but do not dissolve in the solvent from which the solution to be measured is made.† The membrane must be properly chosen so that the solute molecules do not pass through the membrane while the solvent molecules do pass through. It is for this reason that the membrane is said to be semipermeable.

When the device in Fig. 10.7 is filled as shown, the levels in the two chambers adjust and eventually reach equilibrium. Solvent molecules diffuse through the membrane (either out of or into the solution) until there is the same solvent pressure on each side of the membrane. Notice carefully that the solute molecules do not pass through the semipermeable membrane. The system reaches equilibrium only when the solvent pressures are the same on the two sides. Let us call the pressure due to the solvent molecules at the membrane P_s.

FIGURE 10.7

The osmotic pressure of the solute in the very dilute solution is *hdg*, where *d* is the density of the solution or solvent (the two densities are essentially the same).

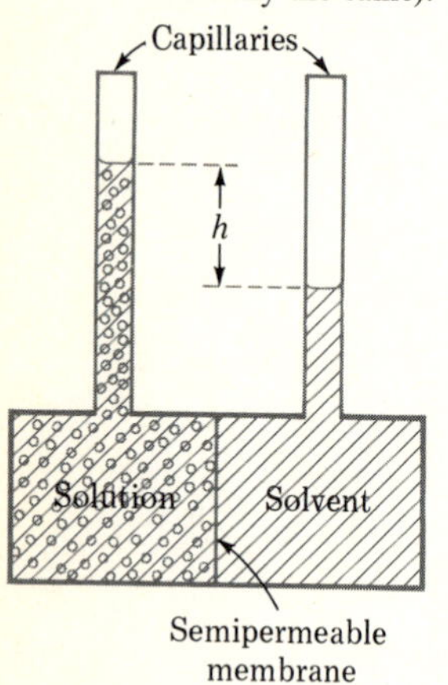

But the total pressure at the left of the membrane is due to the solvent plus the solute "gas" molecules. Let us denote this solute "gas" pressure by π. Then, Dalton's law tells us that the total pressure on the left is

$$\text{Pressure on left} = \pi + P_s$$

However, the pressure on the right is due only to the pure solvent. It, too, is

* Pronounced os-mom′-i-ter.

† A solution is made by dissolving a solid in a liquid, e.g., sugar in water. The solid is called the **solute.** The liquid is called the **solvent.**

P_s once equilibrium is established. Therefore

$$\text{Pressure on right} = P_s$$

DEFINITION

We now see that there is a difference in total pressure on the two sides of the membrane. The difference in pressure between the two sides of the membrane is called the **osmotic pressure.** It is

$$\text{Osmotic pressure} = (\pi + P_s) - P_s = \pi$$

In other words, *the osmotic pressure is simply the pressure of the solute "gas" in the solution.* We can measure it in the osmometer by measuring h shown in Fig. 10.7. Then, since the osmotic pressure is equal to the difference in pressure, we have $\pi = hdg$, where d is the density of the very dilute solution. (In practice, the solution and solvent have negligibly different densities.)

Let us now return to the ideal-gas law. The pressure due to the solute "gas" will be given by the gas law provided the solution is very dilute. Then our conditions for applicability of the gas law will prevail. We can then write

$$\pi V = nRT$$

where n is the number of kilogram moles of solute in the volume V. Or if we represent the molecular mass (or weight) of the solute by M, we have

$$n = \frac{m}{M}$$

where m is the mass of solute in the volume V.

Upon simplification this becomes

$$\pi = \left(\frac{m}{V}\right)\left(\frac{RT}{M}\right)$$

But m/V is the concentration of the solution (in kilograms per cubic meters), and we shall represent it by c. Therefore

$$\pi = \text{osmotic pressure} = \frac{cRT}{M} \qquad (10.5)$$

Osmotic Pressure of Dilute Solution

As we see, *the osmotic pressure is largest for molecules of small molecular weight*. However, for very small molecules, it is difficult to find membranes which will pass the solvent but not the solute.

We can make use of an osmometer and Eq. (10.5) to determine the molecular weight M of a solute molecule. Solving for M, we find

$$M = \frac{c}{\pi}RT$$

If we measure the osmotic pressure π of a solution of known concentration c, the solute M can be found. *In practice, care must be taken to make c small enough for the solute molecules to act like an ideal gas.* It is common practice to use osmometers to measure the molecular masses of molecules in the range from about 5000 to several million.

Illustration 10.11 One of the protein molecules found in blood plasma is albumin. A water solution containing 0.20 g of albumin per 100 cm^3 has an osmotic pressure of 0.74 cm of water at 27°C. From these data, find the molecular mass (weight) of the albumin molecule.

Reasoning We can make use of Eq. (10.5). Since π is 0.74 cm of water,

$$\pi = hdg$$

becomes

$$\pi = (0.74 \times 10^{-2}\ \text{m})(1000\ \text{kg/m}^3)(9.8\ \text{m/s}^2) = 72.5\ \text{N/m}^2$$

Then, from the gas law,

$$M = \left(\frac{c}{\pi}\right)RT$$

But

$$c = \frac{0.20\ \text{g}}{100\ \text{cm}^3} = 2.0\ \text{kg/m}^3$$

and

$$T = 273 + 27 = 300\ \text{K}$$

Therefore, after substituting these values and 8314 J/K for R, we find

$$M = 69{,}000$$

for the molecular mass of albumin.

SUMMARY

Heat energy is related to the KE of a substance. The hotter a substance, the more KE its molecules have. Thermometers are used to measure temperature. Three common temperature scales exist. The Celsius (°C) scale has the freezing point of water at 0°C and the boiling point at 100°C. The Kelvin (or absolute) scale (K) has these two points at 273.15 and 373.15 K. The Fahrenheit (°F) scale takes these two points to be 32 and 212°F. One Celsius degree (or kelvin) is equivalent to $\frac{9}{5}$ Fahrenheit degree.

Many gases obey the ideal-gas law and are called ideal gases. The law is $PV = nRT$. It assumes n moles of gas contained in a volume V at a pressure P. The temperature T must be measured on the Kelvin (absolute) scale. The symbol R represents the gas constant and has a value 8314 J/(kg mol)(K).

A kilogram mole (kg mol) of substance is a mass of substance whose mass in kilograms is equal numerically to the atomic or molecular mass (or weight) of the substance. In a more general sense, it is the mass of substance (in kilograms) which contains 6.02×10^{26} particles of the substance. The number 6.02×10^{26}

particles/kg mol is called Avogadro's number and is represented by N_A. People often use the gram mole (or just mole) which is $\frac{1}{1000}$ kg mol.

The kinetic theory of gases can be combined with the ideal-gas law to show that absolute (Kelvin) temperature is a measure of the translational KE of the gas molecules. Precisely, $T = (2N_A/3R)(\frac{1}{2}m_0v^2)$, where $\frac{1}{2}m_0v^2$ is the average translational KE of a gas molecule. Often, the ratio R/N_A is replaced by k, Boltzmann's constant. Then one has $\frac{1}{2}m_0v^2 = \frac{3}{2}kT$. The average speed of a nitrogen molecule in the air is about 500 m/s, but the molecules have a wide distribution of speeds at any instant.

A gas will behave like an ideal gas provided it obeys two general restrictions. The gas must be far from the conditions under which it will liquefy. The actual volume of the molecules must be small in comparison to the volume available to the gas.

The ideal-gas law is used to define the Kelvin temperature scale. In doing so, the triple point of water is assigned the value 273.16 K.

Tiny dust, smoke, and colloidal particles undergo a zigzag motion called Brownian motion. These particles, if dilute enough, often obey the ideal-gas law. The average translational KE of each particle is $\frac{3}{2}kT$.

Dalton's law of partial pressures states that the total pressure due to a mixture of ideal gases is equal to the sum of the pressures each gas would exert if alone.

The solute molecules in a dilute solution can be assigned a pressure called the osmotic pressure π. If dilute enough, the solute gas obeys the ideal-gas law. In that case, the osmotic pressure $\pi = cRT/M$, where c is the concentration of the solute and M is its molecular mass. This fact is basic to the measurement of molecular masses by use of an osmometer.

MINIMUM LEARNING GOALS

Upon completion of this chapter you should be able to do the following:

1. Define the three common temperature scales and locate the following points on each: absolute zero, ice point, boiling point of water.

2. Change a temperature on one scale to the other two common scales.

3. Draw graphs showing the temperature variation of P and V for an ideal gas. Locate absolute zero on each graph.

4. Write down the ideal-gas law and define each quantity in it. Find P, V, n, or T for an ideal gas when all but one of these quantities is given.

5. Define kilogram mole and gram mole. Explain the relation between the kilogram mole and Avogadro's number. State the value of Avogadro's number.

6. Use Avogadro's number to find the mass of a molecule (or atom) when the molecular weight (mass) of the substance is given.

7. State under what conditions a gas is likely to behave as an ideal gas.

8. Explain the molecular meaning of absolute temperature by reference to the KE of the molecules of an ideal gas.

9. Compute the average translational KE of the molecules of an ideal gas when the gas temperature is given. Sketch a graph showing the distribution of speeds of gas molecules at two different temperatures.

10. Solve problems involving the gas law similar to those in Illustrations 10.6 to 10.10. Distinguish between gauge pressure and absolute pressure.

11. Explain what is meant by Brownian motion and why it occurs.

12. State Dalton's law of partial pressures. Use it to find the resultant pressure due to the mixture of several ideal gases.

13. Describe an osmometer and in the description point out what is meant by osmotic pressure. Explain how osmotic pressure can be used to determine molecular masses.

IMPORTANT TERMS AND PHRASES

You should be able to define or explain each of the following:

Temperature scales: Celsius, Kelvin, Fahrenheit
Ideal gas

Mole, kilogram, and gram
Avogadro's number N_A
$PV = nRT$
Gauge versus absolute pressure
Absolute zero
KE of translation = $\frac{3}{2}kT$
Kinetic theory of gases
Brownian motion
Osmotic pressure

QUESTIONS AND GUESSTIMATES

1. Suppose you were the only survivor of a plane crash on a tiny island in the South Pacific. While waiting for rescue, you decide to construct a thermometer. How might you accomplish this?
2. Sketch a graph showing the variation of the pressure of an ideal gas as a function of $1/V$ as the gas is compressed slowly at constant temperature.
3. We can picture the molecules of an ideal gas to act like tiny balls in continual motion. An ideal gas of colloidal-size particles can also exist. But glass beads and pool balls do not act like an ideal gas. Where (in size) does the dividing line come, and to what is it due?
4. Hydrogen and oxygen gas are sealed off at atmospheric pressure in a very strong glass container containing two electrodes. A spark from the electrodes ignites the gases so that the reaction $2H_2 + O_2 \rightarrow 2H_2O$ results. Will the pressure in the tube be changed after the temperature has come back to its original value (200°C) again? Explain. What if the final temperature is 20°C?
5. In deriving the gas law we assumed perfectly elastic collisions with the wall. This is not a correct assumption, but it makes no difference as long as the walls are at the same temperature as the gas. Explain why.
6. Compare the gravitational PE of a nitrogen molecule which is 1 m above the ground with its KE when the temperature is (*a*) 0°C and (*b*) −270°C.
7. Although the air is mostly composed of N_2 molecules, there is some O_2 present, of course. Do both kinds of molecules travel with the same average speed? What is the exact relation between the speeds?
8. Justify the fact that water molecules form a liquid under standard conditions even though H_2 and O_2 gases do not condense until much lower temperatures are reached.
9. The constitution of the atmosphere changes as we go to extreme heights above the earth. Why should there be a greater proportion of N_2 at high altitudes than at low altitudes compared with the O_2 content?
10. In outer space there is about one molecule per cubic centimeter of volume. Try to devise a method for measuring the pressure of such a thin gas.
11. Making use of Fig. 10.5, estimate the fraction of the nitrogen molecules having speeds in excess of 1250 m/s at 460 and 1700 K.
12. Estimate how large (in volume) a helium-filled balloon must be if it is to be able to lift 500 kg. To start at least, assume the mass of the balloon to be included in the 500 kg. (E)

PROBLEMS

Take atmospheric pressure to be 1.01×10^5 N/m^2 (= 14.7 lb/in^2 = 76 cm Hg = 1 atm).
1. What will Celsius and Kelvin thermometers read when the temperature is (*a*) 77°F? (*b*) −31°F?
2. Normal body temperature is 98.6°F. What temperature is this on the Celsius scale? On the Kelvin scale?
3. According to the handbook, the melting point of mercury is −38.9°C, and its boiling point is 357°C. Change these to Fahrenheit temperatures.
4. A test tube is sealed off at STP.* It is then heated to 300°C. What then is the gas pressure in the tube in atmospheres, in newtons per square meter, in centimeters of mercury, and pounds per square inch?
5. A sealed tank with volume 3.0 ft^3 has an ordinary

* STP stands for "standard temperature and pressure." These are 0°C and 1 atm.

pressure gauge on it which reads 100 lb/in^2. The temperature of the tank is then increased from 27°C to an unknown higher temperature. (*a*) If the gauge now reads 140 lb/in^2, what is the new temperature? (*b*) What was the final pressure in the tank in atmospheres?

6. A tank of compressed oxygen has a volume of 20,000 cm^3 at a gauge pressure of 1500 lb/in^2. What volume will this gas occupy at atmospheric pressure and the same temperature?

7. Air at 27°C and atmospheric pressure in a chamber is suddenly compressed to a volume one-twentieth as large and a pressure of 30 atm. What is the temperature of the gas?

8. A gas at an absolute pressure of 1000 lb/in^2 and 27°C is suddenly expanded into a chamber having 10 times the volume. What is the new pressure of the gas after a few minutes if its temperature is found to be −3°C at that time?

9.* An air bubble at the bottom of a lake 30 ft deep has a volume of 0.23 cm^3. Find its volume just before it reaches the surface. Assume the temperature at the bottom to be 7°C and at the top 17°C.

10. The molecular weight of a typical polyethylene molecule in a piece of polyethylene is about 25,000. Find the mass of this typical molecule. The density of polyethylene is close to 0.95 g/cm^3. How many of these molecules exist in (*a*) 1 g of polyethylene; (*b*) 1 cm^3?

11. Benzene has a density of 879 kg/m^3 and a molecular weight of 78. Find (*a*) the mass of a benzene molecule and (*b*) the number of benzene molecules in 1 cm^3.

12. A 2000-cm^3 tank is filled with nitrogen gas to a gauge pressure of 100 atm. (*a*) How many kilogram moles of nitrogen are in the tank? (*b*) How many kilograms of nitrogen are in the tank? Take the temperature to be 20°C. The molecular weight of N_2 is 28.

13. For many purposes, a pressure of 1×10^{-6} mm Hg is considered to be a reasonably good vacuum. (*a*) At this pressure, find the number of kilogram moles of nitrogen in 1 m^3. (*b*) How many nitrogen molecules exist in 1 cm^3 at this pressure? Assume the air to be pure nitrogen at room temperature. (Nitrogen actually composes about 78 percent of the air.)

14. The actual volume of a nitrogen molecule is about 8×10^{-30} m^3. Find the fraction of the volume actually occupied by nitrogen molecules in nitrogen gas under STP.

15. A piece of Dry Ice, CO_2, is placed in a test tube, which is then sealed off. If the weight of Dry Ice is 0.48 g and the sealed test tube has a volume of 20 cm^3, what is the final pressure in the tube if all the CO_2 vaporizes and reaches thermal equilibrium with the surroundings at 27°C? (Molecular weight of CO_2 is 44.)

16. When a 20-cm^3 test tube was sealed off at very low temperatures, a few droplets of liquid nitrogen were condensed in the tube from the air (boiling point of nitrogen is −210°C). What will the nitrogen pressure in the tube be when the tube is warmed to 27°C if the droplets weigh 0.14 g?

17. Hydrogen gas is to be burned in oxygen to obtain 18.0 g of water. What volume of hydrogen under STP is needed? (Molecular weight of H_2O is $2 + 16 = 18$.)

18. A closed cylinder has an insulated movable piston separating the cylinder into two parts. If equal masses of the same gas are on each side of the cylinder, show that the piston will position so that $V_2 = V_1(T_2/T_1)$, where V_1 and T_1 are the volume and temperature on one side of the piston and V_2 and T_2 apply to the other side.

19.* The temperature of outer space is about 3 K and consists mainly of a gas of single hydrogen atoms. On the average, there is about one such atom per cubic centimeter. (*a*) Find the pressure of this gas in outer space; express your answer in atmospheres. (*b*) Find the average KE of one of the hydrogen atoms. (*c*) What speed will an atom of this energy have? (The atomic weight of a hydrogen atom is 1.)

20. The temperature of the interior of the sun is approximately the same as that in a hydrogen bomb, about 10^8 K. Find the average translational KE of a proton (a hydrogen nucleus) at this temperature and from it compute its average speed. (The atomic mass of a hydrogen atom is 1 and, to good approximation, this same atomic mass applies to the proton.)

21. The density of dry air at STP is 1.29 kg/m^3. Its composition is (by weight) 78% N_2, 21% O_2, 0.9% argon, plus traces of CO_2, helium, and other gases. (*a*) Find the pressure predicted by the gas law for each of the three major components. (*b*) Use Dalton's law to find the air pressure under STP. (The molecular weight of argon gas, 40, is equal to the atomic mass of argon since the gas is monatomic.)

22. Find the osmotic pressure of a 2 g per 100 cm^3 water solution of sugar. The molecular weight of sugar is 342. Assume the temperature to be 27°C. Express your answer in centimeters of water.

23. A solution is made by dissolving 0.250 g of polystyrene in benzene so as to make 100 cm^3 of solution. The osmotic pressure of the solution is measured at 20°C

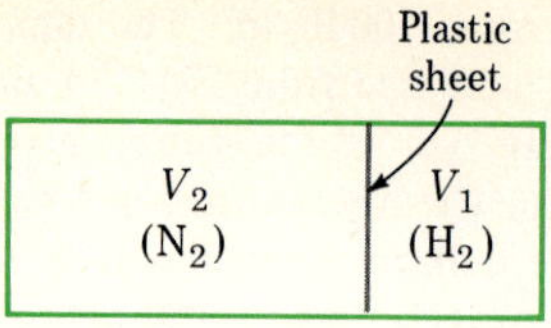

Plastic
sheet

V_2 (N_2) | V_1 (H_2)

FIGURE P10.1

and is found to be 1.67 cm of benzene. Find the molecular weight of the polystyrene. (Density of benzene at 20°C is 879 kg/m^3.)

24.* A thin sheet of plastic separates two containers as shown in Fig. P10.1. Volume 1 contains H_2 gas at STP while V_2 contains N_2 gas at STP. It is also known that $V_2 = 2V_1$. A hole is now made in the plastic. What is the final pressure after equilibrium is achieved? Repeat if the H_2 pressure is 240 cm Hg while the N_2 pressure is 60 cm Hg.

THERMAL PROPERTIES OF MATTER

11

When a substance cools, it loses heat energy. By what mechanism did the heat energy leave the substance? What effect does this loss have upon the atoms of the material? How much heat energy did it lose? The answers to these questions will be discussed in this chapter. In addition we shall investigate such topics as the boiling and freezing of liquids and the mechanisms by which heat is transferred. The knowledge gained from our study of these topics will help us to better understand the nature of heat as well as the behavior of atoms and molecules in matter.

11.1 THERMAL ENERGY, HEAT ENERGY, AND INTERNAL ENERGY

We saw in the last chapter that the Kelvin temperature of an ideal gas has an easily understood meaning: *the average translational KE of a gas molecule is* $\frac{3}{2}kT$, *where k is Boltzmann's constant.* From this we conclude that *the average KE of a molecule in a hot gas is larger than in a cold gas. Most types of gas molecules,* however, *possess energy in addition to translational KE.* For example, a diatomic molecule such as N_2 or O_2 has KE of rotation as well as of translation. One such molecule is shown schematically in Fig. 11.1. For many purposes, the molecule can be thought of as two balls connected by a spring. The spring, which represents the chemical bond between the two atoms, is often distorted in collisions between molecules. As a result, the springlike bond is compressed or stretched. Energy is therefore stored in the distorted bond between the atoms composing the molecule. As we see, even a simple diatomic gas possesses a great deal of energy in addition to its translational KE.

The situation becomes even more complicated as we proceed to more complex molecules. They possess several springlike bonds per molecule, and so the molecules can store even more energy in their distorted bonds. An ideal gas composed of even such complicated molecules conforms to the kinetic theory. As such, the translational KE of its molecules is still given by $\frac{3}{2}kT$, but each molecule has considerable energy in addition.

As the gas molecules collide with each other, the energy of the gas is distributed between translational KE, rotational KE, and energy associated with the "springs" between atoms. We know that the translational KE increases in proportion to T, the absolute temperature of the gas. The rotational and vibrational (spring-associated) energies also increase as T increases. In other words, *T is a measure of the total energy of the gas molecules.* Indeed, it is shown in courses in statistical mechanics that the total energy is proportional to T provided the temperature is not too low.*

These same ideas can be carried over to liquids and solids. A liquid is much like a highly compressed gas. Even though its molecules are too close together for it to behave like an ideal gas, it is still true that T is a measure of its molecular energy. A piece of solid can be likened to a very complex

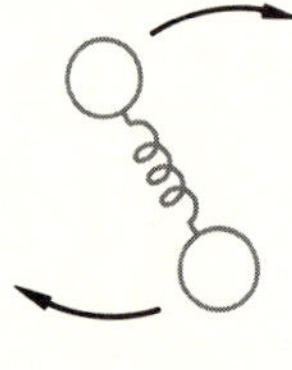

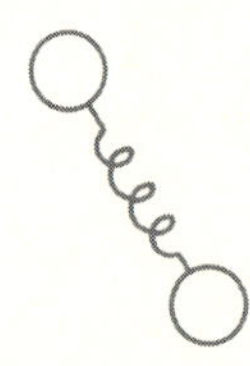

FIGURE 11.1
A gas molecule has rotational energy in addition to KE of translation. It also has energy associated with the springlike bond between its atoms.

*This restriction, that the temperature not be too low, has profound meaning. Departures from the kinetic theory at low temperatures presented a great puzzle for physicists. It was not until the ideas of quantum mechanics were developed that the reasons for these discrepancies became known. We shall elaborate on this in Chap. 26.

molecule. Although its complexity is immense, still there is energy associated with the springlike bonds between its atoms. This energy, too, increases with an increase in T.

As we see *in all forms of matter, the energy associated with the random motion of the molecules and atoms increases with T. This energy associated with the random motion of the atoms and molecules is often called* **thermal energy.** We should point out at once that thermal energy is a colloquial term. It has no precise scientific definition. When we wish to be quantitative, we shall use the terms heat energy and internal energy.

DEFINITION

Internal Energy

The **internal energy** U of a substance is defined as follows: *the total of all kinds of energy possessed by the atoms and other particles constituting a substance is called the* **internal energy** *of the substance.* As we see, U contains the kinetic, potential, chemical, electrical, nuclear, and all other forms of energy possessed by the molecules of a substance.

To define the quantity we call heat energy, we must first consider what happens when two substances at different temperatures are placed in contact. The situation is shown in Fig. 11.2. If the temperature of object A is higher than that of object B, and if the two objects are composed of the same substance, the average energy of a single molecule in A is larger than in B. As a result, when the objects are placed in contact, the molecules in A lose energy to those in B and A is thereby cooled and B warmed.

DEFINITION

The energy which is transferred (or flows) from a high temperature object to a lower temperature object because of the temperature difference is called **heat energy.** *It is represented by the symbol ΔQ.* Experiment shows that the flow of heat energy will cease when the temperatures of the two objects become equal. This is true even if the objects are made from different substances. We conclude that

Heat Energy

> Heat energy flows spontaneously from hot objects to cooler objects but not vice versa.

In the next section we shall define the practical units by means of which heat energy is measured.

11.2 UNITS OF HEAT ENERGY

Historically, the use of heat energy far preceded our understanding of the nature of heat. Consequently, the measurement of heat was done in a purely

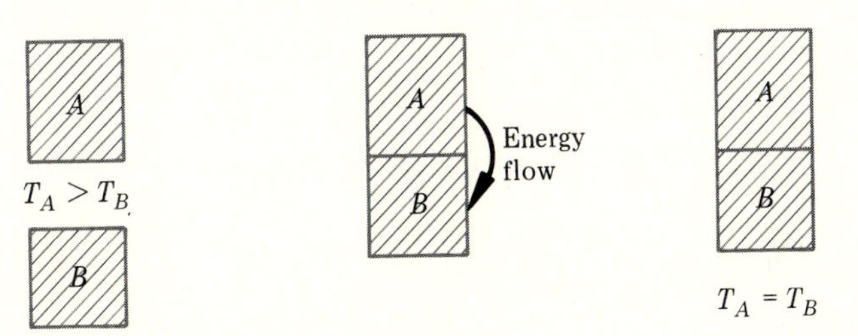

FIGURE 11.2
Heat energy flows from a hot object to a cooler one until their temperatures become equal.

practical way, and the units chosen for it were selected because of their utility. Since water is involved in most uses of heat, it is not surprising that the quantity and units of heat were chosen in terms of an experiment involving water.

Heat Units

The two common units of heat are the **calorie** (cal), *and the* **British thermal unit** (Btu), defined today as follows:

$$1 \text{ cal} = 4.184 \text{ J} \qquad \text{and} \qquad 1 \text{ Btu} = 1054 \text{ J}$$

These definitions conform to the *original definitions* of these units which were stated in the following way: *one calorie is the amount of heat required to raise the temperature of one gram of water from 14.5 to 15.5°C; one Btu is the amount of heat required to raise the temperature of one pound of water from 63 to 64°F.* From either of these definitions we find

$$1 \text{ Btu} = 252 \text{ cal}$$

Nutritionists use still another unit for heat energy. It is also called the calorie, but more properly it should be called the kilocalorie (or large calorie) since

$$1 \text{ nutritionist's calorie} = 1000 \text{ cal}$$

For example, when health scientists tell us our daily diet should contain about 2000 cal of food energy, they are actually speaking in terms of 2000 kcal.

11.3 SPECIFIC HEAT CAPACITY

In order to increase the temperature of an object one must increase the thermal energy of its molecules. We can do this by letting heat flow into the object from a hotter object. Similarly, if we wish to cool an object, we can allow heat energy to flow from the object to a still cooler object. To be able to describe such processes as cooling and heating quantitatively, we must know how much energy is required to change the temperature of an object. *The quantity of heat which must flow into or out of unit mass of a substance to change its temperature by one degree is called the* **specific heat capacity** *of the substance.*

DEFINITION

We represent the specific heat capacity by c. Its definition can be written as an equation. When a quantity ΔQ of heat flows into a mass m of substance, its temperature will increase by an amount ΔT. Then, by definition,

Specific Heat Capacity

$$\text{Specific heat capacity} = c = \frac{\Delta Q}{m\,\Delta T}$$

After clearing fractions, we have

$$\Delta Q = cm\,\Delta T \tag{11.1}$$

The units of c are, typically calories per gram per Celsius degree. A somewhat similar definition for c is used in the British units system, but in that case, 1 lb of substance is considered and the temperature change is measured in Fahrenheit degrees. The appropriate units are Btu per pound per Fahrenheit degree.

Because of the different complexities of substances, each substance has its own unique specific heat capacity. Moreover, the heat requirements for a 1-degree temperature change vary slightly with temperature. This variation is usually slight in a limited temperature range, and so it is often ignored. Values for c appropriate to many substances are given in Tables 11.1 and 11.2. Notice that the values apply to both calories per gram per Celsius degree and Btu per pound per Fahrenheit degree. From the definition of c, can you show why these two values should be equal?

Illustration 11.1 How much heat is given off as 20 g of water cools from 90 to 30°C?

Reasoning The specific heat capacity of water is 1.00 cal/(g)(°C). In the present case, $\Delta T = -60°\text{C}$ and so, from Eq. (11.1),

$$\Delta Q = cm\,\Delta T = [1.00\ \text{cal/(g)(°C)}](20\ \text{g})(-60°\text{C}) = -1200\ \text{cal}$$

The negative sign tells us this is a loss of heat.

TABLE 11.1
SPECIFIC HEAT CAPACITIES, cal/(g)(°C) or Btu/(lb)(°F)

Water	1.000	Aluminum	0.21
Human body	0.83	Glass	0.1–0.2
Ethanol	0.55	Iron	0.11
Paraffin	0.51	Copper	0.093
Ice	0.50	Mercury	0.033
Steam	0.46	Lead	0.031

TABLE 11.2
SPECIFIC HEAT CAPACITIES OF GASES AT 15°C, cal/(g)(°C) or Btu/(lb)(°F)

GAS	c_V	c_P	$\gamma = c_P/c_V$
He	0.75	1.25	1.66
Ar	0.075	0.125	1.67
O_2	0.155	0.218	1.40
N_2	0.177	0.248	1.40
CO_2	0.153	0.199	1.30
H_2O (200°C)	0.359	0.471	1.31
CH_4	0.405	0.53	1.31

Illustration 11.2 A thermos jug contains 300 g of coffee (essentially water) at 90°C. Into it is poured 50 g of milk (also essentially water) at 15°C. What is the final temperature of the coffee?

Reasoning The thermos jug is well insulated, and so we shall assume that no heat flows from the coffee to it. Then the heat lost by the coffee will go into warming the milk. The law of conservation of energy tells us that

$$\text{Heat lost by coffee} = \text{heat gained by milk}$$

Using Eq. (11.1), this equation can be written

$$(cm\,|\Delta T|)_{\text{coffee}} = (cm\,|\Delta T|)_{\text{milk}}$$

In each case, $c = 1.00$ cal/(g)(°C). The mass of coffee is 300 g and of milk is 50 g. If we call t the final temperature of the coffee-milk solution, $|\Delta T|$ for the coffee is $90° - t$ while for the milk it is $t - 15°$. Notice, since we want $|\Delta T|$ to be a positive number, that we write $90° - t$ and not $t - 90°$. Making use of these values, the above equation yields

$$t = 79.3°\text{C}$$

The coffee was cooled 10.7°C by the milk.

Illustration 11.3 An insulated aluminum container "weighing" 20 g contains 150 g of water at 20°C. A 30-g piece of metal is heated to 100°C and then dropped into the water. The final temperature of the water, can, and metal is 25°C. Find the specific heat capacity of the metal.

Reasoning Notice in this case that both the container and the water gain heat. The law of conservation of energy allows us to write

$$\text{Heat gained by can} + \text{heat gained by water} = \text{heat lost by metal}$$

These quantities are as follows.

$$\begin{aligned}\text{Heat gain of can} &= [0.21\ \text{cal/(g)(°C)}](20\ \text{g})(5°\text{C})\\ \text{Heat gain of water} &= [1.00\ \text{cal/(g)(°C)}](150\ \text{g})(5°\text{C})\\ \text{Heat loss of metal} &= c_x(30\ \text{g})(75°\text{C})\end{aligned}$$

Putting these quantities in the above equation yields $c_x = 0.34$ cal/(g)(°C).

11.4 c_V AND c_P FOR GASES

In Table 11.2 *we list two values for the specific heat capacity of gases, c_P and c_V.* The subscripts on these quantities tell us whether the pressure or the volume of the gas was maintained constant during measurement. You will notice from the table that c_V, *the specific heat at constant volume, is less than c_P, the specific heat at constant pressure.* The reason for this is easily understood from the following considerations.

When a gas is heated in a constant-volume container, the heat energy added to the gas must all appear as additional energy of the gas molecules. It is therefore all used to increase the temperature of the gas. But this is not the case if the gas can expand while being heated. To see this, refer to Fig. 11.3. The container is closed on top by a movable piston. The weighted piston exerts an unchanging pressure on the gas. If heat is added to the gas, the gas will expand by lifting the piston. However, the pressure of the gas, equal to the pressure applied by the piston, remains unchanged. But in lifting the piston, the gas does work. This work is done at the expense of the heat energy which flows into the gas. Therefore, not all the heat added to the gas is used to heat the gas. Some of it is used to do work as the piston is lifted by the gas.

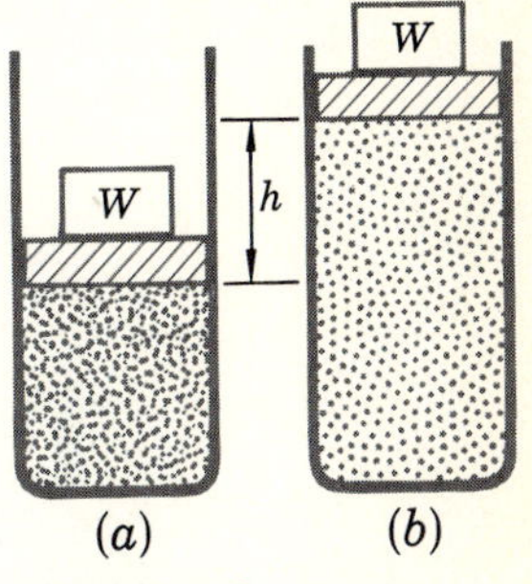

FIGURE 11.3
When the gas in the chamber is heated, the piston rises, thereby maintaining constant pressure. Hence the heat energy must do work in lifting the piston as well as in heating the gas.

We see from this that *more heat will be needed to raise the temperature of a gas under constant pressure than under constant volume. In the first case, some of the heat energy is used in doing work. Because of this, c_V is smaller than c_P. The difference in these two values is equal to the work done on the piston as unit mass of the gas is heated 1 degree under constant pressure.* In a special note on page 254 we evaluate $c_P - c_V$ for an ideal gas and find it to be R/M, where R is the gas constant and M is the molecular weight of the gas.

11.5 HEAT OF VAPORIZATION AND THE BOILING OF LIQUIDS

As we have seen, when heat energy is added to a substance, the internal energy of the substance is increased. This increase in internal energy is usually accompanied by an increase in temperature. However, *a temperature increase does not always occur even though the internal energy of the substance is increased. At the temperature where a phase change occurs in the substance, the temperature remains constant until the phase change is completed.* For example, at the temperature where a crystalline substance melts or where a liquid boils, a phase change is occurring. In the one case, a crystalline phase is changing to a liquid phase. In the other, a liquid phase is changing to a vapor phase. *Even though heat must be added to accomplish the phase change, no temperature increase occurs for the substance.*

When a substance changes from one phase to another, its internal energy is changed. Water molecules in steam have more internal energy than they had when condensed. When the molecules are frozen into ice, they have even less internal energy. Let us now investigate what happens when a substance changes phase. We begin our discussion by considering the phase change from liquid to vapor.

THE QUANTITY $c_P - c_V$ FOR AN IDEAL GAS

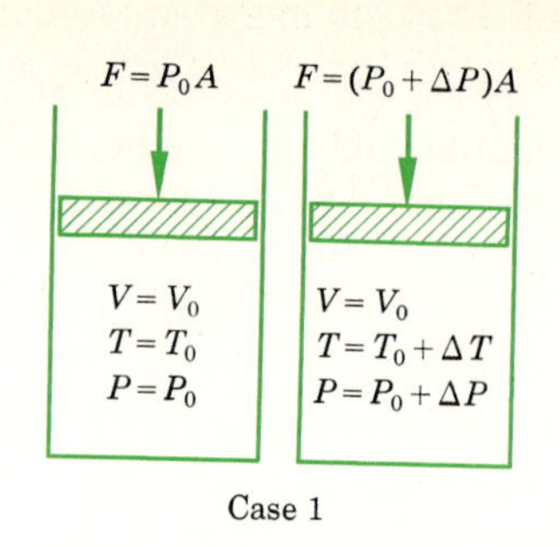

Case 1

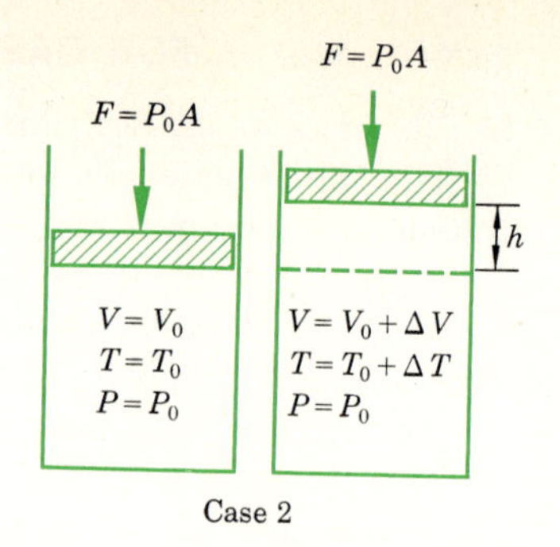

Case 2

The quantity of heat needed to raise the temperature of a gas an amount ΔT depends upon how the gas is heated. Two extreme cases are important: (1) the gas is heated in a container having constant volume, or (2) the gas is heated in a container the volume of which is changed in such a way as to maintain the pressure constant. These two alternatives are illustrated in the figure. Let us examine these two situations to find the heat energy needed to raise the temperature the same amount ΔT in each case.

When heat energy is supplied to an ideal gas, energy may be given to the molecules as internal energy. In some instances, part of the heat energy is used doing work against forces used to confine the gas. If the volume of the gas is held constant as in case 1 shown in the figure, no work is done against the external force holding the piston in place, since no movement of the piston occurs. Therefore, all the heat energy ΔQ_V given to the gas was used to raise the internal energy of the molecules. Calling the specific heat capacity in this case c_V, we have

$$\Delta Q_V = mc_V \Delta T$$

where m is the mass of gas in the container and the subscript V indicates that the volume is held constant.

In case 2, a larger amount of heat energy will be needed to raise the temperature by the same amount ΔT. This follows from the fact that the KE of the molecules must be raised by the same amount as in case 1, since the temperature raises the same amount in each case. But, in addition, an amount of work Fh is done against the force on the piston as the piston rises through the distance h.

We therefore have

$$\Delta Q_P = mc_P \Delta T = \text{increase in energy of molecules} + \text{work done against piston}$$

where the subscript P indicates that the pressure is maintained constant. Since the increase in energy of the molecules is the same in both cases, it is $mc_V \Delta T$. Replacing ΔQ_P by $mc_P \Delta T$ and substituting for the quantities on the right-hand side of the above equation, we find

$$mc_P \Delta T = mc_V \Delta T + Fh$$

However, the force on the piston F is merely the pressure times the area A of the piston, and so $Fh = P_0 Ah$. But Ah is the increase in volume of the container ΔV caused by the lifting of the piston, and so the above equation may be rewritten as

$$(c_V - c_P)m\,\Delta T = P_0\,\Delta V$$

This relation may be simplified by using Eq. (10.2), the ideal-gas law, which was, for the initial container in case 2,

$$P_0 V_0 = nRT_0$$

After heating, in case 2, the equation becomes

$$P_0(V_0 + \Delta V) = nR(T_0 + \Delta T)$$

Subtraction of the first of these two equations from the second yields

$$P_0\,\Delta V = nR\,\Delta T$$

This value for $P_0\,\Delta V$ may be substituted in the above equation involving $c_P - c_V$ to give

$$(c_P - c_V)m\,\Delta T = nR\,\Delta T$$

Upon canceling like factors we find

$$c_P - c_V = \frac{nR}{m}$$

But n, the number of kilogram moles of gas is related to the molecular weight M and the mass of gas m in the container by $n = m/M$. Therefore we can replace n/m by $1/M$ to obtain

$$c_P - c_V = \frac{R}{M}$$

As you will see in Prob. 13 at the end of this chapter, this theoretical result agrees well with the data in Table 11.2.

A liquid placed in an open dish will slowly evaporate. This is a result of the fact that not all the molecules of the liquid have the same energy. As pointed out in the last chapter, *some molecules of a gas have energy far in excess of the average, and others have very little. This is also true in a liquid. The very-highest-energy molecules, if at the surface of the liquid, may actually escape from the other molecules and leave the liquid. This is the process of* **evaporation.**

The molecules which evaporate are the highest-energy molecules. They carry energy away from the liquid, and so evaporation leads to a decrease in the average internal energy of the remaining molecules. Since the temperature of the liquid is merely a measure of this internal energy, it is clear that evaporation will lead to a cooling of the liquid. This process actually forms the basis for many cooling systems.

DEFINITION

Heat of Vaporization

The heat required, i.e., *the energy required, to tear unit mass of molecules loose from each other to change them from the liquid phase to the vapor phase is called the* **heat of vaporization** *of the liquid.* Since the molecules of a liquid are less tightly bound together at a high than at a low temperature, the energy required to tear a molecule loose decreases with increasing temperature. For example, the heat of vaporization of water is 590 cal/g at 10°C and 539 cal/g at 100°C. That is to say, 539 cal of heat energy is required to tear 1 g of water molecules apart at 100°C. The reverse statement is also true. *When a vapor condenses to liquid, the heat of vaporization is given off in the process.* Typical values of heats of vaporization are given in Table 11.3.

Most often, the heat of vaporization is quoted for the normal boiling temperature of the liquid. It is common experience that liquids evaporate most readily when this temperature is reached. In fact, bubbles form within the liquid, and we use this as a common means for telling when a liquid has reached its boiling temperature. Let us now examine the meaning of this temperature in terms of the vaporization of the molecules.

Vapor Pressure

Suppose that a liquid is placed in a closed container from which the air has been removed. Some of the liquid molecules will evaporate into the space above the liquid, as shown in Fig. 11.4. Of course, the reverse process is possible. From time to time, a vapor molecule will hit the surface of the liquid and stick to the liquid. As the number of molecules in the vapor increases, *a condition will be reached at which the number of molecules leaving the liquid will equal the number returning to the liquid from the vapor.* The number of molecules in the vapor will therefore remain constant at this point if the temperature of the system is not changed. *The vapor under these conditions is said to be* **saturated.** *We define the pressure of the molecules in the vapor under this condition to be the* **vapor pressure** *of the liquid. The vapor pressure will increase as the temperature is raised,* since molecules will evaporate more readily at the higher temperatures. This variation of vapor pressure with temperature for water is shown in Table 11.4. Notice that even ice has an appreciable vapor pressure.

FIGURE 11.4

When the vapor is saturated, equal numbers of molecules of the liquid evaporate from the surface and condense from the vapor in a given length of time.

TABLE 11.3

HEATS OF VAPORIZATION AND FUSION

SUBSTANCE	MELTING TEMP, °C	BOILING TEMP, °C	HEAT OF FUSION, cal/g	HEAT OF VAPORIZATION, cal/g
Lead	327		5.9	
Water	0	100	80	539
Mercury	− 39	357	2.8	65
Ethanol	−114	+ 78	25	204
Nitrogen	−210	−196	6.1	48
Oxygen	−219	−183	3.3	51

TABLE 11.4
VAPOR PRESSURE OF WATER AND ICE

TEMPERATURE, °C	VAPOR PRESSURE,* mm Hg, or Torr	TEMPERATURE, °C	VAPOR PRESSURE,* mm Hg, or Torr
−90	0.000070	90	526
−50	0.030	94	611
−10	1.95	99	733
0	4.58	100	760
10	9.21	110	1,075
30	31.8	150	3,570
60	149.4	200	11,650

*The pressure unit Torr is defined by 1 Torr = 1 mm Hg. It is named after the scientist Torricelli.

Once in a while, within the body of a liquid, a group of molecules will attain enough energy to tear apart the liquid and form a small bubble. A much exaggerated case is shown in Fig. 11.5. At low temperatures the vapor pressure within the bubble will be much smaller than the pressure of the atmosphere above the liquid. Hence, the bubble will collapse under the external air pressure before it has had a chance to grow to observable size. However, as the temperature of the liquid is raised, a point will be reached where the vapor pressure within the bubble equals the air pressure on the top of the liquid. The bubble will no longer collapse. Instead, it will grow and rise to the surface of the liquid. This process, happening at many places within the liquid, gives rise to the phenomenon of boiling.

Clearly, *a liquid boils at a temperature where the vapor pressure of the liquid just equals the external pressure on the liquid.* In science and industry, a vacuum is often applied when distilling liquids. The reduced external pressure allows the liquid to boil at a lower temperature and decreases the possibility of chemical reaction within it.

As the vapor of a boiling liquid escapes, the highly energetic molecules in the vapor carry energy away from the liquid. If the liquid is to continue boiling, heat must be continuously supplied to compensate for this energy loss. *If one tries to heat a liquid above its boiling temperature at the particular external pressure to which it is subjected, it will only boil more vigorously. Its temperature will not rise further until all of it has evaporated.* An exception occurs if bubbles fail to form, in which case the liquid will superheat. Since bubbles first form most easily on impurities such as dust particles or air bubbles, it is not uncommon for a very clean, gas-free liquid to superheat. When it does begin to bubble, it may do so with nearly explosive force.

The boiling points and heats of vaporization for liquids are usually given for the boiling point under standard pressure, 760 mm Hg (or Torr). Under these conditions, the heat of vaporization of water is 539 cal/g, and the boiling point is 100°C, of course. It is interesting to notice that atmospheric pressure in the Rocky Mountain region is much lower than 760 Torr at the more than mile-high altitudes of such cities as Denver, Cheyenne, and Laramie. At these altitudes the pressure is near 600 Torr, and water boils at about 94°C. Why is a pressure cooker a near necessity under such conditions?

FIGURE 11.5
The boiling temperature is the temperature at which the vapor pressure in the bubble equals the external pressure on the liquid. (The size of the bubble is exaggerated.)

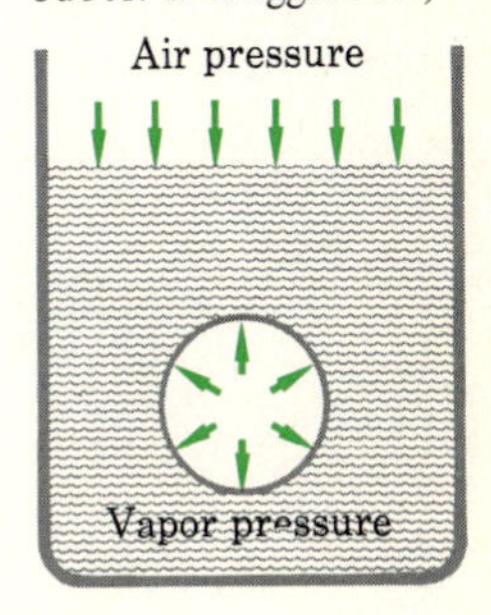

11.6 HEAT OF FUSION AND MELTING

Ice crystals melt at 0°C under standard pressure (76 cm Hg). Before melting, the water molecules are ordered in a crystalline lattice. They are held in place by rather strong intermolecular forces. To melt the crystal, one must tear the molecules out of this tight arrangement and cause them to disorder. This process requires energy, which is usually supplied by heat.

We therefore find that a crystalline material, when heated, begins to melt at a certain temperature. As heat is slowly added to the crystalline-liquid mixture, the temperature remains constant until all the crystals have melted. The substance has a definite melting temperature, and a definite amount of heat must be furnished to melt the crystals at this temperature.

DEFINITION

Heat of Fusion

One defines the **heat of fusion** *to be the amount of heat energy required to melt unit mass of the crystalline material. This same amount of heat must be given off to the surroundings when unit mass of the liquid crystallizes.* The heat of fusion of water to ice is 80 cal/g. Values for other materials are listed in Table 11.3. It is seen that the hydrogen-bonded materials, water and ethanol, have much higher heats of fusion and vaporization than the others. Why?

A few other effects should be mentioned in connection with melting phenomena. First, some plastics such as polyethylene are partly crystalline. That is to say, only a portion of the solid is in an ordered state, the remainder being liquid. The two parts are intimately mixed together, and so the material has intermediate physical properties. These crystalline regions differ in degree of perfection. The less perfect crystals melt at a lower temperature than the larger, more perfect crystals. Hence such materials exhibit a melting range, often 20 or more Celsius degrees wide. The quoted melting point for these materials is usually taken as the temperature at which all the crystallinity has disappeared.

Other plastics, such as polystyrene, are clear, hard, glassy materials. These materials are merely very viscous liquids, as explained in Chap. 9. The molecules are not well ordered in the solid. Although they soften at a fairly definite temperature, no heat of fusion is involved in this process as there would be for the melting of a crystalline material.

Freezing points of liquids can be altered somewhat by applying large pressures to the systems. Those materials which contract upon freezing have their melting points raised by increased pressure. Most materials behave in this way. A few materials, such as water, expand when they freeze. Increased pressure decreases the freezing point of such substances. The pressure of an ice skater's skate upon the ice can cause the ice below it to melt. In this case, the skater is actually skating on ice lubricated with a thin film of water.

11.7 CALORIMETRY

A large number of situations involving heat interchange can be elucidated by the application of the following simple equation:

$$\text{Heat lost} = \text{heat gained}$$

In general, such situations involve the interchange of heat within an insulated vessel, a calorimeter, which effectively isolates the system from its surroundings. *The above equation is simply a statement of the law of conservation of energy*. We shall illustrate its use by several examples.

To handle the examples which follow, we shall also make use of the following facts.

1 When a mass m of substance is heated (or cooled) through a temperature ΔT, it gains (or loses) a quantity of heat given by Eq. (11.1), namely,

$$mc\,\Delta T$$

(We assume that no phase change occurred in this temperature range.)

2 When a mass m of substance is melted (or crystallized), it gains (or loses) an amount of heat

$$mH_f$$

where H_f is the heat of fusion.

3 When a mass m of substance is vaporized (or condensed), it gains (or loses) an amount of heat

$$mH_v$$

where H_v is the heat of vaporization.

Illustration 11.4 A 100-g piece of lead is heated to a temperature of 100°C and then dropped into a cavity in a large block of ice at 0°C. How much ice will melt?

Reasoning Heat lost by lead = heat used to melt ice

$$m_{\mathrm{Pb}}c_{\mathrm{Pb}}\,\Delta T = m_w H_{fw}$$
$$(100\text{ g})[0.031\text{ cal/(g)(°C)}](100 - 0\text{°C}) = (m_w)(80\text{ cal/g})$$
$$m_w = 3.9\text{ g}$$

The various constants were obtained from Tables 11.1 and 11.3.

Illustration 11.5 If 20 g of ice in a 10-g copper calorimeter can is originally at −30°C, how much steam at 100°C must be condensed in this can if the ice is to be changed to water and heated to 40°C?

Reasoning

Heat lost by steam = heat gained by ice and can

The steam must first be condensed to water at 100°C, and then the water must be cooled to the final temperature of 40°C. Hence

$$\begin{aligned}\text{Heat lost} &= mH_v + mc\,\Delta T \\ &= (m)(540) + (m)(1)(100 - 40) = 600m\end{aligned}$$

The copper can need be heated only from −30 to +40°C. On the other hand, the ice must first warm to 0°C, then melt to water, and then be heated to 40°C. Therefore

$$\begin{aligned}&\text{Heat gained} \\ &\quad = (m_{\text{Cu}}c_{\text{Cu}})(70) + (m_w c_{\text{ice}})(30) + m_w H_{fw} + (m_w c_w)(40) \\ &\quad = (10)(0.093)(70) + (20)(0.50)(30) + (20)(80) + (20)(1)(40) \\ &\quad = 65 + 300 + 1600 + 800 \\ &\quad = 2765\ \text{cal}\end{aligned}$$

The units have been omitted from these equations to conserve space. You should supply them and check the units of the answer. Equating heat gained to the heat lost yields the mass of steam required,

$$m = \tfrac{2765}{600} = 4.6\ \text{g}$$

Notice how little steam is required to accomplish this. It is clear that the steam lost most of its heat during the process of condensing, namely, 2500 of the total 2765 cal. Can you see from this why, under certain conditions, hot steam can cause much more severe burns than hot water?

Illustration 11.6 A 10-g lead bullet is traveling at a speed of 100 m/s when it strikes and embeds itself in a wooden block. By about how much will its temperature rise on impact?

Reasoning Assume that all the KE of the bullet is changed into heat residing in the bullet. (Can you justify this assumption?) Then, the heat gained by the bullet will equal the KE loss of the bullet.

$$\begin{aligned}\text{KE of bullet} &= \tfrac{1}{2}mv^2 \\ &= (\tfrac{1}{2})(0.010\ \text{kg})(100\ \text{m/s})^2 = 50\ \text{J} = 12\ \text{cal}\end{aligned}$$

Notice the units conversion from joules to calories

$$\begin{aligned}\text{Heat gained} &= m_{\text{Pb}}c_{\text{Pb}}\,\Delta T \\ 12\ \text{cal} &= (10\ \text{g})[0.031\ \text{cal/(g)(°C)}]\,\Delta T \\ \Delta T &= 39\text{°C}\end{aligned}$$

Hence, if the original temperature of the bullet were 20°C, its final temperature would be 59°C. If the bullet had been traveling at a speed of 600 m/s, we would find ΔT obtained in the above way to be 36 times as large, or the final temperature would be about 1430°C. Of course, the bullet would melt before this temperature was reached, and so the above computation would no longer be correct. How could the computation be carried out in this latter case?

Illustration 11.7 When nutritionists state that 1 kg of bread has a food value of 2600 cal, they mean that if the dried bread is burned in pure oxygen it will give off 2600 kcal of heat energy. (Basically, the body generates heat from food in a somewhat similar chemical reaction.) Estimate how much heat energy a human body gives off each day.

Reasoning Depending upon the person, his or her nutritional calorie intake each day is in the range of 2000 to 3000 "calories." Since these are actually kilocalories, the process of metabolism within the person's body will generate of the order of 2×10^6 cal of heat. Because the body temperature remains nearly constant, the body must lose this energy as it is generated. The methods by which this energy is generated and released within the body are complex and are discussed in texts on biochemistry and biophysics. The air we breathe out and the evaporation of perspiration from the skin are well-known mechanism for cooling the body, but others are important as well.

11.8 THERMAL EXPANSION

As we have seen, the temperature of a substance is a measure of the internal energy of its molecules. *As the temperature of a liquid or solid is raised, the molecules,* having greater energy, *will generally vibrate through larger distances. This increased amplitude of vibration of a given molecule will force its neighbor molecules to remain at a greater average distance from it. Hence, the solid or liquid will expand.* Although there are some notable exceptions to this rule over small temperature ranges (for example, water contracts* in going from 0 to 4°C), it is generally true that substances expand with increasing temperature provided that a phase change does not occur.

Clearly, the thermal expansion of the metal in a building or a bridge can be a matter of considerable practical importance. If provision were not made for thermal expansion, railway tracks and concrete highways would buckle under the action of the hot summer sun. Many of us have lived or worked in buildings where lengthening steam pipes in the heating system give rise to noticeable effects. For these reasons, and many more, it is necessary to know exactly how a material expands with temperature. To this end, a constant of

*In water, hydrogen bonding binds the molecules into groups of several molecules each in a definite structure even above the melting point of ice. As the temperature increases, these groups break up, causing a new, more compact arrangement of the molecules.

linear thermal expansion α (alpha) and a constant of volume thermal expansion γ (gamma) are defined and tabulated.

DEFINITION

The **coefficient of linear thermal expansion** α *is defined to be the increase in length per unit length of a material for a temperature change of one degree.* Written as an equation, this definition is

$$\alpha = \frac{\Delta L/L}{\Delta T}$$

In other words, if a bar of length L expands in length an amount ΔL when the temperature is raised by an amount ΔT, the value of α is specified by the above equation. Notice that the length units cancel out and so the units of α are reciprocal degrees, that is, $1\,°C^{-1}$ or $1\,°F^{-1}$. A few typical values for α are given in Table 11.5.

As an example, if a brass rod is L m long, it will lengthen an amount ΔL given by the definition of α to be

$$\frac{\Delta L}{L} = \alpha\,\Delta T \qquad (11.2)$$

for a temperature change ΔT. Taking L to be 1.00 m and the temperature change to be 50°C, one has (see Table 11.5 for α)

$$\Delta L = (1.00\text{ m})(19 \times 10^{-6}\,°C^{-1})(50\,°C) = 0.00095\text{ m}$$

Since this change in length is really very small, the value of L used to determine ΔL is not sufficiently temperature-dependent to cause worry about

TABLE 11.5

COEFFICIENTS OF THERMAL EXPANSION PER CELSIUS DEGREE AT 20°C

SUBSTANCE	α,* $\times 10^6$	γ,* $\times 10^6$
Diamond	1.2	3.5
Glass (heat-resistant)	~3	~9
Glass (soft)	~9	~27
Iron	12	36
Brick and concrete	~10	~30
Brass	19	57
Aluminum	25	75
Mercury		182
Rubber	~80	~240
Glycerin		500
Gasoline		~950
Methanol		1200
Benzene		1240
Acetone		1490

*This notation at the top of the column means that all values of α and γ have been multiplied by 10^6. Therefore, α for iron is $12 \times 10^{-6}\,°C^{-1}$.

its temperature of measurement. Actually α varies somewhat with temperature, and for very precise work one should use the value appropriate to a given temperature range. In actual practice, however, this complication is very seldom of importance.

DEFINITION

The **volume-thermal-expansion coefficient** for a substance is defined in a manner analogous to the linear coefficient. It *is the relative change in volume per unit change in temperature,* or, as an equation,

$$\gamma = \frac{\Delta V/V}{\Delta T}$$

which yields at once

$$\Delta V = \gamma V \Delta T \tag{11.3}$$

The units of γ are reciprocal degrees. As an example of its use, suppose that 100 cm³ of benzene at 20°C is heated to 25°C. Its volume will change by an amount (see Table 11.5 for γ)

$$\Delta V = (1.24 \times 10^{-3}\,°\mathrm{C}^{-1})(100\ \mathrm{cm}^3)(5\,°\mathrm{C}) = 0.62\ \mathrm{cm}^3$$

This is a 0.6 percent change in volume and is an appreciable change in V for many purposes. It is necessary therefore to stipulate the temperature at which V should be measured if the γ coefficients in Table 11.5 are to apply. Those given there are for $T = 20°\mathrm{C}$. Of course, for small temperature changes not too far removed from 20°C, one can compute ΔV to fairly good precision using V measured anywhere in this small temperature range.

An examination of Table 11.5 shows that the linear-expansion coefficient for solids is essentially one-third the volume-expansion coefficient. This is a general rule for most solids that are isotropic, i.e., the same in all directions. It is left as an exercise for the student to show why this rule follows from the definitions of α and γ.

Illustration 11.8 A slab of concrete in a highway is 20 yd long. How much longer will it be at 126°F than it is at 0°F?

Reasoning We have

$$\Delta L = \alpha L \Delta T$$

From our tabulated values, $\alpha \approx 10 \times 10^{-6}\,°\mathrm{C}^{-1}$. To use this, we must have temperatures in Celsius degrees. We have $\Delta T = 126°\mathrm{F} = (\frac{5}{9})(126) = 70°\mathrm{C}$. So

$$\Delta L = (10^{-5})(70)(20\ \mathrm{yd}) = 14 \times 10^{-3}\ \mathrm{yd} = 0.5\ \mathrm{in}$$

An expansion crack of this magnitude would be required, or the concrete might buckle.

Illustration 11.9 Late one evening an automobile owner had his gasoline tank filled. The temperature of the gasoline was 68°F. The tank held 16 gal. He then parked the car. By the time he returned the next day, the hot sun had warmed the gasoline to 131°F. How much gas had overflowed from the tank?

Reasoning Let us first change the temperatures to the Celsius scale. They are 20 and 55°C. Since the initial temperature was 20°C, no difficulty arises in connection with the γ values listed in Table 11.5. From the definition of γ

$$\Delta V = \gamma V \Delta T = (0.95 \times 10^{-3})(16)(35) = 0.53 \text{ gal}$$

Even if the initial temperature were 55°C, the error involved in using the volume at that temperature to compute ΔV would amount only to about 3 percent, since ΔV is only about 3 percent of the total volume. However, if the initial volume had been specified at a temperature T other than 20°C, and high precision was needed, one would first find the volume at 20°C by use of the relation

$$\Delta V = \gamma V_{20}(T - 20)$$

where

$$\Delta V = V - V_{20}$$

and V is the measured volume at the temperature in question. We have neglected the expansion of the gasoline tank itself. Is this a serious error?

11.9 TRANSFER OF HEAT: CONDUCTION

As we have seen, much of the thermal energy in a substance is KE of motion of the molecules of the material. *When heat energy is transmitted from the hot end of a metal rod to its cold end,* as shown in Fig. 11.6, the process of heat transmission is quite obvious. The molecules at the warm end are vibrating with high energy. They collide with the cooler molecules on their right and give them more energy. These in turn hit the slower molecules on the right side of them, giving them more energy. Hence, *the heat is transmitted down the rod by means of molecular collisions. This mechanism of heat transfer is called* **conduction.**

Heat Conduction

The rapidity with which heat energy flows down a bar depends upon the material of which the bar is made. We all know that metal conducts heat

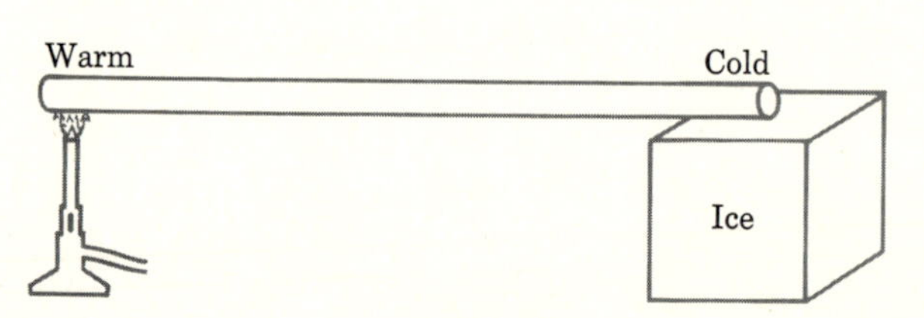

FIGURE 11.6

Heat moves from the hot end of the metal rod to the cold end by conduction.

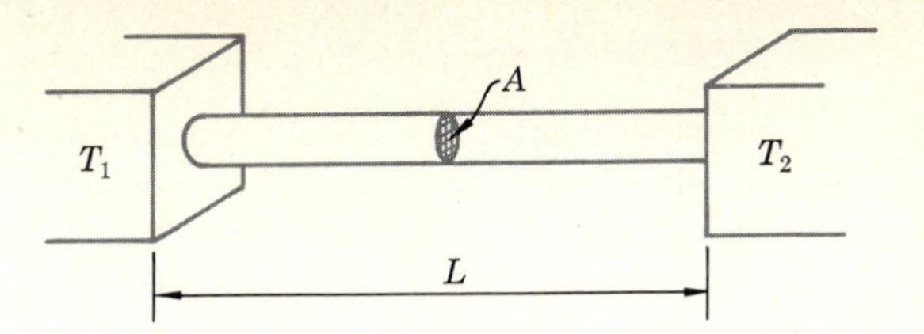

FIGURE 11.7

The rate of heat flow down the rod is proportional to $T_1 - T_2$ and to A, and inversely proportional to L.

better than wood or glass. A rather general relation states that *good electrical conductors are good heat conductors.* In particular, metals are good conductors of heat because the valence electrons in the metal move about freely and, in so doing, carry kinetic energy (heat) along with them.

To express the flow of heat in mathematical terms, we imagine the experiment illustrated in Fig. 11.7. A bar having cross-sectional area A and length L is connected between two constant-temperature devices, as shown, with $T_1 > T_2$. Heat energy will flow down the bar. Let us now ask how much heat ΔQ will flow down the bar in time t. (Assume the sides of the bar to be insulated so that heat cannot flow radially from it.)

One would suppose that, the larger A was, the more heat would flow. Hence $\Delta Q \propto A$. Also, if $T_1 = T_2$, no heat would flow. As a first guess, one might say that $\Delta Q \propto T_1 - T_2$. If the bar were very long, the heat flow would probably be decreased and so one might suppose that $\Delta Q \propto 1/L$. Of course, the longer the time t, the greater the amount of heat that will flow. All these suppositions turn out to be correct. Experiment shows that

$$\Delta Q = \frac{\lambda(T_1 - T_2)At}{L} \tag{11.4}$$

where λ *(lambda) is a proportionality constant called the* **heat conductivity** *for the material* of the bar. Clearly, λ *will be large for a good conductor* and small for a poor conductor. Some typical values are given in Table 11.6. Notice that the units one must use in Eq. (11.4) are stated in the table of heat conductivities.

Illustration 11.10 A brass bar, $A = 2.0\ \text{cm}^2$ and 1.00 m long, is placed with one end in boiling water and the other end on a cake of ice. How much ice will be melted by the heat conducted down the bar in 10 min?

Reasoning First we find the heat conducted down the bar in 10 min. In proper units we have $A = 2.0\ \text{cm}^2$, $L = 100$ cm, $T_1 - T_2 = 100°\text{C}$, $t = 600$ s, and $\lambda = 0.20$ cal/(cm)(s)(°C). Substitution in Eq. (11.4) gives

$$\Delta Q = 240\ \text{cal}$$

But 80 cal is needed to melt 1 g of ice. Hence, 240/80 or 3.0 g of ice would be melted in this time.

TABLE 11.6

HEAT CONDUCTIVITIES, cal/(cm)(s)(°C)

MATERIAL	λ
Silver	1.0
Copper	1.0
Aluminum	0.50
Brass	0.20
Glass	$\sim 20 \times 10^{-4}$
Asbestos paper	$\sim 5 \times 10^{-4}$
Rubber	$\sim 5 \times 10^{-4}$

11.10 TRANSFER OF HEAT: CONVECTION

A simple experiment devised to illustrate **convection** is shown in Fig. 11.8. When the glass tube illustrated is filled with water, a little colored dye placed near the neck remains nearly motionless in the position shown in part *a*. However, if the tube is heated as shown in part *b*, the liquid begins to flow counterclockwise around the tube, carrying the dye with it as illustrated.

The reason for this motion is quite simple. A heated liquid or gas expands, and so the water, in the lower right corner of the tube at *A*, expands when heated. It is now lighter than the rest of the liquid. The heavier column of liquid, on the left, will no longer be supported by the lighter column, on the right. It falls, pushing the water along in the tube. Hence the liquid on the right flows upward. It cools as it moves along, and by the time it reaches the left side, it is cooler and denser than when it was at point *A*. In summary, heated liquid at *A* will rise to *B*. In so doing, it will carry heat with it, and so *heat is moved from A to B by the actual motion of the liquid from A to B. This means of heat transfer is called* **convection.**

Heat Convection

Notice that *conduction does not involve the motion of the molecules over large distances. Heat is transferred from molecule to molecule by collision. In convection, however, the molecules of the transferring material actually move along with the heat.* Only liquids and gases can transfer heat by convection, since it is only in these materials that the molecules can move over large distances.

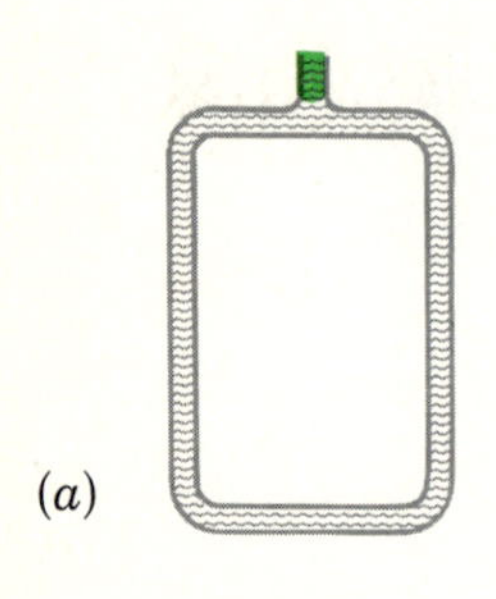

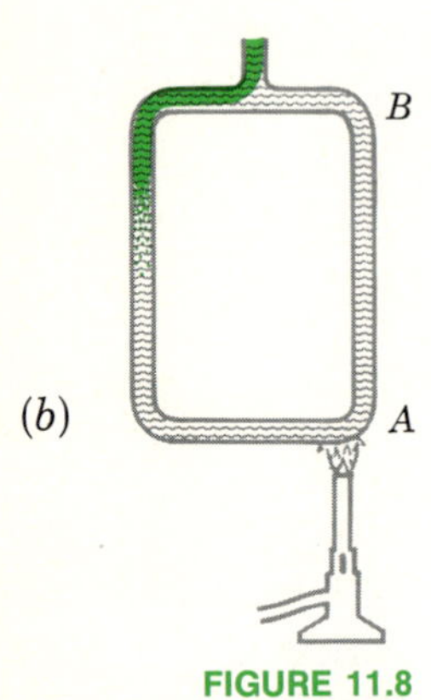

FIGURE 11.8
When the heat is applied to the liquid in the tube, the dye in the liquid shows that the liquid circulates counterclockwise. Heat is carried along by the circulating liquid, giving rise to heat convection.

Many homes are heated by air-convection methods. Even in the case of heating systems without fans, the circulatory movement of the air is appreciable. For example, when one stands over a hot-air register above an air furnace, the rush of hot air from the register is often quite noticeable. Proper design of such convection systems must allow the cool air to return to the furnace much as the cool liquid circulates back to point *A* in Fig. 11.8*b*. Clearly, the purpose of the cold-air registers in such heating systems is to return the cool air to the furnace.

Weather phenomena are partly the result of convective air currents: thermal air-circulatory currents near the edges of mountain ranges are particularly interesting. Quite large effects are noticed at various fixed times of day as the cool air from the mountains flows down and causes the warm air on the nearby plains to rise. The Gulf Stream is another interesting example of a large-scale transfer of heat by convection.

11.11 TRANSFER OF HEAT: RADIATION

We know that the sun warms the earth. It is in fact our major source of heat. We can easily see that the heat from the sun is not transferred to us by either conduction or convection. There are essentially no molecules in the vast reaches of space between us and the sun. Hence vibrational transfer by conduction and circulatory transfer by convection are impossible. We have here a case of *heat transfer through vacuum,* i.e., through nothing. This method of heat transfer *is called* **radiation.**

Heat Radiation

For many years, even to the first decade of this century, most scientists could not conceive of heat and light transfer from the sun through *nothing*. They therefore postulated that all space was filled by a "luminiferous ether." It was not until Einstein showed (in 1905) that such a concept as a mechanical ether was neither useful nor verifiable that it was eventually discarded. However, even before that time, the concept of an ether had run into formidable conceptual difficulties. We no longer consider the ether concept either necessary or convenient. As we shall see in our study of electromagnetic radiation, an understanding of radiation can be achieved without it.

As we have intimated, there is a direct relation between heat radiation, light radiation, and electromagnetic (radio) radiation. We shall in fact see in later chapters that these phenomena are essentially all the same.

11.12 LAWS OF COOLING

Newton showed that a convenient and simple law is obeyed in the cooling of bodies which are not too hot. If the temperature of a body is T_1 and the temperature of its surroundings is T_0, he found by experiment that the heat ΔQ lost by the body in time t is

$$\Delta Q = (\text{const})(T_1 - T_0)t \tag{11.5}$$

Newton's Law of Cooling

This is called **Newton's law of cooling.** We now know it is only approximately true and then only if $T_1 - T_0$ is not too large. It represents the combined effects of conduction, convection, and radiation.

In heat loss by radiation alone, it was first shown experimentally and later by theory that a hot object should radiate heat according to the equation

$$\Delta Q = (\text{const})(T_1^4 - T_0^4)t \tag{11.6}$$

Stefan's Law

This is known as **Stefan's law.** The temperatures involved are absolute temperatures, T_1 being that of the hot object and T_0 being that of its surroundings. *For very small temperature differences, Stefan's law reduces to Newton's law of cooling.**

Furthermore, it was shown that a blackbody (one which does not reflect light) will radiate heat much better than a highly reflective body. As a general rule, *a good heat absorber is a good heat radiator.*

11.13 HUMIDITY

Everyone knows that on a day when the humidity is high, the air contains a great deal of water vapor. Humidity is a measure of the water content of the

*It is left as an exercise for the student to show that this is true. To do so, you will need to use the fact that $T_1^4 - T_0^4 = (T_1^2 - T_0^2)(T_1^2 + T_0^2) \approx 2T_0^2(T_1^2 - T_0^2)$ for $T_1 \approx T_0$.

Relative Humidity

air. Precisely, *the relative humidity* (RH) *is defined to be the ratio of the mass of water vapor per unit volume in the air to the mass per unit volume required to produce saturation at the same temperature.* As pointed out earlier in this chapter, when a saturated vapor is in contact with a liquid, identical numbers of molecules leave and return to the surface of the liquid in a given length of time. Hence, at saturation, no net evaporation occurs. If the vapor is more than saturated, i.e., **supersaturated,** droplets will condense from the vapor and fog or rain will result.

Since the pressure of an ideal gas is proportional to the number of molecules in it, the definition of relative humidity is often couched in terms of pressures rather than masses. Water vapor is very nearly an ideal gas, and so the two definitions are nearly identical. In equation form

$$\text{Relative humidity} = \frac{m}{m_s} \approx \frac{P}{P_s} \tag{11.7}$$

where m and P are the mass per unit volume and pressure of the water vapor in the air and m_s and P_s are the respective values for saturated vapor. Some data for saturated water vapor at various temperatures are given in Table 11.7.

According to Table 11.7, saturated air at 68°F contains 17.1 g/m^3 of water. Suppose that the air actually did contain 17.1 g/m^3 of water vapor. If the air were at any temperature above 68°F,it could contain still more vapor. However, if this amount of water vapor were present and the air cooled below 68°F (as it would, perhaps, after the sun went down), it would become supersaturated as soon as the temperature dropped below 68°F. At that temperature and below, water droplets would begin to fall out of the air in the form of fog, dew, or rain. *We term the temperature at which the air just becomes saturated the* **dew point.**

DEFINITION
Dew Point

The dew point of the air is a useful quantity. Suppose that the temperature of the air on a certain day was 89.6°F. On that day the meteorologist at

TABLE 11.7

PROPERTIES OF SATURATED WATER VAPOR

T, °C	*T*, °F	*m*, g/m^3	*P*, cm Hg
−8	+17.6	2.74	0.23
−4	24.8	3.66	0.33
0	32.0	4.84	0.46
4	39.2	6.33	0.61
8	46.4	8.21	0.80
12	53.6	10.57	1.05
16	60.8	13.50	1.36
20	68.0	17.12	1.75
24	75.2	21.54	2.23
28	82.4	26.93	2.82
32	89.6	33.45	3.55
36	96.8	41.82	4.44

the weather bureau cooled some of the air down until fog or dew began to settle out of it. Suppose that she found the dew point to be 60.8°F. She now knows from Table 11.7 that the air contains 13.50 g/m^3 of water vapor, since this is the value for saturated vapor at 60.8°F. However, since the actual air temperature is 89.6°F, she knows that saturated air at this temperature holds 33.45 g/m^3 of water. From this she computes the relative humidity to be

$$\mathrm{RH} = \frac{m}{m_s} = \frac{13.50}{33.45} \approx 0.40$$

We usually multiply this answer by 100 and say that the relative humidity is 40 percent.

The relative humidity can be measured in other ways besides determining the dew point. One common method, the **wet-bulb–dry-bulb method,** makes use of the fact that liquids cause a cooling effect when they evaporate (because the heat of vaporization is used) and that when the vapor is saturated, no evaporation occurs. Hence, if the reading of a dry thermometer is compared with that for a thermometer having a wet cloth around its bulb, the wet-bulb thermometer will usually read cooler than the dry-bulb. The difference in temperatures is a direct measure of relative humidity; the lower the relative humidity, the greater the difference. Tables have been compiled relating relative humidity to this temperature difference so that one need only read the two thermometers to determine the relative humidity.

High relative humidity in summer often causes us discomfort. This results from the fact that we perspire when we are hot and the evaporation of the perspiration cools us. However, if the relative humidity is 100 percent, no evaporation, and therefore no cooling, will result. It is for this reason that one does not feel the heat nearly so much in a hot, dry climate as in a hot, moist climate.

SUMMARY

When the temperature of an object is raised, its internal energy is increased. The internal energy U of the object is the total energy resident in the object. It includes the kinetic, vibrational, chemical, nuclear, and all other forms of energy possessed by the particles composing the object.

The energy which is transferred from a high-temperature object to a lower-temperature object because of the temperature difference is called heat energy. We represent it as ΔQ. Heat energy flows spontaneously from hot objects to cold objects but not vice versa.

The commonly used units of heat energy are the calorie and Btu. They are related to the joule through 1 cal = 4.184 J and 1 Btu = 1054 J. Moreover, 1 Btu = 252 cal. Nutritionists make use of the kilocalorie and call it a calorie.

By definition, the specific heat capacity c of a substance is the heat energy needed to raise the temperature of unit mass of the substance by one degree. As an equation, $\Delta Q = cm\,\Delta T$, where ΔQ is the heat needed to raise the temperature of a mass m by an amount ΔT.

In dealing with gases, two specific heats are usually given. When the temperature change occurs at constant volume, the specific heat capacity is c_V. At constant pressure, the value is c_P. c_P is always larger than c_V since heat must be furnished to do work in expanding the volume in the case of constant pressure.

The energy needed to vaporize unit mass of substance is called the heat of vaporization H_v. An equal amount of energy is given off when unit mass of the vapor is condensed. In a similar manner, the energy needed to melt unit mass of crystalline substance is called the heat of fusion H_f. An equal amount of heat is given off when unit mass crystallizes from the molten state.

When a liquid is maintained at constant temperature in contact with its vapor, the equilibrium pressure of the vapor is called the saturated vapor pressure. Its value increases with temperature. As the temperature of a liquid is increased, its saturated vapor pressure increases until, at high enough temperature, it equals the external pressure on the liquid. At that temperature, vapor bubbles form and grow in the liquid. The liquid is then said to boil.

If an object of original length L is subjected to a temperature change ΔT, the length will change by an amount ΔL given by $\Delta L/L = \alpha\,\Delta T$. The quantity α is the coefficient for linear thermal expansion. Similarly, an object of volume V undergoes a volume change ΔV when a temperature change ΔT takes place. A coefficient for volume thermal expansion γ is defined by $\Delta V/V = \gamma\,\Delta T$.

Heat can be transferred by three mechanisms. In conduction, the energy is passed from atom to atom (or other particle) by collision. No long-range motion of the particles take place. In convection, high-energy molecules flow as a current and carry heat energy along with them. The particles move over large distances, distances comparable to that over which the heat is transported. In radiation, energy is transmitted through vacuum. This is the mode of transport of energy in a light or radio-wave beam.

Relative humidity is the ratio of the mass of water vapor per unit volume in the air to the mass per unit volume required to produce saturation at the same temperature. Air can hold more water per unit volume at high than at low temperature. If air is cooled far enough, a temperature is reached at which the air is saturated. This temperature is called the dew point. At temperatures below this, some form of precipitation will occur.

MINIMUM LEARNING GOALS

Upon completion of this chapter you should be able to do the following:

1. Distinguish between the terms thermal energy, internal energy, and heat energy by explaining the meaning of each.

2. State the direction of heat flow when two objects of known temperature are placed in contact.

3. Give the meaning of the units calories and Btu (*a*) by reference to their values in joules and (*b*) by reference to the heat energy needed to heat water. Explain the relation of the calorie to the nutritionist's calorie.

4. Define the specific heat capacity both in words and by equation. Give its value for water.

5. Use the equation $\Delta Q = cm\,\Delta T$ to solve simple problems involving the heating and cooling of objects.

6. Explain why c should be larger for a substance composed of complex molecules than it is for a monatomic gas.

7. Explain why c_V differs from c_P for a gas.

8. Define the following terms and relate them to each other: saturated vapor pressure, boiling point, heat of vaporization, evaporation, cooling due to evaporation.

9. Tell qualitatively how the temperature of a crystalline substance changes as a function of time as it is slowly heated to the melting point, melted, heated further, and then vaporized. In so doing, point out the effects of heat of fusion and heat of vaporization.

10. Describe what happens to the boiling point of a liquid as the external pressure on it is changed.

11. Solve simple problems in calorimetry such as those in Illustrations 11.4 to 11.7.

12. Compute the thermal expansion of a rod or a volume when the appropriate thermal-expansion coefficient is given. Or, given the results of an expansion experiment, compute the coefficient.

13. Determine how much heat flows through a slab of material when the temperatures of the two faces of the slab are given. Assume the heat conductivity of the material is known.

14. List the three processes by which heat is transferred and explain each.

15. Define the terms relative humidity and dew point. Calculate the RH when the dew point (or water-vapor concentration) and temperature of the air are given. Assume a table such as Table 11.7 is available to you. Explain why the RH is an important factor in cooling the body by evaporation or perspiration.

IMPORTANT TERMS AND PHRASES

You should be able to define or explain each of the following:

Internal energy U
Heat energy ΔQ
Specific heat capacity c
$\Delta Q = cm\,\Delta T$
Calorie, Btu, nutritionist's calorie
c_P is larger than c_V
Saturated vapor pressure
Evaporation and boiling
Heats of vaporization and fusion
Boiling under reduced pressure
$\Delta L/L = \alpha\,\Delta T$ and $\Delta V/V = \gamma\,\Delta T$
Heat conduction, convection, radiation
Newton's law of cooling
Relative humidity
Dew point
Supersaturated vapor

QUESTIONS AND GUESSTIMATES

1. Although argon gas molecules (Ar) and oxygen molecules (O_2) have roughly the same molecular mass, it takes about twice as much energy to heat oxygen as argon. By reference to their internal energies, explain why this result is not unexpected.

2. In an experiment, a student is given a thermos jug containing an unknown substance at temperature T_1. A quantity ΔQ of heat is added by adding hot water. After equilibrium is again established, the temperature is still T_1. The student concludes that the specific heat of the material in the thermos jug is infinite. Explain why the experiment implies $c = \infty$. What is the probable explanation of these experimental results?

3. Can heat be added to something without its temperature changing? What if the "something" is a gas? A liquid? A solid? Explain.

4. Explain why c_P is larger than c_V for a gas.

5. A particular type of wax melts at 60°C. Describe an experiment by which one could determine the heat of fusion of the wax.

6. How would one go about computing the heat needed to raise the temperature of a mixture of known weights of two gases by 10°C?

7. It is possible to make water boil furiously by cooling a flask of water which has been sealed off when boiling at 100°C. Explain why.

8. Liquid chlorine will boil at 30°C if it is subjected to a pressure of 8.60 atm. Where should one look to find data of this sort? Be specific.

9. In what ways would the earth be different if water contracted as it froze?

10. Why should a gas cool more when it expands an amount ΔV by pushing out a piston than when it expands into a vacuum chamber of size ΔV?

11. Why does a piece of steel feel colder than a piece of wood at the same temperature?

12. What would happen if the earth were to be covered by a dense layer of smog so that the rays from the sun could not reach its surface?

13. Estimate how much the temperature of a human body would rise in 1 day if it retained the approximately 2000 large calories (kilocalories) acquired in food in one day. The value for c for a person is about 0.83 cal/(g)(°C). (E)

14. It is well known that a room filled with people becomes very warm unless it is properly ventilated. Assuming that a person gives off heat equivalent to the person's food energy in a steady way throughout the day, estimate how much the temperature of your classroom would rise in 1 h if there were no heat loss out of the room. (E)

15. About how much water would have to evaporate from the skin of an average size man to cool his body by 1 Celsius degree? How does this fit in with what you have heard about the effect of perspiration on the body? [$c_{body} \approx 0.83$ cal/(g)(°C).] (E)

16. If ice is subjected to high pressure, its melting point is decreased below 0°C. To a rough approximation, the melting temperature decreases by about 5 Celsius degrees for each additional 6000 N/cm^2 of applied pressure. Estimate the melting temperature of ice beneath an ice skater's skate. (E)

PROBLEMS

1. (*a*) How many calories of heat are needed to change the temperature of 10 g of lead from 20 to 100°C? (*b*) How many Btu is this?

2. (*a*) If 20 g of copper is cooled from 80 to 30°C, how many calories of heat are given off? (*b*) How many Btu is this?

3. (*a*) How many Btu are required to heat 3.0 lb of lead at 70°F to its melting point, about 590°F? (*b*) How many calories is this?

4. A 70-kg person consumes about 2500 nutritionist's calories, that is, 2.5×10^6 cal, of food per day. If all of this food energy was changed to heat and none of the heat escaped, what would be the temperature rise of the person's body?

5. Fifty people are sitting in a rectangular room which has dimensions 3.5 by 10 by 10 m^3. On the average, each person consumes 2500 nutritionist's calories per day and loses one-twenty-fourth of this amount of heat each hour. By how much do they raise the temperature of the air in the room in 1 h? For simplicity, assume the air to be N_2 at a density of 1.29 kg/m^3. Neglect the volume occupied by the people. Assume the air to be stagnant with no heat loss to the walls or windows of the room and to be maintained at constant volume.

6. How many calories of heat are needed to change 30 g of ice at -5°C to water at 20°C?

7. How much ice at 0°C is needed to cool 250 g of water at 25°C to a temperature of 0°C?

8. An 18-g ice cube (at 0°C) is dropped into a glass containing 150 g of Coke at 25°C. If negligible heat exchange occurs with the glass, what will be the final temperature of the Coke after the ice melts?

9. Molten lead at 327°C is poured into a cavity of a block of ice at 0°C. How much ice will be melted by 40 g of lead? Assume temperature equilibrium with the block is achieved.

10. How much perspiration must evaporate from a 5.0-kg baby to reduce its temperature by 2°C? The heat of vaporization for water at body temperature is about 580 cal/g.

11. The average energy reaching us from the sun each second is 0.134 J/cm^2. Most of this energy is absorbed as it passes through the earth's atmosphere. Assume that 0.10 percent of it strikes the surface of a lake and is used to evaporate its water. How much water would evaporate from 1 m^2 in a day? Use $H_v = 590$ cal/g.

12. When a volatile liquid such as alcohol or ether evaporates from your skin, noticeable cooling occurs. Suppose 0.020 g of dichloroethane ($H_v \approx 85$ cal/g) evaporates from a 1 cm^3 area of your skin and cools a surface layer 0.035 cm thick. (*a*) How much will the temperature of the skin be lowered? Assume c for the skin to be 0.75 cal/(g)(°C) and its density to be 0.95 g/cm^3. Neglect the fact that the skin may undergo a phase change. Also, consider the value of c to be negligibly small for dichloroethane. (*b*) Why does not the temperature of the skin decrease as much as you calculate?

13. Earlier it was shown that $c_P - c_V = R/M$ for an ideal gas. Compute the quantity $(c_P - c_V)M$ for each of the real gases in Table 11.2 from the data in the table. For comparison, $R = 8314$ J/(kg mol)(K). The values of M for the gases (in the order listed in the Table) are 4.0, 40, 32, 28, 44, 18, 16.

14.* Gas is confined in a vertical cylindrical container by a piston having a weight of 10 N much like the situation shown in Fig. 11.3. When the system is at 20°C, the piston rests at a certain height in the cylinder. After heating to 100°C, the piston has risen 20 cm. How much more heat is required to heat the gas in the container from 20 to 100°C under constant pressure than under constant volume? Assume the gas to be ideal.

15. If 100 g of lead shot at 100°C is dropped into 50 g of water at 20°C contained in a copper calorimeter can of 50 g mass, what is the resulting temperature?

16. To 100 g of water contained in a 50-g copper calorimeter at 35°C is added 20 g of ice at -10°C. What is the final temperature?

17. How much steam at 100°C is needed to change 40 g of ice at -10°C to water at 30°C if the ice is in a 50-g copper calorimeter?

18. A 70-g calorimeter can [$c = 0.20$ cal/(g)(°C)] contains 400 g of water and 100 g of ice at equilibrium. To this is added a 300-g piece of metal [$c = 0.10$ cal/(g)(°C)] which has been heated to a high temperature. The final temperature is 10°C. What was the original temperature of the metal?

19. A copper calorimeter has a water equivalent of 5.9 g. That is, in heat exchanges, the calorimeter behaves like 5.9 g of water. It contains 40 g of oil at 50.0°C. When 100 g of lead at 30.0°C is added to this, the final temperature is 48.0°C. What is the specific heat capacity of the oil?

20. Benzene boils at about 80°C. Benzene vapor at 80°C is bubbled into a calorimeter, the water equivalent of which is 20 g (see Prob. 19), containing 100 g of oil [$c = 0.50$ cal/(g)(°C)] at 20°C. The final temperature after 7.0 g of benzene has been condensed is 30°C. What is the value of the heat of vaporization of benzene? [c for benzene is 0.40 cal/(g)(°C).]

21. When 150 g of ice at 0°C is added to 200 g of water in a 100-g aluminum cup, with cup and water being at 30°C, what is the resulting temperature?

22.* A certain 6-g bullet melts at 300°C and has a specific heat capacity of 0.20 cal/(g)(°C) and a heat of fusion of 15 cal/g. (*a*) How much heat is needed to melt the bullet if it is originally at 0°C? (*b*) What is the slowest speed at which the bullet can travel if it is to just melt when suddenly stopped?

23.* A mass m of lead shot rests at the bottom of a closed cardboard cylinder L units long, as shown in Fig. P11.1. When the cylinder is quickly inverted by rotating it about its center, the lead shot falls through the length of the tube. Show that the rise in temperature of the lead after n reversals is given by

$$\Delta T = \frac{ngL}{c_{\text{Pb}}J}$$

where J is the mechanical equivalent of heat, i.e., the number of joules equivalent to 1 cal, and g is the acceleration due to gravity.

24. A certain straight roadway is made of concrete slabs, placed end to end. Each slab is 25 m long. How large an expansion gap should be left between slabs at 20°C if the slabs are to just touch at 45°C?

25. The marks on an aluminum measuring tape were made when the tape was at a temperature of 25°C. What is the percent error due to contraction if the rule is used at −5°C?

26. A salad-dressing jar has an aluminum screw-type lid. At 20°C, the lid fits so tight that it will not screw off. By what fraction will the lid's diameter expand if it is heated to 85°C under the hot-water faucet? Why should you not heat the can lid any longer than necessary?

27. It is common machine shop practice to "shrink-fit" cylindrical rods into holes in wheels, blocks, and plates. Suppose a 2.000-cm-diameter rod is to be fitted into a 1.985-cm-diameter hole in a brass block. How much must the block be heated so that the cool rod will fit into the hole?

28. A heat-resistant glass flask is calibrated to hold exactly 100 cm³ at 20°C. How much will it hold at 30°C? *Hint:* The hollow flask will expand as though it were actually a solid volume.

29. A 100-cm³ volumetric flask is filled with benzene at 30°C. How much more benzene must be added to have it full after it is cooled to 20°C? (Ignore the expansion of the flask.)

30.** An iron beam 10 ft long has its two ends embedded in substantial concrete pillars. If the structure was made at 35°F, what force will be exerted by the beam on the pillars when its temperature is 98°F? The cross-sectional area of the beam is 10 in².

31.** A flask of mercury is sealed off at 20°C and is completely filled with the mercury. Find the pressure within the flask at 100°C. Ignore the expansion of the glass. ($k_{\text{Hg}} = 2.5 \times 10^{-6}$ in²/lb.)

32.** Show that the coefficient for volume thermal expansion in the case of an ideal gas is $1/T$ provided the expansion is done at constant pressure.

33. If the two ends of a brass rod 0.50 m long with a 0.50-cm radius are maintained at 100 and 20°C, how much heat flows down the rod in 1 min? Assume the sides of the rod are insulated.

34. An asbestos sheet 2.0 mm thick is used as a spacer between two brass plates, one at 100°C and the other at 20°C. How much heat flows through 40 cm² of area from one plate to the other in 1 h?

35.* A brass pipe 10 cm in diameter having a 0.25-cm wall thickness carries steam at 100°C through a vat of circulating water at 20°C. How much heat is lost per meter of pipe in a second?

36.** Two brass plates, each 0.50 cm thick, have a rubber spacer sheet which is 0.10 cm thick between them. The outer side of one brass plate is kept at 0°C while the outer side of the other is 100°C. Find the temperatures of the two sides of the rubber spacer.

37.* Show that Stefan's law of cooling reduces to Newton's law of cooling when $(T_1 - T_0)/T_0 \ll 1$.

FIGURE P11.1

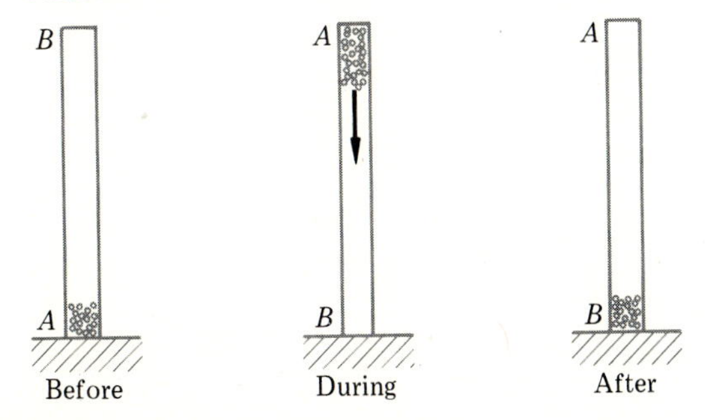

38.* In a closed room with volume 500 m^3, the temperature is 24°C and the humidity is 95 percent. A dehumidifier is turned on to reduce the humidity. Assuming unchanged temperature, what mass of water will be removed in reducing the humidity to 40 percent?

39. On a particular day it is found that water condenses on a water glass when the water temperature is 16°C or lower. The room temperature is 20°C. Find (*a*) the amount of water in unit volume of the air and (*b*) the relative humidity.

40. If the air temperature is 90°F and the relative humidity is 80 percent, how cold must a soft drink bottle be before moisture will condense on it?

41. A person who wears glasses has been outside for some time on a day when the temperature is −4°C. How low must the RH be in a house which the person enters if the glasses are not to fog up? Assume the house temperature is 20°C.

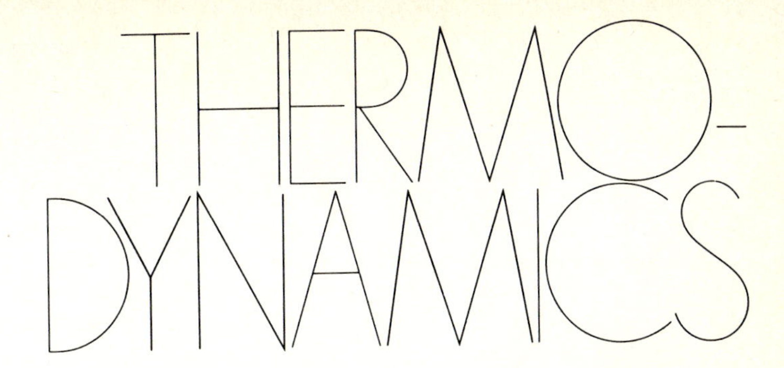

12

Long before the nature of atoms and molecules was known, a powerful way of discussing heat, work, and internal energy had been found. It involves the description of matter in terms of gross, overall properties such as pressure, temperature, volume, and heat flow. This way for discussing the behavior of objects and substances is called **thermodynamics.** Today, even though we understand quite well how atoms and molecules behave, thermodynamics is widely used in all branches of science. In this chapter we shall present an introduction to this important and powerful area of study.

12.1 STATE VARIABLES

In thermodynamics, we most frequently discuss the behavior of a definite group of molecules which we call a **system.** A typical system may be the gas molecules in a container, or those in a solution, or even such a complex system as the molecules in a rubber band. For any very meaningful thermodynamic discussion, the system must be well specified. Only then can we give an unambiguous description of it.

To describe the system, we make use of quantities which apply to the whole system or to some well-defined portion of the system. Typical measurable quantities are pressure, temperature, and volume. In thermodynamics we also make use of such quantities as internal energy, heat, work, and a quantity we shall encounter later called entropy. As the condition of a system changes, these quantities may change. It is important that we know which quantities are suitable in representing the exact condition of the system. Let us now see what they are.

When a container of gas reaches equilibrium, the gas has a definite temperature, pressure, and volume. The ideal-gas law reflects this fact since it tells us that $PV = nRT$. We see that, *given any two of these variables, the other can be calculated and is therefore known. This particular situation, where the gas (the system) has specified values of P, V, and T, is called a* **state** *of the system. Whenever the gas is returned to these same values of P, V, and T, its state will be the same.* Even though each individual molecule within the system may not be doing exactly the same thing whenever the system is brought to this state, the system as a whole still appears the same in macroscopic* measurements.

State Variable

There are certain features of a system which are always the same when the system is in a given state. The variables which describe these features are called **state variables.** For example, P, V, and T are state variables for a system. No matter how the system reaches a particular state, the pressure, volume, and temperature are always the same. For example, the gas in a car tire can be considered to be a system. As long as the tire does not leak, in identical states, the gas in it will have identical values of P, V, and T. The history of the gas in the tire is not important. A given state of the gas always has the same values for P, V, and T, its state variables.

There are quantities which are not state variables. *Work*, for example, *is not a state variable.* The gas in a tire cannot be assigned a definite amount of work. When the gas reaches the same state tomorrow that it has today, there is no assurance that the gas in the tire has done zero net work in the meantime. The "work content" of a gas has no well-defined meaning. Work is not a state variable.

Similarly, *heat is not a state variable.* The gas in a tire can be warmed in many ways—not all of them involving heat flow into the gas. Although the sun may provide heat to a tire and warm it, mechanical action may also change the temperature of a tire. A car tire becomes very warm on a car

*__Macroscopic__ is used to denote measurements involving the average effect of billions or more molecules.

moving at high speeds even though the day may be quite cold. As we see, the temperature of a gas does not necessarily tell us how much heat flowed into the gas to reach the temperature. There is no well-defined meaning to the term "heat content" of a gas. Heat has meaning only when energy flows because of a temperature difference. Heat is not a state variable.

We have also discussed the internal energy of a substance. It consists of all the energy of every type resident in the substance. Since energy is conserved, we can keep a balance sheet of it. A system's internal energy is a well-defined quantity. If you tell me how much internal energy a system has today, I can (in principle at least) tell you how much internal energy it has at a later time. All I need do is measure the energy changes the system undergoes in the meantime. Then, like a good accountant, I can total up the energy balance sheet and tell you how much internal energy the system still possesses. *Internal energy is a state variable.*

12.2 THE FIRST LAW OF THERMODYNAMICS

The law of conservation of energy is one of the two fundamental pedestals upon which all of thermodynamics is based. Indeed, those who first developed the field of thermodynamics were also the earliest investigators of energy conservation. *When the law of conservation of energy is stated in a general way for a system, the statement is called the* **first law of thermodynamics.** Let us now see how the statement is made.

A system in a given state has a definite amount of internal energy U. This energy can be of any and all kinds including, kinetic, potential, chemical, and nuclear energy. The system's internal energy can be changed in only two general ways: (1) heat energy can flow into or out of the system, and (2) the system can do work of some kind against external forces. To summarize this fact in equation form, we define

ΔU = *increase* in internal energy of system
ΔQ = heat which flows *into* system
ΔW = work done *by* the system

Then we have

$$\Delta U = \Delta Q - \Delta W \tag{12.1}$$

First Law of Thermodynamics

This equation is a statement of the **first law of thermodynamics.** *It is also a statement of the law of energy conservation.*

In using the first law we must be very careful about + and − signs. It is a balance-sheet equation, and so debits and credits must be noted with care. By convention, ΔW is the work done *by* the system. It therefore decreases U and hence appears with a negative sign in Eq. (12.1). On the other hand, ΔQ is the heat which flows *into* the system. It therefore increases U and carries a positive sign in Eq. (12.1).

The first law is applicable to all systems, no matter how complex. For example, consider your body as the system. (In order for the system to remain simple, you will not be allowed to eat or excrete.) It loses internal energy as the day goes on. The food you have previously eaten is part of your internal energy. This and other forms of internal energy are used up as the day progresses. Much of the energy is lost as heat flow from your body to the surroundings. A smaller amount is lost as your body, the system, does work on the objects of the world about you. We can write the first law for your body in the following way:

$$-\text{Decrease in internal energy} = -\text{heat lost} - \text{work done}$$

Can you reason out why the signs are all negative? To do so, it is perhaps best to return to the meaning of the first law rather than by plugging directly into Eq. (12.1).

12.3 WORK DONE DURING VOLUME CHANGE

In many practical applications of thermodynamics, the work done by the system is carried out by means of a volume change. It is therefore necessary to know how work and volume change are related. For simplicity, let us consider the system shown in Fig. 12.1. We see there a gas confined to a cylinder by a piston. Let us consider how much work the gas does as it expands and pushes the piston slightly upward.

If we keep the displacement of the piston small the pressure of the gas will not change much. The force exerted on the piston by the gas is given by the definition of pressure to be

$$P = \frac{F}{A} \qquad \text{or} \qquad F = PA$$

where A is the end area of the piston. In the slight displacement Δy, the work done against the piston by the gas is

$$\begin{aligned}\Delta W &= (\text{force} \times \text{distance})(\cos\theta)\\ &= (PA)(\Delta y)(1) = P(A\,\Delta y)\end{aligned}$$

But $A\,\Delta y$ is simply the increase in volume of the gas ΔV. We therefore find

Work Done in Expansion

$$\Delta W = P\,\Delta V \tag{12.2}$$

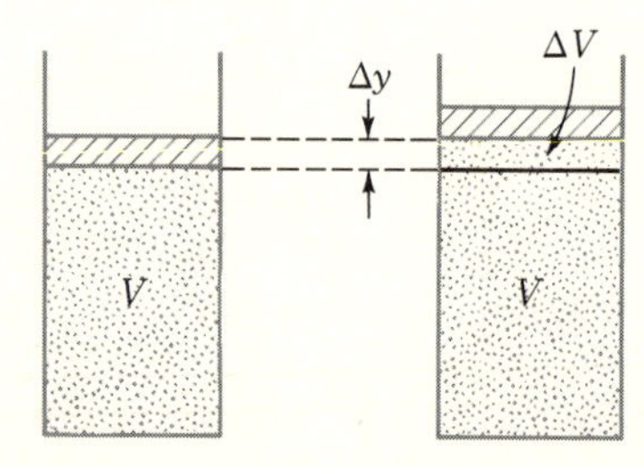

FIGURE 12.1
If the piston has an area A, then $\Delta V = A\,\Delta y$.

Equation (12.2) applies to any system for which P is essentially constant during the volume change ΔV. The work a system does in an expansion ΔV against a pressure P is simply $P\,\Delta V$. If the system contracts, then ΔV is negative. In that case, the system has done negative work. The surroundings have actually done work upon the system.

It is often important to notice that the expansion work of a system can be represented by an area on a particular type of graph called a *P-V* diagram. As a very simple example of such a diagram, consider the case of the thermal expansion of a solid metal object as shown in Fig. 12.2. When the object is heated, it expands. The pressure to which it is subjected is that of the atmosphere, P_A. Of course, P_A remains constant during the expansion. We can represent this expansion as shown in the *P-V* graph in part *b* of the figure.

As we see, the system starts at the point labeled "Cold" in the graph. The object has a volume V_c and the pressure is P_A. During expansion, the system follows the graph line shown. Let us now find the work done by the system. It is, from Eq. (12.2),

$$\Delta W = P\,\Delta V = P_A(V_h - V_c)$$

But the quantity P_A is the height of the shaded rectangle in Fig. 12.2*b* while $V_h - V_c$ is its width. Therefore, $P_A(V_h - V_c)$ is simply the shaded area. We therefore conclude that

Work Related to Area

The expansion work done by a system is equal to the area under its *P-V* curve.

Although this has been derived for an especially simple case, it is true in general. To see this, refer to Fig. 12.3.

We see there a gas confined to a piston. The pressure on the piston can be changed by adding or removing weights. The pressure and volume can also be changed by heating or cooling the gas. Suppose the conditions are varied in such a way that the volume changes from V_A to V_B by the path shown in part *b* of the figure.

Consider the most heavily shaded region in which the pressure is *P*, approximately, and the volume change is ΔV. The work done in this small part of the total expansion is $P\,\Delta V$. But this is equal to the area of the heavily

(*a*)

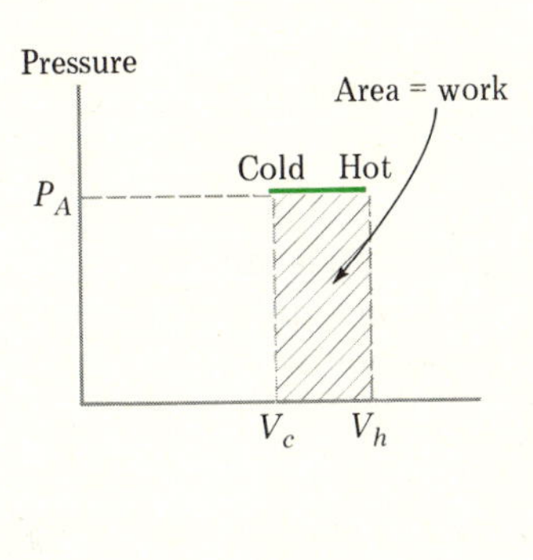

(*b*)

FIGURE 12.2

The work done by the object as it expands against the atmosphere equals the area under the *P-V* curve. (*a*) Object expands as heated; (*b*) *P-V* diagram for the experiment in (*a*).

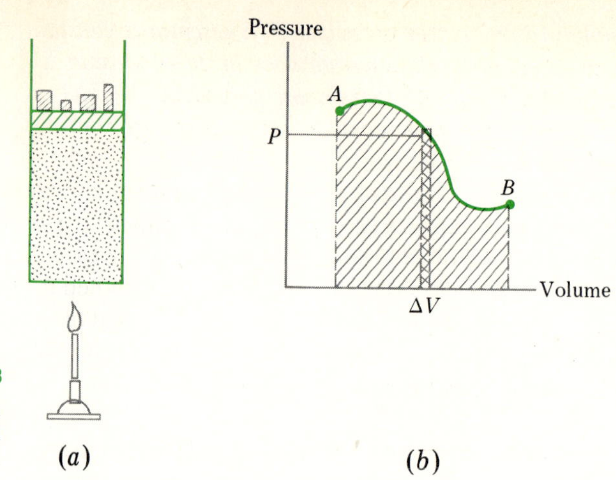

FIGURE 12.3

The work done by the system in going from A to B is equal to the area under the curve.

shaded rectangle. Similarly, the whole expansion from A to B can be thought of as a series of tiny expansions. The work done during each is an area. The total work done is the sum of all these areas and is the area under the curve from A to B. As we see, *the expansion work is always equal to the area under the P-V curve.*

Illustration 12.1 A cube of brass 15 cm on each side is heated from 20 to 300°C. How much work is done by the cube as it expands?

Reasoning The cube expands against atmospheric pressure, about $1 \times 10^5\ \text{N/m}^2$. According to the thermal-expansion equation

$$\Delta V = V\gamma\,\Delta T$$

In our case $V = (0.15)^3\ \text{m}^3$, $\Delta T = 280°\text{C}$, and $\gamma = 5.7 \times 10^{-5}\,°\text{C}^{-1}$. Substitution gives $\Delta V = 5.4 \times 10^{-5}\ \text{m}^3$.

Now we can make use of

$$\Delta W = P\,\Delta V$$

to obtain

$$\Delta W = 5.4\ \text{J}$$

It is of interest to note that the heat energy needed to raise the temperature of this much brass 280 Celsius degrees is about 10^6 cal. Therefore, less than one-millionth of the total heat energy furnished is used in doing expansion work.

12.4 TYPICAL PROCESSES IN GASES

As we have just seen, the P-V curve for a system undergoing change is related to the work done by the system. In drawing such a graph, we assume that the change is slow enough for the pressure and temperature to be uniform at any instant throughout the whole system. Let us now examine several important ways in which a system composed of a gas can undergo change. *An* **isothermal process** *is one during which the temperature remains constant.* Since the temperature of an ideal gas is a measure of its internal energy, an isothermal process is a constant-internal-energy process. For an ideal gas, then, we see that $\Delta U = 0$ during an isothermal process. The first law then tells us that

DEFINITION

Isothermal Process

$$\Delta Q = \Delta U - \Delta W$$

and becomes

$$\Delta Q = -\Delta W \qquad \text{isothermal, ideal gas}$$

To illustrate the meaning of this relation, refer to the situation shown in Fig. 12.4. We see there a container of gas which makes good thermal contact with a heat reservoir. The heat reservoir might be an oven, a cooling bath, or any other constant-temperature device. It will maintain the container of gas at constant temperature provided the piston closing the container is not moved too rapidly.

Suppose weights are slowly added to the piston in Fig. 12.4. The pressure on the gas will slowly increase and the volume will decrease. We show the P-V diagram for this isothermal process in part b of the figure. Its form is given by the ideal-gas law

$$PV = nRT$$

which becomes

$$P = \frac{\text{const}}{V}$$

in this case, where T is maintained constant. This is the equation for the graph line shown in Fig. 12.4.

Suppose now that the gas has been compressed from point A to point B on the graph. If now the force on the piston is slowly reduced, the above relation between P and V will still apply. The system will follow this same graph line as it moves from state B back to state A. A process such as this is

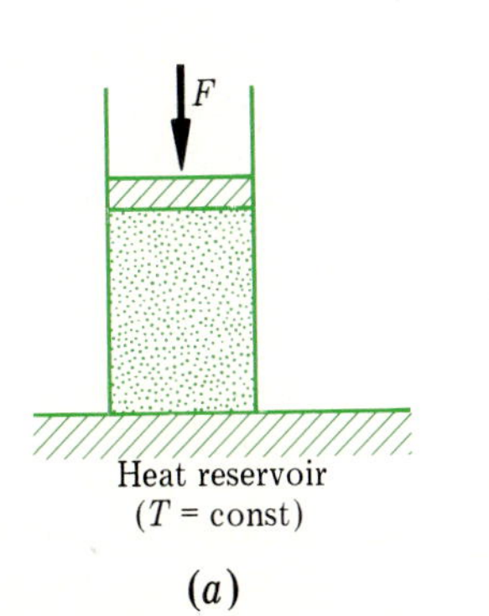

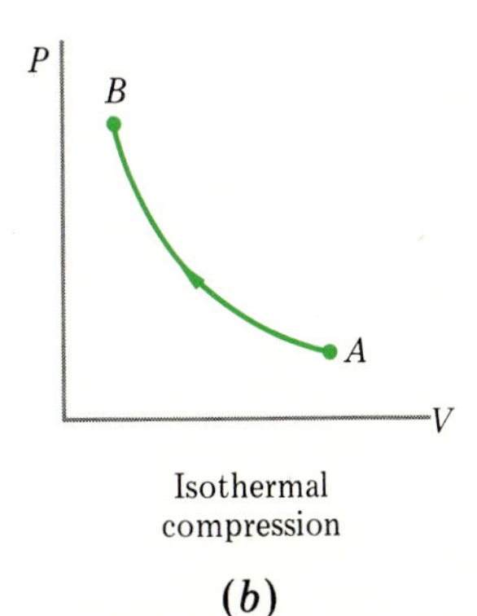

FIGURE 12.4
The P-V diagram for an isothermal compression.

Reversible Process

said to be reversible. Notice that, *in a reversible process, the state variables acquire the same values at all stages of the process independent of the direction in which the process is being carried out.* Not all processes are reversible. For example, *a process which has appreciable friction losses cannot be reversible.* Why?

DEFINITION

An **adiabatic process** *is one in which no heat is lost or gained by the system during the process.* For example, if the system is well insulated from the surroundings, heat transfer is often negligible. Or, if the process is carried out extremely rapidly (such as a very sudden compression of a gas), no appreciable heat will flow into or out of the system in that short time. The process will be adiabatic.

Adiabatic Process

For an adiabatic process, the first law

$$\Delta U = \Delta Q - \Delta W$$

becomes

$$\Delta U = -\Delta W \qquad \text{adiabatic}$$

This relation is not restricted to an ideal gas. It tells us that *if a system does adiabatic work, the internal energy of the system must decrease.* The work is done at the expense of internal energy. But if adiabatic work is done on the system, the internal energy will be increased. The following illustrations show two practical uses of these processes. First, however, we shall examine the adiabatic behavior of an ideal gas in more detail.

In the case of an ideal gas, an adiabatic process is not described in terms of $PV = nRT$ alone. The difficulty is that all three state variables (P, V, and T) change during the process. To find how each changes, we need another equation. It can be found by noticing that the work done on the gas goes completely into increased *internal* energy. This increase in internal energy causes a temperature change in the system. But this same temperature change could have been carried out by adding heat to the system. Hence, a relation between heat, temperature change, and work can be found even for an adiabatic process. For an ideal gas, this line of thought leads to the following result: *if an ideal gas undergoes an adiabatic change from* P_1, V_1, T_1 to P_2, V_2, T_2, then

$$P_1V_1^{\gamma} = P_2V_2^{\gamma} \tag{12.3}$$

where $\gamma = c_P/c_V$ *for the gas.* Typical values for c_P/c_V were listed in Table 11.2.

The P-V diagram graph line for an adiabatic change is given by Eq. (12.3). A typical case is shown in Fig. 12.5. For comparison, the graph line for an isothermal change is shown by the dotted curve. As you might expect, the adiabatic graph line is steeper than the isothermal line at comparable values of P and V.

FIGURE 12.5
Comparison of the adiabatic and isothermal changes for an ideal gas.

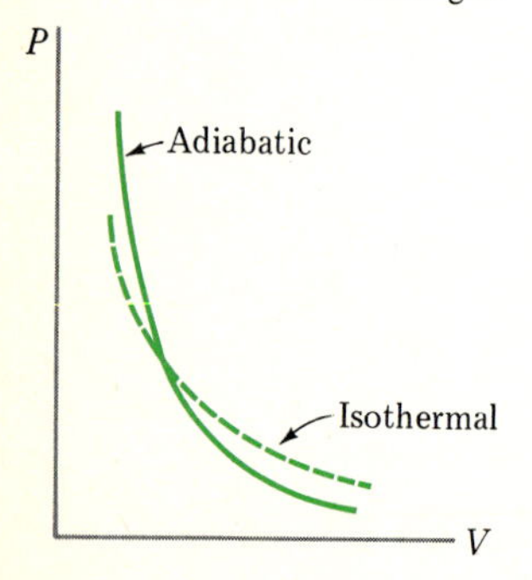

Illustration 12.2 In a diesel engine, the fuel vapor is ignited in the cylinder of the engine by suddenly compressing the gas in the cylinder. Suppose the gas volume is compressed to one-fifteenth its original value. Its initial

pressure is about atmospheric, $1 \times 10^5\ \text{N/m}^2$. If its initial temperature is 27°C, find its final temperature. Assume the gas in the cylinder to be mostly air or nitrogen.

Reasoning This is essentially an adiabatic process since the gas in the cylinder is compressed so rapidly. When we assume the gas to be N_2 and to act like an ideal gas, Table 11.2 gives $\gamma = 1.40$. We know that

$$P_1V_1^\gamma = P_2V_2^\gamma \qquad \text{or} \qquad \frac{P_1}{P_2} = \left(\frac{V_2}{V_1}\right)^\gamma$$

But we are interested in the variation in temperature, not pressure. The ideal-gas law tells us that

$$P_1V_1 = nRT_1 \qquad \text{and} \qquad P_2V_2 = nRT_2$$

Dividing one equation by the other gives, after simplifying,

$$\frac{P_1}{P_2} = \left(\frac{T_1}{T_2}\right)\left(\frac{V_2}{V_1}\right)$$

Substitution of this in the previous equation gives

$$\frac{T_1}{T_2} = \left(\frac{V_2}{V_1}\right)^{\gamma-1}$$

In our case, $V_2/V_1 = \frac{1}{15}$, and so

$$T_2 = T_1(\tfrac{1}{15})^{-0.40}$$

We must use absolute temperatures for T and so $T_1 = 300$ K. Taking logs of both sides of the equation then gives

$$\begin{aligned}\log T_2 &= \log(300) - 0.40(\log 1 - \log 15)\\ &= 2.477 - 0.40(0 - 1.176)\\ &= 2.947\end{aligned}$$

Then, taking antilogs, one finds $T_2 = 886$ K, which is 613°C. Notice how very hot the gas has become because of the adiabatic compression.

Illustration 12.3 As shown in Fig. 12.6*a*, a container is sectioned into two parts with gas at high pressure in one part and vacuum in the other, much larger part. A small hole is opened in the connecting wall so the gas expands into the vacuum chamber. Describe the temperature change of the gas, assuming the process to be adiabatic.

FIGURE 12.6
(*a*) A hole is made in the partition so the gas can expand. (*b*) The expanding gas raises the piston. Under which condition would an ideal gas be cooled most?

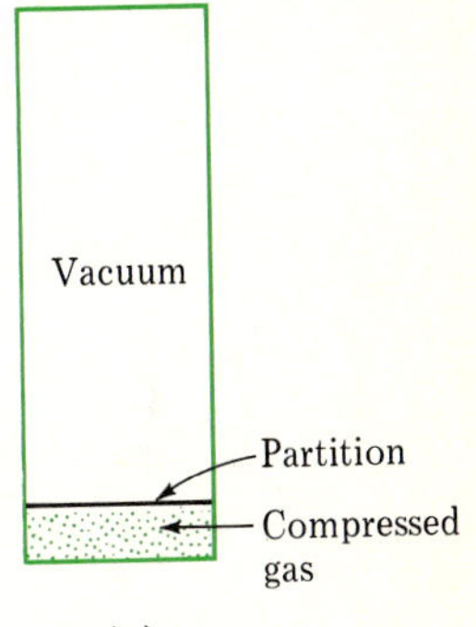

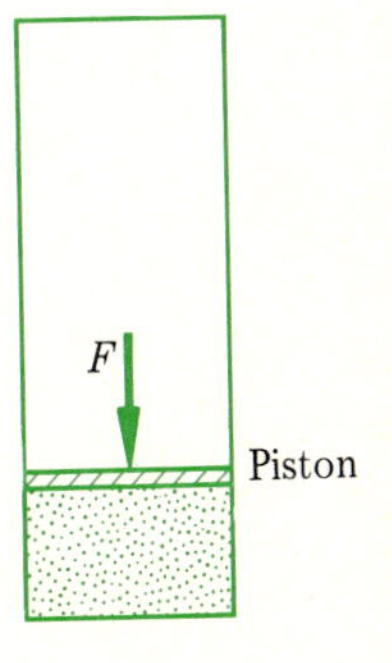

Throttling Process **Reasoning** *This type of process, in which a fluid expands through a small opening or porous disk, is called a* **throttling process.** Since it is adiabatic, we can write, from the first law,

$$\Delta U = -\Delta W$$

where ΔW is the work done by the fluid in the process.

We are asked to deal with the special case of a gas. Suppose first that the gas is ideal. In the expansion shown in Fig. 12.6*a*, the ideal gas does no net work as it expands into the vacuum. This follows because the external pressure resisting the expansion is zero, so $P\,\Delta V = 0$. Since the gas does no work, the fact that $\Delta U = -\Delta W$ tells us that the internal energy of the gas does not change. Since, for an ideal gas, $T \sim U$, we see that T does not change. Therefore, if the expanding gas is ideal, the temperature remains unchanged.

However, the result will be different if the fluid is not an ideal gas. Suppose the compressed material is actually a liquid, e.g., butane, which vaporizes as it expands into the vacuum. Then, energy must be furnished to the molecules to tear the liquid apart, i.e., to supply the heat of vaporization. Since the process is adiabatic, the required energy must come from the internal energy already resident in the fluid. As a consequence, the KE of the molecules is decreased during the expansion process. The temperature of the vapor will therefore be lower than that of the original liquid. (In a sense, this is very much like cooling by evaporation.) You can see that many possibilities exist between the ideal gas, where no cooling occurs, to the volatile liquid, where a great deal of cooling occurs.

In certain cases, even an ideal gas will cool upon adiabatic expansion. For example, suppose the partition is replaced by a movable piston as shown in Fig. 12.6*b*. Then the gas will do work as it expands against the piston. This work will lead to a decrease in the internal energy of the gas, and the gas will cool.

12.5 CYCLIC PROCESSES AND ENERGY CONVERSION

One of the major applications of thermodynamics is to engines and refrigerators. These devices are similar in one important respect: they perform the same operation (or cycle) over and over again. As we shall see, the efficiency of such a device is limited by a law of nature which makes it very difficult to design engines with near 100 percent efficiency.

To begin our discussion, consider the system shown in Fig. 12.7*a*. Suppose the system has the P, V values given by point A of the P-V diagram in part b of the figure. Let us now slowly heat the gas. The volume will then increase from A to B. During this process, the pressure remains constant since the weights on the piston are not changed. Further, heat ΔQ_{AB} is added to the system as indicated. Also, as the gas expands, it does work ΔW_{AB} equal to the shaded area shown.

Next, let us slowly cool the system while simultaneously removing

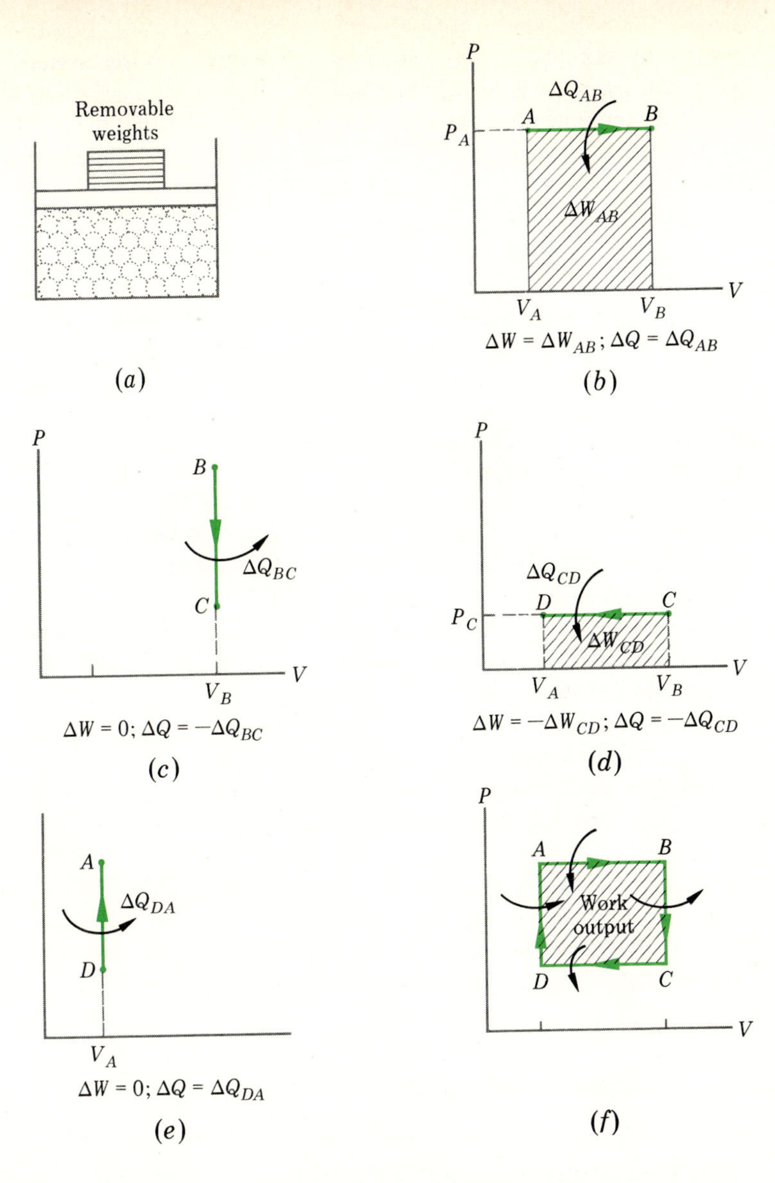

FIGURE 12.7

A simple thermodynamic cycle. The net work done equals the area enclosed.

weights from the piston so as to keep the volume constant. This is part *BC* of the cycle shown in Fig. 12.7*c*. Since the piston does not move, zero work is done in this portion of the cycle. (Notice that the area beneath this portion of the graph line is zero.) But heat is removed from the system, and so $\Delta Q = -\Delta Q_{BC}$, as indicated.

The next step of the cycle is shown in *d*. Now the gas is cooled further at constant pressure. The gas contracts, and so ΔV is negative. For this reason, the system does negative work equal to $-\Delta W_{CD}$, where this is the shaded area shown. Heat is lost by the system in this step, so $\Delta Q = -\Delta Q_{CD}$.

Finally, the cycle is completed by the process shown in *e*. Here the gas is heated while maintaining the volume unchanged by adding weights to the piston. The pressure then increases from *D* to *A*. No work is done by the system during this process. But heat is added to the system, and so $\Delta Q = \Delta Q_{DA}$.

The complete cycle is shown in Fig. 12.7*f*. Notice that the net work done by the system is simply area $\Delta W_{AB} - \Delta W_{CD}$. This is the shaded area in part *f* of the figure. In this case, as well as in all others,

Work/Cycle Is Area Enclosed

The net output work done during a thermodynamic cycle is the area enclosed by the cycle on its *P-V* diagram.

This is an important result because it allows one to find the output work of an engine directly from its *P-V* diagram.

Illustration 12.4 Figure 12.8 is an idealized *P-V* diagram for one cylinder of an ordinary gasoline engine. (This idealized cycle is called the **Otto cycle.**) The air-filled cylinder at *C* is compressed adiabatically to *D*. Gasoline vapor is ignited in the cylinder during portion *DA*, and so heat is added rapidly to the gas with negligible change in volume. The piston then moves out as the hot gas expands adiabatically in portion *AB*. Then, during portion *BC*, heat is released to the exhaust and the original process repeats. Assuming the shaded area to be one-third of the total area of rectangle *AECFA*, find the net output work of this cylinder per cycle.

Reasoning The net output work is equal to the area enclosed by the cycle. Let us first find the rectangular area *AECFA*. It is

$$\text{Area } AECFA = (23 \times 10^5\ \text{N/m}^2)(600 \times 10^{-6}\ \text{m}^3) = 1380\ \text{J}$$

But the shaded area, equal to the net output work, is one-third of this. Therefore, the net output work per cycle is 460 J.

12.6 EFFICIENCY OF AN ENGINE

As the world becomes more concerned about its energy resources, we seek ways of decreasing energy use. One of the most obvious ways is to improve the efficiency of the engines used in our cars and in other vehicles. But we are severely limited in our quest for engine efficiency. The first law of ther-

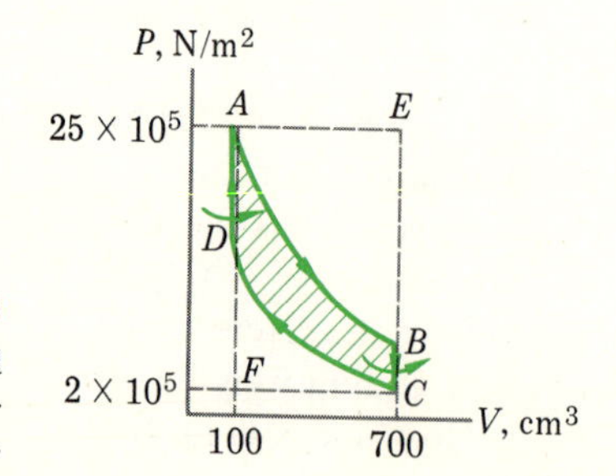

FIGURE 12.8
The Otto cycle, an idealized cycle for the internal-combustion engine.

modynamics places a limit on the efficiency attainable. Let us see what this limit is.

We define the efficiency of an engine in the same way as for any machine. It is defined by

$$\text{Efficiency} = \frac{\text{net output work/cycle}}{\text{input energy/cycle}}$$

The net output work is given by the area enclosed by the engine's P-V cycle. But we shall not use that value for our computation. Instead, we shall compute the efficiency from a direct consideration of the first law of thermodynamics.

Consider any engine which burns fuel for its operation. (An electric motor is actually part of the larger "engine" which includes the power-plant generators.) The burning gasoline, for example, in the internal-combustion engine furnishes heat energy to the system. This constitutes the input energy to the engine. We shall call it ΔQ_{in} and assume it to be taken for one cycle of the engine.

Some of this input heat is always lost from the engine in a nonproductive way. For example, the hot gases of the car's exhaust carry unused energy away from the engine. In addition, frictional energy losses may also be important. We shall denote the total of these exhaust processes ΔQ_{ex}.

Now we invoke the first law, which is really the energy-conservation law. The net output work done by the engine is just the difference between the input and exhaust energies. (This is often represented in diagrammatic form, as shown in Fig. 12.9.) We then have

$$\text{Net output work/cycle} = \Delta Q_{\text{in}} - \Delta Q_{\text{ex}}$$

But the input energy per cycle is simply ΔQ_{in}. We then find from the definition of efficiency that

$$\text{eff} = \frac{\Delta Q_{\text{in}} - \Delta Q_{\text{ex}}}{\Delta Q_{\text{in}}}$$

or

$$\text{eff} = 1 - \frac{\Delta Q_{\text{ex}}}{\Delta Q_{\text{in}}} \tag{12.4}$$

Notice that Eq. (12.4) tells us *the efficiency is always less than unity* (*or 100 percent*) *by the amount* $\Delta Q_{\text{ex}}/\Delta Q_{\text{in}}$. You might then think that all one has

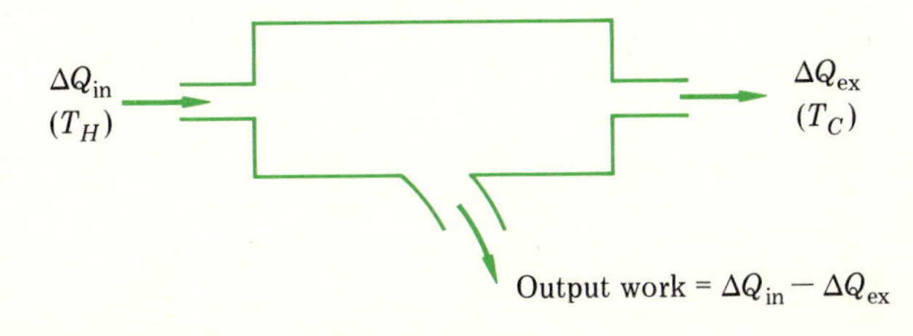

FIGURE 12.9
The output work of the engine is always less than the input by an amount equal to the exhaust heat.

to do is design an engine with very little heat lost in exhaust processes. But this is not as simple as it sounds. The trouble lies in the fact that *work is done by thermal energy only when the heat flows from a hot object to a cooler one. Work is only done when a heat flow exists.*

As a guess, you might think that the work which a given quantity of heat can do would depend upon the temperature difference through which it flows. Then, the higher the temperature difference between the engine's intake and exhaust, the more efficient the engine should be. This turns out to be true in a general sort of way. But the temperature difference of importance turns out to be the difference between the highest temperature and lowest temperature portions of the engine's cycle.

For most real engines, the *P-V* cycle is quite complex. As a result, the engine efficiency depends on exactly how the engine is constructed. Even so, in 1824 Sadi Carnot (pronounced car-no) proved from theory that *one particular engine (and cycle) has the highest possible efficiency. This cycle, called the* **Carnot cycle,** is shown in Fig. 12.10.

Carnot showed that the ratio $\Delta Q_{ex}/\Delta Q_{in}$ in Eq. (12.4) is equal to the ratio of the respective temperatures, T_{ex}/T_{in}. *For the Carnot cycle,* then, we have

Carnot Cycle

$$\text{eff} = 1 - \frac{T_{ex}}{T_{in}} \tag{12.5}$$

Maximum Efficiency

Moreover, *Carnot proved that Eq. (12.5) is the highest possible efficiency for any engine operating between the two temperatures* T_{ex} *and* T_{in}. Usually the exhaust temperature will be no lower than atmospheric temperature, about 300 K. Therefore, to achieve high efficiency, the engine fluid must have a very high temperature during the hottest part of the cycle. Ultimately, this temperature is limited by the material used to construct the engine. In practice, however, there are other limiting factors which make this ultimate consideration unimportant. Typical efficiencies for modern engines are given in Table 12.1. Note that the values are maxima and only approximate.

Illustration 12.5 Estimate the maximum possible efficiency for a steam engine.

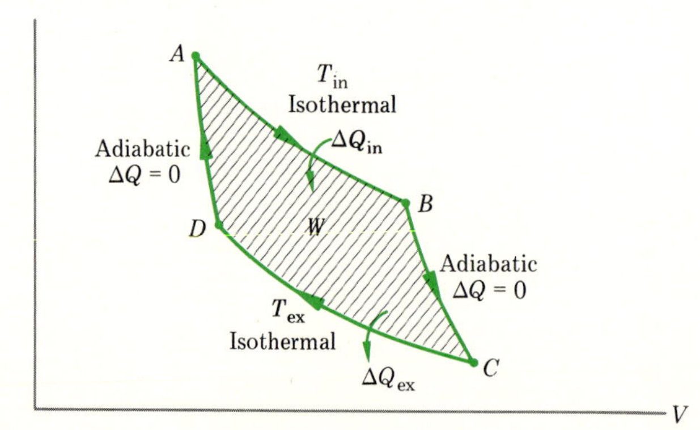

FIGURE 12.10
The Carnot cycle.

TABLE 12.1
TYPICAL MAXIMUM ENGINE EFFICIENCIES

ENGINE	APPROXIMATE EFFICIENCY, %
Steam engine	15
Steam turbine	35
Gasoline engine	30
Diesel engine	40

Reasoning In a steam engine, the hot gas used to drive the piston is steam. Since water boils at 373 K, the pressure of steam goes up rapidly at higher temperatures. As a result, the upper temperature is limited by the strength of the steam boiler. Let us suppose the entering steam has a temperature of $T_{in} = 453$ K (which is 180°C). The exhaust temperature cannot easily be less than ambient. Let us therefore take $T_{ex} = 300$ K (which is 27°C). Then

$$\text{eff} = 1 - \frac{T_{ex}}{T_{in}} = 1 - \frac{300}{453} = 0.34 \text{ or } 34\%$$

In practice, the efficiency might be only one-third of this. Our computation assumed the Carnot cycle and, of course, the actual cycle will be less efficient.

12.7 HEAT PUMPS; REFRIGERATORS

Engines use heat energy to do work. Refrigerators and heat pumps do just the opposite; they use work to transfer heat from a cold region to a warmer one. The energy diagram of a typical compression-type refrigerator is given in Fig. 12.11. Notice that work is used to lift heat energy from a cold region (the inside of the refrigerator) to a warm region outside the refrigerator. Heat would flow in the reverse direction by itself. Work from outside is required to maintain a flow in the direction shown. The first law tells us that

$$\text{Work input} + \text{heat input} = \text{heat output}$$

Many refrigeration systems operate by means of a compressor acting on a fluid. Freon gas* is frequently used as the fluid. It can be liquefied at room

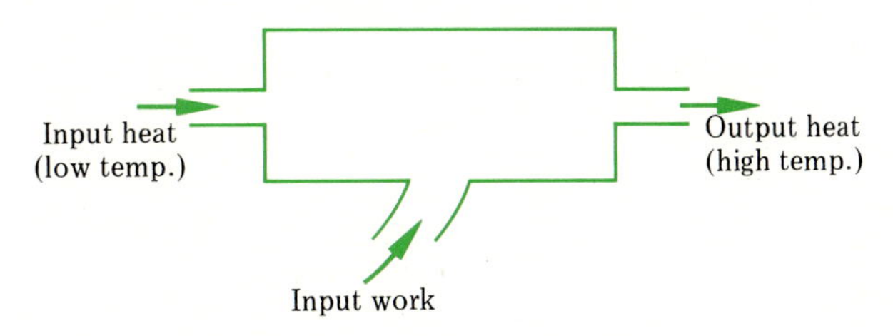

FIGURE 12.11
A heat pump, or refrigerator.

*Freon is a trade name for dichlorodifluoromethane and similar compounds.

temperature by application of moderate pressure. Let us refer to Fig. 12.12 to see the processes used.

The compressor at the top of the diagram is used to compress the Freon gas. Because of the work done on the Freon by the compressor, the gas is heated as it is compressed. This hot, highly compressed gas is then cooled by circulating it through coils over which a fan blows or air can circulate. (When a refrigerator or air conditioner is running, you can feel hot air being blown from it in the uncooled region. This air is carrying heat away from the coils at the right of the diagram.) As the Freon cools somewhat at this high pressure, it condenses to a liquid. This liquid then enters a jet system, which acts as a throttling valve.

You will recall that, in a throttling process, a fluid is suddenly expanded into a region of low pressure. As a result, in the present case, the Freon is greatly cooled as it emerges from the jet as a low-pressure gas. This cold gas is then circulated through coils in the cooling compartment of the refrigerator. Or, in an air-conditioning unit, air is blown into the room across these coils. The cold gas warms somewhat as it cools the region near the cooling coils. It then flows back to the compressor, where the whole process is repeated.

As we see, *work done on the compressor by outside agents* (such as an electric motor), *is used to transport heat from a cold region to a warm region. The heat exhausted is larger than the heat taken up. The heat equivalent of the work done by the compressor is added to the heat transferred from the cold to the warm region.*

Refrigeration units are often rated in tons. This unit is a purely practical unit. A 1-ton refrigeration unit is able to change 1 ton of water at 0°C to ice at the same temperature in a running time of 1 day. Or, in more basic units,

$$1 \text{ ton of refrigeration} = 12{,}000 \text{ Btu/h} = 3513 \text{ J/s} = 840 \text{ cal/s}$$

As we see, a 1-ton refrigeration unit should be able to freeze about 10 g of water per second. In practice, heat flow limitations through the water would greatly reduce this rate.

A heat pump can be used to heat a building. Since it pumps heat from a cold region to a warm region, it can pump heat from the cold outdoors to the interior of a home. In mild climates, many home refrigeration systems are of the reverse-action type. During winter, the refrigeration unit is effectively reversed so that the unit heats instead of cools. Hot air is blown into the

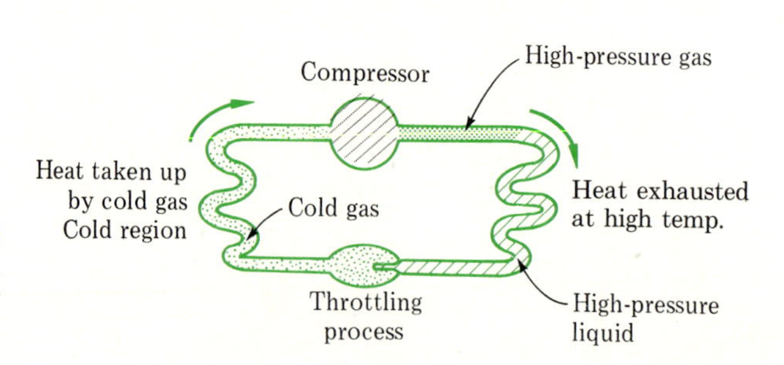

FIGURE 12.12 Some details of a compressor-type heat pump or refrigerator.

house by extraction of heat from outside. As of now, furnaces and electric heat are more economical where the outside temperatures become low for extended periods of time.

As we have seen, heat pumps transport heat in a direction opposite to the direction the heat would normally flow by itself. An outside energy source must do work to accomplish this. The natural tendency in the universe is for heat to flow from hot to cold. We have already seen, in terms of molecules, why this should be. The fact that a preferred direction exists for natural energy flow is basic to the world in which we exist. This effect is summarized by the second law of thermodynamics. In the following sections, we shall examine this law and its consequences.

12.8 THE SECOND LAW OF THERMODYNAMICS

Someone once remarked about the universe that "left to themselves, things go from good to bad to worse." In a very crude sense, this summarizes the second law of thermodynamics. As we have seen, the first law is a statement of energy conservation, but it has nothing to say about the course of events in the universe. Energy is conserved when a stone falls and has its gravitational PE changed to KE. As the stone strikes the ground and comes to rest, its KE is changed to thermal energy. However, a stone resting on the ground never changes the thermal energy in and near it to KE and goes shooting up into the air. The first law does not rule out such a possibility since this reverse process also conserves energy. But the process does not occur.

There are many other processes in the universe which are not ruled out by the first law but which do not occur. For example, heat flows from hot to cold but not from cold to hot. Water evaporates from a saucer, but the vapor in the air does not by itself recondense into the saucer. A dead body decays and turns to dust; but the elements of the earth do not spontaneously form the body in the reverse process. *Nature has a preferred direction for the course of spontaneous events. The second law tells us what that direction is.*

During the history of thermodynamics, the *second law* has been stated in several fully equivalent ways. One of the earliest statements simply summarizes the fact that heat flows naturally from hot to cold:

Second Law of Thermodynamics

> Heat flows spontaneously from a hotter to a colder object but not vice versa. Because of this, it is impossible for a cyclical system to transfer heat from a lower-temperature body to a higher-temperature body indefinitely unless external work is done on the system.

This statement agrees with what we have found out about engines and refrigerators. To operate a refrigerator, an input energy source is required. If heat is to be transferred from a cold region to a warmer region, external input work is needed to accomplish this. And, in the case of an engine, a difference in temperature must exist. The efficiency of the engine is given by $1 - T_{ex}/T_{in}$ and becomes zero if the two temperatures are equal.

As we see, *the second law tells us it is impossible to make use of thermal energy unless the energy can flow to a region of lower temperature.* For example, the waters of the ocean have a huge amount of thermal energy. But we cannot use this energy unless a cooler place is found to which it can flow. As a result, for all usual purposes, the thermal energy resident in the oceans is of no use to us. In the next section, we shall explore the basic reason for this lack of usefulness.

12.9 ORDER VERSUS DISORDER

As any gambler knows, the odds are best for an event happening if the event can occur in many different ways. To illustrate this fact, let us consider a game in which five identical coins are tossed onto a table after being well shaken. There are only six events which can result from such a toss. They are listed as follows:

EVENT	NUMBER OF HEADS UP	NUMBER OF TAILS UP
1	0	5
2	1	4
3	2	3
4	3	2
5	4	1
6	5	0

At first guess you might think that each of these events is equally likely to occur. But that is not correct. The reason is that there is only one way that event 1 or event 6 can occur. For event 1 to occur, all coins without exception must come up tails. For event 6, all coins must come up heads.

However, for event 2 to occur, there are actually five ways in which it could happen. Calling the five coins, A, B, C, D, and E, these ways are as follows:

	Ways for One Head				
WAY	COIN A	B	C	D	E
1	H	T	T	T	T
2	T	H	T	T	T
3	T	T	H	T	T
4	T	T	T	H	T
5	T	T	T	T	H

Because there are five times as many ways that event 2 can happen, event 2 is five times more likely to occur than event 1. Also, event 5 can happen in five different ways. As a result, events 2 and 5 are equally likely to occur. And both these events are five times more likely than events 1 or 6.

In the same way, we can show that events 3 and 4 are equally likely and each can occur in ten different ways. Therefore, 3 and 4 are twice as likely to happen as events 2 and 5, while events 3 and 4 are ten times more likley to happen than events 1 and 6. If you were a gambler, it is obvious which events you should lay your money on if no odds are given.

We can extend this to a situation in which more coins are involved. Suppose 100 coins rather than 5 are tossed. Then, as before, there is only one way in which all the coins can come up heads (or tails). But the number of ways in which other combinations can occur becomes almost unbelievably large. The results are shown in Fig. 12.13. Notice that the number of ways in which 50 heads can come up is about 10×10^{28}. As you can see, the odds against all heads or all tails coming up is so small as to be negligible.

Indeed, the total number of ways for all combinations of heads and tails is about 1×10^{30}. Therefore, the chance that all coins would come up heads is 1 in 10^{30}. If you throw the coins once each 10 s for 10^{22} yr, your chance of all heads coming up once is about 10 percent. For all practical purposes, there is no chance at all of all heads or all tails occurring. As we see from Fig. 12.13, the only really likely occurrence is for nearly equal numbers of heads and tails to occur.

If we consider 10^6 coins instead of 10^2, the situation becomes even more striking. We can summarize all such results in a very simple way. Notice in Fig. 12.13 that the graph line decreases to about one-tenth of its maximum value at the following two numbers of heads: 40 and 60. To give an estimate of the width of the peak, we could say it extends from 50 ± 10. In other words, if you throw 100 coins, the number of heads which should come up is about 50 ± 10. The more general result is as follows:

If one throws N coins, the expected number of heads will be about*

$$\frac{N}{2} \pm \sqrt{N}$$

In the case of 10^6 coins, we should expect $500{,}000 \pm 1000$ heads to come up. Notice how very precise this estimate is. It says the expected number of

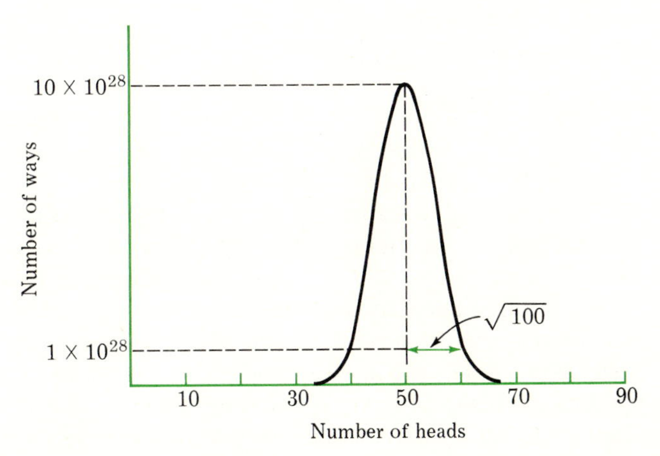

FIGURE 12.13 The number of ways in which the indicated number of heads can come up when 100 coins are tossed.

* Precisely, if N is large, 95.3 percent of the time the number of heads will lie in this range.

heads lies between 501,000 and 499,000, a very narrow range indeed. As you can see, when the number of coins becomes very large, the percentage deviations one will find from the average are very small.

This example with the coins is typical of our universe in general. When things are left to happen by themselves, they occur by chance. As a result, the probability laws applicable to tossed coins applies to these other situations as well. For example, suppose you have a box containing gas molecules, as shown in Fig. 12.14. In the air, there are about 3×10^{19} molecules/cm^3. Let us say the box has 10^{20} molecules in it. We now ask: What are the chances that the molecules will all bunch up in one half of the box?

FIGURE 12.14
What is the likelihood that all the molecules will appear in one side of the box?

From our results with the coins, we can easily clarify this situation. To make the situations similar, call a molecule in the left side of the box a "heads" molecule. Molecules on the right will be "tails" molecules. Our general result from above tells us that the number of "heads" will be about $\frac{1}{2}N \pm \sqrt{N}$. In this case, the number of heads is about

$$50{,}000{,}000{,}000{,}000{,}000{,}000 \pm 10{,}000{,}000{,}000$$

Notice how small the expected deviation is. It is only $10^{10}/5 \times 10^{19}$ or about 1 part in 5 billion. For all practical purposes, the number of molecules in the two halves of the box will be the same. And, of course, there is really no chance at all that all the molecules will, spontaneously, move into one side of the box.

These considerations have fundamental importance for all spontaneous processes. Reasoning from them, we can predict that thermal motion (or other random-type disturbances) causes systems to change from order to disorder. As a crude example of this, consider the case of 100 coins again. Suppose we carefully arrange all of them with heads up. They then have a high degree of order. Now let us give them a type of motion similar to random thermal motion by shaking them. They quickly disorder and never return to their original state of order.

Similarly with the gas molecules in the box of Fig. 12.14. We can give the system order by placing all the molecules in one end of the box. But if we allow them to adjust with spontaneous thermal motion, they become disordered and fill the whole box. Never will they again, spontaneously, return to their original ordered state.

Basic to this discussion are the concepts of order and disorder. We can give a simple method for comparing the disorder of two states. If a state can occur in only one way, it is a highly ordered state. In such a state, each molecule (or other particle) must be placed in a single exact way. In a disordered state, however, there are many possible ways for achieving the state. With these facts in mind, we can relate disorder and probability (or number of ways) of achieving a state. That state which has the highest disorder is the most probable state; it can occur in the largest number of ways. For example, the probability of N coins all coming up heads is very small. This is a state of very low disorder. As we have seen, systems left to themselves move toward states of high disorder.

There are many examples we could give that illustrate behavior of this type. We conclude from them that *in a system composed of many molecules,*

Second Law: Alternate Form

> If a system is allowed to undergo spontaneous change, it will change in such a way that its disorder will increase or, at best, not decrease.

This law of nature, applicable to large numbers of molecules, is an alternate form of the second law of thermodynamics. In the next section we shall see yet another way in which the law can be stated.

12.10 ENTROPY

The implications of order and disorder in a system can be approached in two quite different ways. *Early in the history of thermodynamics, it was recognized that a quantity called* **entropy** *is useful in calculations.* In this approach, entropy is defined in terms of heat flow and absolute temperature. The definition applies to a system which undergoes a reversible change. (By definition, this is a change which occurs in identical ways in both the forward and backward direction. For example, in changing reversibly from state A to state B, the state variables for the system take on definite values at each step during the change. In the reverse process, going from B to A, the system's state variables repeat these steps in reverse order.)

Entropy Change

We represent the change in entropy of a system by ΔS. If, in going from state A to B by a reversible process, an amount of heat ΔQ flows into the system and the system has an absolute temperature T, then, the change in entropy of the system is

$$\Delta S = \frac{\Delta Q}{T} \tag{12.6}$$

As we see, ΔS *is positive* (*the system gains entropy*) *if heat flows into the system*. Notice T is the absolute temperature.

Furthermore, one can show that *entropy is a state variable. Under identical conditions, a system always has the same entropy*. In this respect, it is like P, V, T, and U. Because entropy is a state variable, it proves of value in thermodynamics to make use of it in describing systems and processes.

Long after entropy had been widely used in thermodynamics, the branch of chemistry and physics called statistical mechanics furnished another view of it. *In statistical mechanics, one describes the behavior of gases, liquids, and solids in terms of the statistical behavior of atoms and molecules.* The kinetic theory of gases is a portion of statistical mechanics, for example. In the previous section, we were discussing the behavior of molecules from the standpoint of statistical mechanics.

After years of effort, the work of Boltzmann, Gibbs, and many others led to a union of thermodynamics and statistical mechanics. One of the major results of this union was to provide a meaning for entropy in terms of order and disorder. The end result was as follows.

Suppose a system can achieve a given state in Ω *ways,* where Ω is capital omega. (For a system of 100 coins, $\Omega = 1$ for the all-heads configuration while, for equal heads and tails, $\Omega \approx 10^{29}$.) Then it turns out that *the entropy of the system is*

Entropy Related to Disorder

$$S = k \ln \Omega \tag{12.7}$$

where ln stands for the natural logarithm and k is Boltzmann's constant.

Notice what Eq. (12.7) tells us. If a state of the system can occur in only one way, then $\Omega = 1$. But the logarithm of 1 is zero. So the entropy of such a highly unlikely state is zero. However, if a state can occur in many ways, Ω will be large. The entropy of a highly probable state is therefore large.

Let us now recall that a highly ordered state has a low number of ways it can occur while a disordered state can occur in many ways. Equation (12.7) then tells us that *entropy is a measure of disorder. The more disordered the state of a system is, the larger its entropy will be.* As we see, *entropy is a state variable which measures disorder*. Because of this, we can restate the *second law of thermodynamics* in still another way:

Second Law in Terms of Entropy

If an isolated system undergoes change, it will change in such a way that its entropy will increase or, at best, remain constant.

The following illustration shows how the change in entropy in a process can be computed.

Illustration 12.6 By how much does its entropy change as a 20-g ice cube melts at 0°C?

Reasoning From Eq. (12.6) we have

$$\Delta S = \frac{\Delta Q}{T}$$

Notice that T is the absolute temperature. In our particular case, the temperature $T = 273$ K. Since 80 cal of heat must be added for each gram of ice melted, we have

$$\Delta Q = (20 \text{ g})(80 \text{ cal/g})(4.184 \text{ J/cal}) \approx 6700 \text{ J}$$

Then $$\Delta S = \frac{6700 \text{ J}}{273 \text{ K}} = 24.5 \text{ J/K}$$

Notice that the entropy of the ice increases as it changes to water. The increase is a measure of the increase in disorder of the H_2O molecules.

12.11 HEAT DEATH OF THE UNIVERSE

All spontaneous changes occur in such a way as to increase the disorder of the universe. This is simply a statement of the second law as applied to the universe as a whole. In Chap. 29 you will learn that according to the best knowledge we now have, the whole universe was once confined to a sphere perhaps 10 times larger in diameter than our sun. At that instant, the beginning of time, the universe was an incredibly hot ball of energy. During the billions of years which have since elapsed, the universe has expanded adiabatically with its outer edges traveling outward at speeds near the speed of light.

The laws of thermodynamics apply to this process which has taken perhaps 11 billion years. During that time, heat energy has continuously flowed from hot to cold regions. As it did so, the disorder and entropy of the universe have increased. The thermal energy—indeed, all energy—has become less useful on the average. The temperature of the original fireball has fallen continuously. There are still local hot spots, the positions of the sun and of the stars. But most of the universe has long since cooled far below the temperatures to which we are accustomed. The gas in the vast reaches of space has an average temperature of about 3 K.

We on earth are fortunate. Our nearby sun still floods us with energy. Heat flows to our earth by radiation from the hot sun. We make use of this radiant energy to grow plants. These plants are then used as sources of energy for us and the other creatures on earth. As we have pointed out previously, except for nuclear fuels, the sun is the ultimate source of the energy we use. Notice that the sun's usefulness to us is the result of its very high temperature. As it cools, it will radiate less energy. Over a period of a few billion years, the earth will slowly lose its major energy source.

During all this time, the entropy of the universe will be increasing. The sun and other hot objects will be losing entropy at the rate of $\Delta Q/T_h$, where ΔQ is the heat lost in unit time by the hot object at temperature T_h. Although the cooler portions of the universe will receive this energy, they will be at a lower temperature T_c. As a result, they will gain an entropy $\Delta Q/T_c$ which is larger than $\Delta Q/T_h$. *The entropy and disorder of the universe will continue to increase as hot objects cool and cold objects warm.*

We can picture in our minds a time when everything in the universe has reached the same temperature. Then, no heat flow can occur. The disorder of the universe (and its entropy) will have reached a maximum value. Even though all the original energy the universe once had is still present, the energy will be useless. No plants will grow since there is no hot object to light them. No engines can function since there is no cooler place to which heat can be exhausted. No life will exist anywhere in the universe. *The universe will have undergone what is known as its* **heat death.**

Fortunately for us, this situation will not occur for billions of years. Indeed, we are not certain that it will ever happen. As you will learn in Chap. 29, there is a possibility that the universe will contract and, once again, become a flaming fire ball. But that is another story, and we shall postpone discussion of it until the last chapter in this text.

SUMMARY

Thermodynamics is the branch of science in which the behavior of matter is described in terms of its state variables, chiefly pressure, volume, temperature, internal energy, and entropy. A state variable is any quantity which depends only upon the macroscopic condition of a physical system. Under identical macroscopic conditions, each of the state variables always has the same unique value. Heat and work are not state variables.

The first law of thermodynamics is a statement of the law of conservation of energy. It can be stated in equation form as follows: If heat ΔQ flows *into* a system while the *system does work* ΔW, these two quantities are related to the change in internal energy of the system by

$$\Delta Q = \Delta U - \Delta W$$

When a system subjected to an external pressure P increases its volume by ΔV, it does work equal to $P\,\Delta V$. Therefore $\Delta W = P\,\Delta V$. On a P-V diagram for a system, the work done by the system during a change is equal to the area under the graph line which represents the change. If the system undergoes a cyclic change, the net work done by the system during the cycle is the area enclosed by the cycle on the P-V diagram.

An isothermal process is one carried out without a change in temperature of the system. In an adiabatic change, no heat flow occurs into or out of the system. For an adiabatic process involving a system composed of an ideal gas, the following equation applies:

$$P_1 V_1^{\gamma} = P_2 V_2^{\gamma}$$

where $\gamma = c_P/c_V$ for the gas.

In a throttling process, a fluid is allowed to expand rapidly and adiabatically through a jet or porous disk. Such a process often leads to cooling of the fluid during expansion.

The efficiency of an engine is defined by

$$\text{eff} = \frac{\text{output work}}{\text{input energy}}$$

For a heat engine, this becomes

$$\text{eff} = 1 - \frac{\Delta Q_{ex}}{\Delta Q_{in}}$$

where ΔQ_{ex} is the heat exhausted by the engine in one cycle while ΔQ_{in} is the heat furnished to the engine on each cycle. The Carnot engine has the highest possible efficiency. In it, the engine operates between a high intake temperature T_{in} and a low exhaust temperature T_{ex}. For it

$$\text{eff} = 1 - \frac{T_{ex}}{T_{in}}$$

No engine can be 100 percent efficient.

A refrigerator or heat pump transfers heat from a cold reservoir to a warm reservoir. To do this, energy must be supplied to the device. The first law requires that the heat exhausted to the warm reservoir must equal the sum of the input work and the heat removed from the cold reservoir.

The second law of thermodynamics tells us the direction in which spontaneous change will occur in a system. It can be stated in three equivalent ways:

1 Heat flows spontaneously from a hotter to a cooler object but not vice versa.
2 If a system is allowed to undergo spontaneous change, it will change in such a way that its disorder will increase or remain constant.
3 If an isolated system undergoes change, it will change in such a way that its entropy will increase or remain constant.

The entropy change ΔS of a system can be given in the following ways: if a heat ΔQ flows into a system at temperature T, then

$$\Delta S = \frac{\Delta Q}{T}$$

and if a system can occur in the same state in Ω ways, then the entropy of that state is given by

$$S = k \ln \Omega$$

where k is Boltzmann's constant. The entropy measures the disorder of a system. In the universe as a whole, entropy is constantly increasing.

MINIMUM LEARNING GOALS

Upon completion of this chapter, you should be able to do the following:

1. Explain what is meant by a state variable. Give examples of quantities which are state variables and those which are not.

2. State the first law of thermodynamics.

3. Compute the work done by a system during a given volume change against a known pressure. Use a *P-V* diagram to compute work done by a system during an expansion (or contraction) for which the graph line is given.
4. Define isothermal change and adiabatic change. Apply the first law to each in such a way as to describe the behavior of ΔQ, ΔU, and ΔW.
5. Explain why a gas warms when compressed adiabatically.
6. Describe what is meant by a throttling process and explain why a fluid is often cooled by such a process.
7. Compute the work done per cycle by an engine when its *P-V* cycle diagram is given.
8. Draw on a *P-V* diagram the cycle for a system when the processes carried out during the cycle are told to you in words.
9. Write down the efficiency of a heat engine in terms of (*a*) work or (*b*) heat. Or, for a Carnot engine, in terms of temperature.
10. Explain why the Carnot engine and its cycle are important.
11. Describe the basic ideas of operation for a compressor-type refrigeration unit or heat pump.
12. Give several examples of physical systems which, when left to themselves, become more disordered. Explain why the reverse process is not observed in nature when large numbers of molecules are involved.
13. Use the relation $\Delta S = \Delta Q/T$ to compute the entropy change of a simple system under isothermal conditions.
14. State the second law of thermodynamics in three ways: (*a*) heat flow; (*b*) order and disorder; (*c*) entropy.
15. Explain what is meant by the heat death of the universe.

IMPORTANT TERMS AND PHRASES

You should be able to define or explain each of the following:

State of a system
State variable
First law of thermodynamics: $\Delta Q = \Delta U - \Delta W$
Expansion work: $\Delta W = P\,\Delta V$
P-V diagram
Isothermal process
Adiabatic process
$P_1V_1^{\gamma} = P_2V_2^{\gamma}$
Throttling process
Thermodynamic cycle; area in *P-V* diagram
$\text{eff} = 1 - \Delta Q_{ex}/\Delta Q_{in}$
Carnot engine; $\text{eff} = 1 - T_{ex}/T_{in}$
Heat pump; refrigerator
Second law of thermodynamics
Entropy; $\Delta S = \Delta Q/T$; $S = k \ln \Omega$
Heat death of the universe

QUESTIONS AND GUESSTIMATES

1. Relate the equation $\Delta Q = \Delta U - \Delta W$ to each of the following situations: (*a*) an ice cube is melted by warm water in a thermos jug; (*b*) water in a thermos jug is heated by violent shaking of the jug; (*c*) a person in a deep coma lays in a bed quietly without added medication; (*d*) a hard-boiled egg on a shelf slowly spoils.
2. A container of ideal gas is to be compressed to half its original volume. Under which condition would the work done on the gas be largest, isothermal or adiabatic?
3. I have been told the following story. To test a tire for leaks, the tire was greatly overinflated. As the air was later let out the tire valve, the person doing it suffered a frostbitten finger from exposure to the escaping air. Do you believe the story could be true? Defend your answer.
4. The human body is an incredibly complicated machine. Discuss each of the following for it: (*a*) input energy; (*b*) exhaust energy; (*c*) types of work done by it; (*d*) efficiency.
5. Consider the simple heat engine shown in Fig. P12.1. The heated liquid on the right expands and is lifted by the cooler liquid on the left. As a result, the liquid circulates counterclockwise in the tube. As it does so, it

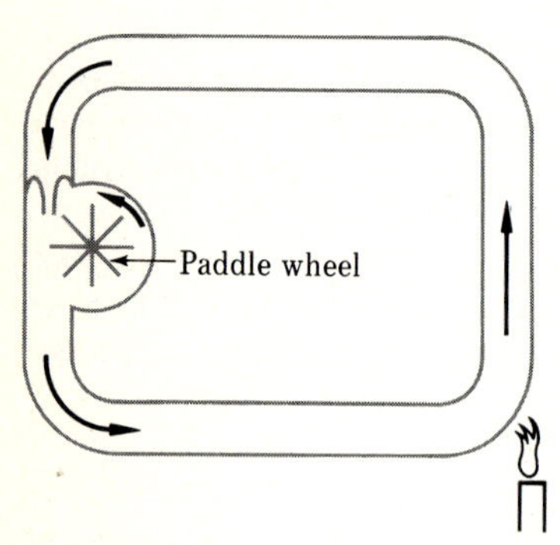

FIGURE P12.1

rotates the paddle wheel, which is then coupled to external devices to do output work. Explain what factors effect the efficiency of this engine. What should be done to make it have its highest efficiency?

6. Body temperature is 98.6°F. Even so, heat is carried away from the body to the environment even when the surrounding temperature is higher than this. Doesn't this contradict the fact that heat only flows from hot to cold? Similarly, a watermelon can be cooled by wrapping it in a wet cloth even on a very hot day. How can you reconcile this with the second law?

7. A baby produces a highly ordered structure as it grows. Each molecule within it is a carefully structured entity, and the molecules are assembled in a highly ordered way. Doesn't this growth and its accompanying high degree of order contradict the second law? Ultimately, what is the energy source for the child? Repeat this question for the case of a growing plant.

8. If you toss two coins a large number of times, what fraction of the times should both coins come up heads? Do the actual experiment and see if your result does approach your predicted value.

9. Each of a pair of dice has six sides labeled from 1 to 6. When the pair is tossed, what is the ratio of the chance that the up sides will total 2 to the chance that they will total 3? What if the total is to be 4 instead of 3?

10. A child wishing to cool off the kitchen of a home opens the refrigerator door and leaves it open. Will this work? Answer the question from both short-term and long-term considerations. Would the situation be any different if an old fashioned ice box was used rather than a refrigerator?

11. Two cylinders sit side by side and are closed by a movable piston. They are identical in all respects except that one contains oxygen (O_2) and the other contains helium (He). Both are now compressed adiabatically to one-fifth their original volume. Which gas will show the largest temperature rise?

PROBLEMS

1. (*a*) By how much does its internal energy change as 15 g of ice at 0°C melts to water at 0°C? Ignore the fact that the ice is less dense than water. (*b*) If you were to include the contraction effect, would your answer be larger or smaller?

2. An ideal gas in a cylinder is slowly compressed to one-third its original volume. During this process, the temperature of the gas remains constant, and the work done in compression is 45 J. (*a*) By how much did the internal energy of the gas change in the process? (*b*) How much heat flowed into the gas?

3. An ideal gas in a cylinder is compressed adiabatically to one-third its original volume. During the process, 45 J of work is done on the gas by the compressing agent. (*a*) By how much did the internal energy of the gas change in the process? (*b*) How much heat flowed into the gas?

4. A gas is carried around the cycle shown in Fig. P12.2. (*a*) Find the work done by the gas in parts *AB*, *BC*, *CD*, and *DE* of the cycle?

5. Figure P12.2 shows a thermodynamic cycle for an ideal gas. The temperature of the gas at point *D* is 20°C.

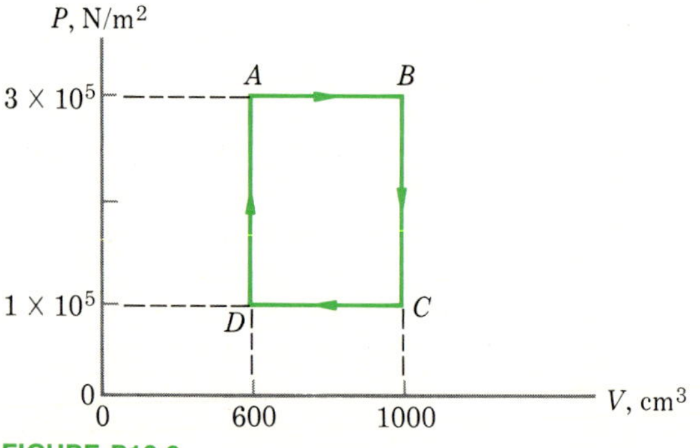

FIGURE P12.2

(a) Find the temperature at points A, B, and C. (b) In carrying through the cycle is heat added in portion AB? BC? CD? DA?

6.* For air, $c_V = 0.177$ cal/(g)(°C). Suppose air is confined to a cylinder by a movable piston under a constant pressure of 3.0 atm. How much heat must be added to the gas if its temperature is to be changed from 27 to 300°C? The mass of air in the cylinder is 20 g, and its original volume is 5860 cm^3. *Hint:* Notice that $mc_V\,\Delta T$ is the internal energy one must add to the gas to change its temperature by ΔT.

7.* A 16,000-cm^3 cylinder is closed at one end by a piston and contains 20 g of air at 30°C. The piston is suddenly pushed in so as to change the gas volume to 1600 cm^3. The compression is adiabatic, and the final temperature of the gas is 500°C. How much work was done in compressing the gas? For air, $c_V = 0.177$ cal/(g)(°C).

8.* Thirty grams of highly compressed air is confined to a cylinder by a piston. Its volume is 2400 cm^3, its pressure is 10×10^5 N/m^2, and its temperature is 35°C. The gas is now expanded adiabatically until its volume is 24,000 cm^3. During the process, 4100 J of work is done by the gas. What is the final temperature of the gas? Assume $c_V = 0.177$ cal/(g)(°C).

9.** Helium gas at 20°C and 1 atm pressure is adiabatically compressed to one-fourth its original volume. What is (a) its final pressure and (b) its final temperature?

10. When gasoline is burned, it gives off 11,000 cal/g, called its heat of combustion. A certain car uses 9.5 kg of gasoline per hour and has an efficiency of 25 percent. What horsepower does the car develop?

11. A 1500-kg car is to accelerate from rest to a speed of 8.0 m/s in 7.0 s. (a) What is the minimum horsepower the engine must deliver to the car if all friction losses are ignored? (b) Using this value and assuming the car uses its fuel with an efficiency of 20 percent, how much gasoline will the car burn in the 7.0 s? Gasoline has a heat of combustion of about 11,000 cal/g; that is, it furnishes this much heat energy for each gram burned.

12. A moderate-sized nuclear power plant might have an output of about 7.0×10^8 W. (a) Assuming the plant to have an overall efficiency of 20 percent, how much energy is consumed by the plant each second? (b) How much heat does it exhaust to the cooling system each second? (c) Repeat (a) and (b) for a fossil-fueled plant having the same output and efficiency.

13. In modern high-pressure steam turbine engines, the steam is heated to about 600°C and exhausted at close to 90°C. What is the highest possible efficiency of any engine which operates between these two temperatures?

14. You are given three pennies which can be tossed to come up heads or tails. If each penny is labeled so we can keep track of it, in how many different ways can the pennies arrange themselves (heads or tails) when they are tossed? What is the chance that all three will come up heads?

15. A pair of dice is thrown. (a) Is the sum of the two up sides of the dice more likely to be 3 than it is 5? (b) In how many ways can the sum be 3? (c) In how many ways can the sum be 5? (d) What is the most likely sum for the dice?

16. When N labeled coins are tossed, there are 2^N different combinations of heads and tails possible. How many combinations are possible for (a) 3 coins, (b) for 5?, (c) for 50? *Hint:* If you don't have a calculator or don't know how to use logs, try a method based upon the idea that $a^{5q} = a^q \cdot a^q \cdot a^q \cdot a^q \cdot a^q$.

17. Making use of the explanation given in Prob. 16, (a) what is the chance that 10 noninteracting ants will all end up in the same half of a box? (b) Repeat for the case where all but one ends up in the same half.

18. About what is the change in entropy of 15 g of water as it is cooled from 20 to 18°C? (Assume T to be nearly constant.)

19.* Five coins can each come up heads or tails. (a) What is the entropy of the configuration where all coins are heads? (b) For the configuration where all but one are heads?

VIBRATORY MOTION

13

In this chapter we return to the mechanics of motion in order to prepare for a study of waves, e.g., waves on a violin string, sound waves in air, and electromagnetic waves traveling through vacuum. As you continue your study of physics, you will find that the motion of waves is at least as important as the motion of material objects. It is therefore important that we understand their behavior. We shall begin this study by discussing the motion of the material objects which generate waves. The following chapter will be devoted to a study of certain types of waves generated by the vibratory motions discussed in this chapter.

13.1 VIBRATING SYSTEMS

Any elastic body or system can be made to vibrate. A few common examples are given in Fig. 13.1. The first three shown are rather simple cases. If the system is displaced to the indicated position and released, the elastic restoring force will cause the system to vibrate.

The fourth example in Fig. 13.1, the drumhead, undergoes a much more complex form of vibration if it is elastically distorted by a blow. Although the fifth system, a pendulum, has no elastic restoring force, we know that it, too, will vibrate. All these systems, together with *all* other *vibrating objects, have one thing in common. If any one of these systems is displaced from its equilibrium position, a restoring force is set up which acts to return the system to its equilibrium position.*

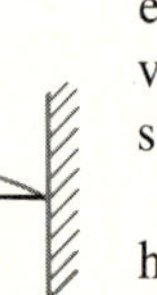

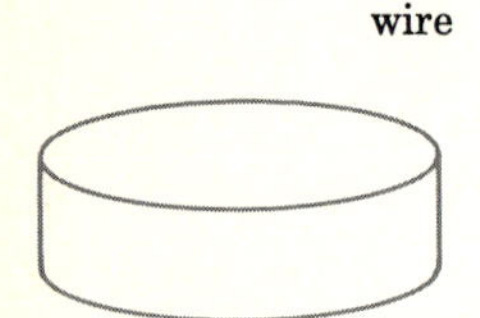

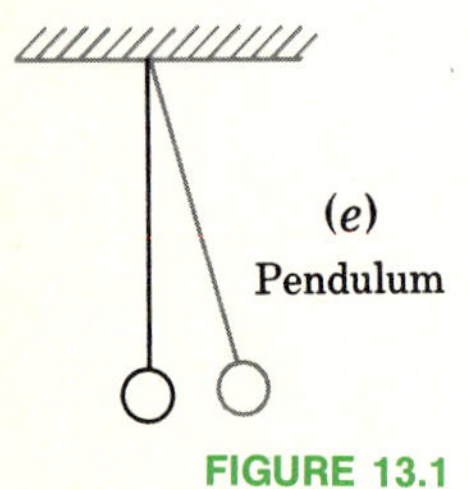

FIGURE 13.1
Typical vibrating systems.

13.2 THE VIBRATING SPRING

One of the simplest vibrating systems is a spring with a mass attached at one end. We shall see that the motion of this device is typical of all simple vibrating motions. In order to remove all complicating features, we shall suppose the mass to rest on a frictionless surface as shown in Fig. 13.2.

The system is shown in its equilibrium position in part *a*. There is no horizontal force acting on the mass in this position. (The pull of gravity down is balanced by the push of the table up, and so the net vertical force on the mass will always be zero.)

Suppose that the spring is now compressed by moving the mass to the position shown in part *b* of the figure. During the process of compressing the spring, we did work upon it, and so we have stored PE in the spring. The compressed spring exerts a force on the mass, tending to drive it back to the $x = 0$ position. If the mass is now allowed to move freely, the spring will keep accelerating it to the right until the position $x = 0$ is reached. The mass will now be moving to the right quite swiftly, and the spring will have lost all the PE stored in it when it was compressed. It is clear that the PE stored in the spring has been given to the mass and now appears as the KE of the moving mass.

Of course the mass will not stop at $x = 0$, since its KE must be lost doing work before it can come to rest. As the mass proceeds to the right of $x = 0$, it begins to stretch the spring and to store energy in it. By the time the mass reaches the position shown in part *c* of Fig. 13.2, it has lost all its KE doing work against the spring. The KE has been changed completely to PE in the stretched spring.

The spring will now accelerate the mass to the left. By the time the mass reaches the $x = 0$ position, all the energy will be in the form of KE once again. The mass will again compress the spring to the position $x = -x_0$, at which point all the KE will be changed to PE stored in the compressed spring. Now the process will repeat itself, and the mass will vibrate back and forth between $x = +x_0$ and $x = -x_0$ forever if there are no friction losses. *This is typical of vibrating systems. Notice that as the mass oscillates back and*

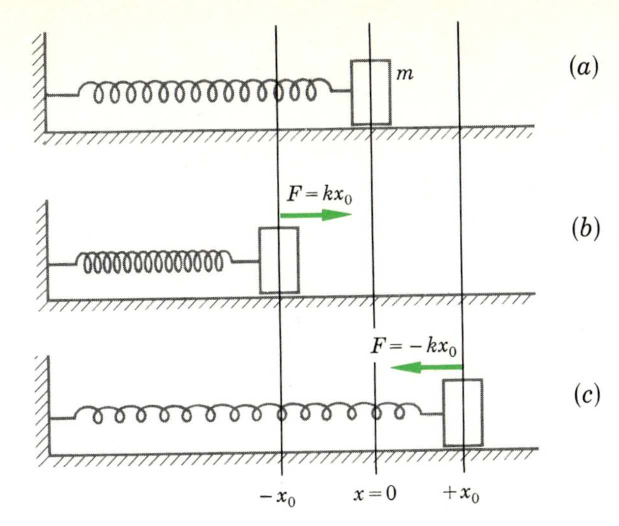

FIGURE 13.2

The mass m is in equilibrium in (a). (b) When the spring is compressed, PE is stored in the spring. Upon release of the system, this PE is changed to KE and finally back to PE when the mass reaches the position shown in (c). Notice that the spring furnishes a restoring force kx.

forth, the energy oscillates back and forth between KE and PE but the total energy must remain constant. Why? momentum must be conserved

Several terms are applied to such vibratory motion:

1 **Amplitude of vibration** is the distance x_0 in Fig. 13.2. It is the maximum displacement of the mass from its equilibrium position.
2 **Period of vibration** τ is the time taken to make one complete oscillation. It is the total time taken for the mass starting at $-x_0$ to move to $+x_0$ and return to $-x_0$ once again. (τ is Greek tau.)
3 The **frequency of vibration** f is the number of complete oscillations the mass makes in unit time. For example, if the period of vibration is 0.10 s, the frequency is 10 cycles per second (cps), which is $1/\tau$. The cps unit is frequently referred to as **hertz** (Hz), named after Heinrich Hertz, who first demonstrated radio-type waves. Both terms will be used interchangeably in this book. (Often ν, Greek nu, is used in place of f.) It is often convenient to recall that the frequency of vibration f is related to the period τ by

Hertz Unit

$$f = \frac{1}{\tau} \tag{13.1}$$

4 The **displacement** x is the distance from the equilibrium position to the mass at any time during the vibration.

13.3 SIMPLE HARMONIC MOTION

To discuss the vibrating spring in a quantitative manner we must decide how the force exerted on the mass varies. *If a spring is not stretched or compressed too far, it obeys Hooke's law,* which we first encountered in Chap. 9. You will recall that in a Hooke's law system, the distortion of a system produces a restoring force proportional to the distortion. For the case shown in Fig. 13.3,

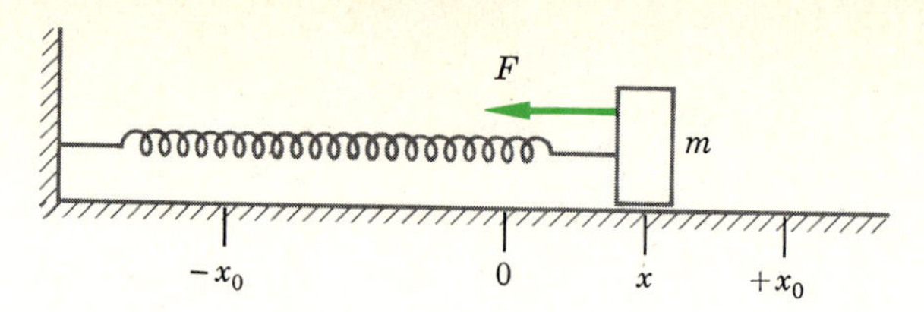

FIGURE 13.3
The Hooke's law force acting on the mass causes it to vibrate with SHM.

this means that the force F with which the distorted spring pulls upon the mass is given by

$$F = -kx \tag{13.2}$$

The negative sign shows the force to be in such a direction as to decrease the distortion. When x is positive, the force on the mass is in the negative direction and is tending to bring the mass back to its equilibrium position. In which direction is the force on m when x is negative?

All systems which obey Hooke's law and in which friction is small undergo similar motion, called **simple harmonic motion** (*SHM*). We shall make use of this fact later when the motion found for the spring-mass system is generalized to other systems. In all these cases, the constant k in Eq. (13.2) is called the **spring constant**. Since it is numerically equal to F/x, it measures the force needed to distort the system unit distance. For a stiff spring, k would be large; it is small for an easily deformed system.

Consider what happens as the spring is stretched from $x = 0$, where $F = 0$, to $x = x_0$, where $F = -kx_0$. Work is done against the spring in the process, and that work can be computed by taking the product of the average force and the distance moved. Clearly, the average force is half the maximum force, and the distance is x_0. Hence, *the energy stored in the spring when it is stretched to its furthest point is just*

Energy Stored in a Spring

$$\text{Maximum PE} = (\tfrac{1}{2}kx_0)(x_0) = \tfrac{1}{2}kx_0^2$$

Indeed, this same reasoning tells us that *a spring which is stretched or compressed a distance x has an energy* $\frac{1}{2}kx^2$ *stored in it*. Knowing this, we can compute the speed of the mass at any position between $-x_0$ and $+x_0$. This follows from the fact that *the total energy of the system, part PE and part KE, must always be constant and equal to the maximum PE, namely,* $\frac{1}{2}kx_0^2$. We have

$$\text{PE} + \text{KE} = \tfrac{1}{2}kx_0^2 \tag{13.3}$$

Substituting gives

$$\tfrac{1}{2}kx^2 + \tfrac{1}{2}mv^2 = \tfrac{1}{2}kx_0^2$$

where m and v pertain to the mass at the end of the spring. Simplifying we have

$$mv^2 = k(x_0^2 - x^2) \tag{13.4}$$

Hence, if the amplitude of vibration x_0 and the spring constant are known, the velocity of the mass can be computed for any position or displacement by using Eq. (13.4).

Of course, the mass is continuously changing its velocity. It travels fastest at $x = 0$, when all the energy is in the form of KE. Let us now observe the character of the acceleration of the mass. This is easily done. At any displacement x, the unbalanced force acting on the mass is a result of the tension in the spring. According to Eq. (13.2), that force is just $-kx$. Using Newton's law, $F = ma$, we have at once for the mass

$$-kx = ma$$

or the acceleration of the mass is just

$$a = -\frac{k}{m}x \tag{13.5}$$

Both Eqs. (13.4) and (13.5) are typical forms for bodies undergoing SHM.

Illustration 13.1 A particular spring stretches 20 cm when a 500-g mass is hung from it. Suppose a 2.0-kg mass is attached to the spring and it is displaced 40 cm from its equilibrium position and released. Find (*a*) the maximum speed of the mass, (*b*) the maximum acceleration of the mass, and (*c*) the speed and acceleration when $x = 10$ cm.

Reasoning Let us first find k, the spring constant. Since 0.50 kg weighs 4.90 N, we have $k = F/x = 24.5$ N/m.

a. The mass will be traveling fastest when it is at its center position, $x = 0$. Since the maximum PE $= \frac{1}{2}kx_0^2$ is

$$\tfrac{1}{2}(24.5 \text{ N/m})(0.16m^2) = 1.96 \text{ J}$$

and since all this energy is kinetic energy at $x = 0$, we find

$$\tfrac{1}{2}mv_{max}^2 = 1.96 \text{ J}$$

Using $m = 2.0$ kg, we find

$$v_{max} = 1.40 \text{ m/s}$$

b. The maximum acceleration occurs when the force kx is greatest, i.e., when $x = x_0$. In that case

$$a_{max} = \frac{F_{max}}{m} = \frac{(24.5 \text{ N/m})(0.40 \text{ m})}{2 \text{ kg}} = 4.9 \text{ m/s}^2$$

c. At $x = 0.10$ m, one has

$$a = \frac{F}{m} = \frac{(24.5\ \text{N/m})(0.10\ \text{m})}{2\ \text{kg}} = 1.22\ \text{m/s}^2$$

Also, since

$$\tfrac{1}{2}kx_0^2 = \tfrac{1}{2}mv^2 + \tfrac{1}{2}kx^2$$

we find

$$1.96\ \text{J} = \tfrac{1}{2}(2\ \text{kg})(v^2) + \tfrac{1}{2}(24.5\ \text{N/m})(0.10)^2$$

from which

$$v = 1.36\ \text{m/s}$$

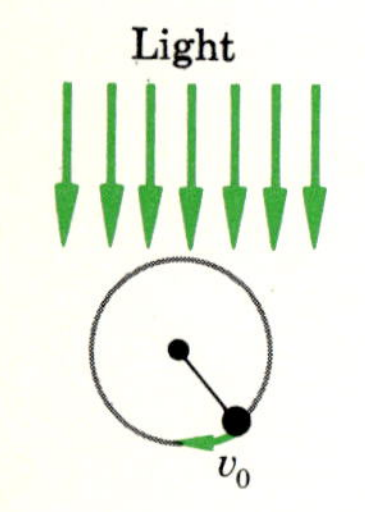

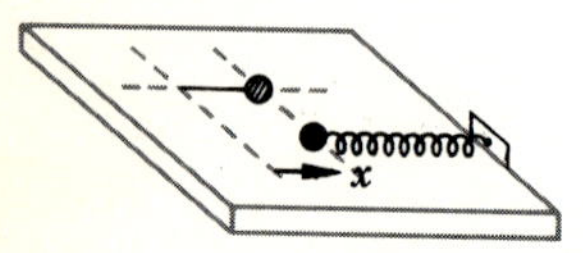

FIGURE 13.4
As the ball rotates around the circle with speed v_0, its shadow moves back and forth in coincidence with the mass at the end of the spring.

13.4 THE PERIOD OF VIBRATION

We have already computed nearly everything concerning harmonic motion. In the last section we showed how the speed and acceleration of the vibrating object can be found. It remains to find the frequency of vibration.

There are several ways of finding the oscillation frequency f. One of the simplest makes use of the similarity between a ball moving along a circular path and a mass vibrating at the end of a spring. The situation is shown in Fig. 13.4. As the ball moves around the circle with constant speed v_0, its shadow on the plane below moves back and forth. The motion of the shadow turns out to be simple harmonic. Indeed, if v_0 is properly chosen, the shadow will move in unison with the vibrating mass at the end of the spring shown there. Let us now prove these statements and then use our results to find the frequency of SHM.

The mass at the end of a spring does not move with constant acceleration. Equation (13.5) says that $a \sim x$, and so the acceleration is changing continuously. For this reason, SHM is not too simple in spite of its name. But by use of the analogy between circular and SHM just mentioned, we can easily arrive at useful results. To do so, let us refer to Fig. 13.5.

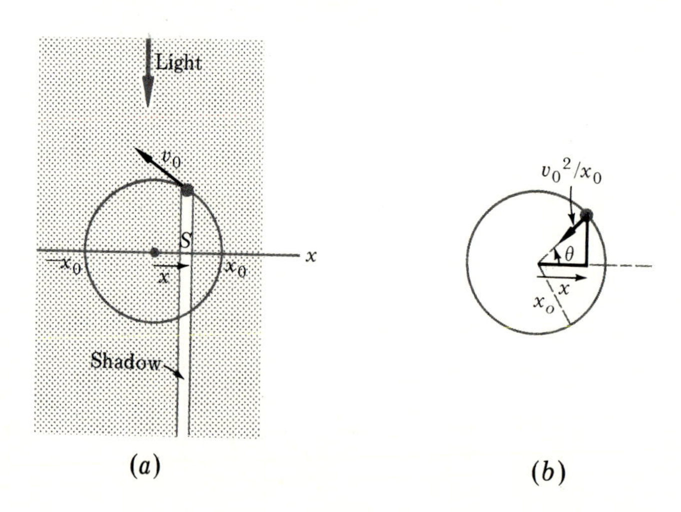

FIGURE 13.5
The ball moves around the circle with constant speed v_0. The x displacement of its shadow moves with SHM with amplitude x_0. For it, $v_0 = x_0\sqrt{k/m}$.

We have learned that a ball moving on a circular path with constant speed v_0 has a centripetal acceleration given by v_0^2/r. This centripetal acceleration is shown in part *b* of the figure. But there, since we choose the radius of the circle equal to x_0, the acceleration is v_0^2/x_0. As the ball moves along the circle, its shadow shown in part *a* will oscillate back and forth along the *x* axis. We claim that the oscillation of the shadow is actually a SHM. To prove this, consider the *x*-directed acceleration of the shadow.

The *x* coordinate of the shadow in part *a* is the same as the coordinate labeled *x* in part *b*. But in part *b*, the *x* acceleration of the ball (and the shadow) is simply the *x* component of the centripetal acceleration. By examination of part *b*, you should convince yourself that the *x* component of the acceleration is

$$\text{Acceleration of shadow} = -\frac{v_0^2}{x_0}\cos\theta$$

The minus sign tells us the acceleration is in the $-x$ direction for the situation shown. In addition, we see from part *b* that $\cos\theta = x/x_0$, and so

$$\text{Acceleration of shadow} = -\left(\frac{v_0^2}{x_0^2}\right)x$$

Notice that this acceleration is very similar to Eq. (13.5) for the acceleration in SHM. We had there that

$$\text{Acceleration in SHM} = -\left(\frac{k}{m}\right)x$$

We can therefore conclude that the motion of the shadow is the same as the motion of a mass at the end of a spring. To make the motions identical, the radius of the circle must be the same as the amplitude of motion. And the speed of the ball in the circle must be chosen so that

Reference Circle

$$\frac{v_0^2}{x_0^2} = \frac{k}{m} \qquad \text{or} \qquad v_0 = x_0\sqrt{\frac{k}{m}} \tag{13.6}$$

You might notice in passing that Eq. (13.6) is a special case of Eq. (13.4). Let us now use this similarity between circular and SHM to obtain an expression for the frequency of the motion.

If we wish to compute the time taken for the vibrating mass on the spring to make one complete vibration, it will be equivalent to computing the time the mass on the circle takes to go around the circle once. The time for one complete trip around the circle of radius *r*, the period, will be

$$\tau = \frac{2\pi r}{v_0} = 2\pi\frac{x_0}{v_0}$$

However, since we know from Eq. (13.6) that

$$v_0 = x_0\sqrt{\frac{k}{m}}$$

the expression for the period becomes

Period of Vibration

$$\tau = 2\pi\sqrt{\frac{m}{k}} \tag{13.7}$$

The frequency of vibration f is just $1/\tau$. Since $F = ma = -kx$ was used in deriving this relation, the units of k should be newtons per meter or pounds per foot.

Equation (13.7) for the period of vibration of a mass m in SHM is a general one. Any system having a mass m vibrating under a Hooke's law force with force constant k will have a period of vibration given by Eq. (13.7). We shall see that it also applies to systems far different from simple springs.

Illustration 13.2 Find the period of vibration for the system of Illustration 13.1.

Reasoning We had in Illustration 13.1 that

$$k = 24.5 \text{ N/m} \qquad \text{and} \qquad m = 2.0 \text{ kg}$$

Applying Eq. (13.7) gives

$$\tau = 2\pi\sqrt{\frac{2.0}{24.5}} \approx 1.8 \text{ s}$$

13.5 SINUSOIDAL VIBRATION: SIMPLE HARMONIC MOTION

We shall now see that a simple mathematical equation can be written for a mass vibrating with SHM. As we have seen, the vibrating shadow in Fig. 13.5 undergoes SHM. The value of x, the displacement, is easily found from part b of the figure. We see that

$$x = x_0 \cos\theta$$

But θ is constantly changing as the ball moves around the reference circle. Since our motion equations state that

$$\theta = \omega t$$

we have at once that

$$x = x_0 \cos \omega t \tag{13.8}$$

where the angular speed ω is measured in radians per second. If we wish to express our answer in terms of f, the frequency in revolutions per second (or, in this case, vibrations per second), we have

$$\omega = 2\pi f \tag{13.9}$$

since 1 rev is equivalent to 2π rad.

Equation (13.8) states that the mass vibrating with simple harmonic motion will do so as the cosine of ωt. To see exactly what this type of vibration entails, consider the experiment illustrated in Fig. 13.6.

A weight is suspended from a spring as shown. If the object is raised to y_0 and released, it will undergo SHM with amplitude, y_0. Behind the vibrating object we shall place a sheet of paper upon which the object marks its position as it vibrates up and down. The paper will be pulled out to the left at constant speed, and a trace of the motion of the object will be plotted upon it. This is shown in the lower part of the figure.

Let us count time from the instant the object was released. This is the end of the trace on the left. We would take this point as $t = 0$. The position of the object at the instant shown in the figure occurs at some later time. Hence, this can be considered a plot of the displacement of the object, y, as a function of the time. According to Eqs. (13.8) and (13.9), the equation of this trace is

$$y = y_0 \cos \omega t = y_0 \cos 2\pi f t = y_0 \cos \frac{2\pi t}{\tau} \tag{13.10}$$

This general type of trace is called a **sinusoidal type of curve,** and the motion

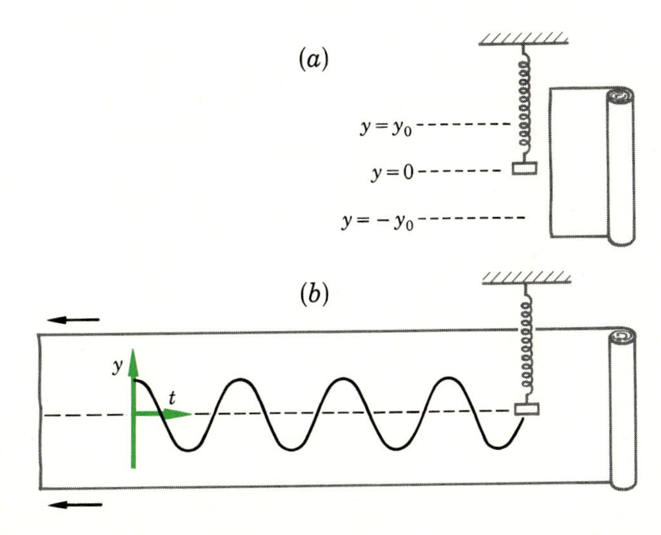

FIGURE 13.6
The vibrating mass traces out a cosine curve as a function of time.

which causes it is called **sinusoidal motion.** Clearly, *SHM and sinusoidal motion are the same.*

13.6 THE SIMPLE PENDULUM

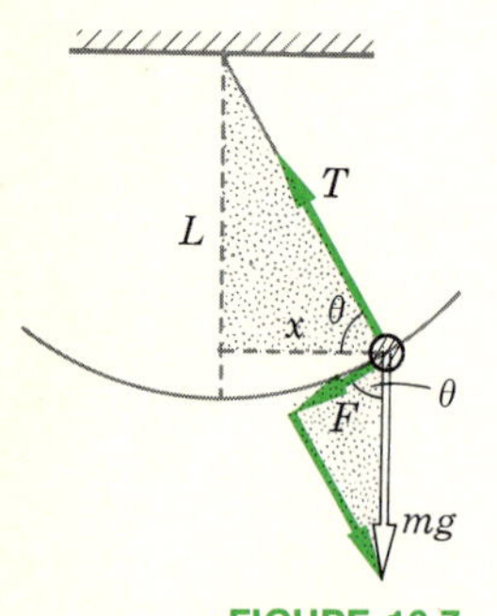

FIGURE 13.7
The two θ's are equal. Notice also that if the swing is not too large, the component of F along x (which is the x-direction restoring force) is very nearly equal to F itself.

We know that a simple pendulum like that shown in Fig. 13.7 vibrates with apparent SHM. If the restoring force on the body is proportional to the displacement, Hooke's law applies and the vibration will be SHM, as we saw previously. Let us now examine the restoring force for the pendulum.

The pendulum ball will move along the arc of the circle shown. Since the tension in the pendulum cord always pulls perpendicularly to this arc, the tension T neither speeds nor slows the motion. It is the weight mg of the pendulum ball which causes the motion. However, only the component of the weight tangential to the circular arc (labeled F) is effective. We have from Fig. 13.7 that

$$F = -mg\cos\theta$$

where the minus sign is used to show that F is a restoring force. But we can also see from the figure that $\cos\theta = x/L$, and so we have

$$F = -\frac{mg}{L}x \qquad (13.11)$$

This is similar in form to Hooke's law,

$$F = -kx$$

except in one detail. In Hooke's law, the displacement x should be in the direction opposite to the force. We see from Fig. 13.7 that this is not true in this case. F and x are not exactly in line. However, if the angle of swing of the pendulum is very small, F and x will be almost exactly in line. Therefore, *the pendulum* does *approximate Hooke's law if it is not swinging too widely. To that approximation, the pendulum will undergo SHM.* From the comparison of Eq. (13.11) with Hooke's law, we see that the spring constant k for the pendulum is just

$$k = \frac{mg}{L}$$

When this value for k is placed in the general formula for the period of a body in SHM, Eq. (13.7) gives

Period of Pendulum

$$\tau = 2\pi\sqrt{\frac{L}{g}} \qquad (13.12)$$

Notice that *the period of a simple pendulum does not depend upon the mass of the bob*. It depends only upon the length of the pendulum and the acceleration due to gravity g. This offers a precise means for measuing g in a simple experiment. If a pendulum of known length is timed so that its period of vibration is known, g can be computed at once from Eq. (13.12). With proper precautions, this method for determining g is extremely accurate.

13.7 FORCED VIBRATIONS

In any real vibrating system, there is always some loss of energy to friction forces. As a result, a vibrating pendulum or mass at the end of a spring vibrates with constantly decreasing amplitude as time goes on. This fact is illustrated in Fig. 13.8. Part *a* shows the vibration one would find in the ideal case with no friction. It is the situation we have been discussing in previous sections. A more realistic situation is shown in *b,* where the vibration is fairly strongly influenced by friction forces. *We say such a system is* **damped** *and the vibration* **damps down** *fairly* quickly.

When the friction forces are very large, the system does not vibrate at all; instead, it simply returns slowly to its equilibrium position, as shown in Fig. 13.8*c*. Such a system is said to be **overdamped.** This situation will exist if the mass at the end of a spring is immersed in a very viscous liquid. The mass does not vibrate at all in such a case. *When the friction forces are just large enough to cause no vibration of the system, we say the system is* **critically damped.**

If any real system is to vibrate for an extended period of time, energy must be added continually in order to replace the energy lost doing friction work. For example, to keep a child swinging at constant amplitude in a swing, one must push the swing from time to time to add energy to the system. This is typical of all vibrating systems. An outside agent must feed energy into the system if the vibration is not to damp down.

Everyone knows that there is a right and a wrong way to push a swing if it is to swing high. One must push with the motion of the swing and not against it. Only in that way can energy be added effectively to the swing. In fact, if one pushes against the motion, the vibration can be brought to a stop since the vibrating object must then do work on the pushing agent. These simple facts have importance in all forced, or **driven,** vibrating systems.

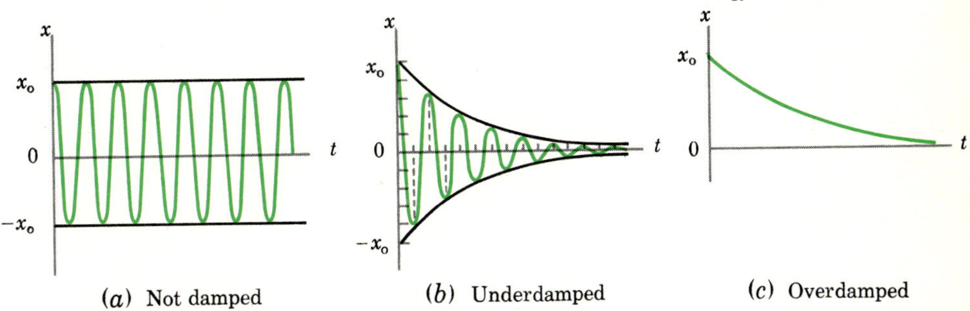

FIGURE 13.8
The free vibration of a system depends upon the extent of energy losses within it.

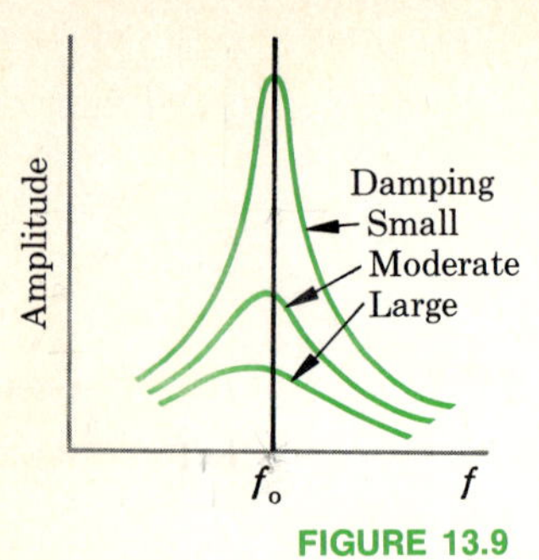

FIGURE 13.9

Response of vibrating systems to a driving force of variable frequency f.

In a driven system, the vibration is usually sustained by a repetitive force acting upon the system. This force has a frequency f, which may or may not be the same as the natural frequency of vibration of the system f_0. Only if $f = f_0$ will the driving agent be most effective in adding energy to the system. At all other frequencies, the driving force will not be quite in step with the motion of the system, and so its action will be less effective in adding energy to it. The variation in the amplitude of a vibrating system with the frequency of the applied force is shown in Fig. 13.9. Notice, as we said above, that *the driving force is most effective when its frequency f equals the natural frequency f_0 of the system. In that case we say the force is in* **resonance** *with the system.* More will be said about f_0, the **resonance frequency** of the system, in the next chapter, where we discuss the resonance vibration of strings.

SUMMARY

In a system which vibrates, energy is constantly interchanged between kinetic and potential. The sum of these two energies is a constant provided dissipative factors are absent. The energy stored in a spring is $\frac{1}{2}kx^2$, where k is the spring constant.

A system subjected to a Hooke's law restoring force undergoes simple harmonic motion (SHM). For a spring-mass system, the acceleration of the mass is given by $F = ma$ to be $a = -(k/m)x$. The velocity of the mass at any displacement x can be found from the energy-balance equation, namely, $\frac{1}{2}kx^2 + \frac{1}{2}mv^2 = \frac{1}{2}kx_0^2$.

When a ball moves along a circle with constant speed, its shadow executes SHM. We can use the ball on a reference circle to compute the period of the SHM. For a mass at the end of a spring, one finds the period to be $2\pi\sqrt{m/k}$.

A simple pendulum approximates SHM provided its angle of swing is small. Its period is given by $2\pi\sqrt{L/g}$.

The amplitude of SHM depends upon the frequency of the driving force. When the driving force has the same frequency as the system's natural frequency, resonance is achieved. Then, the amplitude of vibration is maximum.

MINIMUM LEARNING GOALS

Upon completion of this chapter, you should be able to do the following:

1. Point out the meaning of the following words for a mass vibrating at the end of a spring: amplitude, displacement, frequency, period.
2. Explain how the kinetic, potential, and total energy are related as a system vibrates.
3. Compute the energy stored in a spring in terms of the spring constant and the amount the spring is stretched or compressed.
4. Use the conservation of energy to justify the relation $\frac{1}{2}kx_0^2 = \frac{1}{2}mv^2 + \frac{1}{2}kx^2$. Make use of this relation to find the speed of the mass at the end of a spring.
5. Use $F = ma$ to show that the acceleration of the mass at the end of a spring is given by $a = -(k/m)x$.
6. State the importance of Hooke's law for SHM.
7. Explain how motion in a reference circle is related to SHM.
8. Find the natural frequency and period of vibration for a mass at the end of a spring.
9. Explain why SHM is called sinusoidal motion. Give the equation for a sinusoidal-type curve and explain the quantities in it.
10. Point out what causes the restoring force in the case of a simple pendulum and explain why the motion is only approximately SHM. Give the equation for the period of the motion.
11. Explain the meaning of the following terms in relation to forced vibration: damped, overdamped, critically damped, resonance. Sketch the graph relating vibration amplitude to frequency for a driven vibration.

IMPORTANT TERMS AND PHRASES

You should be able to define or explain each of the following:

$f = 1/\tau$
Amplitude
Displacement
Hertz unit (Hz)
Spring constant k
SHM
$\mathrm{PE} = \frac{1}{2}kx^2$
Reference circle
$\tau = 2\pi\sqrt{m/k}$
Sinusoidal motion
$x = x_0 \cos(2\pi ft)$
$\tau = 2\pi\sqrt{L/g}$
Critical damping
Natural resonance frequency

QUESTIONS AND GUESSTIMATES

1. What basic condition is necessary if a frictionless system is to undergo SHM?

2. A tiny sphere (perhaps a ball bearing) is rolled on a mixing bowl as shown in Fig. P13.1. Explain why vibratory motion occurs in (*a*) but not in (*b*). What causes the restoring force in (*a*)? What does the energy-conservation law tell us about (*a*) if friction forces are negligible?

3. Does a perfectly elastic ball bouncing on a hard floor undergo SHM? Is there any similarity at all? Explain your answer.

4. Do all vibrating bodies without friction losses make periodic complete interchanges between the KE and PE? Explain.

5. A common method used to free a stuck car is to rock it back and forth, with proper shifting of the gears alternately from forward to reverse and back to forward many times over. Discuss the physical basis for this method, paying particular attention to energy transfer.

6. Two equal weights hang together at the end of the same spring, and the system is set vibrating. What happens to the amplitude, frequency, and maximum speed of the end of the spring if one of the weights falls off when (*a*) the spring is at its largest extension and (*b*) the mass is passing through the center position?

7. Will a pendulum clock, properly adjusted at sea level, keep good time in the mile-high city of Denver?

8. Various portions of one's body have characteristic vibration frequencies: a freely swinging arm or leg could be cited. Discuss how these natural frequencies influence how one walks or runs. Estimate the frequencies.

9. Discuss the influence of design characteristics on the suitability and performance of a diving board or trampoline.

10. When a car's wheels are out of balance, the car is likely to vibrate strongly at a certain speed. Explain why.

11. Suppose a pendulum bob consists of a hollow sphere. Would the behavior of the pendulum change if the sphere were filled with water? One-fourth filled?

12. Estimate the natural frequency for up and down vibration of a car on its springs. Under what conditions could this natural frequency become of importance? (E)

13. A glass tube (inner diameter = 1.0 cm) is bent to form a U tube with each side tube 40 cm long and the bottom of the U about 10 cm long (see Fig. P13.2). Estimate the natural up-and-down frequency of vibration for the liquid column when the U is filled to a height of 30 cm with (*a*) water and (*b*) mercury. (E)

FIGURE P13.1

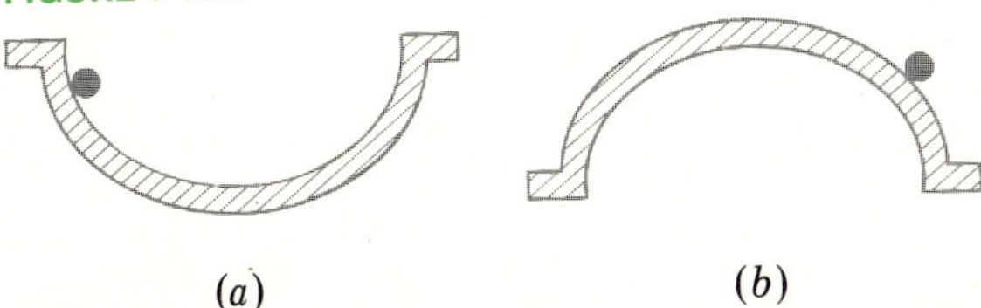

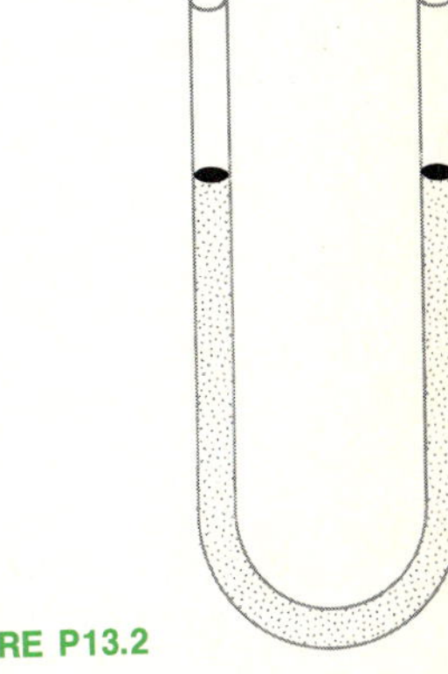

FIGURE P13.2

PROBLEMS

1. As seen in Fig. P13.3, a mass vibrates between two limits, 10.0 and 24.0 cm above the floor. It reaches the lowest point 20 times a minute. What are (*a*) the frequency, (*b*) the period, and (*c*) the amplitude of vibration for this system?

2. A 20-g mass hangs at the end of a vertical spring. When an additional 30 g is added to the spring, the spring stretches 15.0 cm more. (*a*) What is the spring constant for the spring? (*b*) What is the natural period of vibration for the system after the additional 30 g is added?

3. A 2.0-kg mass is hung at the end of a vertical steel wire which has a length of 1.50 m and a cross-sectional area of 8.0×10^{-3} cm^2. (*a*) How much does the wire stretch when the load is added to it? (*b*) What is the natural frequency for vertical vibration of the mass at the end of the wire? (Young's modulus for the wire is 19×10^{10} N/m^2.)

4. Two children notice that they can make a car vibrate up and down by periodically pushing downward on it. The car vibrates through eight complete cycles in 13.0 s. (*a*) Assuming the car to have a mass of 1800 kg, find the spring constant for its suspension system. (*b*) If this value is correct, about how much should the car lower as a 70-kg person enters the car and sits in its front seat?

5. A certain Hooke's law spring stretches 20 cm when a load of 0.35 N is added to it. How much energy is stored in the spring when it is compressed 5.0 cm?

6. A 4.0-lb object is placed at the end of a spring ($k = 3.0$ lb/in), and the spring is then stretched 2 in from its equilibrium position and released. Find (*a*) speed of the weight as it passes through the equilibrium position and (*b*) its acceleration just as it is released.

7. A 200-g mass hangs at the end of a spring ($k = 0.60$ N/m). The mass is now pulled out 3.0 cm from its equilibrium position and released. What is (*a*) the initial acceleration of the mass and (*b*) its speed as it passes through the equilibrium position?

8. A spring-mass system slides on a horizontal, frictionless surface (much like Fig. 13.3). the mass is 50 g. A horizontal force of 0.70 N applied to it stretches the spring 4.0 cm. (*a*) What is the acceleration of the mass when the system is released? (*b*) What is the speed of the mass as it passes through its equilibrium position?

9. As shown in Fig. P13.4, a rotating wheel drives a piston. For the situation shown there, find the following for the motion of the piston: (*a*) frequency; (*b*) amplitude; (*c*) maximum speed; (*d*) speed when the connecting rod is horizontal.

10. For the piston shown in Fig. P13.4, write the piston's equation of motion in the form $x = x_0 \cos \omega t$ with x_0 and ω given numerical values.

11.* For the situation shown in Fig. P13.4, find the magnitude of the acceleration of the piston (*a*) when the connecting rod is horizontal; (*b*) when the radius to the pivot point is vertical; (*c*) when the radius to the pivot point makes an angle θ to the horizontal.

12. A 2.0-kg mass oscillates with SHM at the end of a spring. The amplitude of motion is 30 cm, and the spring has a constant of 400 N/m. Find the speed of the mass when its displacement is (*a*) 30 cm; (*b*) 0 cm; (*c*) 15 cm.

13.** A piston undergoes vertical SHM motion with amplitude of 8.0 cm and frequency *f*. A washer sits freely on top of the piston. At low piston frequencies, the washer moves up and down with the piston. However, at very high frequencies, the washer momentarily floats above the piston as the piston starts its downward motion. (*a*) What is the maximum acceleration of the piston when the washer begins to separate from it? (*b*) What is the lowest frequency at which separation will occur?

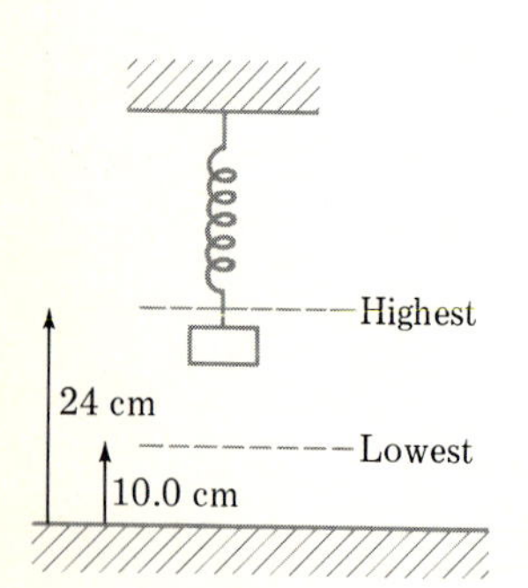

FIGURE P13.3

FIGURE P13.4

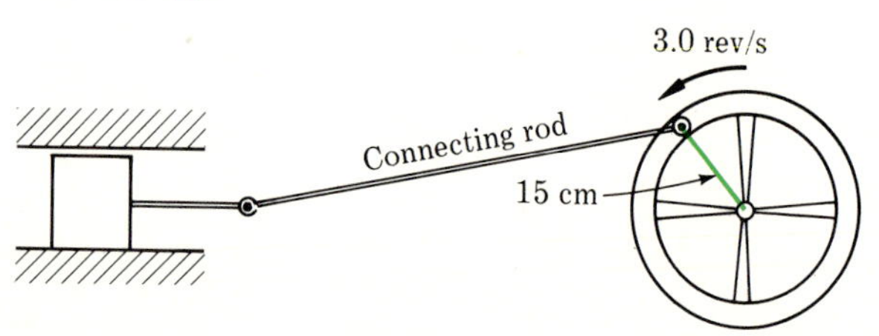

14. A certain simple pendulum has a period of 1.000 s. (*a*) How long is the pendulum? (*b*) If the pendulum bob is a sphere, is the length measured to the top or to the center of the sphere?

15. The length of a certain simple pendulum is 6.0 in. (*a*) With what frequency does it vibrate? (*b*) If it is released at an angle of 37° to the vertical, how fast is the bob going when it passes through the center point?

16. A large clock is controlled by a pendulum. If the clock is taken to the moon, where objects weigh only about one-sixth their weight on earth, how long (in hours) will it take the clock to tick out an hour on its dial?

17. A steel clothesline is strung between two large posts. When a 2-lb object is hung from the center point of the line, the line sags 3.0 in. It sags 6.0 in under a 4-lb weight. If only the 2-lb object is hanging on it and the object is then pulled down 2.0 in more and released, find the (*a*) frequency of vibration and (*b*) the speed of the object as it passes the equilibrium position.

18. Two identical springs hold a mass m on a frictionless table, as shown in Fig. P13.5. Prove that the frequency of vibration of this device is

$$\frac{1}{2\pi}\sqrt{\frac{2g}{d}}$$

if each spring separately stretches d units under the weight of a mass m.

19.** A certain spring has a constant of 15 N/m. The spring is mounted vertically in a tube; a 20-g mass is set on top of the spring and allowed to come to rest. We shall refer to this position as the zero position. By means of a lever, the mass is now pushed 10.0 cm farther down the tube. How high above the zero position will the mass fly when the spring is now released? (Equate the PE in the spring to the PE of the mass. What assumption does this make about the spring?)

20.* A compressed spring with a mass attached to its end is immersed in a container of water at 20.00°C. The heat capacity of the container, spring, mass, and water is 60 cal°C^{-1}. After the spring is released, it vibrates back and forth with decreasing amplitude as a result of friction (or viscous) forces imposed by the liquid. When the system stops vibrating, the temperature is found to be 20.10°C. (*a*) How much energy was stored in the spring? (*b*) If the spring was compressed 8.0 cm, what is the constant of the spring?

21. Show that the maximum speed of a pendulum bob is given by

$$v = \sqrt{2gL(1 - \cos\theta)}$$

if the pendulum starts to swing from an angle of θ to the vertical.

22.* A pendulum is drawn aside to a certain angle and released. When the bob passes the center point, the tension in the string is twice the weight of the bob. Show that the original displacement angle was 60°.

23.* A 10-g bullet is shot with a speed of 1000 cm/s into the 90-g block shown in Fig. P13.6 and becomes embedded in the block. If the block was originally at rest, and if the spring has a constant of 100 N/m, with what amplitude will the spring vibrate after the collision? Ignore friction forces between the block and the table.

24.* If a 2.0-kg mass vibrates at the end of a spring according to the relation $y = 0.20 \cos(9.42t)$ centimeters, find (*a*) the amplitude of vibration, (*b*) the frequency of vibration, (*c*) the spring constant, and (*d*) the maximum speed of the mass.

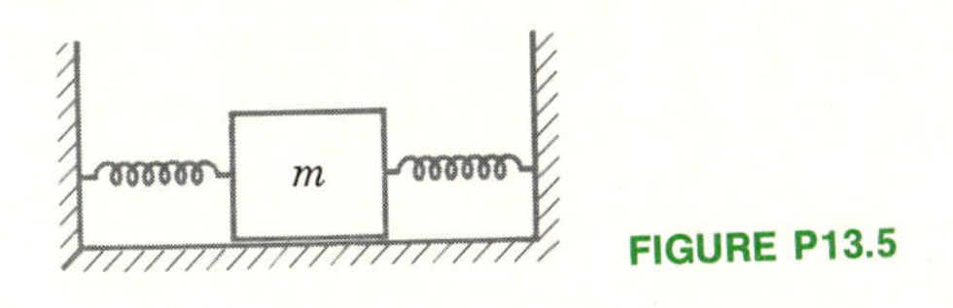

FIGURE P13.5

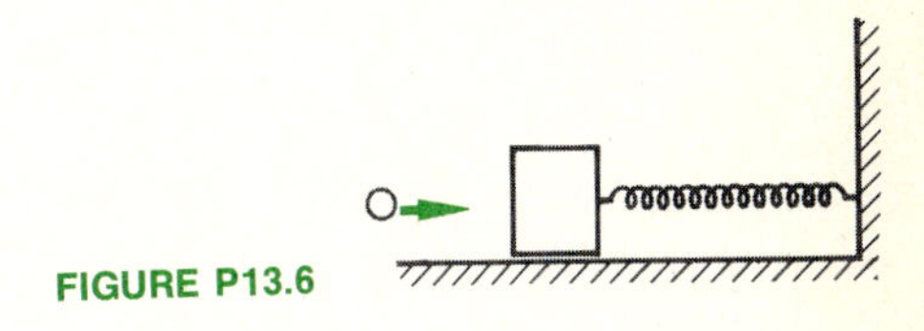

FIGURE P13.6

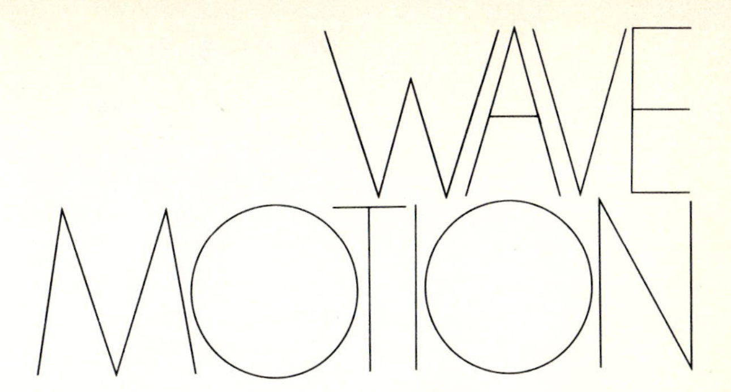

14

The vibrating objects discussed in the last chapter are capable of generating waves. Whether these waves are in air, or metal, or even in vacuum, they have many features in common. In this chapter we shall discuss those principles which are basic to all wave motion. Later chapters will apply these principles to sound, light, and other types of waves.

14.1 IMPORTANCE OF WAVE MOTION

Nearly every vibrating body sends out a wave. We are all aware of the fact that the musical tones given off by a piano or a violin are the result of waves sent out into the air by the vibration of a stretched string. The clarinet also sends out sound waves through the air, but in this case the source of the vibration is the reed blown by the player. In a trumpet, the wave is generated by the vibration of the trumpeter's lips in the mouthpiece of the instrument. There are many other sources of musical sounds—the drum, the triangle, cymbals. Each source consists of a vibrating body which acts in such a way as to generate waves in the air.

Have you ever stopped to think when you are watching TV that everything you see and hear was transmitted to you from the station by waves? The station generated electromagnetic waves by vibrating electrical charge in an antenna. Your TV set responded to these waves and regenerated light and sound waves, which then traveled across the room to you.

If you were to break a bone, the doctor would almost certainly use x-rays, another type of wave, to determine the nature of the fracture. Or if the x-ray picture did not indicate a break, the doctor might suggest treating your strained muscles with another form of wave motion—heat, or infrared, radiation.

Many other types of waves are commonplace to us, but there are other wave phenomena which are not so noticeable. For example, we shall see in Chap. 26 that, to a certain extent, even a baseball may be considered to have wavelike properties.

In this chapter we shall learn the general rules about how waves behave. We shall find that there are two general types of waves into which all mechanical waves can be divided. Within each of these groups there are many different forms of waves. A considerable portion of physics is devoted to the study of their behavior.

14.2 WAVES ON A STRING: TRANSVERSE WAVES

Suppose that a string is tied at one end to a solid wall and at the other to a piece of spring steel, as shown in Fig. 14.1. If the end of the flexible steel rod is struck from below so as to give it an upward velocity, as shown in part *b* of the figure, the end of the rod will vibrate back and forth, as shown in the succeeding parts of Fig. 14.1. We assume it to undergo essentially SHM.

Notice, in part *b*, that the string is pulled upward along with the end of the rod. Its velocity at various places is in the direction of the arrows shown. We see that the string near the rod tends to pull the adjacent string on the right in an upward direction. As the string on the right begins to respond to this pull, it in turn pulls the string adjacent to it upward, and so on. Continuing down through parts *c* to *g* of Fig. 14.1, we see that the disturbance moves continually toward the right. Let us call the velocity with which this disturbance travels along the string v. Later in this section we shall see what its value is in terms of the tension and mass of the string.

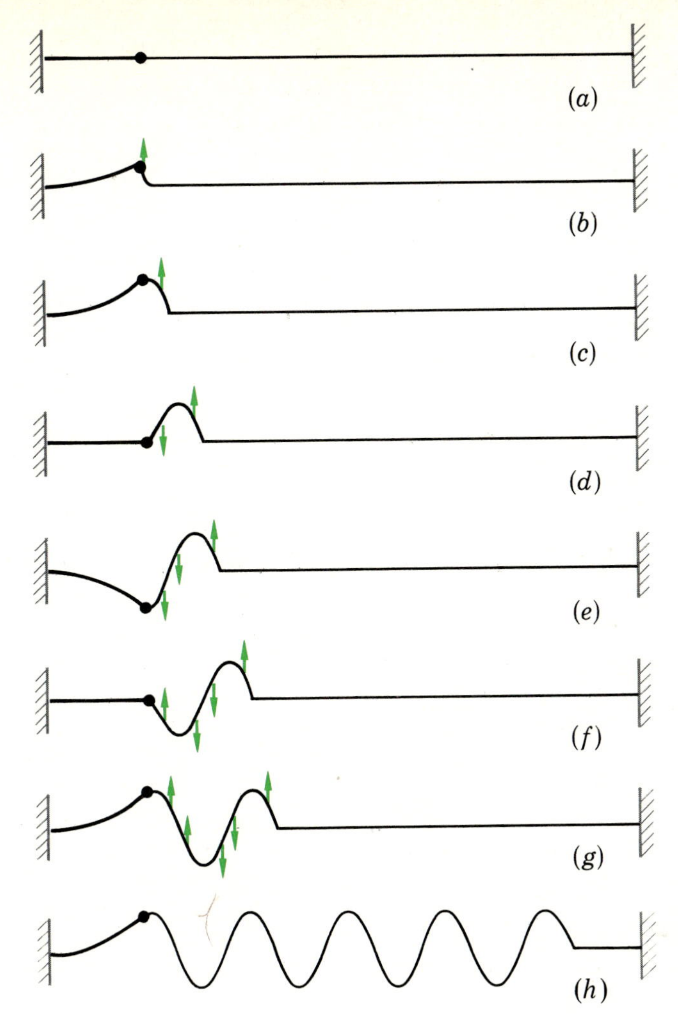

FIGURE 14.1
The vibrating rod sends a wave down the elastic string. Since the motion of the ball is simple harmonic, the wave on the string is sinusoidal.

While the initial disturbance is traveling to the right along the string for the reasons discussed above, the bar continues to vibrate. In so doing, it sends a continually changing disturbance down the string, as shown in Fig. 14.1. Since the source vibrates with SHM, the wave along the string has a sinusoidal shape.

DEFINITION

It will be noticed that only the wave travels to the right in Fig. 14.1. The little pieces of the string simply move up and down. *A wave such as this, where the wave travels in a direction perpendicular to the direction of motion of the particles, is called a* **transverse wave.** This is easily remembered if one recognizes that trans means "across" and that the particles actually travel across the direction of propagation of the wave.

One might well ask: If the particles do not move down the string, what does move in the direction of propagation of the wave? *It is characteristic of waves that they carry energy*. In this particular case, the string had no KE or PE at the start. However, it is clear that in those portions of the wave where

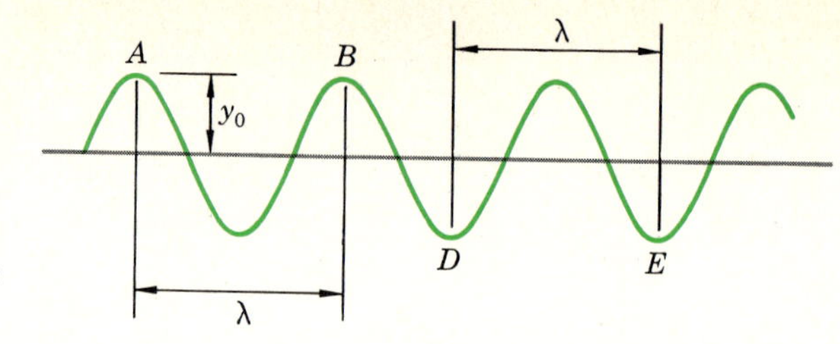

FIGURE 14.2
The amplitude of the wave is y_0, and its wavelength is λ.

the string is moving swiftly, KE is present. This, together with PE, was carried down the string from the vibrating bar. Hence, energy is carried along with the wave in its direction of propagation. The energy travels along with the speed of the wave, v.

The terms used when speaking of waves are easily understood by reference to Fig. 14.2.

Points A and B on the wave are called the **crests** of the wave, while points D and E are called **troughs.** One complete cycle, the portion of a wave generated in one complete vibration of the source, is from crest to crest, i.e., from A to B, or from trough to trough, i.e., from D to E. The horizontal distance between either of these two sets of points is called the **wavelength** of the wave and is represented by the Greek letter lambda (λ). Of course, y_0 is the amplitude of the wave.

There is an important relation between the wavelength λ, the speed of the wave v, and frequency f of any type of wave. To see what it is, let us refer to Fig. 14.3. Notice that the waves travel down the string with speed v as the source sends out an additional wave. It takes a time τ, equal to the period of vibration, for the source to send one additional wavelength down the string. During that time, the waves move a distance λ. Therefore, the motion equation $x = \bar{v}t$ becomes, in this case,

$$\lambda = v\tau$$

Or if we replace τ by $1/f$, this becomes

A Fundamental Relation for Waves

$$\lambda = \frac{v}{f} \tag{14.1}$$

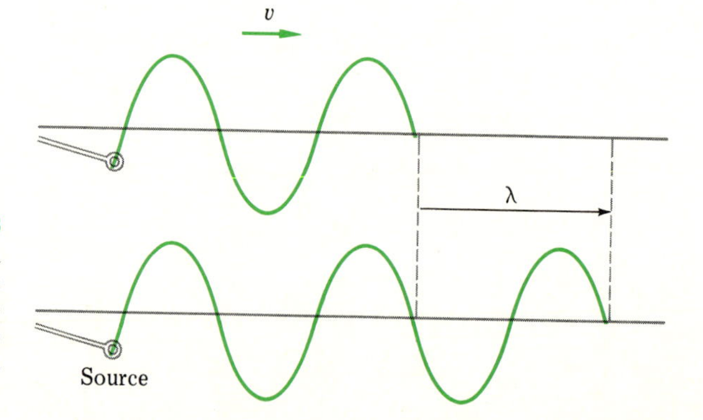

FIGURE 14.3
As the source sends out a complete wave, the wave moves a distance λ along the string. Therefore, the wave travels a distance λ in a time τ.

This extremely important relation is true for all waves. We shall make frequent use of it.

The speed of a wave on a string is given by a particularly simple relation, which we state without derivation. If the tension in the string is T, and if the mass of a length L of the string is m, then the speed of the wave along the string is

$$v = \sqrt{\frac{T}{m/L}} \tag{14.2}$$

Illustration 14.1 A certain guitar string has a mass of 2.0 g and a length of 60 cm. What must the tension in the string be if the speed of a wave on it is to be 300 m/s?

Reasoning Since $v = 300$ m/s, $m = 0.0020$ kg, and $L = 0.60$ m, Eq. (14.2) gives T to be 300 N. Notice that this tension is really quite large since it is equivalent to the weight of about 30 kg pulling on the string.

14.3 REFLECTION OF A WAVE

Until now we have not worried about what happens to a wave when it hits the rigid support on the right. Since the wave represents energy traveling to the right, the energy must either be absorbed by the support or be reflected backward. In practice, some of the energy will be absorbed by the connection at the wall. However, a large fraction of the energy will be reflected if the support is quite solid (or if the end is completely free).* Hence a reflected wave will travel to the left on the string at the same time the original wave is moving to the right. For simplicity, we shall assume that no energy is lost upon reflection.

To study this effect, let us consider only a single wave crest propagating along the string, as illustrated in Fig. 14.4*a*. When this wave reaches the wall, it cannot pull the wall upward in the same way as it would have pulled more string up. Because the wall does not "give," the force at that point is larger than would have been the case if the string had not ended at that point. The wall exerts a downward pull on the string. This pull accelerates the string downward to such an extent that the string's momentum carries it below the zero line. The result is that the pulse is turned upside down as it hits the wall, and the reflected wave appears as shown in part *b* of Fig. 14.4. If the string had been completely free to move up and down at that end, the wave would not have been turned over, although it would still have been reflected (because the energy in the wave could not just disappear at the end of the string!). In summary, *a pulse is inverted by reflection at a fixed end. It is reflected, but not inverted, by a free end.*

FIGURE 14.4

A pulse on a string is inverted when it is reflected from a fixed end.

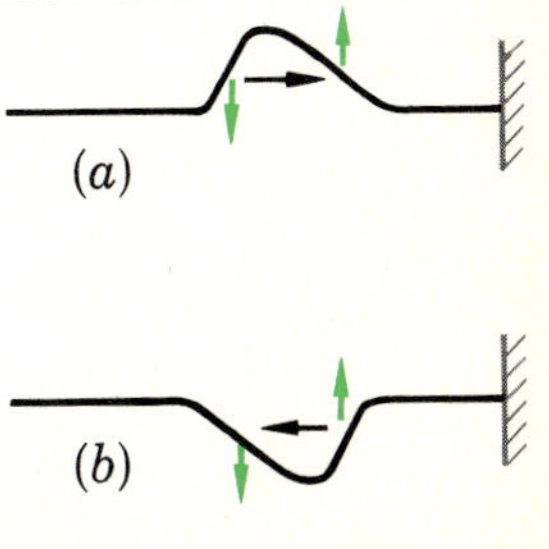

(*a*)

(*b*)

*No work can be done at a fixed end, since the distance moved in the direction of the applied force is zero. Why can no energy loss occur, i.e., work be done, at a perfectly free end?

Next let us consider what happens when a reflected pulse traveling backward along the string meets a second pulse going forward on the string. Suppose two rectangular pulses going in opposite directions meet as shown in Fig. 14.5. The original pulses are shown by dashed lines in the region of overlap. It is found from experience that the string will displace as shown by the full line in this region. As one sees, the string undergoes the vector sum of the individual wave displacements. This is true for all wave systems as long as the displacement is a linear function of the force causing the displacement. One says that the **principle of superposition** applies to such waves. All the waves we shall deal with in this text obey this **principle of superposition:**

Superposition Principle

> A point subjected to two waves simultaneously displaces an amount equal to the vector sum of the individual disturbances.

We are now ready to see what happens when a sinusoidal wave traveling down a string is reflected. You will have to use your imagination a little here, since we shall be drawing pictures of two waves on the same string. These waves, representing displacements of the particles of the string, must be added together to obtain the true behavior of the string.

In Fig. 14.6 we show the two waves traveling to right and left as the dashed curves. The true displacement of the string is obtained by adding the two dashed curves, and the result is shown as the solid curve in the figure. You should add the dashed curves together at a few points to check the displacement of the string. Notice that the two waves exactly cancel each other at the wall, and so the string is not displaced at that point. Furthermore, you can easily see that as time goes on and the waves continue to move, they will always be so matched that they cancel at the wall. This must be true, of course, since the string must remain at rest at that point.

Look, now, at any of the other places where the two waves cancel in Fig. 14.6. It will be noticed that as the waves travel along, they will always cancel at these points as well. Hence we have the important result that *the forward and the reflected wave interfere with each other so as to keep certain points along the string motionless at all times. These points of zero motion are called* **nodes.**

DEFINITION

It is instructive to observe what will happen at the points on the string midway between the nodes. These are the places of large displacement in Fig. 14.6. It will be left as an exercise for you to show that as the two waves

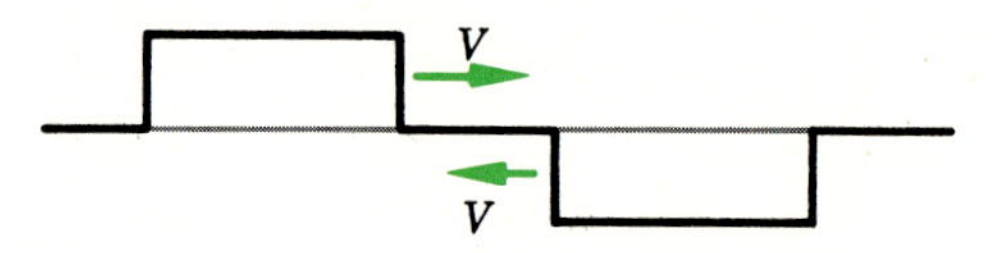

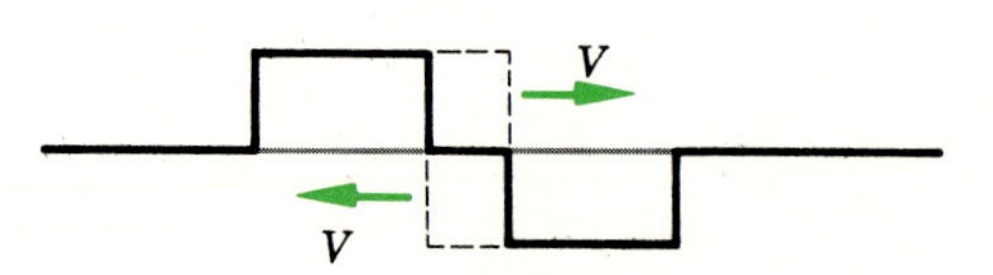

FIGURE 14.5

According to the principle of superposition, two waves on the same string will add as shown when they meet.

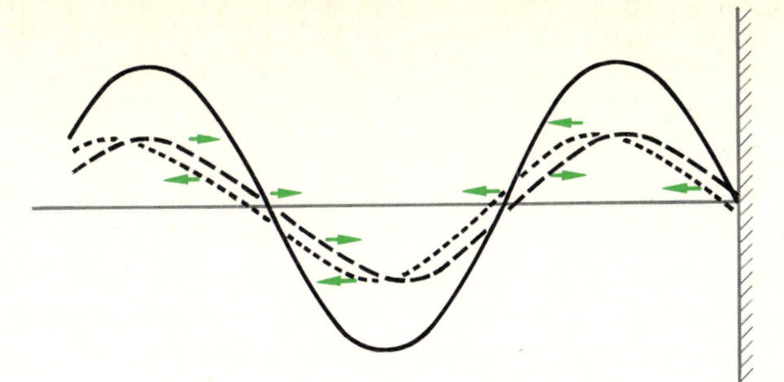

FIGURE 14.6
The incident and reflected waves add together to produce a single wave having a node at the fixed end.

travel along, these intermediate points oscillate back and forth between very large positive and very large negative displacements. This is illustrated in Fig. 14.7. *These points of maximum motion are called* **antinodes.**

DEFINITION

As shown in Fig. 14.7, the string will oscillate back and forth between the two positions shown. The nodes at points P, Q, R, S, and T represent points on the string which remain at rest. No motion of the string occurs at these points. On the other hand, at points A, B, C, and D, the antinodes, the string vibrates back and forth with large amplitude. By comparing Fig. 14.7 with Fig. 14.2, we see that *the distance between adjacent antinodes is* $\lambda/2$. *The distance between adjacent nodes is also* $\lambda/2$.

Distance between Nodes $= \lambda/2$

14.4 RESONANCE

We saw in the last chapter that a vibrating object could be made to vibrate very strongly if the force producing the vibration has the proper frequency. If the force pushes upon the object with a frequency equal to the natural vibration frequency, the force will be in resonance with the object and vibration with very large amplitude will result. Let us review what happens when a force is in resonance with a vibrating system.

As a common example, we know that one does not really have to push very hard on a child in a swing for the child to swing quite high. A small periodic push when the swing is in the proper position will soon build up a large amplitude of oscillation. We can determine why this is true by utilizing some of the concepts we have learned about work.

When one pushes on the swing in the direction in which it is moving,

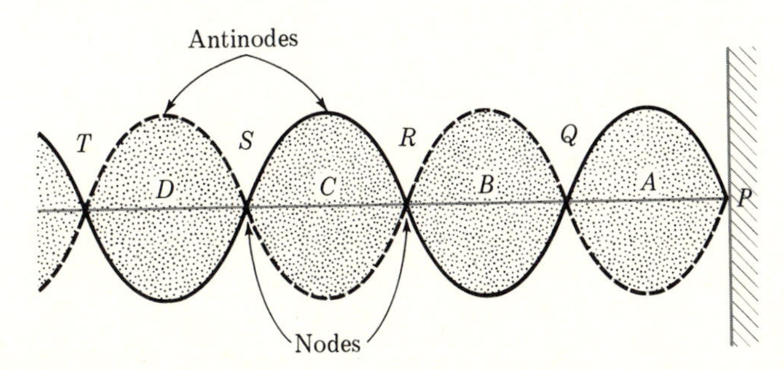

FIGURE 14.7
As the string vibrates to and fro between the limits shown, no motion occurs at the nodes and the greatest motion occurs at the antinodes.

work is done on the swing. The energy thus furnished causes the child to rise somewhat on the next oscillation. When this process is continued over and over again, the child swings higher and higher as energy is put into the system. The limiting amplitude of oscillation occurs when the energy fed in by the person pushing is exactly equal to the frictional energy losses during one complete oscillation.

Conditions for Resonance

Any vibrating system can be excited to wide oscillations by a periodic driving force, as in the case discussed above. Notice, though, that the driving force must be applied at just the proper time if it is to feed energy into the system. (If the driving force pushes on the system when it is moving against the force, the system will do work on the driving mechanism and will lose, rather than gain, energy.) *In the case in which the driving force is just in step with the vibrating system so that the amplitude becomes quite large, the driving force and the system are said to be in* **resonance.** *Two conditions must be satisfied for resonance. The applied force should have the same frequency as the free vibration frequency of the system. In addition, the force must be applied* **in phase** *with the vibration.* That is to say, the force and the vibration must be in step.

Many examples of resonance could be cited. The child in the swing is one. Many clocks have a pendulum which is excited by a mechanism operating in resonance with the pendulum. All musical instruments make use of resonance to produce audible sounds; we shall discuss this example more fully later.

Not all resonances are desirable. It is quite common for a rattle in an automobile to become particularly noisy at a certain speed. The rattling device, a vibrating object of some sort, is in resonance with the vibratory motion of the car at that speed. This type of disturbance becomes very serious in cars which have wheels that are badly out of balance. The thumping action of the wheel against the roadway varies in frequency with the speed of the automobile. It is not surprising that various portions of the car begin severe vibrations when the proper resonance frequency is maintained.

In the next section we shall examine the resonance motions of a vibrating string. We shall see that the string will resonate to driving forces of more than one frequency. Although the computations carried out for the string are very simple examples, it will be seen that much more complicated vibratory systems show the same general features of resonance.

14.5 RESONANCE MOTION OF A STRING: STANDING WAVES

Let us now consider what happens when a vibrator sends waves down an actual string, as shown in Fig. 14.8. If we perform the experiment indicated, interesting results are found. The string shown is attached to the vibrator at one end and is held rigid by the massive pulley at the other. However, the length of the string can be increased by pulling the tuning fork farther to the left, as shown in succeeding parts of the figure.

For most string lengths, the string remains nearly motionless. The

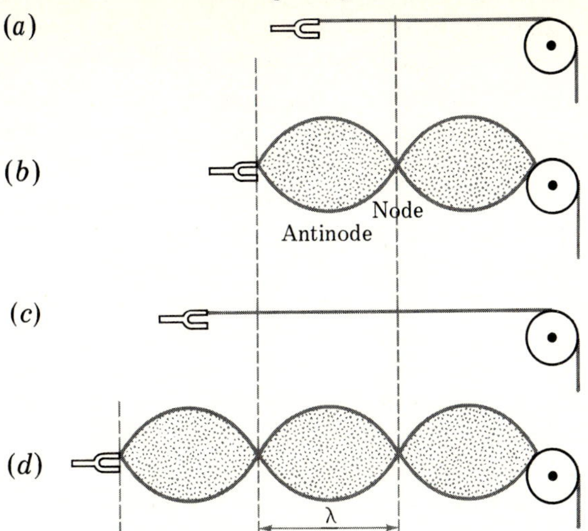

FIGURE 14.8

At what other lengths will the string resonate?

vibrator sends tiny wave pulses down the string, and these are reflected back by the pulley. As the wave pulse goes down the string, it is analogous to a child on a swing moving away from the person pushing the swing. Like the child on a swing, the pulse is reflected and returns to the pusher—the vibrator in this case. Usually the vibrator will not be moving in the precise way needed to reflect the pulse once again and add energy to it. Like a person pushing improperly on the swing, the push will tend to cancel the motion rather than reinforce it. For this reason, the tiny pulses sent down the string by the vibrator do not usually reinforce. As a result, the string remains nearly motionless as shown in part *a*.

Standing Wave

However, a startling result is seen if the string is slowly lengthened. *At some very definite length, the string begins to vibrate widely* as shown in part *b*. *The string vibrates back and forth between the limits shown. Certain points of the string remain motionless, and these are the nodes. Other points, the antinodes, vibrate most widely. We call this pattern consisting of nodes and antinodes a* **standing wave** *on the string.*

It is easy to understand what is happening to cause the standing wave. In this very special situation, the pulse reflected back to the vibrator arrives just in time to be reinforced by the vibrator. It is reflected back down the string by the vibrator, but because the vibrator is exactly in step (or in phase) with it, the pulse is augmented by the vibrator. Consequently, the vibrator keeps adding energy to the pulses as they move back and forth along the string. The wave pulses therefore become very large.

But we have already seen in Sec. 14.3 that the pulses moving to right and left along the string combine to form nodes and antinodes. Therefore the string vibrates in a definite pattern of nodes and antinodes. Moreover, because the sum of these pulses is so very large, the pattern is large and easily visible. As we see, *the large standing wave is a result of the fact that the vibrator pushes in resonance with the wave pulses on the string.*

If now the string in Fig. 14.8b is lengthened further, the reflected pulses will reach the vibrator slightly later than before. They will no longer be in phase with the vibrator and so resonance will no longer occur. As seen in part c of the figure, the string will now remain nearly motionless.

But if the string is lengthened still further, resonance will suddenly occur at the elongation shown in part d. Once again the reflected pulses will be in phase with the vibrator when they reach it, and again large pulses will build up on the string and a large, visible standing wave will result. As we see, the string will resonate to the vibrator at very special conditions. Let us now see what these conditions are in a quantitative form.

Resonance Conditions

You will recall that the distance between nodes is $\lambda/2$. If we examine Fig. 14.8, we see that the vibrator is very close to the position of a node since it cannot move far. Therefore the two ends of the resonating string are at nodes. As a result, the string can only resonate if it is $\lambda/2$ long, or $2(\lambda/2)$ long, or $3(\lambda/2)$ long, and so on. Indeed, *we can state in general that a string fastened firmly at its two ends will resonate only if it is a whole number of half wavelengths long*. For example, in parts a, b, c, and d of Fig. 14.9, the length L of the string is equal to $\lambda/2$, $2(\lambda/2)$, $3(\lambda/2)$, and $4(\lambda/2)$. In general, then, for resonance,

$$L = n\frac{\lambda}{2} \qquad \text{where } n = 1, 2, 3, \ldots \qquad (14.3)$$

Since the wavelength is related to the frequency by Eq. (14.1), we see at once that a string of fixed length will resonate to only certain very special frequencies. In modern terminology, *the resonant frequencies of the string are* **quantized,** *meaning that only certain special resonant frequencies exist.*

FIGURE 14.9

Resonant modes of motion in a string must have nodes at the two fixed ends. These are the four simplest standing waves possible.

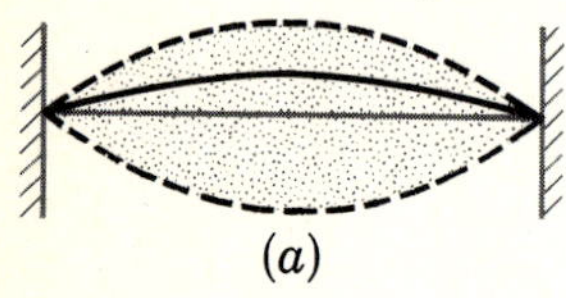

(a)

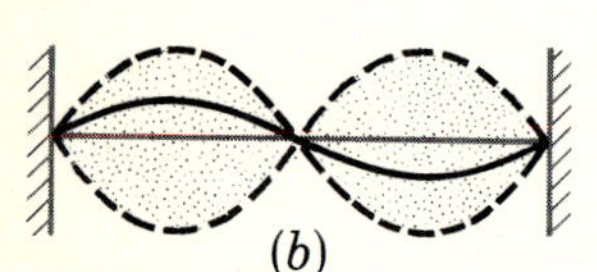

(b)

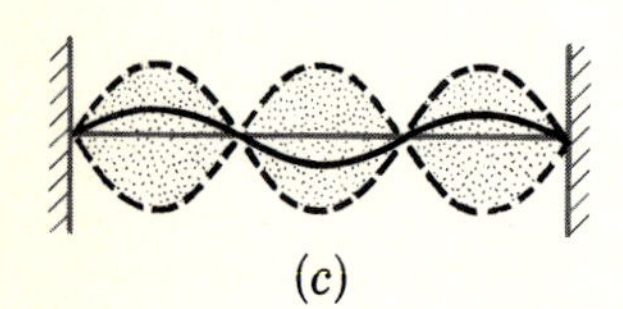

(c)

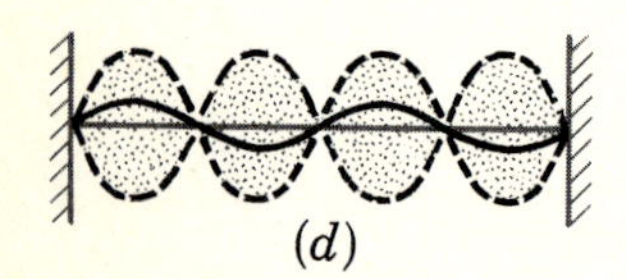

(d)

Illustration 14.2 The speed of a wave on a particular string is 24 m/s. If the string is 6.0 m long, to what driving frequencies will it resonate? Draw a picture of the string for each resonant frequency.

Reasoning The possible resonance wavelengths are given by Eq. (14.3). We have

$$\lambda = \frac{2L}{n} \qquad \lambda_1 = 12\text{ m} \qquad \lambda_2 = 6\text{ m}$$

$$\lambda_3 = 4\text{ m} \qquad \lambda_n = \frac{12}{n}\text{ m}$$

Now we can make use of Eq. (14.1) to find the frequency.

$$f = \frac{v}{\lambda} \qquad f_1 = \tfrac{24}{12} = 2\text{ Hz} \qquad f_2 = \tfrac{24}{6} = 4\text{ Hz}$$

$$f_3 = \tfrac{24}{4} = 6\text{ Hz} \qquad f_n = \frac{24}{12/n} = 2n\text{ Hz}$$

The modes of vibration, i.e., the various standing waves, of the string appear the same as parts a, b, and c of Fig. 14.9 when the frequency is f_1, f_2, and f_3.

14.6 OTHER TRANSVERSE WAVES

We have spent a great deal of space discussing waves on a string because the principles which apply there apply to many other vibrating systems. For example, if a metal bar clamped at its center is struck at its end, as shown in Fig. 14.10, the bar will vibrate. The mode of vibration of the bar is indicated in part b of Fig. 14.10.

The center of the bar must be a node, because it is tightly clamped in place. Since the ends of the bar are not held rigidly, we expect antinodes near them. If we assume the ends to be antinodes, the length of the bar L is one-half wavelength. This follows from the fact that the distance between two successive antinodes is $\lambda/2$. Knowing that $\lambda = 2L$, one could measure the frequency of vibration of the bar and use Eq. (14.1) to compute the speed of a transverse wave in it.

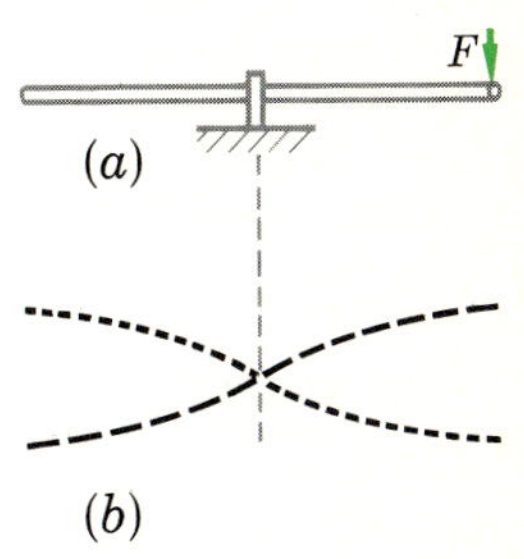

FIGURE 14.10

A transverse standing wave is set up in a bar when it is struck as shown.

If the bar had been clamped as in Fig. 14.11, namely, at a distance $L/4$ from its end, the vibration would have appeared as in part b of that figure. Once again the ends would approximate antinodes, and the clamp point would be a node. In this case

$$\lambda = L$$

while, in the case of Fig. 14.10,

$$\lambda = 2L$$

Since the frequency of vibration is given by the relation

$$f = \frac{v}{\lambda}$$

the respective frequencies of vibration for the bar are

Fig. 14.10: $$f = \frac{v}{2L}$$

Fig. 14.11: $$f = \frac{v}{L}$$

Hence, the bar in Fig. 14.11 would be vibrating with twice the frequency of the bar in Fig. 14.10. If f and L were measured, v could be computed in each case, as was pointed out above.*

FIGURE 14.11

The position of the clamp must be a node.

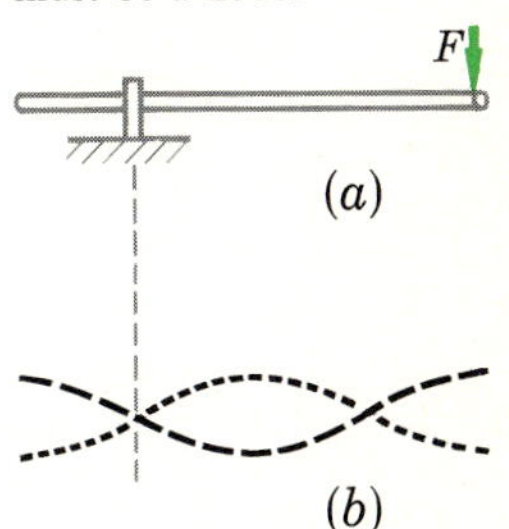

* More detailed computations show that when the finite rigidity and inertia effects are considered, these conclusions are only approximate. In practice the node in Fig. 14.11 occurs at $0.224L$ rather than at $0.250L$.

Many other forms of transverse waves are common. Waves on the surface of water are approximately transverse, since the water particles move vertically up and down while the wave travels horizontally across the water. We shall speak more about these waves when we begin our study of light waves, since there are great similarities between the two.

When a drumhead is struck, transverse waves travel across its surface. Similar waves are set up in a cymbal or, for that matter, any flat sheet of material which is made to vibrate. These waves are more complicated than those on a string, for the wave can travel in any direction in the plane, while in the string it could travel in one direction only. The nodal points in such cases form nodal lines on the plane, lines along which the plate does not vibrate. We shall not discuss this behavior here, but later, in our study of light, we shall encounter a similar situation.

Finally, perhaps the most important type of transverse wave is the electromagnetic wave. This type of wave will be discussed after we have studied the subject of electricity. It will be found there that radio, radar, heat, infrared, light, and ultraviolet waves, as well as x-rays, are all forms of electromagnetic waves. Clearly, from this list, electromagnetic waves are of great practical importance to us.

14.7 LONGITUDINAL WAVES

An interesting experiment can be done with a very long spring placed on a smooth table top and tied at one end. One possible form of this experiment (not the best from a practical standpoint) is illustrated in Fig. 14.12*a* to *d*.

The spring at equilibrium on the table top is shown in part *a*. If it is suddenly compressed as in part *b*, the loops near the end will be compressed before the rest of the spring experiences the disturbance. The compressed loops will exert a force on the loops to the right of them, and so the compression will travel down the spring as indicated. At the fixed end, the compressional energy is reflected; thus the compression is reversed and ends up traveling to the left, as shown in part *d*.

This type of wave is not a transverse wave, since the particles of the spring actually vibrate back and forth in the direction along the spring, the direction in which the wave is propagated. *A compressional wave such as this, where the motion of the particles is along the direction of wave propagation, is called a* **longitudinal wave.**

DEFINITION

A difficulty arises when we attempt to plot a graph of a compressional wave. In a transverse wave, the particle motion was up and down, say, and the displacement on the graph was plotted up and down also. Here we would encounter difficulty plotting the spring displacement in the same direction as that in which it occurs, since distance along the wave and displacement both lie in the horizontal direction. We therefore still plot the magnitude of the displacement vertically, as shown by the black curves in Fig. 14.12. We should always be careful to remember that no motion occurs in this direction in a longitudinal wave. The plot is made in this way merely as a matter of convenience.

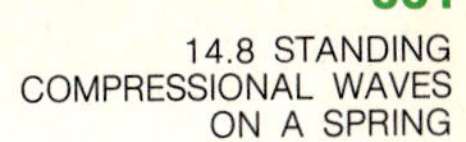

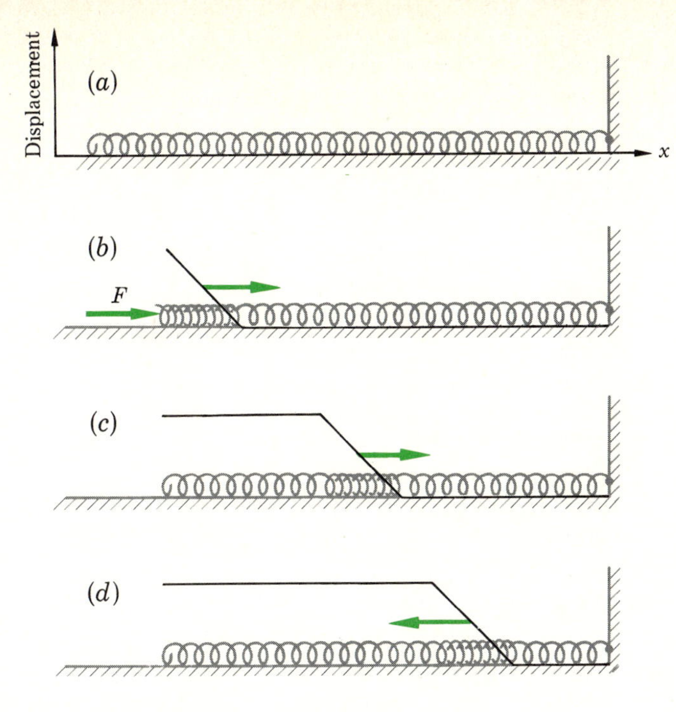

FIGURE 14.12
A longitudinal pulse travels down the spring and is reflected by the wall. The black curves are graphs of the displacement of the spring.

14.8 STANDING COMPRESSIONAL WAVES ON A SPRING

It is clear that *the longitudinal wave on a spring has many features in common with a transverse wave on a string.* If a compressional wave is sent down a spring, the wave and its energy will usually be reflected at the end of the spring. This reflected wave can interfere with the later waves being sent down the spring from the source. *If the proper relation is maintained between the frequency of the driving source oscillating the end of the spring and the various parameters of the spring, resonance will occur.* It is this feature of the spring system which we shall now investigate.

As with resonance on a string, *the position of the driving source will usually be close to a node* (a point of zero motion), since the spring itself at resonance will move much more than the driving source. Also, *if the other end of the spring is fixed solidly to a wall or some other similar object, that end must also be a node. The resonance motion of the spring must then appear as shown in the graphs of Fig. 14.13.*

Notice that although the displacement of the loops of the spring along the spring is plotted vertically, the motion actually occurs in a horizontal direction. For example, point A in Fig. 14.13*a* means that the center point of the spring vibrates back and forth horizontally, with an amplitude indicated by the dashed wave. It is seen in the mode of motion represented in Fig. 14.13*b* that the center point does not move at all. Hence, the center point can be either an antinode or a node, depending upon the frequency which is causing the spring to resonate.

The same relations apply to the vibration along the spring as we found

FIGURE 14.13
Standing waves result from longitudinal vibration of the spring.

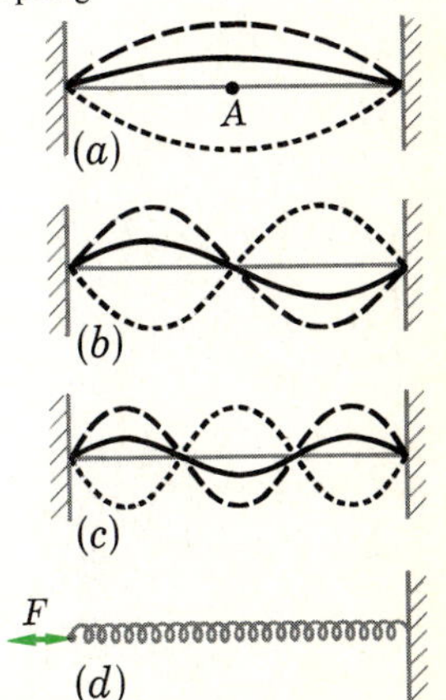

for a wave on a string. From Fig. 14.13, we see that the distance between nodes is still $\lambda/2$. Also, at resonance, in this particular case, the spring must be a whole number of half wavelengths long. That is, at resonance

$$n\frac{\lambda}{2} = L \qquad \text{where } n = 1, 2, \ldots$$

This relation, when combined with the relation between wavelength and frequency, $\lambda = v/f$, tells us at once that the spring resonance frequencies will be

$$f = n\frac{v}{2L} \qquad \text{where } n = 1, 2, \ldots$$

Illustration 14.3 A spring 300 cm long is found to resonate in three segments (nodes at both ends) when the driving frequency is 20 Hz. What is the speed of the wave in the spring?

Reasoning The spring is vibrating as shown in Fig. 14.13*c*. It is seen that in this case

$$3\frac{\lambda}{2} = L$$

or

$$\lambda = 200 \text{ cm}$$

Knowing that $\lambda = v/f$ and since $f = 20$ Hz, we have

$$v = (20 \text{ s}^{-1})(200 \text{ cm}) = 4000 \text{ cm/s}$$

We could, of course, have obtained this result by simply substituting in the relation

$$f = n\frac{v}{2L}$$

given above, using $n = 3$. However, most physicists prefer not to memorize a different relation for each case. They ordinarily use the number of half wavelengths on the total spring to find λ and then the relation $f = v/\lambda$ to find the unknown. As a matter of fact, almost all the situations in resonance which we shall encounter can be described by the use of this relation and an examination of the resonant system. It is not necessary to memorize an equation for each case.

14.9 COMPRESSIONAL WAVES ON A BAR

We have already discussed the transverse vibrations of a bar. It is possible to set up longitudinal vibrations in such a bar as well. There are many possible ways of doing this. One simple method is to strike the end of the bar, as shown in Fig. 14.14*a*.

The blow on the end of the bar sends a compressional wave down the bar. This wave is quite complex in form and is actually a large group of waves of various frequencies. The bar will resonate only to certain frequencies, and hence it selects the proper frequency to which it will resonate.

In this particular case, the bar must have a node at its center and antinodes at its ends. The lowest resonant-frequency mode of motion is shown in Fig. 14.14*b*. Remember that the motion of the bar is longitudinal even though the graph of the motion looks like the similar case for transverse vibration. Clearly in this case the bar is one-half wavelength long. With the length of the bar L known, the wavelength is then also known. If the frequency of vibration of the bar is known as well, the velocity of compressional waves in the material can be computed by using $\lambda = v/f$. (A simple way to measure the frequency of vibration of the bar is to compare the sound given off by the bar with the sound given off by tuning forks of known frequency. More accurate methods will be discussed later.) Some typical values for the speed of compressional waves in various materials are given in Table 14.1.

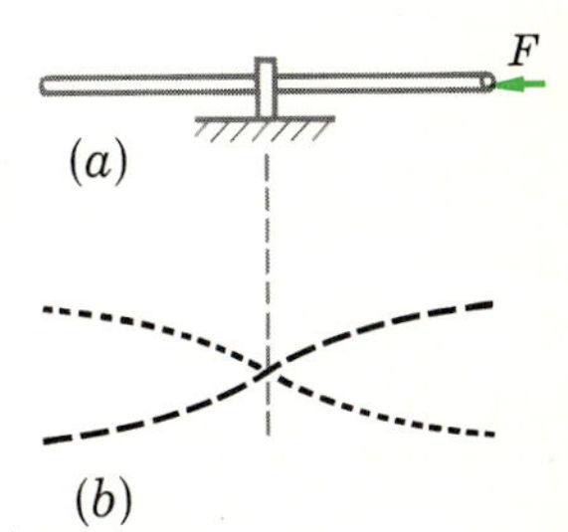

FIGURE 14.14
When the bar is struck as shown, a longitudinal standing wave is set up in it.

Other resonance vibrations of the bar in Fig. 14.14 are also possible. Any vibration which will have a node at the center of the bar and antinodes at the ends will be allowed. You should examine this case to see what the next higher resonant frequency of the bar would be. It will have three nodes along the bar. (Why is the mode of motion with two nodes along the bar not allowed in this case?) Since viscous-energy losses within the vibrating bar are more serious at the higher frequencies, these modes of motion usually die out rather rapidly, i.e., their energy is lost to heat, and the bar ceases to vibrate.

The speed of compression waves in various materials can be related to the density and elasticity of the material. One can show that the speed in liquids is given by

$$v = \sqrt{\frac{E}{d}} \tag{14.4}$$

where d is the mass density of the material and E is the bulk modulus which was defined in Eq. (9.8). Notice that consistent units must always be used in this relation. In the case of rods, the elastic bulk modulus E is replaced by Young's modulus Y.

TABLE 14.1
SPEED OF COMPRESSIONAL WAVES

MATERIAL	SPEED, m/s
Air (0°C)	331
Water (15°C)	1447
Copper	3500
Glass	4000–5500
Wrought iron	4900–5100
Steel	5000

Illustration 14.4 If the metal bar of Fig. 14.14 is 0.925 m long and resonates as shown with a frequency of 2700 cps, find Young's modulus of the metal. The density of the metal is 7.86 g/cm^3.

Reasoning We need first to determine v, so that use can be made of Eq. (14.4). From Fig. 14.14, we see that 0.925 m is equivalent to $\lambda/2$. Hence $\lambda = 1.85$ m. Using $v = \lambda f$ (with the SI units now employed) gives

$$v = (1.85 \text{ m})(2700 \text{ s}^{-1}) = 5000 \text{ m/s}$$

Then from Eq. (14.4)

$$(5.00)(10^3 \text{ m/s}) = \sqrt{\frac{Y}{7.86 \times 10^3 \text{ kg/m}^3}}$$

or

$$Y \approx 2.0 \times 10^{11} \text{ N/m}^2$$

Examination of the Young's moduli given in Table 9.2 indicates that this bar might well have been made of iron.

SUMMARY

An oscillator undergoing SHM motion can send a sinusoidal wave down a string. The distance between two adjacent crests of the wave is the wavelength of the wave λ. It is related to the vibration frequency f and the speed of the wave through $\lambda = v/f$. This is a general relation that applies to all waves.

The principle of superposition of waves states that a particle subjected to two or more wave disturbances responds to the vector sum of the individual disturbances.

Under special circumstances, a string will resonate to a SHM driving force. For a string held rigidly at both ends, resonance occurs if the string's length is $n(\lambda/2)$, where n can be any integer. The string then vibrates with a definite, steady pattern called a standing wave. In the pattern, certain points of the string remain motionless and are called nodes. Other points vibrate most widely and are called antinodes. The distance between adjacent nodes (or antinodes) is always $\lambda/2$ in a standing wave.

When the particles subjected to a wave disturbance move perpendicularly to the direction of propagation, the wave is said to be transverse. Typical transverse waves are waves on a string and electromagnetic waves. But if the particles move along the direction of propagation of the wave, the wave is said to be a compressional or longitudinal wave. Compression waves on a spring or in gases, liquids, and solids are waves of this type.

MINIMUM LEARNING GOALS

Upon completion of this chapter you should be able to do the following:

1. Sketch a sinusoidal wave on a string and point out the following features of it: crest, trough, wavelength, amplitude.
2. Show by means of a sketch what happens to a wave pulse on a string as it strikes a fixed end and a free end.
3. State the relation between λ, v, and f for any wave.
4. Sketch several standing wave forms for a string solidly held at its two ends. Point out the position of nodes and antinodes. Use the number of segments in the standing wave to state the relation between L and $\lambda/2$ for each pattern sketched. Then compute f or v for the string provided L and either f or v are given.
5. Explain the difference between transverse and longitudinal waves and give examples of each.
6. Draw the compressional standing wave resonance forms for the following different situations: (*a*) spring held rigidly at both ends; (*b*) rod held rigidly at one point with that point being either at an end or $0.5L$, or $0.25L$ from one end.
7. Compute the resonance frequency for each situation listed in item 6 provided sufficient data are given.

IMPORTANT TERMS AND PHRASES

You should be able to define or explain each of the following:

Wavelength
Crest; trough
$\lambda = v\tau = v/f$
Transverse wave; longitudinal wave
Resonance; standing wave
Node; antinode
In phase
A segment is $\lambda/2$ long

QUESTIONS AND GUESSTIMATES

1. The two idealized pulses shown in Fig. P14.1 are moving down the string at a speed of 20 m/s. Sketch how the string would look after 0.40 s. Repeat for 0.20 s.
2. Give qualitative physical arguments to justify the fact that the speed of transverse waves on strings should increase as the tension increases but should decrease as the mass per unit length increases.
3. Some stringed musical instruments have strings made of catgut with a thin wire wound around it. What is the purpose of the wire winding?
4. A string fixed at both ends is vibrating with four antinodes, i.e., in four segments. Can one touch the string with a knifeblade without disturbing its vibration? Explain.
5. A variable frequency oscillator sends waves down a string (of length L) whose ends may be considered nodal positions. At a distance of $L/5$ from one end, a tiny clamp holds the string nearly motionless although it still allows wave energy to pass by. Describe the standing waves one will notice on the string as the frequency of the oscillator is slowly increased from a very low value.
6. The little pieces of a string near the antinodes have a great deal of KE. Since the parts of the string near the nodes have very little KE, how did the energy get from the source to the antinodes?
7. At resonance in a string, the reflected wave cancels out the vibration of the incident wave at the nodes. Was energy destroyed? What happened to it?
8. Two identical vibrators are attached at opposite ends of a stretched string. They are adjusted so that when used one at a time, they will cause the string to resonate under transverse motion. When both vibrators are vibrating, will the string resonate? If so, under what conditions?
9. All common metals expand when heated. Try to devise a method for monitoring the temperature of a wire by a vibration-resonance technique. Steel wire lengthens about 0.001 percent for each degree change in temperature. Do you think the vibration method is feasible?
10. Is it possible for two identical waves traveling in the same direction down a string to give rise to a standing wave?
11. From a small height drop a metal bar, glass rod, or a ruler end first onto a solid floor. From the sound given off estimate its fundamental resonance frequency and from this estimate the speed of a compressional (sound) wave in the material from which it is made.
12. A steel guitar string is tuned to 330 Hz. Make an order-of-magnitude estimate of how much the frequency of the string changes when its temperature is lowered 20°C. (E)

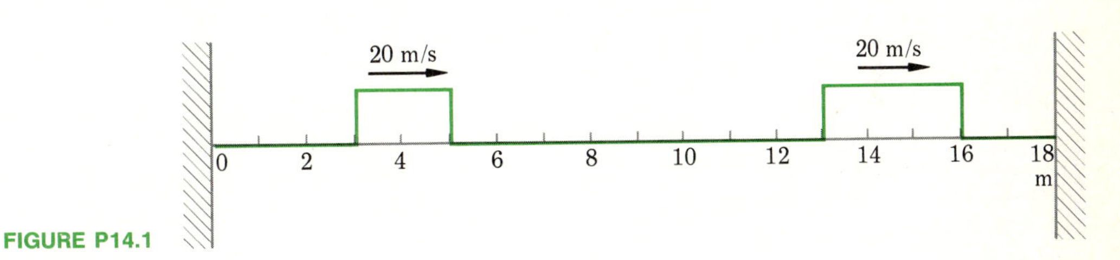

FIGURE P14.1

PROBLEMS

1. Consider the wave pulses shown in Fig. P14.1. They are traveling along the string with speed 20 m/s. How long after the instant shown will the pulses once again look the way they are shown?
2. The wave shown in Fig. P14.2 travels toward the right with a speed of 100 cm/s. (*a*) How many wave crests go by point A each second? (*b*) What is the wavelength of the wave? (*c*) What is the frequency of the wave as computed from $\lambda = v/f$? (*d*) Compare the answers to (*a*) and (*c*).

FIGURE P14.2

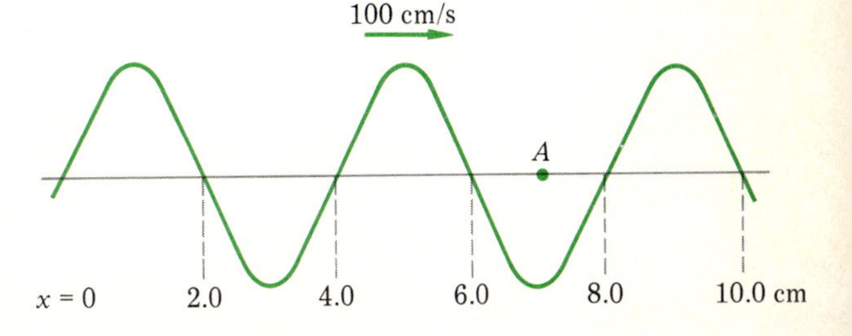

3. As a strong wind blows over a telephone wire stretched between two poles 15 m apart, the wire hums in the wind. The hum has a frequency of 20 Hz, and so it is reasonable to assume that the wire is vibrating with this frequency. Assuming that the two poles are at nodes for the vibrating wire, what is (*a*) the probable wavelength and (*b*) speed of the wave produced on the wire by the wind?

4. A string 2.0 m long "weighs" 4.0 g. It is stretched horizontally by running one end over a pulley and attaching a 1.0-kg mass to it. Find the speed of a transverse wave in the string.

5. How large a weight must be hung on the end of a thread 200 cm long if the speed of transverse waves in it is to be 400 cm/s and if 100 cm of thread "weighs" 0.50 g?

6. When a certain wire stretched between two points 50 cm apart is plucked near its center, it gives off the same tone as a 250-cps tuning fork. Find the wavelength and the speed of the wave in the wire. *Hint:* Since the wire is plucked near its center, it will be vibrating with one antinode and with its ends as nodes.

7. A wire 80 cm long vibrates with four nodes, two of them at the two ends. Find (*a*) the wavelength and (*b*) the speed of the wave in the wire if the wire is being vibrated at 500 Hz.

8. A wire held firmly at both ends vibrates in three segments to a frequency of f_1 and four segments to a frequency f_2. What is the ratio f_1/f_2?

9. Two wires are exactly the same length and are under the same tension. However, the lowest resonance frequency of one is only half that of the other. (*a*) Which wire has the largest mass per unit length, and (*b*) by what factor is it bigger?

10. A certain string held at its two ends will resonate to several frequencies, the lowest of which is 100 Hz. What are the next three higher frequencies to which it resonates?

11. A certain string resonates in three segments for a frequency of 60 Hz. Give four other frequencies to which it will resonate.

12. An iron bar clamped at its center is set into longitudinal vibration by pulling outward on its end. (*a*) If the bar is 100 cm long, find the lowest frequency at which it will resonate. (*b*) Repeat if it is clamped at one end. ($v = 5 \times 10^3$ m/s.)

13. An iron bar 100 cm long is clamped 25 cm from one end. Give the lowest resonance frequency for longitudinal waves in the bar. ($v = 5 \times 10^3$ m/s.)

14. A coil spring is stretched to a length of 4.00 m and set into longitudinal vibration by a vibrator at one end. When the driving frequency is 2.0 Hz, the spring vibrates with five antinodes along its length. What is the speed of compressional waves on the spring? Assume the two ends are nodes.

15. A steel bar 1.00 m long is clamped at one end. What are the first three resonance frequencies for compressional waves? ($v = 5000$ m/s.)

16.* A coil spring 4.0 m long lies on a smooth table, with one end fixed and the other free. The speed of compressional waves in it is 800 cm/s. Give several frequencies with which one could keep hitting the end of the spring if resonating compressional waves are to be sent down the spring.

17.** In Fig. P14.3 is shown a steel wire bent into a loop with a radius of 20 cm. The loop is held rigidly at P and vibrated transversely back and forth by an oscillator at A. Assuming both P and A to be nodes, find the first three resonance frequencies for the loop. ($v = 0.40$ m/s in the steel wire.)

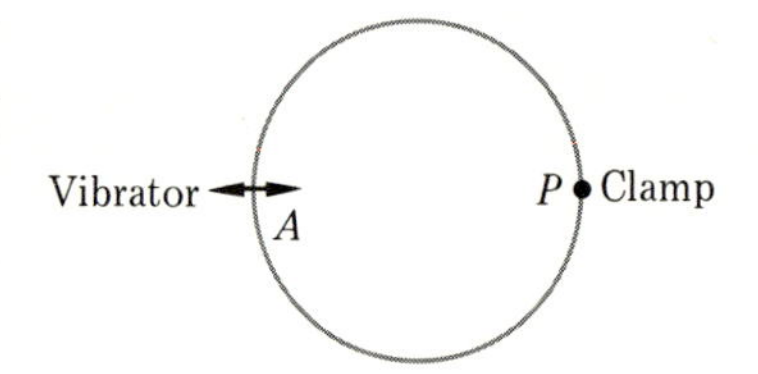

FIGURE P14.3

SOUND

15

The concepts involving wave motion discussed in the last chapter will now be applied to one particular form of wave motion, sound. Not only is a study of sound important in its own right, but in addition it affords us a valuable means for consolidating our knowledge of wave motion in general. We shall find that many of the ideas discussed in connection with sound will also be important in our study of light and other types of wave motion.

15.1 ORIGIN OF SOUND

DEFINITION *Any compressional disturbance traveling through a material in such a way as to set the human eardrum into motion, giving rise to the sensation of hearing, is usually defined as* **sound.**

Generally we think of sound as being a compressional wave in the air, since the ear is usually in contact with the air. However, under water, the sound could reach the ear directly by the vibration of the water against it. In any case, *sound is a material vibration and can occur only where there is a transmitting medium present.* A common example of this is to show that the ringing of a bell cannot be heard if the bell is in a vacuum chamber. The bell itself is vibrating, but there is no material around the bell which can transmit the vibration.

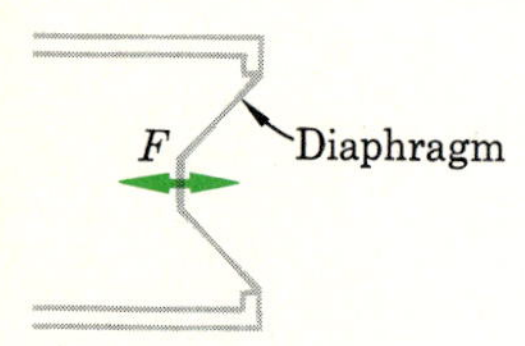

FIGURE 15.1
The flexible loudspeaker diaphragm vibrates back and forth to send out compressions in the air.

Two things are necessary for the generation and propagation of sound. (1) *There must be some object which vibrates in such a way as to send out a compressional wave.* Whether this is the vibrating string and sound box of a violin or the spectacular vibration in the vicinity of an exploding firecracker is only of a secondary importance. Sound is given off from the vibrating object in either case. (2) *There must be material present to transmit the sound.* The violin string transmits its vibrations to the surrounding air, which then carries the vibration through space. A violin string vibrated in vacuum could not be heard, since the vibration could not be propagated away from the string.

15.2 SOUND WAVES IN AIR

Let us now consider the action of a loudspeaker when it is used to generate simple sounds. A simple loudspeaker consists of a cone-shaped sheet of flexible material, with a diaphragm which can be oscillated back and forth, as shown in Fig. 15.1. The oscillatory driving force F is applied at the center of the diaphragm.

When the diaphragm moves to the right, it compresses the air in front of it and a compressional wave travels out through the air. An instant later, the diaphragm is moving to the left, leaving a region of decreased air pressure in its wake, a so-called **rarefaction.** This disturbance, too, travels out from the loudspeaker. Hence a series of pressure waves travel out from the loudspeaker. They consist of alternate regions of high pressure (the compressions) and low pressure (the rarefactions). This situation is illustrated in Fig. 15.2.

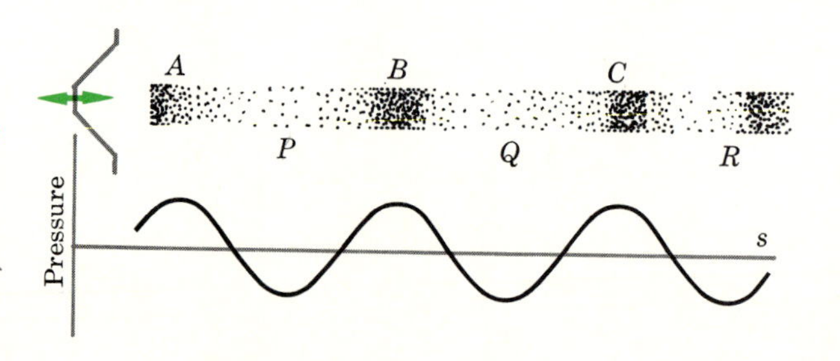

FIGURE 15.2
The sound wave sent out by the loudspeaker consists of alternate high- and low-pressure regions in the air. In practice P changes by only about 0.01 percent or less.

The compressions of the air sent out by the loudspeaker as it moved toward the right are shown as A, B, and C in that figure. When the diaphragm was pulled to the left, a semivacuum was left behind and the rarefactions at P, Q, and R were sent out. A plot of the pressure in the air along this sound wave is given in the lower portion of Fig. 15.2. Notice that the center line is not at zero pressure but is instead at a value equivalent to the average atmospheric pressure. The rarefactions are regions of somewhat reduced pressure. It is assumed in drawing the figure that the diaphragm was oscillating sinusoidally. Even for very loud sounds, the actual pressure variations are only about 0.01 percent of atmospheric pressure.

It is sometimes convenient to speak about the displacement of the air molecules by the sound wave rather than to talk about the pressure in the wave. Certainly the air particles will be moved back and forth in much the same way as the diaphragm moves back and forth. The motion of the air is back and forth along the direction of propagation of the wave, and so this is a longitudinal wave. Moreover, since the air will vibrate in the same way as the diaphragm vibrates, a plot of the particles' displacements will appear much like the graph shown in Fig. 15.2. However, the two graphs would be displaced by one-fourth cycle from each other. Why?

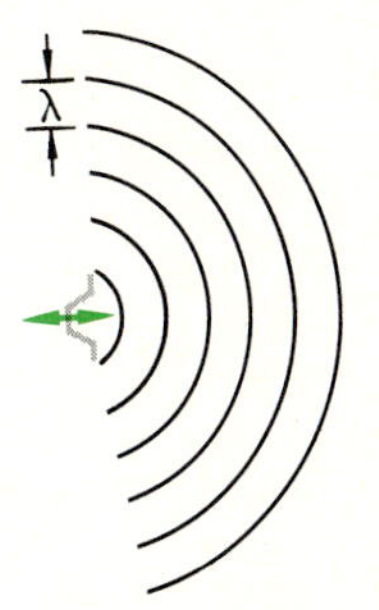

FIGURE 15.3
Crests of waves (wavefronts) spreading out from the loudspeaker are one wavelength apart.

We shall often draw sound waves propagating from a source in the manner pictured in Fig. 15.3. The solid lines represent a particular point on the waves as they spread out and travel through the air. Often these lines (called **wavefronts**) are drawn at the crests of the waves. It is seen that *the distance between* these *wavefronts is the distance from crest to crest on the wave, that is,* λ. Even though most loudspeakers focus the sound waves somewhat in a particular direction, the waves do spread out in many directions from the source, much as is indicated in Fig. 15.3. The waves travel through the air at the speed of sound.

15.3 THE SPEED OF SOUND

Sound waves are compressional waves in matter. We saw in the last chapter that the speed of such waves in fluids is given by Eq. (14.4),

$$v = \sqrt{\frac{E}{d}}$$

You will recall that d is the density of the fluid and the bulk modulus E is given by

$$E = \frac{\Delta P}{\Delta V/V}$$

where ΔV is the change in the volume V caused by a pressure change ΔP. Notice that fluids which are difficult to compress (large E) will have large v. *Sound travels fastest in materials which are least compressible and which have low densities*. We have already listed the speed of sound for various materials in Table 14.1.

Speed of Sound in Air

Of particular practical importance is *the speed of sound in air. At 0°C, its speed is 331 m/s that is, 1086 ft/s.* For temperatures near those normally encountered for air, *the speed of sound increases by about 0.60 m/s for each 1°C rise in temperature.* Also, to a good approximation, the speed of sound in ideal gases is independent of the pressure of the gas.

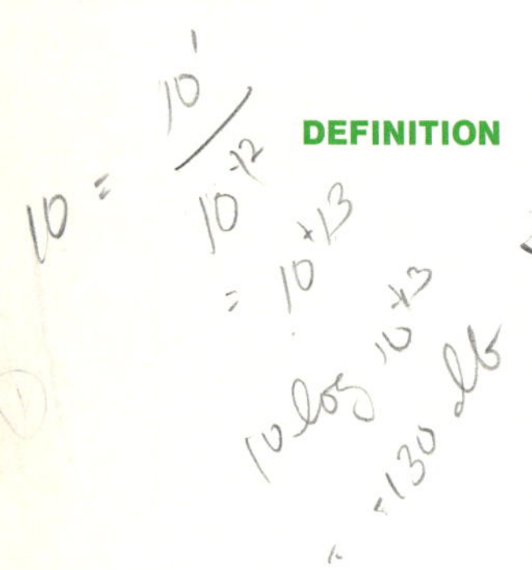

FIGURE 15.4
Sound intensity is measured as the amount of energy flowing through a unit area per second. The area must be perpendicular to the direction of propagation, as shown.

15.4 INTENSITY AND LOUDNESS OF SOUNDS

We saw in the last chapter that the vibrator which sends a wave down a string also sends energy with the wave. Indeed, all waves carry energy along with them. Sound waves are no exception. For example, the loudspeaker in Fig. 15.3 sends out sound-wave energy. The energy flows in the direction of propagation of the wave.

Suppose a sound wave is traveling in the propagation direction shown in Fig. 15.4. We define the intensity of the wave in terms of the energy carried by the wave. To be precise, we erect a unit area perpendicular to the direction of propagation as shown. We then define the intensity of the wave to be the energy carried per second through this unit area by the wave. Since intensity is energy per second, **sound intensity** *is the power passing through a unit area erected perpendicular to the direction of propagation of the wave. The units of sound intensity are typically watts per square meter.* Representative sound intensities are listed in Table 15.1. Notice what a wide range of sound intensities the ear can hear. The ear is a truely remarkable measuring device.

DEFINITION

To better express the way the ear responds to sound, use is frequently made of a sound-intensity scale, the decibel scale, based upon powers of 10, shown in Table 15.2.

Notice that *the lower limit of the sound intensity just audible to the average ear (10^{-12} W/m²) is taken as zero on the decibel scale. Each time the intensity is made 10 times larger, the intensity level in decibels increases by 10 units. It is*

TABLE 15.1
APPROXIMATE SOUND INTENSITIES

TYPE OF SOUND	INTENSITY, W/m²	INTENSITY LEVEL, dB
Pain-producing	1	120
Jackhammer or riveter*	10^{-2}	100
Busy street traffic*	10^{-5}	70
Ordinary conversation*	10^{-6}	60
Average whisper*	10^{-10}	20
Rustle of leaves*	10^{-11}	10
Barely audible sound	10^{-12}	0

*For a person near the source of the sound.

TABLE 15.2
THE DECIBEL* SCALE

INTENSITY, W/m²	INTENSITY LEVEL, dB
10^{-12}	0
10^{-11}	10
10^{-10}	20
10^{-9}	30
⋮	⋮
10^{-1}	110
1	120
10	130

*1 B (bel) = 10 dB and is named after Alexander Graham Bell, the inventor of the telephone.

found that the ear judges sound levels according to the decibel scale. For example, an average person would judge a sound whose intensity is 10^{-10} W/m^2 to be twice as loud as one of intensity 10^{-11} W/m^2.

We can state the definition of the decibel scale in mathematical form as follows:

Decibel Scale

$$\text{Sound level in dB} = 10 \log \frac{I}{I_0} \tag{15.1}$$

where I is the sound intensity in watts per square meter and $I_0 = 10^{-12}$ W/m^2 is an arbitrarily selected reference level. As we see, *the ear responds proportionally to the logarithm of the sound intensity.* For this reason, *it is often referred to as a logarithmic detection device.* It is interesting to note that the eye is also a logarithmic detector, but for light waves rather than sound waves of course.

Illustration 15.1 Find the sound level in decibels of a sound wave which has an intensity of 10^{-5} W/m^2.

Reasoning From Eq. (15.1), we have, after replacing I by 10^{-5} W/m^2 and I_0 by 10^{-12} W/m^2,

$$\text{Sound level in dB} = 10 \log \frac{10^{-5}}{10^{-12}} = 10 \log 10^7 = (10)(7) = 70 \text{ dB}$$

15.5 FREQUENCY RESPONSE OF THE EAR

People vary in their ability to hear sounds. We all know persons whose hearing has been in some way impaired. The sensitivity of their ears had decreased considerably below that of a normal person. However, most people agree fairly well upon the intensity of a sound which is just audible and also upon how loud a sound must be before it causes pain. We can therefore set up average limits of audibility for the normal human ear. The lower limit is the intensity of a just audible sound, while the upper limit is a sound so intense that it hurts the ear.

Frequencies of Audible Sound

Most people cannot hear compressional waves in the air which have frequencies higher than about 20,000 Hz. Waves of higher frequency than this are called **ultrasonic waves**—meaning "beyond" or "above" sound in the sense of frequency. Similarly, we are not able to hear sounds below a certain frequency, about 20 Hz. *The ear is most sensitive in the frequency range near about 3000 Hz.* At frequencies other than this, the sound must be made more intense before it is audible. This variation of sensitivity of the ear with frequency is shown in Fig. 15.5.

It is seen in this figure that *the normal ear is quite insensitive to frequencies above about 15,000 Hz and below about 30 Hz.* Sounds of frequency

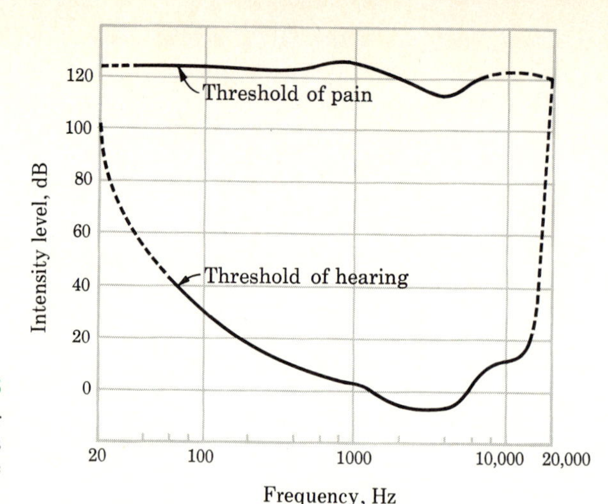

FIGURE 15.5
The normal ear can hear sounds which have intensities above the lower curve in the figure.

outside these limits can be heard only if the sound intensity is very high. Notice that the pain-inducing intensity does not vary much with frequency.

There are some people who have a rather strange impairment of hearing, often unknown to themselves; they are unable to hear sound frequencies above perhaps 6000 Hz. Since most of the sounds we hear consist, partly at least, of frequencies below this, these people are still able to hear sounds which are audible to other people. However, the *quality* of the sounds they hear will be quite unlike those heard by a normal person. Quality and pitch of sound are complex, rather subjective properties of sounds. They are discussed in the next section.

15.6 PITCH AND QUALITY OF SOUND

If a high-quality loudspeaker is driven by an electrical oscillator which generates a sine-wave voltage, the sound wave given off by the loudspeaker will be an almost pure sine wave having the frequency being generated by the oscillator. Anyone who is not tone-deaf will be able to compare the pitch of this sound with another sound. If the frequency of the oscillator driving the loudspeaker diaphragm is increased, the listener will agree at once that the new sound has a higher pitch than the first tone. In such cases pitch and frequency are nearly synonymous.

The above example is not very common, however. If one plucks or bows a violin string, for example, the sound wave given off will not be a pure sine wave. This is readily apparent to anyone who has compared the tone obtained from a violin by an expert with that obtained by a beginner. In one case the tone will be full and melodious, whereas the beginner may obtain rather rasping sounds on the same string. We say that the quality of the tone is different in the two cases.

As we saw in Chap. 14, a string may resonate in more than one way. Typical, simple vibration patterns are shown in Fig. 15.6. These modes of

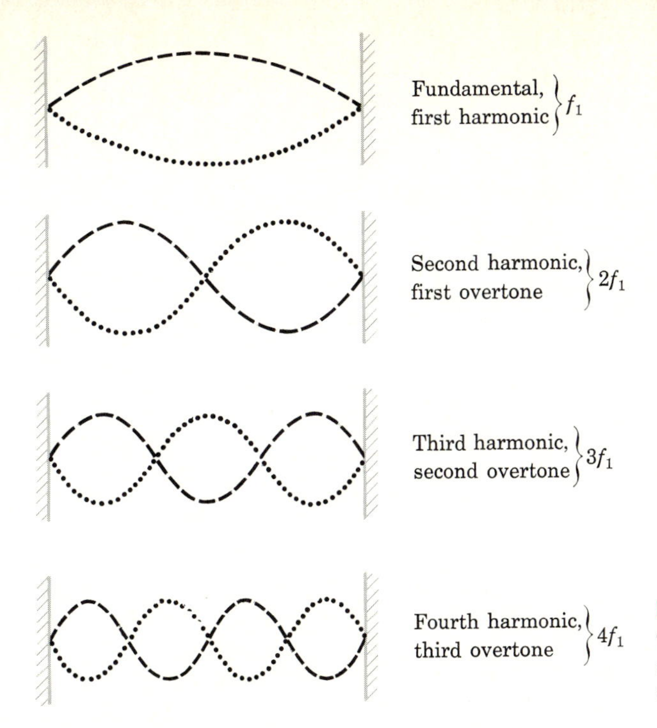

FIGURE 15.6
The four simplest modes of motion for standing waves in a string.

vibration are named as shown in the figure. Since the wavelengths in the cases shown are in ratio $1:\frac{1}{2}:\frac{1}{3}:\frac{1}{4}$ and since $f = v/\lambda$, the vibration frequencies are in the ratio 1:2:3:4, as indicated on the figure.

It is very difficult, however, to cause a string to vibrate exactly as shown in any single pattern of Fig. 15.6. Instead, *if the string is bowed near one end,* as is usually the case, *not only the first harmonic vibration occurs, but several of the other harmonics occur at the same time. The string actually is vibrating in several ways at once.* To find the resulting vibration, one needs to add together the waves for the various harmonic vibrations excited. Proper proportions of each component wave are taken, of course, since some of the component vibrations will be more intense than others.

A typical example for a vibrating violin string is shown in Fig. 15.7. The amplitudes of vibration of the various harmonics are indicated by the length of the vertical bars. In this case, all but the first two harmonics are relatively weak. But, clearly, the tone heard by the ear will be different from either the first or the second harmonic alone.

Also shown in Fig. 15.7 are similar diagrams for the sounds of various other instruments. The piano string shows many more harmonics than the violin string. This is probably the result of the more complex way in which the piano string is set into vibration. The violinist pulls a bow slowly and evenly across the string, whereas the piano string is excited by a hammer-blow.

The quality of a sound or tone is dependent upon the number and type of harmonics occurring in the sound. If all sounds were pure sine waves, much of the variety of sound would be lost. The tone of all human voices would be the

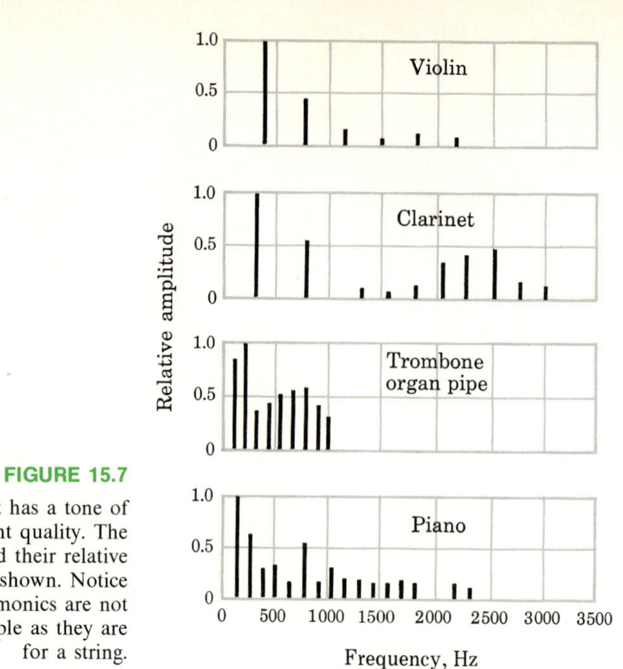

FIGURE 15.7

Each instrument has a tone of different quality. The frequencies and their relative amplitudes are shown. Notice that the harmonics are not always as simple as they are for a string.

same, and a voice could be recognized only by a characteristic frequency or inflection. Much of the beauty of music would be lost if the qualities of all sounds were the same.

In a complex sound such as that shown for the piano or even the clarinet, *the pitch of the sound is not always easily defined. It can no longer be taken as identical to the frequency of the sound, since the sound contains several nearly equal amplitude waves of various frequencies.* In such cases, it is difficult to match tones accurately. For this reason it is not at all uncommon for an inexperienced singer to sing a tone which is twice the frequency of the fundamental tone of a violin and still be unaware that the tones are not the same. Moreover, many listeners would not recognize that the singer was not sounding the desired tone.

15.7 INTERFERENCE OF SOUND WAVES

Suppose that we have a pipe system such as that shown in Fig. 15.8. A pure sine wave is sent in the pipe at the left by a loudspeaker. The sound splits, half the sound intensity going up through section A, while the remaining half goes through the lower section. Each pipe carries half the sound, and this sound is a wave motion in the air, a series of compressions and rarefactions.

Eventually the two waves are reunited at the outlet on the right, at D, where a sound detector such as the ear or a microphone is placed. It is observed that the sound emitted at D can be made loud or very faint, depending upon the position of the sliding pipe EAF. Moreover, as the pipe at A is slowly pulled upward, the sound intensity at D becomes alternately

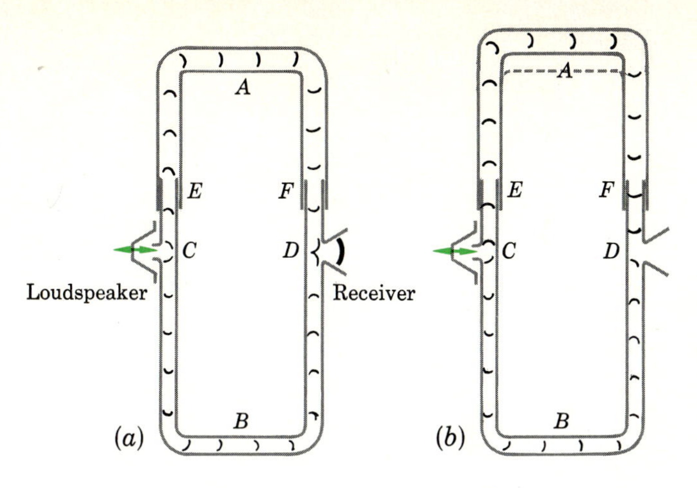

FIGURE 15.8
A sound wave from the loudspeaker is split into two parts. When they are reunited at D, either a loud or a weak sound results, depending upon the path lengths traveled by the two parts.

large and small. We shall now investigate the reasons for this interference phenomenon.

When a compression of the air is caused by a rightward movement of the loudspeaker diaphragm, a region of high pressure starts into the pipe at C. The region of high pressure, the compression, causes compressions to move in both pipes, toward A and toward B. We shall say that the original compression in the entrance pipe at C has split into two equal parts and that one part went up toward A, the other down toward B. Since the compression propagates through the pipe with the speed of sound, the compressions will reach point D simultaneously *provided that the pipe length through A, L_A, is the same as through B, L_B*. They will then reunite at D, giving the original compression, and this will then exit from the pipe system at D.

Of course, the loudspeaker is sending out a pure sound, a sinusoidal wave, consisting of alternate compressions and rarefactions. However, if $L_A = L_B$, the two portions of the original compression will always meet at D. The same is true for the two halves of the rarefactions. Hence, compressions and rarefactions identical to the originals will exist at point D, and the sound will be loud. This fact is represented in Fig. 15.9*a* where the pressure in the A wave at point D is plotted as a function of time and where this is added to the pressure in the B wave at point D to obtain the total pressure at D as a function of time.

On the other hand, suppose that the length along section A of the pipe is somewhat longer than along section B. In that case the half of the compression traveling through A will arrive at point D somewhat later than the half of the compression which traveled through B, since it had farther to travel. If L_A is half a wavelength longer than L_B, the A wave will be one-half wavelength behind the B wave when they meet at D. Hence a compression in the A wave will meet the rarefaction of the B wave at point D. Since the compression will tend to increase the pressure at D and the rarefaction will tend to decrease the pressure at D, the waves will cancel and the pressure will actually remain unchanged. As a result, no sound will be emitted at D. This is illustrated in Fig. 15.9*b*.

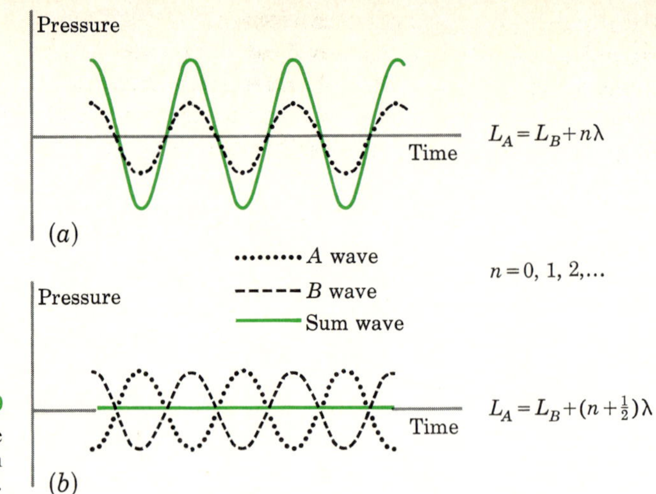

FIGURE 15.9
Waves A and B can reinforce or cancel, depending upon their relative phase.

When one wave is half a wavelength behind another wave, we say that the waves are a half wavelength, or 180°, out of phase with each other. This terminology, 180°, is based upon the fact that a vibration is in many ways akin to a particle moving in a circle, as was pointed out in Chap. 13. One full vibration is taken as equivalent to one full rotation around the circle, i.e., to 360°. Hence, *a retardation of* $\frac{1}{2}\lambda$ *is equivalent to 180°. Similarly, 90° is equivalent to* $\frac{1}{4}\lambda$, *and so on.*

If L_A is now increased still further by pulling the pipe at A upward, the A wave will be held back still more. When L_A is one whole wavelength longer than L_B, a crest of the A wave will again reach D at the same time as a crest of the B wave. Although these crests did not start together at point C, the A crest occurring one compression prior to the B crest, this is of no concern when they reach point D. The wave will appear once again as in Fig. 15.9*a*, and so the original intensity sound will be produced at D.

Conditions for Destructive and Constructive Interference

Similarly, if L_A is increased until its length is $1\frac{1}{2}$ wavelengths longer than L_B, the situation in Fig. 15.9*b* will arise once again. No sound will be heard at D. Moreover, it is clear that no sound will be heard at D whenever $L_A = L_B + (n + \frac{1}{2})\lambda$, where n can be any integer, including zero. *When two waves exactly cancel each other* in this way, *we say that there is complete* **destructive interference.** Clearly, reinforcement, i.e., constructive interference, of the waves occurs whenever $L_A = L_B + n\lambda$, where n is any integer, including zero. Of course, if L_B is greater than L_A, these interferences will also occur.

It is not necessary to have a pipe system such as this to obtain interference. One need only obtain two waves *which are exactly the same in their frequency and shape.* If these waves are combined after traveling different distances, they will interfere with each other. The following Illustration furnishes another example of interference. We shall see later in this text that the interference of light waves is of very great importance.

Illustration 15.2 Two identical sound sources send identical waves toward each other, as shown in Fig. 15.10. They both send out wave crests at the same time, and the sound waves have $\lambda = 70$ cm. Loud sound is heard at the midpoint, point P. But as one proceeds away from P, the sound intensity decreases to near zero at point Q. As one proceeds still further beyond Q, the intensity again increases. How far is point Q from point P?

Reasoning It is easy to see why the sound is large at P. This point is equidistant from the two sources, and so crests from both A and B will arrive there at the same time. The waves from the two sources will therefore reinforce at P.

At Q, the waves from the two sources cancel. This will be the case if the distance $\overline{PQ}$ is $\lambda/4$. In that case, the wave from A must travel $\lambda/4$ beyond P while that from B must travel $\lambda/4$ less than to P. As a result, at Q the wave from B will be $\lambda/4 + \lambda/4 = \frac{1}{2}\lambda$ ahead of the wave from A. At Q, a crest from B will meet a trough from A. Therefore the two waves cancel at Q, and so very little sound is heard.

We conclude that Q is $\lambda/4$ away from P. But λ was given to be 70 cm. Therefore we see that $\overline{PQ} = 17.5$ cm.

15.8 BEATS

When people tune a string on a piano, they do not merely listen to see if the tone of the string is the same as that of the standard tuning fork used for comparison. Instead, they use a much more precise way of judging the accuracy to which the string is adjusted. They listen to **beats** between the sounds of the two vibrating objects. This is a very sensitive method for obtaining agreement of frequency and is widely used for that purpose. We shall now explain the phenomenon of beats between the sounds given off by two nearly tuned vibrating bodies.

Suppose that two vibrating bodies, A and B, vibrate with slightly different frequencies. For example, they might be two loudspeakers as shown in Fig. 15.11. Consider what happens if source A is vibrating with frequency 1000 cps and source B with $f = 999$ cps. At time $t = 0$ we shall say that the loudspeakers are in phase; i.e., both are sending out a compression as shown. If the ear is equidistant from the loudspeakers, the compressions will arrive at the ear together and a large compression will be heard.

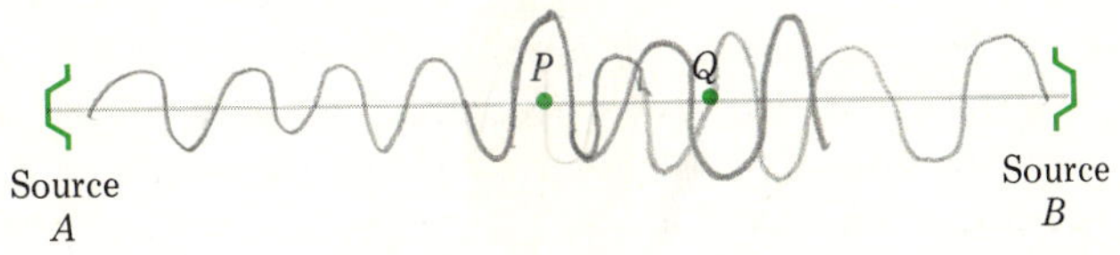

FIGURE 15.10

A loud sound is heard at P, while a very weak sound is heard at Q. Explain how this can be.

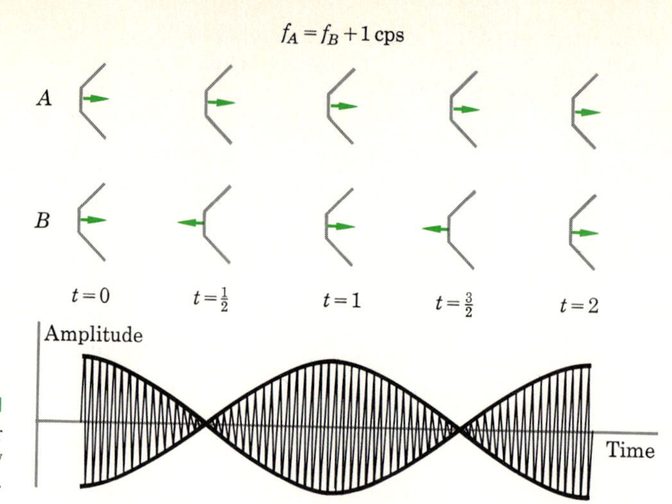

FIGURE 15.11
Beats result when two similar sources vibrate with slightly different frequencies.

As time goes on, loudspeaker B, vibrating at a slightly slower frequency than A, will begin to fall behind. After $\frac{1}{2}$ s, loudspeaker A will have vibrated 500.00 times and will just be sending out a compression, as shown in Fig. 15.11. On the other hand, loudspeaker B will have vibrated only 499.50 times and will be exactly half a cycle behind loudspeaker A. It will be sending out a rarefaction, as is shown. The compression from A will now reach the ear at the same time as the rarefaction from B, and they will exactly cancel each other. Hence, at this instant, no sound will be heard.

As time continues, loudspeaker B will fall still further behind A. After 1 s, B will have vibrated exactly 999 times, while A has vibrated exactly 1000 times. Source B has now fallen exactly one cycle behind source A. Hence they will both once again be sending out compressions together, and a loud sound will be heard.

This process will continue, as shown in Fig. 15.11. At times of 0, 1, 2, 3, etc., s the sources will be in phase, and a loud sound will be heard. However, at times of $\frac{1}{2}$, $1\frac{1}{2}$, $2\frac{1}{2}$, etc., s nothing will be heard, since the sources are 180° out of phase. Hence the ear will hear a series of sound pulses, or beats, one beat each second, as shown by the graph of sound intensity given in Fig. 15.11.

Notice that the frequencies differ by exactly 1 cps in this case and exactly one beat is heard each second. You should carry through the above analysis for the case in which the sources differ by, say, 3 cps. In this case there will be 3 beats per second. In fact, *the number of beats per second equals the difference in frequencies of the two sources*. The piano tuner tries to adjust the tension in a piano string until the beats between the sound from the string and the sound from the tuning fork have become extremely far apart in time.

This phenomenon of beats occurs in any type of vibration which is a combination of vibrations from two sources. It may therefore be used to compare frequencies of vibrations other than sound. As in the case illustrated here, *it is a sensitive means for comparing frequencies.*

15.9 ORGAN PIPES AND VIBRATING AIR COLUMNS

If you hold a vibrating tuning fork over the open end of a glass tube partly filled with water, the sound of the tuning fork can be greatly amplified under certain conditions. While the fork is held as shown in Fig. 15.12, the reservoir R is raised so that the water level will rise in the glass tube. At a certain height of the water, the air in the tube will resonate loudly to the sound being sent into it by the tuning fork. In fact, there are usually several heights at which the tube will resonate.

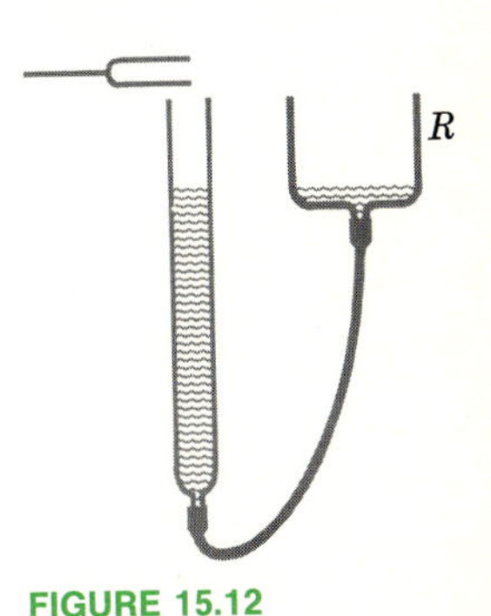

FIGURE 15.12
When the water is at just the right height in the tube, resonance will occur.

The situation here is much like the case of the vibrating string. A wave is sent down through the air in the tube (or in the string); it is reflected at the end and is once again reflected when it reaches the position of the vibrating source. If the tube (or string) is just the proper length, the reflected wave will be reinforced by the vibrating source as it travels down the tube a second time. Hence, the vibrating source builds up a large motion much in the way a swing can be pushed very high with a series of small pushes. This whole general idea of resonance was discussed fully in Chap. 13. We can apply the results found there to the case of a vibrating air column.

The tube of Fig. 15.12 will have a node at the closed end and an antinode near the open end. This follows because the air molecules cannot move at the closed end, for the water at that end will not allow them to move downward. At the open end, the molecules can easily move out into the open space, and so at that point there will be a maximum of motion, i.e., an antinode.*

We see, then, the tube will resonate only to waves which will fit into the tube with a node at one end and an antinode at the other. Several of these waves are shown schematically in Fig. 15.13. Remember that the distance between two nodes or two antinodes is $\lambda/2$. Hence, the distance from a node to an antinode is $\lambda/4$. If we call the length of the pipe L, the length in Fig. 15.13*a* is from a node to an antinode, or $L = \lambda/4$. In part *b*, the tube is three node to antinode lengths long; hence $L = 3(\lambda/4)$, and so on.

FIGURE 15.13
Simple modes of motion for a resonating pipe closed at one end.

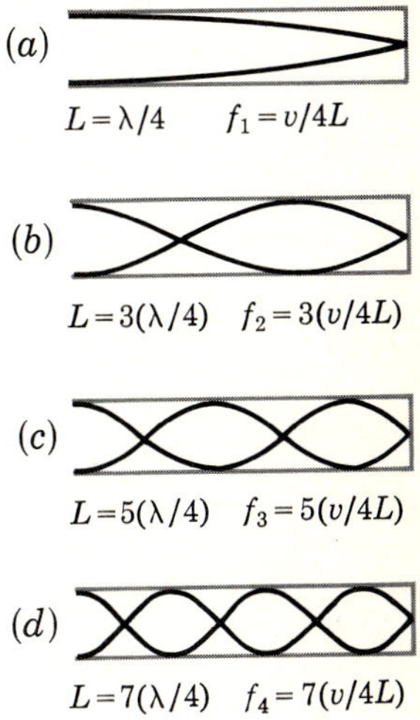

(*a*) $L=\lambda/4 \quad f_1=v/4L$

(*b*) $L=3(\lambda/4) \quad f_2=3(v/4L)$

(*c*) $L=5(\lambda/4) \quad f_3=5(v/4L)$

(*d*) $L=7(\lambda/4) \quad f_4=7(v/4L)$

The frequencies for the resonances shown in Fig. 15.13 can be found from the fact that $f = v/\lambda$. Upon using the values found for λ in terms of L as shown in Fig. 15.13, the resonance frequencies shown are easily computed. Notice that the **first overtone** f_2 is just $3f_1$. The **second overtone** f_3 is $5f_1$, and so on. The customary terminology is to call the frequency $2f_1$ the second harmonic, $3f_1$ the third harmonic, etc. In this case, the tube resonates only to the odd harmonic frequencies. Since the frequency of a given tuning fork is usually known, resonances in a tube such as the one shown in Fig. 15.12 can be used to measure v, the velocity of sound.

Organ pipes are much like the tube mentioned above. However, one excites the tube to vibrate by blowing on a specially shaped cavity at the end of the tube. The operation of the whistle end of the tube is rather complex, and we shall merely say that the blowing process excites many types of

* The antinode is not precisely at the end of the pipe. However, for pipes with radii much smaller than λ, this complication can usually be ignored.

vibrations. The tube resonates to any of the proper vibrations which may be generated at the whistle end.

Closed organ pipes resonate very much like the tube illustrated in Fig. 15.13. An antinode exists near the blowing end, while the closed end is a node. The resonance frequencies are as illustrated in Fig. 15.13. It is possible to obtain two tones at once from an organ pipe by exciting two of the resonances at once. In such a case the quality of the sound changes markedly, but the sound is still heard as a single tone.

Open organ pipes, i.e., the end opposite the blowing end is open, must have antinodes near both ends. The resonant modes and frequencies are obtained in a way similar to that used in obtaining Fig. 15.13. For this case, the situation is as shown in Fig. 15.14. It is seen in this case that all the harmonics are present. The first overtone is $2f_1$, the second is $3f_1$, and so on.

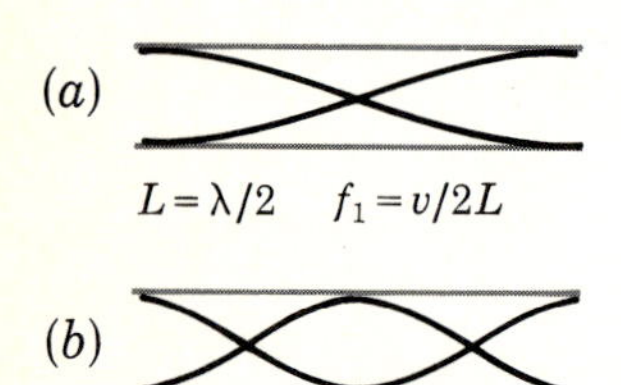

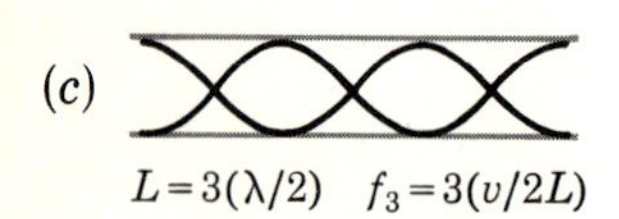

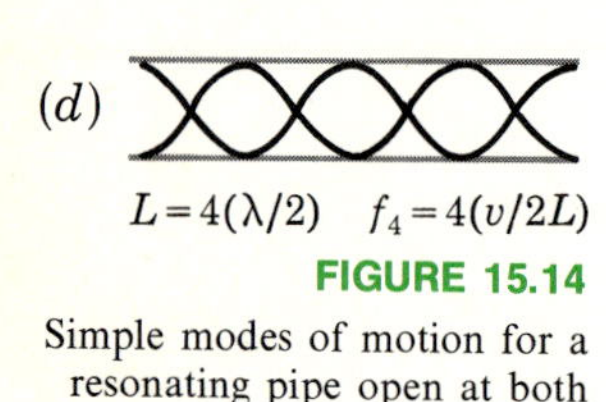

FIGURE 15.14
Simple modes of motion for a resonating pipe open at both ends.

15.10 MUSICAL SCALES AND HARMONIC COMBINATIONS

The ratio of the frequencies of two tones is called a **musical interval.** It is customary to take the ratio of the larger frequency to the smaller so that the interval is a number larger than unity. From this definition, the interval from 800 to 600 cps, namely, $\frac{4}{3}$, is identical to the interval from 1200 to 900 cps, also $\frac{4}{3}$.

Experience shows that most people will say that those tones harmonize which have intervals consisting of the ratio of small whole numbers. For example, tones of 1200 and 600 cps have an interval of $\frac{2}{1}$. This interval is called an **octave,** and these two tones harmonize very well together. On the other hand, tones of 330 and 150 have an interval of $\frac{11}{5}$ and do not harmonize nearly so well as the $\frac{2}{1}$ combination given above.

The natural major musical scale is made up according to the following set of intervals:

$$\text{do}\ \tfrac{9}{8}\ \text{re}\ \tfrac{10}{9}\ \text{mi}\ \tfrac{16}{15}\ \text{fa}\ \tfrac{9}{8}\ \text{sol}\ \tfrac{10}{9}\ \text{la}\ \tfrac{9}{8}\ \text{ti}\ \tfrac{16}{15}\ \text{do}'$$

It is seen that, if do and re are sounded together, the result would be discordant, since the interval is $\frac{9}{8}$, the ratio of two quite large numbers. However, the combinations of others of these tones are much more harmonious. This can be seen by assigning definite frequencies to these tones.

If one calls do 264 cps, the rest of the scale is

do	re	mi	fa	sol	la	ti	do′
264	297	330	352	396	440	495	528

Clearly, the ratio of $\frac{330}{264}$ is $\frac{5}{4}$, and thus do and mi would harmonize fairly well. However, do and fa, being in the ratio of $\frac{4}{3}$, would be better in this respect, and do and sol, being in the ratio of $\frac{3}{2}$, would appear even more melodious. You can easily determine which of these tones will harmonize well together. It is instructive to check your conclusions by using a piano. The above frequencies are chosen such that do is middle C and the others are, in order, D, E, F, G, A, B, and C′.

In spite of these generalizations about pleasing and nonpleasing musical sounds, it must be admitted that a person's musical attitudes depend upon cultural background. The history of music provides us with an interesting example of a changing pattern of acceptability. We need go no further back than the present century to see that musical tastes change. Many of today's accepted composers were first ridiculed because their music was considered hopelessly discordant when played for the first time. In your own experience you are well aware that the older generation often reacts with disgust or dismay to sounds which you consider musical. This, too, is a reflection of the fact that musical taste is undergoing continual change.

Another aspect of musical taste is seen when the music of different cultures is compared. For example, most Americans react with irritation to Arabic music except as a passing novelty. The tone combinations are quite unlike those we use widely, and their music seems both discordant and monotonous to the foreigner. Only after considerable exposure do we begin to appreciate its beauty. It is clear that the psychological aspects of music are far more complicated than the well-understood sound waves which carry the music to our ears.

15.11 DOPPLER EFFECT

If one is standing near the side of the road and a fast-moving car goes by with its horn blowing, a curious phenomenon is observed. Although the pitch of the sound from the horn is steady until the car is nearly even with the point of observation, at the time the car goes by, the pitch suddenly lowers and remains lower as the car recedes down the road. This phenomenon, together with other physical situations in which it occurs, is referred to as the **Doppler effect.**

If one is to believe one's ears, the Doppler effect means that the frequency of the sound striking the eardrum is higher when the car is approaching than when it is receding. In other words, more wave crests hit the ear per second when the car is moving toward us than when it is moving away from us. We shall now show that this is reasonable and that an equation for the effect is readily found.

The physical situation is shown in Fig. 15.15. In the upper part of the

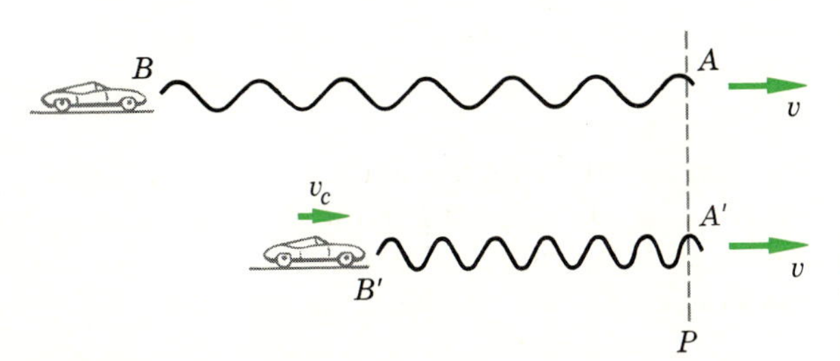

FIGURE 15.15

The Doppler effect results if a sound source is moving.

figure is a stationary car, with the sound wave from its blowing horn traveling away from it. Suppose that the car in the lower portion of the figure is traveling with speed v_c and is blowing an identical horn. Just as it passes the upper car, the wave crest, indicated by A, is pushed out by the upper horn and A' by the lower horn. They both travel to the right with speed v, the velocity of sound, and hence will stay together as shown.

However, because of the motion of the lower car in the figure, the waves are compressed into a smaller region, as indicated. Hence more wave crests will pass by point P each second than in the upper case. As a result, the frequency of sound from the lower car's horn will be larger at P than it would be if the car were standing still. Let us now compute this change in frequency.

Because of the motion of the car, the wavelength of its sound wave is decreased by an amount $v_c\tau$, where v_c is the speed of the car and τ is the period of vibration of its horn. This follows from the fact that the car moves a distance $v_c\tau$ during the time τ between the sending out of wave crests. As a result, the wave crests are closer together by this amount, $v_c\tau$. Calling λ the wavelength of sound from the stationary car, the wavelength of sound for the moving car λ' will be

$$\lambda' = \lambda - v_c\tau$$

This can be simplified by making the substitutions that $\lambda = v/f$ and $\tau = 1/f$, where v is the speed of sound and f is the frequency of vibration of the horn. Also, $\lambda' = v/f'$, where f' is the frequency of sound one hears at P from the moving car. One then rewrites the above equation as

$$\frac{v}{f'} = \frac{v}{f} - \frac{v_c}{f}$$

from which

Doppler Frequency Shift

$$\frac{f'}{f} = \frac{1}{1 - v_c/v} \tag{15.2}$$

where v_c is the speed of the sound source. This can be recast into a form which is useful if $\Delta f = f' - f$ is small. Then,

$$\frac{\Delta f}{f'} = \frac{v_c}{v} \approx \frac{\Delta f}{f}$$

To this approximation, v_c can be considered the velocity of approach of source and observer. It applies to both moving observers and sources.

We see, as we expected, that the larger the ratio v_c/v, the larger the ratio f'/f. When the car is standing still, $v_c = 0$ and so $f' = f$. If v_c is larger than zero, the sound's frequency f' will be higher than the frequency of the horn f. As a typical example, suppose the car is moving at 80 ft/s (55 mi/h). Then, since the speed of sound is about 1000 ft/s, one has $v_c/v = 0.080$ and the ratio of the sound frequency to horn frequency is

$$\frac{f'}{f} = \frac{1}{1 - 0.080} = 1.08$$

If the frequency of vibration of the car horn is 500 Hz, the sound of the moving horn will have a frequency of 540 Hz.

Equation (15.2) also applies to sound from a receding car provided one notices that here v_c will be negative. The receding car stretches out rather than compresses the sound wave, and so the ratio of f'/f will be less than unity. If you follow through the steps used in finding Eq. (15.2) for the receding car, you will see that v_c will be replaced by $-v_c$ in all the equations. Therefore Eq. (15.2) will apply to both cases if v_c is taken as positive for an approaching sound source and negative for one which is receding.

15.12 SHOCK WAVES AND SONIC BOOMS

An interesting situation arises if the speed of the sound source approaches or equals the speed of sound. Then, from Eq. (15.2), we see that the sound frequency f' approaches infinity. This simply means that a nearly infinite number of wave crests reaches the listener in a very short time. We can easily understand this by referring once again to Fig. 15.15.

Suppose the moving car has a speed equal to the speed of sound. Then point B' will coincide with point A'. As a result, all the wave crests between A' and B' will lie upon one another. They, together with the sound source itself, will all pass point P at the same time. All the energy of the sound waves would be compressed into a very small region in front of the sound source. Consequently, this very concentrated region of sound energy, a shock wave, would cause an extremely large sound as it passes point P. Basically this is the origin of the sonic boom which accompanies supersonic aircraft.

When an airplane moves through the air with a speed near that of sound, the noise and air disturbances originating from the plane are built up into a shock wave, which, as we have just seen, is simply a region of very dense sound energy. The exact shape of the shock wave depends upon the speed of the airplane. In general, it covers the surface of a cone, as illustrated in Fig. 15.16. The angle θ of the cone depends upon the ratio of the speed of the plane v_p to the speed of sound v in the following way:

$$\sin \theta = \frac{v}{v_p}$$

As the plane's speed becomes larger with respect to v, the angle of the cone decreases. *The ratio v_p/v is often called the* **Mach number.** In this terminology, a plane traveling at Mach 2 is moving at twice the speed of sound.

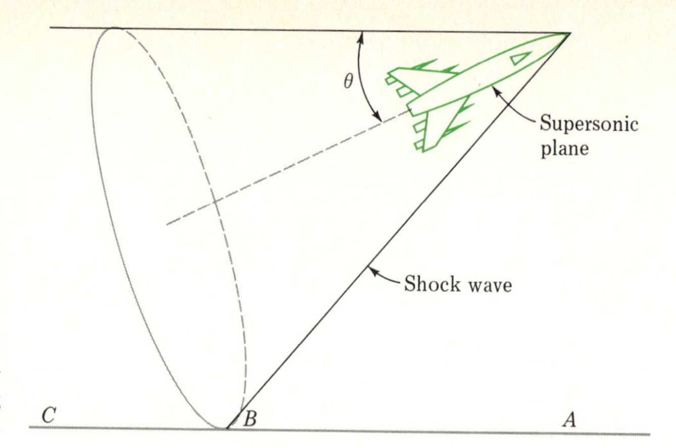

FIGURE 15.16
The sonic boom has already hit point C and is moving through point B toward A.

Since the shock wave is a region of very concentrated sound energy, it can cause severe damage when it strikes something. *The familiar sonic boom is the result of the conical shock-wave surface passing over the earth.* For example, in Fig. 15.16, the sonic boom will soon strike point A; it is currently striking point B; it has already hit point C. Depending upon the intensity of the wave, its effects will be more or less damaging. A flash photograph of the shock wave generated by a projectile in air is shown in Fig. 15.17. Can you show that the speed of this projectile is about 2.1 times the speed of sound in air?

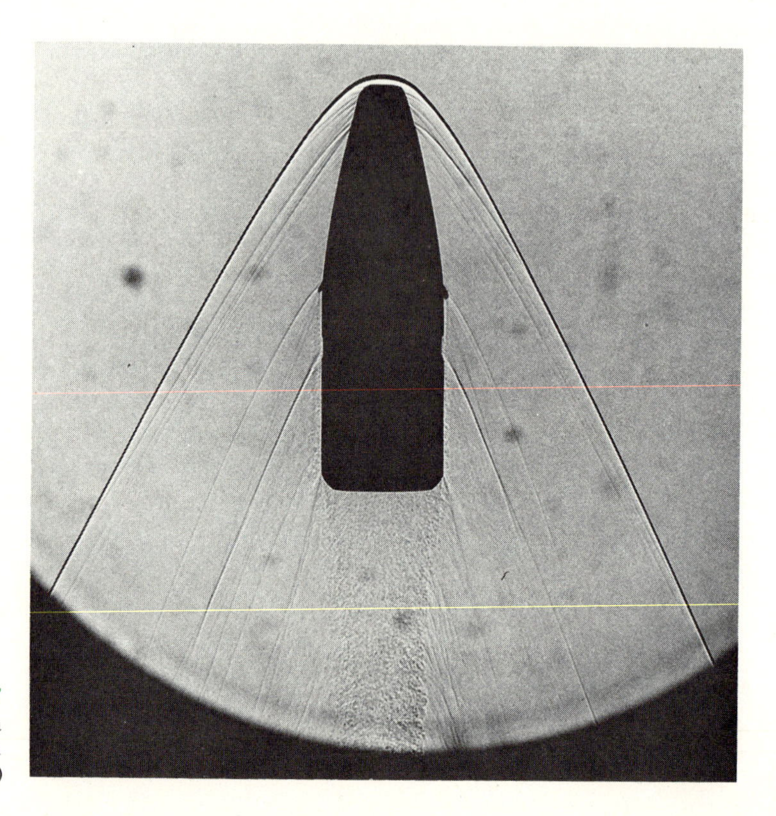

FIGURE 15.17
Shock waves generated by a projectile shooting through air. (*Hewlett-Packard Co.*)

SUMMARY

Sound waves are compressional waves in a fluid or solid. When they strike the eardrum, they give rise to the sensation of sound. The sound source sends out compressions and rarefactions through the material near it. In air at 0°C, sound travels with a speed of 331 m/s (1086 ft/s). For each 1°C rise in temperature, the speed in air increases by about 0.60 m/s.

The intensity of a wave is the power carried by the wave through a unit area erected perpendicular to the wave's direction of propagation. Typical audible sounds cover an intensity range from about 10^{-12} W/m² to about 1 W/m². Since the ear is a logarithmic detector, it is customary to measure sound intensities on a logarithmic scale. The sound level in decibels is related to I, the intensity in watts per square meter, by $10 \log (I/10^{-12})$.

Most persons can hear sounds with frequencies in the range 20 to 20,000 Hz. Sound waves with frequency higher than 20,000 Hz are called ultrasonic waves. The normal ear is most sensitive to sounds close to 3000 Hz.

Most sounds consist of a mixture of several waves of different frequencies. Depending upon the exact mixture of these waves, the sound will have different quality. It is for this reason that different musical instruments, for example, sound differently even though they are playing the same fundamental frequency.

Waves of the same frequency can be combined to cause interference effects. If the waves are in phase, they interfere constructively and give a loud sound. If they are $\frac{1}{2}\lambda$ (or 180°) out of phase, they cancel each other. When two waves of slightly different frequencies interfere with each other, they give rise to beats. The number of beats per second is equal to the frequency difference.

Sound waves can resonate in pipes and tubes. A motion node exists at a closed end while a motion antinode exists near an open end.

The frequency of sound reaching the ear may change because of the motion of the source and/or the motion of the observer. This is called the Doppler effect. If the source and observer are approaching each other, the frequency of sound is increased. If they are receding, the frequency is lowered.

Shock waves originate when sound sources travel faster than the speed of sound. They consist of very intense wave disturbances and carry large amounts of sound energy in a limited region. When the shock wave strikes an observer, it gives rise to what is called a sonic boom.

MINIMUM LEARNING GOALS

Upon completion of this chapter you should be able to do the following:

1. Explain what is meant by a sound wave and point out why sound cannot travel through vacuum.

2. Recall the speed of sound in air at 0°C and compute its value at temperatures near room temperature.

3. Define the intensity of sound. State why the decibel scale is advantageous and, given the sound intensity in watts per square meter, find the intensity level in decibels.

4. Sketch a graph similar to Fig. 15.5 and interpret it in words.

5. Explain the concept of quality of sound and point out why it is different from frequency.

6. Combine two waves of the same frequency and amplitude but of different phase so as to obtain destructive interference and/or constructive interference. Apply your understanding to solving simple problems like that of Illustration 15.2.

7. Use the phenomenon of beats to find the difference in frequency between two sound sources.

8. Find the resonance frequencies of sound in given pipes.

9. Explain what is meant by the Doppler effect and compute the frequency shift noticed for an approaching sound source.

10. Explain how a shock wave originates and why it gives rise to a sonic boom.

IMPORTANT TERMS AND PHRASES

You should be able to define or explain each of the following:

Sound wave
Wavefront

Sound intensity
Intensity level in decibels
The ear is a logarithmic device
Quality of sound
Ultrasonic waves
In phase and out of phase
Phase difference of 0 and 180°
Beats
Overtones versus harmonics
Musical interval
Doppler effect
Shock wave and sonic boom
Mach number

QUESTIONS AND GUESSTIMATES

1. Describe clearly why a bell ringing inside a vacuum chamber cannot be heard on the outside.
2. Would you expect a sound heard under water to have the same frequency as when heard in air if the sources vibrate identically? Explain.
3. When a deep-voiced man inhales hydrogen and then speaks, his voice sounds high-pitched. Why? (v_H = 1270 m/s.)
4. Suppose that a few pipes of a pipe organ were mounted close to a hot steam pipe. Would this affect the performance of the organ? Explain.
5. A siren can be made by drilling equally spaced holes on a circle concentric to the axis of a solid metal plate or disk. When the disk is rotated while a jet of air is blowing against it near the holes, a sirenlike tone is given off. Explain how this gives the sensation of sound to the ear, and state what factors influence the pitch and quality of the tone.
6. It has been claimed that a certain singer could shatter a wineglass by singing a particular note. Could this be true? Explain.
7. When a firecracker explodes in a large room such as a gymnasium, the sound persists for some time and then dies out. Explain what happens to the sound energy given off by the explosion.
8. The reverberation time of a room is the time taken for a sound, for example, the tone from an organ pipe, to die out after the sound source is shut off. Explain why the reverberation time is shorter when the windows are open. What other means can be used to control the reverberation time?
9. There is on the market a device which uses intense ultrasonic waves in water to wash dirt loose from cloth and other objects. Explain how this works, and list its advantages and disadvantages.
10. One can make a soft-drink bottle give off a sound by blowing with one's lips placed properly at the top of the bottle. Estimate the frequencies to which the bottle will resonate. Check your estimate by using tuning forks or a calibrated oscillator and loudspeaker. (E)
11. When making underwater investigations, it is common to use **sonar** to "see" underwater. Pulses of sound are sent out through the water, and the pulses reflected back to the sending point are observed. Explain how one can make inferences about the surroundings by such a method. In what ways is sonar similar to radar? Bats use a form of sonar for their flights through pitch-black caves. How does this work?
12. Suppose on some distant planet there exist humanoids whose hearing mechanisms are designed as follows. From the outside, their heads look like our own. However, a 1-cm-diameter hard-surfaced cylindrical hole passes through the head from ear to ear. At the midpoint of the channel a thin circular membrane acts like a drumhead separating the two halves of the channel. These beings experience the sensation of sound when this drumhead vibrates. What can you infer about their hearing abilities and the ways they will communicate orally with each other?

PROBLEMS

Unless otherwise stated, use 340 m/s for the speed of sound in air.
1. The sound of a lightning flash is heard 6.0 s after the flash. Assuming the light to travel much more swiftly than the sound, how far away was the lightning?
2. A mountain climber notices that it takes 4.0 s for her

voice to be echoed by a distant mountain face. How far away is the mountain face?

3. The speed of sound in water is 1500 m/s. From this fact, what is the bulk compressibility of water?

4. Bats locate objects in the dark by sending out ultrasonic pulses and noting how they are reflected. If the ultrasound has a frequency of 7×10^4 Hz, what is the wavelength sent out by the bat?

5. By what percent will the speed of sound in air change if the air temperature is raised from 0 to 16.5°C?

6. A certain loudspeaker has a circular opening with a diameter of 15 cm. Assume that the sound it emits is uniform and outward through this entire opening. If the sound intensity at the opening is 10^{-4} W/m², how much power is being radiated as sound by the loudspeaker?

7.* A beam of sound has an intensity of 2×10^{-5} W/m². What is the intensity level in decibels?

8.* About how many times more intense will the normal ear judge a sound of 10^{-6} W/m² to be than one of 10^{-9} W/m²?

9.* If the sound intensity is 10^{-4} W/m² at a distance of 2.0 m from a source, and if the intensity decreases inversely as the square of the distance from the source, (*a*) what is the intensity at 20.0 m from the source? (*b*) What is the ratio of the sound levels at the two places?

10. As shown in Fig. P15.1, a tiny loudspeaker is placed in an air-filled tube which is bent into a circle. If the circumference of the circle is 12.0 m, what are the lowest three frequencies of vibration of the loudspeaker for which intense sound will exist in the pipe?

11. The two loudspeakers shown in Fig. P15.2 send out identical sound waves of wavelength 34 cm. The speakers vibrate in phase. (*a*) Give at least three values of x for which loudness will be heard at point P. (*b*) Repeat for minimum sound heard at P.

12. Consider the harmonics shown for the violin string in Fig. 15.7. Assume that the fundamental vibration has a frequency of 380 Hz. From a consideration of the resonances of a string held rigidly at its two ends, what should be its next five resonance frequencies? Are these about the same as those shown in the figure?

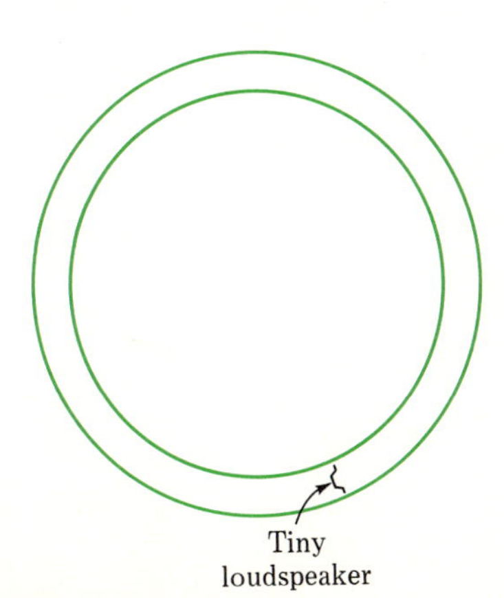

FIGURE P15.1

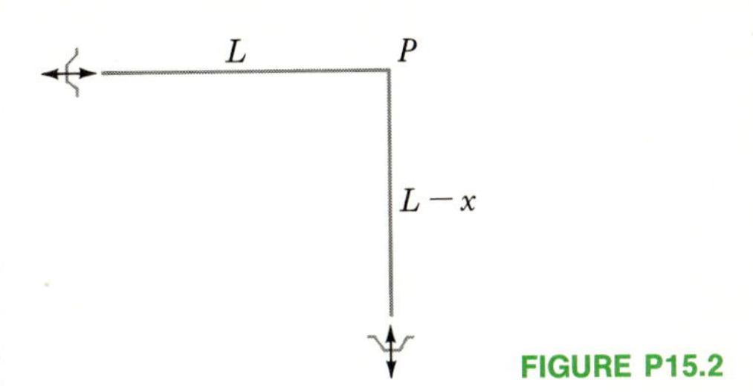

FIGURE P15.2

13. A pipe open at one end and closed at the other is 0.75 m long. To what three lowest frequencies of sound will it resonate? Draw the wave within the tube for each frequency.

14. To what three lowest frequencies will a pipe open at both ends resonate? The pipe is 0.90 m long. Draw the wave within the tube at each frequency.

15. In an experiment like that shown in Fig. 15.12 resonances are found to occur when the water is 31.00 and 41.00 cm high in the tube. If no resonances occur in between, find the frequency of the tuning fork.

16. A man wishes to find how far down the water level is in the iron pipe leading into an old well. Being blessed with perfect pitch, he merely hums musical sounds at the mouth of the pipe and notices that the lowest-frequency resonance is 80 Hz. About how far from the top is the water level?

17. The Lincoln Tunnel under the Hudson River in New York is about 2600 m long. To what sound frequencies will it resonate? What, if any, practical importance do you think this has?

18. At 0°C a certain organ pipe resonates to a frequency of 500 cps. (*a*) What is its resonant frequency at 30°C? It would harmonize well with a 250-cps pipe at 0°C. (*b*) Will it still harmonize when both pipes are at 30°C? Explain your answer.

19. How fast must a car be coming toward you if its horn is to appear to be 10 percent higher in frequency than when standing still?

20. A sound source vibrates at 100 cps and is receding from an observer with a speed of 18 m/s. If the speed of sound is 332 m/s, what frequency does the observer hear?

21. An unknown frequency is compared with a standard 2200-cps tuning fork and found to give rise to 2.0 beats per second. (*a*) What are the possible values of the unknown? (*b*) How could you determine the true value if you had a wad of chewing gum handy?

22. A violinist compares the tone given off by his violin string with the corresponding note on a piano and notices a beat every 3 s between the two sounds. (*a*) By how many vibrations per second do the two instruments differ? (*b*) How could the violinist decide which instrument was higher in pitch?

23. The speed of sound in hydrogen is about 1270 m/s. If an organ pipe which resonates in its fundamental to 800 cps in air is filled with hydrogen, what will be its fundamental resonance frequency?

24.* A tube open at one end and closed at the other is 40 cm long. It will resonate under the action of a 210-cps sound source provided the source is receding from it. How fast must the source be moving? (Use $v = 320$ m/s.)

25.** Show that the frequency of the sound f' heard by a man approaching with speed v_m a fixed horn of frequency f is given by

$$f' = \frac{v + v_m}{v} f \qquad \text{or} \qquad \frac{\Delta f}{f} = \frac{v_m}{v}$$

26.** The device shown in Fig. P15.3, **Kundt's tube,** is sometimes used to measure the speed of sound in metal rods. When the rod R is stroked lengthwise with a rosined cloth, it resonates with a longitudinal motion such that the disk end is an antinode and the support is a node. The vibration of the disk D at the end of the rod sends sound waves of the same frequency down the glass tube. When the tube is adjusted to a position of resonance, such as the position shown, cork dust in the tube indicates clearly the position of nodes and antinodes in the air. If the separation between antinodes is found to be 10.0 cm and the rod is 100.0 cm long, find the speed of sound in the metal rod.

FIGURE P15.3

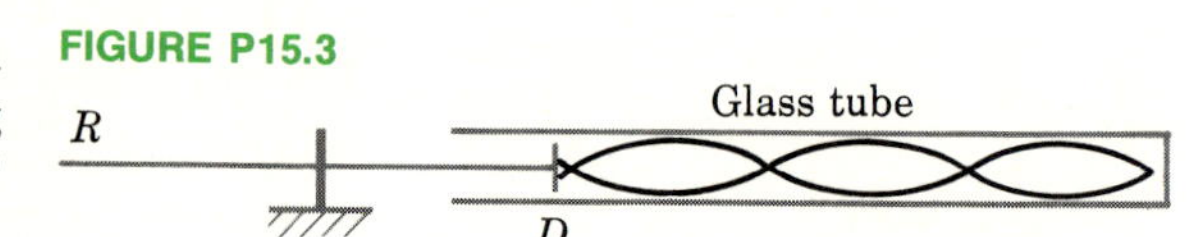

ELECTRIC CHARGES AT REST: ELECTROSTATICS

16

We begin our study of electricity with electrostatics, the physics of electric charges at rest. The attractions and repulsions between charges, electric fields, and the concept of electric potential difference will be discussed. It will be seen that electrical concepts are closely related to the principles we learned in mechanics. The law of conservation of energy will prove to be of great importance in electricity as well as in the previously studied fields of mechanics and heat. After completing our study of electrostatics, we shall be prepared to investigate the behavior of moving charges, and this will be done in the chapters which follow.

16.1 ATOMS AS THE SOURCE OF CHARGE*

An atom is composed of a tiny positively charged nucleus about which there are negatively charged particles called **electrons.** This is illustrated in a schematic way for an atom of carbon in Fig. 16.1. You will recall from courses in chemistry that all atoms are electrically neutral. That is to say, the quantity of positive charge on the nucleus exactly equals the total charge of the electrons about the nucleus. In the case of the carbon atom shown, if $-e$ is the charge on each electron, the charge on the nucleus will be exactly $+6e$. We shall postpone a detailed discussion of the atom to a later chapter and merely make use of its electrical constitution here.

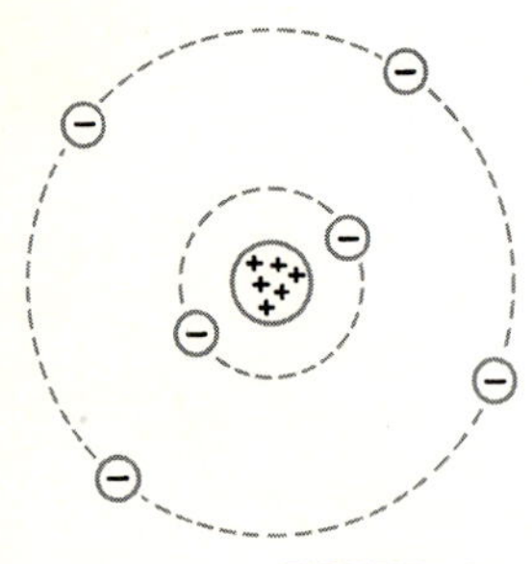

FIGURE 16.1

A schematic representation of a carbon atom. The negative charges on its six electrons are balanced by the positive nuclear charge. (The nucleus and electrons are much smaller than shown.)

It appears that the whole universe is nearly, if not completely, neutral from an electrical standpoint. The earth has very little, if any, excess of either positive or negative charge. This is so nearly true that for nearly all practical purposes the earth can be considered to have no net charge. The vast majority of the charges on and in the earth reside as constituents of atoms. When free negative or positive charges are found, they are usually assumed to have come from the tearing apart of an atom.

Actually, it is not difficult at all to remove an electron from an atom—under certain circumstances. For example, if a bar of ebonite (hard rubber) is rubbed with animal fur, some of the electrons from the atoms in the material of the fur are rubbed off onto the ebonite rod. (The reason for this charge transfer taking place is not simply explained. It is covered in courses dealing with solid-state physics.) Hence, the ebonite rod acquires a net excess of electrons. When the rod is rubbed against a metal object, some of the excess electrons transfer to the metal, as illustrated in Fig. 16.2.

Two Types of Charge

Similarly, if a glass rod is rubbed with silk, some of the electrons leave the atoms in the glass and give rise to an excess of electrons on the silk. Of course, now the glass rod has an excess of positive charges. If the glass rod is touched to a neutral metal ball, electrons will leave some of the atoms of the metal and replace those electrons lost by the atoms in the glass. As a result, the metal ball acquires a net positive charge. Many other materials give rise to a separation of charge when rubbed together. The ones we have described were originally used to define positive and negative charge before the existence of the electron was even known.

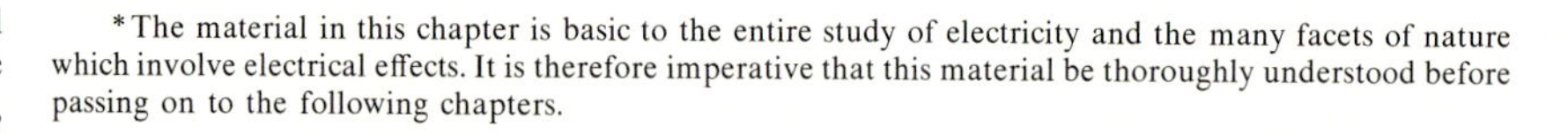

*The material in this chapter is basic to the entire study of electricity and the many facets of nature which involve electrical effects. It is therefore imperative that this material be thoroughly understood before passing on to the following chapters.

FIGURE 16.2

When the negatively charged ebonite rod touches the uncharged metal ball, electrons are conducted off the rod onto the ball.

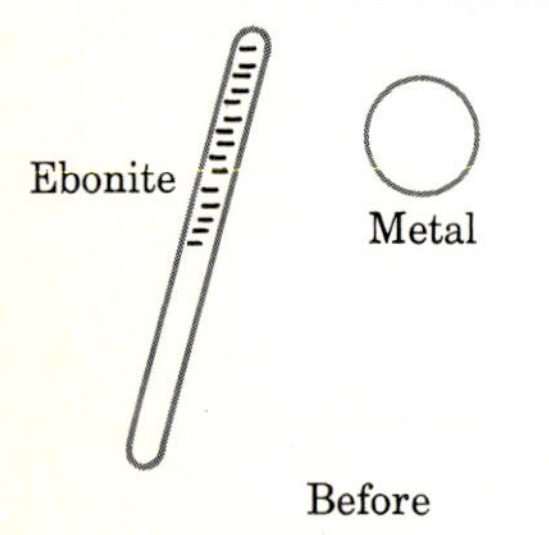

Before

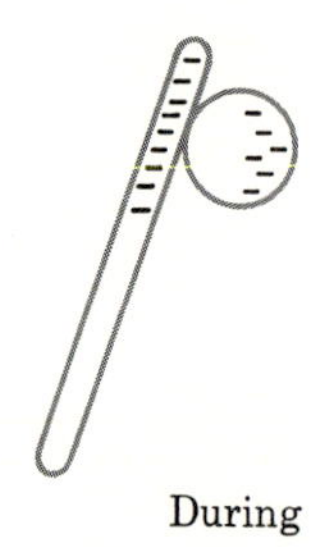

During

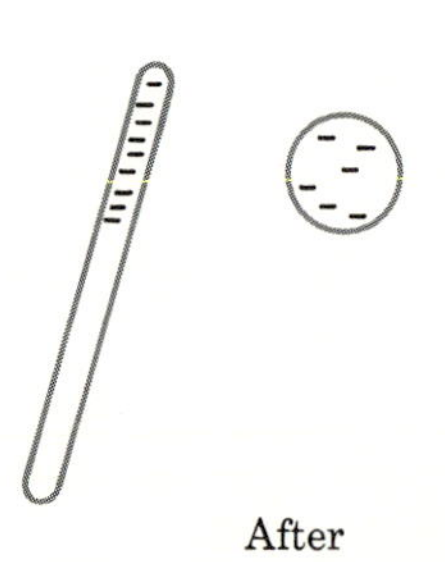

After

16.2 FORCES BETWEEN CHARGES

Now that we know how to obtain charged bodies, it is possible to examine the forces between charges. One of the simplest ways of doing this is to make use of tiny pith balls. These are very light balls, often covered by a metallic paint. It is a simple matter to charge these balls by means of glass and ebonite rods. If the balls are suspended by light threads, four interesting experiments can be performed. They are illustrated in Fig. 16.3*a* to *d*.

From the experimental results shown in Fig. 16.3 we can conclude the following:

1 Like charges repel each other; i.e., two positive charges will repel each other, as will two negative charges.
2 Unlike charges attract each other; i.e., positive charges attract negative charges, and vice versa.
3 The magnitude of electrical forces between two charged bodies often exceeds the gravitational attraction between the bodies. (How do the experiments illustrated show this?)

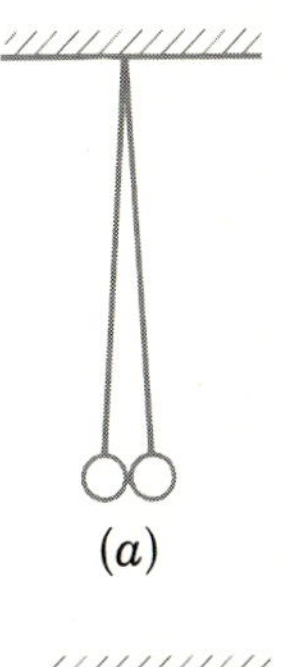

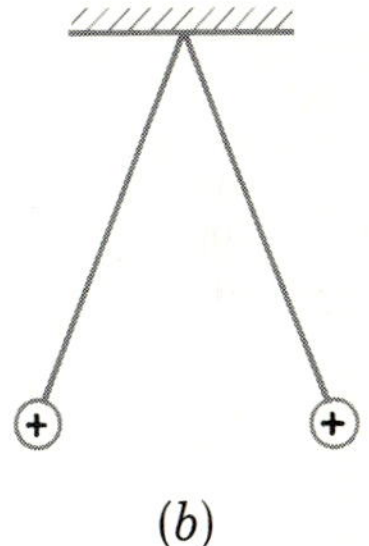

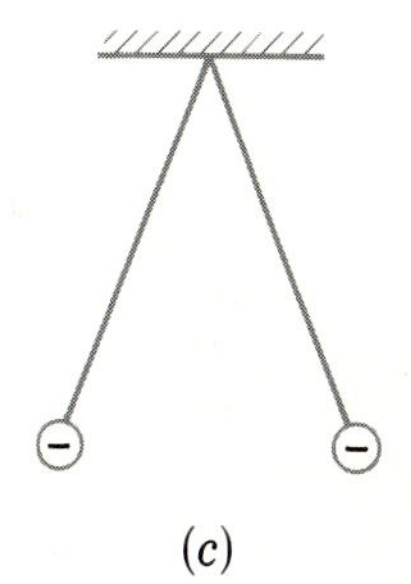

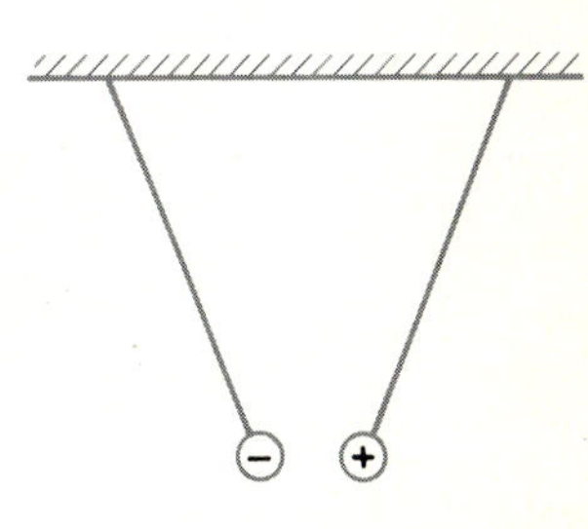

FIGURE 16.3
Charged pith balls show that like charges repel each other while unlike charges attract each other.

16.3 INSULATORS AND CONDUCTORS

Although all materials are made of atoms, and even though all atoms are made of electrons and nuclei, we are well aware that the electrical properties of substances vary widely. *There are two basic groups into which all substances can be divided according to electrical properties. They are the conductors and the nonconductors, or insulators.**

In the second group, insulators, the electrons of any given atom are bound tightly to that atom and cannot wander through the material. Hence, even if an excess of charge is placed near the end of a rod made from an insulator, the electrons in the adjacent atoms cannot move under the attraction or repulsion of the nearby excess charge. They are strongly bound to their atoms and cannot get free, and so no large motion of charge occurs in the rod.

Conductors behave quite differently. In these substances the electrons near the outer portion of the atom, called **valence electrons,** are so close to neighboring atoms that it is difficult to determine exactly which electron belongs to which atom. Under these conditions, the valence electrons in the atoms move about through the metal and may be looked upon, for some purposes at least, as an electron gas contained within the confines of the piece of metal. Even in metals, though, most of the electrons are tightly held in the atom, and only the electrons near the outside of the atom are at all free to move through the solid. Usually only one, two, or three electrons associated with each atom, the valence electrons, can be considered at all free.

You should realize that the precise formulation of the motion of charges in solids is a very complex problem. It can be handled well only by use of

* Materials which are intermediate between these two groups are called **semiconductors.** Some people prefer to class them as a separate group.

quantum mechanics. In fact, the exact behavior of electrons in solids is still an active field for research today, particularly for the substances which are intermediate between conductors and nonconductors, the so-called semiconductors.

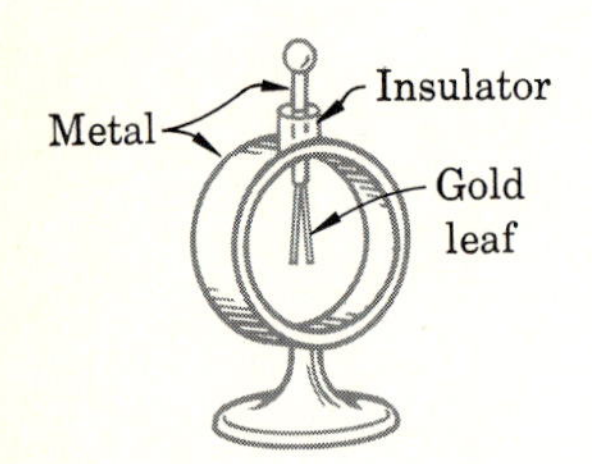

FIGURE 16.4
One type of gold-leaf electroscope. The central portion, consisting of the metal sphere, rod, and leaves of gold foil, is insulated from the case.

16.4 THE ELECTROSCOPE

The electroscope is a simple device used for measuring charges of small magnitude. It is shown in Fig. 16.4. A metal rod from which are suspended two very thin leaves of gold foil is held inside a metal case by the aid of an insulator which keeps the rod from touching the case. Two faces of the case are covered with glass, so that the disposition of the leaves can be seen.

Suppose that some negative charge is placed on the central rod by touching to it a charged piece of ebonite. The charge is confined entirely to the rod and leaves, since they are insulated from other objects. Because like charges repel each other, the negative charges will distribute themselves more or less uniformly over the rod and gold leaves. However, the gold leaves, being free to bend and being repelled by the charges on each other, take up the position shown in Fig. 16.5*a*.

If a negatively charged ball is now brought close to the ball of the electroscope as shown in part *b* of Fig. 16.5, many of the negative charges on the upper part of the electroscope will be repelled down the rod, causing the leaves to stand out further. An opposite effect is observed if a positively charged ball is brought close to the electroscope (Fig. 16.5*c*). In addition, a ball having no net charge does not disturb the electroscope to any great extent. With this apparatus the sign of a charge and its rough magnitude can be determined. You should convince yourself that a similar procedure can be followed if the electroscope is charged positively.

16.5 CHARGING BY CONDUCTION AND BY INDUCTION

There are two general ways for placing charge on a metal object by use of a second object which is already charged. As a concrete example, consider the

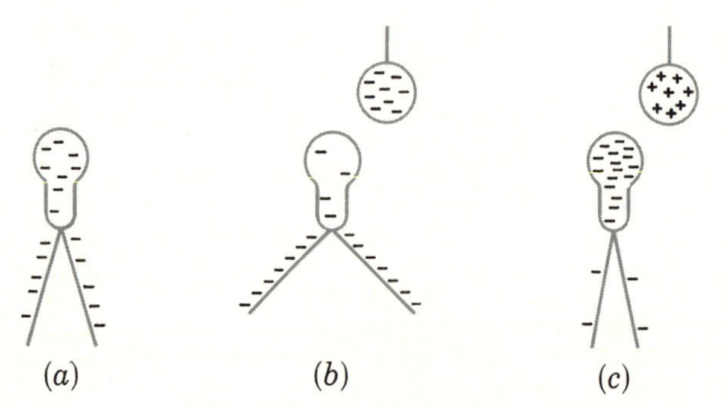

FIGURE 16.5
The charged electroscope is used to determine the sign and approximate magnitude of charge on an object.

problem of charging a metal ball by use of a negatively charged ebonite rod. One can charge the ball by simply touching it to the rod. Upon contact, some of the excess negative charge on the rod moves off onto the ball. This process, shown in Fig. 16.6, is called **charging by conduction.**

Charging by Conduction

The same rod can be used in a different way to charge the ball. This is shown in Fig. 16.7. In this process, charging the ball **by induction,** the rod is not touched to the ball at all. Notice that when the rod is brought close, some of the electrons in the metal are repelled to the right side of the ball, leaving a positive charge on the left side. Since no charge has been added to or subtracted from the ball, it is still neutral, of course. Now, suppose that the ball is touched by another object such as one's finger. (We say in that case that the ball has been **grounded,** and the symbol —|ı is used to show this.*) The negative charges can get still further away from the negative rod if they flow to the object touching the ball. Once this has happened, the ball will no longer be neutral, since it has lost some of its negative charge. After the object touching the ball has been removed, the negative rod can be removed and the ball will have a positive charge. (Why must the touching object be removed *before* the rod?)

Charging by Induction

Upon comparing Figs. 16.6 and 16.7, you can see that an ebonite rod can be used to charge a metal object negatively by *conduction* but that it charges the same object positively by *induction*. You may find it interesting to work out the similar diagrams using a positively charged glass rod. The charges will be reversed in that case.

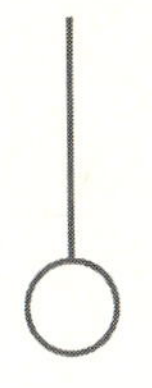

Before

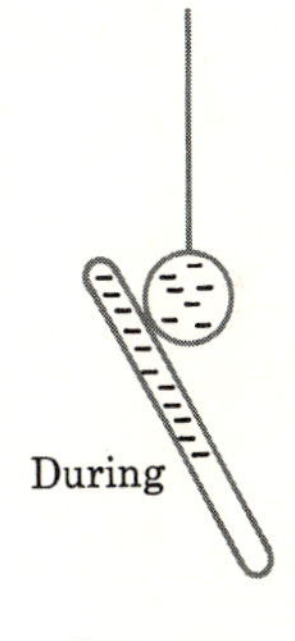

During

After

FIGURE 16.6
A metal ball can be charged by conduction. Note that the charge is shared and therefore the ball and the rod have like charges.

16.6 THE FARADAY ICE-PAIL EXPERIMENT

In 1843 Michael Faraday first carried out a simple but highly instructive experiment. He attached a metal ice pail to an electroscope, as shown in Fig. 16.8*a*. A metal ball suspended from a thread was given a positive charge. The charged ball was then lowered into the pail (without touching it), at which time the leaves of the electroscope were found to diverge. Moreover, the ball could be moved from place to place within the pail, and the leaves remained

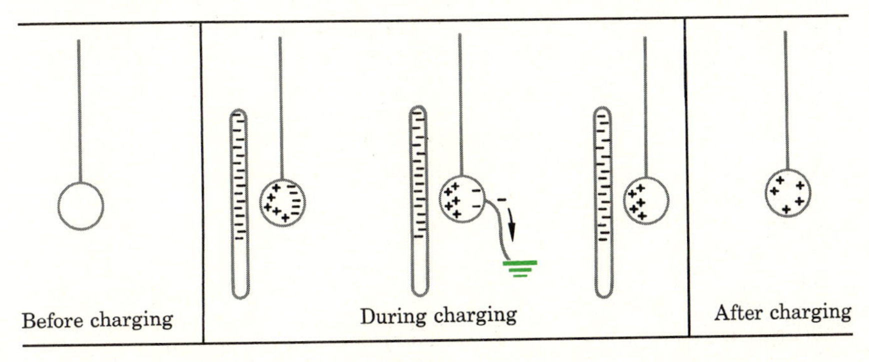

FIGURE 16.7
The method for giving a metal sphere a positive charge by induction. Note that the rod and balls have unlike charges.

* Usually, to ground an object, we attach it to a water pipe or some other object entering the earth. Any excess charge can then flow to the ground, i.e., the earth.

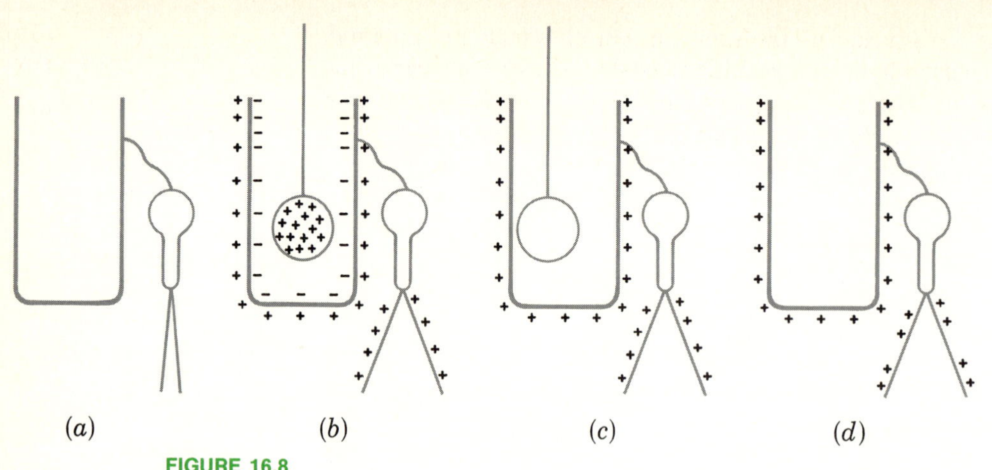

FIGURE 16.8
Faraday's ice-pail experiment.

in the same divergent position. Only when the ball was removed would the leaves return to their original position.

Faraday further noticed that if the charged metal ball, when inside the pail, was touched to the inside of the pail, the electroscope leaves remained in the same divergent position they had maintained before the ball was touched. Now, however, when the ball was removed from the pail, the leaves remained divergent. When the ball was brought close to a second electroscope, it was found to be no longer charged. Apparently, upon touching the inside of the pail, the excess charge on the ball had been completely neutralized. Since the leaves of the electroscope attached to the outside of the pail did not move when the ball touched the pail, Faraday concluded that the inner surface of the pail had just enough charge on it to neutralize the ball.

From these experiments we can draw the following conclusions, illustrated in Fig. 16.8*b* to *d*.

1. A charged metal object suspended inside a neutral metal container induces an equal and opposite charge on the inside of the container.
2. When the charged metal object is touched to the inside of the container, the induced charge exactly neutralizes the excess charge on the object.
3. If a charged object is placed within a metal container, an equal charge of the same sign will be forced to the outer surface of the container.
4. When a path for conduction is provided, all the charge on a metal object will reside on its outer surface.

These are important facts concerning electric charges on metal. We shall interpret them more fully after we have examined Coulomb's law and the concept of electric fields.

16.7 COULOMB'S LAW

The mathematical law by which like charges repel and unlike charges attract was formulated in 1785 by Charles Augustin de Coulomb (1736–1806) and *is called* **Coulomb's law.** By means of a very sensitive balance, similar to that used by Cavendish in gravitation, he was able to measure accurately the force between two small charged balls. A model of the situation is shown in Fig. 16.9. Two small balls, much smaller than shown, with distance r between centers carry charges $+q_1$ and $-q_2$. After a number of experiments Coulomb concluded that the force on sphere 1 varied in proportion to the product of their charges and inversely as the square of their distance apart. In symbols,

$$F \propto \frac{q_1 q_2}{r^2}$$

or

$$F = (\text{const}) \left(\frac{q_1 q_2}{r^2}\right) \tag{16.1}$$

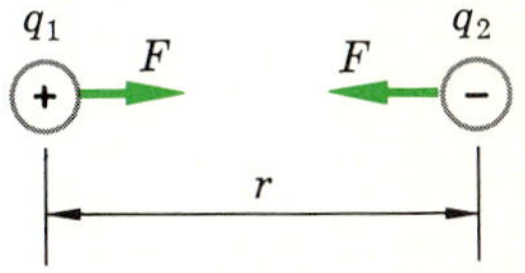

q_1 F F q_2 + − r

FIGURE 16.9
The two unequal, unlike charges attract each other with equal force.

and the force was in the direction shown in Fig. 16.9. According to Newton's law of action and reaction, the force on sphere 2 was identical in magnitude but oppositely directed. Notice that Coulomb's law applies only to point charges. If the charges extend over a large region, the distance r between them is not easily defined.

The constant in Coulomb's law depends upon the units used for the various physical quantities in the equation. For many years, a system of units which made the constant unity was employed. Unfortunately, that system of units was not the same as most people use in practical electrical work, nor did it correspond to the SI. To overcome these difficulties would require us to use a somewhat more complicated proportionality constant in Coulomb's law. This complication is far outweighed by the fact that a more complicated proportionality constant would allow us to use both the SI units and the common (or "practical") electrical units. Following that course, we make use of a unit of charge, the coulomb (C), which will be defined precisely later in terms of forces between electrical currents. We then have the following form for **Coulomb's law,**

Coulomb's Law

$$F = k\frac{q_1 q_2}{r^2} \tag{16.2}$$

where F is in newtons, r is in meters, q is in coulombs, and $k = 8.9874 \times 10^9\ \text{N} \cdot \text{m}^2/\text{C}^2$. For most computations in this text we shall take k to be $9.0 \times 10^9\ \text{N} \cdot \text{m}^2/\text{C}^2$.

Smallest Charge

In order to obtain some feeling for the size of the charge unit, the coulomb, the following facts should be noted. *The magnitude of the charge on the electron and proton is* a very small fraction of a coulomb, 1.60219×10^{-19} C. *No charge smaller than this has ever been found. No charge has ever been found which was not an integer multiple of this basic charge quantum. We therefore believe that all charges consist of multiples of this basic value.* However, small charged pith balls and the like may have excess charges of the order of a microcoulomb (10^{-6} C, or 1 μC) and so the balls carry perhaps a million million charge quanta. We represent the magnitude of the charge quantum by e,

$$e = 1.60219 \times 10^{-19}\ \text{C}$$

The charge on the electron is $-e$, negative, while the proton's charge is $+e$.

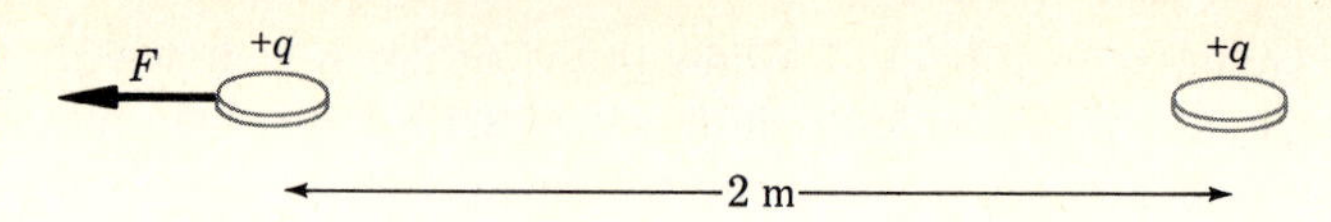

FIGURE 16.10
As shown in Illustration 16.1, only a tiny fraction of the electrons need be removed from a penny to give rise to large electrical forces.

Illustration 16.1 A copper penny "weighs" about 3 g and contains about 3×10^{22} copper atoms. Suppose two pennies are 2.0 m apart and carry equal charges q. (*a*) How large must q be if the force of repulsion on one due to the other is equal to the weight of a penny? (*b*) How many electrons must be removed from a penny to give it this charge? (*c*) What fraction of the atoms have lost electrons in such a case?

Reasoning *a.* The situation is shown in Fig. 16.10. Each penny has a mass of 3 g, or 3×10^{-3} kg. The weight of a penny is mg and therefore 0.0294 N. Coulomb's law, $F = kq_1q_2/r^2$, gives the repulsion force on one of the pennies as

$$F = (9 \times 10^9 \text{ N} \cdot \text{m}^2/\text{C}^2)\frac{q^2}{(2.0 \text{ m})^2}$$

with $F = 0.0294$ N in this case. Solving for q gives $q = 3.6 \times 10^{-6}$ C.

b. Since each electron removed from a penny leaves an unbalanced charge of 1.60×10^{-19} C behind, the number of electrons removed must be

$$\text{Number of electrons} = \frac{3.6 \times 10^{-6} \text{ C}}{1.60 \times 10^{-19} \text{ C}} = 2.3 \times 10^{13}$$

c. But there are about 3×10^{22} atoms in the penny. So the fraction which have lost a single electron is

$$\text{Fraction} = \frac{2.3 \times 10^{13}}{3 \times 10^{22}} = 7.7 \times 10^{-10}$$

Notice how very small a fraction of the electrons must be removed to give an object a sizable charge. Notice also that charges as small as 10^{-6} C give rise to easily measurable forces between ordinary-sized objects.

Illustration 16.2 Find the force on the center charge in Fig. 16.11.

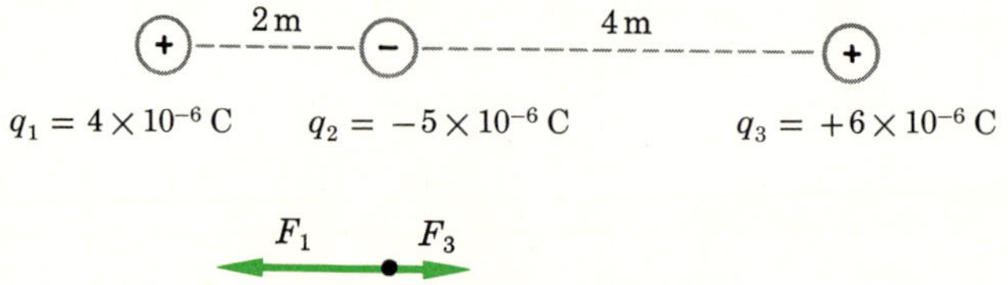

2 m 4 m

$q_1 = 4 \times 10^{-6}$ C $q_2 = -5 \times 10^{-6}$ C $q_3 = +6 \times 10^{-6}$ C

F_1 F_3

FIGURE 16.11
The central charge is attracted to q_1 by the force F_1 and to q_3 by the force F_3.

Reasoning Clearly, charge q_2 will be attracted to q_1 by a force F_1. It is also attracted to charge q_3 by F_3. Using Coulomb's law, we have

$$F_1 = (9 \times 10^9 \text{ N} \cdot \text{m}^2/\text{C}^2) \frac{(4 \times 10^{-6} \text{ C})(5 \times 10^{-6} \text{ C})}{(2 \text{ m})^2} = 0.0450 \text{ N}$$

$$F_3 = (9 \times 10^9 \text{ N} \cdot \text{m}^2/\text{C}^2) \frac{(5 \times 10^{-6} \text{ C})(6 \times 10^{-6} \text{ C})}{(4 \text{ m})^2} = 0.0169 \text{ N}$$

The total force on q_2 is to the left and is

$$F = F_1 - F_3 = 0.0281 \text{ N}$$

Notice that the signs of the charges are not employed in the above use of Coulomb's law, since these are implicit in the direction of the forces. This enables us to compute the magnitude of the force by using Coulomb's law and indicate by a vector drawing which direction is positive or negative.

Illustration 16.3 Find the force on the $+10$-μC charge of Fig. 16.12.

Reasoning The forces on the charge in question are drawn. Using Coulomb's law, $F = kq_1q_2/r^2$, we can find easily that

$$F_1 = 6 \text{ N} \qquad \text{and} \qquad F_2 = 18 \text{ N}$$

As shown in the figure, these two forces act at right angles. We can therefore use the pythagorean theorem to find the resultant force on the 10-μC charge.

$$F = \sqrt{F_1^2 + F_2^2} = 19 \text{ N} \qquad \text{and} \qquad \tan\theta = \frac{F_2}{F_1} = 3.00$$

from which

$$\theta = 71.5°$$

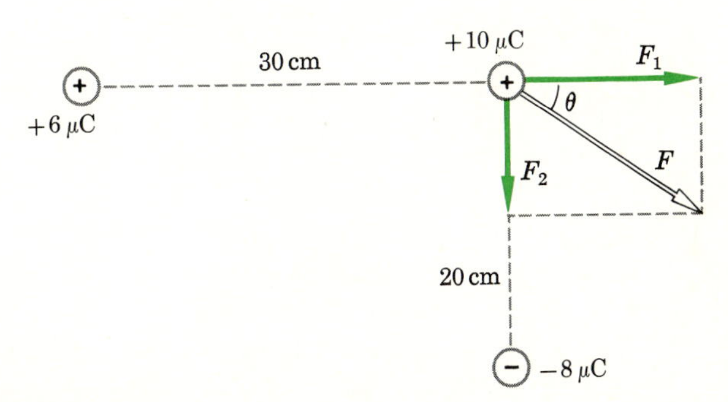

FIGURE 16.12

To find the resultant force on the $+10$-μC charge, we must add the two separate forces exerted by the other two charges upon it.

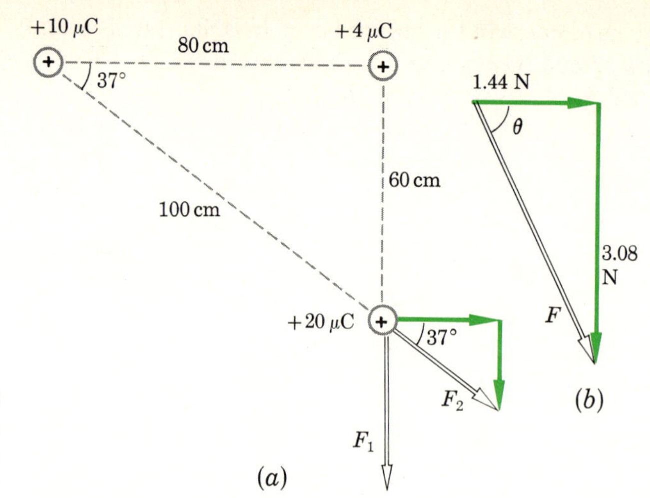

FIGURE 16.13
The vector forces acting upon the 20-μC charge produce the resultant F shown in part b.

Illustration 16.4 Find the force on the +20-μC charge in Fig. 16.13.

Reasoning From Coulomb's law, one finds

$$F_1 = 2.0\ \text{N} \qquad \text{and} \qquad F_2 = 1.8\ \text{N}$$

Resolving F_2 into components, one has $F_{2x} = F_2 \cos 37° = 1.44$ N and $F_{2y} = 1.08$ N, as shown. The total y force on the charge is $F_y = 2.0 + 1.08 = 3.08$ N downward, while the total x force is 1.44 N to the right. They are shown in Fig. 16.13b. Clearly, the resultant force is

$$F = \sqrt{(1.44)^2 + (3.08)^2} = 3.4\ \text{N} \qquad \text{and} \qquad \tan\theta = \frac{3.08}{1.44} = 2.14$$

from which

$$\theta \approx 65°$$

16.8 THE ELECTRIC FIELD

We shall find it convenient to discuss electrical forces in terms of a concept called the **electric field.** It serves much the same purpose in electricity as the concept of a gravitational field serves in mechanics. Before discussing this new concept in detail, let us first review the more familiar situation of the gravitational field.

It is common for people to say that a stone when released falls to the earth because of the downward force of the earth's gravitational field. By this they mean that the stone experiences a gravitational force and that this force causes it to accelerate. When we go far away from the earth out into space, the earth's gravitational attraction for objects becomes quite small. We say that the earth's gravitational field is very weak at large distances from the earth. A gravitational field is said to exist in a region of space where a gravitational force exists on a body. If the gravitational field is strong, the gravitational force on a given mass is strong.

It is convenient to draw pictures of gravitational fields. The gravitational field of the earth is shown in Fig. 16.14. We interpret this picture in the following way: If an object is placed at point A in the figure, it will experience a force directed toward the center of the earth. The lines of force drawn in Fig. 16.14 show the direction of the earth's gravitational field, i.e., the direction of the gravitational force on an object.

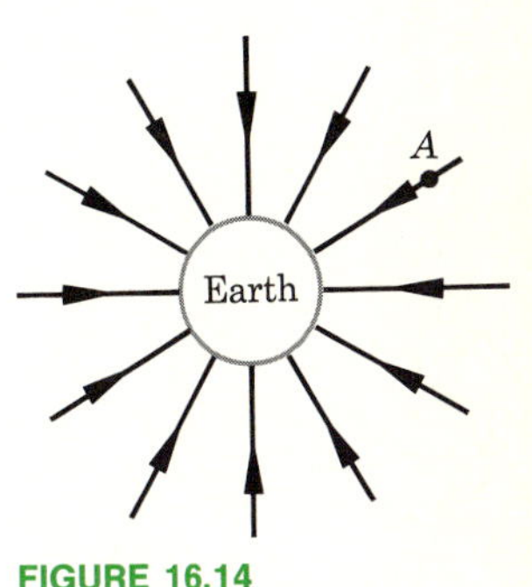

FIGURE 16.14
The gravitational field of the earth is directed radially inward and becomes stronger as one approaches the earth.

Actually, of course, Fig. 16.14 should be drawn in three dimensions, with lines of force directed from all sides in toward the center of the earth. One notices that not only do the lines of force represent the direction of the force but they also indicate the magnitude of the force. Where the field, i.e., gravitational force, is strong, the lines of force are closely spaced. Far away from the earth, where the lines are widely separated, the gravitational field is weak. It is shown in more advanced texts that this is not just qualitatively true but is exactly true for both the gravitational field and the electric field.

DEFINITION

The electric field is defined in a manner analogous to the gravitational field. *It is universally agreed to define the direction of the electric field as the direction of the electric force on a tiny,* **positive test charge.** *A test charge is a fictitious charge endowed with a very special quality: the test charge exerts no forces on nearby charges and therefore does not disturb the charges in the vicinity.* In practice, we can approximate such a fictitious charge by using a tiny object which contains a very small charge. It will therefore disturb the nearby charges by only a negligible amount.

FIGURE 16.15
The electric field is directed radially inward toward a small negative charge and radially outward from a small positive charge.

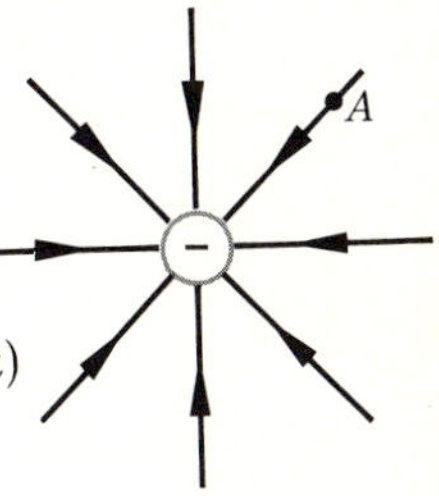

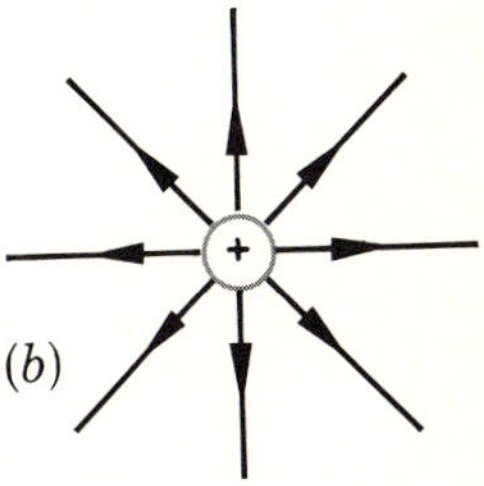

Suppose a positive test charge is placed at point A in Fig. 16.15*a*. It will be attracted by the negative charge along the radial line shown. We define the direction of the electric field at A to be in the direction of the force on the positive test charge. As we see, *the electric field near a negative charge is directed radially inward toward the charge.* We represent the electric field by the field lines shown.

The electric field about a positive charge is shown in Fig. 16.15*b*. Notice that the direction of the arrows on the electric field lines is now away from the charge, since the positive charge repels our positive test charge. The lines which we have drawn in these figures representing the force on the positive test charge are called **electric lines of force.** We may generalize as follows: *lines of force originate on and come out of positive charges, whereas they are directed toward and end on negative charges.*

We wish also to define the electric field strength in a quantitative way. As stated above, the direction of the field at a point is the direction of the force on a positive test charge placed at the point in question. *The magnitude or strength of the electric field is taken to be the electric force* **F** *on the tiny test charge divided by the magnitude of the test charge q. Upon using the symbol* **E** *for electric field strength,* this definition may be written in equation form as

$$\mathbf{E} = \frac{\mathbf{F}}{q} \tag{16.3}$$

The units of E are clearly newtons per coulomb. If q were 1 C, E and F would be numerically equal. For this reason we shall frequently state that *the* **electric field strength** *is the force experienced by a unit positive test charge.* However, we should always realize that in order to measure the field we would use a charge much smaller than 1 C so as not to disturb the other charges present.

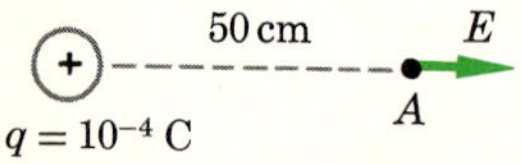

50 cm E

$q = 10^{-4}$ C A

FIGURE 16.16

To find the electric field E at point A, we must compute the force which a unit positive charge would experience if placed at that point.

Illustration 16.5 Find the electric field strength at a distance of 50 cm from a positive charge of 10^{-4} C.

Reasoning We wish to find the field at point A in Fig. 16.16. Since it is the force on a unit positive charge, its direction is to the right, as shown by the arrow. To find the magnitude of the electric field we make use of Coulomb's law (mentally placing a test charge q_t at A). We have

$$F = (9 \times 10^9 \text{ N} \cdot \text{m}^2/\text{C}^2)\left(\frac{q_1 q_2}{r^2}\right)$$

In our case, $q_1 = 10^{-4}$ C, $q_2 = q_t$, and $r = 0.50$ m. Since $F/q_t = E$, we have at once

$$E = (9 \times 10^9)\left[\frac{10^{-4}}{(0.50)^2}\right] = 3.6 \times 10^6 \text{ N/C}$$

You should show that the answer really does have these units.

Illustration 16.6 Find the electric field strength at point B in Fig. 16.17.

Reasoning Mentally we place a positive test charge q_t at point B and

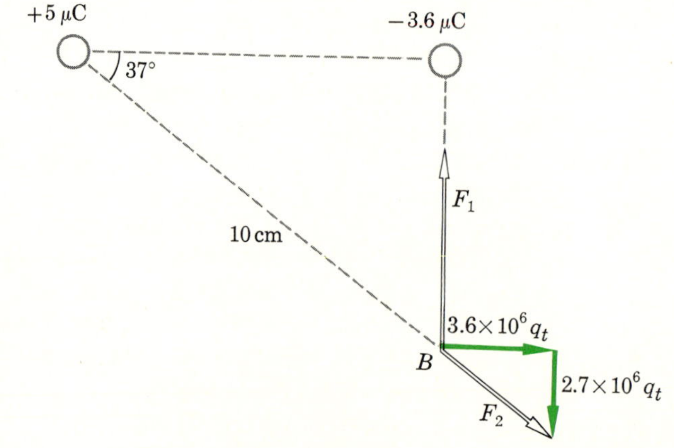

FIGURE 16.17

To find the field at B, we must find the forces F_1 and F_2 exerted upon a positive test charge q_t placed at that point.

compute the force on it, using Coulomb's law. Making use of the vectors shown,

$$F_1 = (9 \times 10^9\ \mathrm{N \cdot m^2/C^2}) \left(\frac{3.6 \times 10^{-6}\ \mathrm{C}}{0.0036\ \mathrm{m^2}} \right) q_t = 9 \times 10^6 q_t \text{ newtons}$$

so

$$E_1 = (9 \times 10^6)\ \mathrm{N/C}$$

Similarly,

$$E_2 = 4.5 \times 10^6\ \mathrm{N/C}$$

From this, after taking components,

$$E_y = 9 \times 10^6 - 2.7 \times 10^6 = 6.3 \times 10^6\ \mathrm{N/C} \qquad E_x = 3.6 \times 10^6\ \mathrm{N/C}$$

and

$$E^2 = E_x^2 + E_y^2 \qquad E = 7.3 \times 10^6\ \mathrm{N/C}$$

16.9 THE ELECTRIC FIELD IN VARIOUS SYSTEMS

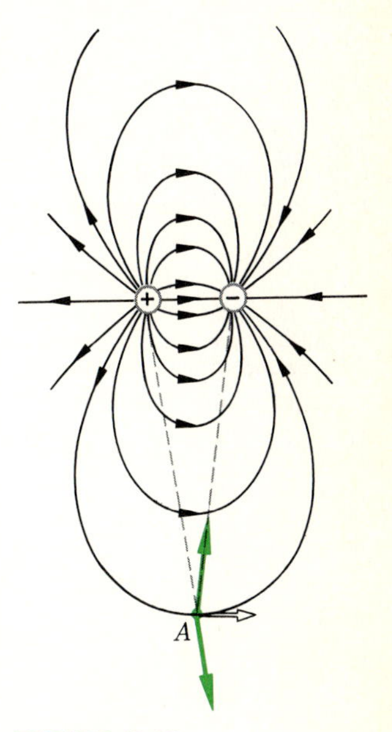

FIGURE 16.18
The electric field lines originate on the positive charge and end on the negative charge.

A great deal of insight into a problem can be obtained by examining the pertinent electric-field drawing. To illustrate this point, let us examine several electric fields and the charge distributions giving rise to them. In doing so we recall that we defined the electric field as the force on a unit positive test charge. Lines of force must always be drawn out of positive charges and into negative charges.

Consider the electric field about two equal charges, one positive, the other negative, as shown in Fig. 16.18. The electric field will be similar about each, except that it will be directed into the negative charge and out of the positive charge. This field is shown in Fig. 16.18. You should examine several points in the figure to convince yourself that a positive test charge actually would experience a force in the directions indicated by the force lines in the figure. To see how this is done, consider point A of Fig. 16.18. A positive test charge at point A will be repelled by the positive charge and attracted by the negative charge. The attractive force equals the repulsive force because the test charge is as close to the positive charge as it is to the negative charge. The resultant of these two forces is along the line of force at point A.

A similar situation, the field in the neighborhood of two like charges, is shown in Fig. 16.19. It is seen in this figure, as well as in Fig. 16.18, that where the lines of force are close together the field is strongest. This is a convenient way of estimating the strength of the electric field.

FIGURE 16.19
Notice how the lines of force about two like charges seem to repel each other. Why must this be so?

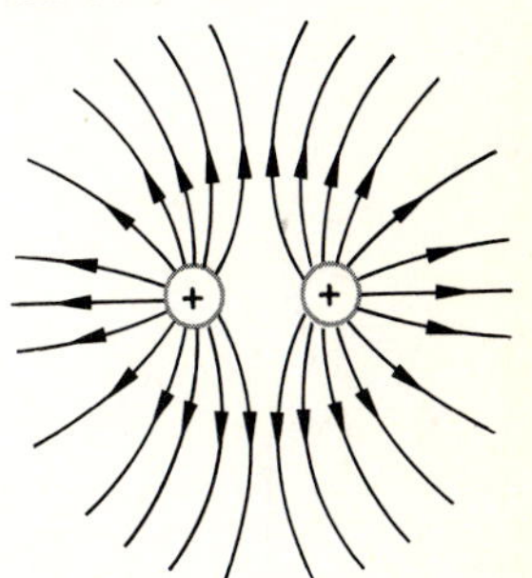

Suppose that a positive *uniformly* charged ball is held above a large metal plate. If the metal plate is attached to the ground, charge may run on or off it to the essentially inexhaustible capacity of the earth. The situation is shown in Fig. 16.20. We see at once that the electrons in the metal plate will be attracted by the positive charge. Although they cannot leave the plate, they congregate on the plate just below the positive charge. Negative charge will flow onto the plate from the ground in order to replace the electrons induced to take up the position near the center of the plate.

The positive test charge used to determine the direction of the field will be pulled and pushed down toward the plate in the central region. You can readily verify that the lines of force shown in Fig. 16.20 actually do represent the force on a positive test charge. Similarly, the density of the lines indicates the field strength. As one would expect, the field is strongest close to the positive charge.

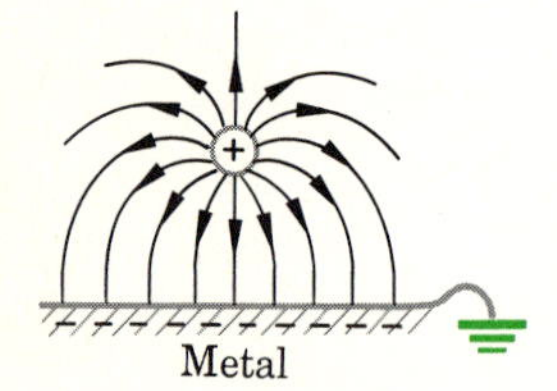

FIGURE 16.20
The positive charge attracts negative charges to the top of the metal plate. Why are the field lines perpendicular to the plate at its surface?

Two rather subtle but important features of Fig. 16.20 should be pointed out. First, the lines of force do not penetrate into the metal plate. All of them stop at the surface. This would indicate that the electric field is zero within the metal; i.e., there is no force on a charge in the metal. This is readily understood from the following reasoning: We are considering only electrostatic conditions, i.e., no continuing flow of charge. If there were a field within the metal, it would exert a force on the free electrons in the metal. The electrons would then move, and a current would flow. But no currents are flowing under electrostatic conditions, and hence the electric field must be zero inside a metal.

The second point has to do with the fact that the lines of force are perpendicular to the metal surface. First we shall point out that the lines of force represent the force on a positive charge. A negative charge would be forced in the opposite direction. In Fig. 16.20, the negatively charged electrons at the surface of the metal experience a force trying to pull them upward. This is the result of the attraction by the positive charge suspended above the metal. However, the electrons are held within the body of the metal and cannot come loose under this force.

Suppose that the force were not exactly perpendicular to the metal. There would be a component of the force along the surface of the metal. Since the electrons of the metal can move in this direction, they would flow under the action of this force and a current would flow. However, since this is an electrostatic situation, no current is flowing. Hence, there can be no component of the field parallel to the metal surface. We conclude that the field is directed perpendicular to the metal surface.

These last facts will be used again later. They are quite general in any electrostatic problem. For this reason we shall state them once again. Under electrostatic conditions:

1 The electric field within a metal is zero.
2 The electric field just outside the surface of a metal is perpendicular to the surface.

Since these facts were proved by reference to the movement of free electrons within the metal, the proofs will not hold true for insulators, since they contain no free electrons. Hence the above rules apply only to metals and not to insulators.

FIGURE 16.21
A charge $-Q_1$ is induced on the inner surface of the metal. Is this true even in the absence of the high degree of symmetry assumed here?

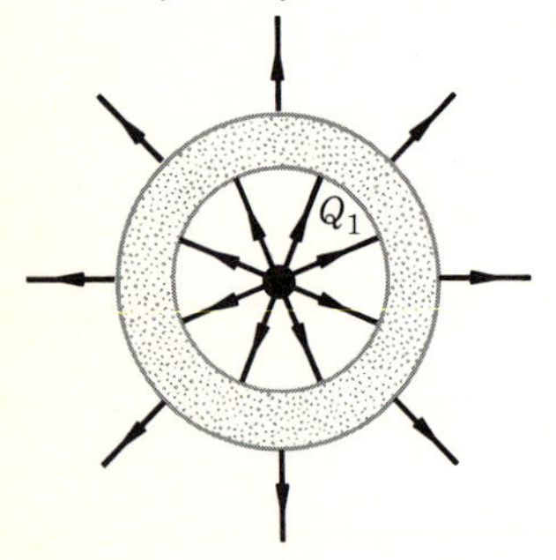

Illustration 16.7 Show by means of a diagram using lines of force that a charge suspended within a metal cavity induces an equal and opposite charge on the interior surface.

Reasoning Let us suppose the metal object to be as shown in Fig. 16.21. Lines of force come out of the positive charge Q_1 suspended in the cavity. Since the field within the metal must be zero in electrostatic situations such as this, the lines from Q_1 must stop when they hit the metal surface. They must therefore be as shown. However, this means the inner surface must possess a negative charge since lines of force go into and terminate only upon negative charges. Moreover, since the number of lines originating on Q_1 is equal to the number terminating on the negative charge, we infer that the charge on the inner surface is $-Q_1$.

If the metal object was neutral before Q_1 was placed in the cavity, and if no charge was allowed to run on or off the object, the object must still have no net charge. Therefore, a charge $+Q_1$, must exist on the outer portions of the object. Can you prove, using lines of force, that the charge must be on the outer surface as indicated? Can you show that the results obtained here are independent of the shape of the object and cavity?

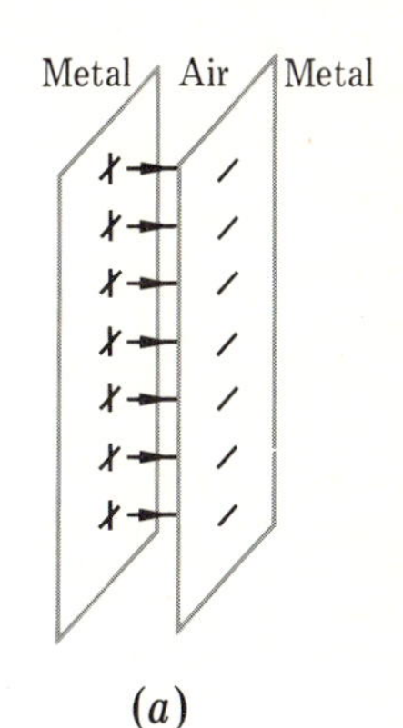

(*a*)

16.10 ELECTRICAL POTENTIAL ENERGY: POTENTIAL DIFFERENCE

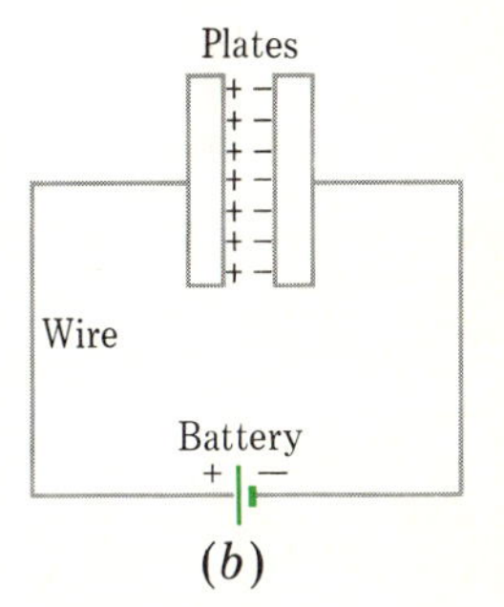

(*b*)

When discussing the movement of an object from place to place within a gravitational field, we made use of the concept of gravitational PE. It will be recalled that the difference in PE between two points was defined to be the work done in moving the object from one point to the other. Moreover, the path along which the object is moved from the one point to the other point was of no consequence; the work done, and therefore the potential energy difference, was the same for all paths. The gravitational field was said to be a *conservative* field since it possessed this property.

In electricity we have the same situation. Consider the electric field between two charged plates. (The plates are charged by connecting them to opposite ends of a battery, as we shall see later. The positive end of the battery places a positive charge on the plate to which it is attached. A negative charge is acquired by the plate connected to the negative side of the battery.) Two such plates are shown in Fig. 16.22. We shall take the plates to have lateral dimensions tremendously larger than the separation between plates. Hence the effect of the edges of the plates will be negligible in the central portion of the system. A close view of a portion of the two plates is shown in part *a* of Fig. 16.22, while the whole circuit is shown in Fig. 16.22*b*.

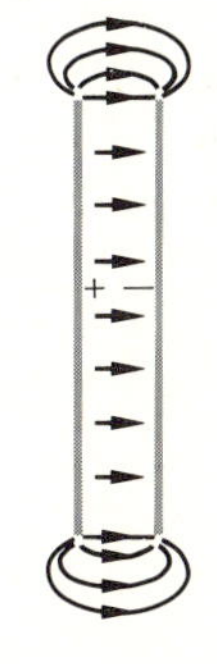

(*c*)

FIGURE 16.22
When a battery is connected to two metal plates, as in (*b*), an electric field is caused to exist between the plates, as shown in (*a*) and (*c*).

If the effects of the edges of the plate are ignored, a positive test charge placed between the plates will experience a force away from the positive plate and toward the negative plate. The electric field E is uniform and in the direction shown in Fig. 16.22*a* and *c*. Near the edges of the plates, the field diagram will appear somewhat as shown in Fig. 16.22*c*. A parallel-plate device of this general sort is called a **capacitor** or **condenser.**

Suppose now that a unit positive charge, an imaginary test charge, is placed between the capacitor plates. It will experience a force E to the right. (The force is E because the electric field is by definition the force per *unit positive test charge.*) Work must be done to move the test charge from place to place within the field. Referring to Fig. 16.23, let us compute the work done in moving the unit positive test charge from A to several other positions between the plates.

In order to move the unit positive charge from A to B in Fig. 16.23, one must pull on it with a force E to balance the force of the electric field. Hence the work done in carrying the unit test charge from A to B is (force) × (distance), or (E) (d). *We define this work to be the potential difference between point A and B.* In general, *the potential difference from point A to point B is the work done in carrying a unit positive test charge from A to B.* Using the symbol V_A for the potential at point A, one has *for the special case of a region in which* **E** *is constant,*

DEFINITION

Potential Difference in Uniform Field

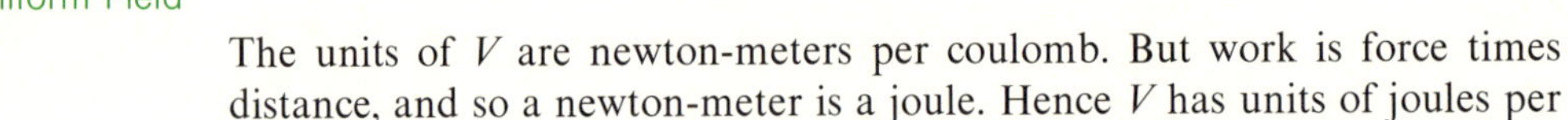

$$V_B - V_A = Ed \qquad \text{constant field} \tag{16.4}$$

The units of V are newton-meters per coulomb. But work is force times distance, and so a newton-meter is a joule. Hence V has units of joules per coulomb. This unit is called the **volt** (abbreviated V). *Potential difference is measured in volts.*

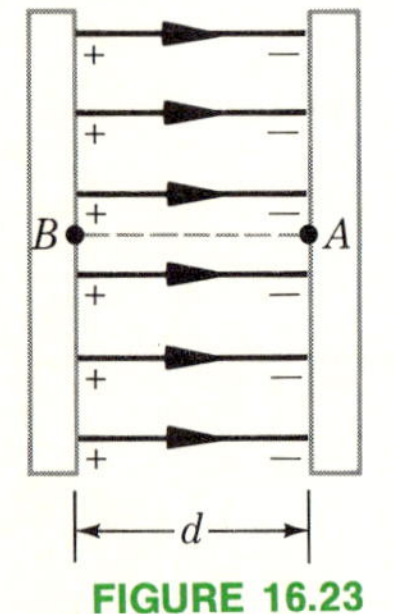

FIGURE 16.23

Since the work needed to carry a unit positive charge from A to B is Ed, the potential difference between the plates is Ed.

It is customary to say that a point such as B is at a higher potential than point A. That is to say, the unit positive test charge had to be pulled from A to B. Hence, by analogy to the gravitational case, B is uphill from A, in other words, at a higher potential than A. One also notices that a positive charge released from point B will be accelerated toward A. The electric field E exerts a force upon the charge, and the charge will fall under the action of that force. A close analogy exists between a body falling from a place of high potential to a lower potential under the action of a gravitational field and the falling of a charge under the action of an electric field.

Let us emphasize, once again, that *potential difference between two points A and B, $V_B - V_A$, is the work done in carrying a unit positive charge from point A to point B.* This work is expressed in units of joules per coulomb or, more commonly, volts. If not a unit charge but a charge q is to be moved from point A to point B, the work required would be

Work Related to Potential Difference

$$\text{Work} = (V_B - V_A)q \tag{16.5}$$

This work is stored in the system as electrical potential energy.

Refer to Fig. 16.24 on the next page. How much work must we do to move a unit positive test charge from A to B? To hold the test charge in place, we must exert a force to the left on it. This force is needed to counterbalance the effect of the electric field on the test charge. But if we move the charge from A to B, our balancing force does no work since the direction of motion is perpendicular to the force. Indeed, we see that no work is needed to move the test charge in a direction perpendicular to the electric field. Therefore, no potential difference exists between points A and B in Fig. 16.24. In fact, it should be clear that all points on the dashed line through A and B are at the same potential. We call this line of constant potential an **equipotential line.** Moreover, the plane which lies through this line and is parallel to the plate is

Equipotentials

a constant potential plane, an **equipotential plane.** *No work is done in moving a charge along an equipotential line or equipotential plane* since such motion is always perpendicular to the lines of force, i.e., the electric field. Conversely, *the lines of force are always perpendicular to the equipotential lines.*

There are an infinite number of equipotential lines and planes between the plates in Fig. 16.24: It is not difficult to show that no total work is done upon a charge in moving it along a path which begins and ends on the same equipotential line or plane. Can you show this? The same reasoning can be used to prove that the work done in carrying a charge from point A to a point such as C in Fig. 16.24 is independent of the path taken. We therefore conclude that *the static electric field is a conservative field. The electric potential difference between two points is a constant not dependent upon the path used for its computation.*

Conservative Field

Illustration 16.8 A 20-V battery is used to charge the condenser plates of Fig. 16.24. How large is the electric field between them? The separation of the plates is 0.10 cm. (The battery causes the potential difference between the plates to be 20 V.)

Reasoning By definition, the fact that the potential difference between the plates is 20 V means that 20 J of work is required to carry a 1-C test charge from the one plate to the other. Since the force on the unit test charge is the field strength E, and since E is constant, we have

$$\frac{\text{Work}}{\text{Charge}} = \left(\frac{\text{force}}{\text{charge}}\right)(d)$$

$$20\text{ V} = (E)(0.10 \times 10^{-2}\text{ m})$$

from which

$$E = 20{,}000\text{ V/m}$$

The units volts per meter are equivalent to the units used previously, newtons per coulomb. Can you show this?

16.11 ABSOLUTE POTENTIALS

So far we have been concerned only with differences in potential because, as in gravitational potential, the choice of a position for zero PE is merely a matter of convenience. The gravitational PE can be measured with respect to any point we choose: the table top, the ground, the top of a building, or wherever. Similarly, in electrical-PE problems, the zero-PE location is a matter of choice. In electrical-circuit theory, a particular wire in the circuit may be attached to the ground, e.g., soldered to a water pipe. This point would usually be taken to have zero PE. In other cases, however, a different zero for electrical potential is frequently taken.

When dealing with atoms and molecules, we frequently specify the zero

FIGURE 16.24

Points A and B are on an equipotential line.

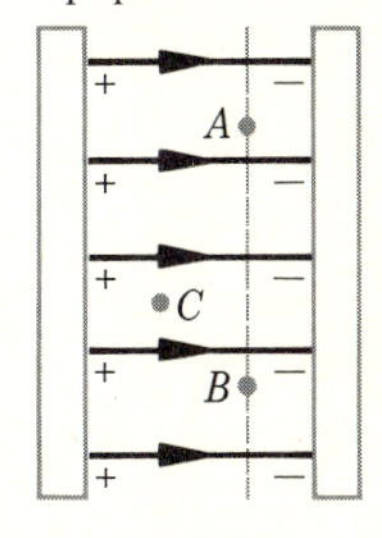

of electrical PE in a different way. To illustrate this method of approach, let us refer to Fig. 16.25, where we see a charged sphere carrying charge $+Q$. This sphere might represent a proton, for example. (The proton is a hydrogen atom nucleus and carries a charge $e = +1.60 \times 10^{-19}$ C.) It is customary in such cases to take the zero of electrical PE to be at an infinite distance away from the charge. Let us now see the consequences of such a choice.

DEFINITION
Absolute Potential

Consider the point B in Fig. 16.25 which is at a radius r from the center of the charge. *We define the* **absolute potential** *at B to be the work done in carrying a unit positive test charge from infinity up to B*. In effect, what we are doing is as follows. So far, we have discussed situations in terms of potential differences, $V_B - V_A$. Now, however, we specify that point A is to be taken at infinity. Further, we specify that the potential at infinity is to be taken as zero so that $V_A = 0$. Then V_B becomes what we refer to as the absolute potential at B. Notice carefully that *when we speak of the absolute potential at a point, we are really speaking about the potential difference between infinity and that point*.

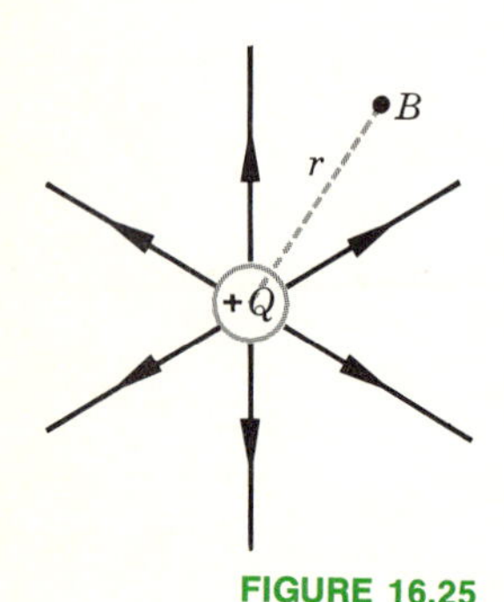

FIGURE 16.25
We define the absolute potential at B to be the work done in carrying the unit positive test charge from infinity up to B.

A particularly simple expression can be found for the absolute potential outside a single, isolated charge such as the one shown in Fig. 16.25. We need to compute the work done in carrying our unit test charge from infinity up to the point B, for example. This computation is not as simple as finding the potential difference between two parallel plates. In the parallel-plate case, E is constant, and so the force on the test charge is constant. Hence the potential difference was simply Ed.

But in the present case, E becomes stronger as we come closer to B. At infinity, E is zero. Clearly, the value of E changes, and so does the force on our test charge as we bring it in from infinity. In spite of that fact, the work done in bringing the test charge from infinity to point B is still easily found using calculus.

We can only state the result here. *The absolute potential V, at point B of Fig. 16.25, due to the charge $+Q$,* that is, the work needed to carry a unit positive charge from infinity up to point B, is

Point Charge Potential

$$V = k\frac{Q}{r}$$

or

$$V = (9 \times 10^9 \text{ N}\cdot\text{m}^2/\text{C}^2)\left(\frac{Q}{r}\right) \tag{16.6}$$

This equation, like Coulomb's law, *applies only to point charges and uniformly charged spheres*. In other cases, the distance r is not usually well defined.

Illustration 16.9 Compute the absolute potential at point B in Fig. 16.25 if $r = 50$ cm and $Q = 5 \times 10^{-6}$ C.

Reasoning From Eq. (16.6) we have at once

$$V = (9 \times 10^9 \text{ N}\cdot\text{m}^2/\text{C}^2)\left(\frac{5 \times 10^{-6}\text{ C}}{0.50 \text{ m}}\right) = 90{,}000 \text{ V}$$

Verify the units given with the answer.

Illustration 16.10 Compute the absolute potential at point B in the vicinity of the three charges shown in Fig. 16.26.

Reasoning When a unit positive test charge is carried from infinity up to point B, work is done against three separate forces. It is done against the repulsions of the two positive charges, and the negative charge actually tends to pull the positive test charge in toward B. Hence the work done consists of two positive amounts of work, done against the repulsions of the two positive charges, and a negative amount of work, the work done *on* the test charge by the attraction of the negative charge. These are each found by use of Eq. (16.6).

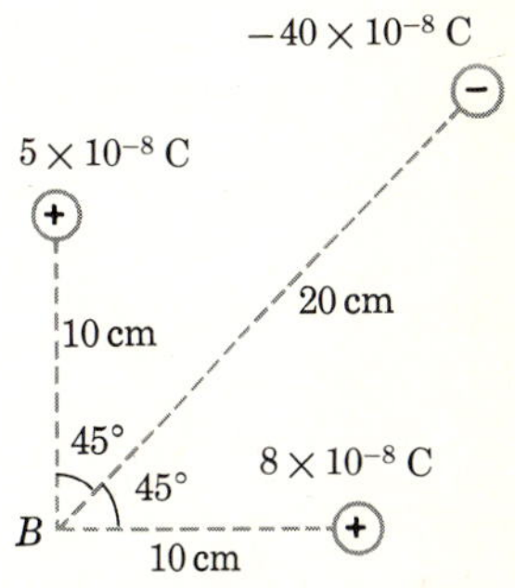

FIGURE 16.26

Find the absolute potential at point B.

Due to $+5 \times 10^{-8}$: $\quad V_1 = (9 \times 10^9)\left(\dfrac{5 \times 10^{-8}}{0.10}\right) = 4500 \text{ V}$

Due to $+8 \times 10^{-8}$: $\quad V_2 = (9 \times 10^9)\left(\dfrac{8 \times 10^{-8}}{0.10}\right) = 7200 \text{ V}$

Due to -40×10^{-8}: $\quad V_3 = -(9 \times 10^9)\left(\dfrac{40 \times 10^{-8}}{0.20}\right) = -18{,}000 \text{ V}$

Unlike forces and fields, work and potential are scalars, not vectors. Hence, being numbers without direction, merely plus or minus, they may be added directly. The result for the absolute potential at point B is $V_B = V_1 + V_2 + V_3$ so

$$V_B = -6300 \text{ V}$$

Since the result is negative, we conclude that the unit positive charge actually has less potential energy at B than it did at infinity. This must mean that the effect of the attraction of the negative charge for it actually was stronger than the repulsion of the two positive charges.

Illustration 16.11 A hydrogen atom consists of an electron in orbit about a proton as pictured schematically in Fig. 16.27. The electron is at a radius $r = 0.53 \times 10^{-10}$ m from the center of the proton. Find the work done in pulling the electron from its orbit out to infinity; i.e., in tearing the electron loose from the atom.

FIGURE 16.27

In the hydrogen atom, the electron is often pictured in orbit about the proton.

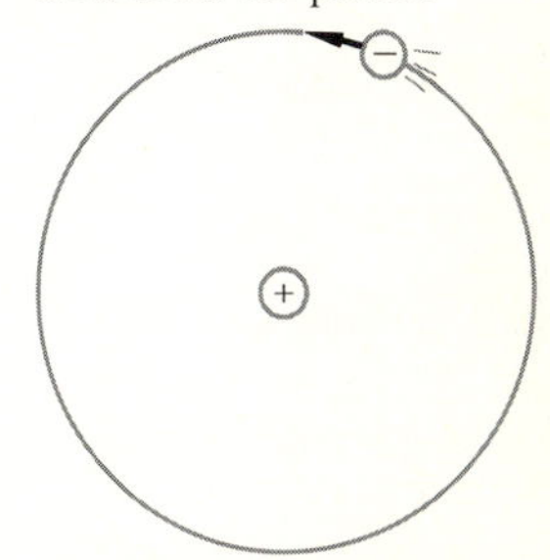

Reasoning Let us first consider an isolated proton. The situation is similar to that discussed in connection with Fig. 16.25. We recall that the absolute potential at point B is given by Eq. (16.6) to be $V = kQ/r$. This is, by definition, the potential difference between infinity and point B.

In the present case, B will be a point on the electron's orbit so $r = 0.53 \times 10^{-10}$ m. Also, Q is the charge on the proton, 1.60×10^{-19} C. Substituting these values gives

$$V = k\frac{Q}{r} = 27.2 \text{ V}$$

This is the potential difference between the electron's orbit position and infinity.

If we are to pull the electron out to infinity, we must carry it through this potential difference. But Eq. (16.5) tells us that the work done in such a case is

$$\text{Work} = |\text{potential difference}| \cdot |q|$$

where q is the charge on the electron.

In this case this becomes

$$\text{Work} = (27.2\ \text{V})(1.60 \times 10^{-19}\ \text{C}) = 4.3 \times 10^{-18}\ \text{J}$$

where use has been made of the fact that the charge on the electron is -1.60×10^{-19} C.

As we see, 4.3×10^{-18} J of work must be done to tear the electron from the hydrogen atom, i.e., to ionize the atom. However, the electron originally had KE in its orbit and so it helps do this work. It turns out that this KE is half as large as the total work which must be done. As a result, an external agent can cause a hydrogen atom to ionize by furnishing only half of the required 4.3×10^{-18} J.

SUMMARY

There are two kinds of charges designated + and −. The electron carries a charge $-e = -1.60 \times 10^{-19}$ C while the charge on a proton is $e = 1.60 \times 10^{-19}$ C. Like charges repel each other and unlike charges attract each other.

The force on a point charge q_1 at a distance r from a point charge q_2 is given by Coulomb's law to be

$$F = k\frac{q_1 q_2}{r^2}$$

where q_1 and q_2 are measured in coulombs and $k \approx 9 \times 10^9$ in SI units.

The electric field strength **E** at a point is the force experienced by a unit positive test charge placed at that point. Its units are newtons per coulomb or volts per meter. Electric fields are represented by field lines which are tangent to **E.** These lines emerge from positive charges and end on negative charges. The field is strongest where the lines are closest together.

Under electrostatic conditions, the field in the body of a metal is zero. The field lines end on the metal surface and are perpendicular to the surface. All the excess charge on a metal object resides on the surface of the object.

The static electric field is a conservative field. By definition, the potential difference from point A to point B is the work done in carrying unit positive test charge from A to B. Its units are joules per coulomb, or volts, and it is represented by $V_B - V_A$. If the line connecting B and A is along the direction of a constant field E, then the potential difference has a magnitude Ed, where d is the distance from A to B.

Equipotential lines, surfaces, and volumes consist of points which are all at the same potential. The electric field lines always intersect equipotential lines and surfaces at right angles.

The potential difference between any point and infinity is called the absolute potential at the point. In this terminology, points at infinity are assigned a value of zero electric potential. The absolute potential at a distance r from the center of a point charge q is kq/r.

MINIMUM LEARNING GOALS

Upon completion of this chapter you should be able to do the following:

1. State what the two kinds of charge are and which kind exists on the proton and on the electron. Describe the force which one charge exerts upon another charge.

2. Distinguish between insulators, conductors, and semiconductors.

3. Explain how an object can be charged by conduction and by induction. Describe qualitatively how charges on a given metal object redistribute when a charged object is brought nearby.

4. State the conclusions one can draw from the Faraday ice-pail experiment.

5. Use Coulomb's law to find the force on a given charge in the vicinity of other point charges.

6. Explain what information can be gleaned from an electric field diagram. In so doing, define E, explain how to draw field lines, and tell how the lines behave near point charges and near and in metals.

7. Use the relation $F = qE$ in simple situations.

8. Sketch the electric field lines in the vicinity of simple charged objects.

9. Define what is meant by a test charge and use it to define the electric potential difference between two points.

10. Compute the potential difference between two specified points in a given constant electric field. Show the equipotential surfaces and lines in simple situations involving specified charged objects.

11. Use the relation $W = (V_B - V_A)q$ in simple, specified situations.

12. Explain in words what is meant by the absolute potential at a point. Give its value at a specified distance from a point charge. Find the absolute potential at a point due to several specified point charges near the point.

IMPORTANT TERMS AND PHRASES

You should be able to define or explain each of the following:

$e = 1.602 \times 10^{-19}$ C
Insulator versus conductor; semiconductor
Electroscope
Electrical ground
Induced charge
Faraday ice-pail experiment
Coulomb's law
Excess charges are a small fraction of the total charge
Electric field and E
Electric potential difference
The static electric field is conservative
Equipotentials
Absolute potential at a point

QUESTIONS AND GUESSTIMATES

1. A plastic comb or ruler can be given an electrostatic charge by rubbing it vigorously with a dry cloth. How could you determine the sign of the charge on the comb or ruler?

2. A plastic comb or pencil charged by the method indicated in item 1 will attract tiny bits of paper. Explain why. (*Hint:* The usually slightly moist paper conducts electricity slightly.) Why does a phonograph record wiped with a dry cloth usually attract dust and lint?

3. Sometimes your whole body becomes charged by walking across a deep carpet or by sliding across a plastic car seat. If you then extend your finger toward a metal door knob or handle, a spark will jump from it. Explain.

4. How can you charge a metal object positively by use of a positively charged plastic comb? Using a negatively charged plastic pen?

5. Two points A and B are at the same potential. Does this necessarily mean that no work is done in carrying a positive test charge from one point to the other? Does it mean that no force will have to be exerted to carry the test charge from one point to the other? Explain.

6. Can two equipotential surfaces intersect? Explain.
7. Prove that all points in a piece of metal are at the same potential if no current is flowing in the metal.
8. The absolute potential midway between two equal but oppositely charged point charges is zero. Can you find an obvious path along which no work would be done in carrying a positive test charge from infinity up to this point? Explain.
9. Starting from the fact that a piece of metal which has no current flowing in it is an equipotential body, prove that the electric field inside a hollow piece of metal is zero.
10. Draw the electric field between a highly charged cloud and a lightning rod mounted on a building. Why must the rod be securely grounded?
11. A metal cube is hung from an insulating string. If it is charged positively, approximately how is the charge distributed on it?
12. If V is zero at a point, must E be zero there as well?
13. What can be said about E in a region where V is constant?
14. Prove that all points of a metal object are at the same potential under electrostatic conditions.

PROBLEMS

1. Two point charges, $+6 \times 10^{-4}$ and $+2 \times 10^{-5}$ C, are separated by a distance of 0.30 m. Find the force on either of them.
2. A point charge 8×10^{-6} C is placed on the x axis at $x = 0$ while an unknown point charge is placed on the axis at $x = 50$ cm. The force on the unknown charge is 300 N in the $-x$ direction. Find the sign and magnitude of the unknown charge.
3. In the Bohr model of the hydrogen atom, an electron circles the nucleus (a proton) at a radius of 0.53×10^{-10} m. (*a*) How large a force does the proton exert on the electron? (*b*) This force supplies the centripetal force which holds the electron in orbit. How fast is the electron moving? (The charge on each particle has magnitude e. The electron $m = 9.1 \times 10^{-31}$ kg.)
4. The following three point charges are placed on the x axis: $+4\,\mu$C at $x = 0$, $+2\,\mu$C at $x = 30$ cm, and $+6\,\mu$C at $x = 60$ cm. Find (*a*) the force on the 2-μC charge; (*b*) the force on the 6-μC charge.
5. As shown in Fig. P16.1, four point charges are placed on the four corners of a square. If $q_1 = q_2 = q_3 = q_4 = 5\,\mu$C and $a = 30$ cm, find the magnitude and direction of the force on q_3.
6. In a certain region of space, a tiny pith ball with charge -3.0×10^{-12} C experiences a force of 5.0×10^{-7} N in the $+x$ direction. What is the electric field magnitude and direction in this region?
7. Two charges are placed on the x axis: $+6\,\mu$C at $x = 60$ cm and $-3\,\mu$C at $x = 0$. Find the electric field at (*a*) $x = 30$ cm and (*b*) $x = 100$ cm.
8. Find the electric field strength at the center of the square in Fig. P16.1 if (*a*) $q_1 = q_2 = q_3 = q_4$; (*b*) $q_1 = q_2 = 2\,\mu$C and $q_3 = q_4 = -4\,\mu$C and $b = 50$ cm.
9. If $q_1 = q_3 = -6.4\,\mu$C and $q_2 = 2\,\mu$C in Fig. P16.2, find the electric field strength at point A.
10.** (*a*) In Fig. P16.2, if $q_1 = q_2 = +6.4\,\mu$C, how large must q_3 be if the field at A is to be directed along a line parallel to the line of the three charges? (*b*) How large is the field in that case?
11. The potential difference from A to B is $+60$ V. How much work is needed to carry a (*a*) proton ($q = e$) from A to B and (*b*) an electron ($q = -e$) from A to B?
12. Two parallel metal plates are 0.20 cm apart and are

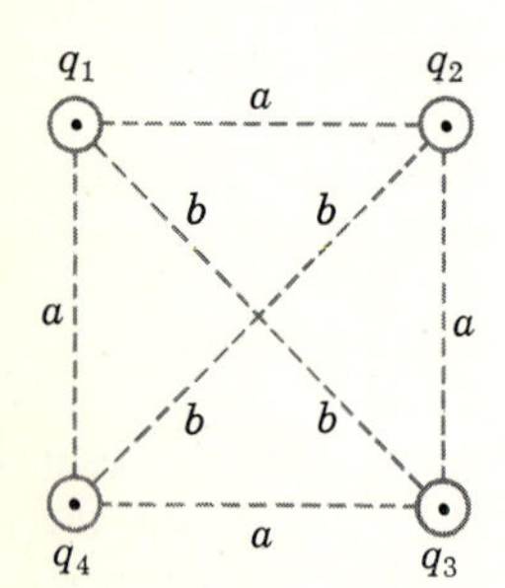

FIGURE P16.1

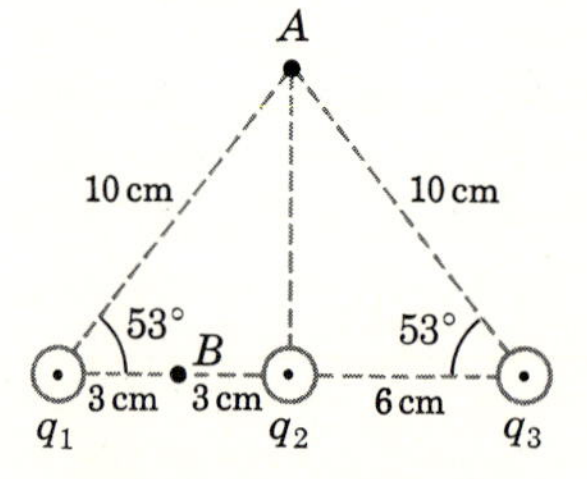

FIGURE P16.2

connected to a 6-V battery. Find the field between the plates.

13. Two parallel metal plates are connected to a 6-V battery. If the field between the plates is 300 V/m, (*a*) how far are the plates separated? (*b*) What would be the magnitude of the force on an electron between the plates? Charge on the electron is -1.6×10^{-19} C.

14. Two charges +6 and $-3\ \mu$C are 60 cm apart. Find the absolute potential at a point halfway between them. How much work is required to carry a 3-μC charge from infinity up to that point?

15. What is the absolute potential at the center of the square in Fig. P16.1 (*a*) if $q_1 = q_2 = q_3 = q_4 = 2\ \mu$C and (*b*) if $q_1 = q_2 = q_3 = 2\ \mu$C and $q_4 = -6\ \mu$C? Take $b = 30$ cm.

16. In Fig. P16.2, if $q_1 = +2\ \mu$C, $q_2 = -8\ \mu$C, and $q_3 = +4\ \mu$C, find the absolute potential at point A.

17. In Fig. P16.2, (*a*) how large is the potential difference between point A and B if $q_1 = +3\ \mu$C, $q_2 = -6\ \mu$C, and $q_3 = -9\ \mu$C? (*b*) What are V_A and V_B?

18.* In Fig. P16.3 is shown a particle between two parallel oppositely charged plates. Assume the particle to be an electron ($q = -e$, $m = 9.1 \times 10^{-31}$ kg). (*a*) How large must the electric field be between the plates if the particle is to remain motionless under the combined effects of gravity and the electric field? (*b*) Should the top plate be positive or negative? (*c*) If the distance between the plates is 2.00 mm, what is the required potential difference between the plates?

19.* The particle between the plates in Fig. P16.3 has a mass of 4×10^{-12} g. It can be held stationary against the pull of gravity by making the potential difference between the two plates 245 V with the upper plate negative. The distance between the plates is 2.00 mm. Find (*a*) the electric field strength between the plates and (*b*) the sign and magnitude of the charge on the particle.

20.* The parallel metal plates shown in Fig. P16.4 are separated by a distance of 10 cm, and the voltage difference between them is 28 V. A small pith ball of 0.60 g mass hangs by a thread from the upper plate. What is the tension in the thread if the ball carries a charge of 20 μC? Two answers are possible. Find both.

21.** A point charge of $+2\ \mu$C is placed on a straight line defined by a +27- and a +3-μC charge which are 100 cm apart. (*a*) Where must the 2-μC charge be placed if the resultant force on it is to be zero? (*b*) Is it in stable or unstable equilibrium?

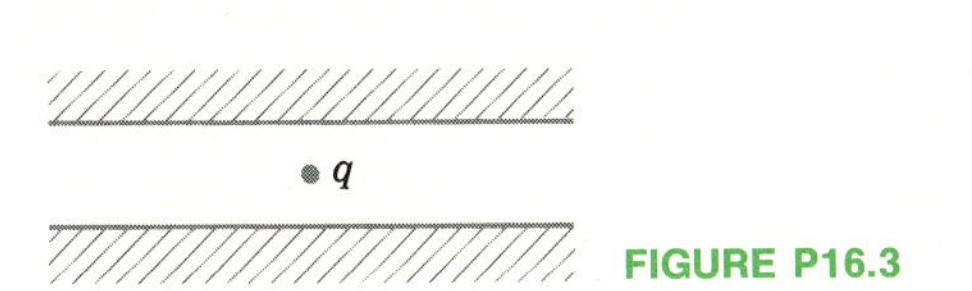

FIGURE P16.3

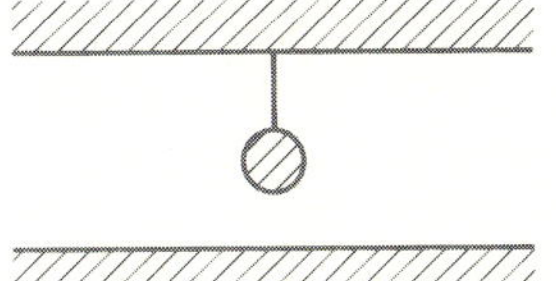
FIGURE P16.4

CIRCUIT ELEMENTS AND THEIR BEHAVIOR

17

In this chapter we shall learn how a potential difference gives rise to current in wires. Three basic electrical devices will be introduced, the battery, the resistor, and the capacitor. The laws governing circuits containing these devices will be discussed as a preparation for the study of direct-current circuits, which will be investigated in the next chapter.

17.1 BATTERIES AS A SOURCE OF POTENTIAL DIFFERENCE

The basic purpose of a battery in most electrical circuits is to supply the energy needed to operate the circuit. Although most batteries are essentially chemical devices, other types of batteries are now becoming prominent. The ordinary lead-cell battery, the common automobile battery, makes use of a chemical reaction to supply energy. This is likewise true of the "dry cell," which is not dry inside in spite of its name. Perhaps you have heard of "solar" cells, which are used to supply electrical energy to devices in space. These batteries operate on quite different principles and transform light and heat energy directly into electrical energy. Other types of nonchemical batteries are in the process of development.

Symbol for a Battery

For most purposes in this book, we shall ignore the internal workings of the battery except to consider it a source of energy and potential difference. The basic property of all batteries is that they provide two terminals (actually metal posts on the battery box) which are at different potentials. *The symbol ordinarily used for a battery is* $\overset{+}{-}|\overset{-}{\vdash}$ *where the long line, marked +, is the positive terminal of the battery and the short line, marked −, is the negative terminal.* Usually the + and − symbols are left off, and the reader is expected to know that the long side is positive. (In passing we might note that the terminals on batteries are usually stamped + and −, or, in some cases, the positive terminal is merely painted red.)

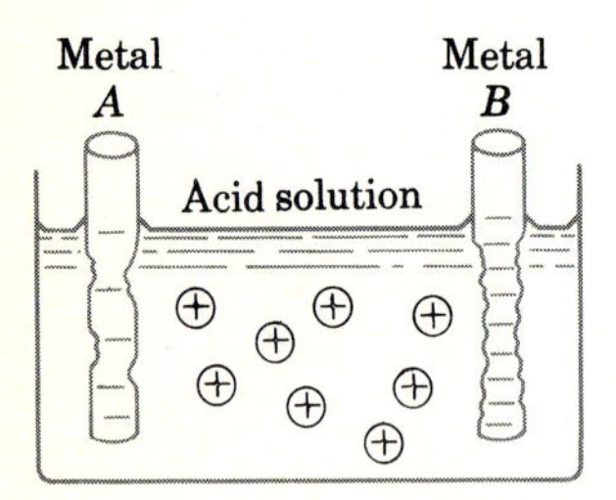

FIGURE 17.1

Positive ions go into the solution, leaving electrons behind on the electrodes. In the case shown, metal B has lost more ions per unit of volume than metal A. Which electrode is at the higher potential?

A simple chemical battery (or cell) can be made by immersing two dissimilar metal rods in a dilute acid, as shown in Fig. 17.1. Most metals dissolve at least slightly in the acid. When the metal dissolves, each atom of the metal leaves at least one electron behind on the rod and enters the solution as a positive ion. The electrode (or rod) from which the ion came is charged negatively by this process. Eventually, it becomes so negative that equal numbers of positive ions are attracted back to the negative electrode, so that the net number of ions leaving the electrode becomes zero. Since the electrode has an excess of negative charge, a potential difference exists between it and the solution.

In Fig. 17.1 the potential difference between the electrode and solution is greater for metal B than for metal A, since B dissolves more extensively than A. Since both electrodes are negative, they are both at a lower potential than the solution. However, metal B is at a lower potential than A, and so, going from electrode A to electrode B, one goes from a higher to a lower potential. This potential difference is called the **electromotive force** (a misnomer), or emf, of the battery. Therefore rod A will be the positive terminal of the battery, and rod B will be the negative terminal. *The emf of a battery will be represented by* $\mathcal{E}$. Do not confuse it with the symbol E used for electric field. Instead, *the* **emf** $\mathcal{E}$ *is the potential difference between the terminals of a battery when the battery is isolated electrically.*

DEFINITION

17.2 CHARGE MOTION IN AN ELECTRIC FIELD

We stated in the last section that the major purpose of a battery is to supply energy. Let us now investigate a few examples which show how a battery supplies energy to charges. Consider first the situation shown in Fig. 17.2*a*. We see there a battery connected to two metal plates. The situation is also shown in the shorthand diagram in Fig. 17.2*b*. As indicated there, we shall assume the two plates to be in vacuum.

The battery places excess charge on the two plates. As indicated, plate A is positive since it is connected to the positive side of the battery. This transfer of charge to the plates requires only a tiny fraction of a second. After that, the situation is electrostatic. You will recall that under electrostatic conditions metal solids are equipotential volumes. For that reason, the wire from C to A and the plate at A are all at the same potential. We say that they are all at the same electrical level. Similarly, points D and B as well as plate B are all at identical levels. But point D is 1.5 V lower than point C. Therefore, plate B must be 1.5 V lower than plate A. The potential difference from B to A is 1.5 V, A being at the higher level.

Suppose now that a positive charge is released from plate A as shown in Fig. 17.2*b*. It will be repelled by plate A and attracted by plate B. As a result, it will accelerate from plate A to plate B. We see, in effect, the positive charge "falls" from the high-potential plate A to the low-potential plate B.

Positive charges fall from high potentials to lower potentials.

Let us now see how much energy the charge acquires as it falls from A to B.

You will recall from the last chapter that work is needed to carry a positive charge "uphill" through a potential difference V. In the present case, plate A is at a potential V volts higher than plate B, where $V = 1.5$ V in this case. To carry a charge q from B to A requires work in the amount qV. (This follows from the fact that V is the work needed to carry *unit* positive charge from B to A.)

But when a charge q is released at plate A and falls to B, it gains back energy equal to this work. We therefore see that

When a charge q falls through a potential difference V, it gains energy in the amount qV.

Let us now make use of these facts in a few examples.

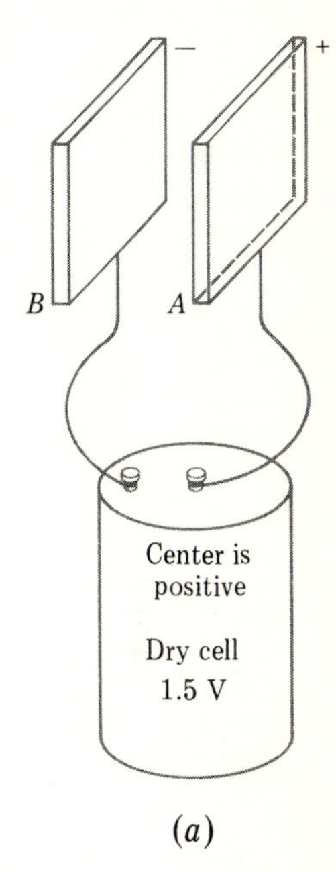

(*a*)

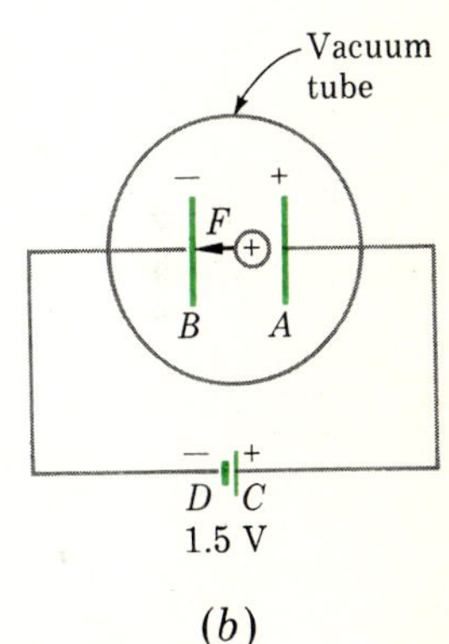

(*b*)

FIGURE 17.2
The potential difference from B to A is 1.5 V, the emf of the battery.

Illustration 17.1 The positive charge in Fig. 17.2*b* is a proton ($m = 1.67 \times 10^{-27}$ kg, $q = +e$). If it is released at plate A, find its speed just before it strikes plate B.

Reasoning The charge falls through a potential difference $V = 1.50$ V. It therefore gains energy in the amount qV, where $q = 1.6 \times 10^{-19}$ C. This energy will appear as an increase in the KE of the proton. Since the initial speed of the proton was zero, we see that

$$\text{Increase in KE} = \tfrac{1}{2}mv_f^2 - \tfrac{1}{2}mv_0^2 = \tfrac{1}{2}mv_f^2$$

But since this increase is equal to qV,

$$qV = \tfrac{1}{2}mv_f^2$$

from which

$$(1.6 \times 10^{-19}\ \mathrm{C})(1.5\ \mathrm{V}) = \tfrac{1}{2}(1.67 \times 10^{-27}\ \mathrm{kg})v_f^2$$

Solving gives $v_f = 1.7 \times 10^4$ m/s. Notice how very fast the proton is moving even after having fallen through such a small potential difference.

Illustration 17.2 Many electronic devices give visual displays of electrical signals. The picture on a TV tube is a familiar example. To create the picture, an electron gun shoots a beam of electrons at the fluorescent material on the end of the tube. As the beam sweeps repeatedly across the tube face, it traces out the picture one sees. A highly simplified electron gun is shown in Fig. 17.3. Electrons boil out of a white-hot filament within a metal enclosure. Many of these electrons wander out of the enclosure through a tiny hole as shown. They then "see" that the enclosure is charged negative and the plate on the right is positive. As a result, they accelerate toward the plate. By the time they reach the plate they are moving at high speed, and some electrons shoot through the tiny hole in the plate. As a result, a pencillike beam of electrons shoots from this device. (In practice, the gun has a more complicated structure than shown.)

Electron Gun

In a typical TV set, the potential difference through which the charges accelerate is of the order of 20,000 V. Find the speed of the electrons which shoot from the gun.

Reasoning Notice in this case that the charge falls from − to +. The electron, being negative, behaves oppositely to a positive charge. In either case, however, the change in KE of the particle has a magnitude $|qV|$. We have, since $q = -1.6 \times 10^{-19}$ C and $V = 2 \times 10^4$ V, that

$$|\text{Change in KE}| = (1.6 \times 10^{-19}\ \mathrm{C})(2 \times 10^4\ \mathrm{V}) = 3.2 \times 10^{-15}\ \mathrm{J}$$

But the electron had negligible initial KE, and so its final KE $= \tfrac{1}{2}mv_f^2$ is equal to 3.2×10^{-15} J. The mass of an electron is 9.1×10^{-31} kg, and so we

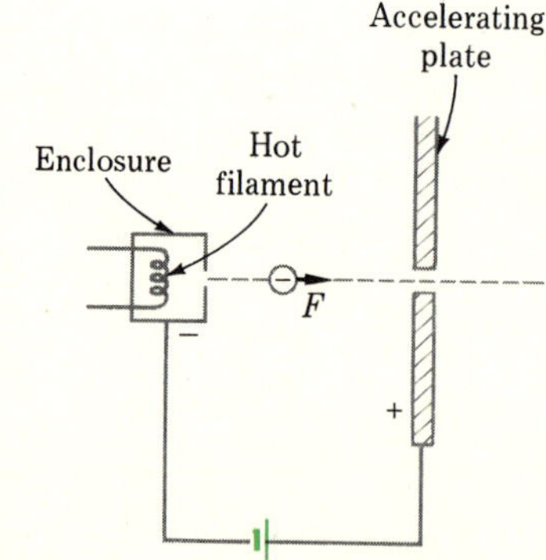

FIGURE 17.3
The electron gun shoots a beam of electrons toward the right. In a typical TV tube, the gun is considerably more complicated than the one shown. The beam travels through vacuum and eventually strikes a fluorescent screen, which would be far to the right.

have

$$v_f = \sqrt{\frac{3.2 \times 10^{-15}\ \text{J}}{\frac{1}{2}(9.1 \times 10^{-31}\ \text{kg})}} = 8.4 \times 10^7\ \text{m/s}$$

As we see, this result is approaching the speed of light, 30×10^7 m/s. In such cases, our result is not strictly correct because the mass of the particle is no longer equal to its rest mass. (This topic is discussed more fully in Chap. 26.) When the situation is analyzed exactly by use of the theory of relativity, one finds $v_f = 8.1 \times 10^7$ m/s. As we see, it is easy to accelerate an electron to speeds which are large enough to show relativistic effects.

17.3 THE ELECTRONVOLT ENERGY UNIT

In our studies of atomic physics the energy of electrons and similar particles will be needed frequently. In most cases, these energies will be acquired by movement of the charged particle through an electrical potential difference. For example, to remove the single electron from a hydrogen atom, one must furnish it with an energy equivalent to that which the electron would acquire in falling through a potential difference of 13.6 V. In other words, the energy required to ionize the hydrogen atom, i.e., to tear its electron loose, is Vq, where V is 13.6 V and q is the electronic charge, 1.6×10^{-19} C. Upon evaluation, this turns out to be 2.18×10^{-18} J.

It is inconvenient to multiply V by q when stating the energy of a particle which has fallen through a potential difference. We therefore *define a new energy unit called the* **electronvolt** (eV). *By definition*

$$1\ \text{eV} = 1.60 \times 10^{-19}\ \text{J}$$

The eV Unit

where 1.60×10^{-19} C is the charge quantum. Since the energy of a charge q which has fallen through a potential difference V is Vq joules, one has

$$\text{Energy in eV} = \frac{\text{(potential difference)(charge)}}{\text{charge quantum}}$$

As typical examples, if an electron falls through a potential difference of 2 million volts (MV), its energy is 2×10^6 eV, or 2 MeV, where MeV stands for mega (or million) electronvolts. However, an α particle (which has a charge of 3.2×10^{-19}, or two charge quanta) will have an energy of 4 MeV when it falls through a potential difference of 2×10^6 V. Of course the electronvolt is not a proper SI unit, and so we must always change it to joules (by multiplying by 1.6×10^{-19}) before using it in fundamental formulas.

17.4 ELECTRICAL CURRENT

As we have seen in our previous discussion, most of the definitions in electricity are made in terms of positive, not negative, charge. This is reputed to have come about in the following way. In the mid-1700s, when electricity was still a curiosity, experimental investigations of it were carried out chiefly as a hobby. One such investigator was Benjamin Franklin. He made several discoveries involving electricity, and he designed one of the first lightning rods. In addition, he was an early proponent of the idea that charge flows from a battery through a wire. He assumed that the charge which flows in a wire is positive charge.

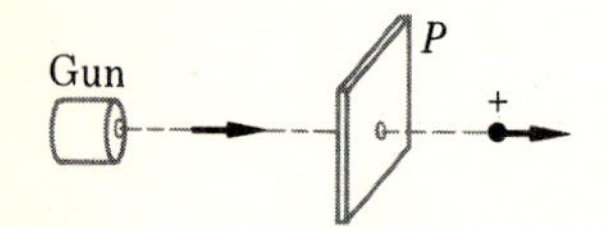

FIGURE 17.4

The beam of moving charges passes through the hole in the plate P. If a charge ΔQ passes through the hole in time Δt, the current is $\Delta Q/\Delta t$.

It was not until about a century later that Franklin's choice was found to be wrong. We now know that electrons, negative charges, flow in wires. Even so, it is surprisingly difficult to prove this fact directly. In nearly all experiments, a flow of negative charge in one direction is equivalent to a flow of positive charge in the other. Because of this fact, there is no overwhelming reason to revise the definitions. We therefore still give preference to positive charge in stating definitions of electrical quantities. Let us now see how one defines the flow rate for moving charges.

If we refer to Fig. 17.4, we see a beam of positive charge being shot toward the right from a positive charge gun. The beam shoots through a hole in the plate P. We wish now to define a quantity which measures how much charge passes through this hole each second. This quantity will be called the electrical current carried by the beam, and we shall designate it by the symbol I. *If a charge ΔQ is carried past a given point by the beam in a time Δt, then the* **current** *carried by the beam is*

DEFINITION

$$I = \frac{\Delta Q}{\Delta t} \tag{17.1}$$

Electrical Current

The units of electrical current, coulombs per second, are given the name **amperes** (A). Notice that if the moving charge is positive, ΔQ and I will be positive. But if the moving charge is negative, ΔQ and I are negative. For this reason

> A flow of negative charge results in a current in a direction opposite to the flow direction.

Although beams of charge are quite common (in TV sets, for example), we are even more familiar with charge flow in wires. To see how current is defined in this case, refer to Fig. 17.5. In a metal, there are many free electrons. Under the action of an electric field, they move through the metal. Such a movement of charge is shown in the figure. However, in keeping with our emphasis on positive charge, we couch our definition in terms of positive charge. *If ΔQ is the quantity of charge passing through a given cross section of*

DEFINITION

FIGURE 17.5

The current in the wire in amperes is defined to be the quantity of positive charge in coulombs flowing through a cross section such as A in 1 s.

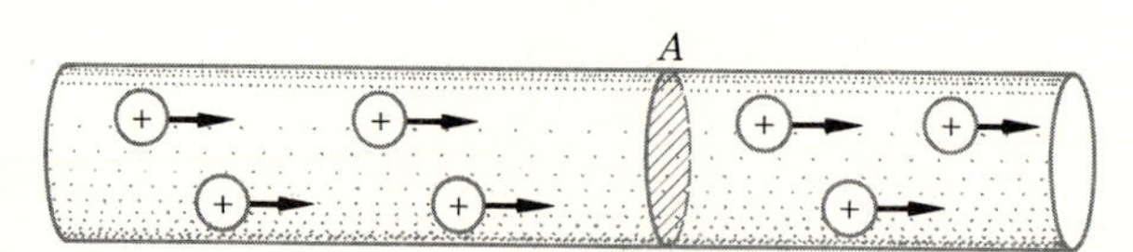

a wire (such as A in the figure) *in time Δt, the current in the wire is*

$$I = \frac{\Delta Q}{\Delta t} \tag{17.1}$$

Current in a Wire

It is measured in amperes (A). As with the charge beam, the current is opposite in direction to the flow of negative charges. It should be noticed that I is the quantity of charge which passes a given point in a wire each second provided the current is steady.

Illustration 17.3 A current of 3.2 A is flowing in a wire. How many electrons pass a given point on the wire in a second?

Reasoning If we knew the total charge passing the point in 1 s, we could divide that by the charge on the electron, 1.6×10^{-19} C, and thereby obtain the number of electrons. Since I is the number of coulombs passing a point on the wire in 1 s, we have at once that 3.2 C of charge flows past the point in 1 s. But each electron carries a charge 1.6×10^{-19} C, and so the number of electrons passing per second is

$$\text{Number} = \frac{3.2}{1.6 \times 10^{-19}} = 2 \times 10^{19} \text{ electrons}$$

This is a tremendously large number, of course. We therefore conclude that any attempt to observe the action of a single electron moving along the wire will be fruitless. There are cases where the effects of individual electrons can be observed, however. For example, the Geiger counter and Wilson cloud chamber are capable of detecting individual electrons. These devices will be discussed in a later chapter.

17.5 A SIMPLE ELECTRIC CIRCUIT

Let us begin our study of electric circuits by examining the simple circuit shown in Fig. 17.6. We see there a flashlight bulb connected to a battery. Current flows from the battery, through the bulb's filament, and back into the battery. Although this current consists of electrons flowing out of the negative terminal N and into the positive terminal P, the direction of the current is opposite to the direction of electron flow. This follows from our definition of current since the current's direction is defined in terms of positive, not negative, charge flow. As indicated on the diagram, the current I flows out of the positive terminal of the battery P, through the filament from A to B, and back into the battery at the negative terminal N.

We know that the filament of the flashlight bulb glows white hot in a situation like this. Clearly, the bulb is being furnished energy by the battery. The charge flow through the filament generates heat, and the filament is

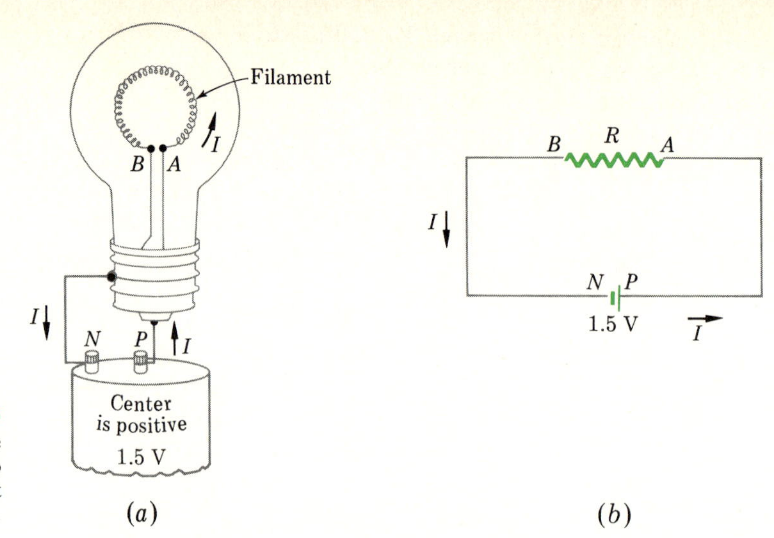

FIGURE 17.6
A simple electric circuit. The filament of the flashlight bulb glows white hot as current flows through it.

thereby made white hot. Apparently the battery gives energy to the charges and causes them to flow through the filament. As they pass through, they generate heat in a manner somewhat analogous to the way heat is generated by friction in a mechanical system.

In the circuit of Fig. 17.6*a*, only the filament from *A* to *B* becomes hot. The current flowing through the other wires (the connecting wires from *P* to *A* and from *B* to *N*) causes negligible heating. This is a result of two factors: (1) the filament wire is much thinner than the others, and (2) the material of the filament offers more resistance to the flow of charges through it. We say that the filament has a high electrical resistance, much higher than the electrical resistance of the other wires. For many purposes, the heat energy generated in the connecting wires is negligible. In such cases, the connecting wires can be assumed to have negligible resistance.

The circuit shown in Fig. 17.6*a* is customarily drawn as shown in part *b*.

DEFINITION

Any wire in which a current generates appreciable heat is called a **resistor.** In the present case, the filament is a resistor. *We represent all resistors by the*

Symbol for Resistance

symbol -ΛΛΛ-. This is the symbol used in *b* to represent the filament. The essentially resistanceless connecting wires are shown by the connecting lines to the battery. Notice also that the current flows out of the positive terminal of the battery and into its negative terminal.

Certain features of the circuit shown in Fig. 17.6 can be clarified by comparing it to an analogous water system. For example, consider the two systems shown in Fig. 17.7. Part *a* is a water system consisting of a water pump which causes water to flow through a pipe system. The pipes and pump are completely filled with water. As a result, if the pump pushes a little water into the pipe at *P*, an equal quantity must flow back into the pump at *N*. Moreover, a similar quantity of water must pass point *A* and flow into the narrow pipe. Of course, this same quantity of water must exit from the narrow pipe at *B*. As we see, the quantity of water flowing past points *A*, *B*, *N*, and *P* each second must be the same. Since the quantity of water passing a point each second is the current, we see that I, the water current in this case, is the same at all points in the water circuit.

Similar considerations apply to the electric circuit shown in part *b*. The wires are "filled" with charges (electrons), and these act like the water molecules in the water pipes. They act like an incompressible fluid which flows like the water in the pipes of part *a*. (This fact is reflected in the terminology of electricians who sometimes refer to the "juice" in the wires.) We can conclude at once that the current I in the electrical circuit is the same at all points of this simple circuit.

There is another important point to notice about the circuits of Fig. 17.7. As we have said, when a little water leaves the pump at P, an equal amount enters at N. Similarly in the electric circuit. When a little charge flows out of the battery at P, an equal amount of charge enters at N. This occurs almost instantaneously. Notice in particular that the charge which leaves P is not the same charge which enters at N. All the charges in the wire simply move along the wires a little. Equal amounts enter and leave the two ends.

It is of interest to examine the energy in these two circuits. Work must be done in the water circuit to push the water through the thin, curling pipe. This pipe offers much, much more resistance to the water flow than the larger pipes. Therefore, we can usually ignore the resistance effects of the larger pipes. The work done in pushing the water through the narrow pipe is lost doing viscous (or friction) work. An amount of heat equal to this work is generated in the resistance pipe. As we see, the energy supplied to the water by the pump is lost to friction work in the resistance pipe. The energy supplied by the pump appears as heat in the resistance pipe.

A similar situation occurs in the electric circuit. The battery supplies energy to the charges and pushes them through the resistance wire. This energy is lost to heat as the charges pass through the resistor. As with the water circuit, usually only negligible heat is generated in the portions of the circuit from P to A and from B to N. These connecting wires offer negligible resistance to charge flow.

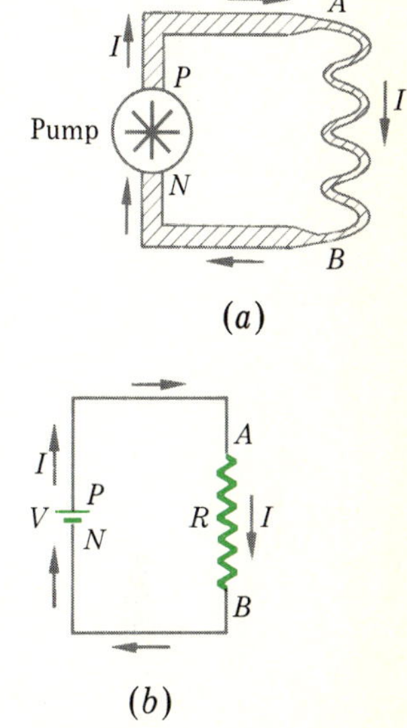

FIGURE 17.7

The water system in (*a*) is analogous to the electrical system in (*b*). Notice that the pump and the battery are the energy sources.

17.6 OHM'S LAW

Let us now examine this simple circuit so as to obtain a quantitative relation between the current, the battery voltage, and the resistance effect of the wire. We redraw the circuit in Fig. 17.8. Notice that the part of the circuit from P to A is designated "high potential" while the part from B to N is "low potential." The reason for this is as follows.

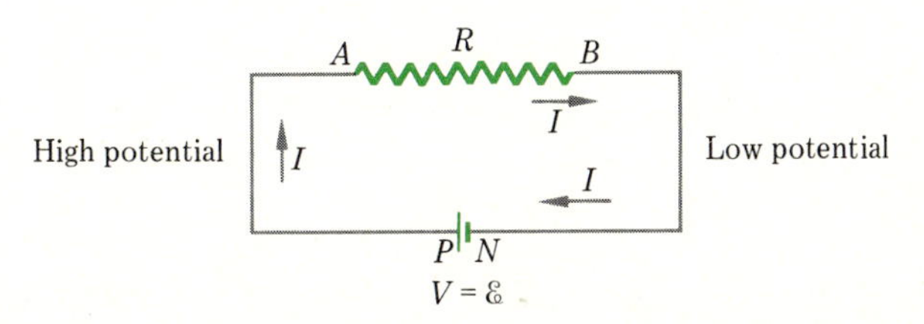

FIGURE 17.8

Current always flows from high to low potential through a resistance. The emf of the resistanceless battery is $\mathcal{E}$.

Consider a positive charge at point P of the circuit. It is at the same potential as the positive terminal of the battery, namely, V volts above point N. As the charge moves along the resistanceless wire from P to A, it loses no energy since negligible heat is generated in this wire. Therefore, when it reaches point A, it is still at the same potential as it was at P. The whole portion of the circuit from P to A is at the same potential. We say that this portion of the circuit is all at the same electrical level.

However, point N is at a potential V volts lower than P. Since the whole circuit from N to B is at the same electrical level, we conclude that this portion of the circuit is V volts lower than the section from P to A. As we see, the potential difference from A to B is V volts. And since positive charges move from points of high potential to points of lower potential, it is not surprising that the current (a flow of positive charge) is directed from A to B. Indeed,

> Current always flows through a resistor from its high-potential end to its lower-potential end.

We are now experienced enough to accept as reasonable an important fact first discovered by Georg Simon Ohm (1787–1854). He found by experiment that the current which flows through a resistor is often directly proportional to the potential difference between its two ends. For example, in Fig. 17.8 he found that

$$I \propto V$$

where V is the potential difference between the two ends of the resistor. For example, if he doubled the battery voltage, the current would double.

DEFINITION

The other factor which influences the current is the magnitude of the resistance effect of the resistor. We call this resistance effect the **resistance** R of the resistor. It is defined in the following way: *If a potential difference V across a resistor causes a current I to flow through it, the resistor's* **resistance** *is*

$$R = \frac{V}{I}$$

The unit of resistance, a volt per ampere, is designated the ohm (Ω). This defining equation for resistance is often written as

Ohm's Law

$$V = IR \tag{17.2}$$

and *is called* **Ohm's law.** However, Ohm also implied that R does not change as the potential across the resistor is changed. This is often true, but not always, as we shall see later. In spite of this, Eq. (17.2) is always true provided one understands R to be defined by it.

Illustration 17.4 Suppose the light bulb in Fig. 17.6 draws a current of 0.25 A when connected across a 1.5-V battery as shown. What is the resistance of the bulb under these conditions?

Reasoning We can make direct use of Ohm's law, Eq. (17.2). The potential difference across the bulb V is 1.5 V. We are told that I through it is 0.25 A. As a result,

$$V = IR$$

becomes

$$R = \frac{1.5\ \text{V}}{0.25\ \text{A}} = 6.0\ \Omega$$

The resistance of the hot bulb is $6\ \Omega$. We shall see later that the bulb's resistance is considerably lower when it is not white hot.

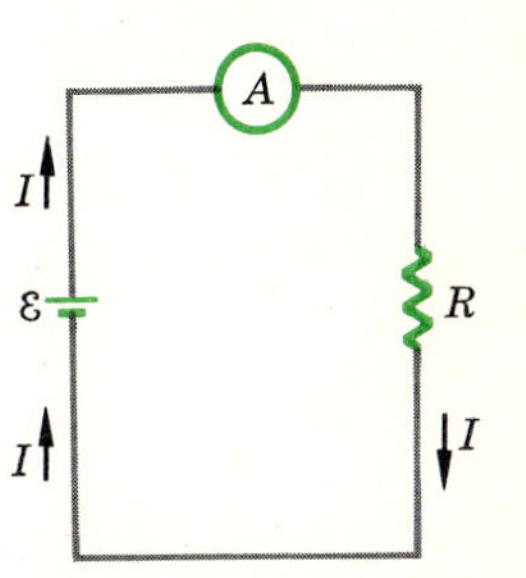

FIGURE 17.9

One connects an ammeter in series with the line to measure the current passing through a wire. The resistance of the ammeter must be very low. Why?

17.7 PLACEMENT OF METERS IN CIRCUITS

As we have indicated before, the connecting wires in a circuit have low resistance. For example, an ordinary house wire has a resistance of order $10^{-3}\ \Omega$ per meter length. As we shall see, this resistance is usually negligible.

When we draw a circuit, we indicate resistances by the symbol -ᴧᴧᴧ-, as pointed out previously. For example, in the circuits of Figs. 17.6 and 17.8, all wires except the one indicated by this symbol are assumed to have negligible resistance. No difference of potential can exist across a resistanceless wire since Ohm's law predicts that

$$V = IR$$

If $R = 0$, then V also must be zero.

Frequently one wishes to measure how much current flows in a wire. For this purpose use is made of an ammeter. The internal mechanism of this meter will be discussed in Chap. 20. At present we shall state only that, like a water flowmeter, it registers the quantity of charge flowing through it per second, i.e., the current.

Purpose of Ammeter

In Fig. 17.9, the same current flows everywhere in the circuit. (To see this, consider the equivalent water-pipe system. If there is no leak in the system, the same amount of water must flow through each pipe.) Hence, an ammeter placed in the circuit in the way shown in Fig. 17.9 will read the current flowing in any wire of the circuit. Notice that all the current flows through the meter. It is said to be **in series** with the circuit.

Purpose of Voltmeter

From time to time it is necessary to measure the potential difference between two points. A voltmeter is used for this purpose. For example, if one wishes to measure the potential difference between the ends of the resistor of Fig. 17.9 (the so-called **potential drop** across the resistor), a voltmeter would be connected as shown in Fig. 17.10. You should compare Fig. 17.10 carefully with Fig. 17.9. *The perfect voltmeter has infinite resistance,* and so no current flows through it. Hence, it does not disturb the circuit when connected as in Fig. 17.10.

On the other hand, *a perfect ammeter has zero resistance,* and so it does not disturb the circuit when connected as shown in Fig. 17.9. A bad mistake is sometimes made by students when they inadvertently connect the ammeter where the voltmeter is connected in Fig. 17.10. Since the potential difference between points a and b is $\mathcal{E}$, the current through the ammeter will be*

$$I = \frac{V}{R_{\text{am}}} = \frac{\mathcal{E}}{0} \to \infty$$

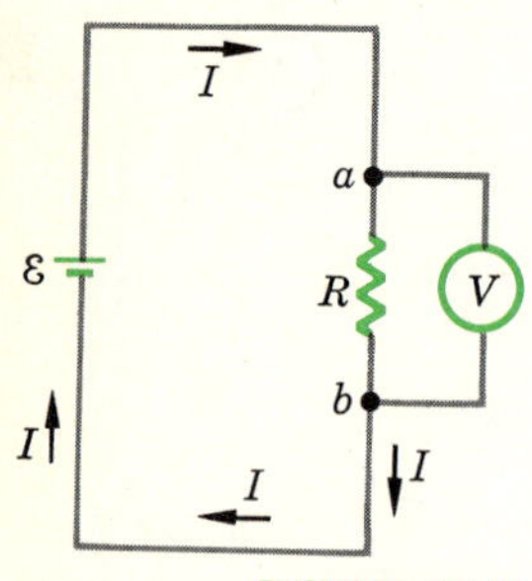

FIGURE 17.10
To measure the voltage drop from point a to point b, one connects a voltmeter as shown. A voltmeter must have a very high resistance. Why?

and the ammeter will be destroyed. An ammeter, since it has a very low resistance, is unable to limit the current flowing through it. Very large currents heat the wires inside the meter, and the internal mechanism burns up.

17.8 RESISTIVITY AND ITS TEMPERATURE DEPENDENCE

Not all materials are good conductors of electricity. There are even differences between the abilities of the various metals to conduct current. For this reason we neeed a quantity which will tell us exactly how good a conductor a material is. Or we could, if we wished, describe the material by stating how large its resistance is. We shall actually use this latter procedure.

Suppose that we have a cylindrical piece of wire like that shown in Fig. 17.11. The resistance of this length L of wire depends upon the length as well as the cross-sectional area of the wire. Since a wire twice as long would be expected to have twice as much resistance, the resistance of the wire is proportional to L. Moreover, the larger the cross-sectional area A, the less the resistance should be. Therefore, R should vary inversely as A. In equation form

$$R = \rho \frac{L}{A} \tag{17.3}$$

Resistivity

The proportionality constant ρ (Greek rho) is a property of the material from which the wire is made. It *is called the* **resistivity** *of the material.* If it is large, the material is a poor conductor. Solving for ρ, we find

$$\rho = R\frac{A}{L}$$

and so its units are ohm-meters or ohms times a length unit. Resistivities of various materials are listed in Table 17.1. Notice that copper and silver are the two best conductors listed. It is obvious why most wire is made from copper.

FIGURE 17.11
The resistance of a uniform wire varies directly as L and inversely to A.

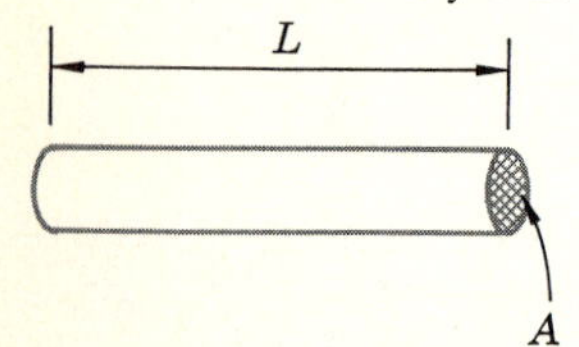

Sometimes a quantity called the **conductivity** is used to describe the electrical properties of metals and other conductors. It is equivalent to $1/\rho$, the reciprocal of the resistivity.

It will be noticed that the resistivity is quoted in Table 17.1 for a

*Recall that the emf of a battery is represented by $\mathcal{E}$.

TABLE 17.1
RESISTIVITIES AND THEIR TEMPERATURE COEFFICIENTS

MATERIAL	RESISTIVITY ρ AT 20°C, Ωm	α AT 20°C, °C^{-1}
Silver	1.6×10^{-8}	4.1×10^{-3}
Copper	1.7×10^{-8}	3.9×10^{-3}
Aluminum	2.8×10^{-8}	4.0×10^{-3}
Tungsten	5.6×10^{-8}	4.5×10^{-3}
Iron	10×10^{-8}	6.5×10^{-3}
Graphite (carbon)	3500×10^{-8}	-0.5×10^{-3}

particular temperature, 20°C. This is necessary because the resistance of wires changes rather markedly with temperature. Over a limited temperature range the resistivity may be represented by an equation of the form

$$\frac{\rho - \rho_{20}}{\rho_{20}} = \alpha_{20}(t - 20°) \qquad (17.4)$$

Variation of Resistance with Temperature

or

$$\frac{R - R_{20}}{R_{20}} = \alpha_{20}(t - 20°)$$

where t is the Celsius temperature.

In both these equations, the resistivity (and resistance) at 20°C, ρ_{20} (and R_{20}), are used as the reference values. The quotient given in each case is the fractional change in resistivity (or resistance). The temperature coefficient of resistance α is a measure of how much the resistance changes with temperature. If α is zero, the resistance is constant. Typical values for α are listed in Table 17.1.

It is seen that the value of α used in Eq. (17.4) must be measured at the same temperature as the reference temperature for the resistance. Sometimes other reference temperatures than 20°C are used, in which case Eq. (17.4) would be altered accordingly. The variation of the resistivity of copper with temperature is shown in Fig. 17.12. Notice that the changes in resistance with temperature are actually quite large.

If an equation such as Eq. (17.4) were strictly correct, the graph in Fig. 17.12 would be a straight line. Actually, the graph is slightly curved. Hence, Eq. (17.4) is only approximately correct for large temperature intervals.

FIGURE 17.12

The actual resistivity of copper is shown as the solid curve. A straight line placed through the data would appear as the broken curve.

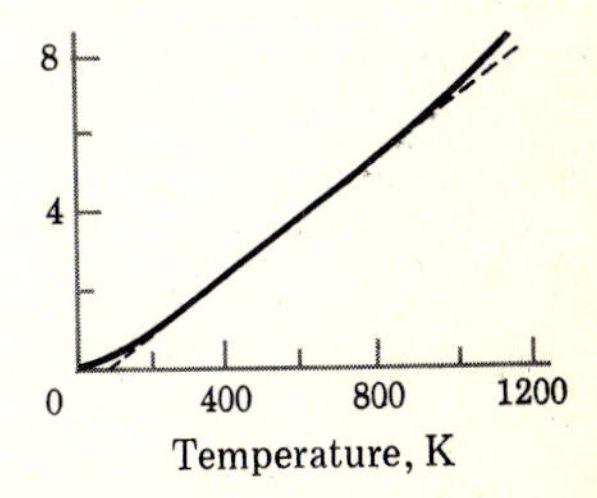

Illustration 17.5 A rectangular bar of iron is 2 by 2 cm in cross section and is 40 cm long. How large is its resistance?

Reasoning We know that

$$R = \rho \frac{L}{A}$$

From Table 17.1 we find $\rho = 1.0 \times 10^{-7}\,\Omega \cdot \text{m}$, and so

$$R = (1 \times 10^{-7})\left[\frac{0.40}{(0.02)(0.02)}\right] = 1 \times 10^{-4}\,\Omega \qquad \text{at } 20^\circ\text{C}$$

where you should carry through the units to verify the units of the answer.

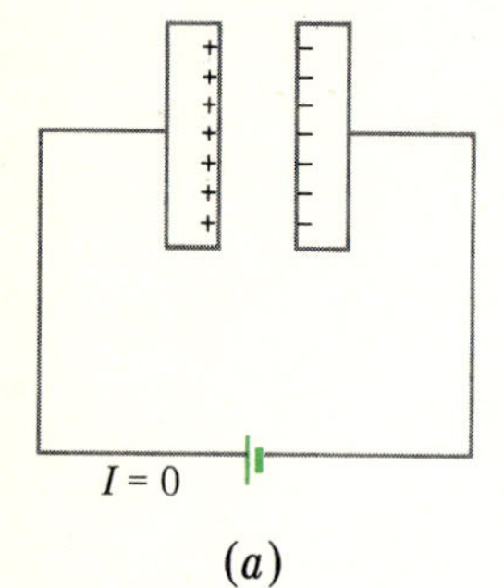

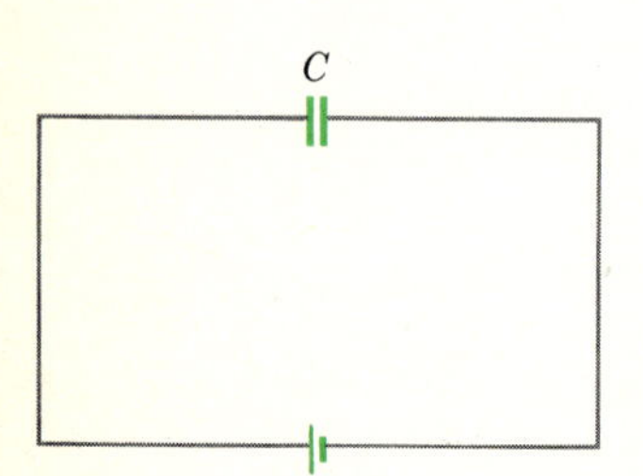

FIGURE 17.13 Equal and opposite charges reside on the inner faces of the capacitor plate. Notice the symbol used for a capacitor in (*b*).

Illustration 17.6 At a temperature of 520°C what is the resistance of the bar in the previous illustration?

Reasoning Making use of the data in Table 17.1 as well as the results of the previous example, we have

$$\frac{R - R_{20}}{R_{20}} = \alpha_{20}(t - 20)$$

$$\frac{R - 1 \times 10^{-4}}{1 \times 10^{-4}} = (6.5 \times 10^{-3})(520 - 20)$$

from which

$$R = 4.3 \times 10^{-4}\,\Omega$$

The resistance of the bar is 4.3 times higher at this temperature than it was at 20°C.

17.9 CAPACITORS (CONDENSERS)

One of the simplest types of capacitor consists of two parallel metal plates. Such a device connected to a battery is shown in Fig. 17.13. The positive terminal of the battery places a positive charge on one plate, and the negative terminal charges the other plate negatively. These charges attract each other, and so they reside on the inner surfaces of the plates, as shown. Such a device is capable of storing charge. As shown in part *b*, *the symbol used for* a capacitor *is* ⊣⊢ *or sometimes* ⊣(⊢.

Symbol for a Capacitor

We now are interested in how much charge Q resides on one of the plates, say the positive one. (Of course, since the system is electrically neutral, an equal negative charge exists on the other plate.) The amount of charge on the plate will depend upon several factors. First, the larger the voltage of the battery, the more charge it will put on the capacitor. Hence, if V is the voltage across the capacitor,

$$Q \propto V$$

SUPERCONDUCTORS

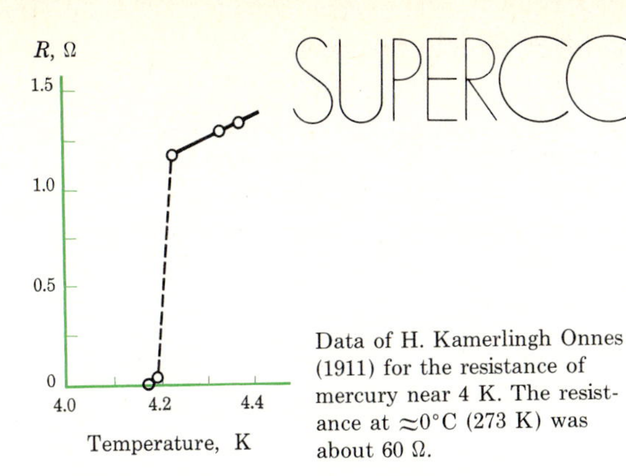

Data of H. Kamerlingh Onnes (1911) for the resistance of mercury near 4 K. The resistance at ≈0°C (273 K) was about 60 Ω.

Ordinary conductors of electricity become better conductors as the temperature is lowered. This was already known in 1835 from the measurements of Heinrich Lenz. There was some speculation in later years that perhaps the resistance of a metal would actually be zero at absolute zero, i.e., at −273°C, or 0 K. However, no one was able to carry resistance measurements to very low temperatures until much later than 1835, because no means existed for cooling objects to such low temperatures.

The attainment of very low temperatures received considerable impetus in 1883, when Wroblewski and Olzewski succeeded in liquefying air. With liquid air as a cooling agent, it was possible to carry out experiments at the boiling point of liquid oxygen (−183°C) and liquid nitrogen (−196°C). Soon thereafter, in 1898, James Dewar (1842–1923) succeeded in liquefying hydrogen, which has a boiling point of −263°C, or 10 K. There remained only one other gas which had not been liquefied, helium. The liquefaction of helium was finally achieved in 1908 by Heike Kamerlingh Onnes (1853–1926), and it was found that liquid helium boiled at 4.2 K.

Using his liquefied helium to achieve lower temperatures than had been possible before, Kamerlingh Onnes set out to measure the resistance of metals at very low temperatures. The measurement technique was quite simple in principle, although quite difficult in practice because of the difficulty of maintaining the temperature constant. Since resistance R is defined by Ohm's law, Kamerlingh Onnes needed only to measure the voltage drop V across a given cylinder of the metal when a definite current I flowed through it. Then, $R = V/I$. For his measurements, he used pure mercury as his metal. (It solidifies at −39°C, and so it was a solid.) When he carried out the requisite measurements in 1911, he obtained the astonishing data shown in the graph.

Although the metal was steadily approaching a very low resistance as the temperature was being lowered, at 4.2+ K the resistance suddenly decreased to zero. The mercury had become what we now refer to as a **superconductor.** Subsequent experiments indicated that a current would flow essentially forever in a ring of the superconductor, even though no source of emf was maintained. As far as experiment has been able to tell, the resistance of the superconductor is essentially zero.

Kamerlingh Onnes succeeded later in showing that lead, tin, and indium also become superconductors at 7.2, 3.7, and 3.4 K, respectively. Since that time, many other superconductors have been found. Most interesting has been the fact that certain alloys become superconductors at rather high temperatures. For example, the compound Nb_3Sn becomes superconducting at 17.9 K. Since no heat is generated when a current flows through a metal having no resistance, superconductors are now being used where this property is of great importance, e.g., in large electromagnets. The theoretical reasons for superconductivity in terms of atomic structure have been clarified only recently, especially since 1950. However, our understanding of superconductivity is still far from complete.

Of course, the larger the plate area, the more charge it should hold. In addition, if the plates are brought closer together, the positive plate will attract more negative charge from the battery to the negative plate. All these factors have to do with the geometry and construction details of the device. They will be a constant of the device. Hence we can write

Charge on a Capacitor

$$Q = CV \tag{17.5}$$

where C is a constant for a given capacitor. We call it the **capacitance** *of the capacitor.* When it is large, the charge on the capacitor will be large for a given voltage across it. Since Q is measured in coulombs and V in volts, the units of capacitance are coulombs per volt. This unit is called the **farad** (F).

If one of the plates has an area A, and if the separation between the plates is d, it is possible to show that *the capacitance of a parallel-plate capacitor is given by*

Parallel-Plate Capacitor

$$C = \frac{\varepsilon_0 A}{d} \quad \text{parallel plates} \tag{17.6}$$

where ε_0 (read "epsilon sub zero") is called the **permittivity** of free space, 8.85×10^{-12} F/m. (The constant ε_0 turns out to be the reciprocal of 4π times the constant k in Coulomb's law.) As an example, suppose that the two plates are square, 10 cm on a side, and separated by a distance of 0.10 mm. Upon changing the length units to meters and substituting in Eq. (17.6), the capacitance is found to be

$$C = 8.85 \times 10^{-10}\ \text{F} = 8.85 \times 10^{-4}\ \mu\text{F}$$

where 1 μF is equal to 10^{-6} F. Since capacitances are usually of the order of 1 μF or less, the latter unit is often used. However, farads must always be used in Eqs. (17.5) and (17.6).

In practice, most parallel-plate capacitors have an insulating sheet of material, called a **dielectric,** *placed between the plates.* This is done for two main reasons. First, the plates can be placed very close together with no fear that they will touch and permit the charges to flow. Many commercial capacitors are formed by taking two thin sheets of metal foil and laying one on top of the other, with a thin film of plastic between them to keep them from touching. The layered sheets are then rolled up into a tight cylinder and

TABLE 17.2
DIELECTRIC CONSTANTS AT 20°C

Air	1.006	Ice (−5°C)	2.9
Paraffin	2.1	Mica	6
Petroleum oil	2.2	Acetone	27
Benzene	2.29	Methyl alcohol	31
Polystyrene	2.6	Water	81

packaged for convenience. The device is essentially a parallel-plate capacitor, but it looks very different from the device shown in Fig. 17.13. Capacitors 1 μF large, a common size, occupy a volume of about 1 cm^3 when made this way.

Dielectrics have a higher dielectric strength, i.e., spark over less readily, than air. For this reason, parallel plates with a dielectric between them can be used at higher voltage differences than air capacitors. Moreover, dielectrics increase the capacitance of a capacitor. *If one has a capacitor with vacuum* (or air to a good approximation) *between its plates, it will have a capacitance given by Eq. (17.6). But if a dielectric is placed between the plates, the capacitance is found to be larger by a factor* k_d *the* **dielectric constant,** *or* **relative permittivity,** *of the dielectric.* Typical dielectric constants are given in Table 17.2. Dielectrics will be discussed in more detail in the next section.

Dielectric Constant

17.10 DIELECTRICS

As pointed out in the previous section, when an insulating material (a dielectric) is placed between the plates of a capacitor; the capacitance of the device is increased. This fact has importance in biophysics and chemical physics as well as from a purely technical viewpoint. From the change in capacitance caused by a given material, one can make inferences regarding the molecular behavior of the material itself. Let us now investigate this subject further.

Many molecules do not have their charges spread out uniformly. Although the molecule as a whole is neutral, the various atoms within it may have (at least partly) lost an electron to other atoms of the same molecule. A few typical examples are shown schematically in Fig. 17.14. Molecules of this type are called **dipolar molecules** since a dipole is by definition two equal and opposite charges separated by a very small distance. The so-called **dipole**

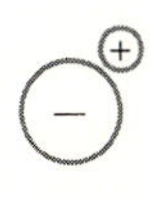

HCl (Hydrogen chloride)

HOH (Water)

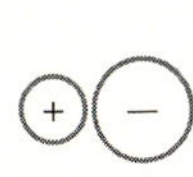

CO (Carbon monoxide)

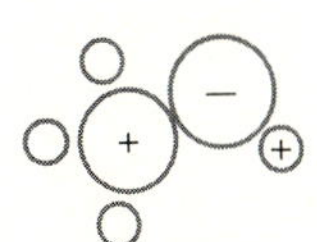

CH_3OH (Methyl alcohol)

FIGURE 17.14
Examples of dipolar molecules.

moment μ of such a molecule is equal to the magnitude of either charge multiplied by the distance of separation. In symbols, $\mu = Qd$.

In a typical molecule, the dipole moment is of the order of 1 D (debye*) which is equivalent to $\frac{1}{3} \times 10^{-29}$ C · m. For example, if the separated charge in the HCl molecule of Fig. 17.14 is taken to be the electronic charge, 1.6×10^{-19} C, then, since the dipole moment of this molecule is 1.03 D,

$$(1.6 \times 10^{-19}\ \text{C})(d) = (1.03)(\tfrac{1}{3} \times 10^{-29})\ \text{C} \cdot \text{m}$$

where d is the charge separation. One then finds $d \approx 0.5 \times 10^{-10}$ m. In many cases, d is smaller than the known separation of the atoms, and so the situation corresponds more to an unequal sharing of charge between the atoms. For HCl the value we have found for d is only about half the known distance between the atom centers.

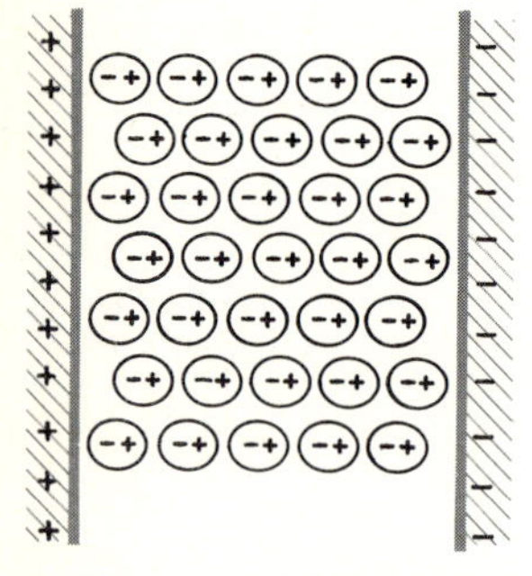

FIGURE 17.15
When a polar dielectric is placed between charged plates, the molecules tend to align in the field. They thereby induce more charge flow onto the plates. The alignment is exaggerated in the figure.

Suppose a dielectric composed of dipolar molecules is placed between two charged plates. The electric field between the plates will rotate the polar molecules so that their negative ends will be closest to the positive plate and vice versa for the positive ends. This is shown schematically in Fig. 17.15. In practice, thermal motion keeps the alignment of the molecules from being even approximately perfect. Notice that, in effect, the negative ends of the molecules have caused a negative charge to be induced on the left side of the dielectric while a positive charge has been induced on the opposite side. This causes a much reduced electric field within the region between the plates, as we shall now see.

In Fig. 17.16*a* we show two charged metal plates. Let us say that the battery which charged them is no longer attached so that charge cannot run on or off the plates. If the electric field between the plates is E_0 and the plate separation is d, the potential difference between the plates is simply $V_0 = E_0 d$. Moreover, the charge on the plates Q_0 is given by

$$Q_0 = C_0 V_0 \qquad \text{or} \qquad Q_0 = C_0 E_0 d \tag{17.7}$$

Now let us suppose a dielectric sheet is placed between the plates as shown in Fig. 17.16*b*. Although we shall assume the dielectric to completely fill the space, we show separations at the edges to simplify the description of what is going on. First, we must notice that the charge on the metal plate is still Q_0 since the battery has been disconnected and no charge can flow on or off the plate. (The dielectric is an insulator.) Therefore the field in the air space between the dielectric and plate is still E_0. However, as indicated previously, a negative charge layer is induced on the surface of the dielectric. Some of the field lines coming from the positive plate end on these negative induced charges. However, since there is less induced charge on the dielectric than the charge on the metal plate, some of the lines do not stop there but continue on through the dielectric. As we see, the field E_d inside the dielectric is smaller then E_0.

* In honor of P. Debye, a famous physical chemist. He received the Nobel prize in chemistry in 1936 for his work with molecules and x-rays.

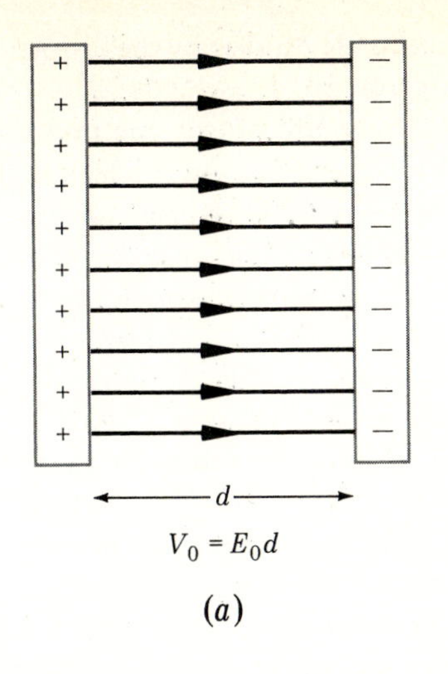

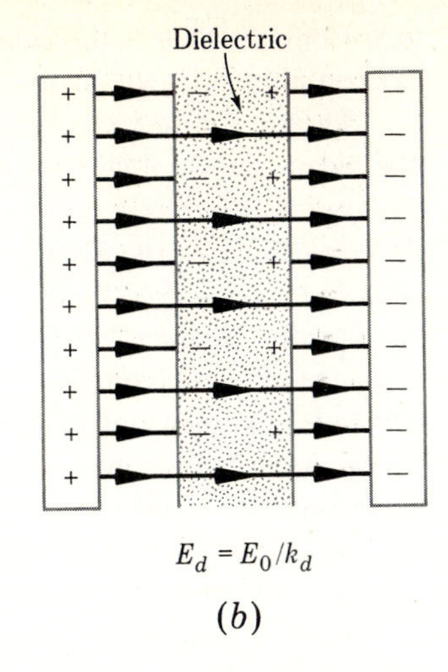

FIGURE 17.16
The dielectric reduces the field between the plates and thereby decreases the potential difference. Can you estimate k_d in the case shown?

If in actuality the dielectric fills the whole region between the metal plates, the field everywhere between the plates will be E_d and the potential difference between the plates will be given by $V_d = E_d d$. Moreover, if we call C_d the capacitance of the parallel-plate device when dielectric is between the plates, we have

$$Q_0 = C_d V_d$$

since the charge Q_0 is still the same. After replacing V_d by $E_d d$, we find

$$Q_0 = C_d E_d d$$

Dividing this relation by Eq. (17.7) and rearranging gives

$$C_d = (C_0)\left(\frac{E_0}{E_d}\right) \tag{17.8}$$

In other words, *the capacitance has been increased in proportion to the ratio of the original electric field to the field within the dielectric.*

It will be recalled that *the dielectric constant k_d of the insulating material was defined equal to the ratio C_d/C_0.* We see from Eq. (17.8) that this means *a material having high dielectric constant will be a material which most effectively reduces the electric field between the plates.* In other words, a high-dielectric-constant material is a material which can have charges induced easily upon its surface. As a rule, those molecules which have the largest dipole moment will have the highest dielectric constants since the induced charge layer will be largest for them. However, water and methyl alcohol are exceptions to this rule; in them the molecules are bound together in more or less "clumps" of molecules by so-called hydrogen bonds. These molecular clumps then act like molecules of much higher dipole moment and give rise to the very large dielectric constants listed for them in Table 17.2. Even nondipolar molecules such as benzene, methane, etc., have noticeable dielectric effects. These molecules are made dipolar by the distorting force of the electric field into which they are placed.

Effect of Dielectric on Forces

The dielectric constants of liquids are of extreme importance in many chemical and biological reactions because the dielectric liquid greatly changes the electrical forces between ions. As we saw, *the electric field in the dielectric between the plates of a capacitor is only* $1/k_d$ *as large as it would be without the dielectric. This same relation holds in more complicated cases.* Two ions in solution exert forces on each other decreased by a factor $1/k_d$ by the presence of the solvent. Since water has a dielectric constant of 80, the force between ions in water is one-eightieth as large as it would be without the water. As a result, the Na^+ and Cl^- ions in NaCl, for example, are rather easily separated in water, and thermal motion is sufficient to cause solution of the NaCl. However, in benzene, $1/k_d = 1/2.3$, and so the forces between the ions are still too large to be overcome by thermal motion. Hence benzene does not dissolve NaCl. Many other similar situations exist in chemical and biological systems where the dielectric nature of the solvent is the controlling factor.

17.11 ENERGY STORED IN A CAPACITOR

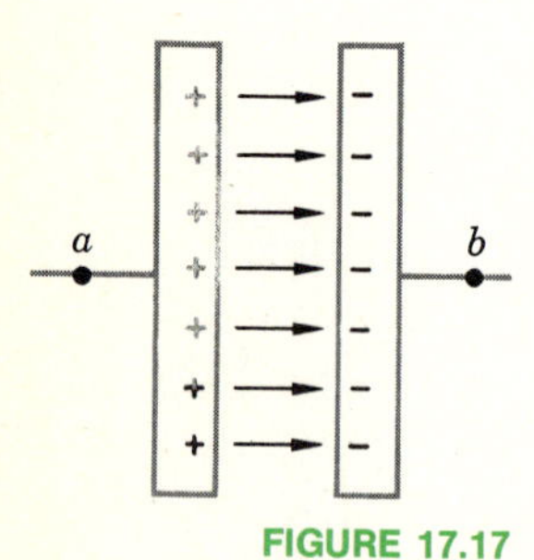

FIGURE 17.17
The energy stored in a capacitor is equal to the work done in charging it, $\frac{1}{2}QV$.

Suppose that the charged capacitor shown in Fig. 17.17 is discharged by connecting a wire from a to b. If the wire is quite thin, it will be found that when the negative charge from the negative plate runs through it to neutralize the positive plate, the wire becomes hot. Since heat is energy, the charged capacitor, which produced the heat energy, must have had energy stored in it. We shall now compute how much energy it had.

Let us consider how much work is required to charge the capacitor. That work will then be equal to the energy stored in the capacitor. When the capacitor is uncharged, no work at all is required to carry a small quantity of positive charge from plate b to plate a. However, as we continue to carry charge across, the field between the plates will become sizable. We shall then need to pull the positive charges across. In fact, if we have charged the capacitor to the point where the potential difference between its plates is V' volts, the work required to carry a charge increment of ΔQ across would be $\Delta Q V'$.

Clearly, the first charge carried across when the capacitor is not charged requires no work, since V' is zero. On the other hand, the final increment of charge carried across from one plate to another is carried through the final potential difference V. The total work done is easily found by using calculus. However, it appears reasonable that the total work done should be equivalent to the work which would be done in carrying the whole charge Q across the average potential difference during the charging process, $\frac{1}{2}V$. This turns out to be a correct assumption. We therefore have that *the energy stored in a capacitor, with charge* Q *and potential difference* V, *is*

Energy of Capacitor

$$\text{Energy} = \tfrac{1}{2}QV \qquad (17.9)$$

By use of the relation $Q = CV$, Eq. (17.8) can be written in the alternate forms

$$\text{Energy} = \tfrac{1}{2}QV = \tfrac{1}{2}CV^2 = \frac{Q^2}{2C}$$

Illustration 17.7 A 2-μF capacitor is charged across a 12-V battery. How much energy is stored in it?

Reasoning We have

$$\text{Energy} = \tfrac{1}{2}CV^2 = (\tfrac{1}{2})(2 \times 10^{-6})(144) = 1.44 \times 10^{-4}\ \text{J}$$

The student should carry the units through this equation to verify the unit of the answer. In so doing, note from $Q = CV$ that 1 F is equivalent to 1 C/V.

17.12 ENERGY STORED IN AN ELECTRIC FIELD

In the last section we saw that the energy stored in a charged capacitor was $\frac{1}{2}CV^2$, where V was the voltage difference across a capacitor having a capacitance C. Although it is not necessary to specify exactly how and where this energy is stored, it is sometimes convenient to think of the energy as being stored in the electric field between the capacitor plates. With this in mind, it would be well to express the equation for the stored energy in terms of the electric field E between the capacitor plates. This can be done by recalling that in the case of a parallel-plate capacitor $V = Ed$, where d is the separation of the plates.

We therefore have for the energy stored in a parallel-plate capacitor

$$\text{Energy} = \tfrac{1}{2}CV^2 = \tfrac{1}{2}CE^2d^2$$

But from Eq. (17.6) the capacitance of a parallel-plate capacitor with plate area A is given by

$$C = \frac{\varepsilon_0 A}{d}$$

provided that the capacitor has vacuum between its plates. If it is filled with a dielectric having a constant k_d, this becomes

$$C = \frac{k_d \varepsilon_0 A}{d}$$

Substituting this value for C in the energy equation yields

$$\text{Energy} = (\tfrac{1}{2}\epsilon_0 k_d E^2)(Ad)$$

But Ad is the volume of the space between the capacitor plates, in other words, the volume in which the constant electric field E exists. Dividing both sides of the equation by the volume gives us an expression for the energy per unit volume, i.e., *the energy which we picture to be stored in unit volume of the region of space where the electric field is E.* We have

$$\text{Energy per unit volume} = \tfrac{1}{2}\epsilon_0 k_d E^2 \qquad (17.10)$$

Notice that the energy stored in unit volume is proportional to the square of the electric field strength. It is often convenient to use Eq. (17.10) as a means for assigning energy to an electric field. Although this expression was derived for a very special case, it is shown in more advanced texts that it has general validity.

SUMMARY

Batteries act as sources of energy for electrical systems. They provide a potential difference through which charges can be made to fall. The potential difference between the terminals of an electrically isolated battery is called the emf ($\mathcal{E}$) of the battery.

When a charge q falls through a potential difference V, it acquires an energy qV. Positive charges, left to themselves, fall from high potentials to low potentials. Negative charges fall in the reverse direction.

Current is defined to have the direction of positive charge flow. If negative charges are moving, the current due to them is defined to flow in the opposite direction. If a charge ΔQ flows past a given point in a time Δt, the current I is given by $I = \Delta Q/\Delta t$. Its units are amperes (A).

When a potential difference V exists between the two ends of a resistor, the current through the resistor is related to V by $V = IR$. The quantity R is the resistance of the resistor, and it is measured in ohms (Ω). Ohm's law not only states that $V = IR$ but also implies that R is a constant, independent of V and I. Current always flows through a resistor from its high-potential end to its low-potential end.

An ammeter measures the current which flows through it. A good ammeter has very little resistance. A voltmeter measures the potential difference between its two terminals. It should have a very high resistance.

The resistivity ρ of a material measures the material's resistance to charge flow. In terms of it, the resistance of a wire L meters long with cross-sectional area A is given by $R = \rho(L/A)$.

The resistance of most materials changes with temperature. If α_{20} is the temperature coefficient of resistance based on a reference temperature of 20°C, then

$$\frac{R - R_{20}}{R_{20}} = \alpha_{20}(t - 20°)$$

where R is the resistance at t degrees if its resistance at 20°C was R_{20}. A similar relation exists with R replaced by ρ.

A capacitor is a device for storing charge. It carries equal but opposite charges on its two plates. If a potential V across a capacitor causes a charge Q on its plates, the capacitance C of the capacitor is given by $C = Q/V$. Its units are farads (F). The capacitance of a parallel-plate capacitor with plate area A and plate separation d is $C = \epsilon_0 A/d$, where ϵ_0 is the permittivity of free space, 8.85×10^{-12} F/m.

Dielectrics usually reduce the electric field in which they are placed. A parallel-plate capacitor with fixed charge Q will have the electric field between its plates reduced by a factor $1/k_d$ when a dielectric fills the region between the plates. The quantity k_d is called the dielectric constant for the dielectric. k_d is about 80 for water and of the order of 2 to 5 for most oils and similar materials. k_d for vacuum is unity.

A charged capacitor has energy $\frac{1}{2}QV$ stored in it. Energy can be assumed stored in an electric field. In a dielectric with constant k_d where the electric field is E, the energy stored per unit volume is $\frac{1}{2}\epsilon_0 k_d E^2$.

MINIMUM LEARNING GOALS

Upon completion of this chapter, you should be able to do the following:

1. State the purpose of a battery in a simple circuit. Explain what is meant by its emf.

2. Find the change in KE of either a positive or negative known charge as it moves through a given potential difference. If its initial speed is given, find its final speed. (Neglect relativistic effects.)

3. Describe a simple electron gun.

4. Give the change in energy in electronvolts of a charge q which falls through a known potential difference. Convert energies in electronvolts to J and vice versa.

5. Make use of the relation $I = \Delta Q/\Delta t$ in simple situations. Define electric current in your own words.

6. Interpret a simple circuit diagram such as the one in Fig. 17.8. State the potential difference between various pairs of points of the circuit.

7. State which end of a resistor is at the higher potential when the direction of current flow through the resistor is given.

8. State the equation form of Ohm's law and explain its meaning. Use it to define resistance and its units.

9. Given a diagram of a circuit, show how an ammeter must be connected in the circuit to measure the current in a given wire.

10. Given a diagram of a circuit, show how a voltmeter must be connected so as to measure the potential difference between two points of the circuit.

11. Compute the resistance of a given piece of wire provided the resistivity of the wire material is known.

12. Find the resistance of a wire at a given temperature when its resistance and temperature coefficient at some reference temperature are known.

13. Draw a diagram of a parallel-plate capacitor and relate C, Q, and V for it. State the units of C.

14. Outline a method by which the dielectric constant of a liquid can be determined if an instrument for measuring capacitance is available.

15. Compute the energy stored in a given capacitor when charged to a known potential difference.

IMPORTANT TERMS AND PHRASES

You should be able to define or explain each of the following:

Emf
Positive charge falls from + to − while negative charge does the reverse
electronvolt
$I = \Delta Q/\Delta t$; ampere
Ohm's law; $V = IR$
Resistivity, $\rho = RA/L$
$(R - R_{20})/R_{20} = \alpha_{20}(t - 20°)$
$Q = CV$; farad
Dielectric constant k_d
Energy $= \frac{1}{2}QV$

QUESTIONS AND GUESSTIMATES

1. Sometimes students insist that current is "used up" as it flows through a resistor. Arguing from the water analogy, how would you convince such a student that current is not lost in a resistor?

2. How do we know which end of a battery is at the higher potential, i.e., positive, in a schematic diagram of a circuit? Of a resistor?

3. Current and voltage are never lost in a circuit. What is lost as the current flows through the circuit?

4. Fluorescent light bulbs are usually much more efficient light emitters than incandescent bulbs. That is, for the same input energy, the fluorescent bulb gives off more light than the incandescent bulb does. Touch each of the two bulbs after it has been lit a few minutes. Explain why the incandescent bulb is a less efficient light emitter.

5. In Fig. P17.1*a* the pump lifts water to the upper reservoir at such a rate that the water level remains constant. The water slowly trickles out of the narrow tube into the lower reservoir. Point out the similarities between this water circuit and the electric circuit shown in part *b*.

6. A parallel-plate capacitor is charged to a fixed charge Q. (The battery is disconnected so charge cannot run on

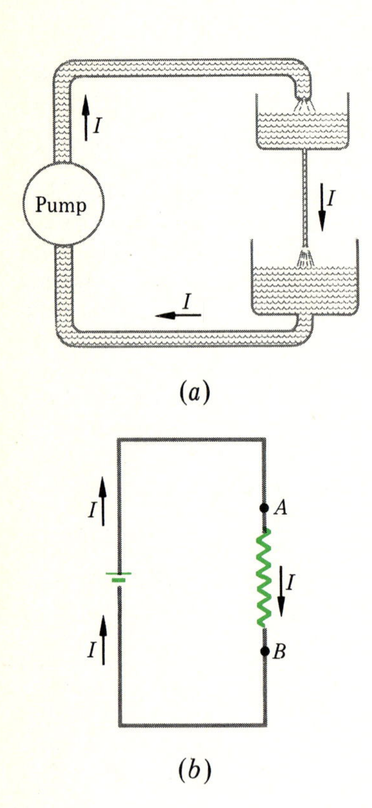

FIGURE P17.1

or off it.) The plates are now pulled somewhat farther apart. In the process, the puller must do work. Why? Does the potential difference between the plates change during the process? What happens to the work done by the puller?

7. There is considerable similarity between the sparking over of an air capacitor and a lightning flash during a thunderstorm. Explain how they are similar.

8. Explain why an ammeter must have a low resistance. What precautions must be taken in handling an ammeter?

9. Why must a voltmeter have a high resistance?

10. If a sheet of metal is placed between two capacitor plates but not touching them, will it appear to have a low or high dielectric constant? Explain.

11. As an electron beam shoots down the length of a TV tube, about how far do the electrons fall because of the effect of gravity? Will a TV set operate if the set is laid on its back so the electron beam must shoot straight upward?

12. Using an ohmmeter (basically a battery in series with a very sensitive ammeter), measure your resistance from one hand to the other. A current of about 0.03 A through the midsection of one's body is sufficient to paralyze the breathing mechanism. About how large a voltage difference between your hands is needed to electrocute you?

PROBLEMS

1. An electron ($m = 9.1 \times 10^{-31}$ kg, $q = -e$) initially at rest falls from the negative filament to the positive plate in a vacuum tube. The terminals of a 12.0-V battery are connected to the filament and plate. (*a*) What is the speed of the electron just before it hits the plate? (*b*) What is its kinetic energy in eV at that time?

2. It is desired to accelerate a proton from rest to a speed of 5.0×10^6 m/s. (*a*) Through how large a potential difference must the proton fall? (*b*) What will be its kinetic energy in electronvolts? ($m_p = 1.67 \times 10^{-27}$ kg, $q = e$.)

3. A proton is shot with a KE of 5000 eV from a negative plate toward a positive plate. The potential difference between the two plates is 1500 V. (*a*) How much KE (in electronvolts) will the proton lose as it shoots to the positive plate? (*b*) What will be its KE (in electronvolts) just before it hits the plate? (*c*) Repeat for an α particle. (The charge on an α particle, a helium nucleus, is $+2e$.)

4. An electron moving with a speed of 5×10^6 m/s is further accelerated through a potential difference of 20 V. What will its new speed be?

5. A current of 2.0 A is drawn from a 6-V battery for 20 min. (*a*) How much charge and (*b*) how many electrons flow from the battery in this time?

6. A certain atom-smashing machine provides a current of 0.0010 A [= 1 mA (milliampere)] through a potential difference of 1 MV. (*a*) How much charge does it provide in $\frac{1}{2}$ h? (*b*) If the current consists of a stream of protons, how many protons are provided in $\frac{1}{2}$ h?

7. An ordinary 60-W light bulb draws a current of 0.50 A when connected across a potential difference of 120 V. What is the resistance of the bulb?

8. How much current flows from a 12-V battery when a 3.0-Ω resistor is connected across its terminals?

9. Number 10 copper wire has a diameter of 2.59 mm. The approximate safe current for this size wire is 30 A. (At higher currents, it becomes too hot.) (*a*) Find the resistance of a 5.0-m length of this wire at 20°C. (*b*) How large a potential difference will exist between the ends of a wire this long when carrying 30 A?

10. The filament of a certain light bulb is 8.0 cm long and has a diameter of 0.150 mm. When the bulb is connected across 120 V, its filament becomes white hot and the bulb draws a current of 0.25 A. (*a*) What is the resistance of the bulb under these conditions? (*b*) What is the resistivity of the filament material at this temperature?

11. An ordinary 60-W tungsten-filament light bulb has a resistance of 240 Ω when lit. The temperature of the filament is about 2000°C under these conditions. If the α value in Table 17.1 is valid even for temperatures this high, what is the resistance of the bulb at 20°C?

12. At 20°C the resistance of a particular tungsten-filament lamp is 30 Ω. When it is lighted, the lamp has a resistance of 240 Ω. About what is the temperature of the filament in the hot lamp?

13.* A graphite resistor is placed in series with a 10-Ω (at 20°C) iron resistor. How large should the resistance of the graphite resistor be in order for the combined resistance to be temperature-independent?

14.** It is desired to make a 100-Ω resistor which is temperature-independent by using a carbon resistor in series with an iron resistor. What should the resistance of each be at 20°C?

15. A typical capacitor used in a radio might have a capacitance of 2.0×10^{-8} F. How much charge exists on the capacitor when it is connected across a 9.0-V battery?

16. Two identical metal plates are placed parallel to each other, one opposite the other. The gap between them is 0.50 mm, and the area of each is 400 cm^2. (*a*) Find the capacitance of the plates if air exists between them. (*b*) If connected to a 12.0-V battery, how much charge would exist on the capacitor? (*c*) Repeat (*a*) and (*b*) if the space between the plates is filled with polystyrene.

17. A certain capacitor has a capacitance of 1.750×10^{-9} F when it has air between its plates. When the capacitor is filled with acetone, what will be its capacitance?

18. How much energy is stored in a 2-μF capacitor which is charged to a potential difference of 300 V?

19.* A certain parallel-plate capacitor can be adjusted so that the gap between the plates can be changed without otherwise disturbing the electrical system. In position *A* the capacitance is 3.0×10^{-9} F, and in position *B* it is 2.7×10^{-9} F. The capacitor is charged by a 12-V battery when in position *A*. The battery is then removed and the capacitor is changed to position *B* without changing the charge on it. (*a*) How much charge is on the capacitor? (*b*) What is the voltage across it in position *B*? (*c*) By how much does its stored energy change in going from *A* to *B*? (*d*) What minimum work must have been done to change the capacitor from *A* to *B*?

DIRECT-CURRENT CIRCUITS

18

In the last chapter we learned about various electrical circuit elements such as batteries and resistances. These elements are connected together in various ways to form practical electric circuits. We shall analyze such circuits in this chapter. It will be found that two simple rules, called Kirchhoff's rules, provide a systematic basis for circuit analysis. Also in this chapter we shall learn about electric power and the basic ideas of electrical safety.

18.1 KIRCHHOFF'S POINT RULE

There are two basic rules which apply to all electric circuits. They are very easy to understand and remember since they are almost obvious. To see what the first rule is, refer to Fig. 18.1.

In part *a* we see a situation where several wires meet at a junction point. Consider point A. The current flowing into the point is I, and the currents flowing out of the point are I_1, I_2, and I_3. What can we say about these currents? We can easily answer this question if we think of the water-pipe analogy. The quantity I would represent a current of so many cubic centimeters of water flowing into point A each second. But if the pipes did not leak, this exact same amount of water must flow out of point A each second. In other words, since point A cannot store water, as much must flow into the point as flows out. In equation form, for this case,

$$\text{Current into } A = \text{current out of } A$$

becomes

$$I = I_1 + I_2 + I_3$$

A similar situation applies to the flow of charge. The current coming into point A must equal the current which flows out of point A. Therefore, in this case, too, $I = I_1 + I_2 + I_3$. Moreover, this same rule must apply to any point in the circuit. At point B, for example,

$$\text{Current into } B = \text{current out of } B$$

becomes

$$I_1 + I_2 + I_3 = I$$

This is identical to the equation found for point A.

We can summarize this result in what is called **Kirchhoff's point rule:**

Point Rule

The sum of all the currents coming into a point must equal the sum of all the currents leaving the point.

We shall find that this simple rule is of very great importance. Can you show that the rule, when applied to Fig. 18.1*b* gives $I_4 + I_7 = I_5 + I_6$?

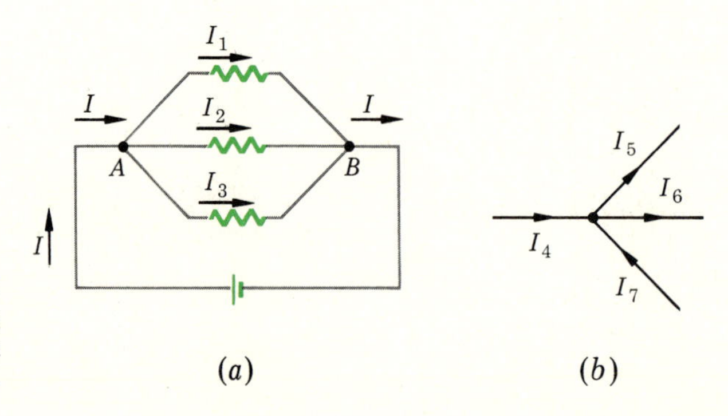

FIGURE 18.1

The point rule tells us that $I = I_1 + I_2 + I_3$ and that $I_4 + I_7 = I_5 + I_6$.

18.2 KIRCHHOFF'S LOOP (OR CIRCUIT) RULE

This second rule applies to circuits in which the current is steady or changing slowly.* In these cases, the electric field near the circuit is essentially electrostatic. Under such conditions, the electric field is a conservative field. By this we mean that the work done in carrying a positive test charge from one point to another is independent of the path followed. Let us see what this tells us about the electric circuit of Fig. 18.2.

Suppose we start at point A in the circuit and carry a positive test charge along it through points B, D, F, G, and back to A. How much total work have we done in carrying the test charge around the circuit and back to the starting point? Since this has been done in a conservative field, the answer is zero.

The situation here is much like a similar situation involving the gravitational field, another conservative field. Suppose you rise in the morning from your bed and return to it at night. The net amount of work you have done for the whole day on your body against the gravitational field is zero. Since the starting and ending point are the same, the gravitational PE of your body is unchanged. The total work done against gravity is zero.

In the case of the electrical circuit of Fig. 18.2, we see that the following is true. If we carry a positive test charge around the circuit and back to the starting point, zero net work is done. This must mean that the charge was carried through an equal amount of voltage rises and voltage drops. The net effect of all these voltage changes (rises taken as positive and drops as negative) is that their algebraic sum is zero. This fact is summarized in **Kirchhoff's loop** (*or circuit*) **rule:**

Loop Rule

> The algebraic sum of the voltage changes around a closed circuit must equal zero.

In other words, if we start at a point on a circuit and move around that circuit so as to come back to the original point, the algebraic sum of the voltage drops and rises which we encounter must add up to zero. We take voltage drops as negative and voltage rises as positive.

As we see, the loop rule is intimately connected with voltage changes. For that reason, let us review how the voltage changes as we move across a resistor, a battery, and a capacitor.

Suppose we move from A to B across the resistor shown in Fig. 18.3. *We*

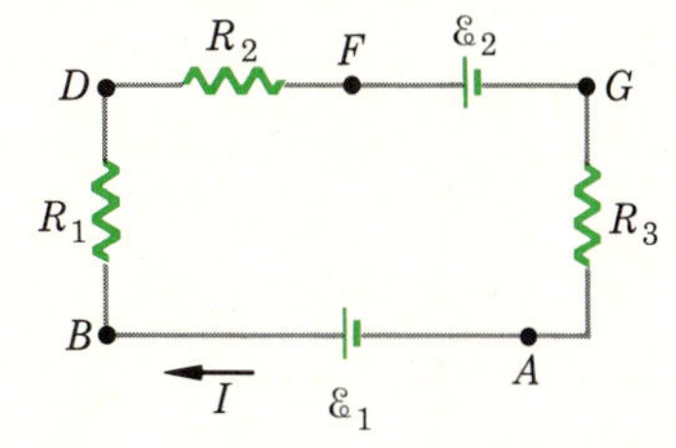

FIGURE 18.2
What does Kirchhoff's loop rule tell us about this circuit?

*As we shall see later, if the current is alternating in the microwave frequency range (10^9 Hz), induced emfs in the circuit loop become important.

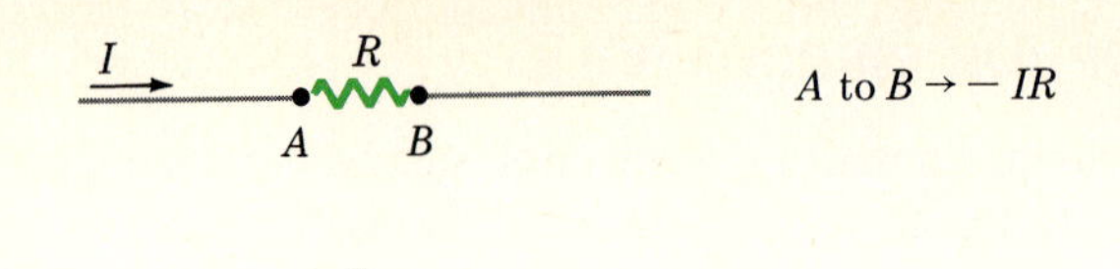

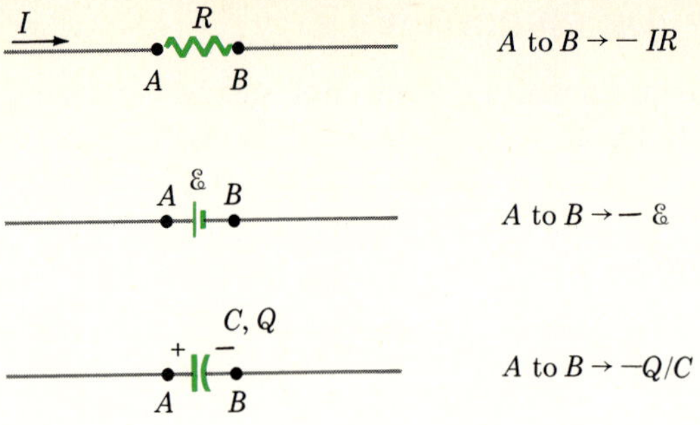

A ℰ B

A to B → − ℰ

C, Q
+ −
A B

A to B → −Q/C

FIGURE 18.3

In each case shown, going from A to B is a voltage drop, a negative voltage change. In the reverse direction, the voltage change would be positive.

know that current always flows from high to low potential through a resistor. Hence point A is at a higher potential than point B. As a result, the voltage change from A to B is a drop in potential. Ohm's law tells us its magnitude is IR. The voltage change in going from A to B is $-IR$, negative because it is a voltage drop.

If we look at the battery in Fig. 18.3, we recall that its symbol tells us the left side is positive. Therefore point A is $\mathcal{E}$ volts higher than point B. Going from A to B, the potential change is $-\mathcal{E}$.

In the case of the capacitor, we must be told which plate is positively charged. According to the diagram, plate A is positive. It therefore is at the higher potential. Since the potential difference across a capacitor is given by $Q = CV$ to be Q/C, we see that the potential change in going from A to B is $-Q/C$.

The potential change in each of these three cases is negative when going from A to B. If we were going from B to A, the change would be positive. Let us now use the loop rule in a few simple circuits before moving on to its more serious applications.

Illustration 18.1 Find the current which flows in the circuit of Fig. 18.4.

Reasoning Let us guess that the current will flow in the direction shown. (You might protest that this is wrong since the 12-V battery will certainly have more effect than the 3-V battery. But one of the nice things about Kirchhoff's rules is that even a poor guesser can use them, as we shall see.) We pick a point such as a and move around the circuit. The voltage changes

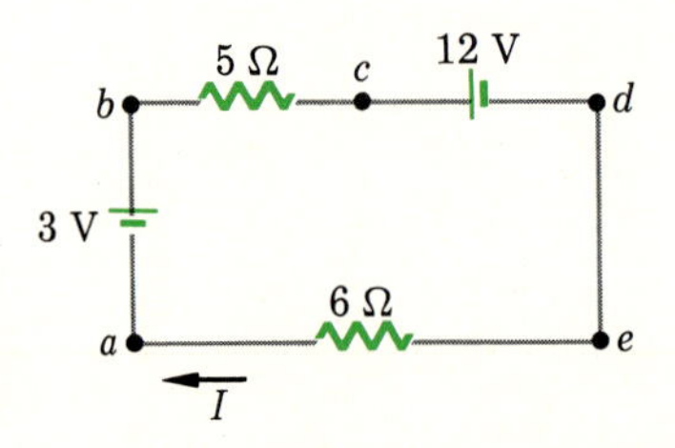

FIGURE 18.4

When we solve this circuit, how will our answer tell us that we have chosen I in the wrong direction?

are as follows, in volts:

$a \to b$:	$+3$
$b \to c$:	$-5I$
$c \to d$:	-12
$e \to a$:	$-6I$

Check them so that you are sure about the signs we have used. The sum of these voltage changes must be zero. Therefore

$$3 - 5I - 12 - 6I = 0$$

Solving for I, we find $I = -\frac{9}{11}$ A. The negative sign for I tells us we guessed its direction wrong. No harm is done.

Suppose we had circled the circuit in the reverse direction. Then our equation would have been

$$+6I + 12 + 5I - 3 = 0$$

from which $I = -\frac{9}{11}$ A, as before.

In solving this circuit, be sure you understand our choice of signs for the voltage changes. Also note that the current is the same at all points in the circuit. Why?

Illustration 18.2 Find the currents which flow in all the wires of the circuit of Fig. 18.5.

Reasoning We assign currents to all the wires and give each a symbol and direction. Once again, we waste little time trying to guess proper direction since our answer will indicate direction.

Starting at a, let us follow the loop $acda$ and write the loop rule. We have, in volts (be sure you understand the signs used)

$$-18I_2 - 9 = 0$$

from which we find at once that

$$I_2 = -0.50 \text{ A}$$

It therefore flows in a direction opposite to that shown.

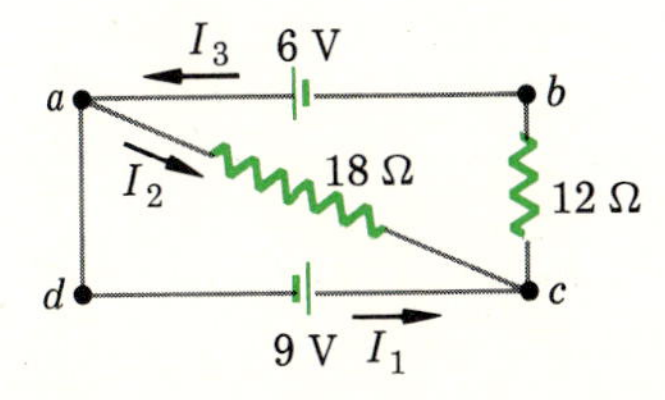

FIGURE 18.5

Find the currents in all three wires.

Now let us move around the loop *abcda*. We have

$$-6 + 12I_3 - 9 = 0$$

from which

$$I_3 = 1.25 \text{ A}$$

We could also write a loop equation for loop *abca*. But no new voltage changes would appear in it. Therefore, this equation would contain no new information and would be redundant.

Instead, we shall write the point rule for point *c*. It is

$$I_1 + I_2 = I_3$$

Substituting, we have, in amperes,

$$I_1 - 0.50 = 1.25$$

Notice that we carry the proper signs for I_2 and I_3. Solving, we find $I_1 = 1.75$ A.

18.3 RESISTORS IN SERIES AND IN PARALLEL

Resistors in Series

There are two configurations of resistor circuits which can be simplified very easily. If we recognize them, we can often greatly decrease the work needed to solve a problem. The first of these two configurations is shown in Fig. 18.6*a*. *Resistors such as these are said to be in series. To move from point A to point B, only one path is possible. That path goes through all the resistors.*

If we connect this combination across a battery as in part *b*, we might guess it would act like a single resistor equal to the sum $R_1 + R_2 + R_3$. In what follows, we shall prove this guess to be correct. What we wish to do is

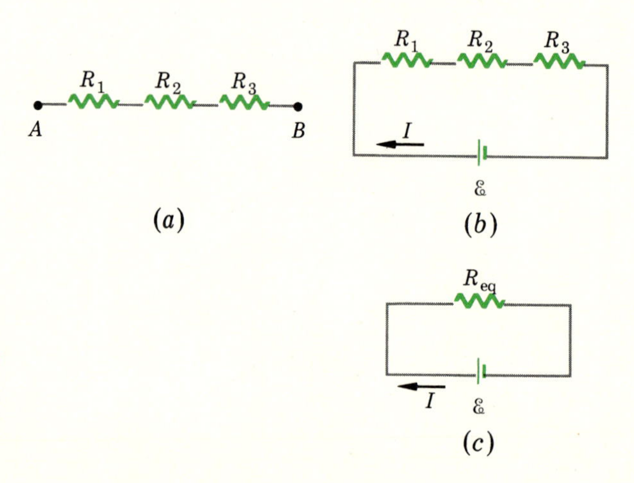

FIGURE 18.6

The three resistors are in series. They are equivalent to $R_{eq} = R_1 + R_2 + R_3$.

find an equivalent resistor R_{eq} in part c of the figure which draws the same current as the combination in part b. To do this, we make use of the loop rule for the circuits of parts b and c.

Going clockwise around the circuit in part b we have the following loop equation:

$$+\mathcal{E} - IR_1 - IR_2 - IR_3 = 0$$

This can be rewritten as

$$\frac{\mathcal{E}}{I} = R_1 + R_2 + R_3$$

Writing a similar loop equation for part c gives

$$+\mathcal{E} - IR_{eq} = 0$$

from which

$$\frac{\mathcal{E}}{I} = R_{eq}$$

From these two equations we see that

$$R_{eq} = R_1 + R_2 + R_3 \qquad \text{series} \tag{18.1}$$

In general, we can say

Equivalent Series Resistance

Several resistances in series are equivalent to a single resistor **equal to their** sum.

Resistors in Parallel

Another common configuration is shown in Fig. 18.7a. *These resistors are said to be in parallel. In a parallel configuration, one end of each resistor is connected to a point A. The other end of each resistor is connected to a point B. When a potential difference is placed across the combination,* as in part b of the figure, *each resistor has this same potential difference across it*. We wish to find the equivalent resistor R_{eq} shown in c which can be used to replace the combination and still draw the same current.

Notice in Fig. 18.7b that the potential difference across each resistor is

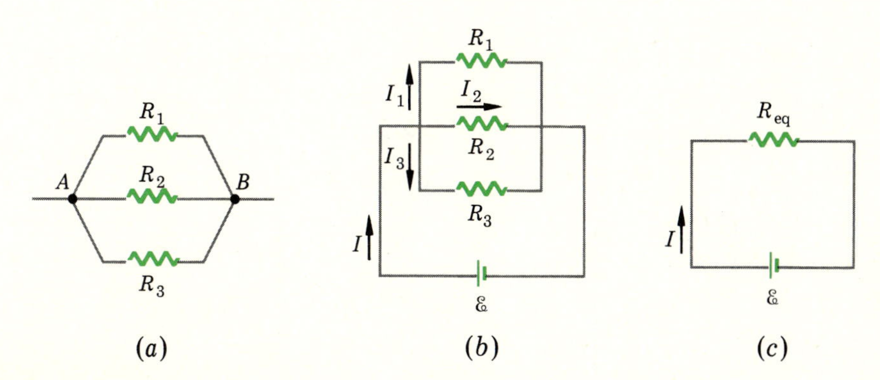

FIGURE 18.7
The three resistors are in parallel. Their equivalent is given by: $1/R_{eq} = 1/R_1 + 1/R_2 + 1/R_3$.

$\mathcal{E}$. Therefore, Ohm's law tells us that

$$I_1 = \frac{\mathcal{E}}{R_1} \qquad I_2 = \frac{\mathcal{E}}{R_2} \qquad I_3 = \frac{\mathcal{E}}{R_3}$$

But the point rule tells us that

$$I = I_1 + I_2 + I_3$$

and so

$$I = \frac{\mathcal{E}}{R_1} + \frac{\mathcal{E}}{R_2} + \frac{\mathcal{E}}{R_3}$$

After rearranging, this becomes

$$\frac{I}{\mathcal{E}} = \frac{1}{R_1} + \frac{1}{R_2} + \frac{1}{R_3}$$

Now let us look at the circuit in part *c*. The loop rule tells us that

$$+\mathcal{E} - IR_{eq} = 0$$

from which

$$\frac{I}{\mathcal{E}} = \frac{1}{R_{eq}}$$

We can now equate these two expressions for $I/\mathcal{E}$ to give

$$\frac{1}{R_{eq}} = \frac{1}{R_1} + \frac{1}{R_2} + \frac{1}{R_3} \qquad \text{parallel} \tag{18.2}$$

In general we can say

Equivalent Parallel Resistance

For several resistors in parallel, the reciprocal of their equivalent resistance is equal to the sum of the reciprocals of the resistances.

Let us now do a few practice examples to see more clearly how these equivalent resistors are obtained and used.

Illustration 18.3 Find the current which flows from the battery in Fig. 18.8*a*.

FIGURE 18.8

The parallel resistors between *b* and *c* are equivalent to 2 Ω, as shown in (*b*). The two series resistors in (*b*) can be combined, as in (*c*).

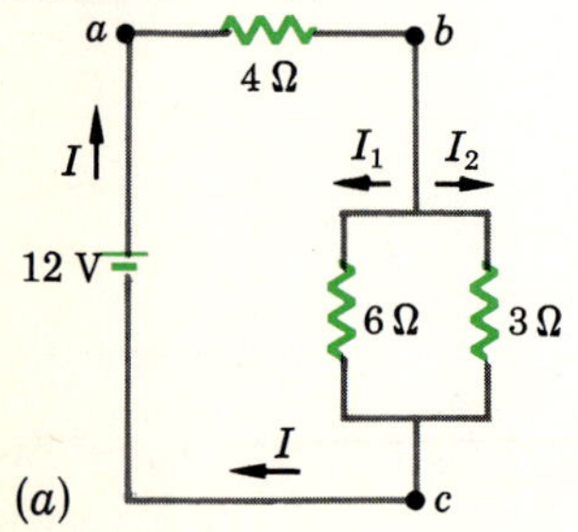

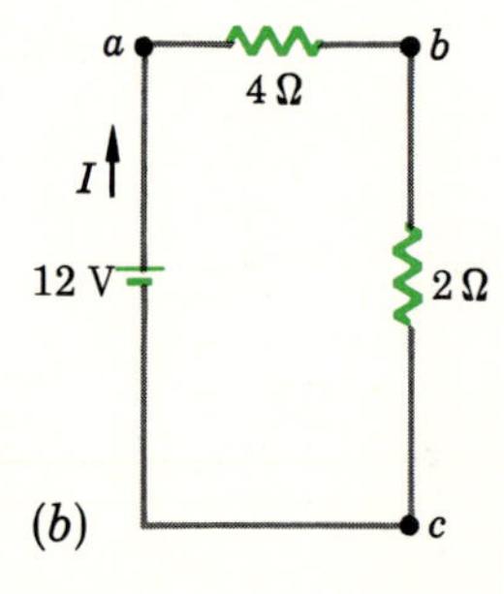

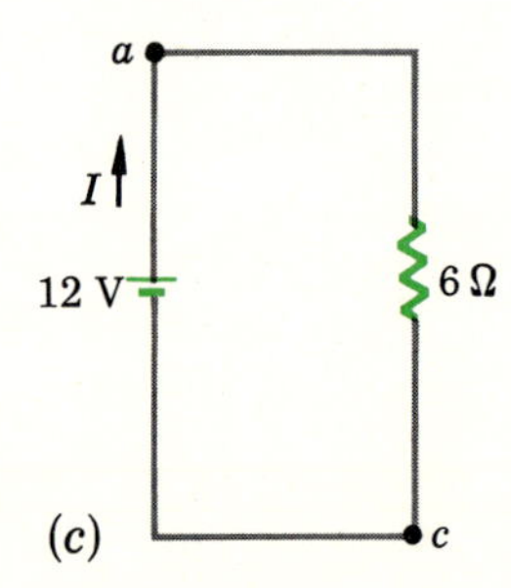

Reasoning We could solve this by use of Kirchhoff's rules. However, it is usually simpler to combine obvious series and parallel resistors before writing the loop equations. The resistances may be reduced as shown in parts *b* and *c* of the figure. Let us first combine the two parallel resistors between points *b* and *c*. We have

$$\frac{1}{R_{bc}} = \frac{1}{6} + \frac{1}{3} = \frac{1}{6} + \frac{2}{6} = \frac{3}{6}$$

or

$$R_{bc} = 2\ \Omega$$

An equivalent circuit is now drawn in Fig. 18.8*b*. The parallel combination has been replaced by its equivalent resistance. We see that the 4- and 2-Ω resistors are connected in series between points *a* and *c*. Their equivalent is

$$R_{ac} = 4 + 2 = 6\ \Omega$$

A new equivalent circuit is now drawn, shown in part *c*. This is a situation to which Ohm's law can be applied. The voltage difference across the 6-Ω resistor is 12 V. Hence we have

$$I = \frac{V}{R} = \frac{12\ \text{V}}{6\ \Omega} = 2\ \text{A}$$

which is the result required.

Illustration 18.4 Find the current which flows from the battery in Fig. 18.9*a*.

Reasoning Here, too, we have two parallel resistors, the 3-Ω and 6-Ω. They can be replaced by a 2-Ω resistor as in part *b*. There we see that the 2-Ω and 4-Ω resistors are in series. So they are equivalent to a 6-Ω resistor. But this equivalent resistor is in parallel with the 6-Ω resistor. Using the reciprocal formula, two 6-Ω resistors in parallel are equivalent to a 3-Ω resistor. This is drawn in part *c*. Finally, the circuit can be reduced to the form shown in part

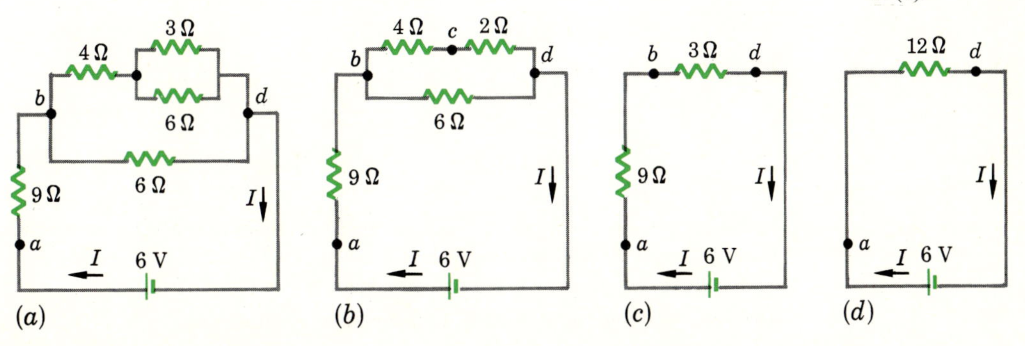

FIGURE 18.9 The complex circuit of (*a*) can be reduced to the simple equivalent circuit shown in (*d*).

d. Using Ohm's law for it, we have

$$I = \frac{V}{R} = \frac{6\ \text{V}}{12\ \Omega} = 0.50\ \text{A}$$

18.4 CAPACITORS IN SERIES AND PARALLEL

Equivalent Capacitance

We shall not examine this situation in detail. Instead, we show the two configurations in Fig. 18.10*a* and *b*. The results for the equivalent capacitors are shown in each case. These results will be obtained in problems at the end of the chapter. Notice that the relations are just opposite to those we found for resistors. *Parallel capacitors add directly while the reciprocals of series capacitances add.*

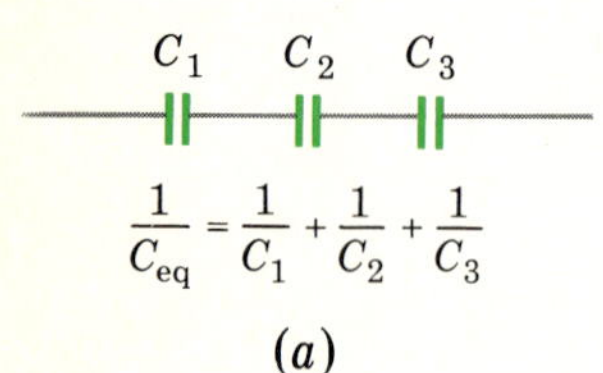

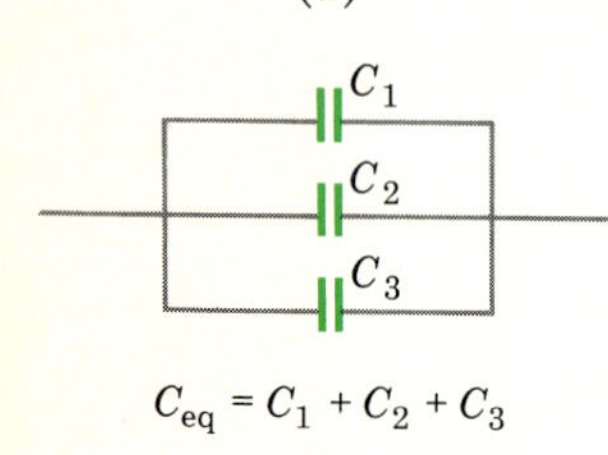

FIGURE 18.10

Reduction of series and parallel capacitances: (*a*) series; (*b*) parallel.

18.5 SOLUTION OF CIRCUIT PROBLEMS

We now have at our disposal the tools needed to solve most direct-current (dc) circuit problems. Before using them in several examples, let us state a few facts we should remember in their use. Although each individual problem will have its own peculiar features, *the following approach is most often useful.*

1 Draw the circuit.
2 Assign a current and direction to each important wire. Be careful to use only one current designation for a given wire even though it may contain several elements. If a branch wire comes off from it, the currents on the two sides of the branch must usually be considered different.
3 Reduce series and parallel resistance systems whenever possible and convenient.
4 Write the loop equations for the remaining circuit. Remember that unless an equation contains a new voltage change, it will be redundant.
5 Write the point equations for the junction points. Unless a new current is used in the equation, it will be redundant.
6 Solve these equations for the unknowns.

Let us now proceed with some examples.

FIGURE 18.11

A circuit easily solved by Kirchhoff's rules.

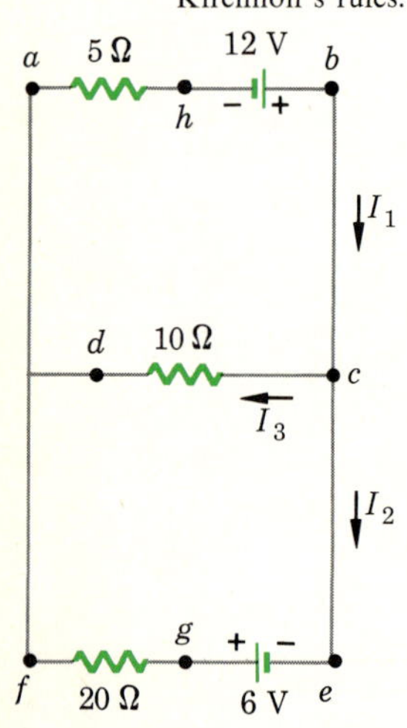

Illustration 18.5 Find the currents in the wires of the circuit shown in Fig. 18.11.

Reasoning Steps 1 and 2 of our procedure are already completed in the diagram. There are no simple resistance systems. (The resistors in *ab* and *ef* are *not* in simple parallel with the center resistor. Notice that the right-hand ends are connected to batteries, not to a common point.)

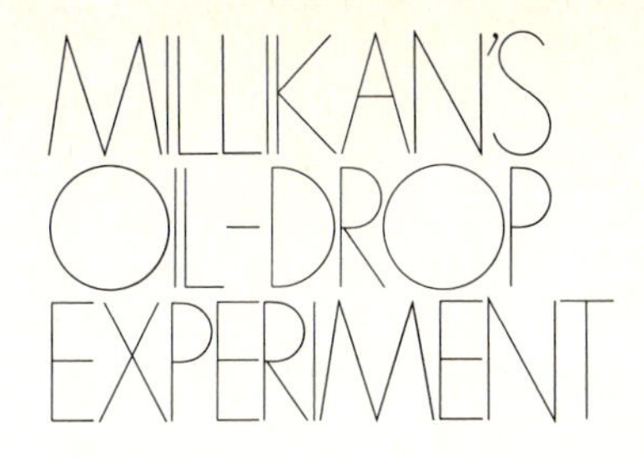

MILLIKAN'S OIL-DROP EXPERIMENT

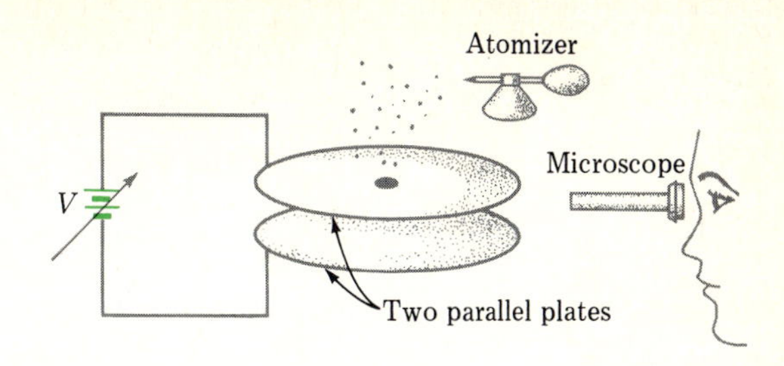

The magnitude of the quantum of charge, the electronic charge e, was first measured accurately by R. A. Millikan and his coworkers (1909–1913). Millikan's experiment has a simplicity and directness which establish without doubt that charge is quantized. His apparatus is illustrated schematically in the figure.

Basically, the apparatus consists of two parallel metal plates (separation $\approx$ 1 mm and area $\approx$ 100 cm^2) with a known variable potential difference between them. Because of the potential difference V, an electric field $E = V/d$ exists between the plates, where d is the plate separation. If now a charged oil droplet (mass m and charge q) finds itself between the plates, a vertical electric force Eq will act on the droplet. By adjusting the potential difference between the plates, this force can be made equal and opposite to the gravitational force mg on the droplet. When this condition is achieved, the resultant force on the droplet is zero and the droplet will remain motionless. Then

$$mg = Eq$$

from which

$$q = \frac{mg}{E}$$

When the electric field E, the mass of the droplet m, and the acceleration due to gravity g are known, the charge q on the droplet can be found.

In practice, Millikan allowed very small oil droplets sprayed from an atomizer to fall through a small hole in the top plate. Many of these droplets contain excess charge generated by friction as the drops are produced in the atomizer. An intense light shining between the plates causes the droplets to sparkle, and so they can be observed by looking through the microscope, as indicated. Provision was also made for changing the charge on the droplet by subjecting the region between the plates to a weak x-ray beam when desired.

The electric field between the plates was calculated from the measured plate separation d and the potential difference between the plates V. To find the mass of the oil particle m, Millikan made measurements of the time taken for the droplet to fall a measured distance through the air in the absence of an electric field. Since the droplets were moving at constant speed—their terminal speed—through the distance chosen, the gravitational force mg was equal to the viscous-friction force between the droplet and the air. Assuming that the viscous force was proportional to the speed of the droplet v, he could write.

$$mg \propto v$$

One of the major sources of error involved in the experiment was associated with the proportionality constant in this relation. To fair accuracy, however, one can use the so-called Stokes' law form for this viscous force and write

$$mg = 6\pi\eta av$$

where η is the viscosity of air and a is the radius of the droplet. Since the radius of the drop is related to its mass through the fact that the density of the oil ρ multiplied by the volume of the drop is equal to the drop's mass,

$$m = \tfrac{4}{3}\pi a^3\rho$$

the mass of the drop can be found from a knowledge of ρ, η, and its terminal speed v.

As shown, when m and E, are known, the charge on the droplet can be found. Millikan carried out thousands of measurements of this general type (slightly modified for higher accuracy). His result for q was always an integer multiple of 1.60×10^{-19} C. He therefore concluded that charge was quantized, the magnitude of the charge quantum being 1.60×10^{-19} C which we designate by e. All charges in nature are found to be integer multiples of this value. In particular, the charge on the electron is $-e$, while that on the proton is $+e$. The currently accepted value of e is 1.6022×10^{-19} C.

Let us now start at point a and write the loop equation for loop $abcda$. It is, in volts,

$$-5I_1 + 12 - 10I_3 = 0 \tag{1}$$

Similarly for loop $dcefd$

$$+10I_3 + 6 - 20I_2 = 0 \tag{2}$$

We have three unknowns, I_1, I_2, and I_3, but only two equations. The loop equation $abcefa$ includes no new voltage changes, and so we ignore it.

To obtain a third equation, we write the point equation for point c. It is

$$I_1 = I_3 + I_2 \tag{3}$$

Substitution of this value for I_1 in Eq. (1) gives

$$-5I_3 - 5I_2 + 12 - 10I_3 = 0$$

which reduces to

$$-5I_2 - 15I_3 + 12 = 0 \tag{4}$$

We now can solve (2) and (4) simultaneously.

or
and $I_2 = I_1 +$

If we knew the potential difference
the relation $Q = CV_{fg}$ to find Q. We can
and moving around the circuit *gdefg*, a
Doing this gives

$$0 + 10 - 6I_1$$
$$V_{fg} = -10 + 1$$

The positive sign shows we have gone u
g, and so the *g* side of the capacitor is p
on the capacitor

$$Q = CV_{fg} = (5 \times 10^{-6}$$

18.6 POWER AND ELECTRICAL H

When a battery causes current to run t
18.15, the battery is furnishing energy.
charge uphill in potential through the
voltage.

We know that the work done in
potential difference V is given by

Work done

If the charge q is transported in a time

Work/s =

But work per second is power, and so

Power =

However, current is defined to be charge
Hence we find that the power furnishe

Power =

This is a general relation for power
operating at a voltage V. In the present i
battery is expended in the resistor. The

Solving (2) for I_3, we find

$$I_3 = 2I_2 - 0.60 \tag{5}$$

Substituting this in (4) gives

$$-5I_2 - 30I_2 + 9 + 12 = 0$$

from which

$$I_2 = 0.60\text{ A}$$

Using this value in (5) gives

$$I_3 = 0.60\text{ A}$$

Now, by use of (3) we find

$$I_1 = 1.20\text{ A}$$

Illustration 18.6 For the circuit shown in Fig. 18.12, find I_1, I_2, and I_3.

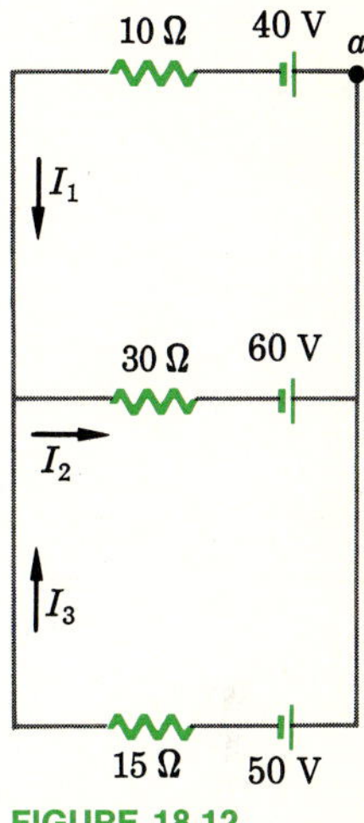

FIGURE 18.12
Find the three unknown currents.

Reasoning Choosing the currents to flow in the directions shown on the figure, and noticing that $I_1 + I_3 = I_2$, we can write the appropriate circuit or loop equations. Taking the loop through the 40- and 60-V batteries, starting at a, and going counterclockwise,

$$-40 - 10I_1 - 30I_2 + 60 = 0 \qquad \text{or} \qquad I_1 + 3I_2 = 2$$

Next from a through the 40- and 50-V batteries gives

$$-40 - 10I_1 + 15(I_2 - I_1) + 50 = 0$$

or

$$2.5I_1 - 1.5I_2 = 1.0$$

After multiplying the second equation by 2 and adding it to the first, we have

$$6I_1 = 4 \qquad \text{or} \qquad I_1 = \tfrac{2}{3}\text{ A}$$

Using this in the first equation gives

$$3I_2 = \tfrac{4}{3} \qquad \text{or} \qquad I_2 = \tfrac{4}{9}\text{ A}$$

And since $I_3 = I_2 - I_1$, we have

$$I_3 = -\tfrac{2}{9}\text{ A}$$

Clearly, I_3 is going in the opposite direction to that drawn on the diagram.

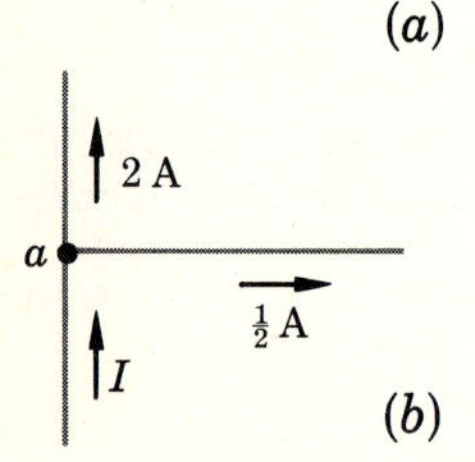

FIGURE 18.13
The ammeter and voltmeter readings are known. We wish to find I, X, and $\mathcal{E}$.

Illustration 1
reads 0.50 A,
ammeter and
resistor is as
direction sho

Reasoning S
tells us that tl
resistor at th
The curr
From the poi

and so

Writing the c
ammeter has
have

from which

The circuit ec

or

Illustration 1
charge on the

Reasoning I
capacitor is cl
solving the re

Writing the lc

or

Using the loo

to show that Eq. (18.3) also is a statement of the power expended in any element through which a current I flows and across whose terminals the voltage is V. *In the case of the resistor,* the application of Ohm's law leads to other forms of the equation,

$$P = VI = I^2R = \frac{V^2}{R} \qquad \text{resistor} \tag{18.4}$$

Recalling that volts are joules per coulomb and that current is coulombs per second, the units of power in this equation are joules per second, or watts.

All of us are aware of the fact that current flowing through a resistor produces heat in the resistance wire. The current through the filament of an incandescent light bulb heats the resistance element to a white heat. An electric stove has a resistance element as its heating unit, as do toasters, flatirons, electric clothes driers, and many other heating devices. The electrical power furnished to a resistor furnishes heat energy. A resistor converts the electrical energy furnished by the battery into heat energy.

The interchange of electrical energy into heat energy allows us to determine the mechanical equivalent of heat by a method different from the one discussed in Chap. 11. We need only measure how much heat is produced by a resistor R carrying a current I. This is conveniently done by placing the resistance wire in a calorimeter containing water. The electrical energy furnished to the water and calorimeter in time t is just Pt or I^2Rt. It is expressed in joules. The heat energy in calories is found from the temperature rise of the calorimeter and its contents. Hence, since the energies must be the same in either units, calories or joules, the conversion between them is easily found. The relation between the units must, of course, be the same as was found by rotation of a paddle wheel in water, namely

$$1 \text{ cal} = 4.184 \text{ J}$$

Illustration 18.9 How much heat (in calories) does a 40-W light bulb generate in 20 min?

Reasoning A 40-W light bulb generates 40 J of heat per second. Hence, in (20)(60) s it will develop

$$\text{Heat} = [(20)(60)](40) = 48{,}000 \text{ J} = 11{,}500 \text{ cal} = 11.5 \text{ kcal}$$

Illustration 18.10 If electrical energy costs 10 cents per kilowatthour, how much does it cost to operate a 700-W drier for 30 min?

Reasoning A kilowatthour (kWh) is the energy unit one obtains if power is measured in kilowatts and time is measured in hours. Hence

$$\text{Energy} = (\text{power})(\text{time}) = (0.700)(0.50)\ \text{kWh} = 0.350\ \text{kWh}$$

The cost of this will be 3.5 cents.

18.7 HOUSE CIRCUITS

We are all familiar with the ordinary electrical circuits which extend throughout our houses. The power company runs at least two wires to each house to provide a potential difference of about 120 V. These lead-in wires to the house usually have a large diameter so that they can carry considerable current without heating up. (The larger the cross-sectional area of the wire, the less its resistance will be. Since heat generated is proportional to I^2R, the low resistance will ensure low heat dissipation.)

In most newer houses, the wires are capable of carrying about 20 A without undue heating. However, *to protect against too large a current, a fuse or circuit breaker is placed in series with the wire. Its purpose is to disconnect the wire* from the voltage source if greater than the allowed current is drawn from it. This procedure automatically disconnects any wire which is accidentally called upon to carry more than the safe current.

A typical house circuit consists of two parallel wires strung through the house from the 120-V source provided by the lead-in wires to the house. This is shown schematically in Fig. 18.16. Each light bulb, appliance, etc., is connected with one terminal to the high-potential wire and the other to the low-potential wire. When the switch to that appliance is closed, current runs through the device from the + to the − wire of the power system. The low-potential wire is usually grounded.

Many 120-V appliances have a third prong on the power plug. This furnishes a connection between a ground wire and the metal frame of the appliance. If, by accident, the high-voltage wire touches the metal frame of the appliance, a direct connection to ground is made. The effect is then the same as connecting the high- and low-voltage wires directly together. A large current then flows through the high-voltage wire to ground. The fuse in the high-voltage wire will then blow. If the ground wire is absent, such a malfunction will leave the whole appliance "floating" at high potential. Anyone touching the metal frame will then suffer a shock.

Let us compute how much current is drawn by the 60-W bulb of Fig. 18.16 when it is turned on by closing the switch. Since power $= VI$, and since $P = 60$ W and $V = 120$ V in this case, we find the current through the bulb to be $I = 0.50$ A. Similarly, when turned on, the stove draws 10 A, the radio draws 0.167 A, and the 120-W bulb draws 1.0 A. If they are all turned on at once, a total of 11.667 A will pass through the fuse. Usually a house circuit would be fused for no less than 15 A, and so no danger exists in this case.

A house which has a large number of electrical appliances requires more than one circuit. Most houses have several separate fused circuits such as the

FIGURE 18.16

The household appliances shown act as resistors. Each is put into operation by closing the appropriate switch.

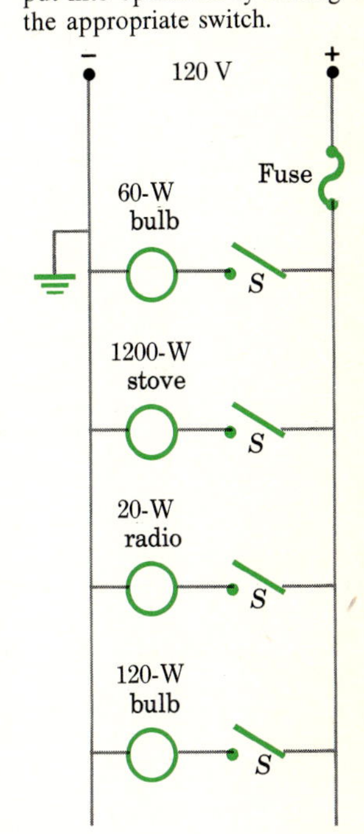

one shown in Fig. 18.16. Each of these starts at the source furnished by the lead-in wires to the house and runs to various portions of the house.

It is interesting to compute the resistance of a light bulb. When the light bulb is cool, its resistance is not too large. However, when it is connected across the rated voltage, usually 120 V, its resistance element becomes white-hot. As discussed in Chap. 17, its resistance will increase considerably when it heats up. When it is hot, it will operate at the wattage stamped on it. Suppose that we have a 60-W 120-V bulb. We know that

$$P = VI = \frac{V^2}{R}$$

or

$$60 = \frac{(120)^2}{R}$$

from which

$$R = 240\ \Omega$$

where you should verify the units of the answer.

18.8 ELECTRICAL SAFETY

Since we use electrical apparatus daily, we should understand the elements of electrical safety. *Electricity can kill a person in two ways: it can cause the muscles of the heart and lungs (or other vital organs) to malfunction or it can cause fatal burns.*

Even a small electric current can seriously disrupt cell functions in that portion of the body through which it flows. When the electric current is 0.001 A or higher, a person can feel the sensation of shock. At currents 10 times larger, 0.01 A, a person is unable to release the electric wire held in his hand because the current causes his hand muscles to contract violently. Currents larger than 0.02 A through the torso paralyze the respiratory muscles and stop breathing. Unless artificial respiration is started at once, the victim will suffocate. Of course, the victim must be freed from the voltage source before he or she can be touched safely; otherwise the rescuer, too, will be in great danger. A current of about 0.1 A passing through the region of the heart will shock the heart muscles into rapid, erratic contractions (ventricular fibrillation) so the heart can no longer function. Finally, currents of 1 A and higher through body tissue cause serious burns.

The important quantity to control in preventing injury is electric current. Voltage is important only because it can cause current to flow. Even though your body can be charged to a potential thousands of volts higher than the metal of an automobile by simply sliding across the car seat, you feel only a harmless shock as you touch the door handle. Your body cannot hold much charge on itself, and so the current flowing through your hand to the door handle is short-lived and the effect on your body cells is negligible.

In some circumstances, the 120-V house circuit is almost certain to cause death. One of the two wires of the circuit is usually attached to the ground, so it is always at the same potential as the water pipes in a house. Suppose a person is soaking in a bathtub; his body is effectively connected to the ground through the water and piping. If his hand accidentally touches the high-potential wire of the house circuit (by touching an exposed wire on a radio or heater, for example), current will flow through his body to the ground. Because of the large, efficient contact his body makes with the ground, the resistance of his body circuit is low. Consequently the current flowing through his body is so large that he will be electrocuted.

Similar situations exist elsewhere. For example, if you accidentally touch an exposed wire while standing on the ground with wet feet, you are in far greater danger than if you are on a dry, insulating surface. The electrical circuit through your body to the ground has a much higher resistance if your feet are dry. Similarly, if you sustain an electrical shock by touching a bare wire or a faulty appliance, the shock is greater if your other hand is touching the faucet on the sink or is in the dishwater.

As you can see from these examples, *the danger from electrical shock can be eliminated by avoiding a current path through the body*. When the voltage is greater than about 50 V, avoid touching any exposed metal portion of the circuit. If a high-voltage wire must be touched, e.g., in case of a power-line accident when help is not immediately available, use a dry stick or some other substantial piece of insulating material to move it. When in doubt about safety, avoid all contacts or close approaches to metal or to the wet earth. *Above all, do not let your body become the connecting link between two points which are at widely different potentials.*

18.9 EMF AND TERMINAL POTENTIAL OF A BATTERY

Probably everyone has noticed at one time or another that the lights on an automobile dim when the motor is first started. The electrical starter used on an automobile draws considerable current from the battery. In so doing, it lowers the potential between the terminals of the battery, and the car lights dim. We shall now investigate this nonconstancy of the terminal potential difference of a battery.

Internal Resistance of Battery

As pointed out in the last chapter, the emf of a battery is generated by the chemical action within it. When no current is being drawn from the battery, the difference in potential between the battery terminals is equal to the emf of the battery. However, *a battery* is a very complex chemical device, and it actually *behaves like an emf in series with a resistor*. An equivalent circuit for a battery is shown in Fig. 18.17.

Notice that when no current is being drawn from the battery, there is no potential drop across the internal battery resistance r. Hence the potential difference between the terminals of the battery is equal to the emf. However, if the battery is connected across a resistor as shown in Fig. 18.18, a current I flows as shown and the terminal potential difference is $\mathcal{E} - Ir$.

For a good 12-V battery the internal resistance would be only of the

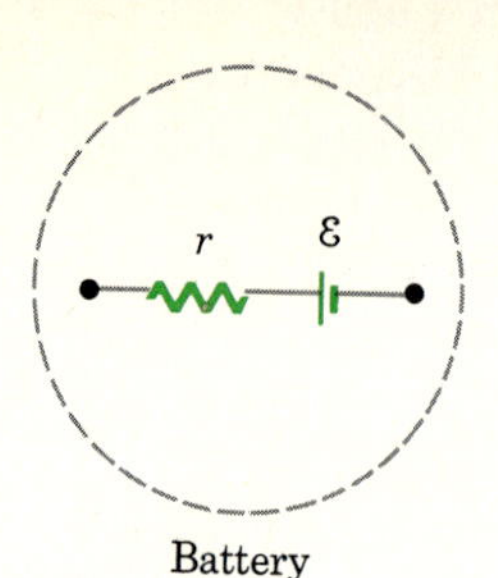

FIGURE 18.17
A battery acts as though it consisted of a pure emf and a resistor in series, as shown.

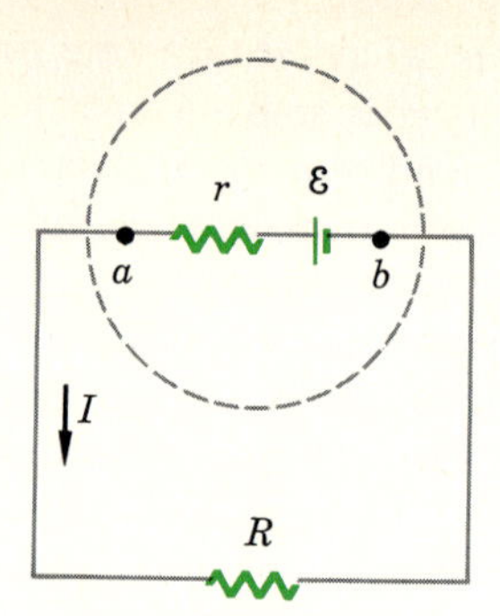

FIGURE 18.18
The terminal voltage V of the battery is $\mathcal{E} - Ir$.

order of 0.01 Ω. If it were connected across a 3-Ω resistor, we would have

$$I = \frac{12}{3 + 0.01} \approx 4 \text{ A}$$

The terminal potential would be the voltage difference between points a and b, which is

$$\text{Terminal potential} = 12 - (4)(0.01) = 11.96 \text{ V}$$

In this case the terminal potential is nearly equal to the emf.

However, as a battery becomes older, its internal resistance increases. If the resistance of the battery in Fig. 18.18 were 1.0 Ω, the current would be

$$I = \tfrac{12}{4} = 3.0 \text{ A}$$

and the terminal potential would be

$$\text{Terminal potential} = 12 - 3.0 = 9.0 \text{ V}$$

It should be clear that when a starter on a car draws 100 A from the battery, the terminal potential of even a new battery will decrease noticeably.

FIGURE 18.19
The 24-V battery is charging the 6-V battery. We find that the terminal potential of a discharging battery is less than its emf while the reverse is true for a battery which is being charged.

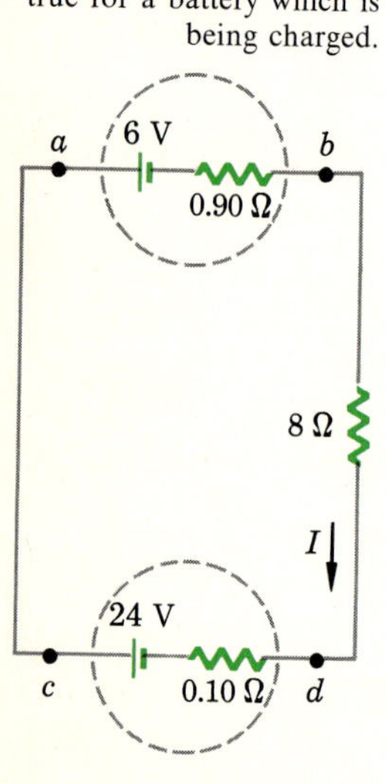

Illustration 18.11 What will be the terminal potential of each of the batteries in Fig. 18.19?

Reasoning The two batteries are opposing each other, and so the effective driving emf in the circuit is 18 V. The current will flow in the direction shown and is

$$I = \tfrac{18}{9} = 2 \text{ A}$$

The potential difference from d to c is

$$V = -0.2 + 24 = 23.8 \text{ V}$$

Hence the terminal potential of the 24-V battery is less than its emf. For the terminal potential of the other battery from b to a we have

$$V = +1.8 + 6 = 7.8 \text{ V}$$

Notice that the terminal potential of this battery is larger than the emf. This is always the case when a battery is being charged as the 6-V battery is in this example.

18.10 THE POTENTIOMETER

If one wishes to measure the potential difference between two points in a circuit, it is sometimes important that the voltage-measuring instrument draw no current, as in trying to measure the emf of a battery which has a high internal resistance. Or suppose that we wished to measure the potential difference between two points on a person's body, as is done in measuring brain waves. The body acts like a battery of extremely high internal resistance. Clearly, no current can be drawn during the measurement if the undisturbed emf is to be measured.

Any voltmeter draws some current from the points to which it is attached. Although electronic voltmeters draw only very small current from the voltage source which they are measuring, ordinary laboratory voltmeters frequently have resistances of only a few hundreds of ohms. Hence, when connected across a few volts, they will draw appreciable current. We require *a simple device for measuring voltage which draws no current in the process. The potentiometer is such a device.*

Suppose you wish to measure an unknown battery of emf denoted as $\mathcal{E}_x$. If you could find a large group of calibrated batteries, you could connect each of the known batteries into a circuit such as the one shown in Fig. 18.20*a*. You could gently tap the switch to see whether the galvanometer would deflect. When a known battery was found for which the galvanometer did not deflect, you would know that the emfs of the two batteries just balanced each other so that no current would flow. Hence you would know that the unknown battery had the same emf as this known battery. The measurement will have been made when no current was being drawn from the unknown, and so the internal resistance of the battery has no effect.

It is impractical to use a large group of known batteries, since they will not usually be available. However, the circuit of Fig. 18.20*b* serves the same purpose and, as we shall see, requires only one standard battery. One merely moves the sliding contact at B along the variable resistor until the voltage drop from A to B is equal to $\mathcal{E}_x$. In effect, the variable resistor provides a variable battery with terminals at A and B. (The part of the circuit below points a and b is frequently referred to as a **potential divider.**) When the

FIGURE 18.20
In (*a*), when no current flows with S closed, we know that $\mathcal{E}_x = \mathcal{E}_{known}$. To obtain a variable $\mathcal{E}_{known}$, the circuit below points a and b in (*b*) is used. The current through G is zero when the voltage drop from A to B equals $\mathcal{E}_x$. This device, a potentiometer, measures the emf of the unknown cell.

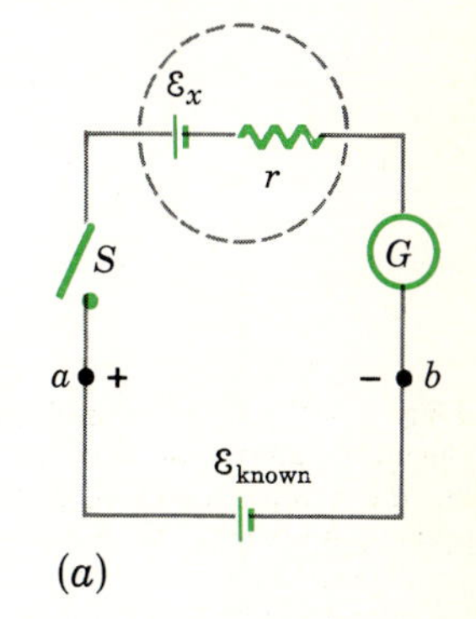

(*a*)

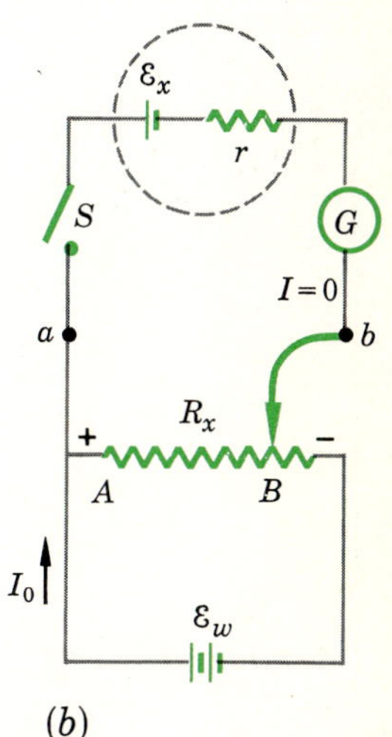

(*b*)

voltage from A to B exactly equals $\mathcal{E}_x$, no current will flow in the upper circuit and one has

$$\mathcal{E}_x = I_0 R_x$$

If $\mathcal{E}_x$ is now replaced by a standard known battery $\mathcal{E}_s$, the potentiometer can again be balanced so that the galvanometer does not deflect. Then

$$\mathcal{E}_s = I_0 R_s$$

Dividing these two equations yields

$$\frac{\mathcal{E}_x}{\mathcal{E}_s} = \frac{R_x}{R_s}$$

In practice, the resistor is calibrated, and so R_x/R_s is known. With the value of the standard cell $\mathcal{E}_s$ known, the unknown emf is easily calculated. (Commonly, the resistor is merely a long, uniform resistance wire. Hence R will be proportional to the length of the wire from A to B. As a result, the ratio R_x/R_s can be replaced by the respective lengths of wire, L_x/L_s.)

Illustration 18.12 By placing electric contacts at various positions on one's head, it is found that potential differences exist which fluctuate in characteristic ways. These potential differences can be recorded by an **electroencephalograph.** Typically these voltage differences are of the order of 5×10^{-4} V and fluctuate in times of the order of 0.10 s. Suppose the resistance of the head between two of these contacts is 10,000 Ω. How large a current can the voltage-recording device draw from the voltage being measured if the voltage read is to be in error by less than 1 percent? How large a resistance must the recording device have?

Reasoning The portion of the person's head between the two electrodes acts as a battery with emf = V_0 and the internal resistance = 10,000 Ω. When a current I is being drawn, the internal resistance causes a voltage error $10{,}000I$. We wish

$$\frac{10{,}000I}{V_0} < 0.01$$

Therefore the recorder cannot draw a current in excess of $0.01\ V_0/10{,}000$, or $10^{-6} V_0$ A. Since the current is caused to flow through the voltage recorder by the voltage source being measured V_0, one can use Ohm's law to find the resistance R_0 of the recorder. Thus

$$V = IR$$

becomes

$$V_0 = (10^{-6}V_0)R_0$$

from which the recorder resistance must be at least

$$R_0 = 10^6\ \Omega$$

Many modern recorders use the potentiometer principle and, in effect, have infinite resistance when a measurement is made.

SUMMARY

Kirchhoff's point rule states that the sum of all the currents coming into a point must equal the sum of all the currents leaving the point. His loop rule states that the algebraic sum of the voltage changes around a closed circuit (or loop) is zero.

Resistances in series add directly; for parallel resistances, the reciprocals add.

Capacitances in parallel add directly; for series capacitances, the reciprocals add.

When a current I flows through a resistance R, the power expended in the resistor is I^2R. This expended power appears as heat energy. If a current I flows through a potential drop V of any kind, the power expended by the current is VI.

The primary safety rule one should follow when dealing with electricity is as follows: never let your body become a connecting link between two points of widely different potential.

A battery acts like an emf $\mathcal{E}$ in series with an internal resistance r. The terminal potential difference of a battery differs from $\mathcal{E}$ by an amount Ir, where I is the current through the battery.

A potentiometer measures voltages without drawing current. As a result, it can be used to measure emf for sources which contain appreciable internal resistance.

MINIMUM LEARNING GOALS

Upon completion of this chapter, you should be able to do the following:

1. State Kirchhoff's point rule and give examples of its application.
2. State Kirchhoff's loop rule and write the loop equation for a series circuit containing batteries and resistors.
3. Reduce a given set of series and parallel resistors to a single equivalent resistor.
4. Reduce a given set of series and parallel capacitors to a single equivalent capacitor.
5. Solve dc circuits which contain batteries, resistances, and capacitances by use of Kirchhoff's rules.
6. Use the power equation, $P = VI$, to find the power loss or gain in a resistor, battery, and capacitor under dc conditions.
7. Sketch a typical house circuit and point out the various elements in it. Compute the current drawn by various portions of a house circuit when the appliances running from it are given.
8. Analyze a given practical situation from the view point of safety.
9. Explain why the terminal potential of a battery is not always equal to the emf of the battery. Find the terminal potential if $\mathcal{E}$, I, and r are known.
10. Sketch the circuit diagram for a potentiometer and explain the function of the device.

IMPORTANT TERMS AND PHRASES

You should be able to define or explain each of the following:

Kirchhoff's point rule
Kirchhoff's loop rule

Series connection: $R_{eq} = R_1 + R_2 + \cdots$ and $\frac{1}{C_{eq}} = \frac{1}{C_1} + \frac{1}{C_2} + \cdots$

Parallel connection: $\frac{1}{R_{eq}} = \frac{1}{R_1} + \frac{1}{R_2} + \cdots$ and $C_{eq} = C_1 + C_2 + \cdots$

$P = VI$

Emf versus terminal potential of a battery

Never become the connecting link

QUESTIONS AND GUESSTIMATES

1. A resistor is connected from point a to point b. How does one tell which it is from a to b, a potential drop or a potential rise? Repeat for a battery and for a capacitor.

2. Explain the following statement: for series resistors, the equivalent is always larger than the largest; for parallel resistors, the equivalent is always smaller than the smallest. What is the similar statement for capacitors?

3. Why would one connect two batteries in series? In parallel?

4. Batteries should never be connected in parallel unless they are nearly identical. Why?

5. Is the energy stored in three identical capacitors charged by a 12-V battery larger when the capacitors are in series or when they are in parallel?

6. Does the energy stored in three capacitors in series depend upon which one is in the middle? Explain.

7. Explain why the charges on several unlike capacitors charged in series are all identical while unlike capacitors in parallel have different charges.

8. When one grasps the two wires leading from the two plates of a charged capacitor, one often feels a shock. The effect is much greater for a 2-μF capacitor than for a 0.02-μF capacitor, even though both are charged to the same potential difference. Why?

9. Birds sit on high-tension wires all the time. Why aren't they electrocuted since sometimes the wires have gaps where there is no insulation on them?

10. If a current of only a small fraction of an ampere flows into one hand and out the other, the person will probably be electrocuted. If the current flows into one hand and out the elbow above the hand, the person can survive even if the current is large enough to burn the flesh. Explain.

11. Mothers frequently worry about their small children playing near electrical outlets. Discuss the various factors which determine how badly shocked the child could be. What would happen if a small child were to cut a lamp cord in two with a pair of noninsulated wire-cutting pliers when the cord is plugged in? Is the child in any danger in this case?

12. Explain why it is much more dangerous to touch an exposed houselight circuit wire if you are in a damp basement than if you are on the second floor.

13. It is extremely dangerous to use a plug-in radio near a bathtub when one is taking a bath. Why? Does the same reasoning apply to a battery-operated radio?

14. For such purposes as electrocardiograms, brain-damage tests, lie detectors, etc., one measures the voltage differences between various portions of a person's body. Why can't one use a simple voltmeter for this purpose? What precautions must one take?

15. Not long ago, car manufacturers changed from a 6-V electrical system to a 12-V system. Why?

16. Estimate the electric-power consumption of a nearby city. How large a current would have to be supplied to the city at 220 V to supply this energy? Explain why very-high-voltage power transmission is of advantage (that is, 10^5 V).

PROBLEMS

Ignore internal resistances of batteries when the resistance is not given.

1. (*a*) Find the equivalent resistance for the circuit of Fig. P18.1. (*b*) Find I if the value of $\mathcal{E}$ is 12.0 V.

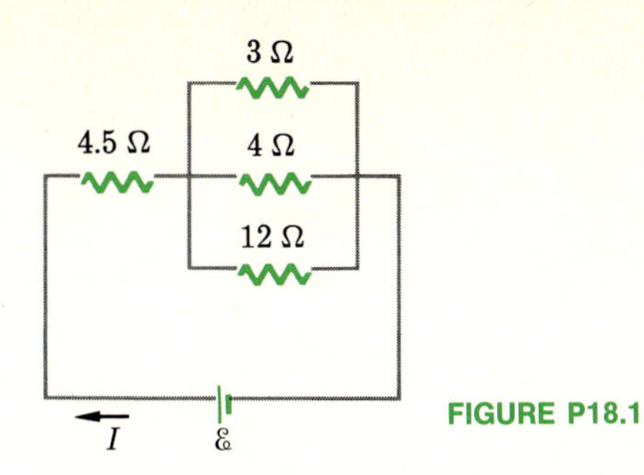

FIGURE P18.1

2. In Fig. P18.2, the value of ℰ is 12.0 V, and each resistor is 2.0 Ω. Find the equivalent resistance and I.
3. Find the equivalent resistance for the circuit in Fig. P18.3 when (*a*) the switch S is open and (*b*) S is closed.
4. Find the equivalent resistance for the circuit shown in Fig. P18.4 if all the resistors are 4 Ω.
5. Each of the capacitors shown in Fig. P18.5 is 2.0 μF. Find the equivalent capacitance of the combination.
6. Find the equivalent capacitance for the combination shown in Fig. P18.6.
7. Find I_1, I_2, and I_3 in the circuit of Fig. P18.7.
8. In Fig. P18.8, the ammeter reads 5 A. Find I_1, I_2, ℰ, and the voltmeter reading.
9.* In the circuit of Fig. P18.9, find I, I_1, and the charge on the 7-μF capacitor. *Hint:* Note that the currents in many of the wires are zero.
10. For the circuit shown in Fig. P18.10, find the (*a*)

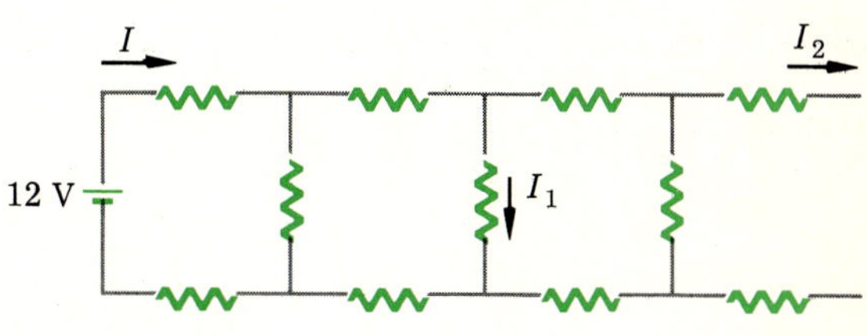

FIGURE P18.4

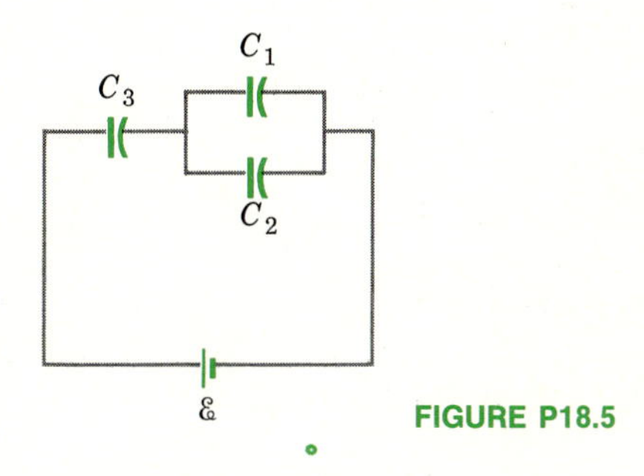

FIGURE P18.5

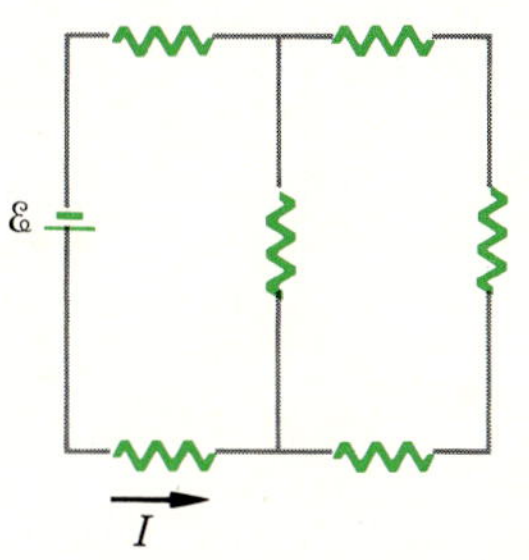

FIGURE P18.2

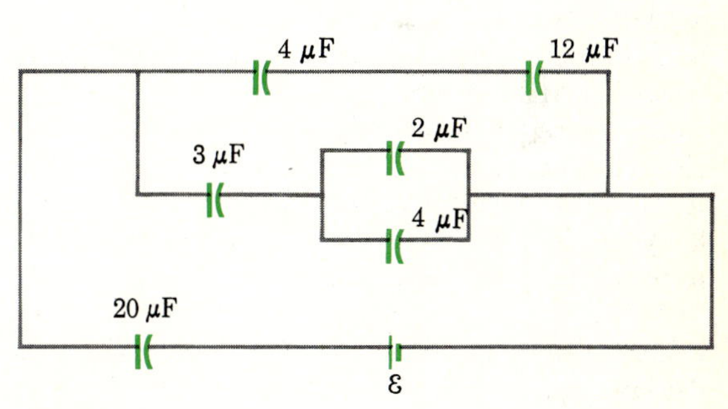

FIGURE P18.6

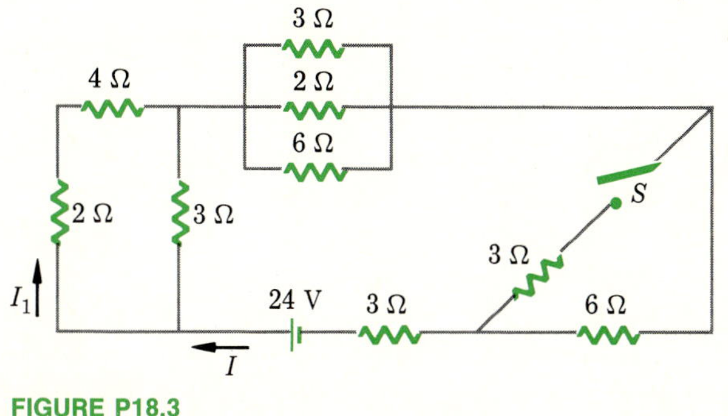

FIGURE P18.3

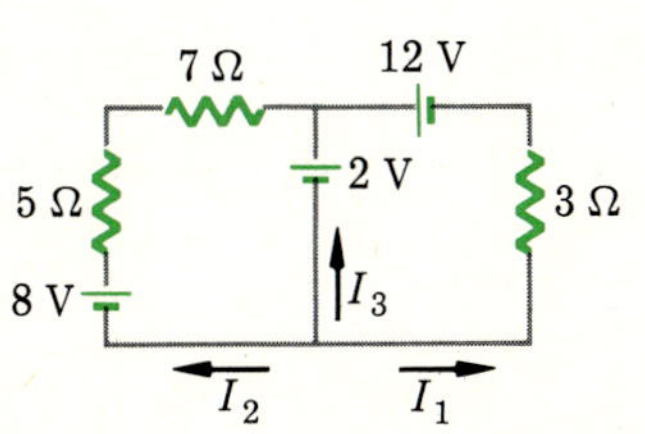

FIGURE P18.7

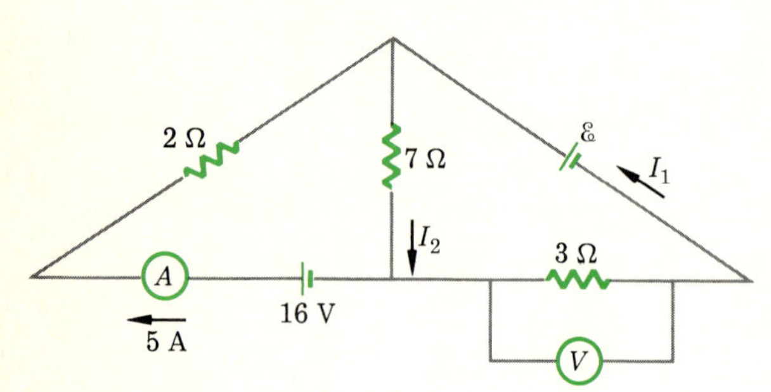

FIGURE P18.8

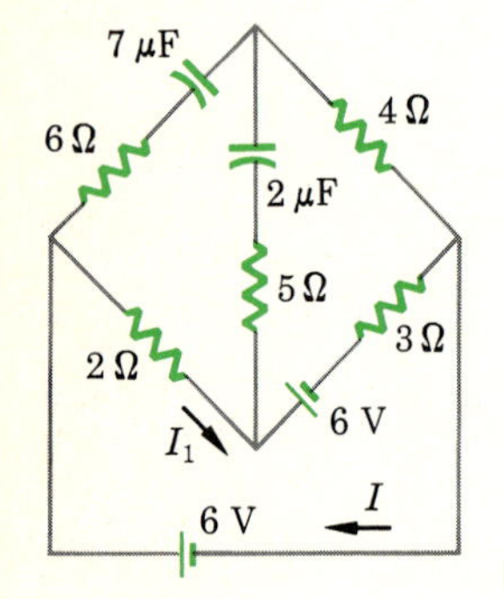

FIGURE P18.9

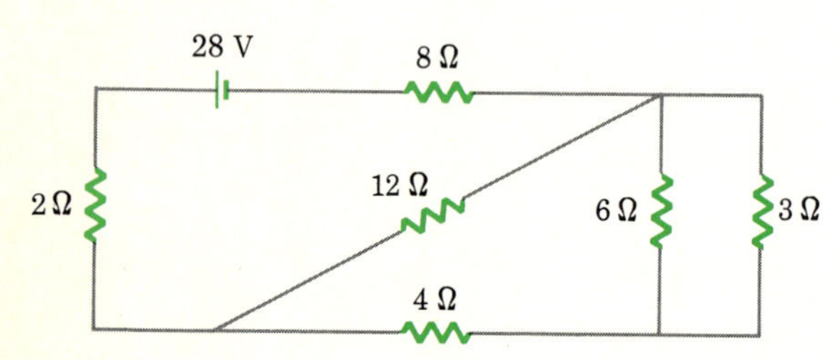

FIGURE P18.10

current through the battery; (*b*) the current through the 12-Ω resistor; (*c*) the power loss in the 8-Ω resistor.

11.* Refer to the circuit of Fig. P18.11. The voltmeter reads 5.0 V, and the ammeter reads 2.0 A with the current flowing in the direction indicated. Find (*a*) the value of R and (*b*) the value of ℰ. (Notice in writing the circuit equation that the voltage drop across R is 5 V.)

12.* In Fig. P18.11, how large must ℰ be if the current through the 6-V battery is to be zero and R is 12 Ω?

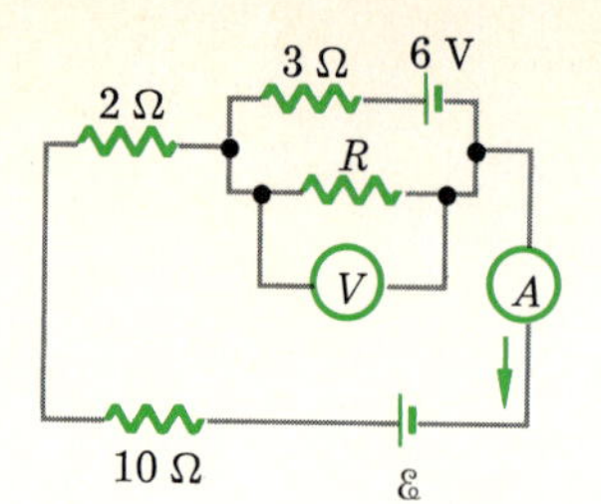

FIGURE P18.11

13.* In Fig. P18.11, what would (*a*) the ammeter and (*b*) the voltmeter read if ℰ was 20 V and R was 6.0 Ω?

14. A particular 120-V house-light circuit has operating on it a 1200-W toaster, one 60-W lamp, and a 600-W soldering iron. The fuse in the circuit blows as soon as another 60-W bulb is turned on. About what size was the fuse?

15. It is planned to operate a 1200-W drier, a washer which requires 360 W, four 60-W bulbs, and a 40-W radio, all from the same 120-V line. For at least how large a current must this line be fused?

16. In Fig. P18.12, the ammeter reads 3.0 A. Find I_1, I_2, and the voltmeter reading.

17. The current through the ammeter in Fig. P18.13 is 3 A in the direction shown. (*a*) Find the unknown emf, X, and the two unknown currents. (*b*) What is the charge on the capacitor?

18. In Fig. P18.14 find, I_1, I_2, ℰ, and the charge on the 2-μF capacitor.

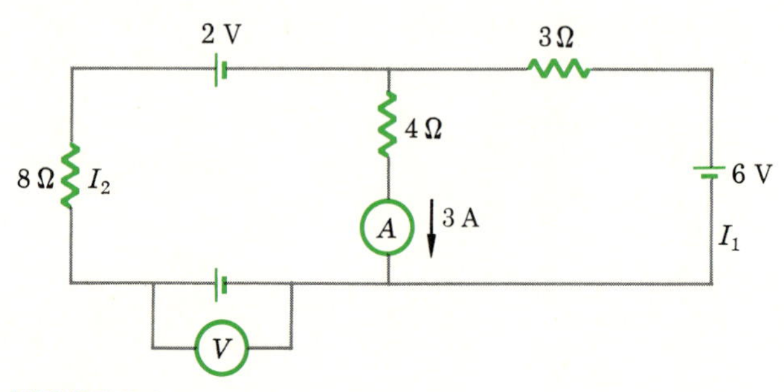

FIGURE P18.12

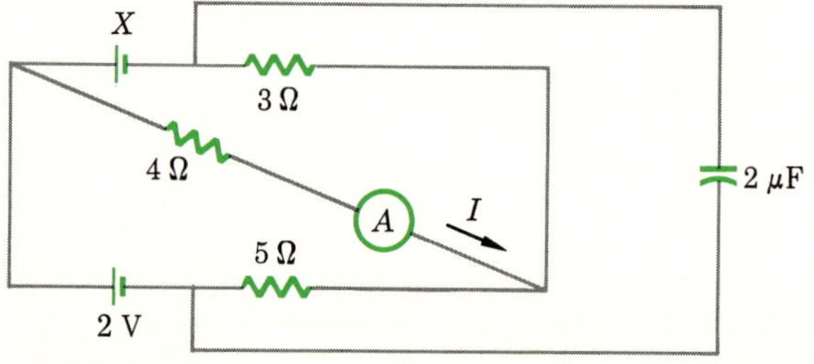

FIGURE P18.13

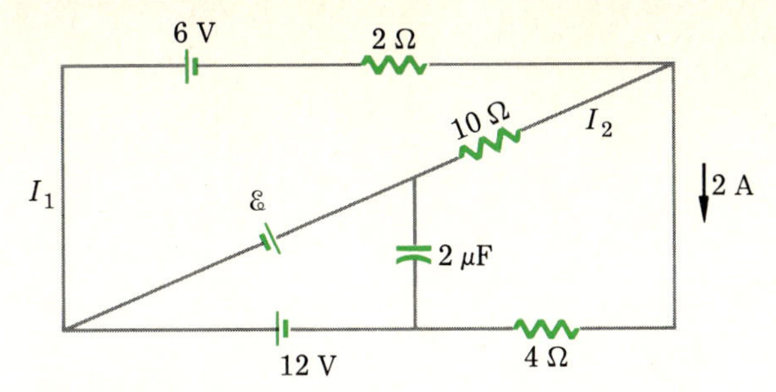

FIGURE P18.14

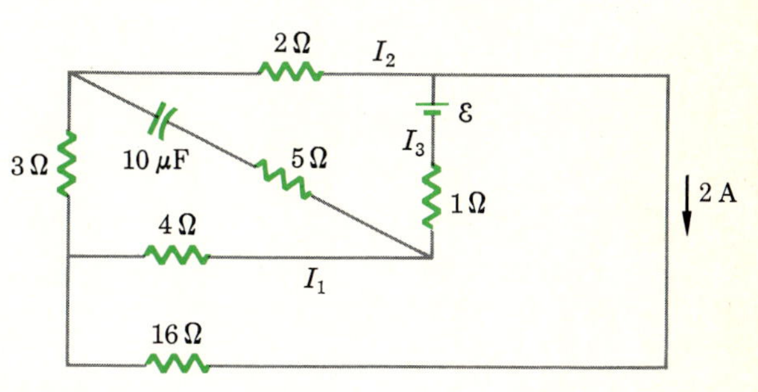

FIGURE P18.16

19. Find the currents I_1, I_2, I_3, and I_4 in the circuit shown in Fig. P18.15.

20.* In Fig. P18.16 find I_1, I_2, I_3, ℰ, and the charge on the 10-μF capacitor.

21.* Two capacitors, 2 and 4 μF, are individually charged to a potential difference of 12 V by connecting them, one at a time, across a battery. After removing them from the battery, they are connected together, the positive plate of one to the positive plate of the other and the negative plate of one to the negative plate of the other. Find (*a*) the potential across each and (*b*) the resultant charge on each. *Hint:* Notice that the potential drop across both will be the same after being connected together.

22.** Repeat Prob. 21 if the capacitors are connected together with the positive plate of one connected to the negative plate of the other.

23.** Each of the resistors shown in Fig. P18.17 has a value of 2 Ω. Find the equivalent resistance between *A* and *B*. *Hint:* From symmetry, many of the wires contain identical currents.

24.* Prove the relation for parallel capacitances given in Fig. 18.10.

25.* Prove the relation for series capacitors given in Fig. 18.10.

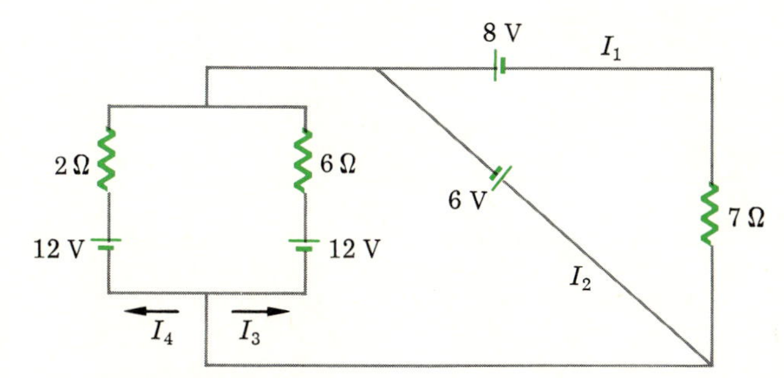

FIGURE P18.15

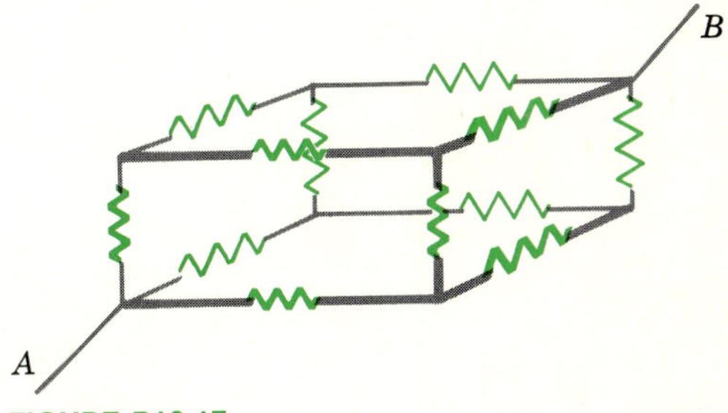

FIGURE P18.17

MAGNETISM

19

All of us, at one time or another, have performed simple experiments with magnets. We know that they have a magnetic field about them and that certain materials experience forces when placed in such a magnetic field. For example, the north pole of one magnet exerts a repelling force against the north pole of a second magnet placed nearby. Conversely, south poles of magnets attract north poles of other magnets. In addition, unmagnetized pieces of certain materials (iron is a common example) are highly attracted by both the south and north poles of magnets.

Although these are facts which many of us learned in grade school, it turns out that the molecular explanation of these facts is not simple. The subject of magnetism today is still an area of active and exciting research. You will see in this and the next chapter that magnetism is of importance to us in many ways. As we proceed to discuss this subject, it will become apparent that magnets and their effects are only a small facet of magnetism.

19.1 PLOTTING MAGNETIC FIELDS

A magnetic field exists in the region about a magnet. As a start, we may describe a magnetic field in terms of its effect upon a compass needle. As you probably know, *a compass needle is merely a small magnet. The end of the needle which points north as a result of the earth's magnetic field is denoted the* **north pole** *of the* needlelike magnet. Furthermore, *we define a magnetic field in such a way that the north pole of the compass needle points in a direction parallel to the field.* With this in mind, together with the fact that like poles repel and unlike poles attract, it is a relatively simple matter to plot magnetic fields.

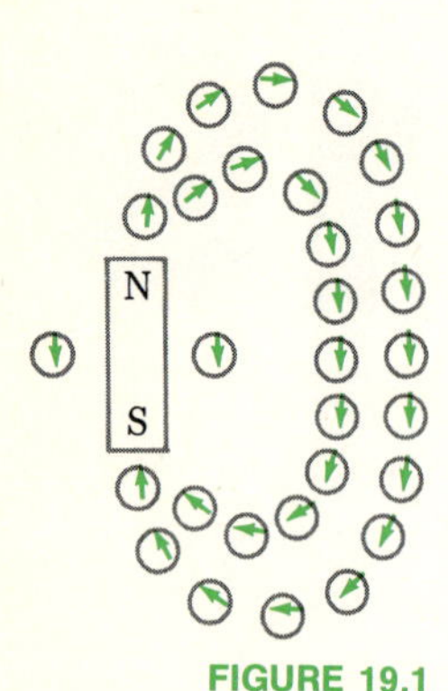

FIGURE 19.1

The direction of the magnetic field in the vicinity of the magnet can be found by using a compass needle.

As a simple example, suppose that we wish to plot the magnetic field about a bar magnet. A compass set in the vicinity of the magnet will align itself in the direction of the magnetic field. This is illustrated in Fig. 19.1, where several compasses are shown. You should satisfy yourself that the north poles of the needles (the arrowheads) will line up, as shown, under the repulsion from the north pole and attraction toward the south pole of the bar magnet. A "picture" of the magnetic field consists in drawing a series of lines about the magnet in such a way that the lines show the direction in which a compass needle would point. This is done for a bar magnet in Fig. 19.2*a*. The fields about other types of magnets are shown in parts *b* and *c* of the same figure. Notice that *the field always points out from the north pole and points into the south pole.* Why? (Remember that the direction of the field at a point is parallel to the direction in which the compass needle points.)

Many important features of a magnetic field can be ascertained from pictures such as those of Fig. 19.2. In particular, *the relative density of the lines in a given figure can be taken as a measure of the field strength,* much as

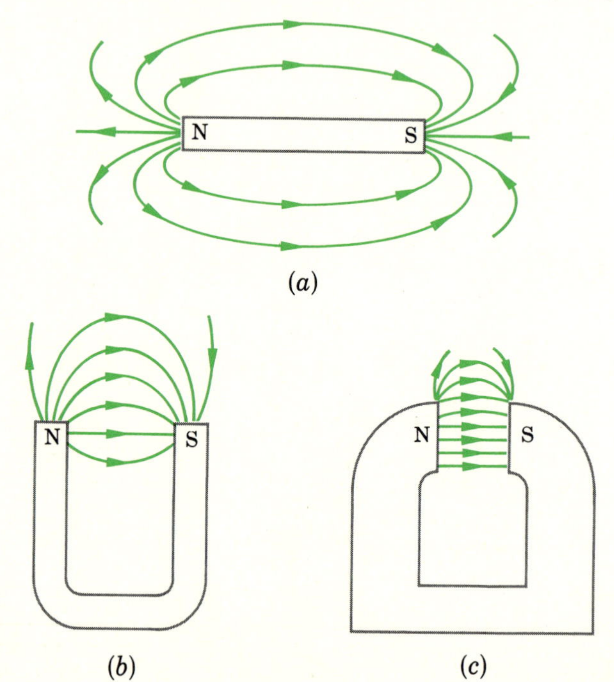

FIGURE 19.2

The magnetic field points away from the north pole of a magnet and into the south pole.

in the case of electric fields. As a result, *the magnetic field lines in a figure not only picture the direction of the field but indicate its relative magnitude as well.*

19.2 MAGNETIC FIELDS OF CURRENTS

Magnets are not the only source of magnetic fields. In 1820, *Hans Christian Oersted discovered* that a current through a wire caused a nearby compass needle to deflect. This indicated *that a current in a wire is capable of generating a magnetic field.* We now know from many other types of experiments that this is indeed the case. Furthermore, there are indications that the magnetic field of a magnet may also be the result of the motion of charges.

Magnetic Field of a Straight Wire

Oersted investigated the nature of the magnetic field about a long, straight wire which was carrying current. His experiment is illustrated in Fig. 19.3. The wire carries a current in the direction indicated. When a compass is placed near the wire, the needle lies with its length tangent to a circle concentric with the wire and the inference is that a magnetic field exists in a circular form about the wire. As is to be expected, the strength of the field is greatest close to the wire. A three-dimensional representation of the magnetic field is shown in Fig. 19.4. In this, as well as later diagrams, the symbol · indicates an arrow coming toward the reader, while × represents an arrow going away from the reader. The symbols are meant to suggest the point and the tail of the arrow.

Right-Hand Rule for Magnetic Field

There is a simple rule for remembering the direction of the magnetic field about a wire, called the **right-hand rule.** If one grasps the wire in one's right hand with the thumb pointing in the direction of the current, the fingers of the hand will circle the wire in the direction of the field. This is shown schematically in Fig. 19.5.

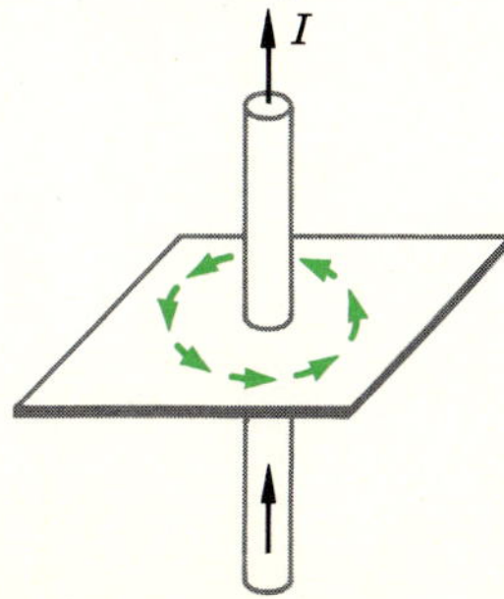

FIGURE 19.3

The magnetic field forms circles concentric to the current-carrying wire.

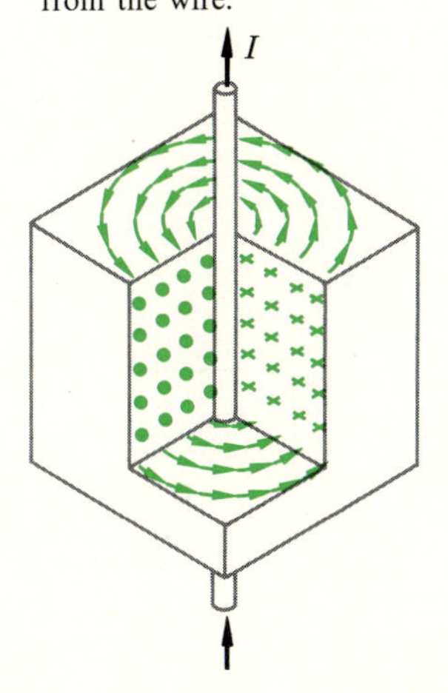

FIGURE 19.4

The magnetic field circles about the long, straight wire. Its magnitude decreases inversely with the distance from the wire.

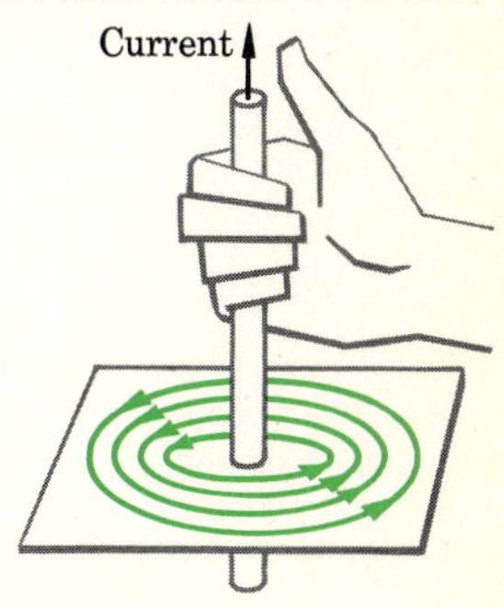

FIGURE 19.5

When one grasps the wire with the right hand in such a way that the thumb points in the direction of the current, the fingers circle the wire in the same fashion as the field.

19.3 FORCE ON A CURRENT IN A MAGNETIC FIELD

Thus far we have discussed only the qualitative features of the magnetic field. We seek now to find some means for precisely measuring it. This can be done by using the fact that *a wire carrying a current experiences a force when placed in a magnetic field.* In a typical experiment to illustrate this phenomenon, a wire is placed between the poles of a magnet, as shown in Fig. 19.6*a*. When a current I is caused to flow through the wire in the direction indicated, the wire experiences a force tending to push it in the direction shown. If the direction of the current is reversed, the force on the wire also reverses. Redrawing the situation of Fig. 19.6*a* in two dimensions as shown in part *b*, we notice that *the line of the wire and a magnetic field line intersecting it determine a plane,* the plane of the page. (Remember, two intersecting straight lines determine a plane.) *The force on the wire is always perpendicular to this plane.* We shall come back to this rule shortly.

The magnitude of the force F on the wire depends on several factors. Resorting to experiment, we can isolate each factor in turn and determine precisely what the interrelations are between them. It is found that the force is directly proportional to both the current I in the wire and the length L of the wire in the magnetic field. When the current is doubled, the force is doubled. We choose to *define the magnetic field in terms of a vector* **B** *which has the direction at a given point identical to the direction of the magnetic field line at that point. The magnitude of* **B** *is taken proportional to the force experienced by a wire which carries a current through the field.* We therefore have that the force on a wire of length L carrying current I in a magnetic field of strength B is given by

Magnetic Field Vector **B**

$$F \propto BIL$$

However, a complication exists. When the wire is directly in line with the field, as shown in Fig. 19.7*a*, the force on it is found to be zero. Further experiments show that the force on the wire is proportional to the component of **B** which is perpendicular to the wire. This component of **B**, namely, $B_\perp$, is shown in Fig. 19.7*b*. *We define our units of field strength B in such a way that the force on the wire is given by*

Magnetic Force on a Current

$$F = B_\perp IL \qquad (19.1)$$

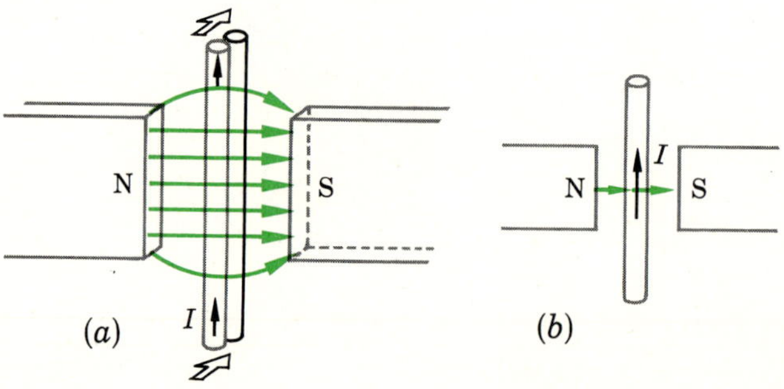

FIGURE 19.6

The magnetic field causes the wire to experience a force. In (*b*) the force is directed into the page.

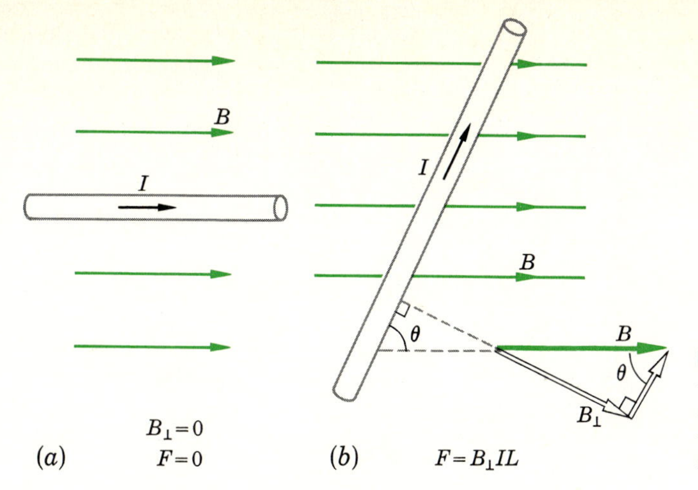

FIGURE 19.7
The force on a current-carrying wire is proportional to the component of **B** which is perpendicular to the wire.

Anyone who has a friend with this name should have no trouble in remembering this equation. Notice that $B_\perp$ is merely $B \sin \theta$, where θ is the angle indicated in Fig. 19.7*b*.

Name and Units for **B**

The unit of B defined by Eq. (19.1) *is webers per square meter* (*Wb/m²*), *also called the tesla* (*T*); we shall use both names interchangeably. From Eq. (19.1) this unit is a N/(A)(m). As usual, F is in newtons, I is in amperes, and L is in meters. We would like to call B the **intensity** of the magnetic field, but that name has been preempted for another quantity. *Several names are used for B, the more acceptable ones being* **flux density** *and* **magnetic induction.** *Colloquially one often refers to it as* **magnetic field strength.**

Another unit often used for B is the gauss (*G*). This unit is much smaller than the weber per square meter. The relation is 10^4 G = 1 Wb/m². *The gauss is not a member of the SI family of units* and must not be used in our equations.

Illustration 19.1 Find the force on the 300-cm length of the wire shown in Fig. 19.7*b* if $\theta = 53°$, the current is 20 A, and B is 2.0 G.

Reasoning The value of $B_\perp$ is $0.8B$. Hence, using Eq. (19.1), we have

$$F = [(2 \times 10^{-4}\ \text{T})(0.8)](20\ \text{A})(3.0\ \text{m}) = 0.0096\ \text{N}$$

Since gauss and centimeters are not proper SI units, they were changed to teslas and meters, respectively. In the next section, we shall learn how to find the direction of the force.

19.4 EXTENSION OF THE RIGHT-HAND RULE

In the previous section it was pointed out that the direction of the force experienced by a wire carrying a current in a magnetic field is perpendicular to the plane defined by the wire and the field. We shall now consider a simple, intuitive extension of the right-hand rule (mentioned in Sec. 19.2) which will allow us to state the direction of the force experienced by the wire. It is purely an intuitive aid for remembering the direction of the force. No real physical significance should be attached to the rule, since it is simply a memory device.

Right-Hand Rule for Magnetic Force

The rule is shown in Fig. 19.8. *Using the right hand, the fingers of the open hand are pointed along the lines of the magnetic field.* The hand is held in such a way that *the thumb points in the general direction in which the current is flowing* in the wire. When this is done, *the force on the wire will be* found to be *in the direction in which the palm of the hand would push.* In the situation shown in Fig. 19.8, the fingers of the right hand point toward the left, the thumb is pointed upward, and the force on the wire is out of the page.

Let there be no confusion on this point. The line of the field vector **B** and the line of the wire together define a plane (the plane of the page in Fig. 19.8). The force on the wire is always perpendicular to this plane. Once you know this, a pure guess allows you a 50 percent chance of obtaining the proper direction for the force. It must be either into or out of a given side of the plane. To find which is the proper alternative, use the rule illustrated in Fig. 19.8. The direction of the force in Fig. 19.8 is toward the reader, i.e., out of the page. Using the same rule, the direction of the force in Fig. 19.7 is into the page.

19.5 FORCES ON MOVING CHARGES

A current is the result of the motion of charged particles, and a wire carrying a current experiences a force in a magnetic field. Is it really necessary that these particles be in a wire in order to experience a force from a magnetic field? It would appear that if the force was exerted directly on the charge carriers, they would experience a force even in open space in a region where there was a magnetic field. This indeed turns out to be the case.

If a positively charged particle such as a hydrogen nucleus, i.e., a proton, is accelerated through a potential difference in an evacuated glass tube, it will travel down the tube and hit the end, as shown in Fig. 19.9*a*. The point of impact P can easily be seen as a bright spot on the glass if the end of the tube has been coated with a fluorescent material. (A television tube produces its

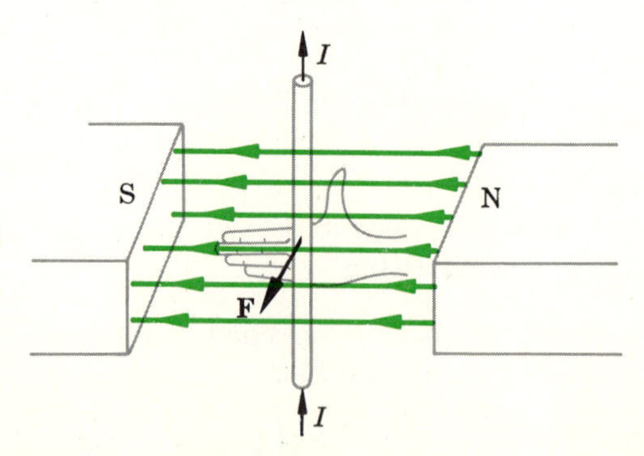

FIGURE 19.8

Using the right-hand rule, the fingers point in the **B** direction, the thumb points in the general direction of I, while the palm of the hand pushes in the direction of **F**.

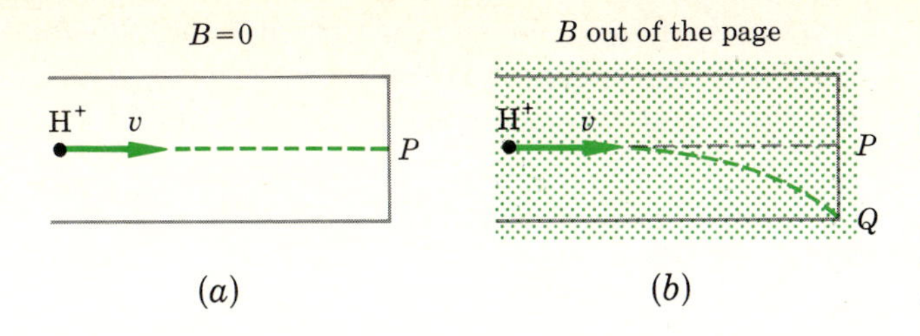

FIGURE 19.9

The proton follows an arc of a circle when deflected in a uniform magnetic field. In (*b*) the dots indicate B to be directed out of the page.

picture when electrons hit the fluorescent screen at its viewing end. We shall consider the positively charged proton rather than the negatively charged electron because we defined current in terms of positive charge flow.)

Suppose, now, that the north pole of a magnet is placed behind the tube so that the field is coming out of the page as shown in part *b*. (As mentioned before, dots represent arrows coming toward you, and crosses represent receding arrows.) It is then found that the bright spot on the end of the tube will move to a position such as Q. The proton has been deflected by the magnetic field. Hence, we have direct experimental evidence for the fact that a charged particle moving in a magnetic field experiences a force.

Moreover, the force on the positively charged particle is perpendicular to both the direction of the field and the direction of motion of the particle. Examination of Fig. 19.9*b* will show that a current traveling to the right will also be deflected downward. This is not surprising in view of the fact that a conventional current is a flow of positive charge.

Magnetic-Force Direction on Moving Charges

If the same experiment is performed by shooting electrons down the tube, it is found that the electrons are deflected upward rather than downward. The magnetic force on a moving, negatively charged particle is in the opposite direction from that on a positive particle. Hence, *no new rules need be learned to determine the direction of forces on moving charges in a magnetic field. We merely consider the moving charge to be a current in the direction of the motion and then use the extension of the right-hand rule to find the direction of the force on the current. If the particle is positive, it will be deflected in the same direction. If negative, it will be deflected in the opposite direction.*

The magnitude of the force on a moving charge q is readily found by a consideration of the equation for the force on a current,

$$F_{\text{wire}} = B_{\perp}IL$$

Consider the wire shown in Fig. 19.10. The current I is traveling to the right, and we can consider the current to be the motion of positively charged particles. Each particle has a charge q and an average speed v. How many charged particles will pass the area at point P per second?

Every free charge within a distance v to the left of P will reach point P in 1 s. (Recall that v is the distance traveled per second.) If there are n free charges in the length v of the wire, then n charges will pass point P in 1 s. But, from the definition of current, it is the charge passing a point per second; we have therefore

$$I = nq$$

FIGURE 19.10

All the charge in the length v will pass through the area at P in 1 s.

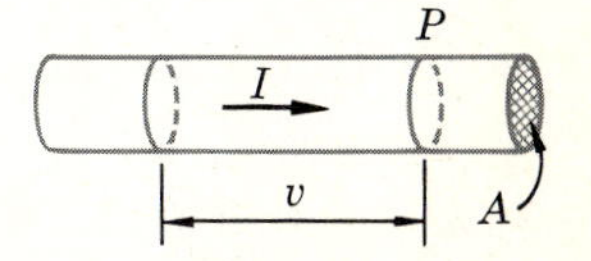

However, the force on these n charges will be just the force on the length of wire which they occupy. The length L in question is just v in magnitude, since the charge travels a distance v per second. Therefore

$$F_{\text{charge}} = \frac{F_{\text{wire}}}{n} = \frac{B_{\perp}IL}{n}$$

or, after substituting for I and L,

Force on Moving Charge

$$F_{\text{charge}} = B_{\perp}qv \tag{19.2}$$

Remember, this force is perpendicular to the direction of the motion as well as to the direction of B. Also remember that $B_{\perp}$ is the component of B perpendicular to v.

Illustration 19.2 A proton traveling at a speed of 10^5 m/s enters a region where there is a magnetic field with $B = 0.020$ T. Further, v and B are perpendicular. Describe the motion of the proton.

Reasoning The situation is as shown in Fig. 19.11. When the particle is at point M, the force is in the direction shown. Hence the proton will deflect upward. But when it reaches a point such as N, the force is still perpendicular to v and is now in the new direction shown. Clearly, the force P is going to force the particle around a circular path of radius r. *A charged particle shot perpendicular to a uniform magnetic field will follow a circular path.* Since the force is always perpendicular to v, *no work is done on the particle* because the motion is never in the direction of the force.

The magnetic force F furnishes the centripetal acceleration needed to hold the proton in a circular path. Our equation for this problem is therefore, from $F = ma$,

$$\text{Magnetic force} = (\text{mass})(\text{centripetal acceleration})$$

$$B_{\perp}qv = \frac{mv^2}{r}$$

FIGURE 19.11
The proton is deflected in a circular path by the uniform magnetic field.

B into the page

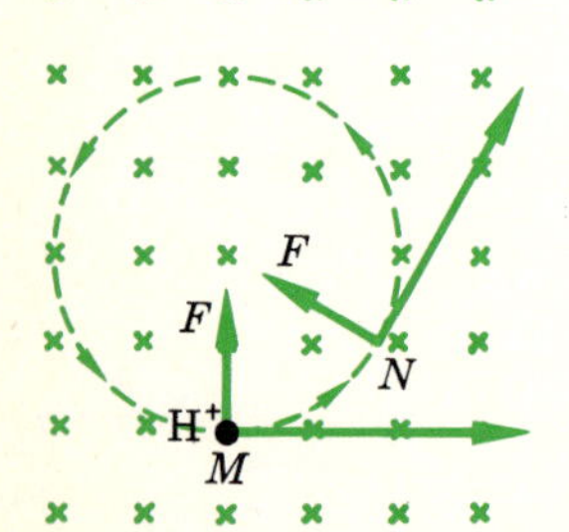

Using the numbers given, we can solve for the radius of the path r. We have

$$r = \frac{mv}{B_{\perp}q} = \frac{(1.67 \times 10^{-27}\ \text{kg})(10^5\ \text{m/s})}{(0.020\ \text{Wb/m}^2)(1.60 \times 10^{-19}\ \text{C})} = 5.2 \times 10^{-2}\ \text{m}$$

or $\quad r = 5.2$ cm

Through how large a voltage had this proton fallen? What would have been its path if it had been an electron?

In the last section we discussed the motion of a charged particle in a magnetic field. One of the most important pieces of information concerning the nature of electrons can be obtained by observing their motion in a magnetic field. Let us illustrate the concepts of the last section by seeing how one can determine the ratio of the charge e to the mass m of the electron.

Suppose that an electron traveling with speed v enters a magnetic field, as pictured in Fig. 19.12. It will describe a circular path, as we have discussed previously. Notice that since the particle is negative, the circle in which the electron moves is not the same as that followed by the positive particle shown in Fig. 19.11. Since the magnetic force Bev supplies the required centripetal acceleration, we can write

FIGURE 19.12

An electron, being negatively charged, is deflected in the direction opposite that in which a positive charge would be deflected.

$$Bev = m\frac{v^2}{r}$$

or

$$\frac{e}{m} = \frac{v}{Br} \tag{19.3}$$

We therefore have a means for determining the ratio e/m by experiment.

Using methods to be described subsequently, the strength of the magnetic field B can be measured, as can the radius of the circular path that the electron describes. If v were known, e/m could be computed. The path of the charged particles can be made visible in a number of ways. One of the simplest is to have air molecules present in the tube, but at a very low pressure. Some of the particles in the beam will collide with the air molecules. As we shall see in Chap. 27, light is given off as a result of such collisions, and so the circular path of the particle beam can be seen as a circular ring of light. The diameter of the ring can then be easily measured.

There are two basic ways of measuring v. One method simply makes use of the fact that a charged particle gains an energy Vq when it falls through a voltage V. Therefore, if we knew the voltage through which the electron was originally accelerated, we could write

$$Ve = \tfrac{1}{2}mv^2 \qquad \text{or} \qquad v = \sqrt{\frac{2Ve}{m}}$$

Substitution of this in Eq. (19.3) gives, after simplification,

$$\frac{e}{m} = \frac{2V}{B^2r^2}$$

Notice that now we need know only the voltage through which the electron was accelerated, in addition to the value of B and r, in order to evaluate the ratio e/m for the electron. The value found is 1.7588×10^{11} C/kg.

A more precise method for measuring velocity makes use of a particle **velocity selector.** This device is shown schematically in Fig. 19.13. Two parallel condenser plates produce a field E which exerts a force qE down-

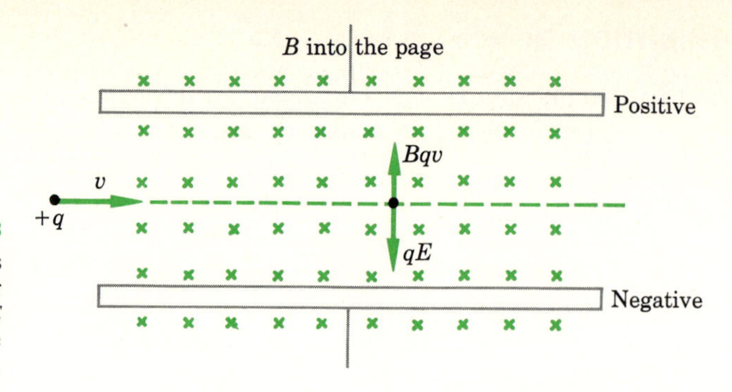

FIGURE 19.13
The velocity selector passes undeflected those particles for which the electric force qE equals the magnetic force Bqv.

ward on a positive particle, as shown. Also, the indicated magnetic field B exerts an upward force on the same positive particle, Bqv. If the voltage across the plates is adjusted properly, the magnitude of E will be such that

$$Bqv = qE$$

and the particle will not deflect. Solving for v yields

$$v = \frac{E}{B}$$

To determine the speed v of the particle one need only adjust the plate voltage V so that the particle travels undeflected through the crossed electric and magnetic fields. With $E = V/d$ known, d being the plate separation, the velocity of the particles can be computed from the above relation. Once v is known, an experiment in which the particle is deflected in a magnetic field alone will permit computation of q/m from Eq. (19.3). Can you show that the above discussion of the velocity selector, using positive charges, applies equally well to electrons?

19.7 THE EARTH'S MAGNETIC FIELD

Experiments such as those outlined in the previous section are complicated by the fact that the earth acts like a very large magnet. Its magnetic field is shown in Fig. 19.14. Only the approximate location of the poles within the earth can be given, but they are indicated roughly on the figure. Notice in particular that the geographic poles, defined by the axis of rotation of the earth, do not coincide with the magnetic poles.

How do we locate the position of the magnetic pole on the north end of the earth? We use a compass, the north pole of which is by definition the end of the needle which points north. Therefore, *the pole at the north end of the earth must actually be the south pole of a magnet, because it attracts the north pole of the compass needle.* This is, of course, purely the result of the fact that we choose to call the end of a magnet which is attracted to the north, a north pole. Although this, the usual definition of north and south poles, causes some confusion, it should cause no difficulty for the careful student.

FIGURE 19.14
The magnetic poles of the earth do not coincide with the poles defined by its axis of rotation. Moreover, the pole near the earth's north pole is actually a south magnetic pole.

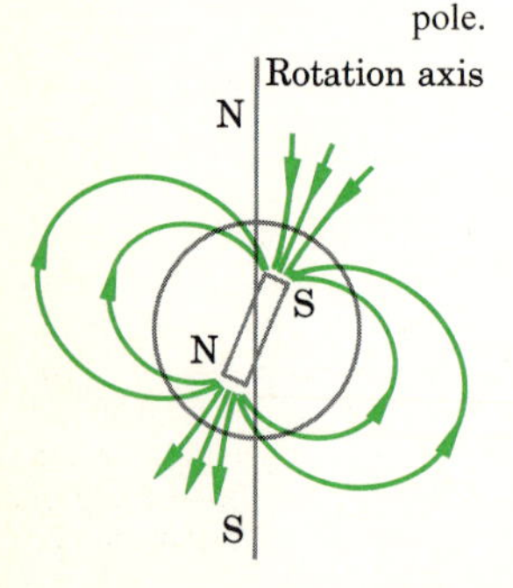

As seen from Fig. 19.14, the earth's magnetic field is parallel to the surface of the earth only near the equator. Near the poles, the field is almost perpendicular to the earth. The quantities needed to specify the field of the earth at a given place are the direction and magnitude of the field. Usually the values given are the magnitude and the angle of dip. The dip angle is illustrated as the angle θ in Fig. 19.15. The value of B is about 10^{-4} Wb/m^2, or 1 G, on the earth's surface.

FIGURE 19.15
We describe the magnetic field at a point on the earth by specifying the flux density B and the dip angle θ.

19.8 LINES OF FLUX AND FLUX DENSITY

In Fig. 19.16*a* we have drawn the magnetic field between the poles of a U-shaped magnet such as that illustrated in part *b* of that figure. You should justify the way in which the field is drawn by recalling that a compass needle aligns itself parallel to the field lines and that the lines emanate from the north pole. As in the case of electric fields, the density of the lines is a measure of the field strength. We shall agree, for computational purposes at least, to draw the lines in the following way.

Relation of Field Lines to B

If the magnetic field has a magnitude B, we shall draw a number of lines equal to B through a unit area perpendicular to the field. Hence, in Fig. 19.16*a*, if the pole pieces had an area A, we would draw BA lines coming out of the upper pole and going into the lower pole (provided that we neglect the fringing effect). *The total number of lines going through an area is called the* **flux** *through the area and is represented by the Greek letter phi* (ϕ). We have, therefore,

DEFINITION

$$\phi = B_\perp A \qquad (19.4)$$

where the notation $B_\perp$ is used to remind us that we must employ the component of B perpendicular to the area.

The reason for this insistence on perpendicularity is easily seen from Fig. 19.17. In part *a* the number of lines going through the area is just $\phi = BA$. However, in part *c* all the lines skim past the area, and none go through it, so $\phi = 0$. The area is parallel to the lines. The intermediate case is more difficult to treat, and at this time it will be stated only that the flux through the area is equal to the product of A and the component of B perpendicular to the area.

Magnetic Flux

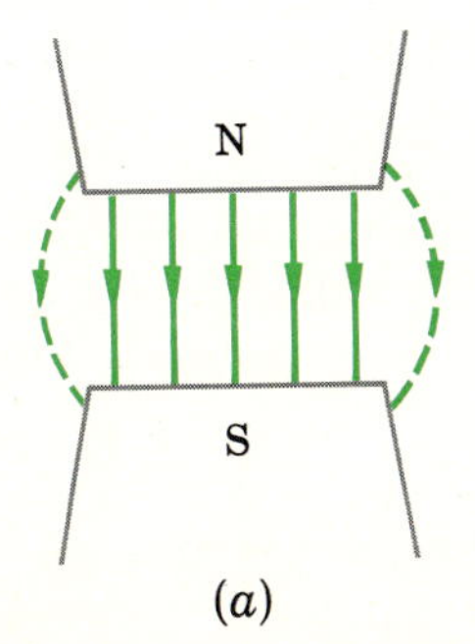

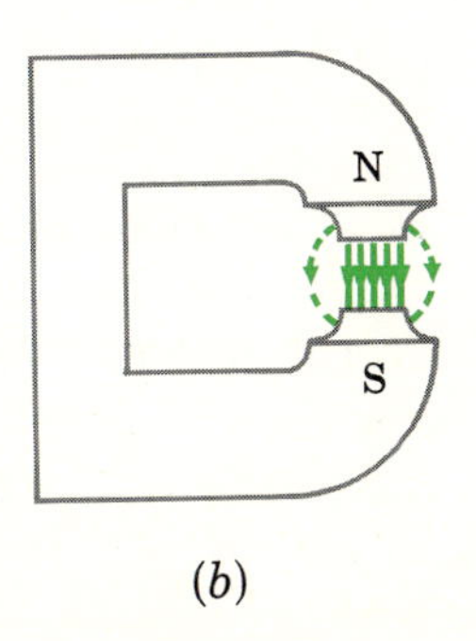

FIGURE 19.16
The magnetic field between the poles of the magnet shown in (*b*) is illustrated in (*a*). Why does the field fringe at the edge of the magnet?

The units of B are webers per square meter, and since flux is B multiplied by an area, we see that the unit of flux is webers. Since B is the flux per unit area, it is frequently called the **flux density.**

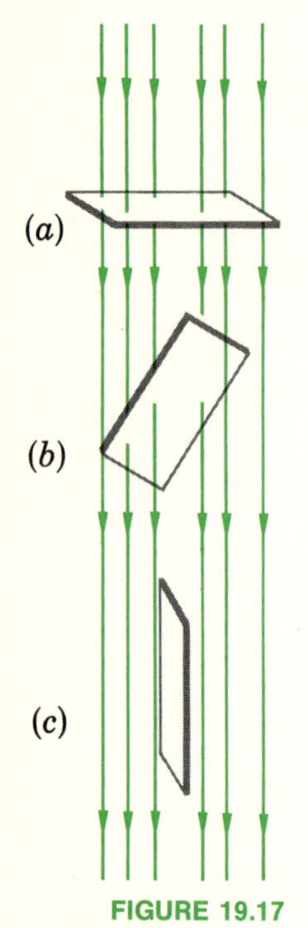

FIGURE 19.17
A maximum amount of flux goes through the area when positioned as shown in (*a*). No flux goes through it when oriented as shown in (*c*).

19.9 AMPÈRE'S LAW AND THE COMPUTATION OF *B*

Although we have pointed out that a magnetic field is generated by a current or flow of charge, we have not yet presented a method for computing B from a knowledge of the current. In general, the computation of B outside a current loop or a more complicated system is a problem of great mathematical difficulty. However, in certain special cases the computation is rather simple, as we shall now show.

The right-hand rule has already provided us with a method for obtaining the qualitative features of the magnetic field outside a current-carrying wire. For example, we know that the magnetic field circles a long, straight wire in the manner shown in Fig. 19.18. Moreover, our intuition tells us that B should become smaller in magnitude as we move farther away from the wire. In order to obtain a quantitative value for B, however, we must make use of a relation first discovered by André Marie Ampère (1775–1836). It may be visualized in the following way.

It was known to Ampère through various experiments that the magnitude of the magnetic field strength B at a distance r from a long wire carrying a current I varied inversely as r and directly with I. In symbols,

$$B \propto \frac{I}{r}$$

The proportionality constant can be determined by experiment and is denoted as $\mu_0/2\pi$ so that we have

$$B = \frac{\mu_0}{2\pi}\frac{I}{r}$$

Permeability of Free Space

for the magnetic induction at point P in Fig. 19.18. μ_0 *is called the* **permeability** *of free space, and its exact value is* $4\pi \times 10^{-7}$ Wb/(A)(m), as will be shown in Chap. 22.

Looking at Fig. 19.18, we notice that the magnitude of B should be constant on a circle symmetric about the wire. This follows from the fact that there is no preferred point on such a circle: Since the wire is circular and is at the center of the circle, each point on the circle is indistinguishable from all other points on the circle. As a result, B should be the same at all points on the circle. If we now multiply B at radius r by the length of the circular path having that same radius (that is, $2\pi r$), we obtain, after using the experimental relation for B given above,

$$(B)(2\pi r) = \mu_0 I$$

Let us state this result in a somewhat different way. We chose a closed path (namely, a circle) which encircled a current I. The product of the length of this closed path and the magnitude of B on the path gave $\mu_0 I$ as the result. We are now tempted to ask whether this result cannot be generalized to a rectangular path instead of a circle. Or perhaps a simple answer might result if we used even such a complicated path as a many-sided polygon which circles a wire. Ampère was able to show that this was indeed the case. He showed that, no matter how complicated the closed path which encircles a wire carrying current I, a simple result could be obtained. One need only break the path up into small-length vectors such that the magnetic-field-strength vector **B** was essentially constant on each given length. Then, if the product of each length times the component of **B** parallel to the length was taken, the sum of all such products was equal to $\mu_0 I$.

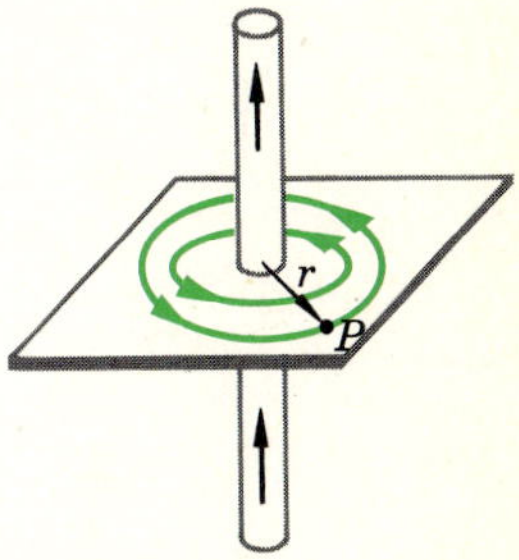

FIGURE 19.18

The magnetic field circles a long, straight wire which is carrying a current.

Ampère's discovery is embodied in what is now known as **Ampère's circuital law.** Before writing the law in mathematical form, let us apply it to the complicated path shown in Fig. 19.19. We have already split the path into a large number of small lengths $l_1, l_2, \ldots, l_{100}$, where we assume the number of lengths to be 100 for the purpose of discussion. Actually, we should probably use much smaller lengths so that **B** could not vary appreciably on the length. We must now take the length l_1 and multiply it by the component of **B** parallel to the length at that particular position. We denote this product as $(lB_\parallel)_1$. Similarly at the second segment, the product is $(lB_\parallel)_2$, and so on. Ampère's circuital law now states that

$$(lB_\parallel)_1 + (lB_\parallel)_2 + (lB_\parallel)_3 + (lB_\parallel)_4 + \cdots + (lB_\parallel)_{100} = \mu_0 I$$

where $+ \cdots +$ represents all the other terms of the sum.

We may write the above equation in symbolic form in the following way:

$$\Sigma(lB_\parallel)_n = \mu_0 I \qquad (19.5)$$

Ampère's Circuital Law

where the symbol $\Sigma(lB_\parallel)_n$ means to take the sum of all the various $lB_\parallel$ products

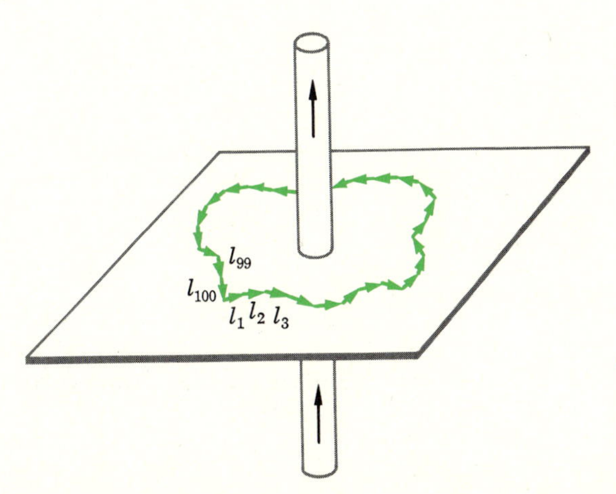

FIGURE 19.19

The closed path or loop formed by the vectors shown is used in connection with Ampère's law.

around the closed path. Equation (19.5) *is the mathematical statement of Ampère's circuital law.* It must be remembered that the elements of length l must be taken small enough for B to be sensibly constant on each element of length.

As an example of the use of this law, let us compute the flux density B outside a long, straight wire. Suppose that we wish to find B at point P in Fig. 19.20. We know from the right-hand rule that B will circle the wire in the direction of the arrows. Furthermore, the magnitude of B everywhere on the circle shown should be constant, since all points on the circle are symmetrical with respect to the wire. For the purpose of using Ampère's law, let us consider the circle of Fig. 19.20 to be composed of a large number, say 1000 short segments or vectors extending around the circle. Writing down Eq. (19.5) gives

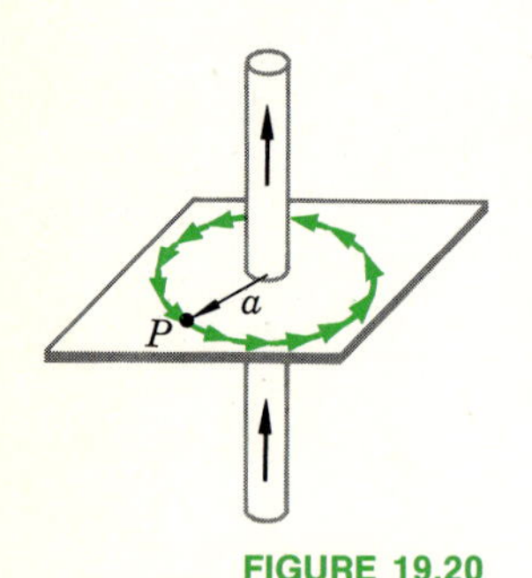

FIGURE 19.20
The vectors form a closed loop concentric to the wire. B is constant on this path.

$$(lB_\parallel)_1 + (lB_\parallel)_2 + (lB_\parallel)_3 + \cdots + (lB_\parallel)_{1000} = \mu_0 I$$

where again $+ \cdots +$ represents all the other terms of the sum not shown.

But we have already seen that B circles the wire in the same way that the little vectors do. In addition, B is constant at each place on the loop. Hence, all the $B_\parallel$ are equal, and their value is the same as the total B at point P, since B circles parallel to the loop. As a result, the above equation becomes, after factoring out $B_\parallel$ or B,

$$B(l_1 + l_2 + l_3 + \cdots + l_{1000}) = \mu_0 I$$

Now the sum of the lengths l_1, etc., is just the distance around the circle, $2\pi a$. Therefore we find

$$(B)(2\pi a) = \mu_0 I$$

B for a Straight Wire

$$B = \frac{\mu_0 I}{2\pi a} \qquad \text{for a long, straight wire} \tag{19.6}$$

which is the result found by experiment, as mentioned earlier.

Illustration 19.3 Find B at a distance of 5 cm from a straight wire carrying a current of 20 A.

Reasoning Using Eq. (19.6),

$$B = \frac{(4\pi \times 10^{-7})(20)}{(2\pi)(0.05)} = 0.80 \times 10^{-4}\ \text{T} = 0.80\ \text{G}$$

Place the units in the above equation, and verify the units of the answer. The field at this distance from the wire is about as strong as the earth's field. With this in mind, can you think of any practical problem Oersted might have had in obtaining the form of the field about a wire?

19.10 MAGNETIC FIELD OF A CIRCULAR LOOP

Suppose that a long wire carrying a current I is bent into the form of a circle such as that shown in Fig. 19.21*a*. The field close to the wire must circle it, just as it did for the straight wire. However, the pattern at some distance from the wires is considerably distorted. This is shown more clearly in Fig. 19.21*b*, where the cross section of the loop is shown. The symbol · indicates a wire carrying a current out of the paper, while × represents a current going into the page. (Recall that the symbols arise from the pointed tip and the feathered tail of an arrow.)

It turns out that the computation of B for this case is extremely complicated except for a point on the axis of the loop. We cannot easily use Ampère's circuital law, because there is no obvious path in Fig. 19.21 on which B is constant or known. Other methods must be used in this case. They involve calculus, and so we shall treat this example only in a qualitative way.

If we have a number of loops of wire wound tightly together, the picture of the field would appear exactly as in Fig. 19.21. However, the loop shown would now be interpreted as being composed of several turns or loops of wire. If the current in the wire were I for both the single loop and the coil with N loops, the field B for the coil would be N times as large as for the single loop.

19.11 THE SOLENOID

A solenoid is a long, cylindrical coil of wire. To show the magnetic field within this device, a solenoid looser than ordinary is illustrated in Fig. 19.22*a*. The cross section of a more common, tightly wound solenoid is shown in Fig. 19.22*b*. Notice how the field within the solenoid is directed straight through it. In addition, the field is much stronger inside the tightly wound, long solenoid than it is on the outside. For many purposes, a long solenoid may be considered to have a field parallel to its axis within the solenoid. The flux lines outside the solenoid have fanned out so far in space that they are very

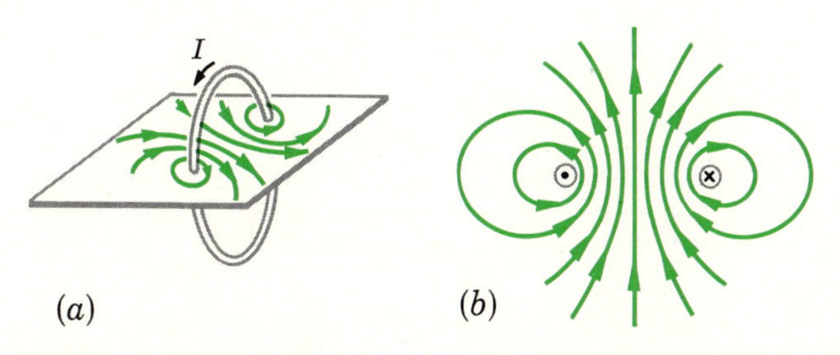

FIGURE 19.21

Two views of the magnetic field about a current-carrying loop.

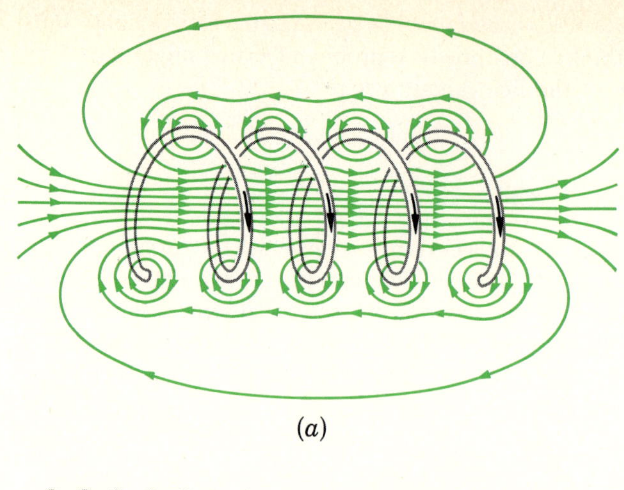

(a)

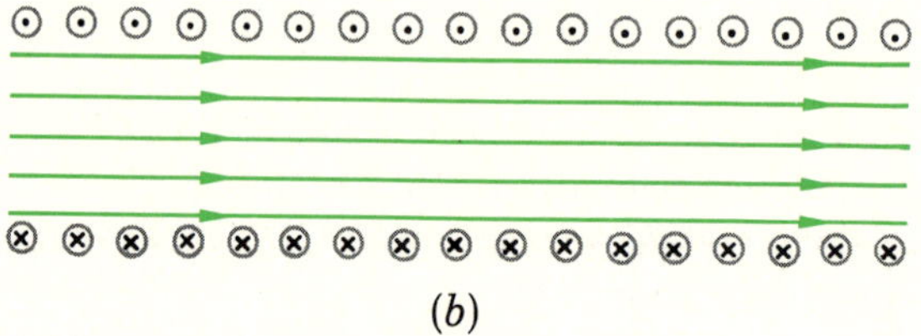

(b)

FIGURE 19.22
The fields are essentially uniform inside the solenoids.

widely separated, indicating that the field is negligibly weak outside the solenoid.

It will be possible for us to compute the value of B within a long solenoid by using Ampère's law. Consider the path, or loop, *abcda* in Fig. 19.23. In this case Ampère's law becomes

$$(lB_{\|})_1 + (lB_{\|})_2 + (lB_{\|})_3 + (lB_{\|})_4 = (\text{current encircled})(\mu_0)$$

Inside the solenoid B is parallel to l_1, and so

$$(lB_{\|})_1 = l_1 B$$

We have already seen that B outside the solenoid is very small provided the solenoid is very long, and so we shall approximate it as zero. Thus

$$(lB_{\|})_3 = 0$$

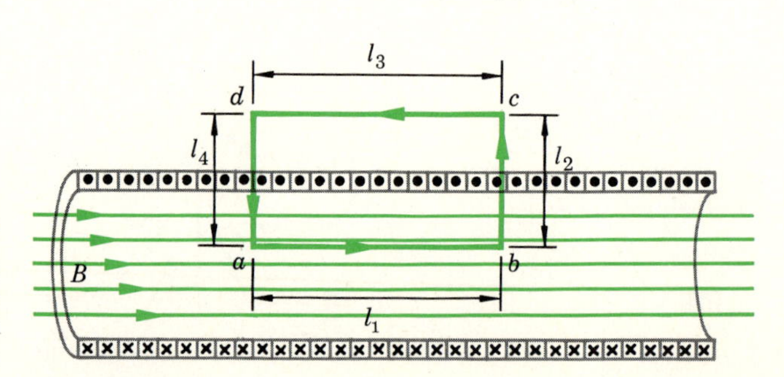

FIGURE 19.23
The value of B within the solenoid can be found by applying Ampère's circuital law to the path *abcda*.

Moreover, B is perpendicular to l_2 and l_4 inside the solenoid and zero outside; so

$$(lB_{\parallel})_2 = (lB_{\parallel})_4 = 0$$

Therefore, Ampère's law in this case reduces to

$$l_1 B = (\mu_0)(\text{current encircled})$$

If there are n loops of wire on unit length of the solenoid, we have encircled nl_1 loops by this path. (In Fig. 19.23 this number is actually 13.) Each loop carries a current I out of the area described by our path, and so the current encircled is just $nl_1 I$. Therefore,

$$l_1 B = \mu_0 n l_1 I$$

$$B = \mu_0 nI \qquad \text{inside a long solenoid} \qquad (19.7)$$

B in a Solenoid

Notice that B is independent of the position within the solenoid, and so the field within a long solenoid is not only parallel to the axis but is uniform as well. This is true, of course, only for points far from the ends.

19.12 THE TOROID

As a final example, we shall consider a toroid. This device is just a solenoid bent into a circle so that it has no ends. *A schematic diagram of a toroid is shown in Fig. 19.24.* Of course the toroid should have many more turns of wire on it than shown. *The magnetic field is confined almost entirely to the inside of the coil and circles the toroid as indicated.*

To compute B in this case, we shall take our closed loop to be the colored circle shown. By symmetry, B will be constant everywhere on the circle, and it will be tangential to the circle. Upon replacing the circle by a series of vectors and remembering that B is constant, Ampère's law gives

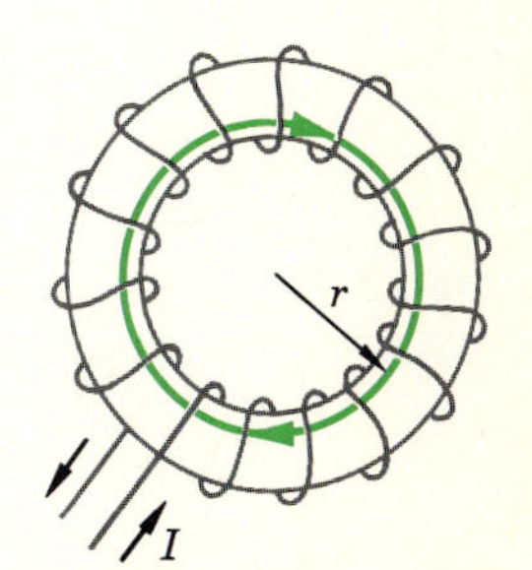

FIGURE 19.24
In a toroid the field circles inside the coil, as shown. It is directed clockwise inside the toroid.

$$(lB_{\parallel})_1 + (lB_{\parallel})_2 + \cdots + (lB_{\parallel})_n = (\mu_0)(\text{current encircled})$$

Since B is constant,

$$B(l_1 + l_2 + \cdots + l_n) = (\mu_0)(\text{current encircled})$$

or

$$(B)(2\pi r) = (\mu_0)(\text{current encircled})$$

We see that the current encircled is NI, where N is the total number of turns of wire on the toroid. Therefore,

$$2\pi r B = \mu_0 NI$$

or

$$B = \frac{\mu_0 NI}{2\pi r} \qquad \text{toroid} \qquad (19.8)$$

B in a Toroid

Notice that in the case of the toroid, B is not uniform within the coil, since it varies with r. However, if the diameter of each loop is small compared with the diameter of the ring, r will not vary much from place to place within the toroid and so B will be nearly constant.

Illustration 19.4 A long solenoid has 50 turns of wire on it for each centimeter of its length. The current in the wire is 0.50 A. The diameter of the solenoid is 2.0 cm. Find B in the solenoid and the flux through the solenoid.

Reasoning To find B, we make use of Eq. (19.7).

$$B = \mu_0 nI = (4\pi \times 10^{-7})(50 \times 100)(0.50)$$
$$= 31.4 \times 10^{-4}\ \text{T} = 31.4\ \text{G}$$

where you should supply the units to the equations. The flux ϕ is the number of lines going through the solenoid. There will be B lines per unit area going down the solenoid, and so $\phi = BA$, where A is the cross-sectional area of the solenoid. Hence

$$\phi = (31.4 \times 10^{-4}\ \text{Wb/m}^2)(\pi \times 10^{-4}\ \text{m}^2) \approx 10^{-6}\ \text{Wb}$$

where use has been made of the fact that $1\ \text{T} = 1\ \text{Wb/m}^2$

Illustration 19.5 Consider the two parallel long, straight wires shown in Fig. 19.25. Find the force on unit length of the lower one resulting from the current in the upper one.

Reasoning The current I_1 causes a magnetic field into the page at the position of I_2. Its magnitude is

$$B = \frac{\mu_0 I_1}{2\pi r} = \frac{\mu_0 I_1}{2\pi b}$$

But since the force on a wire is $B_\perp IL$, and since $I = I_2$ in this case, we have

$$F = \frac{\mu_0 I_1}{2\pi b} I_2 L$$

from which

$$\frac{F}{L} = \frac{\mu_0 I_1 I_2}{2\pi b}$$

FIGURE 19.25
The magnetic field from the upper wire acts on I_2 so as to push the lower wire toward the upper wire.

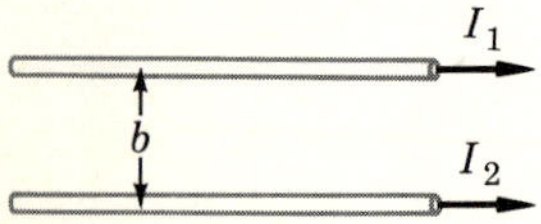

Can you show that the force is in such a direction as to push the lower wire toward the upper one? Notice that when both wires carry the same current, $I_1 = I_2 = I$ can be equated to $(2\pi bF/L\mu_0)^{1/2}$. This provides a way of measuring currents in terms of force and length measurements. We shall see in Sec. 22.12 that this is the basic experiment used to define the ampere.

19.13 AMPÈRE'S THEORY OF MAGNETS

One cannot help noticing the similarity between the magnetic field of a bar magnet and that of a solenoid, as shown in Fig. 19.26*a*. Ampère noticed this too. In fact, he noticed that even the field of a single loop was exactly like what one would expect for a short magnet as shown in Fig. 19.26*b*. He theorized that all magnetic effects, even in bar magnets, are the result of circulating currents. We believe that his ideas are basically correct, although we still do not understand completely why some of the basic particles, the electron, for example, act like small magnets.

We see at once that if Ampère's theory is accepted, it is completely impossible to obtain a north pole by itself. The north pole is merely one side of a current loop (or of a group of current loops), and the other side will always be present as a south pole (see Fig. 19.26). It is for this reason that physicists today speak of the **pole fiction.** That is to say, there is no particle which can be considered to be a magnetic pole. Unlike free positive and negative charges, north and south poles can never be separated from each other completely.*

Ferromagnetic Materials

Actually, of course, Ampère had no idea about the structure of atoms and molecules, even though he believed matter to be made up of atoms, and so he could not say how the current loops were present in a piece of magnetized iron. We know today that *each atom behaves like a little magnet. Some atoms, like iron, cobalt, and nickel* (as well as gadolinium and dyspro-

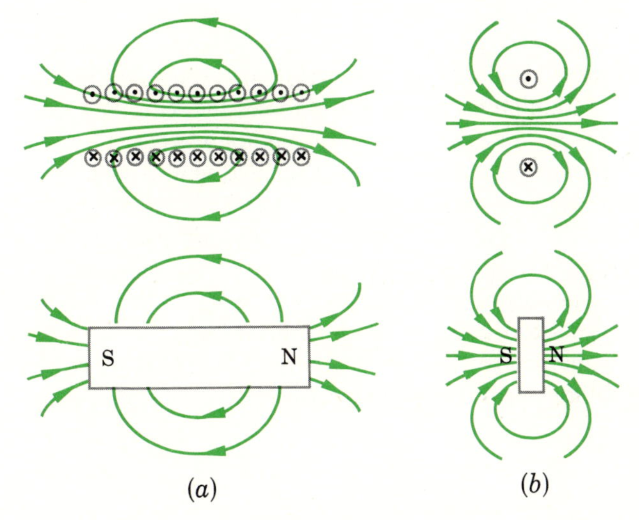

FIGURE 19.26
The magnetic field about a solenoid is very similar to that of a bar magnet having about the same size.

*One recent set of experiments was thought to show that a single pole can exist on a "monopole particle." Unfortunately for those who love excitement, the interpretation given the experiments appears to have been wrong. The search for a monopole particle is continuing.

sium), *cooperate with each other in such a way as to act like quite strong little magnets. These are called the* **ferromagnetic atoms.** Other atoms act like much weaker magnets.

It is often convenient to think of the magnetic nature of the atom as being partly the result of the circulation of the electrons within the atom, thereby acting as a current loop. Moreover, even the electron itself acts like a small bar magnet. One can think of the electron as a charge spinning about its axis and thereby constituting a current loop, and this effect is called the **spin** of the electron. We should point out, however, that most physicists frown upon such detailed pictures of the atom, since there is reason to believe that these pictures cannot be proved experimentally. Moreover, this way of picturing the spin is not completely satisfactory in certain respects, and we are not yet able to reconcile it with the quantitative results of experiment.

We know that if we place a group of tiny magnets close to each other, they will try to arrange themselves so that each south pole is close to a north pole. This is the result of the attractions of unlike poles and the repulsions of like poles. The lowest PE of the system is reached when the magnets are arranged somewhat as shown in Fig. 19.27*a*. Notice that the magnets arranged in this way are equivalent to a large magnet.

However, if the magnets are strongly agitated by shaking the paper or board upon which they rest, they will break loose from each other and end up as shown in Fig. 19.27*b*. Notice now that the individual magnets tend to cancel each other out. No longer do they act as a strong bar magnet.

Curie Temperature

An analogous situation exists with atoms in a solid. Thermal vibrations tend to agitate the system and prevent them from ordering themselves as in Fig. 19.27*a*. Only certain atomic magnets, iron, and the other ferromagnetic materials, can preserve the pattern shown in part *a* at ordinary temperatures. Even these atoms, when heated hot enough, acquire enough thermal energy to break loose and disorient as in part *b*. The temperature at which this happens is quite definite for any type of atom and is called the **Curie temperature.** We should note, however, that the analogy is not complete. There are, in addition to the magnetic forces between the ferromagnetic atoms, other forces, which are much more complex. These forces can be understood only in terms of quantum mechanics, and so we are unable to discuss them further here. They actually play a major role in the aligning of the atomic magnets.

Ferromagnetism and Domains

Most materials have their atomic magnets, if any, randomly oriented as in Fig. 19.27b. The ferromagnetic materials, however, consist of little regions in which the atoms are all aligned as in Fig. 19.27a. Each of these oriented regions is called a **domain.** In an ordinary piece of iron pipe, each domain may contain as many as 10^{16} atoms and consist of a region a small fraction of a millimeter in linear dimension. However, the domains in an unmagnetized

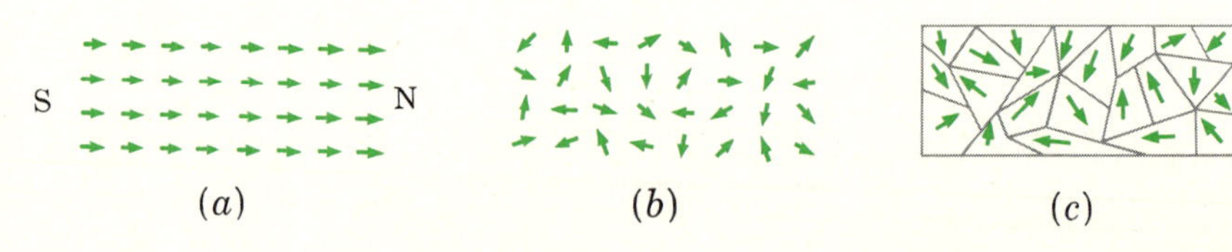

FIGURE 19.27

(*a*) A magnetized piece of iron; (*b*) the disordered, unmagnetized iron shown schematically; (*c*) a more realistic picture of the domains.

piece of iron will be randomly oriented, and so the situation will be much as in Fig. 19.27*b*, where now the arrows represent domains instead of atoms. A more realistic picture of the domains is given in part *c*. If one is to magnetize a bar of iron, the domains within it must be lined up within the material. One can do this in the following way.

Suppose that one starts with the unmagnetized bar of iron shown in Fig. 19.28*a*. A solenoid with a current in it, such as shown in part *b* of the figure, possesses a rather weak magnetic field within itself, as we saw in Illustration 19.4. If, now, the iron is placed in the solenoid, the magnetic field B of the solenoid will exert a force to the right on the north poles of the domains. The south poles are forced to the left. If the friction forces between the domains are not too large, the domains will orient under the action of the field, as shown in part *c* of Fig. 19.28.* The iron is now a bar magnet, with north and south poles. In "soft" iron, the domains are easily oriented, but in "hard" iron the field must be made quite large or the domains must be agitated by heat or mechanical means to allow them to turn into the direction of the field. (The designations hard and soft refer only to the magnetic properties, not to the physical hardness.) It is possible, however, to align the domains nearly perfectly and form a large bar magnet as shown in part *c*.

Once the domains have been aligned, the total flux density B consists of two parts. The original small field of the solenoid is still present. However, the field produced by the bar magnet is usually hundreds of times larger than the field of the solenoid alone. *The combination of a solenoid and a piece of very soft iron is called an* **electromagnet.**

If, now, the current in the solenoid is turned off, the domains in a soft bar of iron will return nearly to the original state shown in part *a*. Thermal motion causes them to disarrange. This is a desirable situation in the case of an electromagnet, because it makes it possible to turn it on or shut it off at will. A hard piece of iron on the other hand will retain most of its alignment and will be a permanent bar magnet as shown in part *d* of Fig. 19.28.

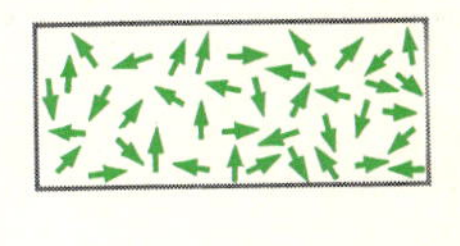

(*a*)

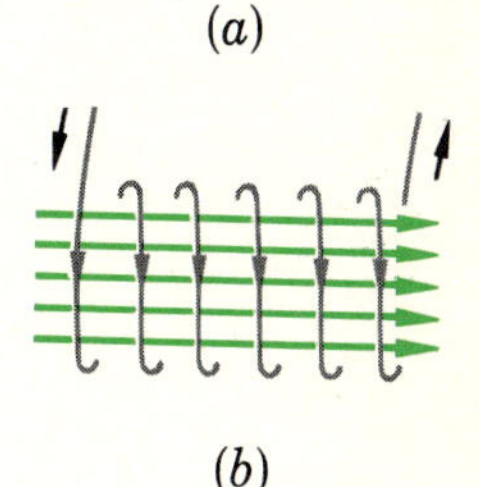

(*b*)

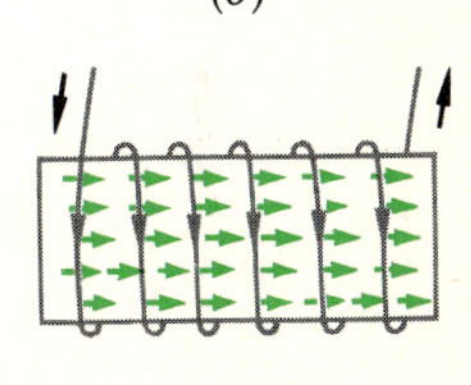

(*c*)

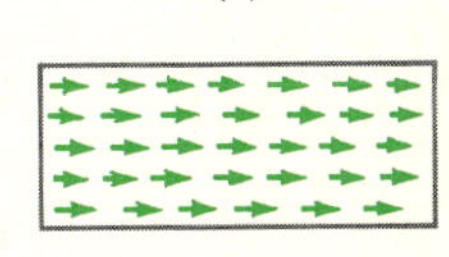

(*d*)

FIGURE 19.28
An unmagnetized piece of iron can be magnetized by using the field of a solenoid to line up the domains.

19.14 THE B-VERSUS-H CURVE

Now consider the toroid illustrated in Fig. 19.29. If it is wound on a wooden core or on some other nonferromagnetic material, the field inside the toroid will be almost entirely the result of the current in the wire. We found in Eq. (19.8) that B in the toroid was

$$B = \mu_0 \frac{NI}{2\pi r} \qquad \text{toroid}$$

where N is the total number of loops on the toroid.

FIGURE 19.29
The value of B within the toroid depends upon the material upon which the coil is wound.

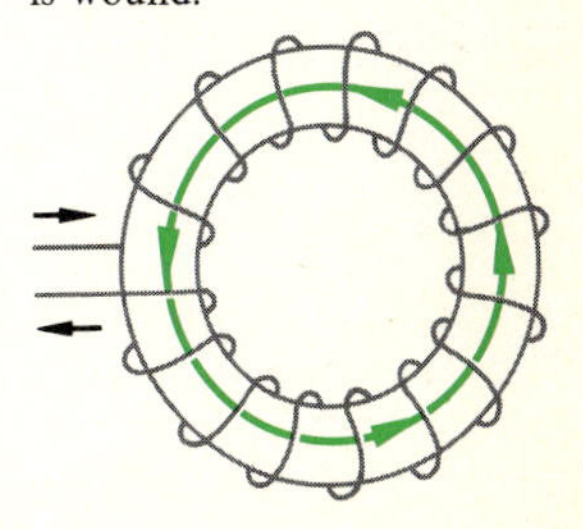

* Actually the domains as a whole rotate comparatively infrequently. Much of the alignment occurs by growth of the properly oriented domains at the expense of the improperly oriented domains. During this process, the domain walls are observed to move. Energy is required for this process, too, and the effects of both processes are essentially the same.

If we were to wind the toroidal coil on an iron core, this field resulting from the current in the wire would tend to magnetize the iron. The aligned magnetic domains would greatly increase B within the toroid, as discussed in the previous section. In this case, Eq. (19.8) will no longer be correct for the field within the ferromagnetic material of the toroid. Actually, Eq. (19.8) should be multiplied by some large factor to give the correct flux density. It is customary to write

$$B = \mu \frac{NI}{2\pi r} \qquad \text{toroid} \tag{19.9}$$

where μ is called the **magnetic permeability** of the material upon which the toroid is wound.

In the case where the toroid is hollow, μ becomes just μ_0, the permeability of free space. All materials except those which are ferromagnetic have values of μ close to μ_0. This is just another way of saying that nonferromagnetic materials do not alter the flux much, because the alignment tendencies of the atomic magnets are not strong enough to cause any great effects. However, μ for ferromagnetic materials is usually hundreds of times larger than μ_0, since they greatly increase B.

Often it is convenient to separate the magnetizing field, which is caused by the current in the wire, from the combined fields of the current and the material of the solenoid. Hence we define the magnetizing field intensity, or **magnetic field intensity,** H to be (within the toroid)

$$H = \frac{B}{\mu}$$

or

$$B = \mu H \qquad \text{simple toroid} \tag{19.10}$$

Notice that Eq. (19.9) yields

$$H = \frac{NI}{2\pi r} \qquad \text{toroid} \tag{19.11}$$

and so H, the magnetic field intensity, is not dependent upon the material upon which the toroid is wound. The definition of H given here is not complete for cases other than a toroid or in empty space. If in effect unbalanced poles at the surfaces or in the interior of iron bars seem to be present, as in the case of a bar magnet, then H is actually defined to be a combination of the effects of currents in wires and the effects of the poles. Complications also arise if the domains are not at equilibrium.

Since the value of B in a hollow toroid is $\mu_0 H$, it is customary to express the magnetic effect of a material by comparing $\mu_0 H$ with the value of B actually found within the toroid. The quantity $\mu_0 H$ can be computed from a knowledge of the dimensions of the toroid and the current through it by means of Eq. (19.11). We shall see later that B in the toroid can be measured quite easily. Hence, we can plot a curve called the B-versus-H curve for a material. (In our system of units it is actually best to plot a B-versus-$\mu_0 H$

curve, although the only difference will be in a factor of μ_0 multiplying each value of H.)

Let us first plot a B-versus-$\mu_0 H$ curve for air. Since air has no appreciable magnetic properties, μ for air is essentially μ_0 and so Eq. (19.10) tells us that $\mu_0 H = B$ for air. The curve of B versus $\mu_0 H$ is therefore as shown in Fig. 19.30. As we saw in Illustration 19.4, the values of B within an ordinary toroid may be of the order of 0.003 Wb/m², provided that it is not wound on a ferromagnetic core. Since $B = \mu_0 H$ in this case, $\mu_0 H$ and B will increase exactly in step. When one is zero, the other is zero. They both reach 0.003 Wb/m² together.

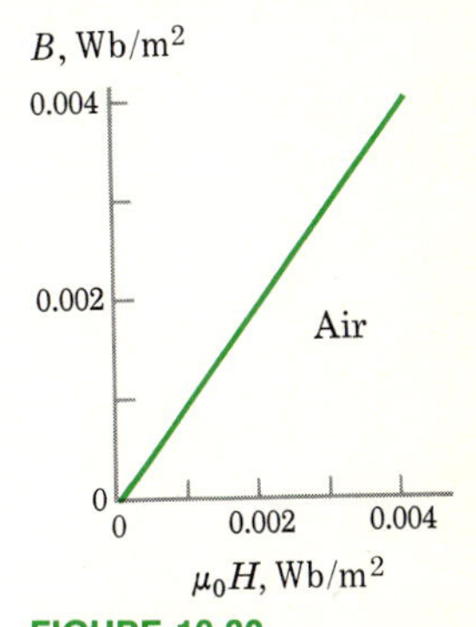

FIGURE 19.30
This B-versus-H curve is for an air-core toroid. Compare it with Fig. 19.31.

Suppose, though, that the toroid is wound on an iron core. Since $\mu_0 H$ does not depend upon the material of the core, being purely the field resulting from the current itself, we see that $\mu_0 H$ will still vary from zero to about 0.003 Wb/m² in this case also. However, we know that the alignment of the domains in the iron core will greatly increase B. At low currents in the wire, the magnetizing field due to the current will not align many of the domains. But as $\mu_0 H$ increases, the domains will continue to line up and B will increase tremendously. This is illustrated in Fig. 19.31.

Notice that the vertical scale, the B scale, is 500 times larger in Fig. 19.31 than it was in Fig. 19.30. If we tried to plot the line of Fig. 19.30 on the graph of Fig. 19.31, the line would be nearly horizontal through zero. The iron core contributes a field very much larger than that contributed by the current in the wire.

It is seen that the B-versus-$\mu_0 H$ curve levels off above point D. This occurs as a result of the fact that nearly all the domains have been lined up at that point. Increasing $\mu_0 H$ beyond that value lines up very few more domains, and so B remains nearly constant. We say in this case that the iron is **saturated;** i.e., its domains are all lined up, and B is essentially as large as it can become.

Since the magnetic permeability of the iron, μ, is defined by the relation

$$B = \mu H$$

its value can be determined from Fig. 19.31. For example, at point D we have known values of B and H, namely, B_D and H_D. One has that

$$\mu_D = \frac{B_D}{H_D}$$

As a result of the fact that the experimental relation between B and H is not a straight line for iron, μ will not be a constant but will vary greatly with H.

FIGURE 19.31
The B-versus-H curve is for an iron-filled toroid. Notice the difference in vertical scale between this graph and that shown in Fig. 19.30.

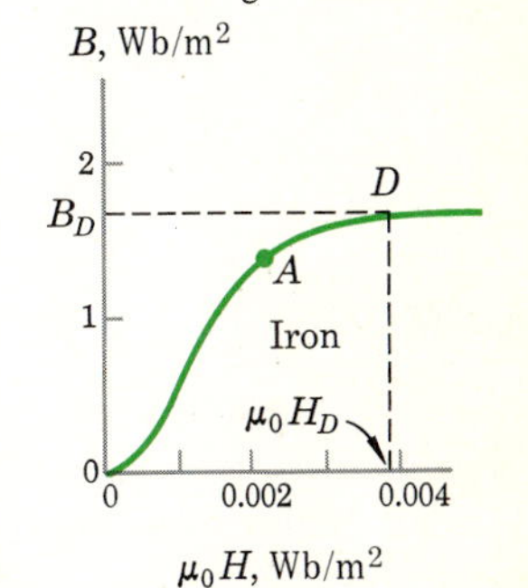

19.15 THE HYSTERESIS CURVE

Let us consider another experiment with an iron-core toroid. Suppose that we have increased the magnetizing field H until the iron is saturated. We shall be at point D in Fig. 19.32. If we now decrease the current in the toroid, $\mu_0 H$ will decrease. However, since there will be more or less friction between the domains, the domains will tend to remain aligned. Hence the value of B will not decrease tremendously and may return to point E on the curve, rather than point C.

If the current is turned off in the toroid, some of the domains will still remain aligned. We shall be at point F on the curve. If the current is reversed in the toroid, the field $\mu_0 H$ will be negative and will actually be able to turn enough of the domains around so that B becomes zero, as shown at point G. As the magnetizing field $\mu_0 H$ is made still more negative, the iron will eventually be saturated in the reverse direction and this is what has happened at point J.

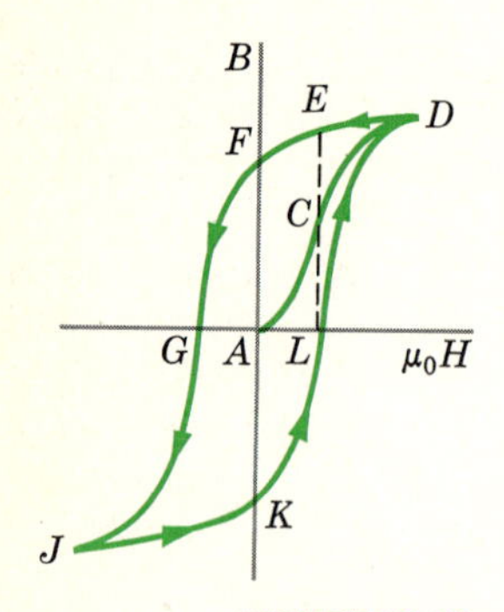

FIGURE 19.32
The hysteresis curve for a ferromagnetic material.

As the current in the toroid is decreased once again to zero, we follow through points J, K, and L back to D on the curve, eventually ending up with the iron magnetized as it originally was. However, in carrying the iron through the cycle D, F, B, J, K, L, D, much energy is lost in doing friction work, in turning the domains around, and in moving domain walls. One can show that the friction and other work done is proportional to the area of this loop, the so-called **hysteresis loop.** This work shows up as heat within the iron.

To make a permanent magnet, it appears that the best material is one having a hysteresis loop or curve such as that shown in Fig. 19.33*a*. This material would be advantageous because it has a high **retentivity,** the length AF, and requires a large **coercive force** AG to demagnetize it. Hence, it would be a strong magnet (much flux through it), and it would be difficult to demagnetize (therefore permanent).

On the other hand, the iron for an electromagnet should have a hysteresis loop such as that in Fig. 19.33*b*. (Why?) We shall see in the next chapter that motor and generator cores should also use iron of this same type. (Is this a hard or soft iron?)

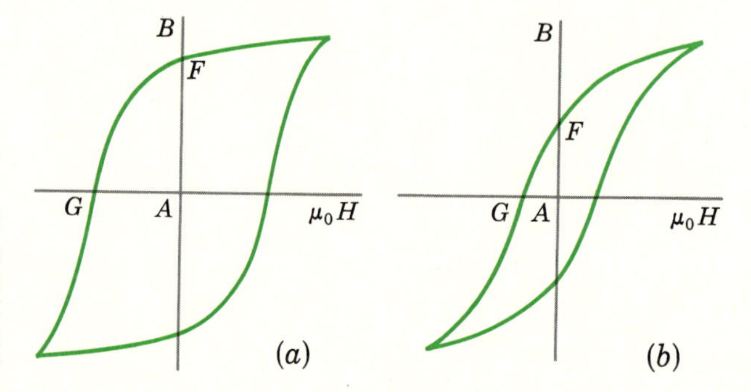

FIGURE 19.33
The material whose hysteresis curve is given in (*a*) would make a better permanent magnet than that whose curve is shown in (*b*). Why?

SUMMARY

A compass needle is a tiny bar magnet. The end that points north is its north pole. Like poles of magnets repel and unlike poles attract. Each magnet always has both a north and a south pole.

Magnetic fields exert forces on magnetic poles. A compass needle aligns itself tangentially to the magnetic field lines. These lines come out of north poles and enter into south poles of magnets.

A straight wire of length L which carries a current I in a magnetic field of strength **B** experiences a force given by $F = B_{\perp}IL$. In this expression, $B_{\perp}$ is the component of **B** perpendicular to the wire. The units of B, the flux density (or magnetic induction), are the webers per square meter, which is the same as the tesla (T). A plane is defined by the intersecting lines of B and L. The force is in a direction perpendicular to this plane and is given by a modification of the right-hand rule.

Moving charges also experience a force in a magnetic field. It is given by $F = qB_{\perp}v$, where $B_{\perp}$ is the component of **B** perpendicular to the velocity **v**. The force direction is the same as on a current composed of the moving particles. A charge which moves with speed v perpendicular to a magnetic field **B** describes a circular path. The radius of the path is found by equating the magnetic force $B_{\perp}qv$ to the centripetal force.

We agree to draw B magnetic field lines through unit area perpendicular to the field. The flux ϕ through an area A is the number of lines which pass through it. It can be computed from $\phi = B_{\perp}A$.

Ampère's circuital law can be used to compute magnetic fields due to currents. The field outside a long straight wire is $\mu_0 I/2\pi r$; the field in a long solenoid is $\mu_0 nI$.

Most materials do not change the magnetic field in which they are placed by much. The ferromagnetic materials, however, greatly increase magnetic fields in which they are placed and have regions of well-aligned atoms called domains. Each domain acts like a magnet. The domains can be aligned by placing the material in a strong magnetic field.

MINIMUM LEARNING GOALS

Upon completion of this chapter, you should be able to do the following:

1. Sketch the magnetic field in the vicinity of (*a*) various magnets, (*b*) a straight wire, (*c*) a loop of wire, (*d*) a solenoid, (*e*) a toroid.

2. Use a compass to determine the magnetic field lines in a given region.

3. Find the magnitude and direction of the force on a given straight wire current in a specified magnetic field.

4. Use $F = B_{\perp}IL$ to find one of the quantities when the others are given.

5. Use $F = qB_{\perp}v$ to find one of the quantities when the others are given. Describe quantitatively the path followed by a particle of known q and v moving perpendicular to a given magnetic field.

6. Determine the flux through a given area when **B** in that region is known.

7. Compute the magnetic field for a long, straight wire and for a solenoid when sufficient data are given.

8. Given a list of common materials, select those which greatly alter the magnetic field into which they are placed.

9. Explain the meaning of a magnetic domain and describe in terms of them what happens when a bar of ferromagnetic material is magnetized or demagnetized.

10. Sketch the magnetization curves for nonmagnetic and for ferromagnetic materials. Explain the basic features of each.

IMPORTANT TERMS AND PHRASES

You should be able to define or explain each of the following:

Like poles repel, unlike poles attract
Magnetic field line
Right-hand rule
$F = B_{\perp}IL$; tesla, webers per square meter, gauss
Flux density, magnetic induction
The magnetic force does no work
Velocity selector
A south pole exists near the earth's north pole
Flux; $\phi = B_{\perp}A$
Ampère's circuital law
$B = \mu_0 I/2\pi r$; $B = \mu_0 nI$
Ampère's theory of magnets; poles come in pairs
Ferromagnetic materials; domains
Magnetization curve
Hysteresis curve

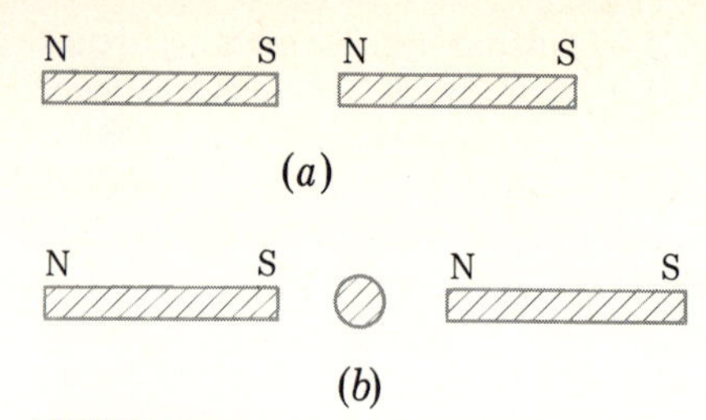

FIGURE P19.1

QUESTIONS AND GUESSTIMATES

1. Sketch the magnetic field for two identical bar magnets placed as shown in Fig. P19.1*a*. Sketch the magnetic field for the situation shown in part *b* if the circular piece of metal is iron. Repeat if the circular piece is brass.

2. A magnet will attract an unmagnetized nail made of iron, and that nail will attract another. Explain why.

3. A circular loop lies on a table. It carries a current in the clockwise direction as seen from above. At the center of the loop, a bar magnet sits vertically with its north pole on the table and its south pole straight above the north pole. Describe the forces on the loop caused by the magnet.

4. Two circular loops lie on a table and are concentric. The larger loop carries a current of 10 A counterclockwise while the smaller carries a current of 5 A clockwise. Describe the forces on each loop.

5. As shown in Fig. P19.2, two high-voltage leads cause a beam of charged particles to shoot to the right through a partially evacuated tube. Their path is shown by a fluorescent screen placed along the length of the tube. When a horseshoe magnet is brought close, the beam deflects as shown. How could you determine the sign of the charge on the particles?

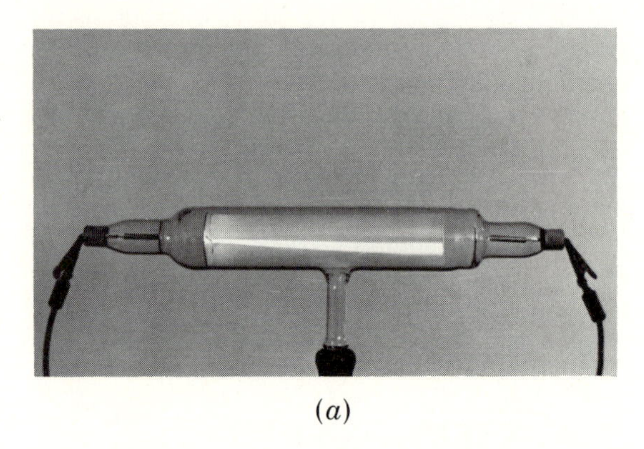
(*a*)

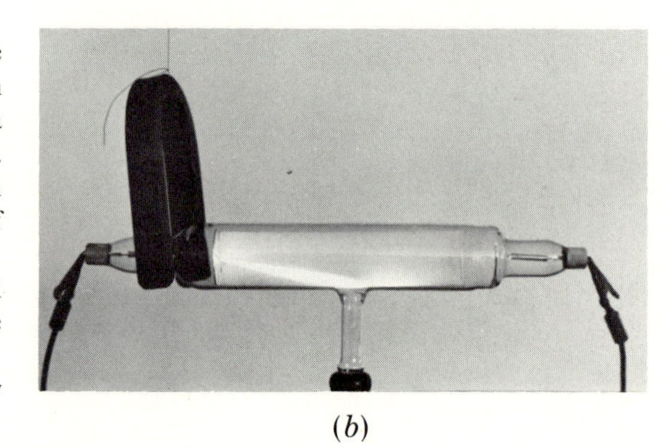
(*b*)

FIGURE P19.2

6. An electron is shot into a long solenoid at a small angle to the solenoid axis. Describe the motion of the electron.

7. Why can't one obtain an isolated north pole by breaking a magnet into two pieces?

8. If properly directed, a wire carrying a current in the earth's magnetic field will experience an upward force. Would it be possible to support an airship by sending a huge current through a metal rod mounted in the ship? Discuss.

9. A positive charged particle from space shoots toward the earth's equator along a radial line perpendicular to the earth's magnetic axis. Describe the motion of the particle. Estimate to within an order of magnitude the speed a proton must have to reach the earth along such a line if one ignores collisions with the air molecules. Repeat for a particle shooting in along the axis.

10. Estimate the effect of the earth's magnetic field on the electron beam in a TV set.

PROBLEMS

1. A power line parallel to the earth's surface carries a current of 20 A straight west. At that point, the earth's magnetic field is 0.80 G parallel to the earth's surface and directed straight north. (*a*) Find the force due to the field on a 15-m length of the wire. (*b*) What is its direction?

2. Repeat Prob. 1 if the wire carries the current at an angle of 30° north of west.

3. In Miami, Florida, the earth's magnetic field is approximately straight north but is directed at an angle of 57° below the horizontal. (The dip angle is 57°.) Its magnitude is about 0.51 G. Find (*a*) the magnitude and (*b*) the direction of the force it causes on a 15-m length of wire carrying a current of 20 A straight upward.

4. A particular power line is carrying a steady current of 50 A at 1000 V parallel to the earth and toward the north. If the earth's magnetic field is 0.60 G at that point and the angle of dip is 53°, (*a*) find the force on a 3-m length of wire. (*b*) What is the direction of the force?

5. A long straight wire carries a current of 12 A straight west along the floor in a house. (*a*) How large is B 2 m above the wire? (*b*) What is its direction?

6. How large is B inside a long solenoid having 1000 turns on its 50-cm length if it carries a current of 3 A?

7. The radius of a narrow toroid is 10 cm, and it has 200 turns of wire on it. How large a current must flow in it if B is to be 0.5 Wb/m^2 inside it?

8. If the toroid of Prob. 7 was wound on an iron core for which $\mu/\mu_0 = 300$, how large would the current have to be to produce $B = 0.50$ T?

9. A particular air-core solenoid is 2.0 m long and has 10,000 turns of wire wound on it. The diameter of its cross section is 3.0 cm. (*a*) If a current of 5.0 A flows in the coil, how large is the magnetic flux density in the central portion of the solenoid? (*b*) Repeat for the case where the solenoid is wound on an iron core which has $\mu = 50\,\mu_0$. (*c*) How much flux passes through the solenoid in part (*a*)?

10. A straight wire carrying a current of 10 A lies along the axis of a long solenoid in which the value of $B = 0.20\ \text{Wb/m}^2$. How large is the force on a 1.0-cm length of the wire?

11. Two parallel long straight wires 20 cm apart carry currents of 20 and 10 A in opposite directions. (*a*) How large is B resulting from the 10-A wire at the position of the other wire? (*b*) Find the magnitude and direction of the force on a 1-m length of the 20-A wire. (*c*) Repeat part (*b*) for the other wire.

12. Find B midway between two long, straight, parallel wires (separation = 20 cm) if they carry currents of 10 and 20 A in (*a*) opposite directions and (*b*) the same direction.

13.* A long straight wire carries a current of 20 A along the axis of a solenoid in which the magnetic field (in the absence of the wire) is 5.0 G. Find the magnitude of the magnetic field at a distance of 0.40 cm from the axis of the solenoid. The solenoid radius is 0.50 cm.

14. Find B at point P in Fig. P19.3.

15.* For the situation shown in Fig. P19.3, where is the magnetic field zero?

16. An electron having a speed of 10^7 cm/s enters a uniform magnetic field ($B = 0.020$ T) in a direction perpendicular to the field. Describe quantitatively the path taken by the electron.

17. A proton falls from rest through a potential difference of 3.34×10^5 V. It then enters a region perpendicular to a magnetic field, $B = 0.50\ \text{Wb/m}^2$. How large is the radius of the circle in which the proton travels? A proton is a hydrogen nucleus. Its charge is $+1.6 \times 10^{-19}$ C, and its mass is 1.67×10^{-27} kg.

18. A proton is accelerated through an unknown potential difference, enters a region perpendicular to a magnetic field of $2.00 \times 10^{-2}\ \text{Wb/m}^2$, and describes a circle having a radius of 30.0 cm. What is the energy of the proton expressed in electronvolts?

19. Describe quantitatively the path of an electron traveling at a speed of 10^7 m/s toward the west parallel to the earth, where $B = 0.5$ G and the angle of dip is 53°.

20.** The magnetic field in a long solenoid is 30 G. An electron is shot into the solenoid at an angle of 10° to the axis of the solenoid with a speed of 5×10^6 m/s. The electron follows a helical path. What is the radius and pitch of the helix?

21.** A proton traveling at 10^5 m/s enters a region of space where there is a uniform magnetic field of 0.100 T. Its velocity makes an angle of 30° with the direction of B. Describe quantitatively the path of the proton. *Hint:* Split the velocity into two components, perpendicular and parallel to the field. The parallel component will not interact with the field. The perpendicular component alone would cause the particle to move in a circle.

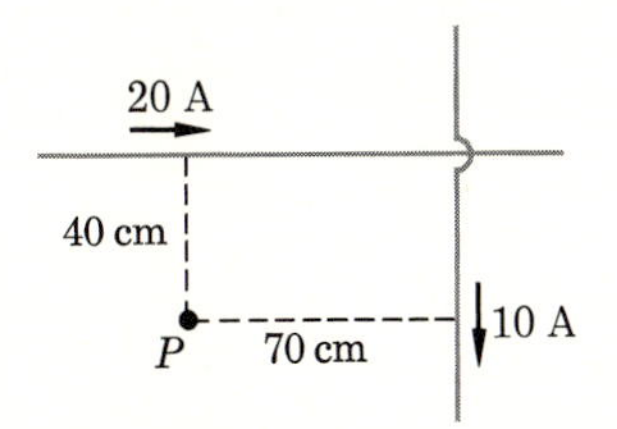

FIGURE P19.3

22. It is found that a particular beam of electrons travels in a straight line through crossed magnetic and electric fields in the condenser situation illustrated in Fig. 19.13. The value of B is 0.050 Wb/m^2, and the plates are separated by 10 cm and have a voltage of 100 V across them. Find (*a*) the speed of the electrons; (*b*) the radius of the circle in which the electrons will travel when the plate voltage is reduced to zero.

23. A narrow cylindrical beam of electrons is shot straight down a TV tube and causes a point of light on the screen. If the current carried by the beam is 0.10 mA, find (*a*) the number of electrons striking the screen each second; (*b*) the value of B at a radius of 2 cm from the beam. (*c*) As viewed by the person watching TV, is B clockwise or counterclockwise about the beam as axis?

24.* When a fast-moving particle such as an electron shoots through superheated liquid hydrogen, bubbles form along the path of the particle. In Fig. P19.4 one sees the paths of several particles in such a "bubble chamber." The paths are curved because of a magnetic field perpendicular to and into the page. Assume it to be 20 G. Is the particle leaving A and moving toward the right positive or negative? Assuming it to be an electron, about what is its speed? (The tracks are actual size. Assume them to be in the plane of the page.)

25.* The particle which starts at B in Fig. P19.4 slows as it moves through the liquid hydrogen. As a result it spirals inward. Assume the same data as in Prob. 24 and assume the particle to be an electron. Find its speed at point C.

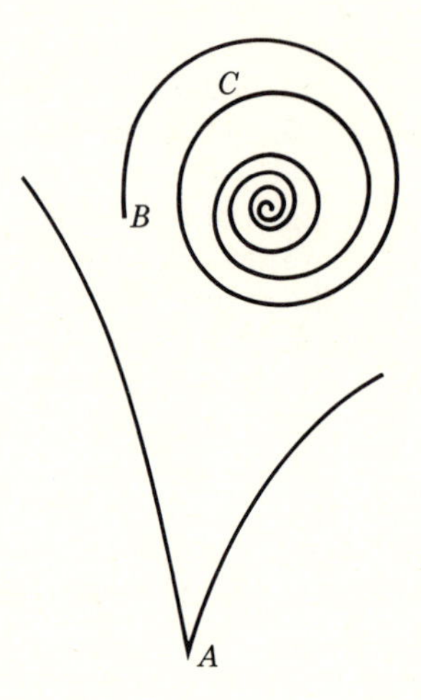

FIGURE P19.4

ELECTRO-MECHANICAL DEVICES

20

In the last chapter we saw that a wire which carries a current through a magnetic field experiences a force. This fact is used in the operation of electrical meters and motors, as well as in many other electrical devices. The operation of some of these devices will be discussed in this chapter. We shall also investigate how a current can be induced to flow in a coil by changing the flux through it. This phenomenon is basic to the understanding of electric generators and the action of inductances. The importance of induced currents and voltages will become apparent in the following chapter when we begin a study of alternating currents and resonating circuits.

20.1 THE GALVANOMETER

Most laboratory ammeters and voltmeters are modified galvanometers. They all operate in basically the same way. Essentially, *they consist of a coil of wire suspended in a magnetic field.* When current flows through the wire, the coil experiences a force because of the interaction of the current with the magnetic field. This force tends to rotate the coil, and *the rotational movement of the coil is used as a measure of the current flowing through it.*

A schematic diagram of the galvanometer coil is shown in Fig. 20.1. Current flows in the upper wire, around the several loops of the coil, and out the bottom wire. The coil is mounted so that it can rotate about line OO' as axis. The permanent magnet produces a field B from left to right across the coil. Since the upper and lower portions of the coil carry current parallel to the magnetic field, $B_{\perp} = 0$ and the force on them is zero.

The right-hand rule for finding the direction of forces on wires tells us that side P of the coil will experience a force out of the page. Side Q will experience a force into the page. Although these forces are essentially equal and opposite, they exert a torque on the coil. As the coil rotates, it twists a spring, which in turn opposes the turning of the coil. The coil will turn until the restoring torque furnished by the spring is equal to the torque caused by the current in the coil. Since the force on the wire, and hence the torque, is proportional to the current I, the angle through which the coil turns against the restoring spring will be proportional to I provided the spring obeys Hooke's law.

There are various ways of observing the angle through which the coil rotates. In the most sensitive galvanometers, a tiny mirror is mounted on the coil, and a beam of light reflected from the mirror moves in proportion to the rotation angle of the coil. A more common system is to mount a needle on the coil. As the coil turns, the needle deflects. The deflection of the needle as a function of current in the coil can be used to measure the current flowing through the meter coil, provided that one has a calibrated scale for the needle deflection. A top view of a device of this nature is shown in Fig. 20.2.

In the meter movement shown in Fig. 20.2 one sees that the pole pieces of the magnet are peculiarly shaped so as to have the magnetic field parallel to the plane of the coil at all times as the coil rotates through the allowed angles, thereby giving a more uniform torque. An iron core is usually provided for the coil in order to increase the magnetic field as well as to

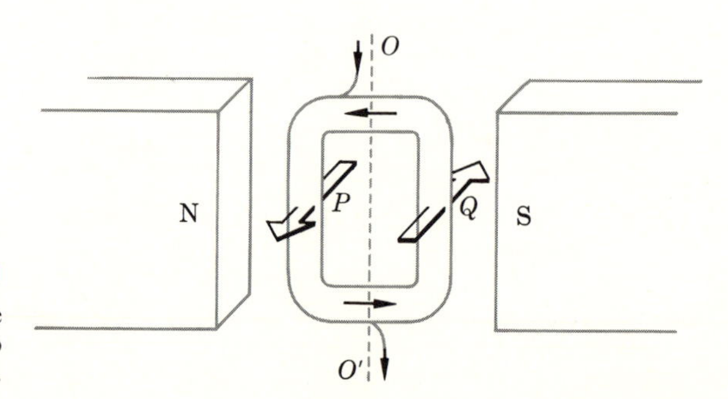

FIGURE 20.1

The current in the galvanometer coil causes it to deflect in the magnetic field.

supplement the shaping of the pole piece. By careful design, the meter scale can be made a nearly linear measure of I, as shown on the diagram.

The sensitivity of a galvanometer movement, i.e., how large a deflection results from a given amount of current, depends upon several factors. Of course, the stiffness of the restoring spring is of primary importance.If the instrument is to be rugged and portable, the spring must not be too delicate. The sensitivity also depends upon the number of turns of wire on the coil. If the number of turns is doubled, the torque on the coil will be doubled as well.

A very sensitive galvanometer will give full-scale deflection if a current of only a fraction of a microampere (10^{-6} A, abbreviated μA) goes through it. Such a highly sensitive galvanometer must have a large number of turns of wire on its coil, and so its resistance might easily be 100 Ω. Even so, a voltage of 10^{-4} V across its terminals would cause a current of 10^{-6} A to flow through it. Hence, it could be used as a very sensitive voltmeter as well as an ammeter or galvanometer. The table-model galvanometers used widely for student work give a full-scale deflection for a current of about 1 mA (10^{-3} A). They have a resistance of about 20 Ω.

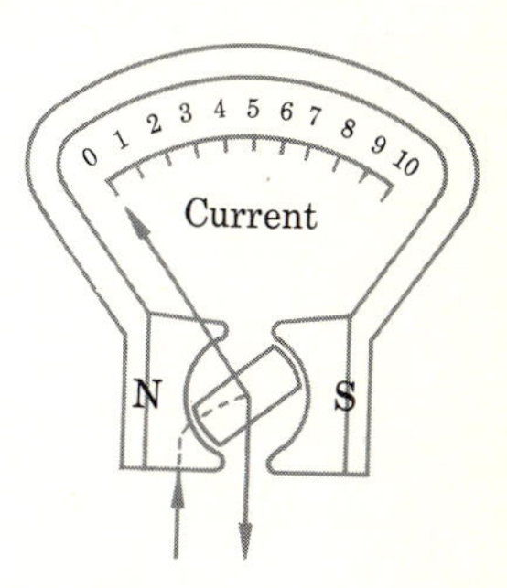

FIGURE 20.2
Schematic diagram of an ammeter.

20.2 AMMETERS

An ammeter is a modified galvanometer movement. If it is to be a very sensitive ammeter, it may be a galvanometer without any modification. However, most movements will give full-scale deflection for a few milliamperes or less, and so they cannot be used directly to measure currents larger than this. Suppose, as a concrete example, that we wish to make an ammeter which will deflect full scale for a current of 2 A. The movement to be used has a resistance of 20 Ω and deflects full scale for a current of 3 mA. (This would be a 60-mV movement; that is, 60×10^{-3} V across its terminals would cause a current of 3 mA to flow through it, deflecting it full scale.)

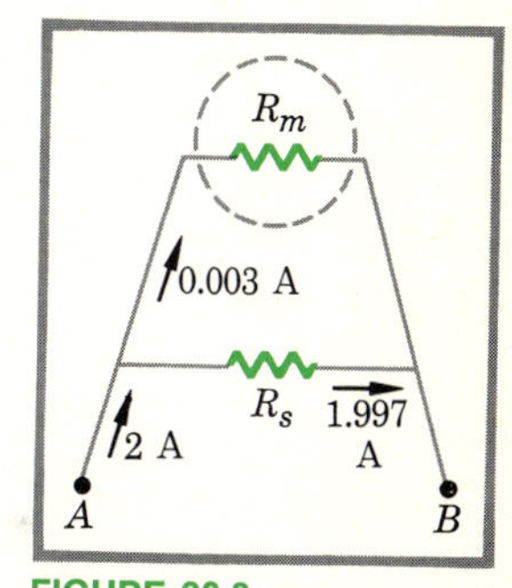

FIGURE 20.3
Only a small portion of the current goes through the movement of an ammeter. Most of it goes through the shunt resistor R_s.

In order to make the desired ammeter from this movement or galvanometer, we must find some way of allowing 2 A to flow through the meter while only 0.003 A flows through the movement. This can be done by the meter design shown in Fig. 20.3. The actual movement or galvanometer is shown as the central circular device having resistance R_m.

Ammeter Construction

When a current of 2 A flows into this meter through terminal A, we want only 0.003 A to flow through the movement, which has a resistance R_m. This means that *we must place a small resistance in parallel with the movement* so that 1.997 A will go through it. *We call this small resistor a* **shunt** resistance, and it is indicated as R_s in Fig. 20.3. To find how large R_s should be, we proceed as follows.

The movement will deflect full scale when there is a potential of 60×10^{-3} V across its terminals from A to B. But under these same conditions, the shunt resistor R_s must carry a current of 1.997 A. Using Ohm's law, we find

$$60 \times 10^{-3} = 1.997 R_s$$

or

$$R_s \approx 0.030\ \Omega$$

Notice that *the shunt resistor is extremely small.* Often it is just a piece of copper wire. Moreover, the resistance of the ammeter will be less than 0.030 Ω, since R_s is in parallel with R_m. This is as it should be, since we do not wish the ammeter to disturb the circuit when it is connected in series to measure the current flowing. (What would happen if this ammeter were accidentally connected across a potential difference of 1 V?)

Frequently one encounters ammeters having more than one range. These meters contain several alternate shunts. The sensitivity of the meter is determined by which shunt is connected across the meter terminals.

20.3 VOLTMETERS

Voltmeter Construction

We can construct a voltmeter from the movement we used in the last section to make an ammeter. The movement deflected full scale for a current of 0.003 A and had a resistance of 20 Ω. Hence, the potential difference across its terminals at full-scale deflection is (20)(0.003), or 60 mV. Let us construct a 90-V voltmeter from this movement.

An appropriate circuit for this purpose is shown in Fig. 20.4. When the potential across the terminals is 90 V, we wish a current of 0.003 A to flow through the meter. To find the resistance R_x which must be placed in series with the movement, we apply Ohm's law to the circuit of Fig. 20.4. We have

$$90 = (0.003)(20 + R_x)$$

Solving for R_x yields

$$R_x = 29{,}980\ \Omega$$

Clearly, this meter possesses a very high resistance. *Since we use a voltmeter by connecting it across a potential difference, we require it to have a high resistance* so that it will not disturb the circuit. The best voltmeters have a much higher resistance than the one we have designed. Our meter is said to have a resistance of 29,980/90 or 333 Ω/V. This does *not* mean that the resistance varies depending upon the scale reading. It is merely the ratio of the resistance of the meter to its maximum scale reading. A similar meter possessing a more sensitive movement would have a higher resistance. (Why?)

FIGURE 20.4

To make a voltmeter from a sensitive movement, we place a large resistor in it.

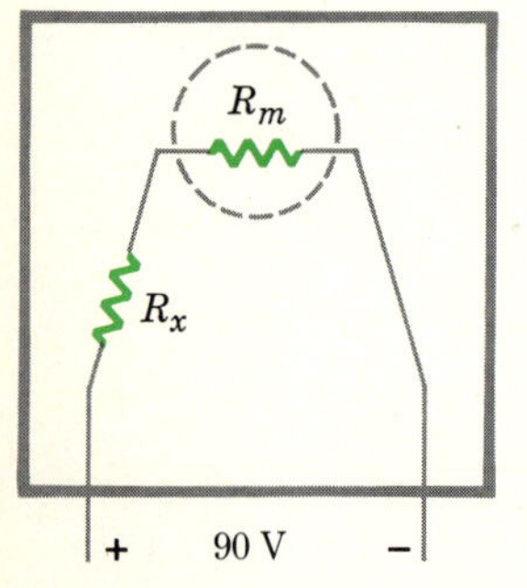

20.4 TORQUE ON A CURRENT LOOP

As described in the previous sections, *the operation of a meter movement depends upon the fact that a current-carrying coil experiences a torque, or turning effect, in a magnetic field.* This is a fact of considerable importance since it is basic to the operation of many devices, including electric motors. For that reason, let us examine quantitatively the torque on such a coil.

The situation of a current-carrying coil in a magnetic field is shown in Fig. 20.5*a*. As indicated, the coil is mounted on an axle and can rotate. Using the right-hand rule, the forces on the various sides are found to be those

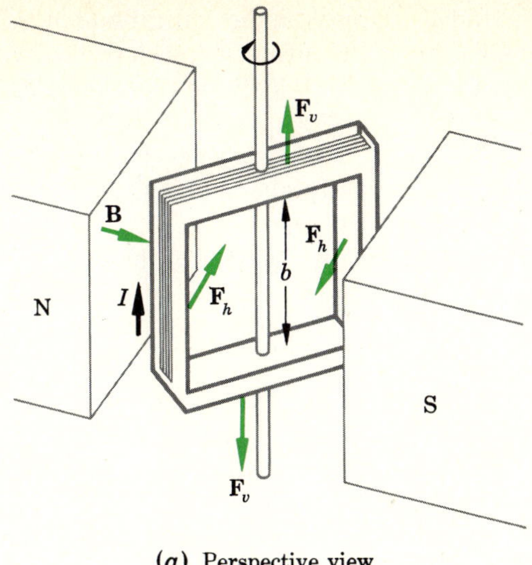

(a) Perspective view

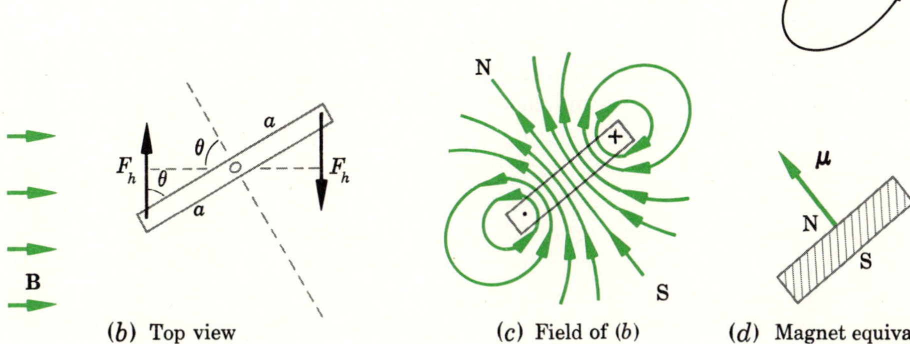

(b) Top view (c) Field of (b) (d) Magnet equivalent

FIGURE 20.5

The coil in (*a*) will experience a torque and for this purpose is equivalent to the short bar magnet shown in (*d*).

illustrated. Notice that only the two forces F_h cause a torque about the axis of rotation. Even these two forces will cause no torque when the plane of the coil is perpendicular to the field of the magnet. Maximum torque occurs when the lines of B skim past the surface of the coil, i.e., when the field lines lie in the plane of the coil, for then the lever arm for F_h is maximum.

To obtain a quantitative expression for the torque on the coil we note that each of the two forces F_h gives a torque

$$(F_h)(\text{lever arm})$$

From Fig. 20.5*b* we see the lever arm to be $a \sin \theta$, where θ is the angle so labeled. Therefore the torque on the coil is

$$\text{Torque} = 2F_h a \sin \theta$$

But F_h is simply the force on the vertical side of the coil. If the vertical side has a length b, and if the current is I, each vertical wire will contribute a force BIb to F_h. But there are N loops on the coil, so $F_h = NBIb$ and the torque becomes

Torque on a Coil

$$\text{Torque} = (2ab)(NI)(B \sin \theta)$$

Notice that $2ab$ is simply the area of the coil. We can therefore write

$$\text{Torque} = (\text{area})(NI)(B \sin \theta) \tag{20.1}$$

Although we have derived Eq. (20.1) for a very special shaped coil, it turns out that the equation is true for all flat coils. Since NI is the current flowing around the coil, we see that the important features of the coil (aside from its orientation) are the area of the coil and the current in it. In view of this, it is customary to define a quantity called the **magnetic moment** of a current loop:

Magnetic Moment

$$\mu = \text{magnetic moment} = (\text{area})(I)$$

There is a definite advantage to thinking of a current loop as a bar magnet characterized by its magnetic moment, as we now shall see.

We have already pointed out that a current loop has a magnetic field similar to that of a bar magnet. This is shown once again in Fig. 20.5*c*. Notice that the coil acts like a short, fat bar magnet with north and south poles as indicated. This similarity is further pointed out in Fig. 20.5*d*. Moreover, if we refer back to parts *a* and *b*, we see that the coil itself experiences a torque similar to that upon its equivalent bar magnet. For example, in part *a* the coil will act like a bar magnet with its north pole on the far side of the coil. Since this side is close to the north pole of the magnet causing the external field, the coil will be rotated in the direction indicated there.

To make full use of this analogy, one gives direction to the magnetic moment μ and makes it a vector. This is shown in Fig. 20.5*d*. We see that $\boldsymbol{\mu}$ is taken perpendicular to the area of the loop and in the direction given by the right-hand rule; i.e., if the fingers circle in the direction of I, the thumb points along $\boldsymbol{\mu}$. As a result, the magnetic moment vector $\boldsymbol{\mu}$ points out of the north pole of the equivalent magnet. This has the following important consequence:

> When a current loop is placed in a magnetic field, it rotates so as to align its magnetic moment vector with the magnetic field vector.

One can appreciate why this is true by recalling that a compass needle is simply a bar magnet and that the field direction is defined to be that direction along which the needle aligns. *We shall find it convenient from time to time to think of a current loop as a magnet with magnetic moment* μ.

The torque on a current loop is of particular importance for electric motors. However, before discussing the operation of motors, it is important to understand the principle behind the operation of a generator. We shall see that this same principle is operative in motors and influences their electric behavior. In the next section we shall see how an electric generator produces a voltage.

Illustration 20.1 Niels Bohr in 1913 postulated the following reasonably successful picture for the hydrogen atom. At its center was a proton; the proton charge is $+1.6 \times 10^{-19}$ C, and its mass is 1.67×10^{-27} kg. The diameter of the proton is of the order of 10^{-15} m. Circling the proton at a radius of 0.53×10^{-10} m is an electron whose charge is -1.6×10^{-19} C and whose mass is only $\frac{1}{1840}$ that of the proton. Bohr concluded that the electron circled the nucleus 6.6×10^{15} times each second. This motion of the charged electron around a circle is equivalent to a current in a loop of wire. Find the magnetic moment of the hydrogen atom resulting from the orbital motion of its electron.

Reasoning By definition, the magnetic moment μ is IA, where A is the area of the loop, namely,

$$A = \pi r^2 = (\pi)(0.53 \times 10^{-10})^2 \text{ m}^2 = 0.88 \times 10^{-20} \text{ m}^2$$

To find I we note that the current in a loop is equal to the charge passing a given point in a second. For the present case, the electron circles the atom 6.6×10^{15} times each second, carrying a charge of 1.6×10^{-19} C past a given point that many times each second. Hence

$$I = (1.6 \times 10^{-19} \text{ C})(6.6 \times 10^{15} \text{ s}^{-1}) = 1.05 \times 10^{-3} \text{ A}$$

One therefore finds the atom to act like a small magnet with magnetic moment

$$\mu = IA = 9.3 \times 10^{-24} \text{ A} \cdot \text{m}^2$$

When techniques became available to carry out the measurement of this magnetic moment, the predicted value was confirmed. However, in addition, the electron *itself* was found to act like a small magnet, and this was pictured as being the result of a spinning motion of the charged electron about an axis through its own center.

20.5 INDUCED EMFS

A simple but fundamental experiment can be performed using the equipment shown in Fig. 20.6*a*. We see there two simple series circuits. One consists of a battery and switch in series with a long wire. The wire is coiled around an iron rod as shown. We call this coil the **primary coil** since it is attached to the battery. A second independent wire is also coiled around the

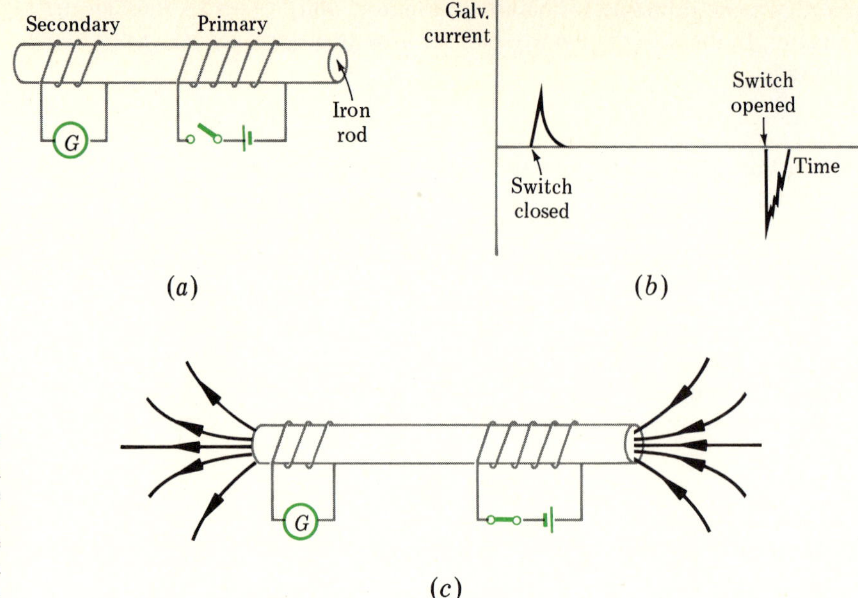

FIGURE 20.6
An induced current exists in the secondary only when the primary current is changing. The current pulses are actually much narrower than shown in (*b*).

rod. This coil is in series with a galvanometer but has no battery in its circuit. It is called the **secondary coil.**

Since there is no battery in the secondary-coil circuit, one might guess that the current through it would always be zero. But a startling fact emerges if the switch in the primary circuit is either suddenly closed or opened. At that exact instant, the galvanometer suddenly deflects and then returns to zero. A current flows in the secondary-coil circuit for a short instant. It is as though the secondary circuit possessed a battery, a source of emf, for just an instant. We say that an **induced emf** exists in the secondary coil.

Figure 20.6*b* shows another feature of this induced current and emf. As we see there, the short-lived current flows in one direction when the switch is pushed closed. But the emf causes a current pulse to flow in the opposite direction when the switch is pulled open. This tells us that the direction of the induced emf depends upon whether the current in the primary coil is increasing or decreasing.

A second, somewhat similar experiment is shown in Fig. 20.7. It involves a bar magnet and a coil in series with a galvanometer. When the magnet is stationary beside the coil, as in parts *a* and *c* of the figure, no current flows in the coil. However, if the magnet is moved relative to the coil, current flows in the coil as indicated in parts *b*, *d*, and *e*. As we see, an induced emf exists in the coil only when the magnet and coil are in relative motion. No induced emf exists when conditions are not changing.

Michael Faraday (1791–1867) discovered and studied these effects. He concluded that an induced emf exists in a coil only if the flux through the coil is changing. For example, refer to Fig. 20.6*c*. Since the field lines follow the iron rod as shown, a considerable flux passes through the secondary coil when the switch is closed. If the switch is now opened, this flux decreases to

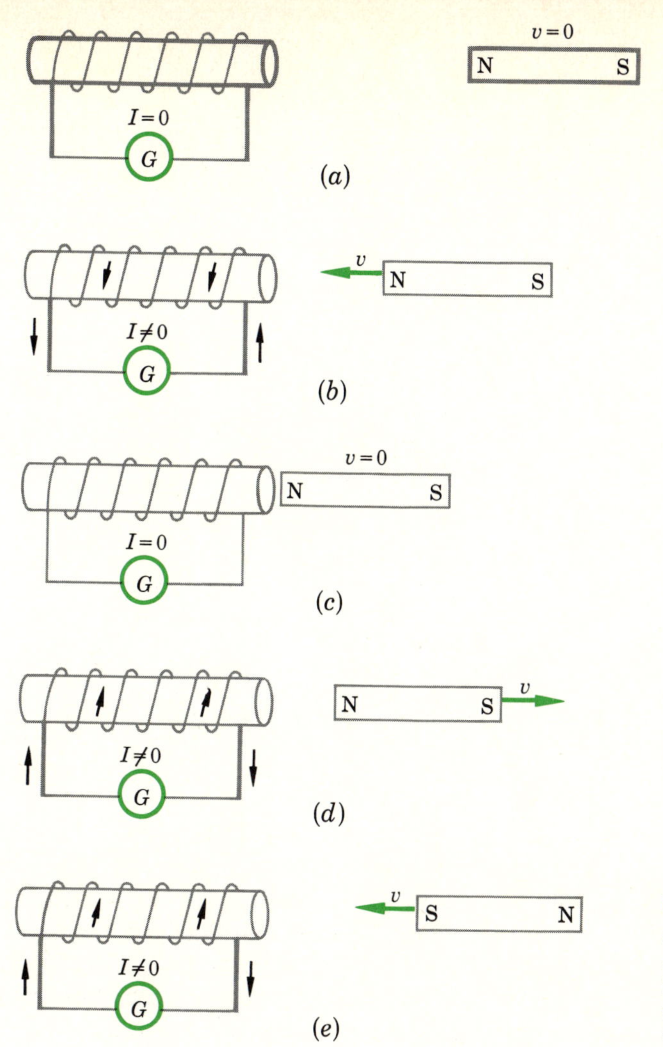

FIGURE 20.7

Current is induced in the coil only when the flux through it is changing. Why does the current flow in the directions shown?

zero. During the time in which the change in flux is occurring, an induced emf exists in the coil. But no induced emf exists when the flux is not changing.

If we refer to Fig. 20.8, we can see that this same situation exists for the coil and magnet. As shown there, the flux through the coil changes as the magnet is moved either closer or farther. During the time when the flux through the coil is changing, an induced emf exists in the coil. Let us now investigate this effect quantitatively.

The direction of the induced emf can be seen by reference to the experimental results in Figs. 20.6 to 20.8. For example, by reference to Fig. 20.8 we can learn how to predict the emf's direction. As we see in part *a* and *b*, the flux is changed in such a way as to increase the flux through the coil toward the left. But, as shown in part *c*, the induced current in the coil sets up a flux in a direction opposite to this. The induced emf is in such a direction as

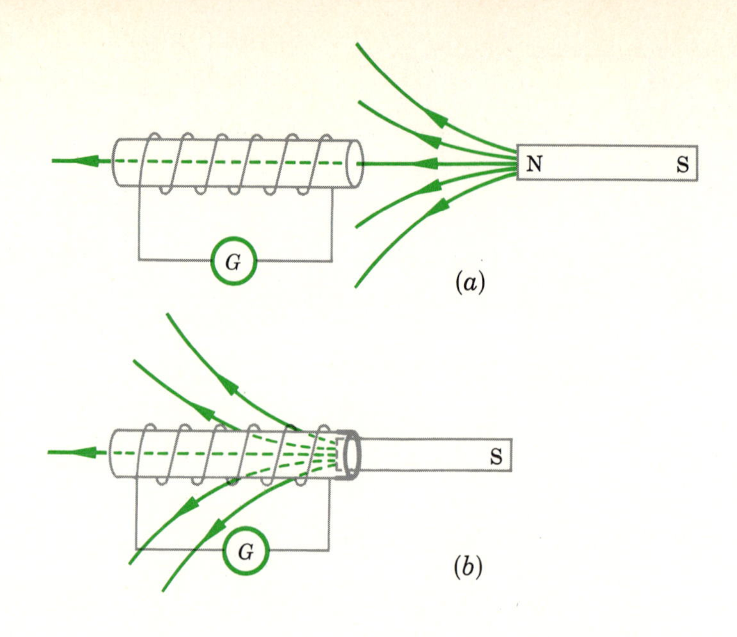

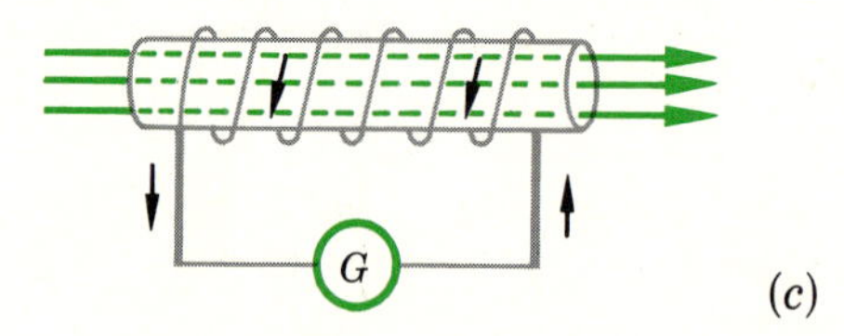

FIGURE 20.8

When the magnet is moved from (*a*) to (*b*), the current in the coil flows in the direction shown in (*c*). Why?

to produce a current, the flux from which will tend to cancel the change in flux through the coil. If you examine Figs. 20.6 and 20.7, you will notice similar behavior. The induced current is in such a direction as to try to cancel the change in flux through the coil. This rule is embodied in **Lenz's law,** which may be stated as follows:

Lenz's Law

> A change in flux through a loop of wire will induce an emf in the loop. The direction of the current produced by the induced emf will be such that the flux generated by the current will tend to counterbalance the original change in flux through the loop.*

In other words, if the flux through a coil is toward the *left* and *increasing,* the induced emf will be in such a direction as to produce flux through the coil toward the *right*. If the flux is toward the *left* and *decreasing,* the induced emf will try to put flux through the coil toward the left. *The induced emf exists only while the flux is changing.*

Faraday discovered the mathematical relation governing the induced

*This is a consequence of the conservation of energy. If the flux generated by the induced current were in such a direction as to augment the change in flux, the induced current could continue to induce more current, without end.

emf in the coil. If the coil has N loops of wire, and if the flux changes by an amount $\Delta\phi = \phi_2 - \phi_1$ in a time Δt, then the average induced emf during this time is

$$\text{emf} = -N\frac{\Delta\phi}{\Delta t} \qquad (20.2)$$

Faraday's Law

This is called **Faraday's law.** The negative sign is purely a formality and reminds us that the induced emf is in such a direction as to oppose the change in flux.

Illustration 20.2 A solenoid having 100 turns of wire has a cross-sectional area of $4.0\ \text{cm}^2$. If a magnet is suddenly brought close to the solenoid, increasing B from zero to 0.50 T in 0.020 s, how large is the average emf generated in the solenoid?

Reasoning It will be recalled that $\phi = B_\perp A$. Hence

$$\phi_1 = 0 \qquad \text{and} \qquad \phi_2 = (0.5)(4 \times 10^{-4}) = 2 \times 10^{-4}\ \text{Wb}$$

We therefore have that

$$\Delta\phi = 2 \times 10^{-4}\ \text{Wb}$$

Faraday's law yields

$$\text{emf} = (100)\left(\frac{2 \times 10^{-4}\ \text{Wb}}{2 \times 10^{-2}\ \text{s}}\right) = 1.0\ \text{V}$$

You should be able to show that since a weber is a newton-meter per ampere, the unit of the answer is volts.

20.6 MUTUAL INDUCTION

Faraday's law for the induced emf in a coil applies to any method for changing the flux through a coil. Suppose that we have two coils placed side by side, as shown in Fig. 20.9. When the switch S is open, both coils have zero flux through them. We have called the coil in the battery circuit the primary coil and the other the secondary coil.

If the switch is suddenly closed, the primary coil will act as an electromagnet and will generate flux in the region near it. Some of the flux from the primary coil will go through the secondary coil. Hence, the flux through the secondary will change when S is suddenly closed. According to Faraday's law, an induced emf will be generated in the secondary for an instant as the

FIGURE 20.9

Why does the current flow from A to B through the resistor at the instant the switch S is just opened?

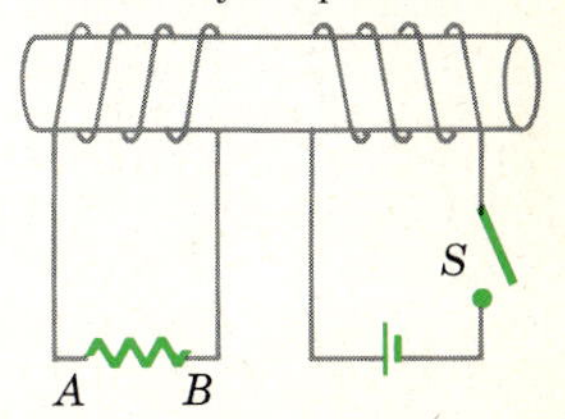

current rises from zero to its final value in the primary. You should be able to show that the direction of the induced current through the resistor in Fig. 20.9 will be from B to A when S is just closed. The current will flow in the opposite direction just as the switch is opened.

The magnitude of the induced emf generated in the secondary will depend upon many geometrical factors. Among these are the number of turns of wire on each coil, how close together the coils are, their cross-sectional area, and so on. (Why?) In addition, since the flux through the secondary will be proportional to the current in the primary, the induced emf in the secondary will be proportional to the rate of change of current in the primary, $\Delta I_p/\Delta t$. We therefore write the following equation for the induced emf in the secondary:

Mutual Inductance

$$\text{emf}_{\text{sec}} = -M\frac{\Delta I_p}{\Delta t} \tag{20.3}$$

As indicated above, *the proportionality constant M contains the effects of the geometry of the two coils. It is called the* **mutual inductance** of the two coils. If the emf is in volts, I in amperes, and t in seconds, *the unit M is defined to be the* **henry** (*H*), or V · s/A. This unit is named after one of the contemporaries of Faraday, Joseph Henry (1797–1878), who actually discovered independently many of the results attributed to Faraday. Unfortunately, Henry's results, obtained while he was in Albany, N.Y., were not widely published, and so they had very little influence on scientific progress at that time.

Illustration 20.3 Two coils of wire wound on an iron core have a mutual inductance of 0.50 H. How large an average emf is generated in the secondary by the primary as the current in the primary is increased from 2.0 to 3.0 A in $\frac{1}{100}$ s?

Reasoning Let us first point out that when two coils are wound on a core of iron, the flux will be much larger than if they are wound on a nonmagnetic core. Hence, the value of M given here is much larger than one would ordinarily have for noniron-core coils.

Making use of Eq. (20.3) we have, using the proper SI units,

$$\text{emf} = (0.50)\left(\frac{1.0}{0.01}\right) = 50 \text{ V}$$

Notice that this emf will exist for only an instant in the secondary. As soon as the current in the primary becomes steady, the flux will no longer be changing and there will be no induced emf.

20.7 SELF-INDUCTANCE

If we believe Faraday's law, we know that any change in flux through a coil will induce an emf in the coil. This means that *when a current through a coil changes, the coil induces an emf in itself.* Therefore, if we consider the coil shown in Fig. 20.10, the flux in it will change from zero to some finite value when the switch is first closed. A flux will be generated by the current, and the flux will be directed toward the left through the coil. By Faraday's law, there will be induced in the coil an emf which tries to produce flux to the right through the coil. Hence, the induced emf must be opposed to the emf of the battery. If the switch is suddenly opened, the induced emf will be aiding, rather than opposing, the battery. (Can you show this?)

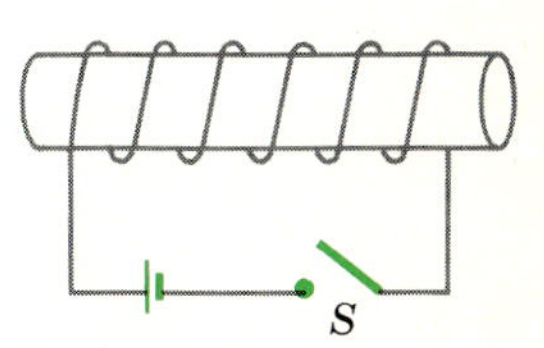

FIGURE 20.10

When the switch is first closed, the coil will induce an emf in itself. Will it aid or oppose the battery?

Here, too, the geometry of the coil as well as the core material will determine how large the induced emf will be. If $\Delta I/\Delta t$ is the rate of change of current through the coil, we can write for the average induced emf

$$\text{emf} = -L\frac{\Delta I}{\Delta t} \tag{20.4}$$

The constant of proportionality L is called the **self-inductance** *of the coil. It has the same units as mutual inductance, namely, henrys.*

Self-Inductance

Obviously, if the coil is wound on an iron core, the flux through it will be much greater than if no magnetic material were present. Hence, *if a large self-inductance is desired, the inductor should be wound on an iron core.* We shall return to the behavior of mutual and self-inductances in later sections. They are of particular importance in alternating-current (ac) circuits, where the current, and thus the flux, is changing continually.

20.8 MOTIONAL EMFS

In certain cases at least, the induction of an emf in a circuit by a changing flux can easily be explained in another way. This second method of treatment gives added insight into the process involved. It is based upon the fact that a moving charge experiences a force when in a magnetic field. We learned that the force F upon a charge q moving with speed v perpendicular to a magnetic field $B_\perp$ is given by

$$F = B_\perp qv \tag{20.5}$$

FIGURE 20.11

The electrons in the wire experience a force directed toward the top end of the wire as the wire moves through the field as shown.

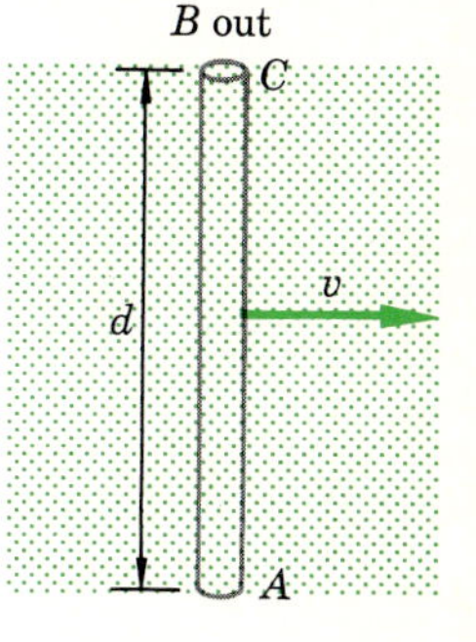

With this fact in mind, consider the motion of a wire, length d, in a magnetic field, as illustrated in Fig. 20.11. Upon recalling that v is in the direction of charge motion and that the motion of these charges is in fact a current, the positive charges will be forced toward the lower end of the wire. Hence, the bottom of the wire will be positively charged, and the top will be negatively charged. There will be an electric field E within the wire.

The definition of E is force per unit charge, or F/q. At equilibrium, the electric field force directed from the bottom to the top of the wire will exactly balance the magnetic force on the positive charges, which is directed from top to bottom. By Eq. (20.5), the magnetic force per unit charge is

$$\frac{F}{q} = B_\perp v$$

while the electric force per unit charge is

$$\frac{F}{q} = E$$

Equating these two expressions yields

$$E = B_\perp v \tag{20.6}$$

We have therefore found that *a wire moving with constant speed v perpendicular to a magnetic field $B_\perp$ develops within it an electric field given by (20.6). This field results in a potential difference between the two ends of the wire.* Since the potential difference is simply the work done in carrying unit charge from C to A, we have

Motion-induced emf

$$\text{Induced emf} = V = Ed = B_\perp vd \tag{20.7}$$

It is possible to derive Eq. (20.7) by Faraday's law as well. To do so, we must have a circuit for which we can write

$$\text{Induced emf} = N\frac{\Delta\phi}{\Delta t}$$

Let us suppose that the moving wire of Fig. 20.11 is connected by sliding contacts to a stationary circuit, as shown in Fig. 20.12. In a time of 1 s, the moving wire will move through a distance v to position MN. As a result, the area of the loop of wire has been increased by an amount vd, the shaded area shown. Consequently, the change in flux through the circuit in time Δt is just

$$\Delta\phi = B_\perp \Delta A = B_\perp v\Delta t\, d$$

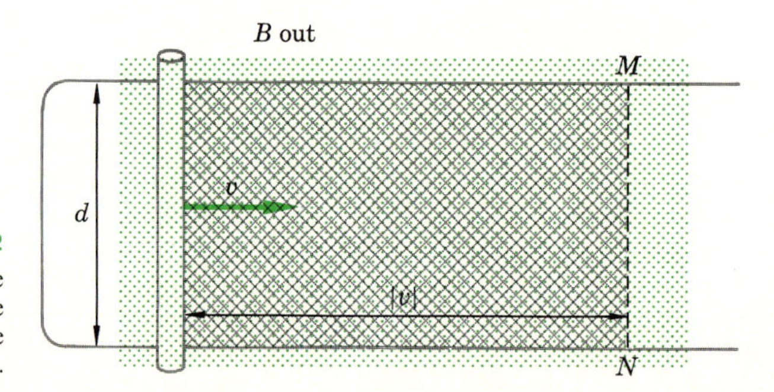

FIGURE 20.12

The emf induced in the wire can be treated as being the result of the change of the flux through the loop.

Using Faraday's law for the induced emf in the loop yields

$$\text{Induced emf} = \frac{\Delta\phi}{\Delta t} = B_{\perp}vd$$

This is exactly the same as Eq. (20.7). We have now shown that the induced emf given by Faraday's law was actually induced entirely in the moving straight wire of the circuit. Hence, the origin of the induced emf is readily understood in terms of the forces on charges moving through a magnetic field. Unfortunately, many situations are not this simple. One need consider only the case of the two mutual-inductance coils of Fig. 20.9 to see that the concepts involved are not always obvious.

You should consider how Eq. (20.7) would have to be modified if the velocity v were not perpendicular to the wire. This is perhaps most easily done by reference to Fig. 20.12. Can you show that d must be replaced by the component of d perpendicular to v in that case?

20.9 THE AC GENERATOR

A generator produces a voltage difference between two terminals by changing the flux through a coil. In theory, the flux could be changed by either moving a magnet with respect to the coil or moving the coil with respect to a magnet. The latter procedure is more easily realized in practice and is the one ordinarily used.

The coil of wire shown schematically in Fig. 20.13 rotates about the axis AA'. A uniform magnetic field perpendicular to AA' is furnished by magnet poles as shown. Notice that one end of the coil is attached to the slip ring R and the other end to R'. These rings rotate with the coil, but contact is made to stationary terminals by means of the brushes b and b', which slide along outside the rings.

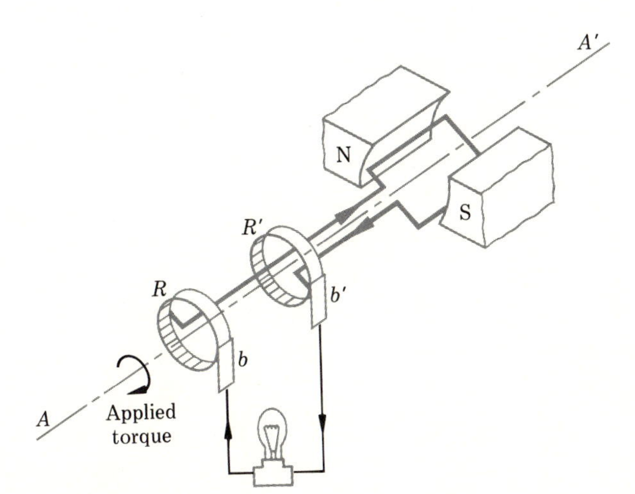

FIGURE 20.13

An alternating emf is produced between terminals b and b' as the coil rotates in an external magnetic field.

To see exactly how the induced emf between the terminals of the coil is generated, let us refer to Fig. 20.14. The coil is assumed to be rotating clockwise. To find the direction of the current flow we make use of Lenz's law.

In Fig. 20.14*a*, the number of lines of flux through the coil is decreasing as the coil rotates from the position shown. Hence, the induced emf in the coil will drive current in the direction shown so as to increase the flux, i.e., counterbalance the change. You should examine each portion of Fig. 20.14 and be able to explain clearly why the current flows in the directions shown. Notice that part *d* is exactly the same as part *a* except that the coil has been turned through 180°. However, the direction of the emf generated in the coil has reversed in going from position *c* to *d*. Therefore, the current in the coil will reverse each time the plane of the coil is perpendicular to the field.

An expression for the emf in the coil can be obtained from a consideration of either the rate of change of flux through the coil or the motion of the wires through the magnetic field. If one is able to use calculus, the flux approach is most convenient. However, we shall use the other approach, since it does not require the use of calculus.

We had in the last section that the induced emf in a wire is Bvd, where all three quantities, B, v, and d, must be perpendicular to each other. If they are not perpendicular, we take their perpendicular components. In Fig. 20.15, the rectangular loop having N turns and dimensions b by a (a being the dimension perpendicular to the page) is rotating clockwise. The net induced emfs in the two sides of the loop parallel to the page in part b are zero. Hence, we shall be concerned only with the induced emf in the wires of the loop which are perpendicular to the page.

Consider the wires of the loop which are perpendicular to the page at the top of Fig. 20.15*b*. The induced emf in each is trying to drive current out of the page. Since B and the wire are already perpendicular, to obtain the emf we need only multiply by the perpendicular component of v, namely, $v \sin \theta$. The induced emf in each of the top wires is

$$B(v \sin \theta)a$$

Clearly, all the top wires will be inducing current to flow in the same direction, and so their effects are additive. Moreover, the N lower wires of the loop will be trying to cause the current to flow in a like direction about the loop. Consequently, the total induced emf will be $2N$ times larger than the

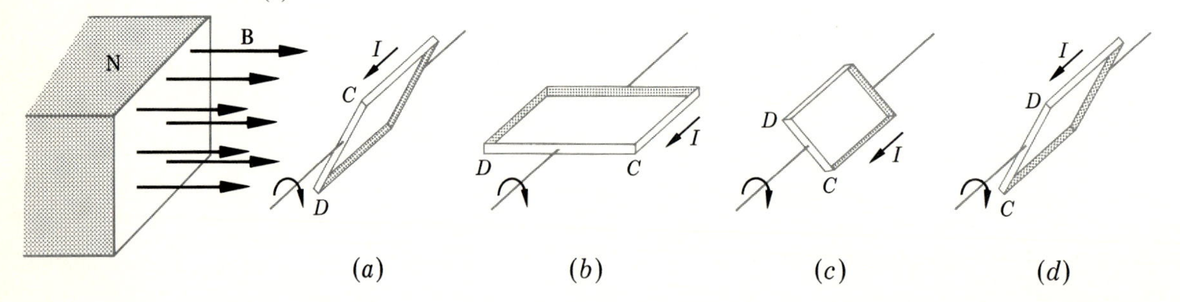

FIGURE 20.14
As the loop rotates in the external magnetic field, the direction of the induced current reverses between the positions shown in (*c*) and (*d*).

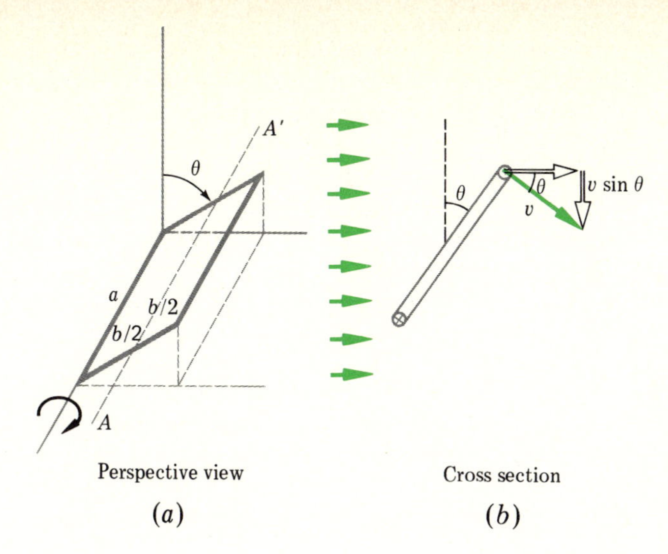

FIGURE 20.15

The induced emf in the rectangular loop is caused by the speed $v \sin \theta$ of the wires perpendicular to the page.

above expression. We therefore find

$$\text{Induced emf} = 2NBav \sin \theta \qquad (20.8)$$

It is more common to express this equation in terms of the rotational velocity ω. We know from $v_T = \omega r$ that $v = \frac{1}{2}b\omega$, and since $\theta = \omega t$, we have

$$\text{emf} = NB\omega ab \sin \omega t$$

Since ab is the area of the loop A, this is equivalent to

$$\text{emf} = NBA\omega \sin \omega t \qquad (20.9)$$

Induced emf in Rotating Coil

Equation (20.9) turns out to be applicable for any flat loop of wire, whether it is rectangular or not. Notice that ω is expressed in radians per second. If we wish it to be in rotations per second f, we must use the fact that $\omega = 2\pi f$.

If you look back to the work on vibrating systems, you will see that Eq. (20.9) is similar to the equation

$$y = y_0 \sin \omega t = y_0 \sin 2\pi f t$$

used there. This equation represented the displacement of a mass vibrating at the end of a spring or some similar device. It told us that the position of the mass oscillated back and forth in a sinusoidal motion between $+y_0$ and $-y_0$. The quantity y_0 was the amplitude of vibration, and f was the number of vibrations per second.

We see, then, that *Eq. (20.9) says that the voltage difference between the two terminals of the rotating coil is oscillating back and forth.* In this case, the polarity of the terminals reverses in an oscillatory fashion. If one terminal is

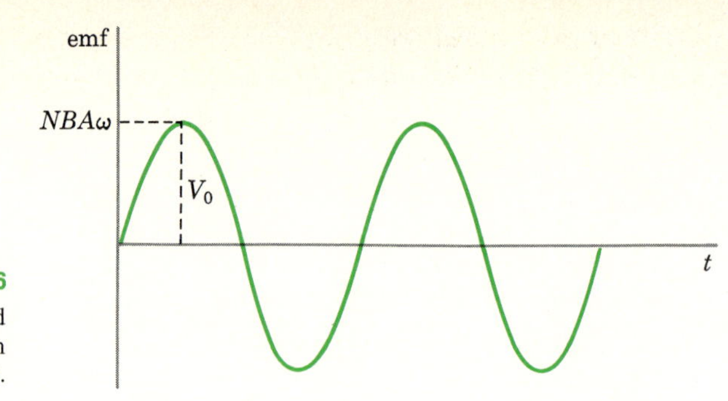

FIGURE 20.16
An alternating emf is induced in a coil rotating in a uniform magnetic field.

at first positive with respect to the other, the polarity soon reverses so that it becomes the negative terminal. A plot of the voltage difference between the terminals as a function of time gives a sinusoidal curve, as shown in Fig. 20.16. Negative values merely indicate that the induced emf in the coil has reversed in direction.

By comparing the vibration equation in Eq. (20.9) we see that the amplitude of vibration y_0 is, in this case, just $NBA\omega$. This value is shown on the graph of Fig. 20.16. Calling this amplitude V_0, we can write Eq. (20.9) as

An Alternating Voltage

$$V = V_0 \sin 2\pi f t \tag{20.10}$$

showing quite clearly that the voltage difference between the terminals is alternating sinusoidally.

From the above considerations, the fact becomes evident that a coil of wire rotating in a magnetic field has an alternating emf generated between its terminals. If such a generator is used as the power source in the simple circuit shown in Fig. 20.17, the current through the resistor will reverse its direction $2f$ times per second. (Notice that the symbol for an alternating-voltage generator is ∼.)

The ac generators used by power companies are usually more complex than the one discussed here, but their basic operation is the same. Mechanical energy to rotate the coil is usually furnished by steam turbines or by waterpower. Let us just briefly consider the conversion of energy in a system such as that shown in Fig. 20.17.

FIGURE 20.17
A simple ac circuit.

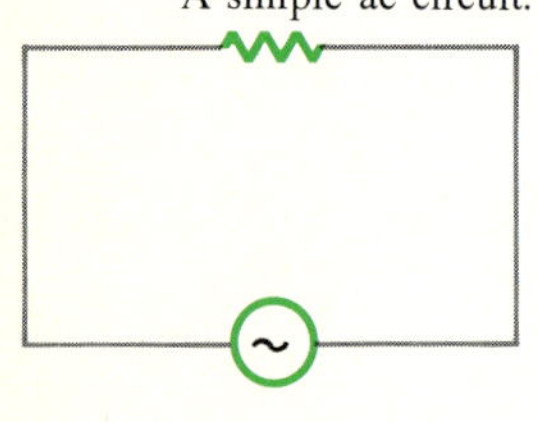

If the circuit is open so that no current can flow in the generator coil, very little force need be exerted to rotate the coil. However, as soon as current is drawn from the generator (the coil), the magnetic field will exert a force on the current-carrying wires of the generator and these forces are in such a direction as to stop the coil from rotating. Hence, the mechanical energy fed into the generator is dependent upon the current drawn from the generator—more current requires more mechanical energy.

At an instant when the voltage of the generator is V, the power being delivered to the resistor of Fig. 20.17 is VI. Clearly, if I is very small, the power consumed by the resistor is small and the mechanical energy needed to operate the generator is small. We therefore see that *the energy needed to operate the generator depends directly upon the energy being drawn from it. The*

Relation of Electrical and Mechanical Energy

mechanical energy is transformed to electrical energy by means of the interactions between magnetic field and charge motion within the coil of the generator.

20.10 MOTORS

A motor is essentially a generator run backward. As with a generator, *a simple motor consists of a coil rotating in the field produced by a magnet.* The simplified diagram of Fig. 20.18 shows the coil to be situated between the pole pieces of a magnet. In practice, this would probably be an electromagnet rather than a permanent magnet. Moreover, the coil itself is wound on a soft iron core to intensify the magnetic field resulting from the current through it. The current-carrying coil itself acts like a bar magnet as discussed earlier; its strength is increased more than a hundredfold by the iron core.

In the figure, the north pole of the coil is repelled by the north pole of the permanent magnet, and if the coil is given a slight, counterclockwise rotation (as viewed along AA' from A to A'), the coil will be made to rotate by the mutual repulsion of the poles. After rotating 180°, the north pole of the coil will be adjacent to the south pole of the permanent magnet and would be retarded from rotating away from that position. However, when it reaches that position, the sliding contacts on the split slip ring slide over the gap and the current flowing through the coil reverses. This in turn reverses the poles of the coil, and so once again the situation shown in Fig. 20.18 is achieved. Repulsion is maintained, and as a result rotation continues.

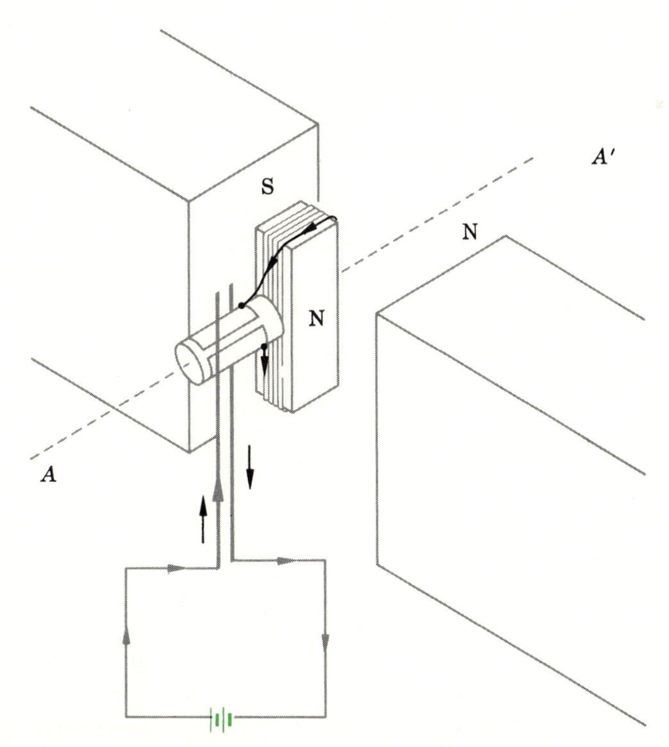

FIGURE 20.18

A simple dc motor. With the slip ring as shown, which way should the motor rotate?

There are various modifications of such a motor. Most motors consist of several loops wound with their planes through AA' but at various angles to each other. Each loop has current flowing through it for only a small portion of a cycle during the time when its orientation to the field is right for obtaining maximum torque. Such a motor gives a much more uniform torque than one could obtain from a single loop.

Some motors use electromagnets, while others make use of permanent magnets to produce the magnetic field. The exact way in which the magnet coils and rotating coil (or armature) are connected differs from motor to motor. Some motors run on both ac and dc voltage, while others run on only one or the other. The student is referred to other texts for a more complete description of these devices.

Counter (or Back) emf

Before leaving the subject of motors, we should point out that *the current through the motor is controlled chiefly by its* **counter emf,** an induced emf which will now be described. (The resistance of a good motor is usually quite low.) When the coil (or armature) rotates in the field of the permanent magnet, an emf is induced in it. The induced emf is in such a direction as to oppose the emf which is sending current through the coil. It is for this reason that it is called a back or counter emf. Since the resistance of a motor is usually quite small, the chief limitation on the current through it is caused by the counter emf. If the motor is overloaded, it will slow down and therefore draw more current from the source. (Why?) This increased current which flows through the overloaded motor may on occasion become large enough to burn it out.

FIGURE 20.19
A motor acts as though it were a resistance in series with a counter emf $\mathcal{E}$.

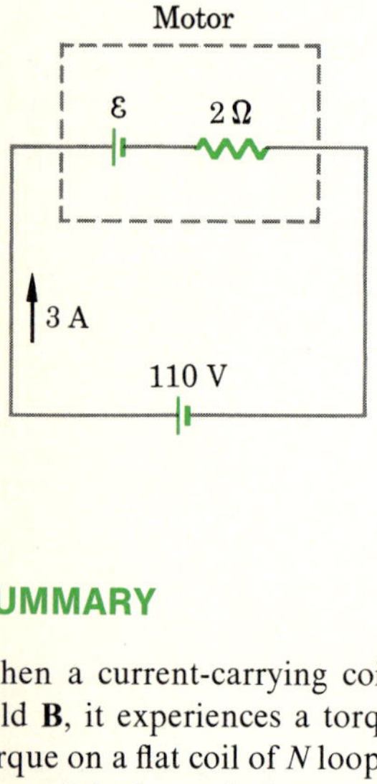

Illustration 20.4 A particular motor has a resistance of 2.0 Ω. It draws 3.0 A when operating normally on a 110-V line. How large is the counter emf it develops?

Reasoning The motor can be thought of as a battery in series with a resistance. Since the battery is to represent the counter emf, it must oppose the operating power source. The situation is shown in Fig. 20.19. Combining the batteries and writing Ohm's law gives

$$110\text{ V} - \mathcal{E} = (3\text{ A})(2\ \Omega)$$
$$\mathcal{E} = 104\text{ V}$$

SUMMARY

When a current-carrying coil is placed in a magnetic field **B**, it experiences a torque. The magnitude of the torque on a flat coil of N loops and area A is $ANIB \sin\theta$, where θ is the angle between **B** and a perpendicular to the area. The magnetic moment of a current loop has a magnitude $\mu = IA$. It is a vector and has the direction given by a modification of the right-hand rule. A magnetic field tends to align the magnetic moment vector along the field lines.

Ammeters and voltmeters make use of a galvanometer movement, which consists of a coil in a magnetic field. The coil rotates in proportion to the current through it. Ammeters have a low-resistance shunt in parallel with the movement. Voltmeters have a high resistance in series with the movement.

When the flux through a coil is changed, an emf is induced in the coil. Its magnitude is $N(\Delta\phi/\Delta t)$, where N is the number of loops on the coil. This is called Faraday's law. The direction of the current produced by the emf is given by Lenz's law. The induced emf is in such a direction that the flux generated by the current will tend to counterbalance the original change in flux through the loop.

If the current in one coil changes at a rate $\Delta I/\Delta t$, it can induce an emf in a nearby coil. The mutual inductance M of the two coils is defined by emf $= -M(\Delta I/\Delta t)$. When a current changes in a coil, an emf is induced in the same coil. This self-induced emf is given by $-L(\Delta I/\Delta t)$, where L is the self-inductance of the coil. Both M and L are measured in henrys (H).

When a wire of length d moves with speed v perpendicular to a magnetic field $\mathbf{B}$, an induced emf $= B_{\perp}vd$ is induced between the ends of the wire. Any time that a wire cuts through magnetic field lines, an induced emf will exist in it.

An ordinary alternating-voltage generator consists of at least one coil rotating in a magnetic field. The changing flux through the coil induces a sinusoidal emf in it. If the coil is rotating with frequency f, then the induced emf is given by $V = V_0 \sin(2\pi ft)$.

Since motors consist of a coil rotating in a magnetic field, they have an emf induced in the coil. This is a back emf which opposes the voltage source which runs the motor. Under normal operation, the current through a motor is mostly limited by this back emf.

MINIMUM LEARNING GOALS

Upon completion of this chapter you should be able to do the following:

1. Explain the major features of a meter movement. Tell how it is used to make an ammeter or a voltmeter.
2. Compute the shunt resistance needed to make a given movement into an ammeter of stated range.
3. Compute the series resistance needed to make a given movement into a voltmeter of stated range.
4. State which way a current-carrying coil will turn when placed in a given position in a magnetic field. Compute the torque on the coil when sufficient data are given.
5. Point out where the effective north and south pole regions are for a current-carrying loop. Explain what is meant by the magnetic moment vector for the current loop.
6. Give the direction of the induced emf in a coil caused by a nearby changing current or moving magnet. Describe qualitatively how the emf behaves as a function of time in simple experiments involving a nearby circuit or magnet. Relate your answers to Lenz's and Faraday's laws.
7. Make quantitative use of Faraday's law in simple situations such as those in Illustrations 20.2 and 20.3.
8. Define mutual and self-inductance. Explain their qualitative features in terms of Faraday's law.
9. Explain qualitatively why a wire cutting magnetic field lines should have an emf generated between its ends. Compute this emf in the case of a wire moving perpendicular to the field lines.
10. Sketch the details of a simple ac generator. Explain how it gives rise to an ac voltage. Sketch a graph of the voltage versus time.
11. Explain the meaning of back emf for a motor and why it depends upon the speed of the motor.

IMPORTANT TERMS AND PHRASES

You should be able to define or explain each of the following:

Meter movement
Shunt resistor
$\tau = ANIB \sin\theta$
Magnetic moment; $\mu = IA$
Faraday's law; emf $= -N(\Delta\phi/\Delta t)$
Lenz's law
Mutual and self-inductance
Motional emf $= B_{\perp}vd$
Ac voltage; $V = V_0 \sin(2\pi ft)$
Counter (or back) emf

QUESTIONS AND GUESSTIMATES

1. What design factors influence how sensitive a galvanometer movement will be?

2. Why does a "good" voltmeter make use of a more sensitive movement than a cheap instrument? In what way is it better?

3. Why does a "good" ammeter make use of a more sensitive movement than a cheap meter? In what respect is it better?

4. An inventor claims that he has an electric generator which runs a motor which in turn keeps the generator running. Additional current from the generator is then used to light bulbs, etc. What do you think of this idea? Why?

5. The law of conservation of energy leads us to conclude that the self-induced emf in a coil is an opposing emf. Explain why. Does this also apply to the induced emf in a motor? Could one devise a perpetual-motion machine if the induced emfs were not in the direction given by Lenz's law?

6. An overloaded motor will frequently blow a fuse before much damage is done. Explain what has happened.

7. Explain why a motor using permanent magnets will operate on direct but not on alternating current, while the same general motor using electromagnets will operate on both.

8. A long, straight wire carries a current along the top of a flat table. A rectangular loop of wire lies on the table as well. If the current in the long wire is shut off, in what direction will the induced current flow in the loop? Draw a diagram for several positions of the loop relative to the wire, showing in what direction the induced current will flow in the loop.

9. How can we ascertain whether the earth is moving through a uniform magnetic field? Could we tell if the earth were moving through the field rather than just carrying the field along with it?

10. A copper ring lies on a table. There is a hole through the table at the center of the ring. If a bar magnet is held vertically by its south pole high above the table and is then released so that it drops through the hole, describe the forces which act upon the magnet.

11. A very long copper pipe is oriented vertically. Describe the motion of a bar magnet dropped lengthwise down the pipe.

12. Is it possible to design an airplane or rocket which would use the interaction between the earth's magnetic field and an electric current as a propulsion mechanism?

13. The electron itself has a magnetic moment. How can this be justified qualitatively in terms of a spinning spherical model for the electron? Extend your argument to the neutron which, although uncharged, also possesses a magnetic moment.

14. A 6-V battery is placed in series with a knife switch, a 2-Ω resistor, and a 0.5-H self-inductance. The time taken for the current to rise to about two-thirds its maximum value after the switch is first closed is $L/R = 0.25$ s. Estimate the induced emf in the inductor when (*a*) the switch is first closed and (*b*) the switch is suddenly opened. (E)

15. Suppose the amount of charge one can place around the outside of a phonograph record by rubbing it with a dry cloth is 1×10^{-6} C. About how large a torque is exerted on the record as it rotates on a turntable in the earth's magnetic field? About how large a torque is needed to lift the record from the turntable? (E)

16. When a series of loops of wire swings past the north pole of a magnet, as shown in Fig. P20.1, **eddy currents** are induced in them. Show the direction of the currents at two different times, and show the force on the loops. Similar eddy currents occur in solid pieces of metal. Explain.

17. Induced currents are used to damp the motion of coils on very sensitive galvanometers. These coils are close to friction-free and swing for a long time if allowed to do so. However, when the two ends of the coil are connected together, the coil will stop swinging immediately. Why? This is called **magnetic damping.**

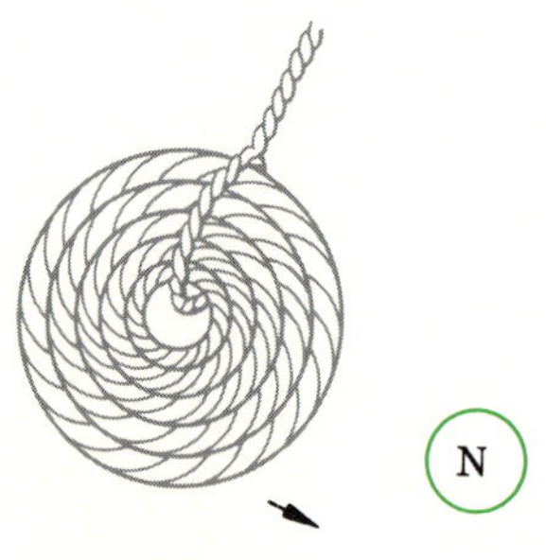

FIGURE P20.1

PROBLEMS

1. A certain galvanometer movement has a resistance of 40 Ω and deflects full scale for a voltage of 100 mV across its terminals. How can it be made into a 2-A ammeter?

2. If a meter movement deflects full scale to a current of 0.010 A and has a resistance of 50 Ω, how can it be made into a 5-A ammeter?

3. How can the meter movement described in Prob. 2 be made into a 20-V voltmeter?

4. How can the galvanometer of Prob. 1 be made into a 50-V voltmeter?

5.* How can the meter movement of Prob. 2 be made into an ammeter having two ranges, 10 and 1.0 A?

6.* How can the meter movement of Prob. 2 be made into a voltmeter having two ranges, 12 and 120 V?

7. A flat coil of wire has 30 loops of wire on it. Each loop has a radius of 2.0 cm. What is the magnetic moment of this coil when a current of 5.0 A flows through it?

8. The coil of Prob. 7 hangs in a magnetic field as shown in Fig. P20.2. (*a*) Find the torque on it. (*b*) Will it tend to rotate so as to increase or to decrease θ? The angle $\theta = 120°$.

9.* A circular coil with a 5.0-cm radius and with 20 loops of wire on it carries a current of 25 A. The coil lies on a horizontal table top. The current in it flows clockwise as seen from above. It is acted upon by the earth's magnetic field, which at that place is directed north and downward with a dip angle of 55°. The value of B is 0.85 G. (*a*) Find the torque on the coil due to B. (*b*) Which part of the coil tries to rise from the table top, its north, east, south or west side?

10.* Figure P20.3 shows a coil of wire (radius b) around a solenoid (radius a). The solenoid is actually much longer than shown. If the magnetic field in the solenoid is changing at a rate of 0.020 T each second, (*a*) find the induced emf in the outer coil. The outer coil has N loops on it while the solenoid has n loops per meter length. (*b*) If a bar of iron for which $\mu = 300\mu_0$ is placed inside the solenoid so as to fill it, what will the emf induced in the outer coil be?

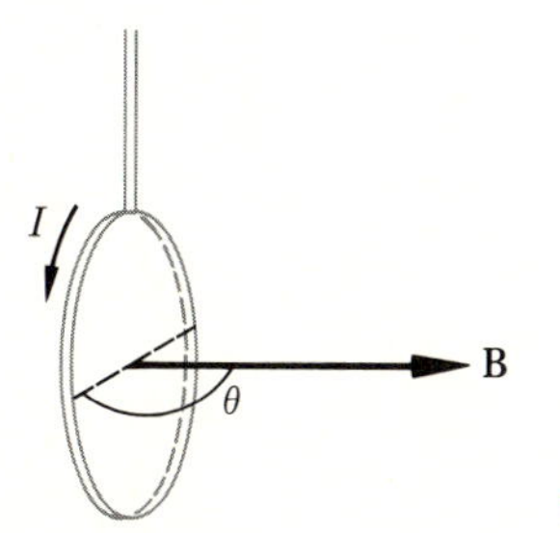

FIGURE P20.2

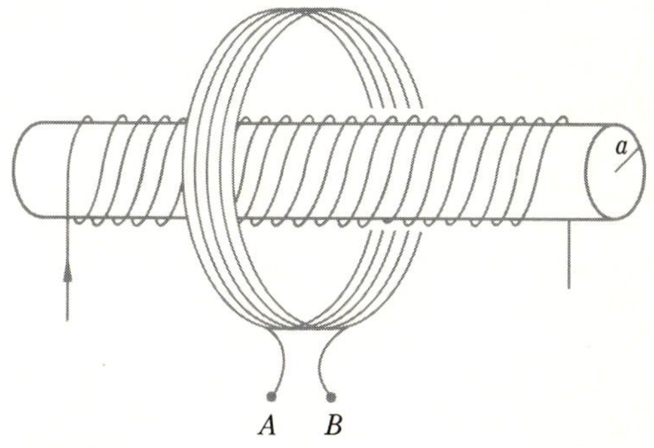

FIGURE P20.3

11.* The current in an air-core solenoid is increasing at a rate of 2.0 A/s. There are 10^6 turns of wire on the solenoid for each meter of its length, and its cross-sectional area is 2×10^{-4} m^2. A secondary coil of 10^4 turns is wound over the solenoid. How large an emf is induced in the secondary?

12. The average field produced within an iron-core toroid is 1.40 T. A secondary coil of 100 turns is wound on the toroid. If the cross-sectional area of the toroid is 0.50 cm^2, how large an average emf is induced in the secondary coil if the current is stopped in the toroid in 0.010 s?

13. A bar magnet can be moved in or out of a 100-turn coil into which it fits snugly. It is found that an average emf of 0.30 V is induced in the coil when the magnet is suddenly brought up and inserted in it in 0.10 s. If the cross-sectional area of the magnet is 2.0 cm^2, find the value of B in it.

14.** Two coils are wound tightly on the same iron core. The cross-sectional area of both is about 4.0 cm^2. When a current of 3.0 A flows in the primary coil, B is 0.20 Wb/cm^2. There are 100 turns on the secondary. (*a*) How large an emf is induced in the secondary if the current in the primary drops uniformly to zero in 0.050 s? (*b*) What is the mutual inductance of the coils?

15.** A long iron-core solenoid with 1000 turns has a cross-sectional area of 4.0 cm^2. When a current of 2.0 A flows in it, $B = 0.50$ T. (*a*) How large an emf is induced in it if the current is turned off in 0.10 s? (*b*) How large is its self-inductance?

16. A metal airplane flies parallel to the ground and toward the west at a speed of 200 m/s. If the downward vertical component of the earth's field is 0.80 G, (*a*) what is the potential difference between the tips of the wings, which are 25 m apart? (*b*) Which wing tip is positive, the north or the south? (*c*) Can this voltage be measured? (*d*) If so, how?

17. An engineer decides to light the lights in a train station by utilizing the emf induced in the axles of the trains running on the tracks. Upon assuming the vertical component of the earth's field to be 0.80×10^{-4} Wb/m^2 and the tracks to be 1.5 m apart, (*a*) how large an emf is produced between the tracks by a train traveling 30 m/s? (*b*) Could this voltage be utilized on the moving train? (*c*) If so, how?

18. If the 200-turn coil in a generator has an area of 500 cm^2 and rotates in a field with $B = 0.60$ Wb/m^2, how fast must the coil be rotating in order to generate a maximum voltage of 150 V?

19. The coil of a motor has a resistance of 5.0 Ω. When the motor is turning at rated speed, it draws a current of 2.0 A from 120 V. (*a*) How large is the counter emf of the motor? (*b*) How much current would it draw if the coil were stopped from rotating?

20. Very large motors take nearly a minute to get up to speed after they are turned on. One such motor has a resistance of 0.50 Ω and draws 8.0 A on 120 V. What resistance (the starting resistance) must be placed in series with the motor if it is not to draw more than 30 A when first turned on? (This resistance is later removed, of course.)

ALTERNATING CURRENTS AND REACTIVE CIRCUITS

21

In the past few chapters we have been concerned mainly with direct currents, i.e., currents which flow continuously in one direction. We saw in the last chapter, however, that a voltage source of alternating polarity is obtained by rotating a coil in a magnetic field. An alternating-voltage source such as this gives rise to alternating currents and these, too, are of great importance. We shall see in this chapter how such currents behave when sent through resistances, capacitances, and inductances.

21.1 CHARGING AND DISCHARGING A CAPACITOR

Let us begin our study of varying current circuits by examining the simple circuit shown in Fig. 21.1*a*. Suppose the switch is open initially and that no charge exists on the capacitor. We wish to know what will happen when the switch is suddenly closed.

The battery will try to send current around the circuit in a clockwise direction. Since there is initially no charge on the capacitor, the current i will be limited only by the resistor R. Therefore, just after the switch is closed (at $t = 0$), the current will be $i_0 = V_0/R$, as shown in part *b*. But as time goes on, the capacitor will become charged. The current to the capacitor will decrease. The current must drop to zero when the capacitor is fully charged. The exact way the current behaves in this circuit is shown in Fig. 21.1*b*.

The curve followed by the current in part *b* is called an **exponential decay curve.** Analysis of the circuit shown in part *a* shows that the current drops to a value of $0.3679i_0$ in a time equal to the product of R and C. (It is an interesting problem in unit manipulation to show that ohm-farads are equivalent to seconds.) *We call the product RC the* **time constant** *of this circuit. It is the time in seconds required for the current to decrease to about* 0.37 *times its initial value.*

RC Time Constant

As the current flows in the circuit, the capacitor charges. When fully charged, the capacitor's charge is $q_0 = CV_0$. The charge q on it as a function of time is shown in part *c* of the figure. Notice that here the time constant measures the time in seconds taken for the capacitor to become about two-thirds charged. As we see, *the time constant is a rough measure of the time taken to charge a capacitor.*

If a charged capacitor C is connected directly across a resistance R, the capacitor will discharge through it. Assuming the initial potential difference between the capacitor terminals to be V_0, the current flowing from the capacitor as it discharges will follow the curve shown in Fig. 21.1*b*. The capacitor discharge current behaves the same as the charging current. It turns out that *the capacitor is about two-thirds discharged in one time constant.* Here, too, *the time constant RC is a rough measure of the time required for the process.*

Illustration 21.1 In most TV sets a capacitor is charged to a potential

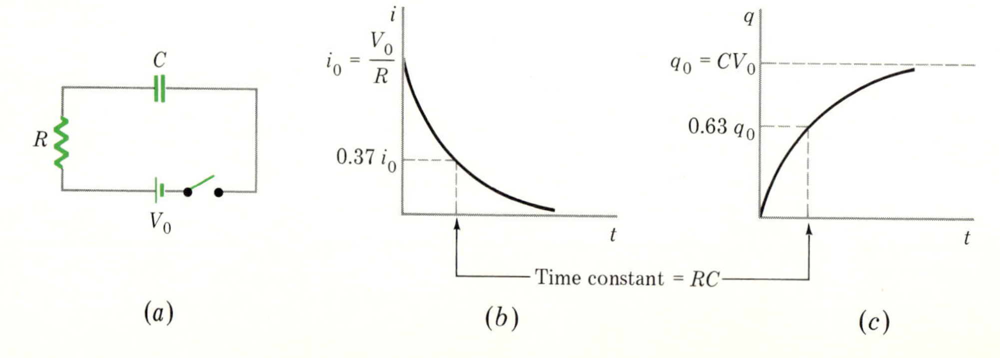

FIGURE 21.1
The time constant *RC* is a convenient measure of the time taken for a capacitor to charge or discharge.

difference of about 20,000 V. As a safety measure, a resistor is connected across its terminals so it will discharge after the set has been turned off. Suppose this so-called bleeder resistor is $10^6\ \Omega$ and $C = 10\ \mu F$. About how long must one wait after turning off the set before it would be safe to touch the capacitor?

Reasoning The time constant for this *RC* circuit is *RC*. Therefore,

$$\text{Time constant} = (10^6\ \Omega)(10^{-5}\ \text{F}) = 10\ \text{s}$$

In 10 s the capacitor would lose about two-thirds of its charge. At a time 10 times this, the capacitor should be essentially discharged. Therefore, one should wait about 100 s before touching the capacitor in this case.

A more quantitative approach to this problem could be had if one knows *a peculiar feature of the exponential-decay-type curve. In such a curve, the quantity decreases by a factor of* 0.3679 $\approx$ 0.37 *during each time constant.* Therefore $i = 0.37i_0$ at $t = RC$. At $t = 2RC$, one has $i = (0.37)(0.37)i_0$. At $t = 3RC$, one has $i = (0.37)^3 i_0$, and so on. After 10 time constants, $i = i_0(0.37)^{10}$, which is $4.5 \times 10^{-5} i_0$. At the time $t = 10RC$, the current and charge have been reduced to 4.5×10^{-5} times their initial values.

21.2 AC QUANTITIES; RMS VALUES

Perhaps the most widely encountered type of varying current circuit is what we refer to as ac (for alternating-current) circuits. You will recall that a coil which rotates in a magnetic field gives rise to a sinusoidal emf. This type of emf causes an alternating current such as that shown in Fig. 21.2. It is this type of voltage and current which power companies furnish to their customers. *All sinusoidal-type currents and voltages have an average value of zero over one or more complete cycles.* Even if the ac current flowing through a wire causes the wire to become white hot (as in an incandescent light bulb), the average current in the wire is zero. This fact may be seen quite easily as we shall now show.

Average Value of AC Quantities

The alternating current in Fig. 21.2 is positive as much as it is negative. To find its average value, we must add together its values at various times

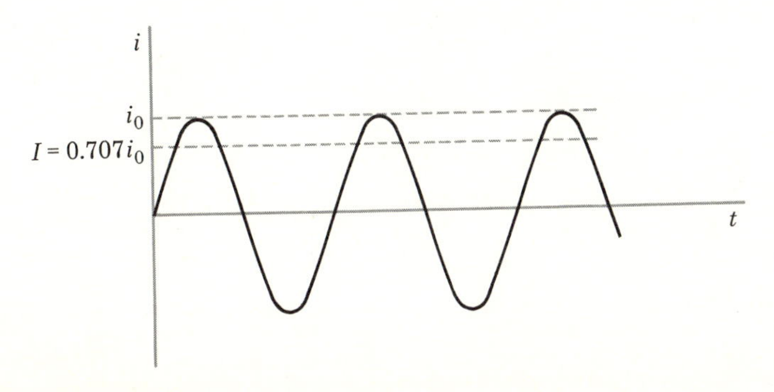

FIGURE 21.2

For an alternating current, the effective, or rms, current is $i_0/\sqrt{2}$.

and divide our sum by the number of values added. Clearly, there will be as many negative as positive values for the current. When added together, they will give zero and so the average current is zero. A similar line of reasoning shows that the average value of an alternating voltage is likewise zero.

Except for such applications as electroplating and battery charging, *an alternating current can do useful work even though its average value is zero.* For example, in using electricity to produce heat in a resistor, as in an electric stove, the fact that the current is reversing periodically is of no importance. We are actually interested only in the heat developed, and, as we have seen before, this is determined by the electrical power delivered to the resistor, i^2R. (Notice that the current is squared in this expression, and so whether it is positive or negative is of no importance.)

Actually, then, *we are more interested in the average power delivered to a resistor than we are in the average current through it.* We therefore wish to find the average value of i^2R. But since R is constant, we really need to know the average value of i^2. The average value of i^2, the **mean-square current,** is the average of a group of positive numbers, since i^2 is always positive. For a sinusoidally varying current (or voltage) the average value of i^2 (or v^2) is one-half its maximum value, $\frac{1}{2}i_0^2$. Or, after taking the square root, the **root-mean-square** (rms) **current** is just

RMS Current and Voltage

$$I \equiv i_{\text{rms}} = \frac{i_0}{\sqrt{2}} = 0.707 i_0$$

where i_0 is the peak current shown in Fig. 21.2. *The rms voltage is*

$$V \equiv v_{\text{rms}} = 0.707 v_0$$

where v_0 is the maximum voltage during the sinusoidal cycle. The factor 0.707 is $1/\sqrt{2}$. *Frequently, the rms values are called* **effective** *values.*

Most ac voltmeters and ammeters read the effective voltage or current. From time to time one encounters a meter calibrated to read the peak voltage v_0 or peak current i_0. Of course, most dc (direct- or steady-current) meters read average values, and so they will not deflect when connected into ac systems. From the way in which the effective, or rms, current is defined, the power loss in a resistor is merely I^2R, where I is the rms value. Of course, in a dc system, the rms current, average current, and instantaneous current are all equal.*

AC Power Loss

21.3 RESISTANCE CIRCUIT

We shall introduce the subject of ac circuits by considering in turn three different circuit elements connected in series with an alternating-voltage

* We shall always use V and I for rms values, the meter readings. The letters v and i will be used to indicate the instantaneous values of voltage and current.

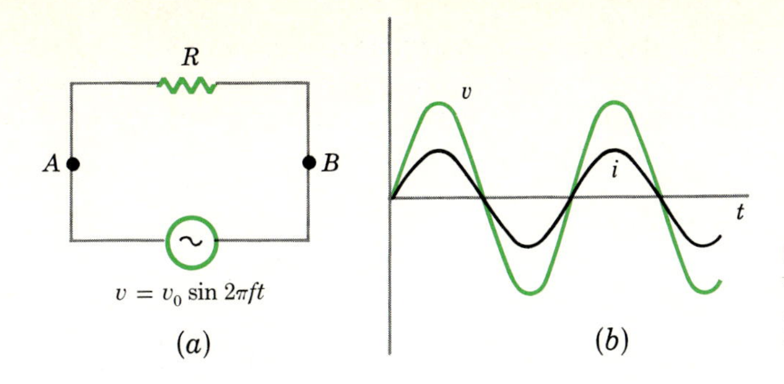

FIGURE 21.3

The current in a resistor is in phase with the voltage across the terminals.

source. First let us consider the simple resistance circuit shown in Fig. 21.3*a*. Ohm's law tells us at once that the voltage difference from A to B is just iR. When the voltage is a maximum, the current also will have its maximum value. When the voltage is zero, the current will be zero as well. This behavior is shown graphically in Fig. 21.3*b*. We say that the current and voltage are *in phase* when they are thus in step.

As outlined in the previous section, the power loss in the resistor is I^2R. In this particular case, where only a resistance is present, $I = V/R$, and so the power loss could also be written as IV, where I and V are the rms meter readings. We shall see in the next sections that there is no average power loss in pure capacitors or inductors. *All power losses in simple ac circuits occur in resistors.*

21.4 CAPACITANCE CIRCUIT

Let us now consider the capacitance circuit shown in Fig. 21.4*a*. We know that the potential from A to B is equal to the voltage of the ac source, $v_0 \sin 2\pi ft$. However, we recall that the potential across a capacitor is given by q/C. Hence we have

$$\frac{q}{C} = v_0 \sin 2\pi ft$$

Since C is a constant, we see that the charge on the capacitor will

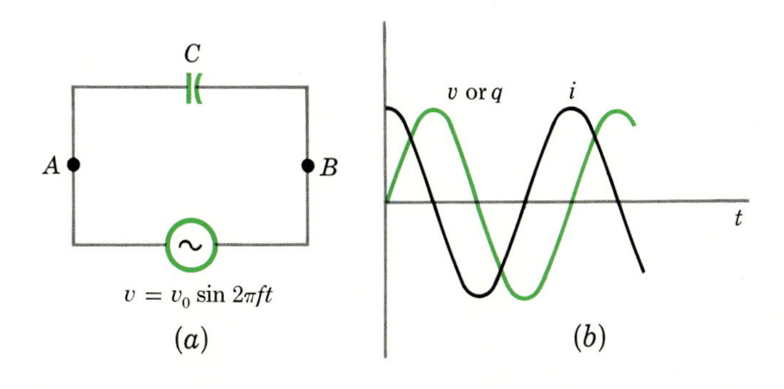

FIGURE 21.4

The voltage across a capacitor reaches its maximum one-quarter cycle later than the current. Both the voltage and the charge on the capacitor are in phase.

oscillate in value in the same way as the voltage of the source. When the source voltage is positive, the capacitor is positively charged. When the voltage is negative, the capacitor is negatively charged. Moreover, the charge reaches its maximum at the same time as the voltage becomes maximum. Hence the graph of voltage versus time will look the same as a graph of charge versus time as shown in Fig. 21.4*b*.

To see how the current in the circuit varies with time, we recall that current is the rate of change of charge. Hence, the rate of change of charge on the capacitor, $\Delta q/\Delta t$, is the current in the circuit. But $\Delta q/\Delta t$ is merely the slope* of the curve one obtains when q is plotted against t. This curve was plotted in Fig. 21.4*b*, and so we need only plot its slope and we shall have a graph of the current in the circuit. This is shown as the black curve in Fig. 21.4*b*.

Notice now that the current is not in step with the voltage. While the current has its maximum at $t = 0$, the voltage does not reach its maximum until $\frac{1}{4}$ cycle later. We say that the *current leads the voltage* by 90° or $\frac{1}{4}$ cycle in such a capacitive circuit. We shall now see what this implies for the power dissipated in the circuit.

During the time interval from e to f in Fig. 21.5, the current is coming out of the positive terminal of the power source (since both v and i are positive), and so the source is furnishing power. However, during the portion fg of the cycle, current is actually going in the direction opposite to which the source would like it to go. The source is being charged rather than discharged. In the portion gh of the cycle, the source is once again furnishing energy, while in section hj it is being charged once more.

In this circuit it is apparent that the power source is being charged, i.e., energy is being fed back into it, as much as it is being discharged, i.e., furnishing energy to the circuit. Hence, the average power drawn from the source is zero. We see, therefore, that *the capacitor,* unlike a resistor, *does not consume energy*. Energy is stored within the capacitor by the source during part of the cycle, but the capacitor returns this energy to the source as it discharges. You may recall that we showed in Chap. 17 that a capacitor could store energy and that the amount stored was $\frac{1}{2}QV$.

FIGURE 21.5
The capacitor is drawing energy from the voltage source in intervals ef, gh, and jk, while during intervals fg and hj it is pushing energy back into the voltage source.

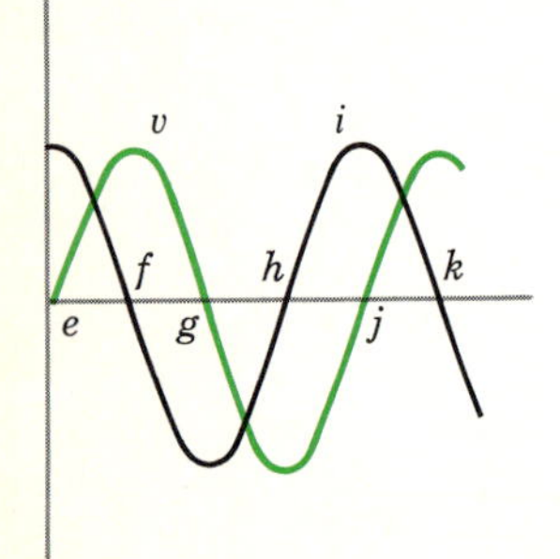

It is also of interest to see how a capacitor behaves as the frequency of the source changes. At very low frequencies very little current will flow in the circuit. When the frequency is zero, i.e., dc conditions, the current is always zero except at the first instant when the capacitor was being charged. If the source frequency is, say, 1 cps, the capacitor will charge and discharge once each second and so some current will flow, but not much.

However, if the frequency of the source is 1 million cycles per second, the capacitor will charge and discharge a million times in 1 s. Since current is the rate of change of charge, the current will now be quite large because the charge is changing rapidly.

---

*The **slope** of a curve is exactly what the word implies. It is the rate at which the curve is rising. If the curve is flat and horizontal, it is not rising at all and so its slope is zero. If the curve is rising rapidly, its slope is large. If the curve is decreasing, i.e., going down, it is, in effect, rising at a negative rate. Its slope is therefore negative.

We therefore see that a capacitor keeps the current extremely small in a circuit such as that of Fig. 21.4*a* when the frequency is low. On the other hand, it does not impede the flow of current at high frequencies. *A capacitor acts much like a large resistance at low frequencies and like a small resistance at high frequencies,* but causes no power loss due to heating. We would like to speak of its "resistance," but this terminology would lead to confusion. We therefore designate its ability to impede the flow of current as its **reactance.** The reactance of a capacitor may be shown by means of calculus to be related to the rms current and voltage by an Ohm's law type equation, namely

$$V = IX_C \tag{21.1}$$

Ohm's Law Form for a Capacitor

where X_C, the reactance of the capacitor in ohms, is given by

$$X_C = \frac{1}{2\pi fC}$$

The expression for the reactance of the capacitor (the capacitive reactance) is reasonable. When the frequency f is small or when the capacity of the capacitor is small, X_c will be large. And as we surmised previously, X_c will be small when the frequency is high.

Illustration 21.2 Suppose in the circuit of Fig. 21.4 that $C = 0.4\ \mu\text{F}$ and that $v = 100 \sin (2\pi ft)$ volts with $f = 20$ Hz. Find the rms current in the circuit. Repeat if $f = 2 \times 10^6$ Hz. (Ordinary power-line frequency is 60 Hz.)

Reasoning We know that $V = IX_C$, where V and I are the rms values. Since $V = 0.707\ v_0$, and since $v_0 = 100$ V, we see that $V = 70.7$ V. But $X_C = 1/2\pi fC$. Substituting $f = 20$ Hz and $C = 0.4 \times 10^{-6}$ F gives $X_C = 19{,}900\ \Omega$. Therefore,

$$I = \frac{V}{X_C} = 0.0036\ \text{A}$$

At a frequency of 2×10^6 Hz, the same procedure yields

$$X_C = 0.199\ \Omega \qquad \text{and} \qquad I = 355\ \text{A}$$

Notice that what we have said is very true; *at high frequencies a capacitance impedes the current much less than at low frequencies.*

21.5 INDUCTANCE CIRCUIT

The behavior of the simple self-inductance circuit shown in Fig. 21.6*a* can be analyzed in a manner similar to that used for the capacitance circuit.* First we notice that the voltage difference between points A and B is equal to the source voltage, $v_0 \sin 2\pi ft$. However, it is also equal to the voltage induced in the inductor by the changing current and flux in the circuit. In the last chapter we found this voltage to be $L\,\Delta i/\Delta t$. Equating these two voltages yields

$$v_0 \sin 2\pi ft = L\frac{\Delta i}{\Delta t}$$

Since L is a constant, we see at once that the source voltage is proportional to the value of the rate of change of current in the circuit. But since the rate of change of current is the slope of the graph of current versus time, we have a way of finding one curve if the other is known. The voltage and current curves in the inductance circuit are shown in Fig. 21.6*b*. Notice that the voltage graph is indeed proportional to the value of the slope of the current graph.

Here, too, the current and voltage are 90° out of phase. In this case, though, the voltage is 90° ahead of the current. We say that *the voltage leads the current by* 90° in this case.

Once again, we can use the same reasoning as in the last section to show that *the inductor consumes no energy on the average.* Although the source stores energy in the inductor during part of the cycle, the inductor gives it back to the source in a later portion of the cycle. It is not difficult to show that the energy stored in an inductor is $\frac{1}{2}Li^2$. You would do well to examine Fig. 21.6*b* and ascertain during which part of the cycle the source is losing energy and during which part energy is being returned to it.

The general behavior of the inductance circuit as the source frequency is changed is also of interest. We know, of course, that the inductor will always try to counterbalance or impede the change in current. In fact, the induced emf in it is $-L\,\Delta i/\Delta t$ and is therefore proportional to the rate of change of

*Notice that the symbol ⌇ is used to represent an inductance coil.

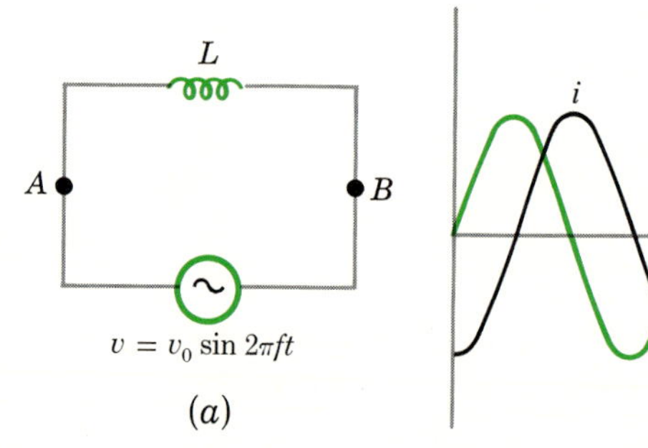

FIGURE 21.6
The voltage across the inductance leads the current through it by 90°, or one-quarter cycle. Notice the symbol used for inductance.

the current. As a result, *when the current is changing very slowly, the inductor will not have much effect. However, at very high frequencies when the current is trying to change rapidly, the impeding effect of the inductor will be very large. We represent this impeding effect by the* **inductive reactance** X_L.

The inductive reactance, X_L, is related to the rms current through it and voltage across it by

$$V = IX_L \tag{21.2}$$

Ohm's Law Form for an Inductance

where $X_L = 2\pi fL$ is the inductive reactance. As we expected, the reactance of the inductor is large at high frequencies and small at low frequencies. It is measured in ohms.

Notice that *capacitors and inductors behave oppositely as a function of frequency. The current-impeding effect of capacitors is large at low frequencies and small at high frequencies. The reverse is true for inductors.* Of course, *the impedance effect of a resistor is independent of frequency.*

Illustration 21.3 Suppose the inductance coil in Fig. 21.6 has a value of 15 mH. The source voltage, as read by an ac meter, is 40 V, and its frequency is 60 Hz. Find the current which flows through it. Repeat for a frequency of 6×10^5 Hz.

Reasoning We make use of the Ohm's law form, $V = IX_L$. In the 60-Hz case, $V = 40$ V and

$$X_L = 2\pi fL = 5.65\ \Omega$$

Therefore, $I = 40\ \text{V}/5.65\ \Omega = 7.1$ A.

At a frequency of 6×10^5 Hz, the value for X_L becomes $5.7 \times 10^4\ \Omega$. Then we find that $I = 7.1 \times 10^{-4}$ A. Notice how very much larger the inductor's impeding effect is at high frequencies than at low frequencies.

21.6 COMBINED *LCR* CIRCUIT

We shall briefly discuss the series circuit illustrated in Fig. 21.7*a*, which contains all three of the elements discussed in previous sections. The equivalent of Ohm's law for this circuit is

$$V = IZ \tag{21.3}$$

where

$$Z = \sqrt{R^2 + (X_L - X_C)^2}$$

Ohm's Law Form for a Series *RCL* Circuit

The quantity Z is called the **impedance** *of the circuit. Its units are ohms.* It is easily seen that Eq. (21.3) reduces to the forms given in the previous sections if all but one impeding element is zero.

FIGURE 21.7
Since the impedance for the circuit shown in (*a*) is $Z = \sqrt{R^2 + (X_L - X_C)^2}$, we may think of Z as being the hypotenuse of the right triangle shown in (*b*). The angle θ is the phase difference between the current and the voltage.

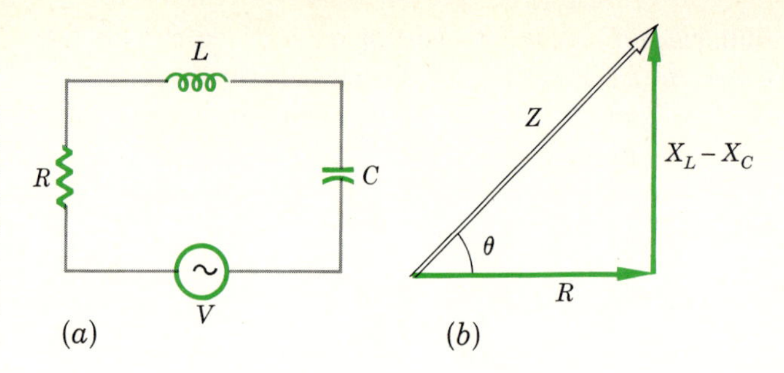

The current in the circuit of Fig. 21.7*a* is not usually in phase with the voltage. It is possible to find the angle θ between the current and voltage by means of a vector diagram. This diagram is shown in Fig. 21.7*b*. If X_C and R are zero, $\theta = 90°$ and so the current and voltage are 90° out of phase, as we saw in the previous section that they should be. You should convince yourself that the vector diagram gives the proper value for θ when both X_L and X_C are zero.

The power consumed by the circuit shown in Fig. 21.7 is I^2R. This follows from our previous discussions of the zero power loss in pure inductors and capacitors. Since the voltage drop across the resistor will not usually be the same as that of the source, it is clear that the power loss in the circuit cannot be equal to VI, as it was for dc circuits. It turns out that the correct expression for the power loss is

Power Loss in an AC Circuit

$$\text{Power} = VI \cos\theta \tag{21.4}$$

where θ is the phase angle between the voltage and the current shown in Fig. 21.7*b*. The factor $\cos\theta$ is often called the **power factor**. Notice that $\theta = 0$ in a pure resistance circuit. For a resistor, the power factor is unity. But for a pure capacitor or inductor, the power factor is zero. Why?

Illustration 21.4 Suppose in Fig. 21.7*a* that the voltage source has an rms value of 50 V and a frequency of 600 Hz. Suppose further that $R = 20\ \Omega$, $C = 10.0\ \mu\text{F}$, and $L = 4.0$ mH. Find (*a*) the current in the circuit and (*b*) the voltmeter readings across R, C, and L individually.

Reasoning We shall make use of $V = IZ$. Let us therefore find X_L and X_C at this frequency, 600 Hz. We have

$$X_L = 2\pi fL = 15.1\ \Omega \qquad \text{and} \qquad X_C = \frac{1}{2\pi fC} = 26.5\ \Omega$$

Then we find that

$$Z = \sqrt{(20)^2 + (15.1 - 26.5)^2} = 23.0\ \Omega$$

Now, using $I = V/Z$, we find $I = 2.17$ A.

To find the voltage drop across R, we use $V_R = IR$ and note that $I = 2.17$ A. Therefore,

$$V_R = 43.4 \text{ V}$$

The voltage drop across the inductance is given by $V_L = IX_L$ to be

$$V_L = (2.17)(15.1) = 32.8 \text{ V}$$

Similarly, we see that

$$V_C = IX_C = 57.5 \text{ V}$$

Notice that the potential difference across the capacitor is larger than the source voltage. This points out *a peculiar feature of ac circuits. The sum of the voltmeter readings around a closed circuit is not zero; voltages don't add properly if the rms voltage readings are used.* This fact is a result of the average character of the rms readings. They do not represent the instantaneous voltages, which can be either positive or negative. These instantaneous voltages do add properly. The rms voltages, however, are always positive by definition. Clearly, they cannot add to give zero. Kirchhoff's loop rule does not apply to them.

21.7 ELECTRICAL RESONANCE

Ac circuits which contain both capacitance and inductance show an important resonance phenomenon. To illustrate this fact, consider the series circuit shown in Fig. 21.8*a*. We know that the current in this circuit, which has no resistance, is given by

$$I = \frac{\text{V}}{Z} = \frac{\text{V}}{X_L - X_C}$$

Notice that *when $X_L = X_C$, the current in the circuit should become infinite.*

It is easy to obtain the condition $X_L - X_C = 0$ because X_L increases with frequency while X_C decreases with frequency. Figure 21.8*b* shows how these quantities vary for the C and L values given in this circuit. We see that the impedance becomes zero at $f = 4500$ Hz in this case. This frequency, *the frequency at which $X_L = X_C$, is called the* **resonance frequency** *of the circuit and we denote it by f_0.* Since $X_L = 2\pi fL$ and $X_C = 1/2\pi fC$, we have at resonance that

LC Resonance Frequency

$$2\pi f_0 L = \frac{1}{2\pi f_0 C}$$

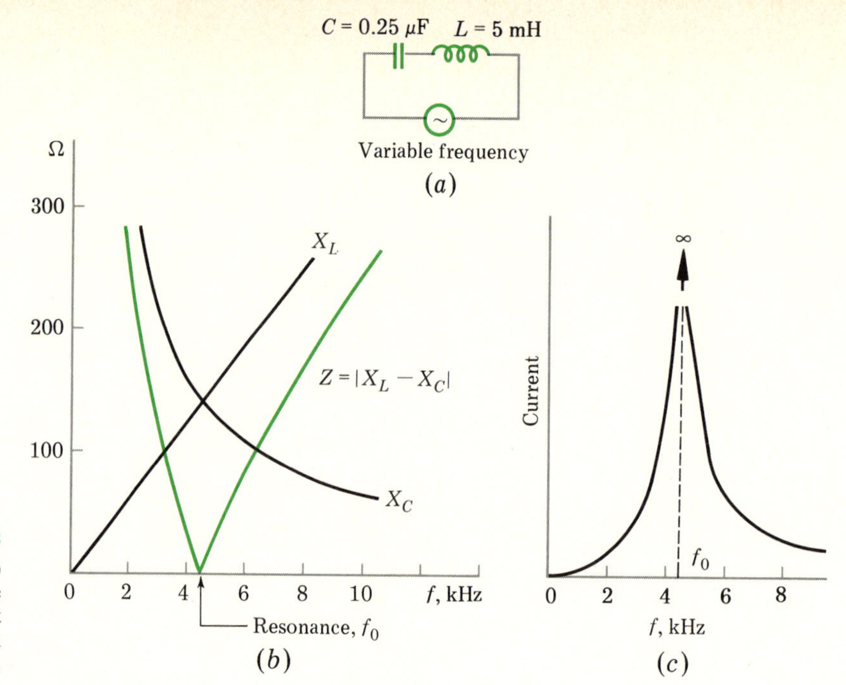

FIGURE 21.8
As the source frequency in (*a*) is changed, X_L and X_C change as shown in (*b*). The current in the circuit varies as shown in (*c*).

from which the resonance frequency is found to be

$$f_0 = \frac{1}{2\pi}\sqrt{\frac{1}{LC}} \tag{21.5}$$

Figure 21.8*c* shows how the current in this circuit varies as the oscillator frequency is changed. (Of course, the voltage of the oscillator must be kept the same for all frequencies.) As we see, *the current peaks sharply at the resonance frequency. In practical circuits the peak would be finite rather than infinite because all wires have some resistance.* Even so, the effect is very dramatic and has important applications, as we shall see in the next chapter.

We can understand electrical resonance better if we recognize that it is much like mechanical resonance. You know that mechanical systems often have a natural frequency at which they vibrate. If pushed with this frequency, they will vibrate widely; they resonate.

A simple *LC* circuit also has a natural frequency of vibration. Let us explore this analogy between resonance in electrical and mechanical systems. For this purpose, consider the *LC* circuit and child on a swing shown in Fig. 21.9. Suppose at the starting instant the current in the circuit is zero and that the child on the swing is at its highest position. The charge on the capacitor is q_0 and an energy $\frac{1}{2}(q_0^2/C)$ is stored in it. By analogy with this, the child on the swing possesses gravitational PE.

We know that in the electrical system the capacitor will begin to discharge through the inductor. The current will rise rather slowly because the

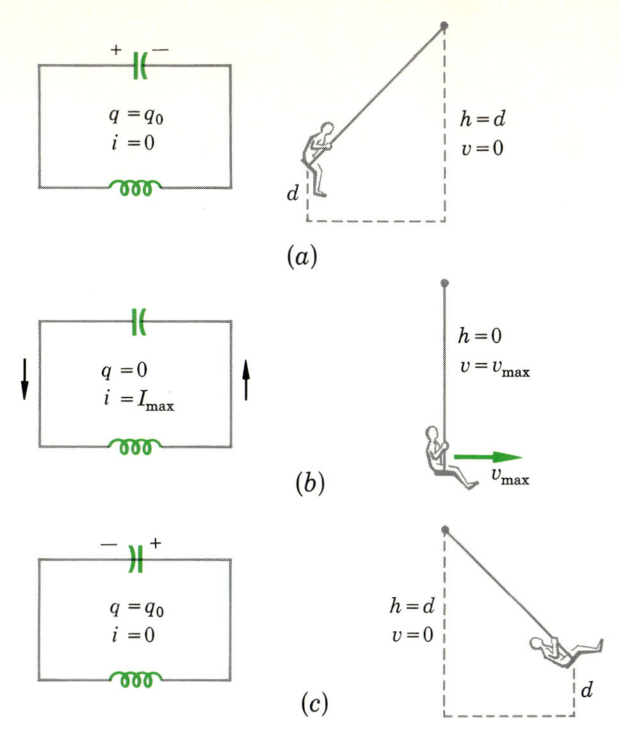

FIGURE 21.9

Just as the energy of the swing continually oscillates between potential and kinetic, the energy of the circuit is alternately stored in the charged capacitor or the current-carrying inductor.

inductor opposes any change in current. Similarly, the child on the swing will begin to pick up speed as the inertia of the swing system is overcome by the accelerating forces acting upon it. Both the child and the capacitor lose their PE. Once the swing has reached the bottom of its path, the PE has all been changed to KE. Similarly with the circuit: once the capacitor has lost all its charge, the current is flowing strongly in the circuit and the original energy is now stored in the inductor. Its value is $\frac{1}{2}Li^2$. This situation is shown in part *b* of Fig. 21.9.

Of course the child on the swing will not stop at the bottom of the path. The system's inertia will keep it moving until it comes to rest in the position shown in part *c* of Fig. 21.9. Now its energy is all PE once again. Much the same thing happens in the electrical circuit. The inductance, having inertia of a sort, opposes any change in current, and so the current does not stop at once. By the time the current finally stops, the capacitor is fully charged again, as in part *c*. These processes repeat over and over again.

Clearly, then, the electrical circuit undergoes an energy interchange much like the child and swing. The child's energy alternates between potential and kinetic, while the energy in the circuit is alternately stored in the capacitor and inductor. Both systems would oscillate back and forth forever if there were no energy losses. In the case of the swing, friction losses eventually damp out the oscillation. In the electrical case, resistive effects cause some of the energy to be lost, and so the oscillation slowly damps down in amplitude.

The analogy can be extended further. Both the child and the circuit possess natural resonance frequencies for their motion. The swing system constitutes a pendulum; we have already computed its natural frequency of vibration in a previous chapter. In the case of the circuit, its natural resonance frequency is the resonant frequency computed in Eq. (21.5).

If we wish to cause the child to swing very high, we must push on the swing at just the proper time and with the same frequency as the resonant frequency of the swing. We have seen that a very large current could be built up in the LC circuit if the oscillator "pushed" on the circuit at its resonant frequency. Hence, even the resonance behavior of the two systems is quite similar. It will be shown in the next chapter that the LC resonant circuit discussed here forms an integral part of any radio or TV receiver.

21.8 THE TRANSFORMER AND POWER TRANSMISSION

A transformer is a mutual inductance designed to change a given ac voltage into a larger or smaller ac voltage. It consists of two coils of wire wound on an iron yoke, somewhat like the system shown in Fig. 21.10. The primary coil has impressed across it an ac voltage v_P. This causes a current to flow through the primary coil, and this current gives rise to flux in the iron, as shown.

Since the flux follows the iron yoke, it passes through the second coil as well. Of course, the flux is changing all the time, because v_P varies sinusoidally. This varying flux through the secondary induces a sinusoidal emf of magnitude v_S in the secondary.

Most transformers have very little resistance in their wires. Hence, the current through the primary is impeded only by the inductance of the primary itself. The behavior of a circuit consisting of an inductor connected directly to an alternating-voltage source was discussed in detail in Sec. 21.5. For that circuit, we equated the driving voltage to the induced emf in the inductor. In the present case we have from Faraday's law

$$v_P = N_P \frac{\Delta\phi}{\Delta t}$$

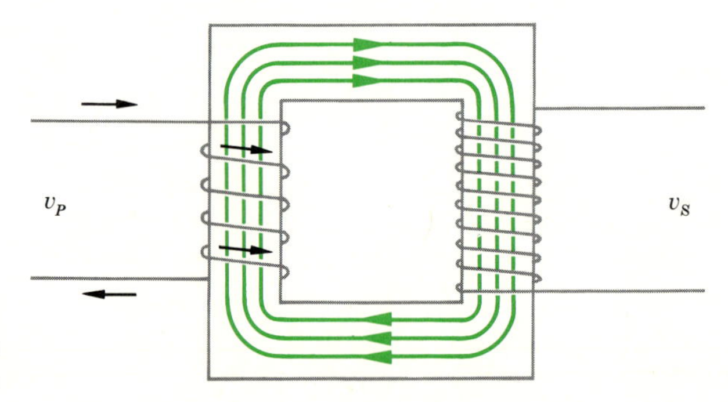

FIGURE 21.10
An iron-core step-up transformer.

where N_P is the number of turns on the primary coil and $\Delta\phi/\Delta t$ is the rate of change of flux through it. However, since the flux follows the iron yoke, all the flux will go through the secondary coil as well. The induced emf in the secondary is therefore

$$v_S = N_S \frac{\Delta\phi}{\Delta t}$$

Division of this expression by the equation for the voltage across the primary coil yields

$$\frac{v_S}{v_P} = \frac{N_S}{N_P}$$

However, the rms values are simply $0.707v_0$, and so we can rewrite this equation in terms of V_S and V_P:

$$\frac{V_S}{V_P} = \frac{N_S}{N_P} \tag{21.6}$$

Transformer Equation

This is the transformer equation, and it tells us how the secondary voltage is related to the primary voltage. The two voltages are in the same ratio as the number of turns on the coils. When N_S is larger than N_P, V_S will be larger than V_P. This is called a **step-up transformer** since the voltage is increased by it. The reverse case is called a **step-down transformer**.

If the secondary circuit is not closed, current cannot flow in it. Hence, there is no power loss in the secondary coil when it is not in use. Moreover, we showed in Sec. 21.5 that there was also no power loss in an inductor which has no resistance. This fact makes it possible for the power company to keep their transformers running throughout a city even when no one is using the electricity they are providing. The transformers themselves consume very little energy.

However, if current is drawn from the secondary, to run a heater, for example, energy is being consumed by the heater. This energy must be fed into the primary of the transformer so that it can be delivered to the secondary. Under these conditions, the loss in power at the secondary causes the primary to act as though it had resistance.

There are many uses for transformers. Every power-line-operated radio and TV set contains one or more of them. It is necessary to transform the 120-V house-line voltage to about 5 V in order to operate the electronic components. In addition, the TV picture tube requires a voltage about 100 times as large as the line voltage. A transformer is needed to provide this.

Another use of transformers has to do with power transmission. Many power companies provide power to cities which are perhaps 100 km from the generators. This proves to be quite a problem. Suppose that in a city of 100,000 people each person is using 120 W of power. This would be the equivalent of one or two lighted light bulbs for each person. The power consumed is (120)(100,000) W, and at a voltage of 120 V we have

$$\begin{aligned} \text{Total power} &= VI \\ (120)(100{,}000) &= 120I \\ I &= 100{,}000 \text{ A} \end{aligned}$$

where we have assumed the power factor to be unity.

Since an ordinary house wire can safely carry only about 30 A without overheating, the power company would need the equivalent of about 3000 of these wires to carry power to the city. Although this is not impossible, the cost of the copper alone would be tremendous. The power companies get around this difficulty quite nicely by noticing that the important quantity is VI and not I alone. They therefore choose to transmit power over long distances at very high voltages. In the above example, if V had been 100,000 V, we would have

$$(120)(100{,}000) = 100{,}000\, I$$

or

$$I = 120 \text{ A}$$

It is for this reason that the power companies use high-tension or high-voltage difference lines to transmit power over large distances.

Of course, they would not dare to have such high voltages wired directly to a house. The danger from electrocution and fire would be tremendous. Instead, they use step-down transformers to convert these high voltages to the normal voltage used in houses in the United States, about 120 V.

Many houses also have 240-V lines. The large appliances such as ironers, driers, stoves, etc., are sometimes run from 240-V lines rather than 120 V. This is for essentially the same reason that the power companies use high voltages. You should be able to explain why these large power-consuming devices are more profitably run on 240 V than on 120 V.

SUMMARY

When a dc source charges a capacitor C through a resistor R, the current decays exponentially. A characteristic time RC, called the time constant, is of importance. It measures the time taken for the capacitor to charge (or discharge) by about 63 percent. In this same time, the charging and discharging currents fall to about 37 percent of their initial values.

The average value of a sinusoidal type current or voltage is zero. Ordinary ac meters read $i_0/\sqrt{2} = 0.707i_0$ and $v_0/\sqrt{2}$, where i_0 and v_0 are the peak values. These meter readings are represented by I and V. They are called the rms or effective values.

When an ac voltage V is impressed across a resistor R, $V = IR$ applies. For a capacitor C across which an ac voltage of frequency f is applied, one has $V = IX_C$ where $X_C = 1/2\pi fC$. In the case of an inductor L with an ac voltage across it, one finds $V = IX_L$ with $X_L = 2\pi fL$. The quantities X_C and X_L are called the capacitive and inductive reactances.

In a series circuit containing an ac source together with inductance, capacitance, and resistance, the relation $V = IZ$ applies. In this relation, the circuit impedance Z is equal to $\sqrt{R^2 + (X_L - X_C)^2}$.

A pure inductor and a pure capacitor consume no average power. Power losses occur only in resistors. The power loss there is I^2R. In a series ac circuit containing

L, C, and R, the power loss is $VI\cos\theta$, where V is the applied voltage. The quantity $\cos\theta$ is called the power factor. The angle θ is given by $\tan\theta = (X_L - X_C)/R$.

An LC circuit resonates at a frequency for which $X_L = X_C$. At the resonance frequency $f_0 = (1/2\pi)\sqrt{1/LC}$ the current in the circuit becomes very large. The less resistance in the circuit, the greater the current at resonance.

Transformers consist of two (or more) coils. The flux from the primary coil links the secondary coil. When ac is used in the primary, an ac voltage is induced in the secondary. The primary and secondary voltages are related to the number of loops on the two coils by $V_S/V_P = N_S/N_P$.

MINIMUM LEARNING GOALS

Upon completion of this chapter you should be able to do the following

1. Sketch the current and charge curves for an RC circuit during charging. Define the time constant for the circuit and relate it to the curves. Explain the significance of the time constant for discharge of the capacitor through a resistor.
2. Sketch a typical ac voltage or current curve. On the sketch, show the peak, average, rms, and effective values. Relate the rms value to the peak value in a quantitative way.
3. State the Ohm's law form which applies to an ac voltage impressed upon a resistor. Sketch the current and voltage curves on the same graph. Compute the average power loss in the resistor if sufficient data are given.
4. Explain why the impeding effect of a capacitor should be higher at low frequencies than at high. Use $V = IX_C$ in simple situations.
5. Sketch the current and voltage curves for a capacitor connected across an ac power source. State the average power loss in the capacitor.
6. Explain why the impeding effect of an inductor should be larger at high frequencies than at low. Use $V = IX_L$ in simple situations.
7. Sketch the current and voltage curves for an inductor connected across an ac source. State the average power loss in the inductor.
8. Use the relation $V = IZ$ for simple problems involving series RCL circuits.
9. By use of $V = IZ$, explain why a resonance frequency exists for an LC circuit. Show how to find the resonance frequency.
10. Explain what a transformer is and state the relation which gives the ratio V_S/V_P. Describe several uses of transformers.

IMPORTANT TERMS AND PHRASES

You should be able to define or explain each of the following:

RC time constant
Ac voltage or current
Average, rms, effective values
$I = i_0/\sqrt{2}$; $V = v_0/\sqrt{2}$
$P = I^2R$; $P = VI\cos\theta$; power factor
Resistance; capacitive reactance; inductive reactance
Impedance
$X_L = X_C$ at resonance; resonance frequency
Step-up and step-down transformers

QUESTIONS AND GUESSTIMATES

1. You are given a 2-μF capacitor, a dry cell, and an extremely sensitive, versatile current-measuring device. How could you use these to measure the resistance of a resistor which is thought to be about $10^8\ \Omega$? Could you do the measurement using an ordinary voltmeter in place of the current meter?

2. In some places low-frequency ac voltage (considerably less than 60 cps) is used. The electric lights operated on this voltage can be seen to flicker rapidly. Explain the cause of this flickering.
3. For which of the following uses would dc and ac voltage be equally acceptable: incandescent light bulbs, electric stove, electrolysis, TV set, fluorescent light, neon-sign transformer, battery charging, toaster, electric clock?
4. Draw an analogy between the vibration of a mass m on a spring and the oscillation of an LC circuit. What quantities in the mechanical system correspond to L and C in the electrical system? Explain.
5. Compare the equation for the resonance frequency of a mass vibrating at the end of a spring with the resonance equation for an LC circuit. What analogy can you draw between them?
6. A dc voltmeter is connected across the terminals of a variable-frequency oscillator. How would the meter behave as the frequency of the oscillating voltage is slowly increased from 0.01 to 100 cps? Explain.
7. Why would it be unwise to use 1000-V ac lines in a home, even though this would be more economical from a wiring standpoint?
8. If by magic you had containers of positive and negative charge which you could ladle out in small portions, how could you use these charges to build up a large oscillation in an LC circuit? If you could ladle only a limited amount at a time, what effect would a slow increase in resistance of the circuit have?
9. The following statement was published in a daily newspaper. "A warning that home electrical appliances can cause fatal injuries has been sounded by City Health Director, J. R. Smith. His comments followed the death of an 18-year old boy who accidentally electrocuted himself by inserting a fork in a toaster. Dr. Smith pointed out that even adults can be killed by such electrical shocks. Ordinary house current is 110 volts but the voltage is increased if the current is grounded, he said." How should the last sentence have been worded?
10. The devices shown in Fig. P21.1 are called **filters.** When an ac voltage is put into the device, the output ac voltage depends upon the frequency of the oscillating voltage. One of these devices lets the input voltage pass through undisturbed if the oscillation frequency is high. The other passes only low-frequency voltages. Explain which is which.
11. A typical electrocardiogram (ECG) graph is shown in Fig. P21.2. This is a graph of the voltage difference between the left leg and left arm. From the graph, estimate the average voltage, rms voltage, and the relation between peak voltage and rms voltage for this waveform. Why doesn't a simple galvanometer deflect in this way when attached to these two points on your body? (E)

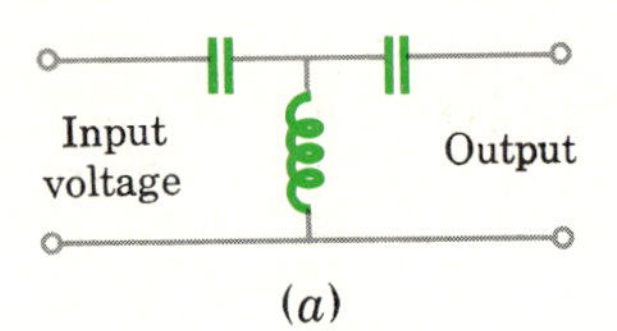

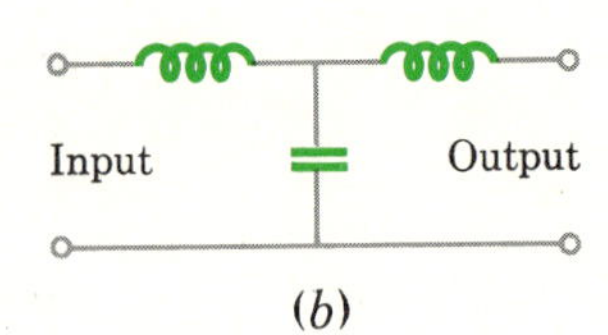

FIGURE P21.1

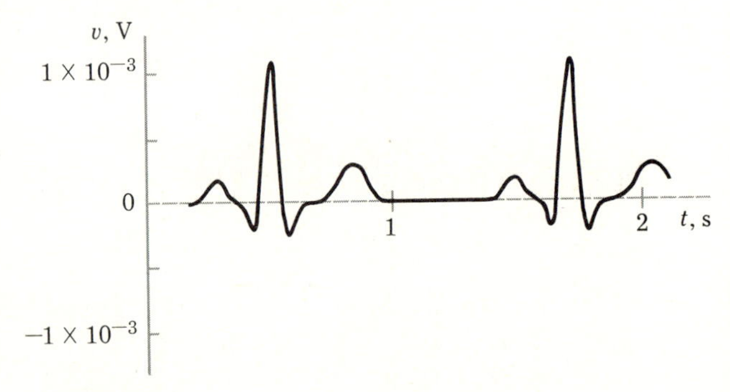

FIGURE P21.2

PROBLEMS

1. A series circuit consists of a 6.0-V battery, a 2×10^6-Ω resistor, a 4.0-μF capacitor, and an open switch. The capacitor is initially uncharged. The switch is now closed. (a) What is the time constant of the circuit? (b) About how long will it take for the capacitor to become two-thirds charged? (c) How much charge will flow onto it in the time calculated in (*b*)? (*d*) About what was the average current which flowed into the capacitor during this time interval?

2. Suppose you measure the resistance of your body between your two hands with an ohmmeter and find it to be 62,000 Ω. A 12.0-μF capacitor has been charged to 9.0 V and disconnected. You now grasp the two terminals of the capacitor with your two hands. (*a*) What is the time constant of the circuit involving your body and the capacitor? (*b*) About what would be the potential difference across the capacitor after $\frac{3}{4}$ s? (*c*) What was the charge on the capacitor when the potential across it was 9 V? (*d*) About what was the average current which flowed through your body in this $\frac{3}{4}$ s?

3. An ordinary ac ammeter reads 2.00 A when connected in an ac circuit. How large does the current actually get in the circuit?

4. An ordinary ac voltmeter reads 120 V when connected across an ac house line. What is the maximum voltage between the lines?

5. A current given by the relation $i = 5 \sin 360t$ amperes flows through a 20-Ω resistor. How much power does it dissipate in the resistor?

6. A voltage $v = 60 \cos 360t$ is impressed across a 20-Ω resistor. How much power is dissipated in the resistor?

7. An rms 120-V source having a frequency of 60 Hz is connected directly across a 10-μF capacitor. How large is the rms current in the circuit?

8. By what factor does the current in a capacitor circuit change if the frequency of the voltage source is made 10,000 times larger without changing the voltage? (The circuit contains only the capacitor and voltage source.)

9. An ac voltage source is connected directly across a resistanceless 0.50-H inductance coil. How large must the voltage be to give a current of 2 A if the frequency is (*a*) 60 Hz, (*b*) 6×10^5 Hz?

10. A 100-V, 180/π-cps voltage is connected across a 20-Ω resistor. (*a*) Find the current drawn from the voltage source. (*b*) Repeat for a frequency of 18,000/π cps. (*c*) How much power is dissipated in each case?

11. Repeat Prob. 10 if the resistance is replaced by a 1.00-μF capacitor.

12. Repeat Prob. 10 if the resistance is replaced by a 0.10-H inductor.

13. A 0.10-H inductor having a resistance of 36 Ω is connected across a 120-V, 180/π-cps source. How much current does it draw?

14. When connected across a 100-V, 60 Hz source, a 0.10-H coil draws a current of 2.0 A. What is the resistance of the coil?

15. (*a*) How large a capacitor must be connected in series with a 0.10-H coil if they are to resonate with a frequency of 60 Hz? (*b*) How large a coil would be needed to resonate at this same frequency with a 1-μF capacitor?

16. A capacitor and inductor are connected in series across a 120-V, 60-Hz source. The inductor has an inductance of $\frac{1}{9}$ H, and its resistance is 20 Ω. The capacitor is 1 μF. (*a*) Find the current in the circuit. (*b*) Repeat for a frequency of 6000 Hz.

17.* (*a*) How large an inductor must be connected in series with a 10-μF capacitor, a 20-Ω resistor, and a 100-V, 60-Hz source if the current is to be 4 A? (*b*) Repeat for a frequency of 6000 Hz.

18. The following elements are connected in series across a 100-V, 200/π-cps voltage source: $R = 10.0$ Ω, $C = 2.50$ μF, $L = 2.50$ H. (*a*) What current is drawn from the source when they are all connected in series across it? (*b*) How much power is being dissipated by the circuit? (*c*) How large is the power factor?

19. Repeat Prob. 18 if the value of L is changed to 5.00 H.

20.* A large coil of wire is wound on a cylindrical piece of wood. Using appropriate meters, one finds that a 10.0-V dc source produces a current of 0.50 A through it. A 70-V, 60-Hz ac source causes a current of 2.00 A to flow through it. What is the inductance of the coil?

21.* An inductance coil draws a current of 0.60 A when connected across a 12-V battery. When connected across a 120-V, 60 Hz source, it draws 3 A. Find (*a*) the power drawn from the ac source and (*b*) the inductance of the coil.

22.* A 60 mH, 10-Ω coil is connected across a 120-V, 60-Hz source. How much power does it dissipate?

23. A manufacturer wishes to have a 2000-W electric heating unit on his stove, in addition to three 1000-W units. Will it be possible to connect this stove to a 120-V, 30-A house line, or should it be designed for use on 240 V?

24. The heaters of some vacuum tubes require a voltage of 6 V. What should be the ratio of primary to secondary turns on a transformer used to operate the tube from a 120-V line? If the transformer was inadvertently connected backward, how large a voltage would be placed across the output terminals?

25. Neon signs require a voltage of about 12,000 V for their operation. What should be the ratio of N_S/N_P for a transformer which would allow it to operate off the 120-V lines?

ELECTRONICS AND ELECTROMAGNETIC WAVES

22

We saw in the last chapter that charge oscillations of definite frequency can be set up in a resonant electrical system composed of capacitance and inductance. Such circuits are widely used in electronic devices employed for the generation and reception of radio waves. The essentials of some of these devices will be discussed in this chapter. Moreover, we shall investigate how radio waves are generated and received by these devices. We shall find that these waves are only a small portion of a wide variety of waves, electromagnetic waves. The nature of electromagnetic radiation will be investigated, and we shall find that this radiation includes such diverse types of wave motion as radio waves, heat radiation, light, and x-rays.

22.1 THERMIONIC EMISSION

There are two major classifications of electronic devices. One encompasses devices which make use of vacuum tubes, and the other encompasses electronic systems which make use of solid-state devices such as transistors. Of course, many electronic instruments make use of both vacuum tubes and solid-state devices. In this section we shall learn about thermionic emission, a phenomenon basic to all vacuum tubes.

To a first approximation we can consider the valence electrons in a metal to be free to move anywhere within the metal. They therefore behave in many respects like gas molecules in a container. In the case of a metal, the metal surface is the container. It is possible to learn a good deal about the behavior of the valence electrons in a metal by treating them as if they were an electron gas, i.e., a gas made up of electrons rather than molecules. We shall use the electron-gas approach to discuss the phenomenon of thermionic emission, the release of electrons from a white-hot metal.

We can easily compute the kinetic energy of the valence electrons within the metal. It will be recalled from Chap. 10 that the gas law can be written in two different forms,

$$PV = \tfrac{2}{3}\nu_0(\tfrac{1}{2}m_0v^2) \qquad \text{and} \qquad PV = \nu_0kT$$

where P is the pressure of ν_0 gas molecules each of mass m_0, confined to a volume V at an absolute temperature T. The quantity $\tfrac{1}{2}m_0v^2$ is the kinetic energy of the gas molecule, and k is Boltzmann's constant, 1.38×10^{-23} J/K. Equating the two expressions yields

$$\tfrac{1}{2}m_0v^2 = \tfrac{3}{2}kT$$

This equation says that the average translational KE of *any* ideal-gas molecule in *any* box is just $\tfrac{3}{2}kT$. It must also be true for the valence-electron gas in a block of metal provided that these electrons can really float freely within the metal. To that approximation we then have the important result that each valence electron in a metal has an average KE equal to $\tfrac{3}{2}kT$. Of course, some of the electrons will have more energy than this, and some will have less. But we can use this figure as a basis for discussion.*

Is there any possibility that the electrons can escape from the metal? In order to answer this question, we must consider what holds the electrons within the metal. A major portion of the force holding the electron to the metal is purely electrostatic in origin. Consider what would happen if the electron tried to get away from the metal block. As shown in Fig. 22.1, the negative electron when just outside the surface of the metal will induce a positive charge on the surface. The positive charge in turn will exert an attractive force on the electron and will try to pull it back to the metal. The

FIGURE 22.1
The electron must overcome the attraction of the induced charges on the surface of the metal if it is to escape from the metal.

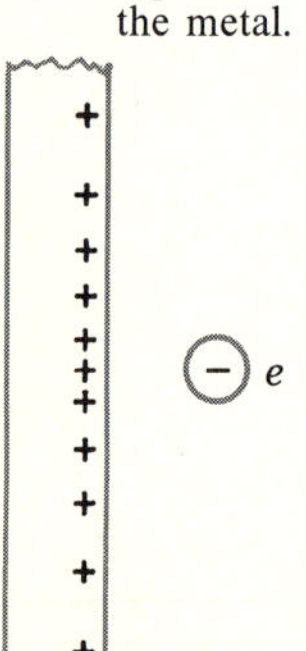

*At low temperatures and low electron energies, quantum effects become important, and the electron-gas approximation is badly in error. However, at high temperatures and high electron energies, this approximation is rather good. We shall be interested here in the high-energy electrons, and so the electron-gas approximation will be valid.

electron will not be able to escape unless it has enough KE to overcome this attraction.

From these considerations we see that a certain amount of work is necessary to tear the electron away from the metal. Unless the electron possesses enough KE to do this work, it cannot escape. *The amount of energy needed to overcome the forces holding the electron within the metal and pull the electron loose is called the* **work function** *of the metal.* We should point out that the energy discussed above is only a portion of the work-function energy. The other energies involved are more difficult to compute, and we cannot discuss them here.

DEFINITION

Work Function

In view of the above discussion, we see that *an electron can escape from a metal provided that it has enough KE.* But since the average KE of an electron is proportional to the temperature of the metal, it is clear that the metal must be heated before the electrons can escape. *For most metals, no appreciable number of electrons can escape from the surface unless the metal is heated red-hot.*

22.2 THE DIODE AND RECTIFICATION

The principle of thermionic emission is basic to the operation of a diode, the simplest type of vacuum tube. This tube consists of two basic elements enclosed in a glass vacuum tube T as illustrated in Fig. 22.2. A piece of hot metal, the cathode, or filament, acts as a source of electrons. Sometimes the cathode is just a fine wire through which a current flows, heating it red-hot. This is the situation pictured. In practice, the cathode or filament voltage $\mathcal{E}_f$ is of the order of 5 V. Some diodes use an indirectly heated cathode, in which case $\mathcal{E}_f$ in Fig. 22.2 would be missing and a separate heating element, not connected electrically to C, would heat the cathode.

In any event, the cathode is heated hot enough for appreciable thermionic emission to occur. The purpose of the other element within the tube, an unheated metal plate P, is to collect the electrons emitted by the hot cathode. Since the tube is evacuated, the electrons from the filament move freely until they collide with the walls of the tube or with the plate. However, because of the voltage difference between the filament and the plate provided by the plate battery $\mathcal{E}_p$, the electrons emitted from the cathode are attracted by the

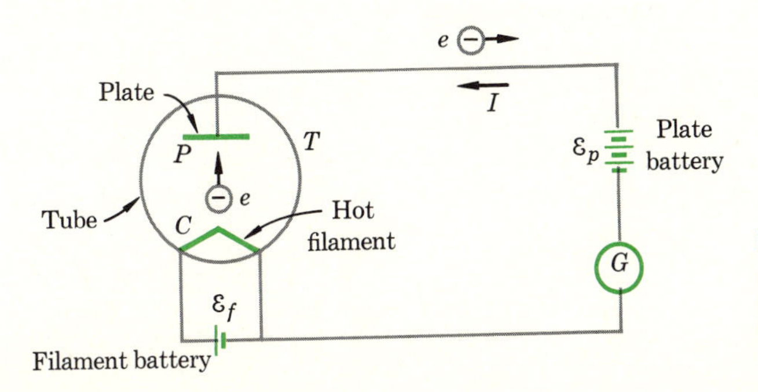

FIGURE 22.2

Electrons are emitted by the hot cathode C and travel through the vacuum tube T to the positive plate P.

plate. In most tubes, if $\mathcal{E}_p$ is made about 200 V or larger, nearly all the electrons emitted by the cathode are collected by the plate.

The diode is most widely used for its rectifying action. As pointed out in the last chapter, it is more convenient in practice to produce and transmit alternating current than direct current. But for numerous purposes, for charging batteries, for electroplating, and especially in electronic devices, one needs direct current. Direct current can be obtained from alternating current by using a diode in the circuit shown in Fig. 22.3. Here the plate battery has been replaced by the ac voltage we wish to rectify.

The voltage $\mathcal{E}_p$ alternates as shown in Fig. 22.3. When $\mathcal{E}_p$ is positive, the plate will be positive and the filament will be negative. Electrons emitted by the hot filament will flow to the plate. This gives rise to a plate current I_p. Since this plate current flows through R_p, the load, or plate, resistor, there will be a voltage difference between a and b, V_{ab}. However, when the voltage $\mathcal{E}_p$ reverses and makes the plate negative with respect to the filament, the electrons from the filament will be repelled by the plate and I_p will drop to zero. For this reason, V_{ab} will be zero when $\mathcal{E}_p$ is negative. The variation of V_{ab} is also shown in Fig. 22.3. It is seen that V_{ab} is never negative, and so we have succeeded in transforming the ac voltage $\mathcal{E}_p$ to a dc voltage V_{ab}.

22.3 THE SEMICONDUCTOR DIODE

As you are probably aware, more and more electronic instruments are being built using semiconducting devices, transistors, and diodes in place of vacuum tubes. These devices are small pieces of crystalline solid which require no heat for their operation. Since a filament as such is not required, the circuitry and power requirements are less than for ordinary vacuum-tube devices. In addition, they can be made very small, and so space can be saved by their use. For example, see the miniaturized circuit shown in Fig. 22.4.

The atoms of silicon or germanium form the basis for many semiconducting devices. Pure silicon and germanium are nonconductors. Both

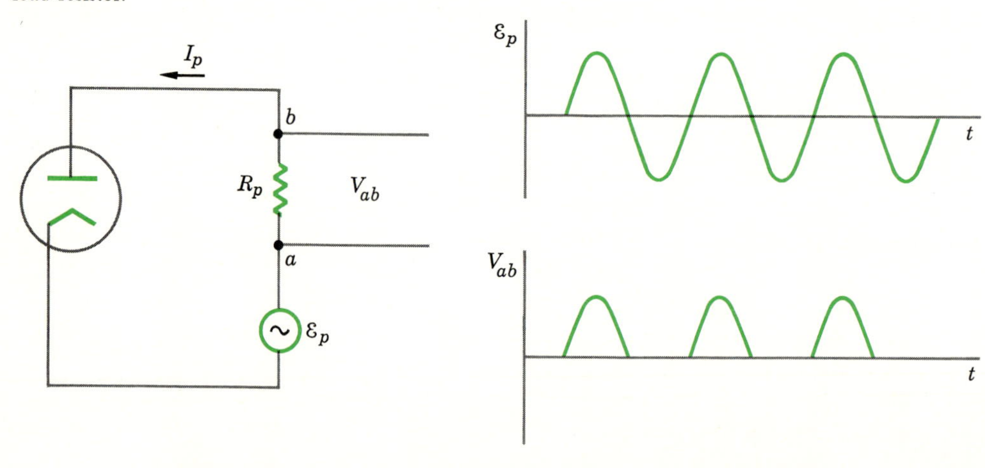

FIGURE 22.3
Although the voltage $\mathcal{E}_p$ is alternating as shown, the diode circuit provides a rectified voltage V_{ab} across the load resistor.

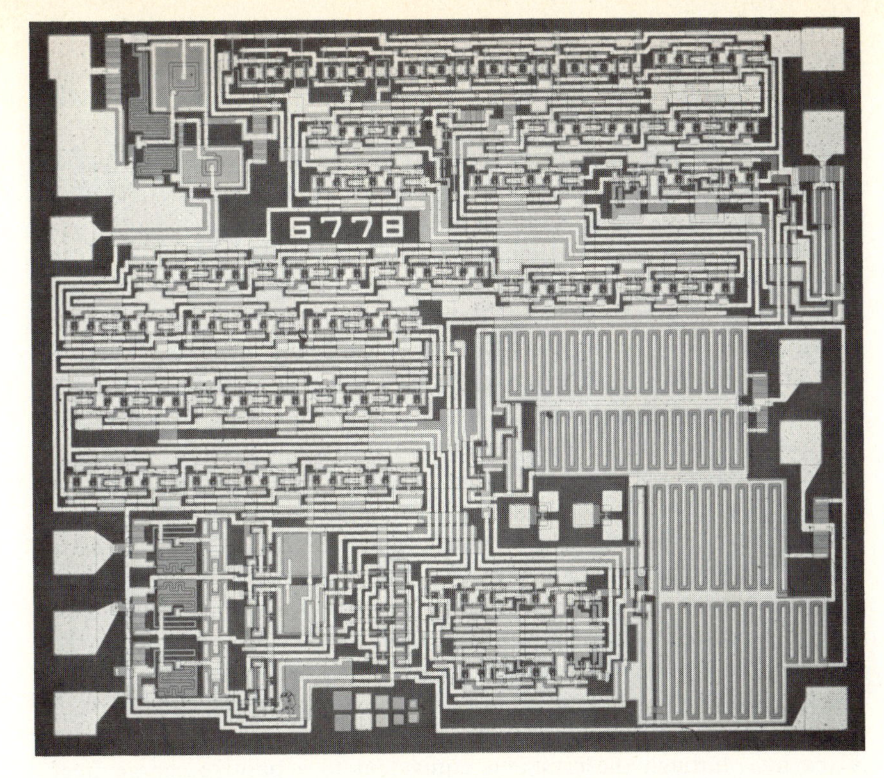

FIGURE 22.4

This miniaturized timing circuit contains the basic electronics for an electric watch. Although it contains several hundred electronic components, its packaged dimensions are about 4 x 5 x 2 millimeters. (*Courtesy of RCA.*)

these atoms have four outer, or valence, electrons. They must lose these four electrons or gain four other electrons in order to have complete electron shells. Hence, these atoms combine with other atoms in such a way that they can effectively gain or lose four electrons. In crystalline germanium and silicon they do this by forming covalent bonds with their neighbors. That is to say, each atom shares its outer four electrons with four other atoms, while sharing four additional electrons from these four atoms, thereby satisfying all. This situation is shown schematically in Fig. 22.5, where the lines between atoms represent a shared electron. Notice that each germanium atom has eight electrons about it, and so its outer shell is filled. This would also be the case for silicon.

FIGURE 22.5

The covalent bonding allows each atom in the lattice to fill its outer shell. None of the electrons is free to move in the lattice.

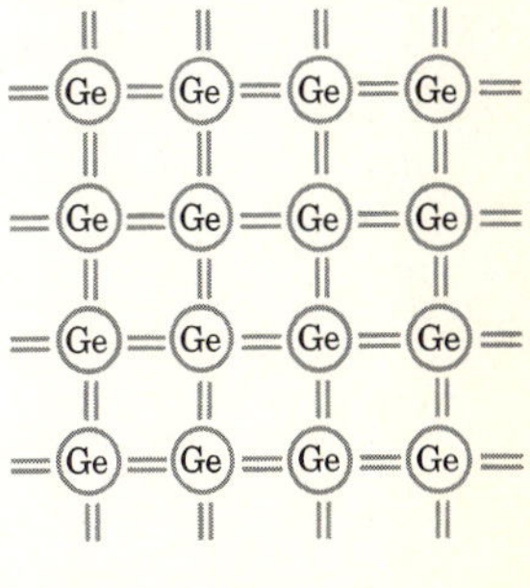

Suppose, now, that we purposely add an extremely small quantity of arsenic atoms as an impurity to the germanium from which we make a crystal. Arsenic is used since it is next to germanium in the periodic table and has essentially the same size as a germanium atom. Therefore, the arsenic impurity atoms fit into the lattice quite well and replace a germanium atom, as shown in Fig. 22.6. Because arsenic is pentavalent rather than tetravalent, it has one too many electrons to fit in the lattice, with the result that the extra electron is not held very tightly by the arsenic atom and escapes easily. Once the electron is loose, it can travel through the crystal rather freely, much like a free electron in metal. Hence, the original nonconducting germanium (nonconducting because the electrons were all tied tightly in the lattice structure) has acquired a few free electrons, one from each arsenic impurity atom. The impure crystal will now be a conductor. Since the number of free electrons is small, it is a poor conductor or semiconductor.

This type of system may also be obtained by adding a few pentavalent impurity atoms to silicon. (Which ones would you expect to be best?) Such impure crystals have an excess of electrons, and it is these electrons which act as charge carriers and carry current through the crystal. This type of crystal is called an ***n*-type semiconductor,** since *the charge carriers are negative.*

n-Type Semiconductor

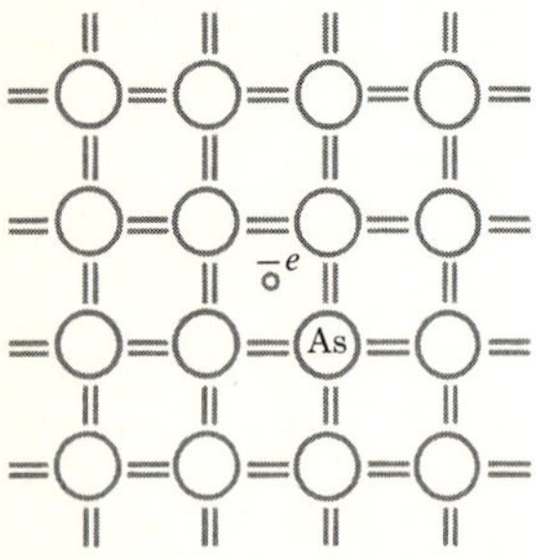

FIGURE 22.6

When a pentavalent arsenic atom replaces a germanium atom in the germanium crystal, its fifth valence electron is not needed to complete the covalent bonding and therefore is relatively free to move in the crystal.

Another type of semiconductor can be made by adding a trivalent atom such as gallium to germanium. In this case, the impurity atom lacks one electron which is needed to make the lattice complete. *This electron vacancy,* shown by the oval at *a* in Fig. 22.7*a*, *is called a* **hole.** Now, however, the electron at *b* in Fig. 22.7*a* can easily slip over to fill the hole. This in turn leaves a hole at *b*. Another electron can slip into this hole, and so on. The hole becomes free from the original impurity atom and wanders about the crystal more or less freely. After a short time, the situation may be as shown in Fig. 22.7*b*, *c*, and *d*.

Notice that the region near the hole in Fig. 22.7*d* has an excess of positive charge and is not neutral. (Actually, it has a deficiency of one electron.) On the other hand, the gallium impurity atom now has one too many electrons and is not neutral either. However, the extra electron near the gallium atom is quite tightly held in the lattice, and so it is not free to move. The hole, though, is still free to move about. As it does so, the site of the excess positive charge moves with it. Hence the more or less free movement of the hole through the crystal is equivalent to a positive charge freely moving through the crystal. This type of semiconducting crystal is called a ***p*-type semiconductor,** since *positively charged holes carry the current.*

p-Type Semiconductor

To make a semiconducting diode rectifier, one combines an *n*- and a *p*-type semiconductor, as shown in Fig. 22.8*a* and *b*. (In practice, very small pieces of semiconducting crystal are placed between metal electrodes. The entire assembly is placed in an insulated container, with leads running to the two metal plates.) Originally, of course, each piece of semiconductor is electrically neutral. In addition, since the metal plates can receive or give up electrons at will, holes and excess electrons can travel relatively freely through the junction to the metal. This is not possible at the semiconductor junction, as we shall now see.

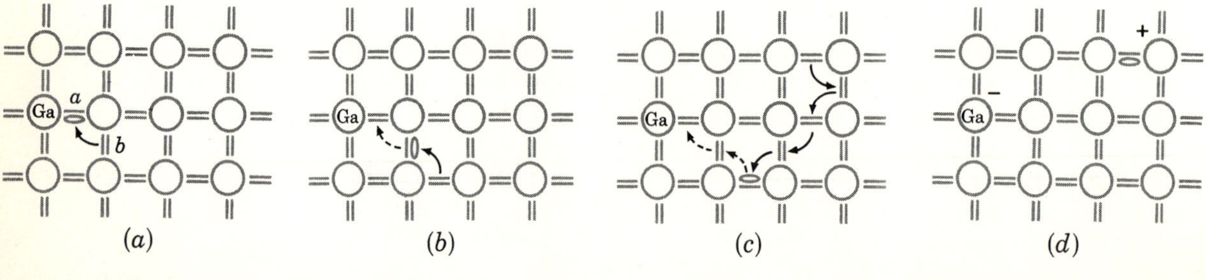

FIGURE 22.7

The trivalent gallium impurity atom in the germanium lattice leaves a vacant bonding site *a* in (*a*). This vacant site can move around as nearby electrons displace it. In (*d*) it has moved essentially free from the gallium atom and is now a nearly free-moving hole. Notice that the fixed gallium atom is now negatively charged while the hole carries a positive charge with it.

Suppose that one applies a voltage to the diode, as shown in Fig. 22.8*a*. The electrons and holes will be made to move in the directions shown. Since we are effectively trying to separate the negative from the positive charge, this will require considerable work. A steady current cannot flow in this

(*a*) (*b*) (*c*) (*d*)

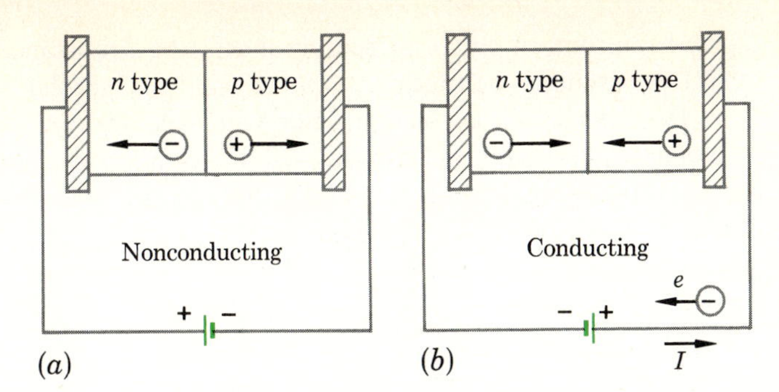

FIGURE 22.8
The diode conducts current in (*b*) but does not pass current when the voltage is reversed, as shown in (*a*).

situation, because it is possible to drain only a few of the electrons and holes from the semiconductors. As a result, the diode will not conduct current in the direction desired by the battery in part *a*. As shown in *b*, a reversed battery brings holes and electrons together and so the diode then conducts.

22.4 APPLICATIONS OF ELECTRONIC DEVICES

In this section we shall present a few important applications of diodes and other simple electronic devices. Electronic technology is so complex that most users of electronic devices cannot be expected to master their detailed construction. Instead, we usually consider the complex of vacuum tubes, diodes, transistors, and other solid-state devices as units designed for specific purposes. For example, the inner workings of an electronic amplifier are usually quite complex, but the amplifier as a unit simply transforms a given input voltage into a much larger output voltage. For a good amplifier, the input and output voltage waves would be of exactly the same shape. As we see, the function and result of use of the amplifier unit are quite easily understood even though the inner workings are very complex.

In order to gain practice in the use of electronic devices as circuit units, we shall examine a few applications of them. As you will see, the effect of the device on an input signal must be known, but the internal workings of the device will not be of major importance for our discussion.

1 *Half-wave rectifier* We have already seen how this device operates. It is shown again in Fig. 22.9. The symbol for a diode is ⊳|. It conducts in the direction of the arrow.

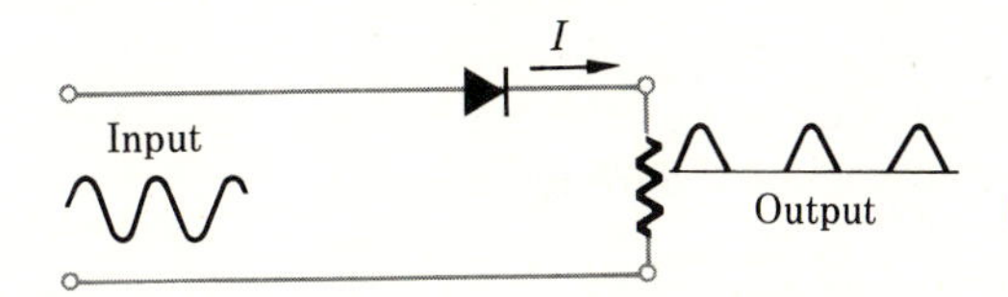

FIGURE 22.9
Half-wave rectifier.

2 *Half-wave rectifier, filtered* As shown in Fig. 22.10, a capacitor C is placed across the output. If no current were being drawn from the output, the capacitor would become fully charged and maintain a constant dc voltage

at the output. In practice, the output acts like a resistance of value R. If R is large enough, the time constant RC will be larger than the period of the voltage wave. Then the capacitor will only drain slightly during each voltage pulse and the output will be nearly steady. (We specify the "steadiness" by the "ripple," the ratio of the voltage variation to the maximum voltage during the cycle.)

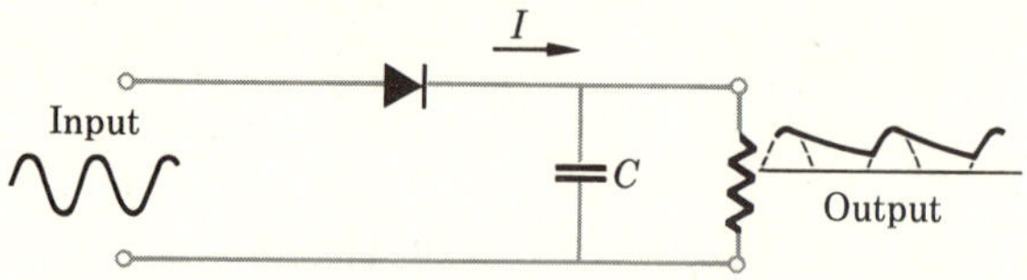

FIGURE 22.10
Filtered half-wave rectifier.

3 *Full-wave rectifier* At the instant shown in Fig. 22.11, the top end of the secondary is positive and the bottom is negative. One-half cycle later, the voltage reverses, but the lower diode will then conduct current to the right. As a result, current is furnished to the output on both halves of the cycle.

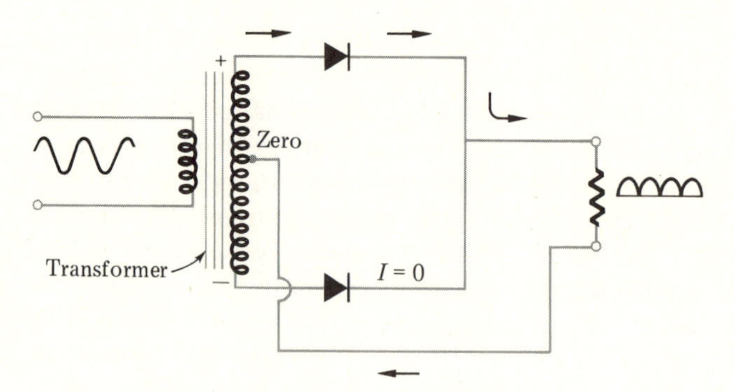

FIGURE 22.11
Full-wave rectifier. The center tap transformer in essence provides two voltage sources which are one-half cycle out of phase.

4 *An x-ray-tube circuit* Refer to Fig. 22.12. Inside the tube, electrons are accelerated through the high voltage between the filament and plate. Their impact with the plate generates x-rays by means of processes we shall discuss in Chap. 27. In high-output tubes, the electron beam heats the plate. To minimize this effect, the plate is rotated past the beam.

As you see, *we can convey information concerning an electronic circuit without showing the details of each device.* We call a diagram like the left part of Fig. 22.12 a **block diagram.** Often in such a diagram only one of the two wires is shown. The second wire is often assumed to be grounded and is omitted. Let us now leave this discussion of electronic devices and turn to the subject of electromagnetic waves.

22.5 GENERATION OF RADIO WAVES

Before beginning our study of the way in which electric and magnetic signals are transmitted through space, let us discuss a few qualitative features of

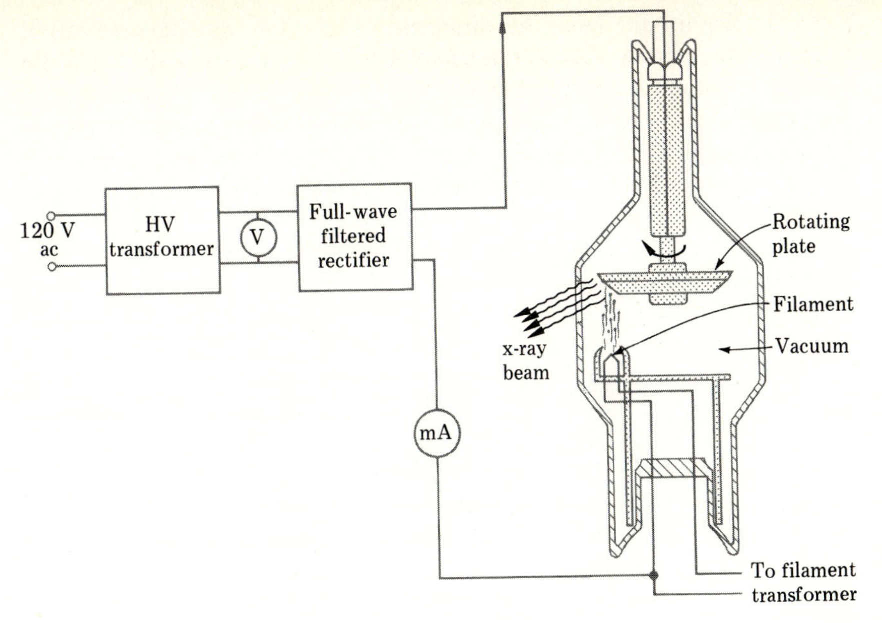

FIGURE 22.12
An x-ray tube circuit.

radio waves. Although these preliminary ideas will be modified and extended in later sections, they will enable us to grasp some of the simpler concepts involved.

Each radio station is assigned a certain radio frequency at which to operate. The frequencies are within the range of about 500,000 to 1,700,000 cps for the usual commercial broadcasting stations. For the purposes of the preliminary discussion, we shall consider a station whose assigned frequency is 1,000,000 cps. This would be marked on most radio dials as 1000 kc/s. The station assigned to this frequency uses an alternating voltage of 1×10^6 cps, and the graph of this voltage as a function of time is shown in Fig. 22.13. This voltage is applied to the primary of a transformer, as shown in Fig. 22.14. The voltage induced across the secondary of the transformer is used to place charge on the transmitting antenna of the radio station.

As shown in Fig. 22.14, the antenna can be considered to be a long piece of wire, and, in our station, the top end of the antenna is charged positively with respect to the lower end 1×10^6 times per second. Since the charging potential alternates, the charge on the antenna is constantly reversing.

Consider, now, what is taking place at the instant when the antenna is charged positive on top and negative on the bottom, as illustrated in Fig. 22.15. The electric field around the wire appears somewhat as shown.

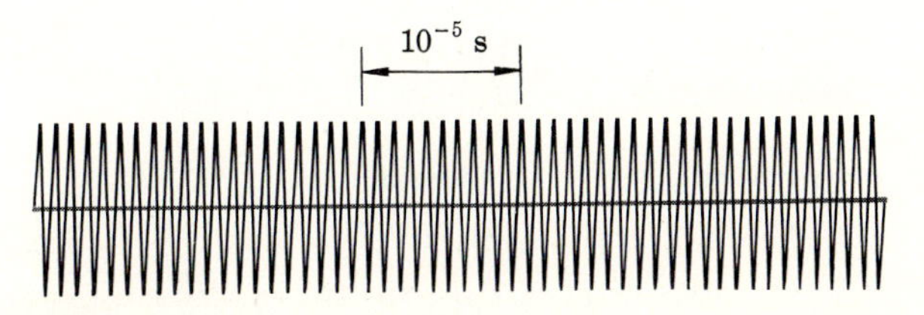

FIGURE 22.13
The alternating voltage used to operate the transmitter discussed in the text is sinusoidal and has a frequency of 10^6 cps.

However, an instant later the direction of the field will have reversed, because the antenna will then be charged oppositely to that shown in the figure. Still later, the charge on the antenna will again be as shown, and the field will also have reversed again. In fact, the number of field reversals per second will equal twice the carrier frequency of our station, 1×10^6.

It is apparent from the above that *an observer some miles away from the radio antenna,* at point A in Fig. 22.15, for example, *will observe that there is an electric field present at that location.* The field results from the charges located on the antenna. *It will be a very weak field, but it will be oscillating as a function of time. Its frequency will be the same as that of the radio station.*

Let us now see how the electric field takes shape throughout the area near the antenna. In Fig. 22.16 we see the electric field pattern develop as it leaves the antenna. Part *a* shows the situation at the instant the antenna is first charged. As time goes on, the charge decreases (part *b*), becomes zero (part *c*), and recharges fully negative (part *d*); finally, in part *e*, the situation is shown after two cycles have passed by after part *c*.

Suppose the antenna of Fig. 22.16 is sending its electric field out over the surface of the earth. We show the right-hand side of Fig. 22.16*d* in Fig. 22.17*a*. If the surface of the earth is along the x axis, then the electric field along the earth is as shown in parts *b* and *c* of the figure. Notice that at points A and C the electric field is directed downward while at B it is upward.

But the electric field generated by the antenna travels out across the

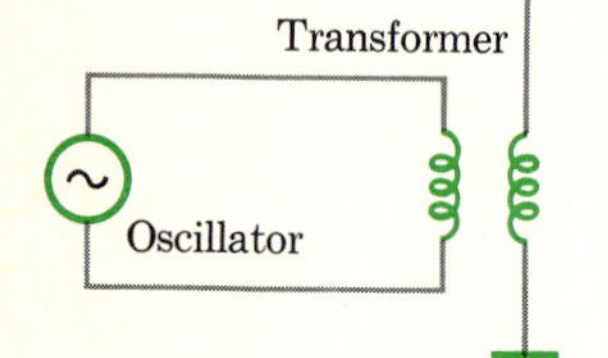

FIGURE 22.14
The high-frequency alternating-voltage source is used to charge the top of the antenna alternately positive and negative.

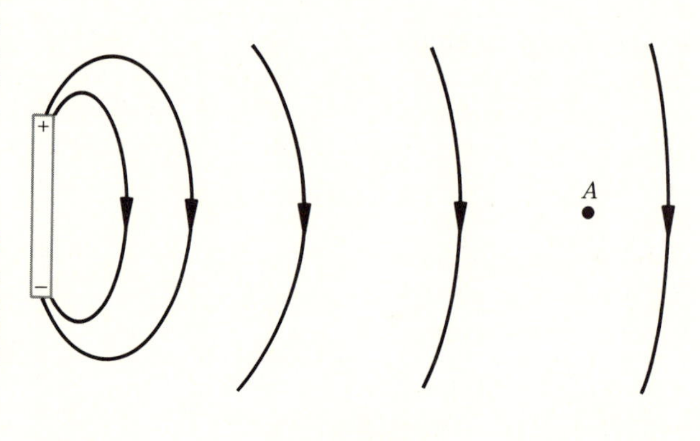

FIGURE 22.15
When the antenna charge oscillates back and forth, the electric field at point A will alternately point up and down.

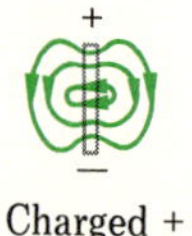

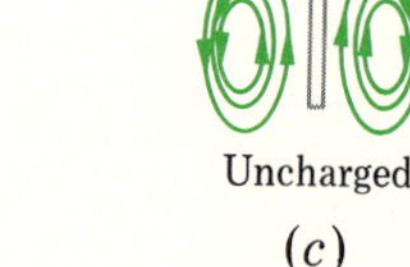

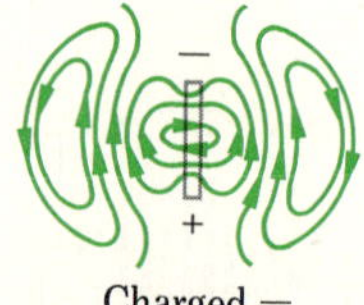

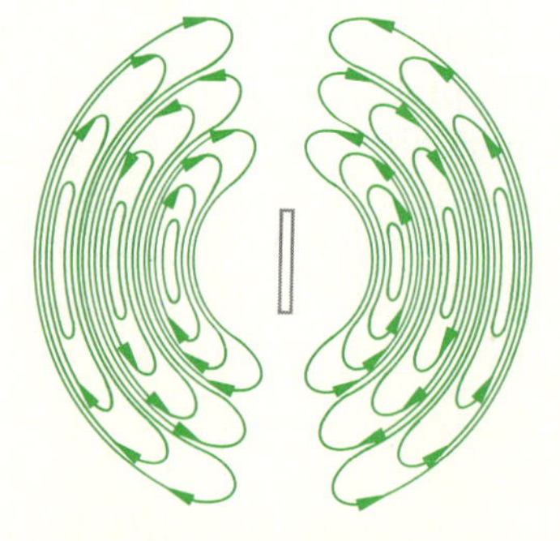

FIGURE 22.16
The electric fields shown move out away from the antenna. As they pass a distant point, the field at that point reverses direction with the same frequency as that of the radio station.

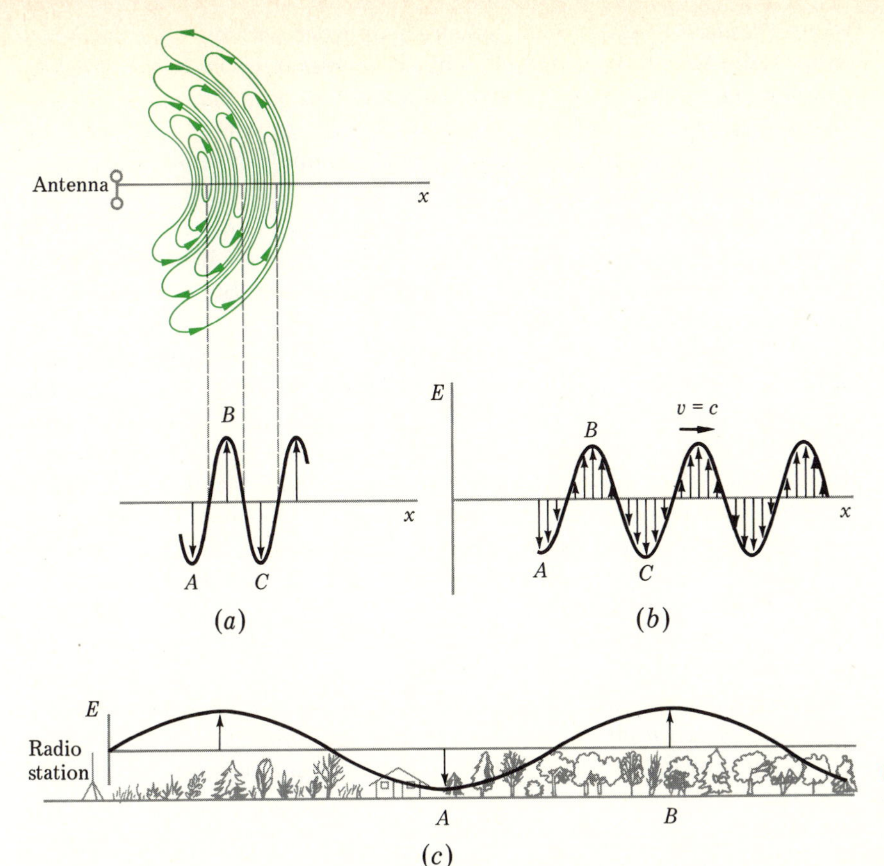

FIGURE 22.17

The electric field wave travels away from the antenna with speed $v = c$.

earth with a certain speed v. As a result, a person at point A on the earth will experience an oscillating electric field. The electric field wave shown in parts a, b and c is moving toward the right. When a crest passes point A, the field will be directed upward. When a trough passes point A, the field will be directed downward.

We see from this that each point on the earth experiences a vibrating electric field. The frequency of this vibration is equal to the vibration frequency of its source, the vibrating charge on the radio-station antenna. *The electric field wave sent out by the source obeys the same relation we have found for all other waves. The wavelength of the wave* λ, *its frequency* f, *and its speed* v *are related by* $\lambda = v/f$. The frequency of the wave sent out by a radio station is of the order of 10^6 Hz. If we know the speed v of the electric field wave, we can find its wavelength. Before discussing the speed of these waves, though, let us consider for a moment another aspect of the signal sent out from the antenna.

$\lambda = v/f$

In addition to the electric field wave, a magnetic field wave is sent out from the antenna. This wave is a result of the fact that when the charge on the antenna changes, a current flows in the antenna wire. As we have seen previously, this current will cause a magnetic field to circle the antenna as shown in Fig. 22.18. Notice that *the magnetic field is in a direction perpendic-*

ular to the electric field. It, too, oscillates in direction with the same frequency as the radio station. We see, then, that *an observer some miles away from the antenna experiences not only an oscillating electric field but also a magnetic field perpendicular to the electric field which oscillates with it. This combined electric and magnetic field is what we call the station's electromagnetic (EM) radio wave.*

Relation of **E** to **B**

We represent this combined EM wave in Fig. 22.19. Notice that *the electric field* **E** *is perpendicular to the magnetic field* **B**. *Moreover, the two waves are in phase.* (This latter point is not obvious but is the result of detailed computation.)

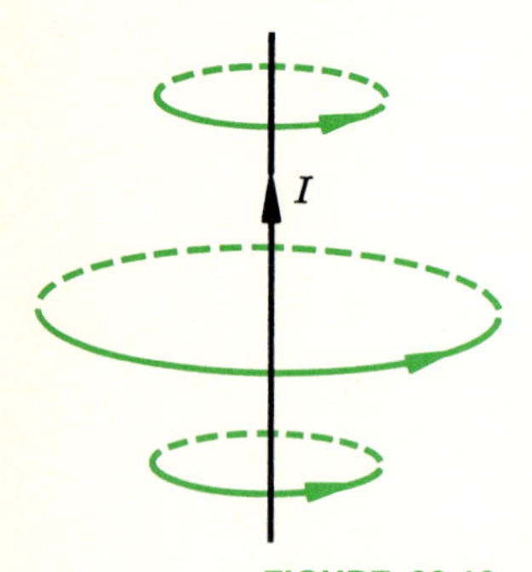

FIGURE 22.18
A magnetic field circles the antenna.

The possible existence of EM radio waves was predicted from theoretical considerations many years before the first radio was built. This prediction was made by a brilliant physicist, James Clerk Maxwell, in 1865. He set down in mathematical form the laws found by Faraday and the other experimenters in the field of electricity. His now famous equations, Maxwell's equations, are four in number. Using these equations, *Maxwell showed that radio-type waves should exist.* Furthermore, *he found theoretically that their speed in vacuum should be* 2.998×10^8 *m/s.* (*For most purposes we shall take this to be* 3×10^8 *m/s and will represent it by* c.)

This predicted speed for radio-type waves was astonishing to the people of Maxwell's time. (Remember, radio-wave generators were not yet invented.) *The speed he found was exactly equal to the measured speed of light. Maxwell inferred from his result that light waves are a form of EM waves.*

Before leaving this section we should point out a fact thus far assumed implicitly: *EM waves can travel through vacuum, empty space. No material is needed for their propagation.* Unlike other types of waves we have studied, they consist of a vibration of a field, not of a material substance.

Illustration 22.1 The oldest radio station in the United States is station KDKA in Pittsburgh, which went on the air in 1920. It operates at a frequency of 1.02×10^6 Hz. What is the wavelength of its EM wave?

Reasoning We know that for any wave $\lambda = v/f$. In our case $v = c = 3 \times 10^8$ m/s, $f = 1.02 \times 10^6$ Hz. Substitution gives $\lambda = 294$ m.

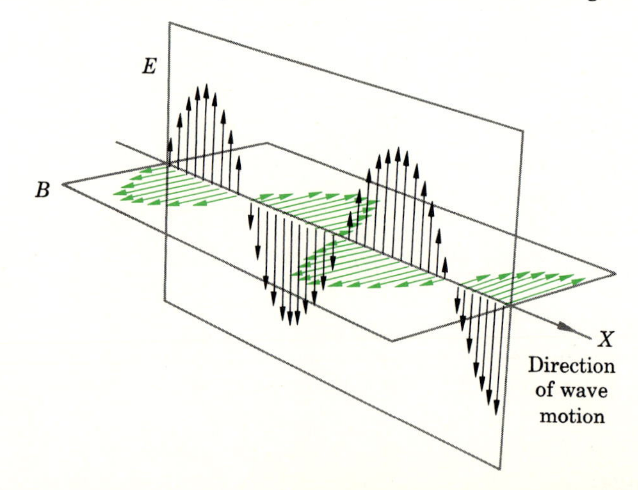

FIGURE 22.19
The oscillating antenna sends out a magnetic field wave perpendicular to the electric field wave. A "snapshot" of this EM wave along a line perpendicular to the center of the antenna is shown.

22.6 THE ELECTROMAGNETIC WAVE SPECTRUM

Speed of EM Waves

As we saw in the previous section, *EM waves travel with a speed* $c = 3 \times 10^8$ *m/s through vacuum. Ordinary radio waves have wavelengths of the order of a few hundred meters long.* Maxwell inferred that light waves are also *EM* waves.

Subsequent investigations have confirmed Maxwell's inference that *light waves are EM waves*. However, as we shall see in later chapters, *the wavelength of light waves is exceedingly small.* Light waves have wavelengths in the neighborhood of 0.0000005 m, or 5×10^{-7} m. The exact wavelength of light depends upon its color, red light being near 6.5×10^{-7} m, while blue light is near 4.3×10^{-7} m.

Anyone noticing this wide difference between the wavelengths of radio and light waves, both electromagnetic, might well wonder whether there was not some form of wave in between these two. This is actually the case. We can of course produce radio waves with wavelengths much shorter than 100 m by using very-high-frequency oscillators. Radar waves (or microwaves) are of this type and can be made with wavelengths as short as 0.01 m. We now know that heat radiation consists of EM waves having λ's intermediate between the values for radar and light. As a matter of fact, short-wavelength heat waves are nothing more than infrared light, as most of you probably already know.

The EM Wave Spectrum

This great span of EM-wave wavelengths extending from radio waves down through light waves is shown in Fig. 22.20. Electromagnetic radiation

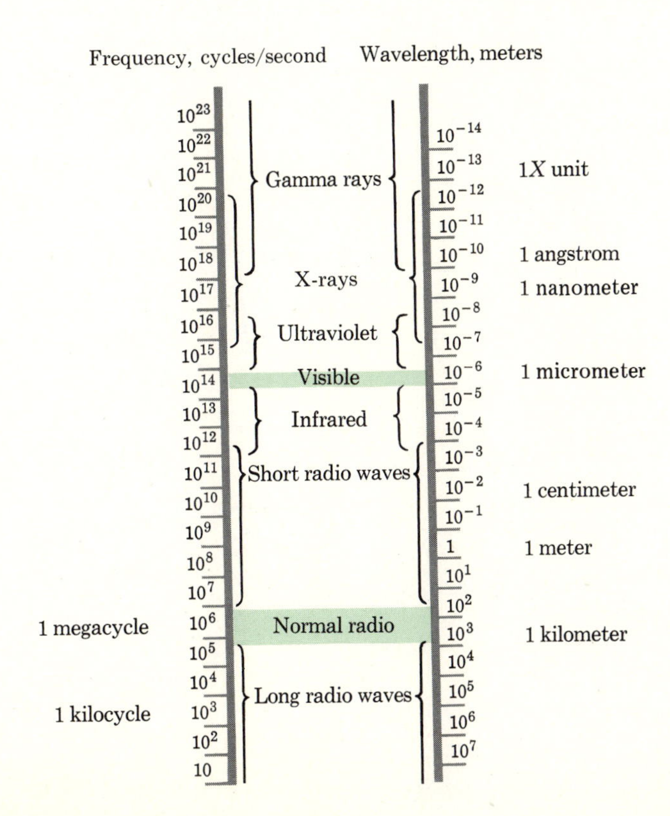

FIGURE 22.20
The electromagnetic spectrum.

does not stop with blue light. There are electromagnetic waves still shorter. We are all familiar with the next shorter waves, ultraviolet radiation. Still shorter than these are x-rays. A type of radiation from radioactive substances, the γ ray, is essentially the same as x-ray radiation but still shorter in wavelength.

We see therefore that *EM waves differ widely depending upon their wavelength.* The next few chapters of this book will be devoted to a very small wavelength range of these radiations, ordinary light. In still later chapters we shall study other types of radiation in the EM spectrum.

22.7 RECEPTION OF RADIO WAVES

As we have seen, a radio transmitter blankets the surrounding area with an electric field which is oscillating at a specified frequency. If one holds a straight piece of wire in this field, the electrons in the wire will move under the action of the oscillating electric field. Of course, the electric field will be very weak at a distance many miles from the transmitter, but a sensitive detecting system can still measure the movement of the charges in the wire as a result of the field. A radio is a device constructed for this purpose.

Suppose that the electric field oscillates vertically, as shown in the upper part of Fig. 22.21. It will separate charges in the vertical radio antenna. For example, when the electric field points upward, the top of the antenna will be charged positive and the lower end negative. This, of course, causes a charge on the condenser C. However, this charge will soon flow out of the condenser, because one-half cycle later the field will be pointing downward and the charge on the antenna and condenser must be reversed. A half cycle later still, the antenna will charge the capacitor as it was originally. From this we

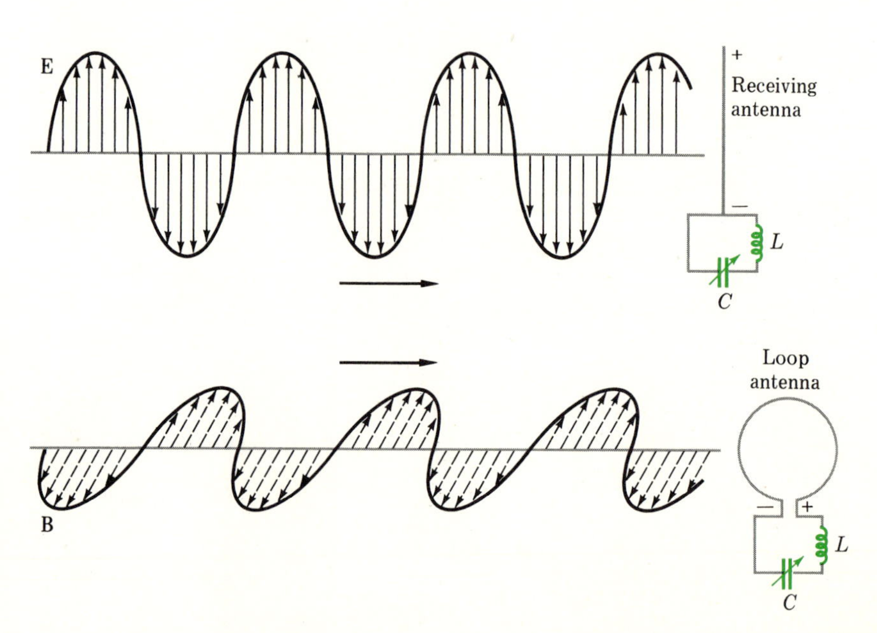

FIGURE 22.21
The radio's antenna acts like a voltage source since the radio wave passing by it induces potential differences in it.

see that the antenna is taking energy from the radio wave and giving it to the capacitor.

Frequency-Selection Method

If one adjusts the value of the capacitor so that the natural frequency of the LC circuit is the same as that of the radio station, the circuit will resonate under the driving action of the antenna. Consequently, the LC circuit will build up a large response to the action of radio waves to which it is tuned. Thus, one is able to tune the radio to receive a particular station. The resonant circuit will then show an oscillatory voltage across C which is nearly exactly proportional to the output voltage of the radio station. It will consist of a carrier wave modulated with the effect of the station's microphone system. However, the voltages across C will be extremely small except for the very nearest stations. For this reason most radios amplify this voltage by rather complex electronic circuits.

Most portable radios used today do not make use of the oscillating electric field. Instead, they receive the signal transmitted by the oscillating magnetic field portion of the EM wave. As shown in the lower part of Fig. 22.21, *a loop antenna is used to detect the magnetic portion of the wave.* As the wave passes by the loop, its changing flux induces an emf in the loop. This oscillating emf is used to drive the resonant circuit, as in the previous case. The loop antenna usually consists of a coil wound on a rod made of a ferromagnetic material, a ferrite. If you look inside a small portable radio, this rod is usually quite visible.

A loop antenna often shows directional properties. If the radio is oriented so that the loop is related to the field as shown in Fig. 22.21, reception is best. Why? But if it is rotated 90°, the reception will be minimum. This property of a loop antenna can be used to locate the source of the EM waves. How?

22.8 THE RELATION BETWEEN E AND B IN ELECTROMAGNETIC WAVES*

In this section we shall use Faraday's law of induced emfs to obtain a relation between the electric field intensity $\boldsymbol{E}$ and magnetic field intensity $\boldsymbol{B}$ in an electromagnetic wave. Let us confine our attention to the magnetic-field portion of the wave propagating along the x axis. A "snapshot" of this wave at a given instant is shown graphically in Fig. 22.22*a* and in cross section in part *b*. We know from the previous sections that this wave is traveling through space—in our case along the x axis to the right. As the magnetic field moves along to the right, it causes a change in magnetic flux through a loop such as the one shown in part *c* of the figure. The loop is assumed to have infinite resistance so that no current can flow in it. We shall now calculate the induced emf in this loop by two methods and thereby obtain information concerning the EM wave.

Consider what happens to the flux through the loop in Fig. 22.22*c*

---

*Sections 22.8 through 22.10 may be omitted without loss of continuity.

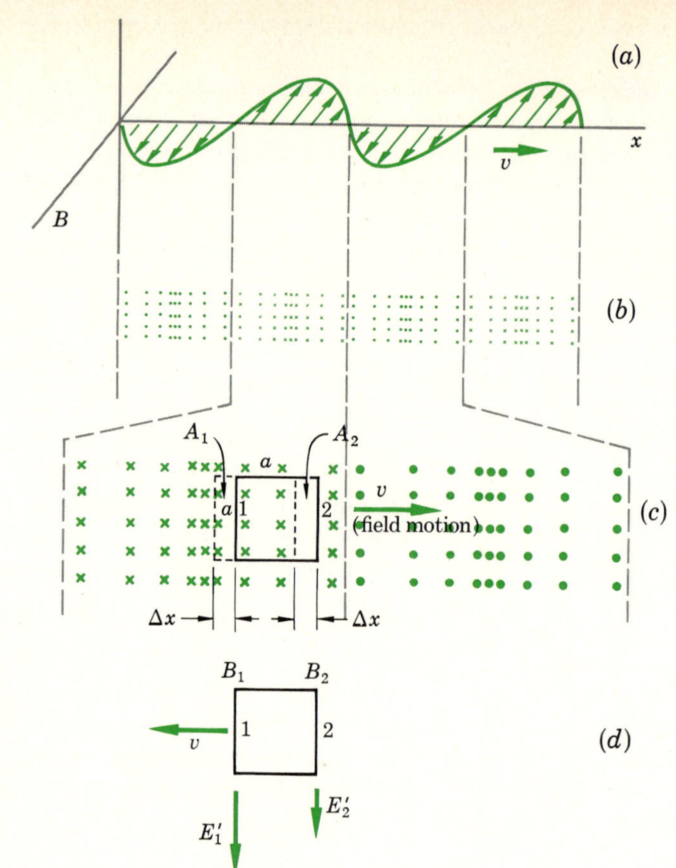

FIGURE 22.22
The magnetic portion of the electromagnetic wave, shown in (*a*) and (*b*) at a certain instant, travels to the right with speed v. In (*c*) the loop is displaced to the left along the wave as the wave moves to the right. This induces electric fields E_1' and E_2', as shown in (*d*).

during the small time Δt that it takes for the wave to travel the small distance Δx. (We indicate this displacement in the figure by showing the new position of the loop *on the wave* as the displaced area within the dotted lines. Actually, of course, the loop stays in one place, and the wave moves.) Since the distance Δx is actually much shorter than illustrated, the value of B on the rectangular area A_1 is very nearly B_1 and it is B_2 on A_2. As the wave moved, the flux through area A_2, namely, B_2A_2, was lost from the loop but the flux through A_1 was gained. We therefore have

$$\Delta\phi = \text{flux gain through loop} = B_1A_1 - B_2A_2 \qquad (22.1)$$

However, A_1 and A_2 are both equal to $a\,\Delta x$. Since Δx is merely the distance traveled by the wave in time Δt, we have that Δx is given by $v\,\Delta t$, where v is the speed of the wave. Substituting this value in Eq. (22.1) gives

$$\Delta\phi = (B_1 - B_2)v\,\Delta t\,a \qquad \text{or} \qquad \frac{\Delta\phi}{\Delta t} = av(B_1 - B_2)$$

According to Faraday's law this is simply the induced emf in the loop. Hence,

$$\text{emf} = av(B_1 - B_2) \tag{22.2}$$

You can easily verify that the emf is counterclockwise in the loop.

Let us now compute the emf in another way. Suppose that we were actually traveling through space with the EM wave. The wave would then be at rest with respect to us, i.e., in our reference frame, and the magnetic field would be stationary and unchanging. However, the loop would appear to be traveling to the left in Fig. 22.22*c* with a speed v. As we learned in Chap. 20, wires 1 and 2 will be found to have a voltage difference between their ends because of their motion through the magnetic field. The positive charges in the wires will experience a force directed toward the lower end of the wire because of their motion in the field, and so the induced emf will be directed toward the lower end of the wires. From Eq. (20.7), the induced emf in a wire of length a is Bva. Since the two induced emfs in wires 1 and 2 are opposite to each other in the loop, the net induced counterclockwise emf in the loop is

$$\text{emf} = av(B_1 - B_2) \tag{22.3}$$

which is identical to the result of Eq. (22.2).

The value of the computation leading to Eq. (22.3) is seen when one considers the fact that the induced emfs in wires 1 and 2 are the result of forces exerted on the positive charges of the wires. Since the emf or voltage difference across the wire is the force per unit positive charge, that is, E, multiplied by the length of the wire, we have

$$aE_1' = avB_1 \qquad \text{and} \qquad aE_2' = avB_2$$

or

$$E_1' = vB_1 \qquad \text{and} \qquad E_2' = vB_2 \tag{22.4}$$

where the directions of E_1' and E_2' are shown in Fig. 22.22*d*. We therefore see that a magnetic field B moving with speed v gives rise to an electric field E' which is perpendicular to it. The magnitudes of E' and B are proportional and are given by $E' = vB$.

The original B wave of Fig. 22.22*a* is redrawn in Fig. 22.23. Also, the induced electric field E' is drawn on the same figure by using the results we

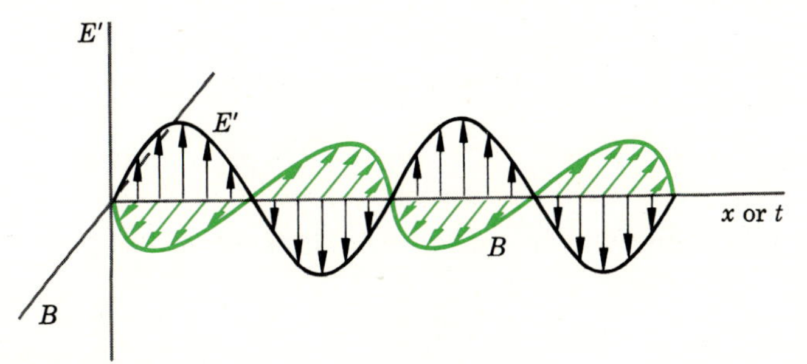

FIGURE 22.23
The magnetic portion of the EM wave traveling to the right generates an electric field wave E' as indicated.

have just found. Since $E' = vB$, both fields must reach their maximum values together and be zero together. Although they are perpendicular to each other, they are exactly in phase. The similarity between this figure and that for the EM wave shown in Fig. 22.20 is apparent. We are therefore led to suspect that the electric-field portion of an EM wave is, in a sense at least, generated by the magnetic-field portion of the wave as it travels through space. This will be given further confirmation by the results to be obtained in the next sections.

22.9 MAGNETIC FIELDS INDUCED BY CHANGING ELECTRIC FIELDS

We saw in the last section that the magnetic-field portion of an EM wave generates an electric field which travels along with it. This is another example of Faraday's induction law, which states that a varying magnetic field can give rise to an emf, and this implies that an electric field is generated in the process. Since a varying magnetic field gives rise to an electric field, we might well inquire whether a varying electric field gives rise to a magnetic field. That would mean that the varying electric field would act like a current, since all magnetic fields we have thus far encountered are generated by charge motion. Let us investigate the effect of a changing electric field to see how it might behave like a current.

A physical situation which combines electric fields and currents in a straightforward way is shown in Fig. 22.24. The long, straight wire shown there is carrying a current I to the capacitor and charging up the capacitor. Of course, this current cannot continue for long, since the current must decrease to zero as the capacitor becomes fully charged. However, during the charging process, the current at some instant is I, and we shall discuss what is happening at that instant. We shall ignore the fringing effects at the edge of the capacitor and assume E, the electric field, to be uniform between the plates.

FIGURE 22.24

The current I, charging a capacitor as shown, results in an equivalent current through the space between the plates. This equivalent current is proportional to the rate of change of electric flux passing between the plates.

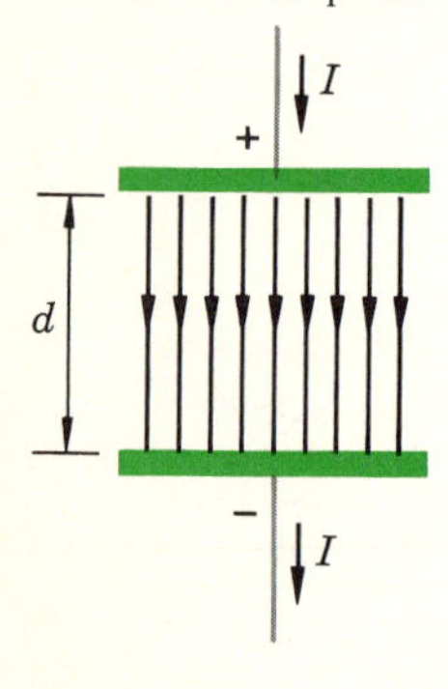

Since the voltage difference between the plates is just Ed, as we saw in Chap. 16, and since the charge Q on a capacitor of value C is given by $Q = CV$, we have at a particular instant

$$Q = CEd$$

At a time Δt later, the charge will have become $Q + \Delta Q$, and the field will have increased to $E + \Delta E$. Therefore

$$Q + \Delta Q = C(E + \Delta E)d$$

Subtracting the first of these equations from the second gives

$$\Delta Q = C\,\Delta E\,d$$

If we divide each side of this equation by the time taken for the change,

Δt, and replace C by its value given in Chap. 17, namely, $\varepsilon_0(A/d)$, where A is the area of a plate and ε_0, the permittivity of free space, is $8.85 \times 10^{-12}\ C^2/(N)(m^2)$, we find

$$\frac{\Delta Q}{\Delta t} = \varepsilon_0 A \frac{\Delta E}{\Delta t} \tag{22.5}$$

Since E may be thought of as the number of electric field lines per unit area in the same way that B is the number of magnetic field lines through unit area, the quantity AE is very often referred to as the **flux** *of* E through the area. It is designated as ϕ_E. The subscript E is necessary so that we shall not confuse the electric flux with the magnetic flux $\phi = BA$. In any event, $A\,\Delta E$ is the change in electric flux through the area of the capacitor, and so Eq. (22.5) can be written as

$$I = \varepsilon_0 \frac{\Delta \phi_E}{\Delta t} \tag{22.6}$$

Equivalence of Changing Electric Flux and Current

where use has been made of the fact that current is $\Delta Q/\Delta t$.

Equation (22.6) indicates that a changing electric flux due to a changing electric field is equivalent to a real current traveling in a wire. Because of this, we infer the following: a changing electric field will give rise to an equivalent current, which we can obtain from Eq. (22.6). This equivalent current will in turn generate a magnetic field, just as any real current would do. To find this magnetic field, we may make use of Ampère's circuital law, which was discussed in Chap. 19. We found that if we choose a closed path of total length $l = l_1 + l_2 + l_3 + \cdots + l_n$, the following equation applies:

$$\Sigma(B_\parallel l)_n = (\text{all the current encircled by closed path})(\mu_0)$$

The path has to be made up of lengths $l_1, l_2, \ldots, l_n$ such that $B_\parallel$ is essentially constant on each length. The left side of this equation says to take the component of B parallel to each length, multiply that component by the length, and add all such products for all the lengths which constitute the loop, or closed path. This sum is then equal to all the current which flows through the loop, multiplied by μ_0. From what has been said in regard to equivalent currents resulting from changing electric fields, we infer that the equivalent currents must also be included when writing down the current encircled by the loops.

The supposition that a changing electric flux acts exactly like a current with magnitude given by Eq. (22.6) flowing in the same direction constitutes one of the basic assumptions introduced by Maxwell in writing his celebrated equations. His justification for making the assumption was quite simple: only by making the assumption could he reconcile the results predicted by his equations with the results of experiment. Although he could give intuitive justifications for it, such as the one we have given, in the last analysis the assumption can be justified only by testing it against experiment. The experimental result which weighed most heavily in the justification of this equivalent-current concept was the measured speed of light. It can be predicted exactly by the use of the equivalent-current concept. We shall show one way of doing this in the next section.

22.10 THE SPEED OF ELECTROMAGNETIC WAVES

Let us now consider the motion through space of the electric-field portion of an EM wave. We shall follow much the same procedure that we used in Sec. 22.9 in applying the Faraday induction law to the magnetic-field portion of the wave. It will be recalled that we found there that the magnetic field B generated an electric field E' and that they were related by the equation

$$E' = vB \tag{22.7}$$

where v is the speed with which the wave travels.

In Fig. 22.25*a* we show graphically the electric field at a particular instant along the x axis. We show it in a horizontal instead of a vertical plane so that we can represent it more easily in the other portions of this figure. The

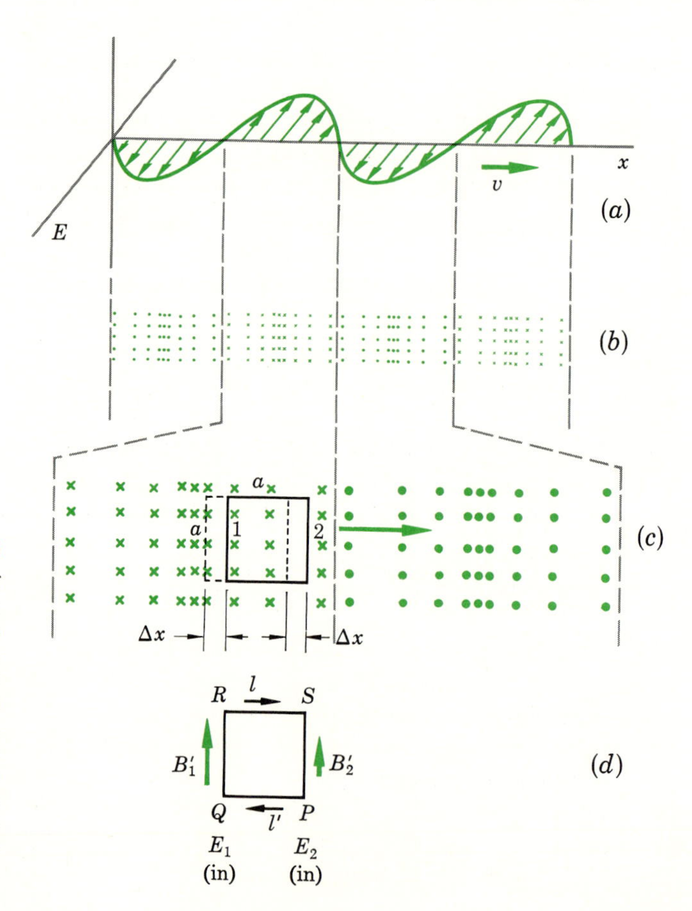

FIGURE 22.25

The electric-field portion of the electromagnetic wave shown in (*a*) and (*b*) travels to the right across the loop shown in (*c*). As a result, the loop is displaced to the left on the wave, as indicated. The change in electric flux through the loop gives rise to an equivalent current. This current generates magnetic fields in the vicinity of the loop.

wave is assumed to travel with speed v to the right. A cross section of the electric field wave is shown in part b and, enlarged, in part c. We wish to consider the change in electric flux through the rectangular loop shown in part c as the wave moves to the right across it. Interpreting this change in electric flux as an equivalent current, we shall apply Ampère's circuital law to the loop in an effort to find the magnetic field.

The wave traveling to the right with speed v will move a distance $v\,\Delta t$ during a time Δt. After a time Δt the loop will be displaced to the left a distance $\Delta x = v\,\Delta t$ on the wave, as shown by the dotted position in part c of the figure. In the process, an amount of flux $(a\,\Delta x)E_2$ is lost from the right side of the loop and a flux $(a\,\Delta x)E_1$ is gained at the left side. We therefore have that the change of flux is

$$\Delta\phi_E = (a\,\Delta x)E_1 - (a\,\Delta x)E_2 = (av\,\Delta t)(E_1 - E_2)$$

where we have replaced Δx by $v\,\Delta t$. The equivalent current will be given by

$$I = \varepsilon_0 \frac{\Delta\phi_E}{\Delta t}$$

or

$$I = \varepsilon_0 av(E_1 - E_2) \tag{22.8}$$

This will be the current encircled by the loop and the current to be used in Ampère's circuital law.

Referring to part d of Fig. 22.25, let us add the products $(B_\parallel l)_n$, starting at Q and moving clockwise through Q, R, S, P, and back to Q. Since E is constant at all points along the line of which QR is a part, the induced magnetic field along it will be constant and we call it B'_1. Similarly along SP the magnetic field will be constant, B'_2. Since side PQ and side RS of the loop are subject to identical electric fields, the values of B along them should be identical. Notice, however, that as we add length vectors proceeding around the loop in a clockwise direction, the vectors composing side RS will be oppositely directed to those on side PQ. Since $\mathbf{B}'$ is in the same direction on the upper side as on the lower side, if $\mathbf{B}'$ is parallel to $\mathbf{l}$ on the upper side it will be antiparallel to $\mathbf{l}'$ on the corresponding point of the lower side. Therefore in writing Ampère's circuital law, each $(B_\parallel l)_n$ on the upper side of the loop will be canceled by an equal and opposite value from the lower side of the loop.

Since the values of $(B_\parallel l)_n$ from the upper side of the loop cancel those from the lower side, the sum in Ampère's circuital law reduces to $B'_1 a - B'_2 a$. Equating this to μ_0 times the current encircled gives

$$B'_1 a - B'_2 a = \varepsilon_0 \mu_0 av(E_1 - E_2)$$

After solving for $E_1 - E_2$ this becomes

$$E_1 - E_2 = \frac{B'_1}{\varepsilon_0\mu_0 v} - \frac{B'_2}{\varepsilon_0\mu_0 v} \tag{22.9}$$

If the values of the electric field, E_1 and E_2, had been zero, i.e., if there had really been no field, then there would be no equivalent current and B_1' and B_2' would be zero. Furthermore, if the loop had been taken very long so that side QR was in a region where an electric field existed while side SP was in a region where E_2 was zero, then both B_2' and E_2 would be zero. We would therefore have, from Eq. (22.9),

$$E_1 = \frac{B_1'}{\varepsilon_0 \mu_0 v}$$

But since side QR could have been taken to be anywhere, this relation is a general one for any point, not just for the point where E is E_1. We are thereby justified in dropping the subscripts in the equation and writing in general

$$E = \frac{B'}{\varepsilon_0 \mu_0 v} \tag{22.10}$$

Equation (22.10) tells us that the moving electric-field portion of the EM wave generates a magnetic field. The changing electric field flux is equivalent to a current, and this equivalent current generates the magnetic field. Referring again to Fig. 22.25*c* and *d*, the electric flux increases into the page as the wave moves to the right (or as the loop moves to the left). This is equivalent to a current into the page. Our right-hand rule tells us that the magnetic field must act predominantly clockwise around the loop. Since E_1 is larger than E_2, Eq. (22.10) tells us that B_1' must be larger than B_2', as we have indicated them to be in the figure. The direction of B_1' and B_2' must therefore be upward, since B' must act clockwise around the loop. If we compare the relative orientation of B' to E_1, it is apparent that the relative orientations are the same as those shown in Fig. 22.23 between B and E'. (Recall that in drawing Fig. 22.25 we rotated the plane of the E vectors perpendicular to the plane of the page for ease in drawing. In making the comparison, mentally rotate the lower end of the diagrams 90° out of the page.) Hence, Fig. 22.23 also is applicable to the relationship we have found between E and B'.

The comparison between the present case and that treated earlier is clearly seen by reference to the resulting equations for the two cases. We found previously that the magnetic portion of the electromagnetic wave B generated an electric field E' given by

$$E' = vB \tag{22.7}$$

We now find that the electric-field portion E generates a magnetic field B' given by

$$E = \frac{1}{\varepsilon_0 \mu_0 v} B' \tag{22.10}$$

In both cases, E' and B as well as E and B' are in phase, and the relative directions of the fields are identical. This must mean that B' is identical to B

in regard to direction and phase. Similarly, E' and E are identical in direction and phase. The obvious question to ask is: Are E and E' identically the same and are B and B' identical as well? We have already shown their identity in direction and phase. If their magnitudes are the same, then the answer to our questions would be "yes."

Let us *assume* for the moment that E and E' as well as B and B' *are* the same. Division of Eq. (22.7) by Eq. (22.10) would then give

$$\frac{E}{E'} = \frac{1}{\varepsilon_0 \mu_0 v^2} \frac{B'}{B}$$

After canceling E with E' and B with B', we find

$$1 = \frac{1}{\varepsilon_0 \mu_0 v^2}$$

Solving for v gives

$$v = \sqrt{\frac{1}{\varepsilon_0 \mu_0}} \qquad (22.11)$$

Placing in the experimentally determined values $\mu_0 = 4\pi \times 10^{-7}\ \mathrm{N \cdot s^2/C^2}$ and $\varepsilon_0 = 8.85 \times 10^{-12}\ \mathrm{C^2/(N)(m^2)}$, we find

$$v = 2.998 \times 10^8\ \mathrm{m/s}$$

which is equal to the measured speed of light!*

This is a truly amazing result. We have used the Faraday induction law together with Maxwell's concept of an equivalent current to find the speed of EM waves, and that speed has turned out to be the measured speed of light. In doing so we made the assumption that the electric field generated by the changing magnetic field was actually the electric portion of the EM wave. Similarly, the changing electric-field portion of the EM wave induced a magnetic field which we have identified with the magnetic portion of the EM wave. In a sense, then, we have assumed that the EM wave regenerates itself as it travels through space. This assumption has led us to conclude that the wave travels with the speed of light. Many experiments have now shown this conclusion to be correct, thereby lending credence to our suppositions.

As we have already mentioned, Maxwell was the first to discover this interrelationship between light and electromagnetism. He concluded that light was an EM wave, a conclusion amply verified during the years since then. Although radio waves had not yet been discovered by the time that Maxwell died in 1879, Heinrich Hertz (1857–1894) succeeded in generating EM waves other than light in 1887. These waves were later shown to travel with the speed of light, and so Maxwell's prediction was fully justified.

---

* We shall see in the next section that the coulomb and ampere are now defined by assuming this to be exactly true.

22.11 DEFINITION OF THE ELECTRICAL UNITS

In the previous sections we learned the following important facts:

1 EM waves travel through vacuum at a speed $c = 1/\sqrt{\varepsilon_0\mu_0}$. This has a numerical value of 2.998×10^8 m/s.
2 In an EM wave in vacuum, **E** and **B** are mutually perpendicular. They are related through $E = cB$.

We are now prepared to state precisely the definitions of the quantities we have been using in our study of electricity. The delay has been necessitated by the fact that the velocity of light as expressed in terms of ε_0 and μ_0 forms the basis for these definitions. Now that we have shown that *the velocity c of EM radiation in vacuum is* $\sqrt{1/\varepsilon_0\mu_0}$, we can make use of this fact to define values for ε_0 and μ_0.

Values of μ_0 and ε_0

The permeability of free space μ_0 *is arbitrarily defined to be* $4\pi \times 10^{-7}\ N \cdot s^2/C^2$. Using this value, *we can define* ε_0 *in terms of the measured speed of light c.* Thus, since

$$\mu_0 = 4\pi \times 10^{-7}\ \mathrm{N \cdot s^2/C^2}$$

and

$$c^2 = 1/\varepsilon_0\mu_0 = (2.998 \times 10^8\ \mathrm{m/s})^2$$

we have

$$\varepsilon_0 = 8.85 \times 10^{-12}\ \mathrm{C^2/(N)(m^2)}$$

In order to define the unit of current, the ampere, we make use of the fact that a wire carrying a current experiences a force when placed in a magnetic field. If we consider two long, parallel, straight wires through which the same current I flows, the wires will experience forces because each is in the magnetic field of the other. When the separation of the wires is d, the magnetic field at one, because of the current of the other, is simply

$$B = \frac{\mu_0 I}{2\pi d}$$

as we found in Eq. (19.6). The field will be perpendicular to the second wire, and so the force on a length L of the wire will be

$$F = BIL$$

or, after substituting for B and dividing through by L,

$$\frac{F}{L} = \frac{\mu_0 I^2}{2\pi d}$$

Since the force on unit length of the wire, F/L, can be measured, together with the separation of the wires d, the current I can be evaluated in terms of known quantities. When F is measured in newtons, and when μ_0 is given the value $4\pi \times 10^{-7}\ \mathrm{N \cdot s^2/C^2}$, the current is in the unit we define to be the ampere. Hence the ampere is defined directly in terms of force and length, both fundamental units.

Electrical Units

The coulomb of charge is defined to be the charge carried through a cross section of a wire in one second when the current in the wire is one ampere. As a result, the definition of the coulomb is based directly upon the same measurement used to define the ampere.

To define the unit of flux density, B, we make use of the relation

$$F = B_{\perp}IL$$

If a 1-m length of wire carries a current of 1 A perpendicular to a magnetic field and the force on that wire is 1 N, then the value of the flux density B is, by definition, 1 T (or $\mathrm{Wb/m^2}$), or $1\ \mathrm{N \cdot s/(C)(m)}$.

Definition of the other quantities used in electricity has already been made in terms of the quantities defined above, together with force, length, and time units. We shall not repeat them all here. However, it should be pointed out that we have succeeded in defining all the electrical units in terms of definite experiments involving the measurement of forces, lengths, and times. As a result, anyone who is able to duplicate our units for these three basic quantities will be able to duplicate our electrical units as well.

SUMMARY

When substances are heated white hot, electrons "evaporate" from them in a process called thermionic emission. The most energetic electrons in the material have enough thermal energy to supply the work-function energy, the energy needed to tear an electron loose from the material.

Both the thermionic and solid-state diode conduct current in only one direction. As a result, they can be used to change ac voltages and currents into dc voltages and currents. Solid-state electronic devices make use of semiconducting materials. In n-type materials, electrons constitute the current carriers. In p-type materials, the current carriers are electron vacancies called holes. They act like positive-charge carriers.

When an oscillator causes reversing charges to appear on the transmitting antenna of a radio station, an electric and magnetic field wave is sent out from the antenna. This electromagnetic (EM) wave consists of two mutually perpendicular waves, one electric, the other magnetic. The two waves are in phase. They travel through empty space with the speed of light, $c = 2.998 \times 10^8$ m/s.

There are many types of EM waves. They differ in wavelength. Radio waves have λ of the order of hundreds of meters. At successively shorter wavelengths are radar, infrared (heat), visible light, ultraviolet light, x-rays, and γ rays. They cover a wavelength range down to less than 10^{-10} m. In all EM waves, $E = cB$ in vacuum.

Electrical units are defined in terms of the speed of light. By definition, $\mu_0 = 4\pi \times 10^{-7}\ \mathrm{N \cdot s^2/C^2}$. Since the speed of light is given by $1/\sqrt{\varepsilon_0 \mu_0}$, this fact is used to define ε_0. All other electrical units can then be defined in terms of the units of mass, length, and time.

MINIMUM LEARNING GOALS

Upon completion of this chapter you should be able to do the following:

1. Give a qualitative explanation why thermionic emission depends upon temperature.

2. Explain how a vacuum-tube diode operates and describe its use as a rectifier.

3. Describe how an n-type semiconductor can be made from silicon. Repeat for a p-type semiconductor.

4. Explain qualitatively why a p-n type junction shows rectification. Give the symbol for a solid-state diode and show which direction it passes current.

5. Draw a circuit for a half-wave rectifier and explain

how its output can be smoothed. Draw a circuit for a full-wave rectifier and explain its principle of operation.
6. Sketch the E and B fields in an EM wave. Compute the wavelength of a radio wave when the frequency of the radio station is given.
7. Arrange a list of various types of EM waves in order of decreasing wavelength. State which type of wave a given wavelength belongs to.
8. Describe the two ways in which radio waves can be detected by a radio. Explain the function of an LC circuit in the radio. Explain how a particular station is selected by the radio.
9. Give the speed of EM waves in vacuum.

IMPORTANT TERMS AND PHRASES

You should be able to define or explain each of the following:
Thermionic emission
Vacuum-tube diode
p-type and n-type semiconductors
Solid-state diode
Half-wave and full-wave rectifiers
EM wave
EM wave spectrum
Speed of EM waves c

QUESTIONS AND GUESSTIMATES

1. All other things being equal, which type of material would be preferable for the filament of a vacuum tube, one with large work function or one with small work function?
2. The circuit shown in Fig. P22.1 is a full-wave rectifier (often referred to as a bridge-type rectifier). Examine it and explain why the current is rectified and flows during both halves of the cycle.
3. Some radio stations have their transmitting antenna vertical while others have theirs horizontal. Describe and compare the EM waves generated by these two types of antennas. In particular, how are **B** and **E** directed relative to the earth's surface?
4. Refer to the previous question. If you open up a transistor radio, you can see how its coil antenna is mounted. How could you use the radio to tell whether a distant station's antenna is vertical or horizontal.
5. Electromagnetic waves from most of the radio stations in the world are passing through the region around you. How does a radio or TV set select the particular station you want to listen to? When you turn the dial on a radio, what is happening inside to select the various stations?

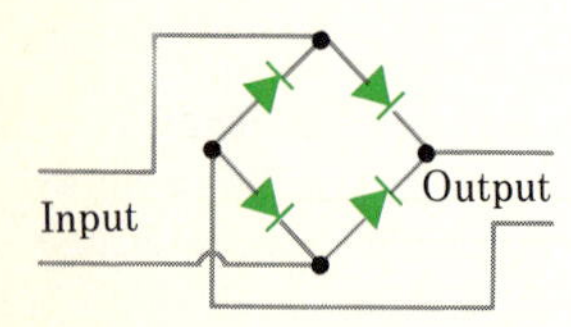

FIGURE P22.1

6. There are two types of radio and TV receiving antennas in use; one picks up the electric part of the EM wave, and the other picks up the magnetic. Examine a pocket transistor radio or a table radio and see which method is used. Is it possible to use both?
7. From time to time in the movies or on TV one sees the good guys trying to locate a clandestine radio transmitter by driving through the neighborhood with a device which has a slowly rotating coil on top. Explain how the device works.
8. It is claimed that in the vicinity of a very powerful radio-transmitting antenna, one can sometimes see sparks jumping along a wire farm fence. What do you think of this claim?
9. In microwave ovens, foods and utensils are subjected to very-high-frequency radar (EM) waves. If a spoon is left in such an oven, it becomes very hot. What heats it? Can you explain the heating action in terms of the electric part of the wave? The magnetic? How are nonmetallic substances heated in the oven? Will a glass dish heat up in such an oven?
10. There is some doubt about the safety of human exposure to intense radio and microwaves. Why would one expect the danger to depend upon the frequency of the waves? Which would you expect to present the most danger (if any), radio or microwaves?

PROBLEMS

1. The temperature of the white-hot filament in a light bulb or vacuum tube is of the order of 2500 K. (*a*) What is the average KE of a free electron in the filament? (*b*) Compare this with the work function of tungsten, about 4.5 eV.

2. Assuming the valence electrons in a metal to be free, what is the average speed of an electron in a white-hot piece of metal at 2500 K?

3.** Assuming one free electron per atom, (*a*) find the pressure of the electron gas inside a block of copper at 300 K. (*b*) Compare this with standard atmospheric pressure. (Density of copper = 8.92 g/cm^3 and atomic mass = 63.5.)

4. A half-wave rectifier system is connected to a 50,000-Ω resistor as its load. It is rectifying 60-Hz ac voltage. (*a*) What magnitude filter capacitor must be used (as in Fig. 22.10) if the time constant of the filter-load system is to be 10 times as large as the "no-current" time of the rectifier? (*b*) Why is it difficult to obtain a smooth dc voltage from a 60-Hz rectifier system if large currents are to be drawn from it?

5. If, in Fig. 22.10, the output voltage has a maximum value of 12.0 V and a minimum value of 9.6 V, what is the value of the ripple for the rectifier system?

6. Radio station WJR in Detroit operates on a frequency of 760 kc/s. (*a*) What is the wavelength of the wave it sends out? (*b*) How long does it take a wave crest sent out from the station to reach the moon, 3.84×10^8 m away?

7. As your heart beats, small potential differences are set up between various parts of your body. (These are measured by electrocardiograms and similar medical tests.) Suppose your heart beats 75 times each minute. (*a*) What is the frequency of the EM wave sent out by your body because of this consequent oscillating voltage and charge on your body? (*b*) What is the wavelength of the very faint EM wave your body sends out?

8. One way to provide heat to muscles and other portions of the body is by means of diathermy. In this, radar waves are sent into the body much like a microwave oven sends waves into the material to be heated. The oscillating electric field of the wave causes dipolar molecules and ions to move back and forth, thereby generating friction-type heat. The oscillating magnetic field of the wave induces emfs which cause Joule heating. Standard diathermy frequencies are 900 and 2560 MHz. (*a*) What are the wavelengths of EM radiation in air which arise from these frequencies? (*b*) Can you see any problem in using diathermy on the face of someone whose teeth have metal fillings?

9. An explosion occurs at a distance of 5.0 km from an observer. How long after the observer sees the explosion will the sound from the explosion be heard? (Take the speed of sound to be 340 m/s.)

10. A wire 10 m long is oriented in the direction of the electric field vector in an electromagnetic wave. (*a*) If the voltage difference between the ends of the wire is to have a maximum value of 10^{-4} V, how large must the maximum electric field of the wave be? (*b*) What is the maximum value of the magnetic field in the same wave? (This would be a rather weak signal.)

11.* A rectangular loop antenna measures 30 by 10 cm. It has 200 turns of wire on it. About how large an emf can be generated between the terminals of the loop by an electromagnetic wave of frequency 10^6 cps which has a maximum electric field of 10^{-5} V/m?

12.** Two circular plates each of area A placed one above the other act as a parallel-plate capacitor. Their separation is d. Show that B between the plates is given by $\mu_0 Ir/2A$ when a current I is flowing into the capacitor. The distance r is the radial distance measured from the line of centers of the plates to the point being considered. *Hint:* Apply Ampère's circuital law to a circle of radius r through the point where B is to be found. Use the equivalent current through it.

13. The negative electron circles the positive nucleus with a frequency of 3.6×10^{15} Hz in Bohr's picture of the hydrogen atom. If this system obeyed the laws studied in this chapter, it should act like an antenna radiating waves of this frequency. (*a*) Find the wavelength of the waves and (*b*) state in which portion of the EM spectrum they should be found.

THE PROPERTIES OF LIGHT

23

In this and the next few chapters, we shall be concerned primarily with a very small portion of the entire electromagnetic spectrum. This portion is composed of the small range of wavelengths to which the eye is sensitive, the wavelengths referred to as **light.** However, even though our primary concern will be with light, much of what we learn will be applicable to all EM radiation.

23.1 THE CONCEPT OF LIGHT

Even in ancient times the properties of light were a source of wonder and a stimulus to experimentation. Its nature has always been a subject of great speculation. In Newton's time, scientific investigations into the properties and nature of light were conducted by nearly all the scientists of the day. Newton himself derived a great deal of his fame from his experiments with light.

In spite of this wide interest in light, its very nature remained in dispute even until the first decade of the present century. During Newton's time, and for many years later, there was disagreement whether a light beam was a stream of corpuscles or a wave of some sort. Newton was a great proponent of the corpuscular theory, and, because of his prestige, many others were inclined to this view as well. In 1670 Christian Huygens, Newton's contemporary, was able to explain many of the properties of light by considering it to be wavelike in nature. Both these ideas concerning the nature of light had their supporters.

It was not until 1803, when Thomas Young (and a little later Augustin Fresnel) presented evidence to show that light beams could interfere with each other much like sound waves, that the wave theory became almost universally accepted. At about this time the speed of light in water was measured. The observed speed was less in water than in air. This contradicted the corpuscular theory and supported the wave theory. Hence, by 1865, when Maxwell found theoretically that EM waves should travel with the speed of light, the idea of light waves was fairly well accepted.

One would think, then, that by 1900 the nature of light was reasonably well understood. However, at that time we still knew very little about the emission of light by atoms. It was not until about 1913 that Bohr gave the first reasonably correct interpretation for the mechanism of light emission. His concepts were greatly modified, and it was not until about 1930 that the emission of light could be said to be well understood. In addition, Einstein showed in 1905 that at least one property of light, the photoelectric effect, which will be discussed in Chap. 26, was best explained by considering light to act as quanta or particles. This concept has been expanded through the years until we now consider light to possess a sort of dual personality, part wavelike and part particlelike. More will be said about these and other developments in later chapters.

It is clear that the subject of light has a long and varied scientific history. We expect that in years to come our understanding of the nature of light will continue to grow. For the next few chapters, however, it will be sufficient to concentrate upon the aspects of light evident from its EM character. Other characteristics of light involving its particle nature and dealing with its atomic origin will be discussed in later portions of this book.

Wavelengths of Light

The wavelength of visible light waves can be measured by methods to be discussed in Chap. 25. These wavelengths turn out to lie in the range 4×10^{-5} to 7×10^{-5} cm. The position of the various colors on the wavelength scale is shown in Fig. 23.1 and also on the color plate insert. Notice too that three units of length measurement are introduced in Fig. 23.1.

OPTICAL SPECTRA

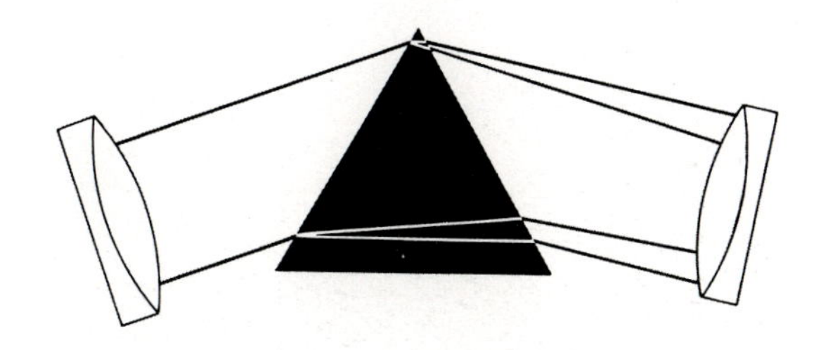

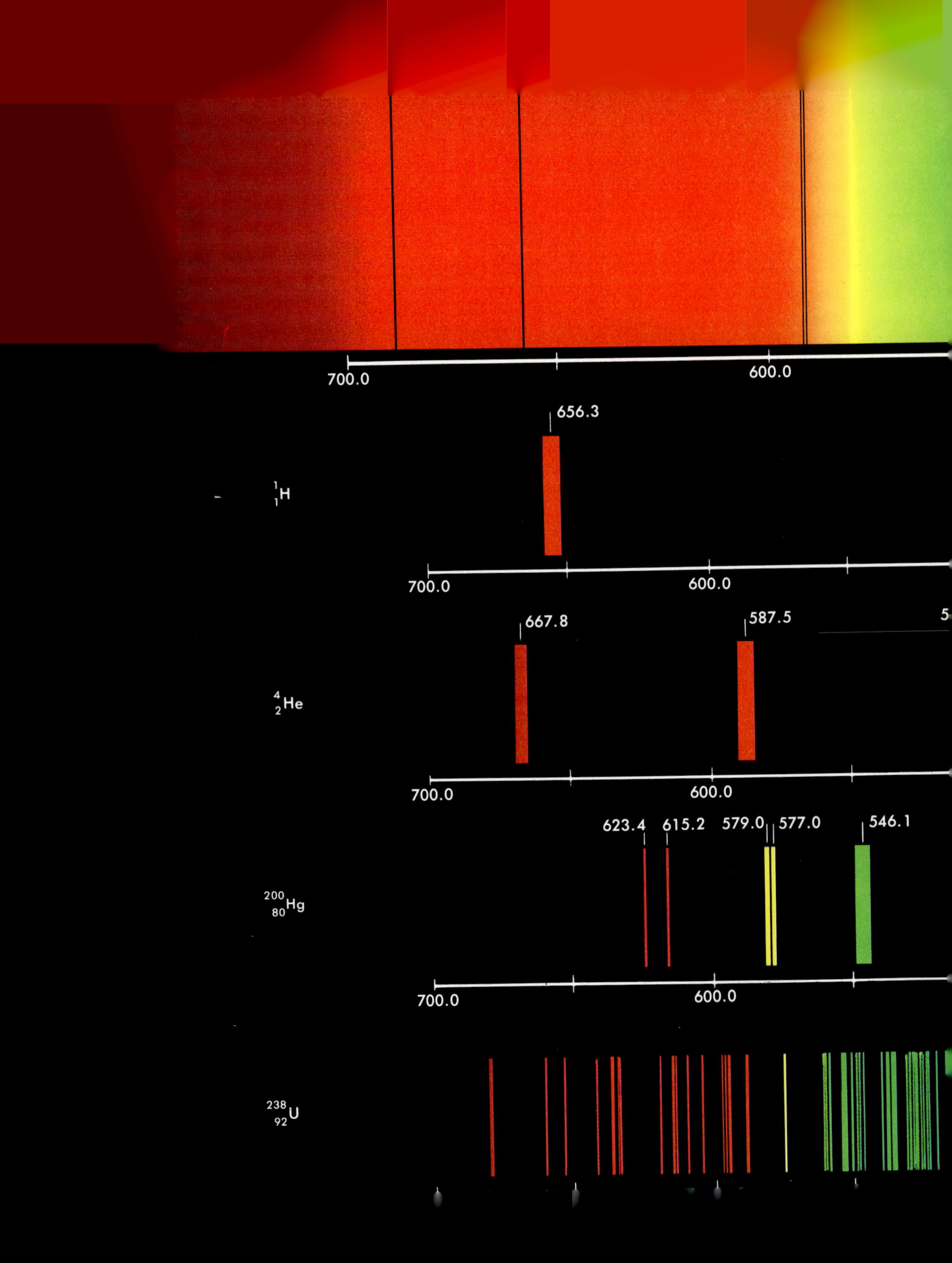

700.0
600.0
656.3
$^{1}_{1}H$
700.0
600.0
667.8
587.5
$^{4}_{2}He$
700.0
600.0
623.4
615.2
579.0
577.0
546.1
$^{200}_{80}Hg$
700.0
600.0
$^{238}_{92}U$

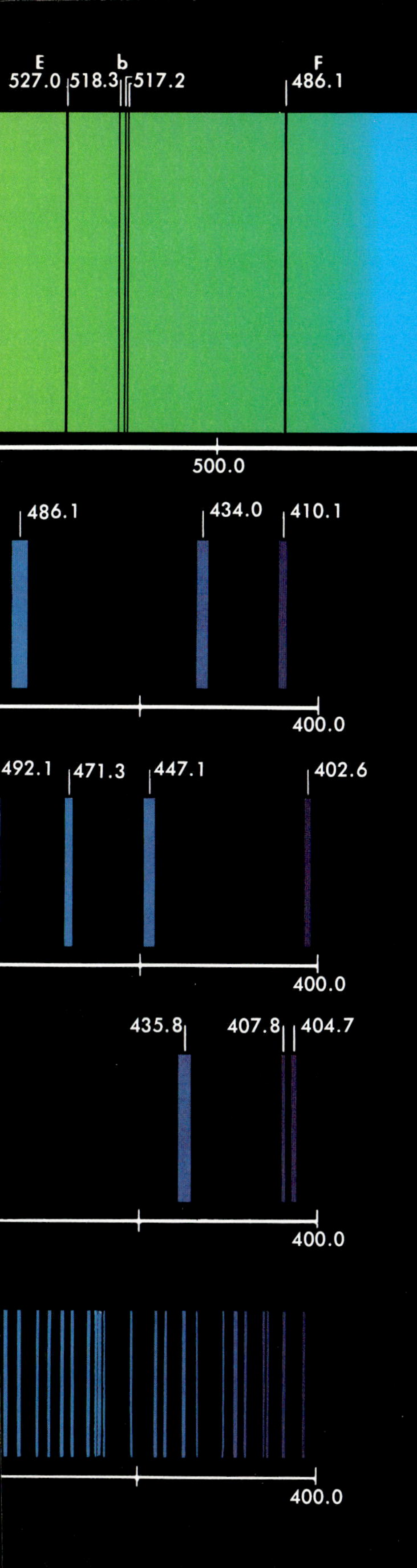

Such diverse and fundamental information on the nature of matter as the composition of distant stars and the structure of atoms and molecules has been obtained by analysis of the light emitted from substances heated to incandescence.

In the SPECTROSCOPE, such light, passed through a slit and a prism, is broken up into its component wavelengths, which are observed as colored lines (i.e., light of different energies) characteristic of the differences between the various electron energy levels of the atoms. This EMISSION SPECTRUM is CONTINUOUS when the images of the wavelengths are uninterruptedly overlapping; it is a LINE SPECTRUM when only certain specific wavelengths are emitted, as shown here for the elements hydrogen, helium, mercury, and uranium.

On the solar spectrum across the top of this plate appears a series of dark lines—FRAUNHOFER LINES—forming an ABSORPTION SPECTRUM. Some of the light from the intensely hot interior of the sun is absorbed by the cooler gases of its outer layers as the light energies raise the atoms in the cooler layers to higher energy states; bright lines are not, therefore, seen for these changes.

The spectra are calibrated in nanometers (1 nm = 10^{-9} m); the letters are arbitrary designations introduced by Fraunhofer for lines important in spectroscopy.

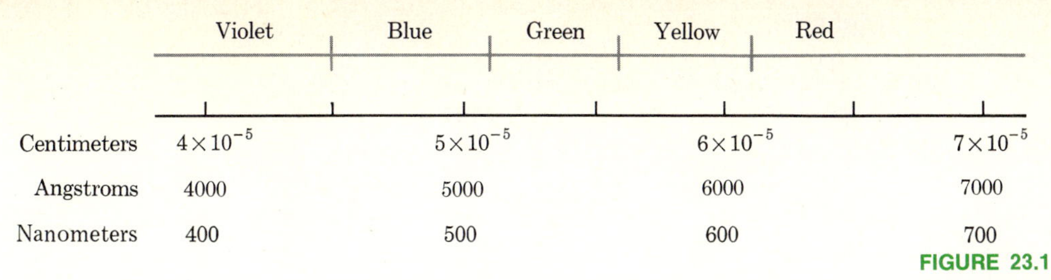

FIGURE 23.1

The correspondence between wavelengths and color shown here is only approximate. Colors such as blue-green and orange occupy the intermediate regions. (See the color plate.)

Commonly, light wavelengths are measured in

Angstroms (Å): $1\ \text{Å} = 10^{-8}\ \text{cm} = 10^{-10}\ \text{m}$
Micrometers (μm): $1\ \mu\text{m} = 10^{-4}\ \text{cm} = 10^{-6}\ \text{m}$
Nanometers (nm): $1\ \text{nm} = 10^{-7}\ \text{cm} = 10^{-9}\ \text{m}$

Light waves are electromagnetic in nature, consisting of an electric field perpendicular to, and in phase with, a magnetic field, as discussed in the last chapter. The electric field for a wave propagating in one direction in space is shown in Fig. 23.2. It is assumed that the wave is moving along the x axis to the right. Notice that the vibrating electric field is perpendicular to the x axis. Hence, *light waves are transverse waves,* since the vibration is perpendicular to the direction of propagation. As such, they will have many properties in common with waves on a string or on the surface of water, since these, too, are transverse waves.

23.2 THE SPEED OF LIGHT

There are several methods for determining the speed of light. We shall mention here only one of these general methods, the method which has been most widely used in high-precision work. It was used by Michelson in the early 1930s to measure the speed of light between two mountain tops in California. The same general method had been used previously over much smaller distances to measure the speed of light in materials other than air.

The method uses the apparatus shown in simplified form in Fig. 23.3. A beam of light from the source is reflected from one side of a cube M having

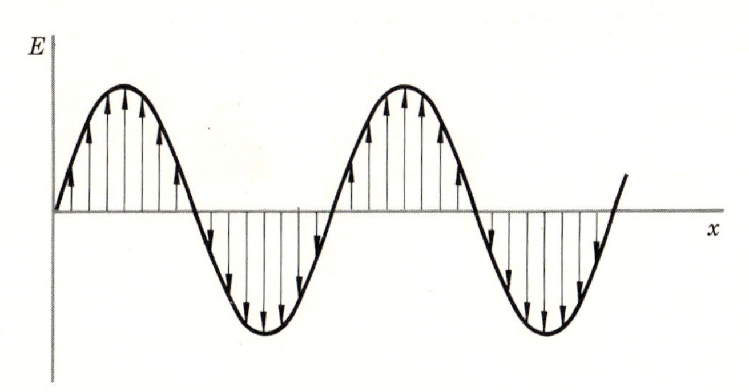

FIGURE 23.2

The electric field of the electromagnetic wave vibrates perpendicular to the direction of propagation. Hence the wave is transverse.

TABLE 23.1
REFRACTIVE INDICES ($\lambda = 5890$ Å)

MATERIAL	$c/v = n$	MATERIAL	$c/v = n$
Air*	1.0003	Crown glass	1.52
Water	1.33	Sodium chloride	1.53
Ethanol	1.36	Polystyrene	1.59
Acetone	1.36	Carbon disulfide	1.63
Fused quartz	1.46	Flint glass	1.66
Benzene	1.50	Methylene iodide	1.74
Lucite or Plexiglas	1.51	Diamond	2.42

* Normal temperature and pressure.

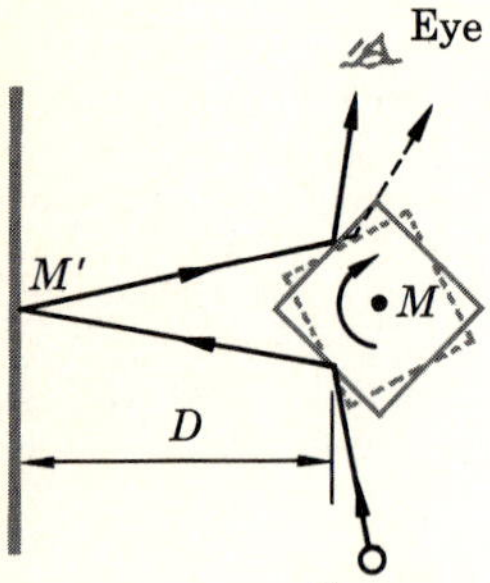

FIGURE 23.3
If the mirror is rotating at just the proper speed, the beam will be reflected into the eye of the observer. In practice the distance D is much larger than shown.

mirrored surfaces on four sides. The beam of light is then reflected from mirror M', back to the cube, where it is reflected again as shown. If the cube is at just the right position, the beam of light will enter an observer's eye in the position indicated. (Actually the observer looks through a telescope system placed where the eye is shown and sees an image of the source.)

Suppose, however, that the cube is rotating about an axis through its center, perpendicular to the page. When it is in the position indicated by the heavy lines in Fig. 23.3, the beam is reflected to mirror M', as shown. However, by the time the beam returns to the cube from M', the cube will perhaps have rotated to the dashed position and the light beam will not be reflected into the observer's eye. The cube must rotate through $\frac{1}{4}$ rev during the time needed for the beam to travel to M' and back if the beam is to be properly reflected, for only then will the cube reflect the beam into the eye of the observer.

The measurement technique is to speed the rotation of the cube until the reflected beam is properly reflected. At that speed of rotation we know that the time taken for $\frac{1}{4}$ rotation of the cube is equal to the time the light takes to travel a distance $2D$. One need know only the speed of rotation of the cube and D in order to compute the speed of the light. The value at present accepted for the speed of light in vacuum is

Speed of Light

$$c = 2.997925 \times 10^8 \pm 0.000010 \times 10^8 \text{ m/s}$$

The speed of light in air is only about 0.03 percent less than this value.

Light travels fastest through vacuum. Its speed in other materials is always less than c. Moreover, the speed in materials other than vacuum depends upon the wavelength of light as well as the constitution of the material. We have listed in Table 23.1 the ratio of the speed of light in vacuum to that in various materials. The wavelength used for the table is $\lambda = 5890$ Å. This is the wavelength of the yellow light given off by a sodium-vapor lamp.

23.3 REFLECTION OF LIGHT

When a stone is dropped into a large, still pond of water, a set of circular waves moves out from the point where the stone hit the water. We are all familiar with this situation, but let us look at it in some detail. The circular waves, or **wavefronts,** are shown in Fig. 23.4. They travel outward from the center in the directions shown by the arrows. These arrows, in the direction the wavefronts travel, are called **rays.** Notice that *the rays are always perpendicular to the wavefronts.* Hence we can specify the motion of a wave either by use of rays or by drawing the wave itself. Both methods are of value.

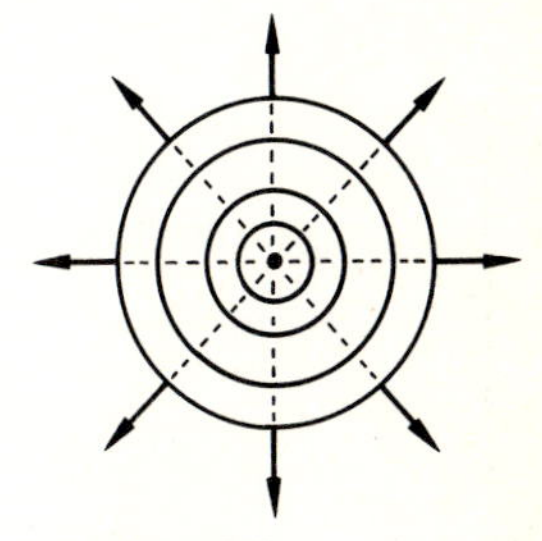

FIGURE 23.4
The rays are perpendicular to the wavefronts and show the direction of motion of the wave.

If we consider the case where we are far removed from the source of the wave, the wavefronts become nearly straight lines (or flat planes if we are dealing with three dimensions). This is illustrated in Fig. 23.5. As we can infer from that figure, at distances far removed from the source, the waves are plane waves, and the rays are parallel. It is often convenient to use plane waves in computations, and when the source is far removed, or when a suitable lens is used, the waves are very nearly of this sort.

Suppose that a series of plane water waves is incident upon a flat wall, as shown in Fig. 23.6*a*. The velocity of the wave can be thought of as being split into two components, one perpendicular to the wall, the other parallel to the wall. Upon striking the wall, the perpendicular component of the velocity will be reversed in direction, and the wave will be reflected back upward as shown in Fig. 23.6*b*. If the incident wavefront *CD* had continued in its original direction, it would have been in position *CD* in Fig. 23.6*b*. Instead, it was reflected upward, and so the original point *C* on the wavefront actually is at point *A* above the surface with $\overline{AB} = \overline{BC}$. Notice also the direction of motion of the wave after reflection.

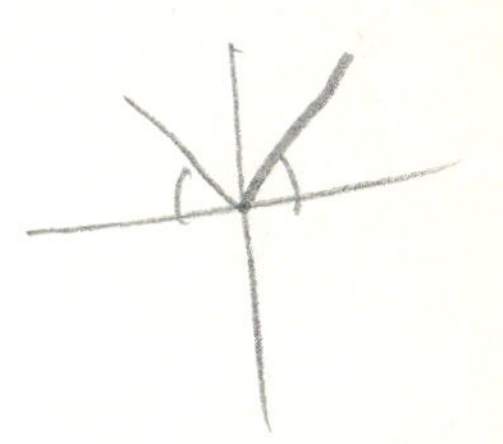

In Fig. 23.6*b*, the fact that $\overline{AB} = \overline{BC}$ tells us at once that all the four angles indicated as θ in the figure are equal. (Why?) Moreover, since the rays are perpendicular to the wavefronts, the angles made by the rays with the normal to the surface are also θ, as shown in part *c* of the figure. (Why?) This leads us to conclude that the angle of incidence i is equal to the angle of reflection r.

The fact that a water wave is reflected in such a way that the angle of incidence equals the angle of reflection is of general validity. We could have used the same reasoning to show that light waves would also be reflected in this way. Notice that the only basic assumption made was that the velocity component perpendicular to the surface was reversed upon reflection and the other component was unchanged. Our result will be true for any type of wave for which this assumption is true. Measurements on light and other forms of electromagnetic radiation confirm our deduction. We may therefore formu-

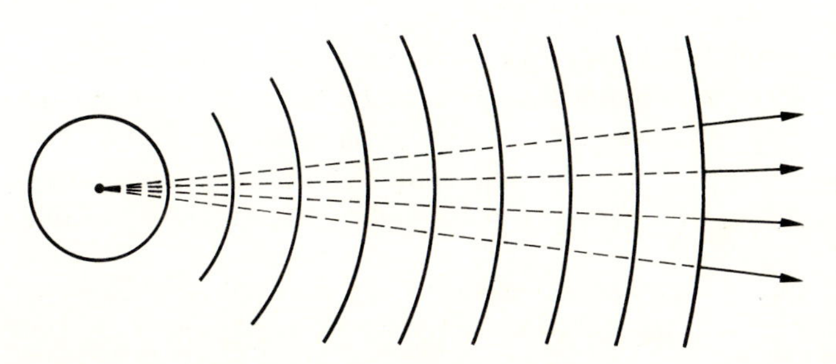

FIGURE 23.5
As the spherical wave spreads out farther from the source, it becomes more like a plane wave and the rays become more nearly parallel.

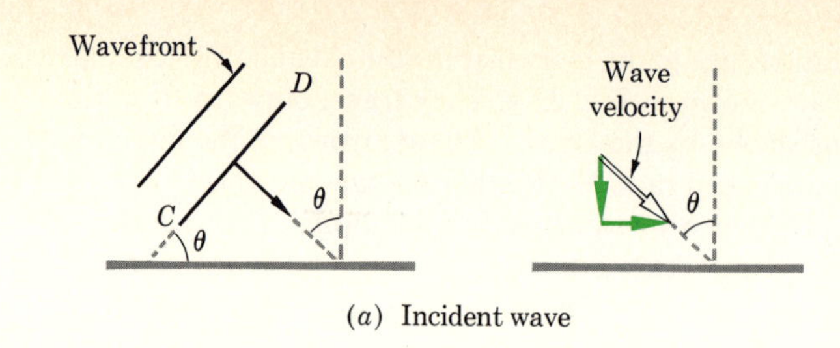

(a) Incident wave

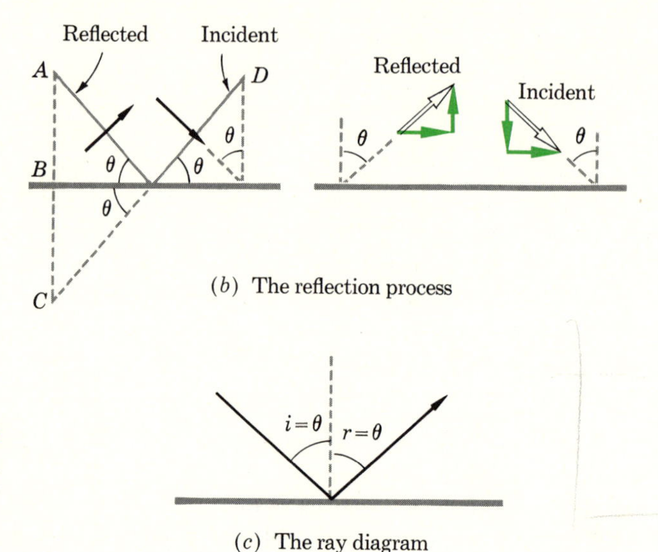

(b) The reflection process

(c) The ray diagram

FIGURE 23.6
The incident wave is reflected in such a way that the angle of incidence *i* equals the angle of reflection *r*.

late the following rule:

Law of Reflection

The angle of incidence equals the angle of reflection.

23.4 REFRACTION OF LIGHT: SNELL'S LAW

Refraction

When a beam of light enters water from air, its path bends, as shown in Fig. 23.7. *This change in direction of a ray as it passes from one material to another is called* **refraction.** The angle θ_1 is, of course, the angle of incidence, and angle θ_2 is called the angle of refraction. Some of the light beam hitting the water surface will also be reflected, as shown by the dashed ray in Fig. 23.7, but we ignore this in this section.

In order to find a relation between θ_1 and θ_2, it is convenient to consider the motion of the wavefronts in a plane wave. The situation is shown in Fig. 23.8. We shall say that the light has a speed v_1 in the upper material and a speed v_2 in the lower material, with v_1 being greater than v_2. (If the upper material is air, $v_1 = c = 3 \times 10^8$ m/s.) You should be able to show that the angles labeled θ_1 and θ_2 are the same in the two parts of the figure. We assume that the wavefront ABC has moved to position $A'B'C'$ after a time t.

Hence we have in Fig. 23.8*b*

$$d = v_1 t \qquad \text{and} \qquad l = v_2 t$$

which yields, after division of one equation by the other,

$$\frac{d}{l} = \frac{v_1}{v_2}$$

Moreover, from Fig. 23.8*b* we see that

$$\frac{d}{BB'} = \sin\theta_1 \qquad \text{and} \qquad \frac{l}{BB'} = \sin\theta_2$$

which, after division of one by the other, gives the relation

$$\frac{d}{l} = \frac{\sin\theta_1}{\sin\theta_2}$$

But since $d/l = v_1/v_2$, the relation becomes

$$\frac{\sin\theta_1}{\sin\theta_2} = \frac{v_1}{v_2} \tag{23.1}$$

This relation is one form of **Snell's law.** In its more common form it is usually written for the case where material 1 is vacuum. In that case $v_1 = c$, and

$$\frac{\sin\theta_1}{\sin\theta_2} = \frac{c}{v_2} = n \tag{23.2}$$

DEFINITION

The quantity n is called the **absolute index of refraction** of material 2 (or, more commonly, the index of refraction of material 2). *The absolute index of refraction of a material is the ratio of the speed of light in vacuum to the speed of light in the material.* We have already listed this quantity for various materials in Table 23.1. Notice that the absolute index of refraction is always

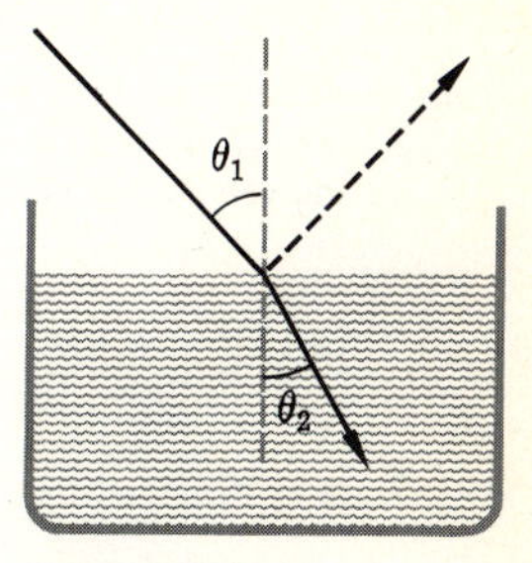

FIGURE 23.7
When a ray of light passes from an optically less dense material to an optically more dense material (air to water, for example), the ray is refracted toward the normal to the surface.

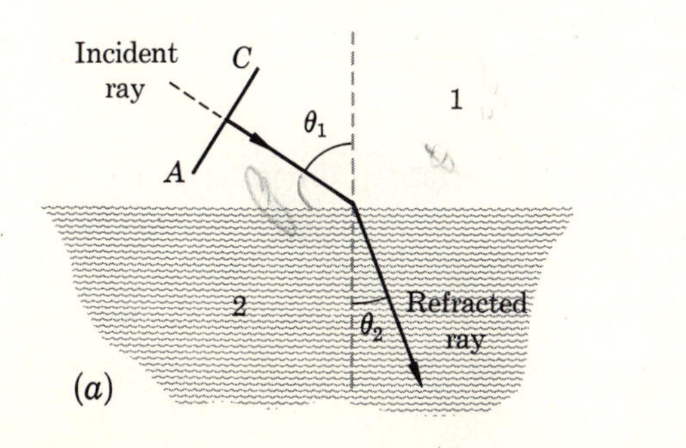

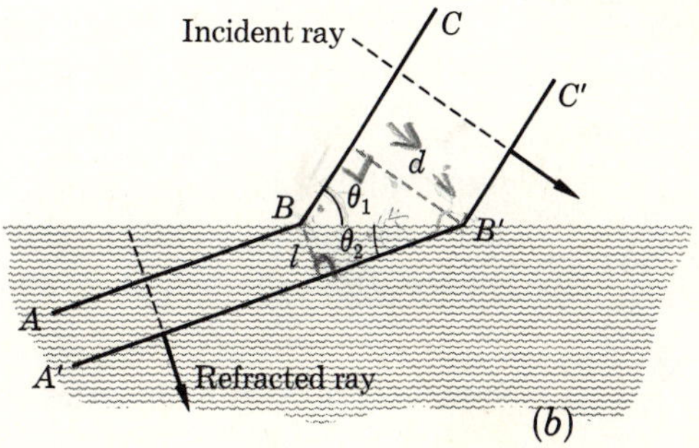

FIGURE 23.8
Since the wave travels slower in the lower material than it does in the upper material, θ_2 is smaller than θ_1.

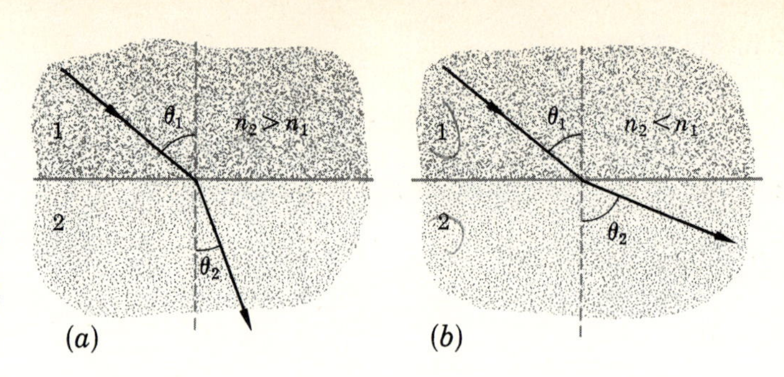

FIGURE 23.9

If $n_2 > n_1$, the beam bends toward the normal, while if $n_2 < n_1$, the reverse is true.

larger than unity or equal to it. Of course, the index of refraction of vacuum would be c/c, or unity.

Frequently the ratio v_1/v_2 in Eq. (23.1) is called the **relative index of refraction.** This terminology arises because $n_1 = c/v_1$ and $n_2 = c/v_2$, so that one has $v_1/v_2 = n_2/n_1$. Hence, Eq. (23.1) can be written

Snell's Law

$$n_1 \sin \theta_1 = n_2 \sin \theta_2 \tag{23.3}$$

which we shall refer to as **Snell's law.** If material 1 is air, this relation reduces to Eq. (23.2), since $n_1 = 1$.

We see that if n_2 is greater than n_1, $\sin \theta_1$ is larger than $\sin \theta_2$. In this case θ_1 would be larger than θ_2. This is the instance illustrated in Fig. 23.9*a* and is the most common one. Sometimes, however, we are interested in the reverse case, where n_2 is smaller than n_1. This would be applicable to a beam of light going from glass to air, for example. Under these circumstances Eq. (23.3) predicts that θ_2 is larger than θ_1, as shown in Fig. 23.9*b*.

Illustration 23.1 A diver beneath the surface of the ocean shines a bright searchlight up at an angle of 37° to the vertical. At what angle does the light emerge into the air?

Reasoning The situation is shown in Fig. 23.10. Notice that material 1 is water and material 2 is air. Applying Snell's law and using $n_1 = 1.33$ and $n_2 = 1.00$ gives

$$1.33 \sin 37° = 1.00 \sin \theta$$

$$\sin \theta = 0.80 \qquad \text{or} \qquad \theta = 53°$$

FIGURE 23.10

The underwater searchlight sends out a beam which bends away from the normal as it passes into the air.

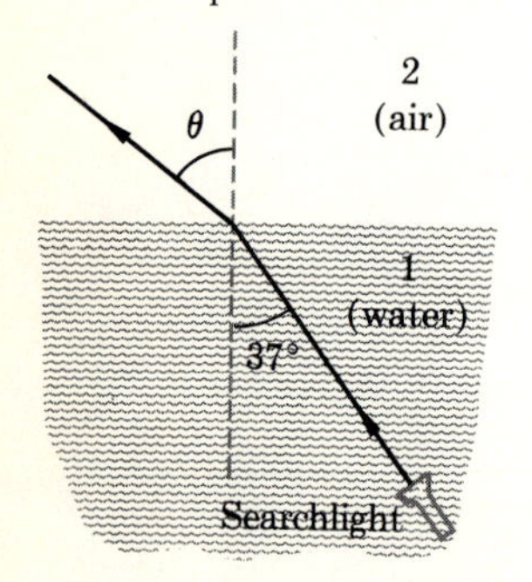

Illustration 23.2 Light is incident upon the surface of the water in a level, flat-bottomed glass dish, as shown in Fig. 23.11. At what angle will the light emerge from the bottom of the dish?

Reasoning At the air-water surface we have

$$1.00 \sin \theta_1 = n_w \sin \theta_2$$

At the water-glass interface we have

$$n_w \sin \theta_2 = n_g \sin \theta_3$$

Since quantities equal to the same thing are equal to each other, we find

$$1.00 \sin \theta_1 = n_g \sin \theta_3$$

Notice that this equation is exactly the relation one would find if the water were not present and the light went directly into the glass from the air.

Proceeding, at the lower glass-air surface we have

$$n_g \sin \theta_3 = 1.00 \sin \theta_4$$

Combining this with the previous equation gives

$$\sin \theta_1 = \sin \theta_4$$

from which $\theta_1 = \theta_4$.

This important result shows that *a uniform layer of transparent material does not change the direction of a beam of light*. The beam is usually slightly displaced sideways, however. (Why?)

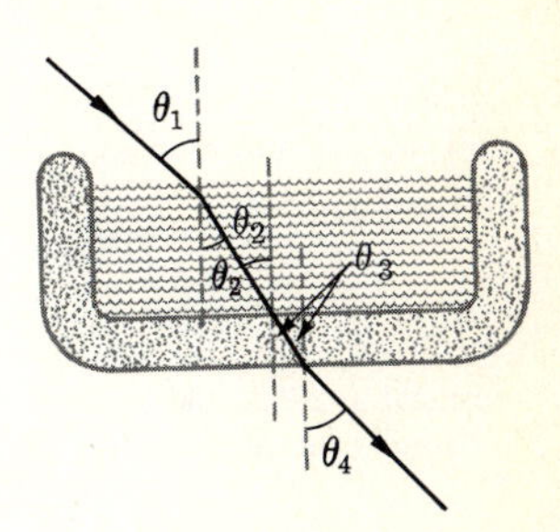

Note: $\theta_2 > \theta_3$

FIGURE 23.11

The direction of travel of a light beam is not altered by a parallel plate of transparent material.

23.5 TOTAL INTERNAL REFLECTION

Aside from their romantic value, diamonds owe a great deal of their beauty to the phenomenon of total internal reflection. It is this property which causes diamonds to sparkle in all directions. What happens is that a beam of light becomes trapped within the diamond. When it finally does emerge, the ray of light can be emitted in any one of many directions. Hence the crystal gives off light (or sparkles) in random directions. There are many other cases where total internal reflection is of importance. We shall now investigate this type of behavior of light in some detail.

Consider a light source O below the surface of a lake, as shown in Fig. 23.12*a*. Snell's law tells us that

$$\sin \theta_2 = \frac{n_1}{n_2} \sin \theta_1$$

and since we are assuming n_1 to be larger than n_2, we have that θ_2 is larger than θ_1, as shown. Notice that ray OC is bent nearly parallel to the water surface as it emerges. Clearly, the critical case, shown in Fig. 23.12*b*, occurs when $\theta_2 = 90°$ and the emerging beam just skims the surface. Thus, when $\theta_2 = 90°$,

$$\sin 90° = \frac{n_1}{n_2} \sin \theta_c$$

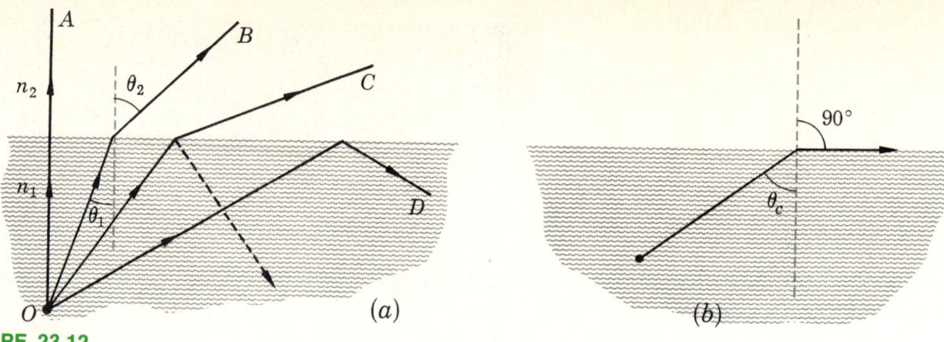

FIGURE 23.12
When θ_1 is greater than the critical angle θ_c, the beam is totally internally reflected.

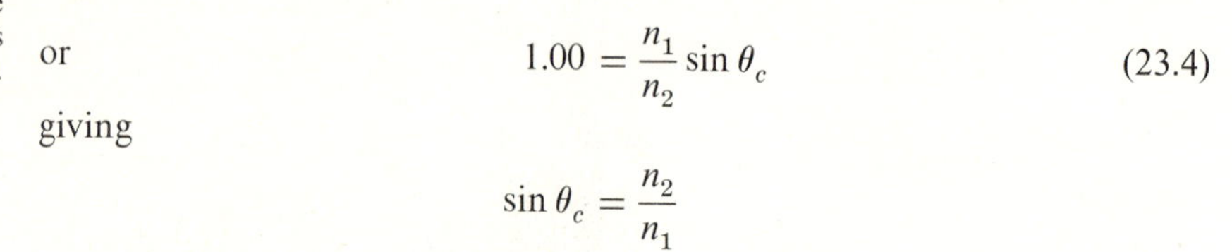

or

$$1.00 = \frac{n_1}{n_2} \sin \theta_c \tag{23.4}$$

giving

$$\sin \theta_c = \frac{n_2}{n_1}$$

If θ_1 is larger than θ_c, the critical angle, Snell's law would tell us that $\sin \theta_2$ is greater than 1.00. However, the sine of 90° is 1.00, and this is the largest that the sine can become. Hence Snell's law tells us that it is impossible to find the refracted beam if θ_1 is greater than θ_c. This is easily understood by reference to Fig. 23.12*b*. When $\theta_1 = \theta_c$, the refracted ray is just barely getting out of the water. If θ_1 is larger than θ_c, the beam will not leave the water: it will all be reflected as shown by the beam *OD* in the figure.

Total Internal Reflection

We see, therefore, that *a beam of light trying to travel from an optically dense material* (a material of high index of refraction) *to an optically less dense material will be totally internally reflected if the angle of incidence exceeds the critical angle given by Eq. (23.4)*. Notice that total internal reflection can occur only if the beam is going from the water to air and not if it is going from air to water. In most cases we are concerned with air as the second material, and so $n_2 = 1.00$. For those cases, $\sin \theta_c = 1/n_1$. Typical critical angles are 49° for water, 42° for crown glass, and 24° for diamond. Since diamond has such a high index of refraction, the critical angle is quite small. Hence the beam of light within a diamond must hit the surface nearly straight on if it is to emerge into the air. The jeweler cuts the crystal in such a way that once a beam of light gets inside it, the chance that it will strike a surface at an angle of 24° or less is quite small. As a result, the trapped beam reflects many times within the crystal before it is able to escape.

Total internal reflection makes it possible to "pipe" light around corners. By using a gently curved rod of glass, light which enters one end is totally internally reflected around the curve, as shown in Fig. 23.13. By using very narrow glass fibers in a bundle, the composite picture of an object can be piped from place to place; such a device is called a **light pipe.** See Fig. 23.13*b* for an example of this use.

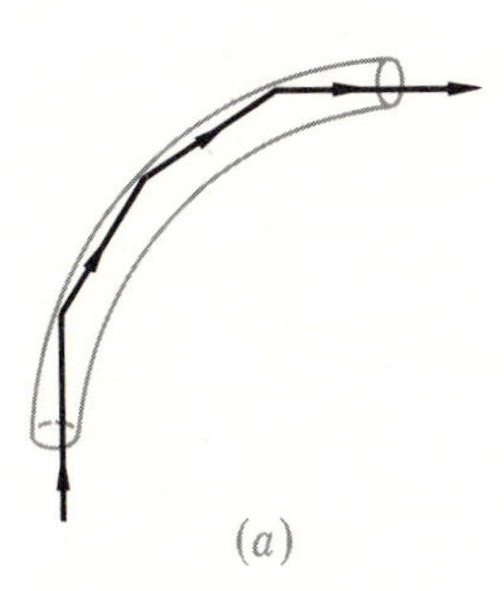

(a)

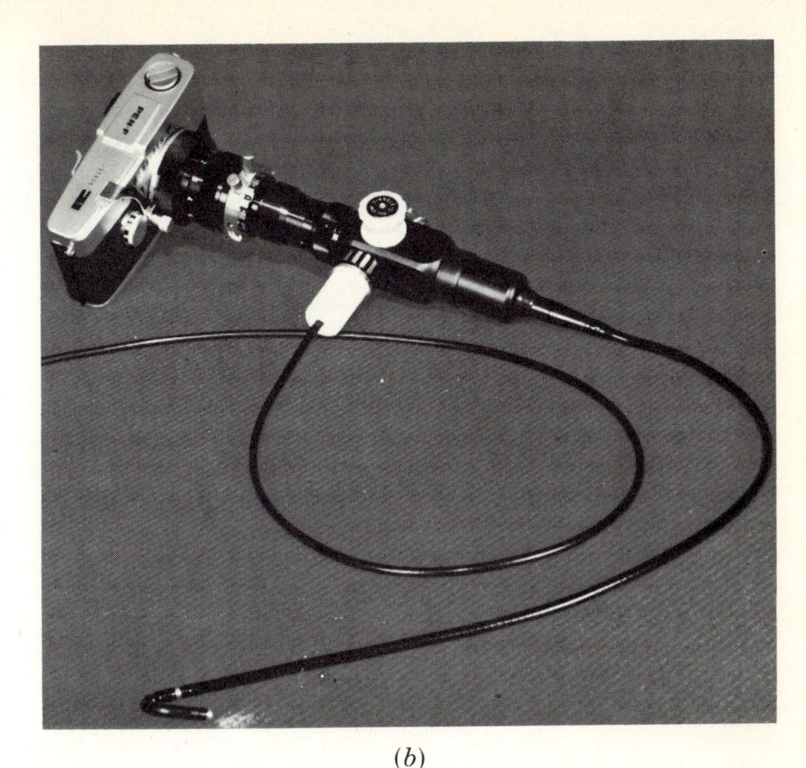

(b)

FIGURE 23.13

Light is caused to follow a glass fiber by total internal reflection. (b) A glass-fiber gastroscope attached to a camera. A light source outside the picture at the left supplies light to the fiber bundle at the bottom. This light pipe is inserted through the throat to the stomach. Light reflected from the stomach wall is reflected back up through the central fibers of the bundle and forms an image on the film of the camera. Visual observation is also possible. (*American Optical Corp., Fiber Optics Division*)

23.6 THE PLANE MIRROR

Now that we understand the phenomena of reflection and refraction, we are prepared to consider the important topics of lenses and mirrors. First we shall consider how a plane mirror forms an image.

Every day you look at yourself in a plane mirror. You see the image of your face in front of you. Actually, if you stop to examine exactly what you are doing, you perceive that the image of your face is behind the surface of the mirror. In fact, it appears to be about as far behind the mirror as your face is in front of the mirror. Let us now examine such a reflection in a plane mirror in order to understand clearly why the image is seen as it is.

Suppose that you place an object in front of a mirror, as shown in Fig. 23.14*a*. You wish to find where the eye shown in the figure will see the image of the object to be. When you see the tip of the object, your eye sees a ray of light which was emitted or reflected by the object tip. If you see the image of the object tip in the mirror, you really see light from the object tip which was reflected by the mirror, as shown in Fig. 23.14*a*. It is apparent from this same figure that you will see the object as being off in the general direction from which the ray came. Hence you know that the image of the object tip will be somewhere along the line *EM* or its extension.

To see exactly where the image of the object will be, refer to Fig. 23.14*b*. The two reflected rays shown appear to the eye to have come from the single point *I*. It is at this point that the image of the object tip will appear to be. In fact, the same reasoning may be used to show that the total image of the

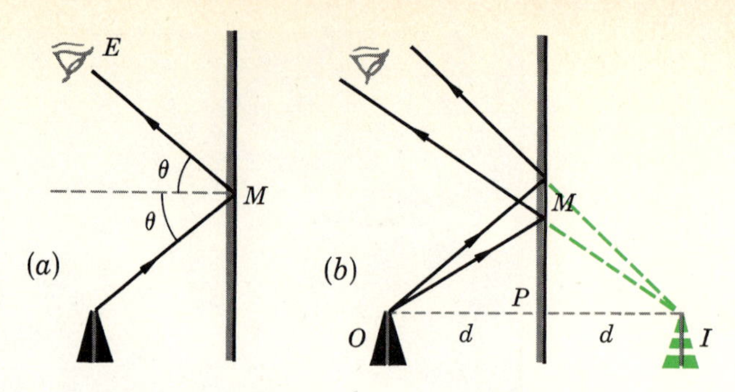

FIGURE 23.14
The image formed by the plane mirror is as far behind the mirror as the object is in front of it.

object will appear as shown at I. From the geometry of the triangles OMP and IMP, it is clear that if the object is a distance d in front of the mirror, the image is a distance d behind the mirror.

Virtual (or Imaginary) Image

This type of image, one through which the observed rays do not actually pass, is called a **virtual** *or* **imaginary image.** In other words, the rays reaching the eye do not really come from the point where we see the image. There is no possibility whatsoever that a sheet of paper placed at I behind the mirror would have a lighted object appear on it. The mind merely imagines that the light comes from I. It is always true, of course, that the image of an actual object seen by reflection in a plane mirror is a virtual image. The image is always exactly as far behind the mirror as the object is in front.

23.7 THE FOCUS OF A CONCAVE SPHERICAL MIRROR

FIGURE 23.15
A spherical mirror M is a portion of a hollow sphere. The radius of the mirror is R, its center is point C, and its principal axis is line $\overline{PA}$.

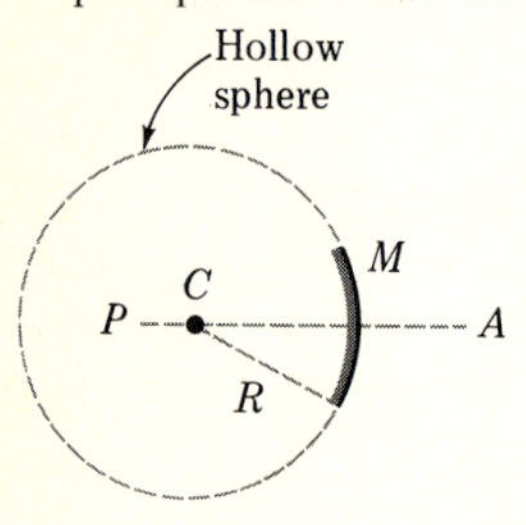

Although flat, or plane, mirrors are used by all of us, spherical mirrors are more or less of a curiosity to most people. They are widely used for technical purposes, however, and we shall describe their operation. A spherical mirror is actually a portion of the surface of a hollow sphere, as shown in Fig. 23.15. If the sphere reflects from its inner surface, it will reflect light as shown in Fig. 23.16*a*. Such a mirror is called a **concave spherical mirror.** In the event that the mirror reflects from its outside surface, it will reflect light as shown in Fig. 23.16*b*. This type of mirror is called a **convex spherical mirror.**

Notice that in drawing Fig. 23.16 we have assumed that the light came from a distant source so that the rays are parallel and the wavefronts are straight or plane. (Recall that a water wave loses its curvature as it goes farther and farther from the source of the wave.) The figure indicates that the parallel rays traveling along the principal axis of the mirror (defined in the legend of Fig. 23.15) are all reflected to or from a point F. This is approximately correct, as we shall soon show. *This point to which the light from a distant object is reflected by a concave mirror is called the* **focus** (*or* **focal point**) *of the mirror.* If we were to reflect light from the sun by a concave mirror, the light would be reflected into a tiny spot (an image of the sun) very near the focus of the mirror. This follows because the light waves from a source as distant as the sun should be nearly plane by the time they reach the earth.

DEFINITION

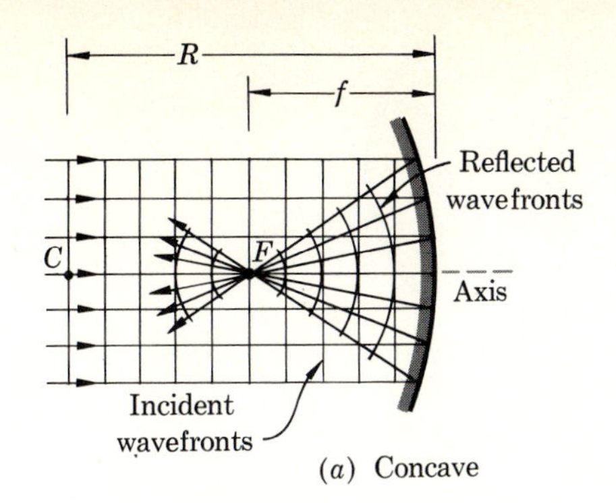

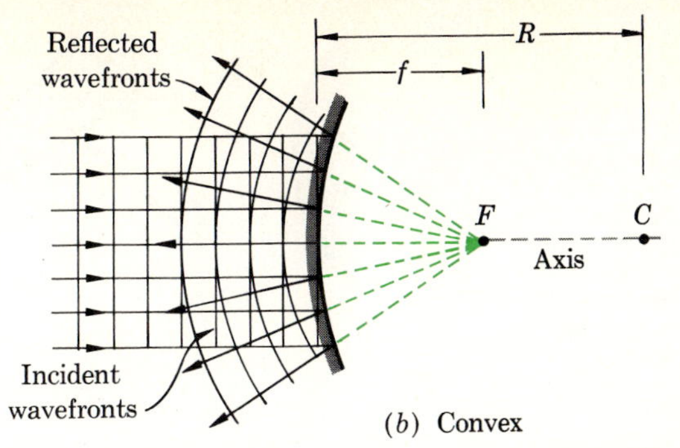

FIGURE 23.16
The reflected light from a concave mirror is converged to the focus F, while the light reflected from a convex mirror appears to diverge from the focus.

The rays of light from the distant sun are so nearly parallel that we can consider them so.

Until now we have asked the student to accept the above assertions without proof. However, it is a simple matter to demonstrate exactly how three special types of the infinite number of rays from a source will be reflected by a concave mirror. Once this is known, it becomes quite easy to ascertain the position of an image formed by such a mirror. We shall now consider how the concave mirror reflects these three important rays.

First, consider the ray shown in Fig. 23.17. It is parallel to the axis of the mirror, striking it at A. Let us draw a radius CA from the center of the sphere of which the mirror is a part. Then, since CA is a radius of the sphere, it is therefore perpendicular to the mirror surface at A. The reflected ray makes the same angle to the perpendicular at A as the incident ray does, since the angle of incidence at any smooth reflecting surface equals the angle of reflection. Moreover, since the incoming ray is parallel to the axis CB, the angle at C must also be i, as indicated.

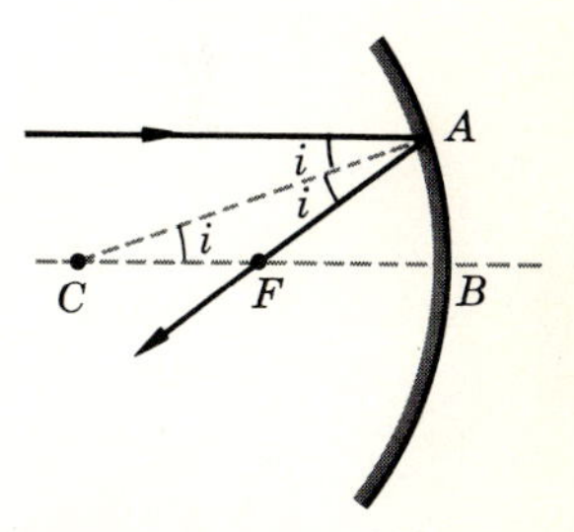

FIGURE 23.17
A ray parallel to the principal axis of a concave mirror is reflected back through the focal point.

The triangle CFA is isosceles, and CF must therefore be equal to FA. But if the angle i is quite small, the length FA is nearly the same as FB. Hence any ray parallel to the axis and not too far from it will be reflected through a point F which is halfway between the mirror surface B and the center of the sphere of which the mirror is a part. In other words, if the mirror has a radius of curvature R, rays parallel to the axis of the mirror will be reflected through the point F, which is a distance $\overline{FB} = \frac{1}{2}R$ from the mirror. This point, *the point to which parallel rays are focused, is the focus of the mirror, and the length* $\overline{FB}$ *is called the* **focal length** f *of the mirror. We have* $f = \frac{1}{2}R$.

Focal Length f

Of course, it is not completely true that all rays parallel to the axis are focused exactly at the focal point. An approximation was made in arriving at this result. If the length AB in Fig. 23.17 is only a small fraction of the sphere diameter, the approximation is fairly good. However, you should draw the case of a ray reflected at an angle $i \approx 90°$ to see that, in that case, the ray will not be reflected through the focus. It is for this reason that useful spherical mirrors are only a very small portion of a sphere in size. Larger spherical mirrors give fuzzy images. Parabolic mirrors do not suffer this disadvantage but they are more expensive to make.

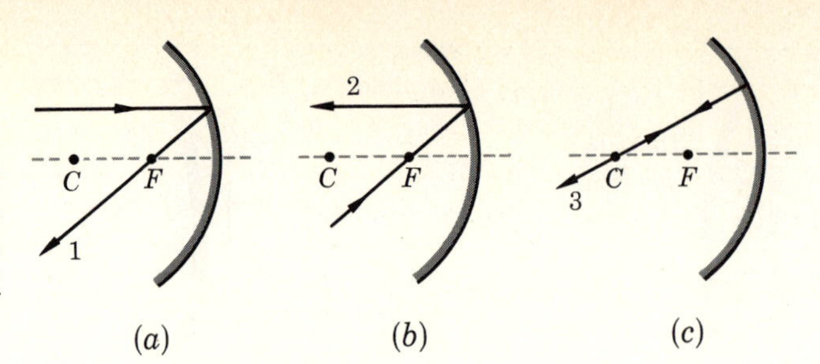

FIGURE 23.18

The three rays shown are easily drawn with a straightedge provided that the center C and the focus F of the mirror are known.

23.8 THE THREE REFLECTED RAYS AND IMAGE FORMATION

We now know that a ray parallel and near to the axis of a spherical mirror may be considered to reflect through the focal point. There are two other rays which are also easily traced. All three of the rays are shown in Fig. 23.18*a* to *c*.

Ray 1 in part *a* has already been discussed. Ray 2 in part *b* is just the reverse of ray 1, and the same geometric arguments used for ray 1 apply to ray 2 as well. Ray 3 in part *c* passes through the center of curvature of the mirror and strikes the mirror. Since it is traveling along a radius of the mirror surface, it hits the mirror perpendicularly. Hence it will be reflected straight back upon itself, as shown. We therefore have the following rules *for concave mirrors:*

A ray parallel to the mirror axis is reflected through the focus.

A ray through the focus is reflected parallel to the mirror axis.

A ray through the center of curvature is reflected back through the center of curvature.

These rules can now be used to locate the position of images.

Suppose that we wish to find the image formed by the mirror of the object O shown in Fig. 23.19. Let us say that the object is a light bulb. The

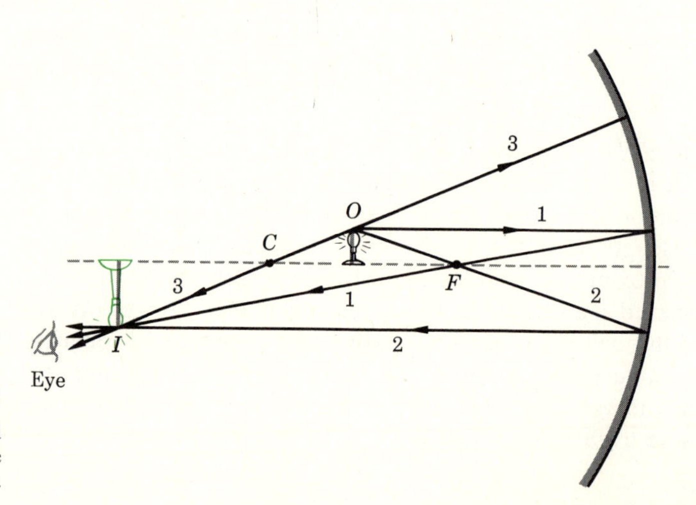

FIGURE 23.19

A real image I is formed from the object O. Trace the three rays from the object.

bulb will emit light in all directions, but we know how to treat only three of the millions of rays we could draw coming from it. These three rays travel as shown in Fig. 23.19. They are exactly the rays described by our rules, and you should trace each in turn to see that it is properly drawn. Once the positions of C and F are known, only a straightedge is needed to draw these rays.

DEFINITION

If you now place your eye as shown, the three rays will appear to come from point I. You will actually see the light bulb to be at I, and this is called the **image** of the object at O. Moreover, since the rays of light actually do converge on point I and pass through it, a sheet of paper placed at I will show a lighted picture of the original bulb. This is a **real image:** *at a real image, the light actually passes through the point to reproduce the object.* Notice how this differs from the imaginary, or virtual, image found for the plane mirror.

Real Image

Suppose that the bulb was on the tip of the post represented in Fig. 23.19. Rays of light will emanate from it as well as from the bulb. (For example, light from the bulb might strike the post and be reflected from it. This light would appear to be emitted from the post, and it will act as if it were emitted rather than reflected light.) We could treat each little portion of the post as a new light source and find its image. You should trace a few rays to show yourself that the image of the post lies along the image shown at I. From now on, we shall locate only a single point of an image and immediately draw in the rest of the image in this way.

Consider the situation shown in Fig. 23.20. Once again we draw our three rays. Now, however, ray 2 does not go through the focus on its way to the mirror, since the object is inside the focal length. However, it still appears to come from the focus and is reflected parallel to the axis, as always. Notice that rays 1, 2, and 3 appear to come from the image I, as shown. Hence, the eye sees the image to be behind the mirror in this case. Notice that the image is virtual (imaginary), erect (right side up), and magnified (larger than the object).

23.9 THE MIRROR EQUATION

To derive a mathematical equation which describes the location of the image, let us refer to the example shown in Fig. 23.21. The ray ABE in part a is not

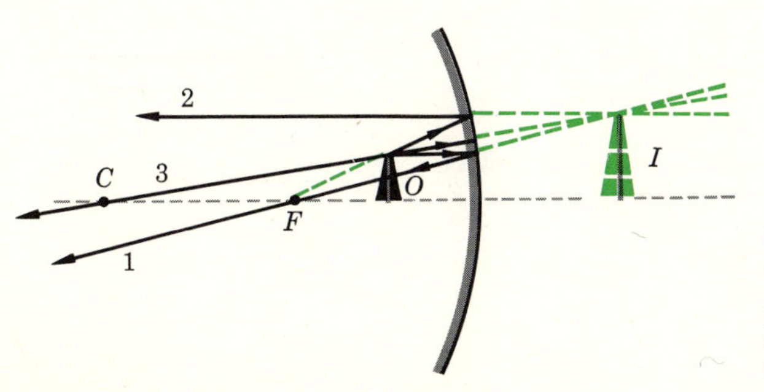

FIGURE 23.20

The three rays appear to come from the virtual image I. Notice especially rays 2 and 3.

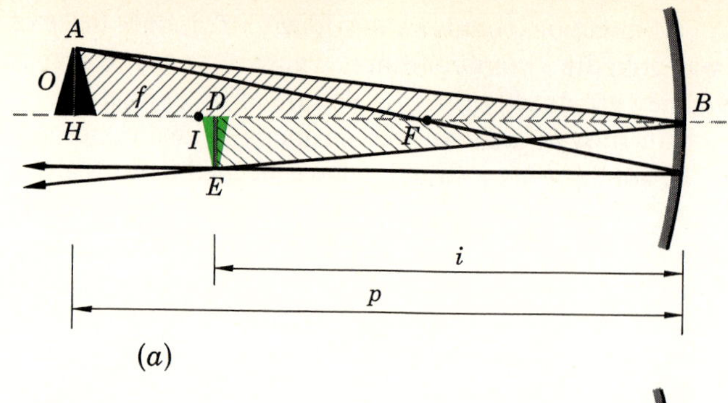

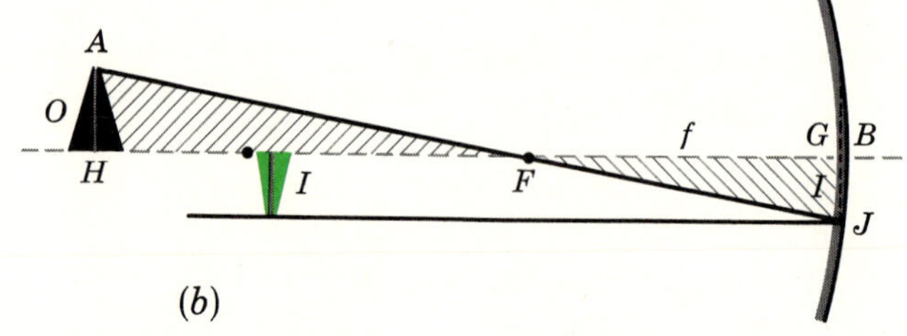

FIGURE 23.21
The shaded triangles are similar in each drawing. We assume that the distance $\overline{FG}$ in part *b* is negligibly different from $\overline{FB}$.

a usual one. However, it is reflected in such a way that angle *ABH* equals angle *FBE*. For this reason, the shaded triangles in Fig. 23.21*a* are similar triangles. Taking the ratios of corresponding sides gives

$$\frac{O}{I} = \frac{p}{i}$$

In Fig. 23.21*b*, the shaded triangles are also similar. Hence

$$\frac{O}{I} = \frac{\overline{HF}}{\overline{FG}}$$

But $\overline{HF}$ is just $p - f$, and $\overline{FG}$ is nearly f. To this approximation we have

$$\frac{O}{I} = \frac{p - f}{f}$$

Equating this to the expression found in part *a* of the figure gives

$$\frac{p}{i} = \frac{p - f}{f}$$

After dividing this equation by p and rearranging, we have

The Mirror Equation

$$\frac{1}{p} + \frac{1}{i} = \frac{1}{f} \qquad (23.5)$$

where $f = \frac{1}{2}R$ and R is the radius of curvature of the mirror.

Equation (23.5) is the mirror equation. It allows us to compute the distance i of the image from the mirror surface, provided that the object distance from the mirror surface p and the focal length f are known. To compute the relative heights of the object and image, we note that $O/I = p/i$, as found above. *The magnification produced by the mirror is defined to be the ratio of the image height to the object height.* Therefore,

DEFINITION

$$\text{Magnification} = \frac{I}{O} = \frac{i}{p} \tag{23.6}$$

Magnification

Illustration 23.3 An object 2 cm high is placed at a distance of 30 cm from a concave mirror having a radius of curvature of 10 cm. Find the position and size of the image.

Reasoning The mirror relation applies, with $p = 30$ cm and $f = \frac{10}{2} = 5$ cm. Hence, with all distances in centimeters,

$$\frac{1}{30} + \frac{1}{i} = \frac{1}{5} \qquad \text{or} \qquad \frac{1}{i} = \frac{6}{30} - \frac{1}{30} = \frac{5}{30}$$

and

$$i = 6 \text{ cm}$$

The image will be on the same side of the mirror as the object, i.e., on the silvered side, and since the light will actually be passing through it, the image is real. Its size is found from Eq. (23.6) to be

$$\frac{I}{2} = \frac{6}{30} \qquad \text{or} \qquad I = \tfrac{2}{5} \text{ cm high}$$

It is always wise to check the algebraic solution by drawing the appropriate ray diagram.

Illustration 23.4 An object is placed 5.0 cm in front of a concave mirror having a 10-cm focal length. Find the location of the image.

Reasoning If we refer to Fig. 23.20, we see that this example should give an image on the wrong side of the mirror, i.e., the unsilvered back. Our answer should make this evident. Using the mirror equation, taking all distances in centimeters,

$$\frac{1}{5} + \frac{1}{i} = \frac{1}{10}$$

gives

$$\frac{1}{i} = \frac{1}{10} - \frac{2}{10} = -\frac{1}{10} \qquad \text{or} \qquad i = -10 \text{ cm}$$

Notice that i is negative. Hence, when images are behind, i.e., on the back of, the mirror, the image distance is negative. This is not surprising in view of the fact that i was taken positive for the reverse case. Clearly, the image must be virtual in this event. An image behind the mirror is, of course, virtual. It does not appear on a screen placed there.

23.10 CONVEX MIRRORS

A convex spherical mirror is a portion of a sphere which has a reflective coating on the outside surface. This surface is convex when viewed from outside the sphere, and hence the mirror is called a convex mirror. In Fig. 23.16 we illustrated the behavior of parallel rays reflected from such a mirror. They appear to diverge from a point behind the mirror. This point is said to be the **focal point** for the convex mirror. *Parallel rays incident on a convex mirror are reflected as though they came from the focal point.* To prove that the parallel rays behave in this way, we proceed in much the same way as we did for the concave mirror.

Focal Point of Convex Mirror

Referring to Fig. 23.22, we see from the law of reflection and the geometry involved that several angles are equal, as shown. The triangle AFC is isosceles so that $\overline{AF} = \overline{FC}$. If the length $\overline{AB}$ is small compared with the radius of curvature of the mirror, $\overline{AF}$ is nearly equal to $\overline{BF}$. Consequently, $\overline{BF}$ nearly equals $\overline{FC}$, and so here, too, the focal point may be considered to be midway between the mirror and the center of curvature of the mirror.

We are therefore able to write rules for drawing three rays in the case of the *convex mirror:*

A ray parallel to the axis is reflected as though it came from the focal point.

A ray heading toward the focal point is reflected parallel to the axis.

A ray heading toward the center of curvature is reflected back on itself.

These three rays are illustrated in Fig. 23.23. You should trace them to see that they conform to the rules stated above. Notice that all three reflected rays appear to come from the image I behind the mirror. As you see, the image is virtual, upright, and diminished in size.

The algebraic relation used in locating the image for a convex mirror

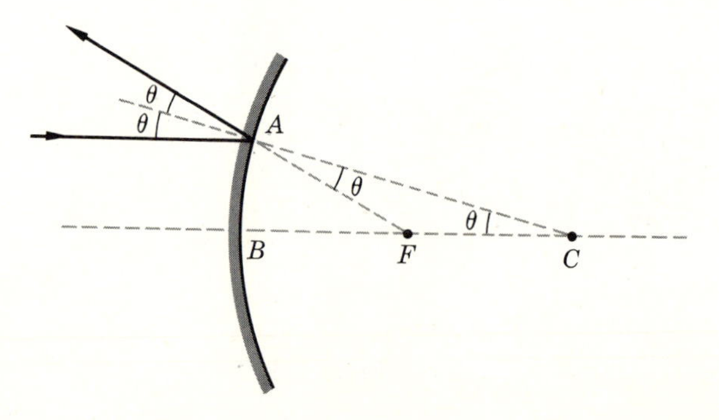

FIGURE 23.22
A ray parallel to the axis is reflected as though it came from the focal point.

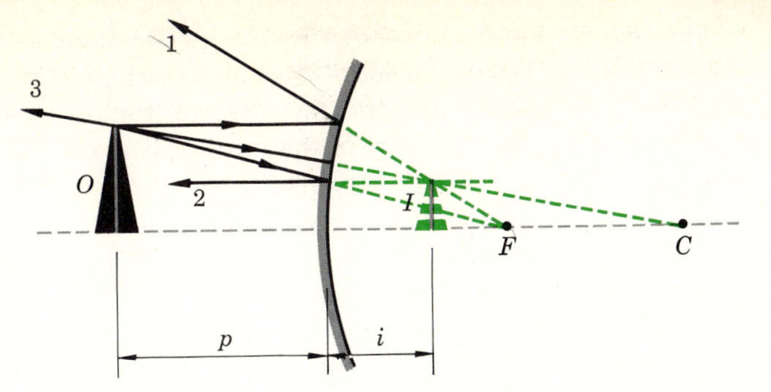

FIGURE 23.23 You should be able to draw the three rays shown for any situation involving a convex mirror.

can be obtained by reference to Fig. 23.24*a* and *b*. You should be able to show that triangle ABH is similar to EBD in part *a*. In part *b*, triangle JFG is similar to triangle EFD. This being true, the following equations are found, as in the case of the concave mirror:

$$\frac{O}{I} = \frac{p}{i} \qquad \text{and} \qquad \frac{O}{I} = \frac{f}{f - i}$$

In writing these, the distance $\overline{BG}$ has been considered negligibly small.

Equating the two expressions, inverting, dividing by i, and rearranging yields

$$\frac{1}{p} - \frac{1}{i} = -\frac{1}{f}$$

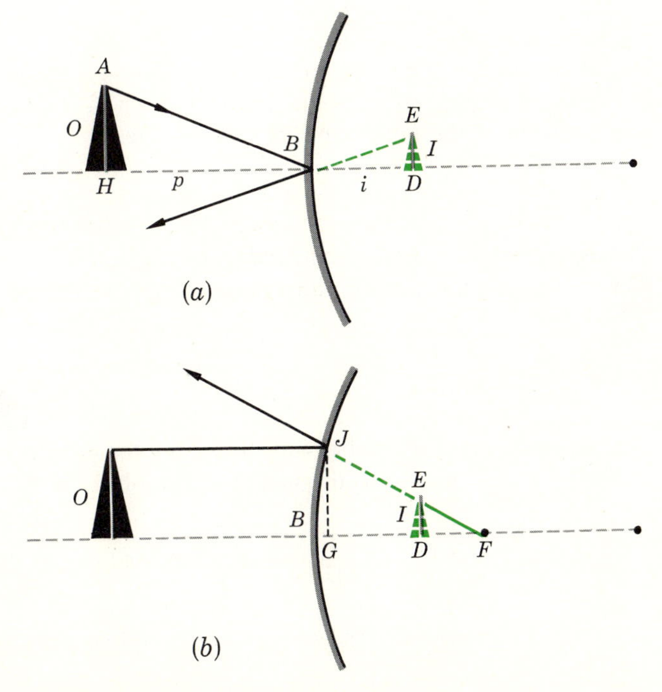

FIGURE 23.24 Triangles HBA and DBE are similar, as are DFE and GFJ. We make the assumption that the length $\overline{FG}$ is essentially the same as $\overline{FB}$.

Notice that, except for signs, this equation is the same as Eq. (23.5) for the concave mirror. The difference in signs alerts us to the fact that the image in this case is behind the mirror rather than in front of it. Also, the negative focal-length term is the result of the mirror being convex rather than concave.

Rather than remembering two mirror equations, we shall now set up rules which will allow us to use Eq. (23.5) even for convex mirrors. If we agree always to call image distances behind mirrors, i.e., virtual-image distances, negative, the negative sign may be omitted from the i term in the convex-mirror equation. Moreover, if we always say that the focal length of a convex mirror is negative, the other negative sign may be omitted as well. Hence for *all* mirrors we can write

Mirror Equation and Its Sign Conventions

$$\frac{1}{p} + \frac{1}{i} = \frac{1}{f} \quad \text{mirrors} \tag{23.5}$$

where we agree that:

1. Object distances are positive if the object is on the reflecting side of the mirror, negative otherwise.
2. Image distances are positive if the image is on the reflecting side of the mirror, negative otherwise.
3. The focal length of a concave mirror is positive, and it is negative for a convex mirror.

In addition, we can use Eq. (23.6) for the magnification in either case without use of any negative signs whatsoever.

Illustration 23.5 A convex mirror with a 100-cm radius of curvature is used to reflect the light from an object placed 75 cm in front of the mirror. Find the location of the image and its relative size.

Reasoning Since the mirror is convex, $f = -R/2 = -50$ cm. Using the mirror equation and taking all distances in centimeters, we have

$$\frac{1}{75} + \frac{1}{i} = -\frac{1}{50} \qquad \text{or} \qquad i = -30 \text{ cm}$$

The negative sign tells us the image is behind the mirror, and it is, of course, virtual. Its size relative to the height of the object is

$$\frac{I}{O} = \frac{i}{p} = \frac{30}{75} = 0.40$$

It would be wise to check this solution graphically to see that no mistake has been made.

23.11 THE FOCAL POINTS OF A LENS

A properly constructed lens is capable of focusing a beam of parallel light into a small region at a focal point. The mechanism by which this is done is illustrated in Fig. 23.25*a* and *b*. We recall that a light wave travels more slowly in glass than in air. Hence, the central portion of the incident plane wave shown in Fig. 23.25*a* is found to have fallen behind the outer portions of the wave because it traveled a greater distance in the glass. The emergent wave is therefore curved as shown. Since the rays, i.e., the directions of travel of the light, are perpendicular to the wavefronts, the light is converged toward point F on the axis.

Although we shall not prove it here, the various rays will converge to a single point, as shown, if the surfaces of the lens are portions of spheres. However, this is only an approximation, and it becomes a poor one if the surfaces are highly curved, in other words, if they constitute more than a very small portion of a sphere. Moreover we assume the lens to be relatively thin. In these circumstances, the rays parallel to the axis are converged nearly to a point. *The point to which parallel rays are converged by a converging lens is called the* **focal point** *of the lens*.

DEFINITION

There is a simple way to determine the position of the focal point of a lens such as the one shown in Fig. 23.25*a*. If the light from the sun is passed through it, an image of the sun is formed. Since the sun is very far away, the waves from it are essentially flat and the rays are parallel. Hence, the place where the image of the sun is formed is the focal point of the lens.

A different type of lens is shown in Fig. 23.25*b*. Notice that since this lens is thinner in the middle than on the edges, the outer portion of the wave will fall behind. Now the emergent wave is spherical in form but diverging from the lens. Under the same general restrictions as stated for the other type of lens, this diverging wave appears to come from the point F in the figure. *This point, from which the original parallel beam appears to diverge, is called the* **focal point of the diverging lens.**

DEFINITION

We see then that two general types of lens are possible. Converging lenses, which are thickest in the middle, cause a parallel beam of light to converge to the focal point. Diverging lenses are thickest near the edges and cause a parallel beam of light to spread out (or diverge) as though from a point, the focal point. Even if the lenses were turned around, they would still behave in this way.* Hence each lens has two focal points, one on each side

* When lenses are so thick that the focal length and thickness are comparable, this statement must be qualified.

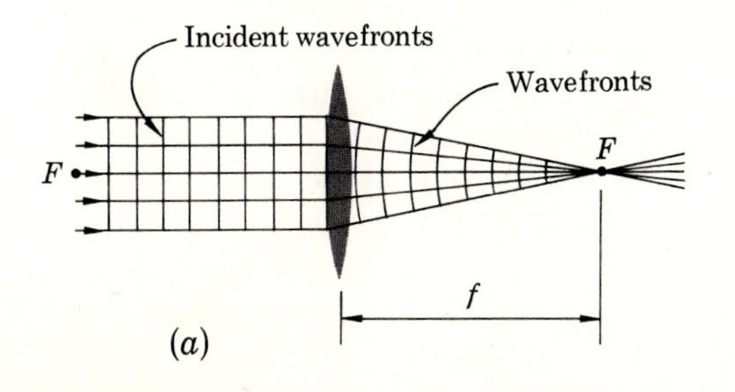

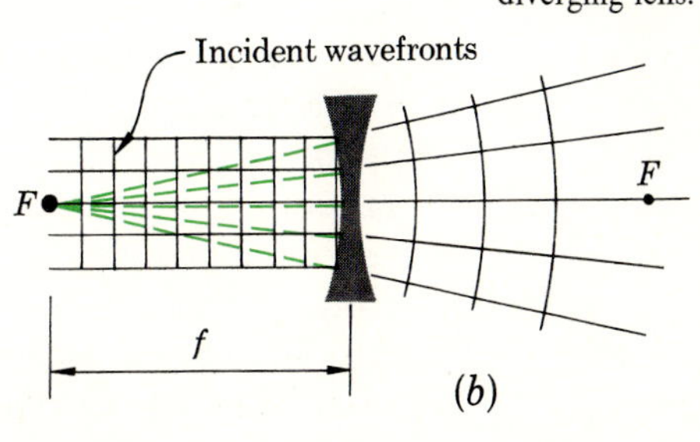

FIGURE 23.25
Parallel rays are converged to the focal point by the converging lens. They are diverged and appear to come from the focal point in a diverging lens.

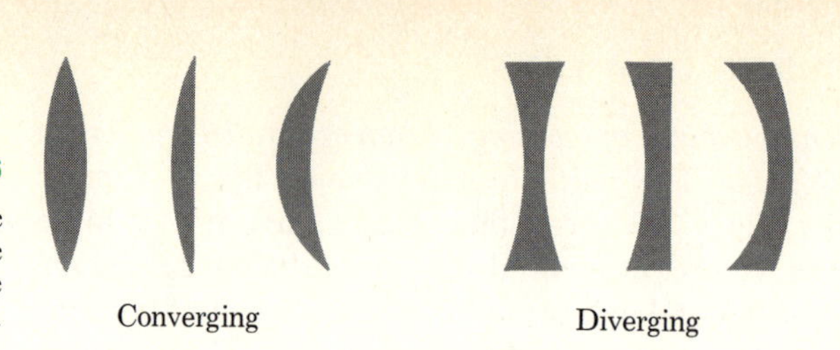

FIGURE 23.26

The converging lenses are thickest at the center; the reverse is true for the diverging lenses.

Lens Focal Length

of it, and both are at the same distance from its center. This distance *from the center of the lens to the focal point is called the* **focal length f of the lens.** Several different shapes of common lenses are shown in Fig. 23.26. Notice that the converging lenses are all thickest at the center.

Suppose that a glass lens is immersed in a liquid having an index of refraction which is the same as that of the glass. In this case, the wave will travel with the same speed in the glass and in the liquid. For this reason, the wave will not be disturbed by passing through the lens, and the lens will not cause it to focus. Thus, the focal length of a lens depends upon the medium in which it is used. Most commonly this medium is air. However, this is not always the case. The focal length of a lens immersed in a medium other than air is given by

$$f_{\text{medium}} = \frac{f_{\text{air}} n_m (n_g - 1)}{n_g - n_m}$$

where n_m is the refractive index of the medium and n_g is the index of refraction of the glass. Can you show that this relation is reasonable for the limiting cases of $n_m \to n_g$ and $n_m \to 1.00$?

23.12 RAY DIAGRAMS FOR THIN LENSES

We saw in the last section that a ray traveling parallel to the lens axis is bent by the lens. It is converged to the focus by a converging lens and diverged from the focus by a diverging lens. These rays are shown as ray 1 in Fig. 23.27*a* and *b*.

The second ray is just the reverse of ray 1. For the convex lens, it comes from the source, through the other focal point, and is converged parallel to the axis by the lens. Notice that it is exactly the same as ray 1 traveling in the reverse direction. In the diverging case, ray 2 heads toward the other focal point but never reaches it. The lens diverges it parallel to the axis as shown. It, too, is ray 1 traced in the reverse direction.

The third ray comes from the source and goes straight through the center of the lens without deflection. It is easy to see why it behaves this way by referring to Fig. 23.28. Notice that the rays of light which go through the center of the lens enter and leave the lens at surfaces which are parallel to each other. Hence, the ray behaves as though it had gone through a flat plate of glass. It will be recalled that a ray of light is not deviated in direction by a flat plate which has parallel faces. Therefore, the rays of light passing through the lens center will proceed undeviated.

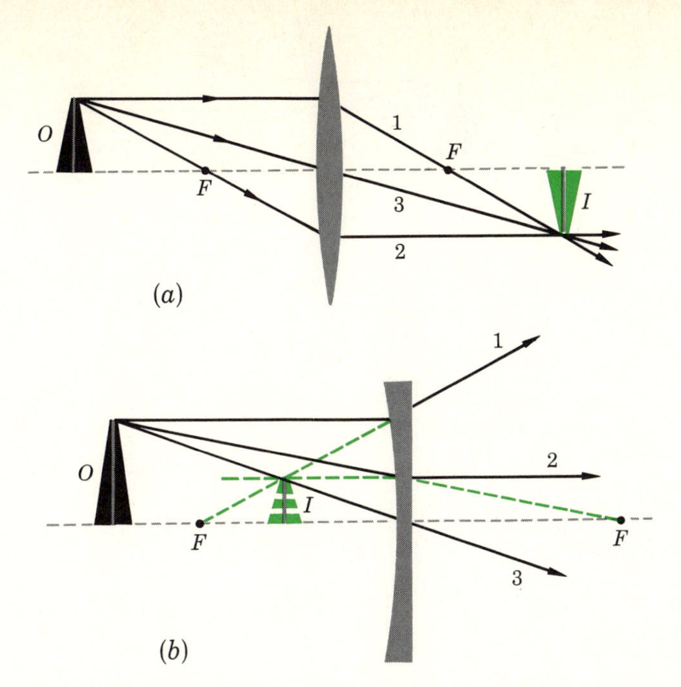

FIGURE 23.27

Graphical location of the images formed by thin lenses consists of drawing the three rays shown.

We are now able to draw three of the many rays of light which pass through a lens. Any two of these allow us to locate the image of an object. Two examples of this construction have already been given in Fig. 23.27. Notice that the image in part *a* is real, since the three rays actually converge at the image. A screen placed there would catch the light and show an image of the object. In part *b*, however, the image is virtual, or imaginary, since the three rays merely appear to come from the image position. A screen placed at that position does not show an image, because the three rays actually do not meet at that place.

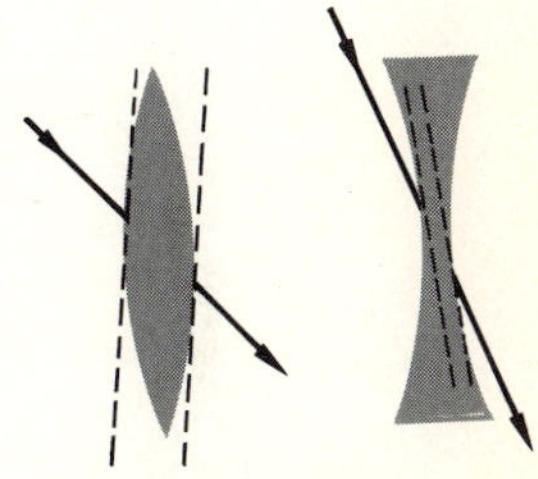

FIGURE 23.28

The ray passing through the center of the lens essentially passes through a flat plate and is therefore not deviated. A small displacement of the ray occurs, but this is not shown in the figure. Why is it negligible for a thin lens?

Illustration 23.6 A convex lens of focal length 10.0 cm is used to form an image of an object placed 5.0 cm in front of it. Draw a ray diagram to locate the image.

Reasoning The appropriate ray diagram is shown in Fig. 23.29. We notice

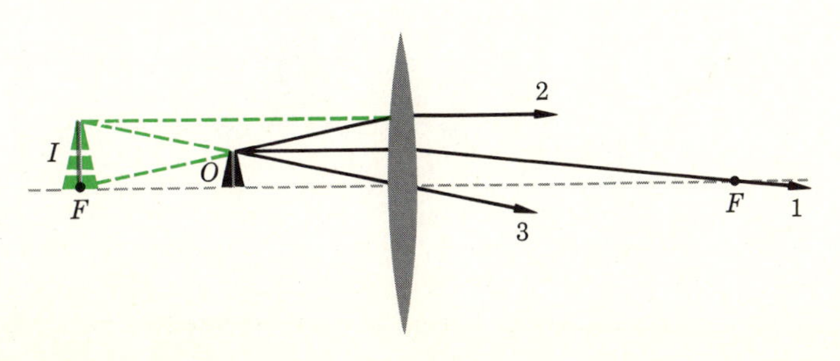

FIGURE 23.29

Virtual images are formed by convex lenses when the object is inside the focal point.

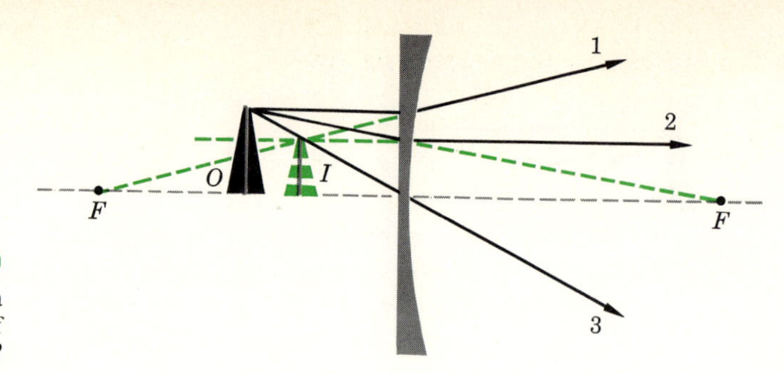

FIGURE 23.30
Is the image formed by a diverging lens always virtual if the object is real?

that the eye will assume that the three rays come from the image position indicated. As we see, the image is virtual, erect, and enlarged.

Illustration 23.7 A concave lens of focal length 10.0 cm is used to form an image of an object placed 5.0 cm in front of the lens. Find the image position by means of a ray diagram.

Reasoning The appropriate ray diagram is shown in Fig. 23.30. Here, too, the image is virtual. It is erect and diminished in size.

23.13 THE THIN-LENS FORMULA

Consider the image formed by the converging lens shown in Fig. 23.31*a* and *b*. In part *a*, triangles *ABH* and *EBD* are similar, and so we can write

$$\frac{I}{O} = \frac{i}{p}$$

From the two similar triangles *JFB* and *EDF* shown in part *b*, we also have

$$\frac{I}{O} = \frac{i - f}{f}$$

FIGURE 23.31
Triangles *ABH* and *EBD* are similar, as are *JFB* and *EDF*.

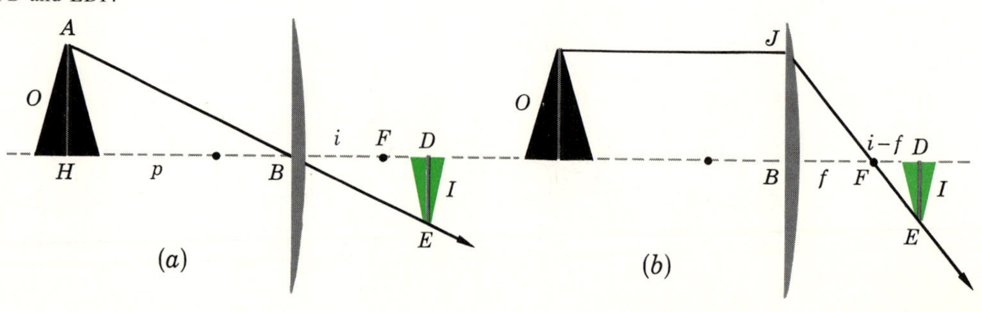

Equating these two expressions and simplifying yields

$$\frac{1}{p} + \frac{1}{i} = \frac{1}{f} \tag{23.5}$$

This relation is exactly the same as the mirror equation (23.5). We have obtained it by taking p and i positive when the object and image are in their normal positions. That is, the object is on the side from which the light comes, and the image is on the side to which the light is going. Also, Eq. (23.6) for the magnification applies to lenses as well, since one of the above relations is identical to it.

We can derive the relation applicable to concave lenses by referring to the sets of similar triangles illustrated in Fig. 23.32*a* and *b*. We have

$$\frac{I}{O} = \frac{i}{p} \qquad \text{and} \qquad \frac{I}{O} = \frac{f - i}{f}$$

Equating and simplifying, we find that

$$\frac{1}{p} - \frac{1}{i} = -\frac{1}{f}$$

In order to make this relation appear the same as Eq. (23.5), we must take both i and f negative. If we then agree to say that concave (diverging) lenses have negative focal lengths, the negative sign preceding the $1/f$ term may be omitted. Moreover, if we agree to call image distances negative when the image lies on the side of the lens from which the light came, the negative sign on the $1/i$ term may be omitted. In that case, Eq. (23.5) will apply to all lenses and mirrors. Notice also that Eq. (23.6) for the magnification applies to all mirrors and lenses.

Let us briefly review our sign conventions in using Eq. (23.5),

Lens Equation and Its Sign Conventions

$$\frac{1}{p} + \frac{1}{i} = \frac{1}{f} \tag{23.5}$$

1 The object distance is always positive if the object is on the side of the lens or mirror from which the light is coming.

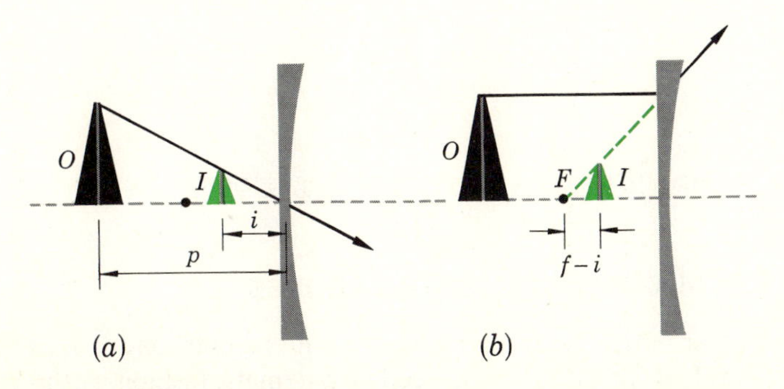

FIGURE 23.32

A consideration of the similar triangles shown leads to the lens equation.

2 The image distance is positive:
  a In the case of mirrors if the image is on the side of the mirror where the light is, i.e., in front of the mirror.
  b In the case of lenses if the image is on the side of the lens to which the transmitted light goes.

3 The focal length of converging mirrors and lenses is positive, while the focal length of convex mirrors and diverging lenses is negative.

Notice that the natural or normal positions for the object and image are positive. It is normal for the object to be on the side from which the light comes. We usually accept an image formed by a lens as normal when it is on the side of the lens to which the light is going. On the other hand, for mirrors we usually think of the image as normal when it is on the same side as the object, since the light is reflected, not transmitted. Of course, converging lenses and mirrors usually seem more normal to us than diverging lenses and mirrors do. Hence, the normal things are positive, while the others must carry a negative sign.

Before leaving this section we should point out that several different sign conventions are sometimes used. Negative signs are often employed in the lens and mirror equations. There seems to be no general agreement which system is the better. When using any other form of the lens equation than the one given here, you must always abide by the sign rules pertinent to that equation. We, of course, shall always use Eq. (23.5) and the rules we have just stated.

Illustration 23.8 A diverging lens with a 20-cm focal length forms an image of a 3.0-cm-high object placed 40 cm from it. Find the image position and size.

Reasoning From Eq. (23.5) we have, using all lengths in centimeters,

$$\frac{1}{40} + \frac{1}{i} = -\frac{1}{20} \qquad \text{or} \qquad i = -\tfrac{40}{3} \text{ cm}$$

The image is on the "wrong" side of the lens, i.e., on the same side as the light is coming from. It is imaginary, of course, as a ray diagram will easily show. To find the image size, we use Eq. (23.6).

$$\frac{I}{O} = \frac{i}{p} \qquad \text{or} \qquad I = \frac{(3)(\frac{40}{3})}{40} = 1 \text{ cm}$$

Notice that we do not carry the signs along when finding the image size.

23.14 COMBINATIONS OF LENSES

Most optical instruments contain more than one lens. These systems are easily dealt with if we proceed in systematic fashion. Let us consider the final

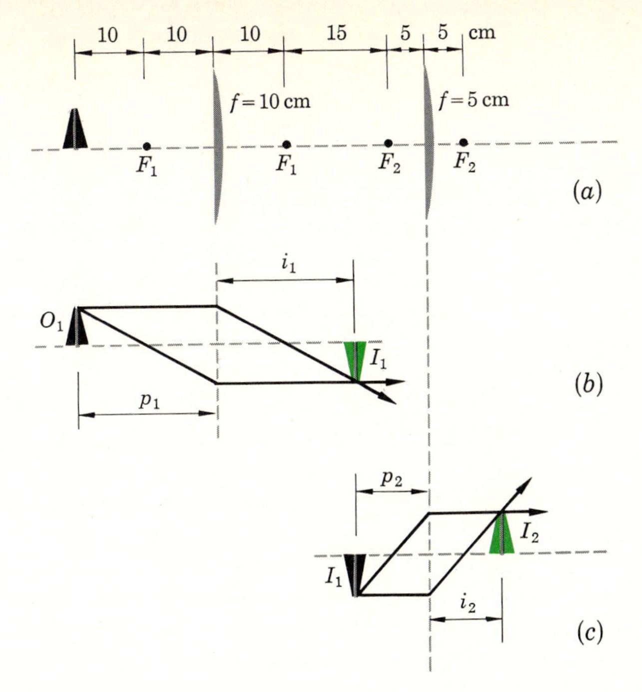

FIGURE 23.33 In order to find the image cast by a lens combination, we consider each lens in turn, as in (*b*) and (*c*).

image formed by the two lenses shown in Fig. 23.33*a*. The object is 20 cm from the first lens, which in turn is 30 cm from the second lens. Both lenses are converging. To solve this by a ray diagram as well as by formula, we proceed as in Fig. 23.33*b* and *c*.

First, *ignore the second lens completely, and find the image of the original object cast by the first lens.* The ray diagram locates the image I_1 in part *b* of the figure. By formula we have, using centimeters,

$$\frac{1}{20} + \frac{1}{i_1} = \frac{1}{10} \qquad \text{or} \qquad i_1 = 20 \text{ cm}$$

where i_1 is shown in the figure.

We now use the image formed by the first lens as an object for the second. This is possible, since an eye or any other instrument placed to the right of the image will see the image as though there were really an object at that place. Ignoring the first lens completely and using I_1 as the object for the second lens, we draw the ray diagram in Fig. 23.33*c*. The final image is located at I_2, or, in equation form,

$$\frac{1}{p_2} + \frac{1}{i_2} = \frac{1}{f_2} \qquad \text{and} \qquad \frac{1}{30 - 20} + \frac{1}{i_2} = \frac{1}{5}$$

or

$$i_2 = 10 \text{ cm}$$

where i_2 is shown in the figure. Clearly, the final image formed by the two lenses in combination, I_2, is real and erect.

To find the size of the final image in terms of O_1, the height of the object, we apply Eq. (23.6) twice. For the situation of Fig. 23.33*b* we have

$$\frac{I_1}{O_1} = \frac{20}{20} \qquad \text{so} \qquad I_1 = O_1$$

Now, using Fig. 23.33*c* and remembering that the height of I_1 is the same as O_1 and that I_1 is really the object for the second lens,

$$\frac{I_2}{I_1} = \frac{i_2}{p_2} \qquad \text{or} \qquad I_2 = (O_1)(\tfrac{10}{10}) = O_1$$

Hence, we find the unusual situation where the image is the same height as the original object.

SUMMARY

Light consists of EM waves in the visible wavelength range, 4×10^{-7} to 7×10^{-7} m. The speed of light in vacuum is $c = 2.998 \times 10^8$ m/s. It travels less fast in all other media. We characterize a material optically by its index of refraction n. It is the ratio of the speed in vacuum to the speed in the material. For air, $n = 1.0003$; for water it is 1.33.

When light reflects from a smooth surface, the angle of incidence equals the angle of reflection. When light passes from one material (index of refraction n_1) to another (index n_2), it refracts and obeys Snell's law: $n_1 \sin\theta_1 = n_2 \sin\theta_2$. If $n_2 > n_1$, the ray bends toward the normal. If $n_2 < n_1$, θ_1 is smaller than θ_2. This gives rise to total internal reflection for $\theta_2 = 90°$. The critical angle for total internal reflection is given by $\sin\theta_c = n_2/n_1$.

Real images are places where light rays focus to give optical duplicates of an object. Such images can be displayed on a screen placed at the image position. Virtual (or imaginary) images are places from which the rays only appear to come. A screen placed there will not display the image.

Plane mirrors give rise to virtual images of objects placed in front of them. The image is the same distance behind the mirror as the object is in front. The image is virtual, upright, and the same size as the object.

Parallel rays are converged to the focal point by a converging mirror or lens. Parallel rays appear to diverge from the focal point when reflected by a convex mirror or after passing through a diverging lens. The object distance p is related to the image distance i and the focal length f by $1/p + 1/i = 1/f$. Sign conventions for use of this relation are given in Sec. 23.13. Also, the ratio of image length to object length (the magnification) is given by i/p.

Images for mirrors and lenses can be found graphically by tracing three rays. These rays are described in Secs. 23.8 and 23.10 for mirrors and Sec. 23.12 for lenses.

MINIMUM LEARNING GOALS

Upon completion of this chapter you should be able to do the following:

1. Give the approximate wavelength limits of the visible spectrum and arrange a list of colors in order of their wavelengths.

2. State the speed of light in vacuum. Compute n for a material when the speed of light in it is given or vice versa.

3. Explain why a distant source gives rise to parallel rays. Distinguish between wavefront and ray; give the relation of one to the other.

4. Draw the reflected ray when the incident ray on a smooth surface is given.

5. Explain the meaning of refraction and be able to use Snell's law for a refracted beam.

6. Using a diagram, show why total internal reflection

occurs only if $n_2 < n_1$. Use Snell's law to find the critical angle for total internal reflection. List a few uses of this phenomenon.
7. State whether a given image is real or virtual.
8. Use ray diagrams to locate images for both single mirrors and single lenses.
9. Use the lens and mirror equation to obtain p, i, or f if two of the three are given or described to you. Relate f to R for a mirror. State the sign for f in any given case.
10. Find the size of an image when the object size is given or vice versa.
11. Tell whether a lens is diverging or converging in air when its shape is given to you.
12. Explain how the focal point and focal length of a concave mirror and converging lens can be obtained by experiment.

IMPORTANT TERMS AND PHRASES

You should be able to define or explain each of the following:
Visible spectrum
$c = 3 \times 10^8$ m/s
Index of refraction
Wavefront; ray; parallel light
Incidence angle = reflection angle
Refraction; Snell's law
A parallel plate does not deviate a beam
Total internal reflection; critical angle
Virtual versus real image
Focal point; focal length; $f = R/2$
Three rays for image location
Lens and mirror equation; sign conventions
$I/O = i/p$
Convex, concave, diverging, converging

QUESTIONS AND GUESSTIMATES

1. Consider a concave mirror and an object at infinity. Where is the image formed? Is it upright or inverted? Is it larger or smaller than the object? Answer these questions as the object is slowly moved in toward the mirror. In particular, note the positions where any of the answers change.
2. Repeat Question 1 for a convex mirror.
3. Repeat Question 1 for a converging lens.
4. Repeat Question 1 for a diverging lens.
5. Reflection of light from a smooth mirror is called **specular reflection.** (The word specular is derived from the Latin word for mirror.) When reflection occurs from a rough surface, it is called **diffuse reflection.** Describe the nature of this latter type of reflection, using diagrams to illustrate.
6. An object under water does not appear to be as far below the surface as it really is when one looks at it from above the water. Explain why.
7. Explain, using a wavefront diagram, why a lens can be either converging or diverging depending upon the material in which it is embedded.
8. Can an empty water glass focus a beam of light? A full water glass? Is it possible to start a fire by accident if a bowl of water is set in a sunlit window?
9. A spherical air bubble in a piece of glass acts like a small lens. Explain. Is it converging or diverging?
10. How can one determine the focal length of a converging lens? Of a diverging lens?
11. Repeat Question 10 for mirrors.
12. Two plane mirrors are placed together so that they form a right angle. An object is then placed between them. How many images are formed? Repeat for an angle between the mirrors of 30°.
13. About how much longer does it take for a pulse of light from the moon to reach the earth because of the presence of air than if there were a vacuum above the earth? (E)
14. A "solar furnace" can be constructed by using a concave mirror to focus the sun's rays on a small region, the furnace region. How would you expect the temperature of the furnace to vary with area of the mirror and focal length of the mirror? (E)
15. Newton believed light consisted of a stream of particles and that the "light corpuscles" were strongly attracted by the water surface as light went from air to water. How would this lead to the observed refraction effect? Why did the observed speed of light negate this idea?

16. In various science museums (as well as in some unexpected places), a room is so designed that a person can whisper at one particular point in the room and be heard clearly at a certain distant point. How must the room be constructed so as to achieve this effect? (Sound waves can be focused by reflection just as light waves are.)

PROBLEMS

1. A girl has an earring in her left ear. She looks at herself in a plane mirror and sees an earring in the ear of her image. Is it in the right or the left ear of the image?
2. Suppose length D in Fig. 23.3 is 10 m. Give the two lowest possible speeds of rotation of the drum shown for which the beam will be properly reflected.
3. The index of refraction of a certain transparent plastic is 1.48. What is the speed of light in the plastic?
4. A radar beam is used to locate an airplane. The time taken for a radar pulse to travel from the transmitter to plane and back to transmitter is 5.0×10^{-4} s. How far away is the plane?
5. A radar pulse reflected by the moon requires 2.6 s for the round trip from earth to moon and back. How far away is the moon?
6. How far will a beam of light travel in water in the time it takes to travel 1 cm through air?
7. Light at an incidence angle of 37° enters a flat glass plate ($n = 1.50$). (*a*) What is the angle of refraction inside the glass? (*b*) After the beam leaves the plate, what is the angle between it and the beam incident on the plate?
8. The index of refraction of glass is different for different wavelengths. Flint glass has an index of refraction 1.650 for blue light ($\lambda = 4300$ Å) and 1.615 for red light ($\lambda = 6800$ Å). A beam of light consisting of these two colors is shone into flint glass at an incidence angle of 50°. Find the angles between the two color beams in the glass.
9. A beam of light is incident from air at an angle of 53° on a layer of water floating on a layer of carbon disulfide. Find the angle the beam makes in each liquid.
10. At what angle must a fish look to see an insect sitting on the shore at the water's edge? Assume the fish to be below the surface of a still lake.
11. If a beam of light is traveling inside a solid cube in a plane parallel to the base, it will keep traveling around inside the cube forever provided that it strikes each surface at an incident angle of 45°. (*a*) What must the refractive index of the cube be if the beam is to be totally internally reflected at each surface? (*b*) What practical difficulties appear in such a scheme?
12. A concave mirror having a 10-cm radius of curvature forms an image of an object 2 cm high that is placed 20 cm in front of the mirror. Find (*a*) the position and (*b*) the size of the image. (*c*) Is it real or virtual? (*d*) Is it erect or inverted? Repeat for object distances of (*e*) 10, (*f*) 8, and (*g*) 4 cm. (Check with a ray diagram.)
13. Find the (*a*) position, (*b*) size, and (*c*) nature of the image formed of a 3-cm-high object placed 50 cm in front of a convex mirror having a 20-cm radius of curvature. Repeat for object distances of (*d*) 20 and (*e*) 5 cm. (Check with a ray diagram.)
14. If an object is placed 40 cm in front of a convex mirror having a focal length of 10 cm, (*a*) where is the image formed? (*b*) What is the magnification? (*c*) Is the image real or virtual? (Check with a ray diagram.)
15.* A concave mirror having a 200-cm radius of curvature is used to form a real image of an object. (*a*) Where must the object be placed if the image distance is to equal the object distance? (*b*) Are the object and image superimposed? (*c*) Compare the object and image size. (Check with a ray diagram.)
16.** (*a*) Where must the object be placed if the image formed by a convex mirror is to be half as far from the mirror as the object? (*b*) How large is the magnification in this case?
17.* Given a concave mirror having a 20-cm focal length, (*a*) where must an object be placed if the image is to be real and twice the size of the object? (*b*) Repeat for the case where the image is to be virtual.
18.* A virtual image is formed by a convex mirror having a 10-cm focal length. (*a*) Where must the object be placed if the image is to be half the size of the object? (*b*) Is it possible to obtain a virtual image larger than the object for this type of object?
19. A converging lens, focal length 20 cm, is to form an

image of a 3-cm-high object. Find the position, size, and nature of the image for the following object distances: (*a*) 100 cm, (*b*) 40 cm, and (*c*) 10 cm.

20. Find the position, size, and nature of the image formed by a diverging lens having a focal length of −30 cm if the 5-cm-high object is at distances of (*a*) 90 cm, (*b*) 60 cm, and (*c*) 20 cm.

21.* (*a*) Where must an object be placed with respect to a converging lens if the image is to be the same size as the object? (*b*) Is the image real or virtual? Express your answer in terms of f.

22.** If a diverging lens is to be used to form an image which is half the size of the object, where must the object be placed?

23.* Two identical converging lenses, $f = 30$ cm, are 90 cm apart. (*a*) Find the final image position for an object placed 120 cm in front of the first lens. (*b*) How large is the magnification of the system?

24.* An object is placed 12 cm in front of a converging lens ($f = 8$ cm). At a distance of 36 cm beyond the first lens is a diverging lens having $f = -6$ cm. (*a*) Find the position and magnification of the final image. (*b*) Is it real or virtual? (*c*) Is it erect or inverted?

25.* A converging lens, $f = 20$ cm, is followed by a diverging lens, $f = -30$ cm, their separation being 30 cm. If an object is placed 10 cm in front of the first lens, find (*a*) the image position and (*b*) its magnification.

26.** An object is placed 40 cm in front of a converging lens, $f = 20$ cm, which in turn is 50 cm in front of a plane mirror. Find all the images formed by this system.

27.** An 8-cm-focal-length diverging lens is placed 16 cm to the left of a concave spherical mirror having a radius of 20 cm. If an object is placed 8 cm to the left of the lens, find all the images formed by the system.

28. A crown-glass lens has a focal length of 20 cm in air. What is its focal length when it is submerged in water?

OPTICAL DEVICES

24

Now that we understand lenses and mirrors, we are in a position to discuss some common optical devices. We shall discuss the operation of such widely different devices as the human eye and the telescope, as well as the microscope and other instruments which are of importance to us. In so doing not only shall we obtain practice in the use of lenses and mirrors, but we shall also become more intelligent operators of the optical devices we are frequently called upon to use.

24.1 THE EYE

The most familiar of all optical devices, the eye, is also one of the most complicated. Although the actual lens system used in it is not very complex, the associated interpretive equipment is as complex as man himself. The exact way in which the image formed on the retina of the eye is transformed into our sensation of sight is a problem still challenging biophysicists. It will be necessary for us to restrict our discussion to the lens system and the formation of the image on the retina of the eye.

A simplified diagram of the eye is shown in Fig. 24.1. As you probably already know, the cornea is a protective cover for the eye, the iris diaphragm controls the amount of light entering, and the retina is a sensitive surface which transmits the image formed on it to the brain. When a ray of light enters the eye, it is refracted at the cornea. Lesser refraction effects occur in the pupil and lens, since the refractive indices of the cornea, pupil, lens, and fluid portions of the eye are quite similar.

For the normal relaxed eye, these combined refraction effects form an image of distant objects on the retina. Hence, the focal length of the eye is about the distance from the retina to the lens. If you draw ray diagrams for a converging lens, you will find that the image distance increases as the object is brought closer to the lens. In the eye, however, the image distance must always be such that the image is formed at the retina. This will be true only if the eye lens is made more converging as the object viewed is brought closer to it. The muscles of the eye alter the shape of the rather deformable eye lens so as to make it thicker (more converging) when viewing close objects.

Myopia (Nearsightedness)

Normally, one is able to relax the eye lens to the state where a distant object is focused on the retina. A person unable to do this is said to be **nearsighted,** or **myopic.** *The myopic eye is able to focus only objects which are less than a certain distance from the eye. This distance is called the* **far point** *of the eye.* Since the eye remains too converging to allow proper focus for very distant objects, a myopic person must be fitted with spectacles having a diverging lens in order to see objects far away clearly. We have illustrated this effect in Fig. 24.2*a*, where the dashed lines indicate the position to which distant objects would be focused without the spectacle lens.

Hyperopia (Farsightedness)

People who have **hyperopia,** *i.e.,* **farsighted** *people, are able to relax the eye to see distant objects but are unable to make the eye lens fat enough to focus nearby objects onto the retina. The normal person cannot make the lens*

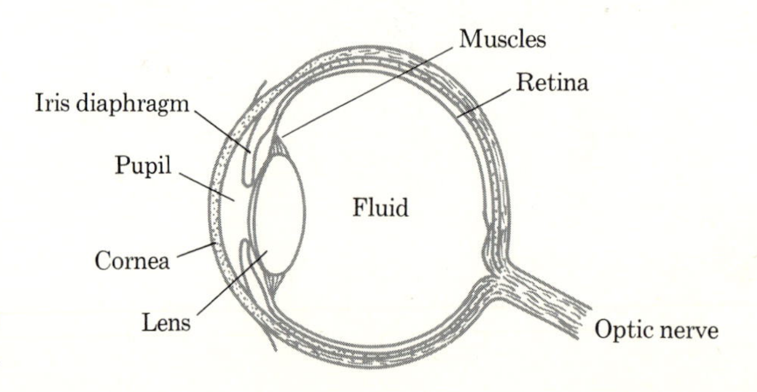

FIGURE 24.1
Diagram of the human eye.

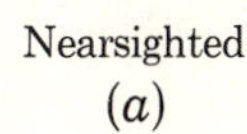

Nearsighted
(*a*)

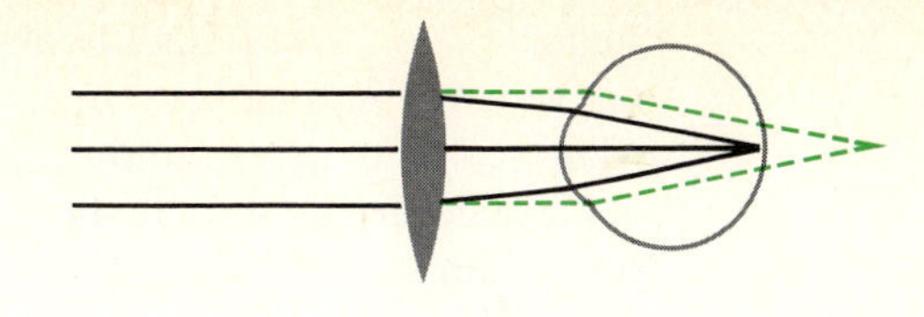

Farsighted
(*b*)

FIGURE 24.2

A diverging lens is used to correct myopia, whereas a converging lens is needed to correct hyperopia. The diagrams are exaggerated.

converging enough to see objects closer than about 25 cm, the normal **near point** *of the eye.* A farsighted person must wear a converging spectacle lens in order to aid the eye to bring nearby objects to focus on the retina. This situation is shown in Fig. 24.2*b*.

If the eye has very little ability to alter the shape of the lens, we say that the eye has lost its **accommodation.** Such an eye may be able to focus neither very distant nor very close objects. The use of bifocal spectacles allows one to look through diverging lenses when looking straight ahead or converging lenses when looking down. Some people actually have three types of lenses built into a single spectacle lens.

Astigmatism

Another common type of eye defect is **astigmatism.** When a person with this type of eye defect views the test pattern shown in Fig. 24.3*a*, some of the lines appear darker than others. This is caused by a nonspherical shape of the eye lens. As you know, a spherical lens focuses parallel rays at a single point,

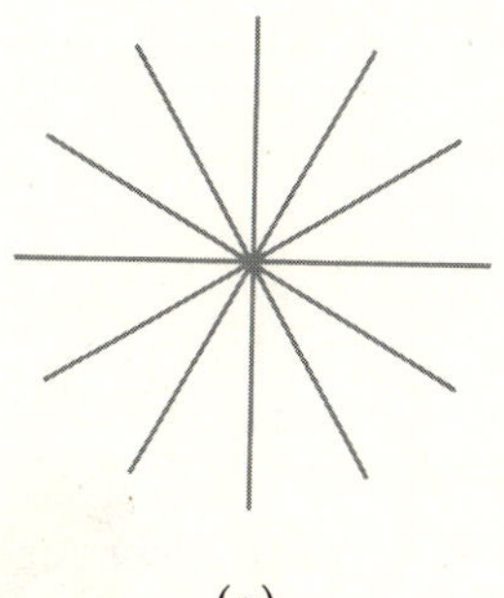

(*a*)

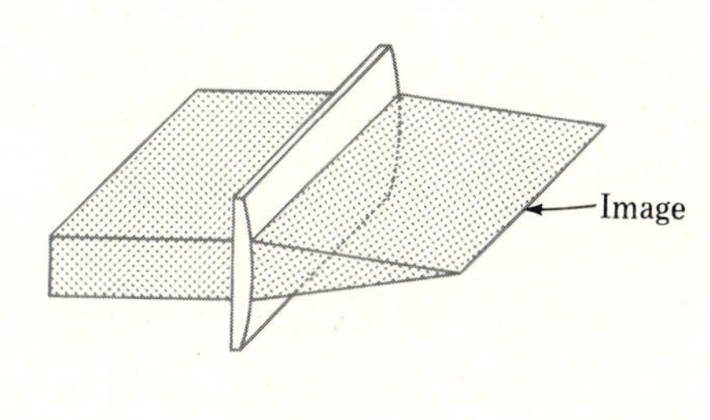

(*b*)

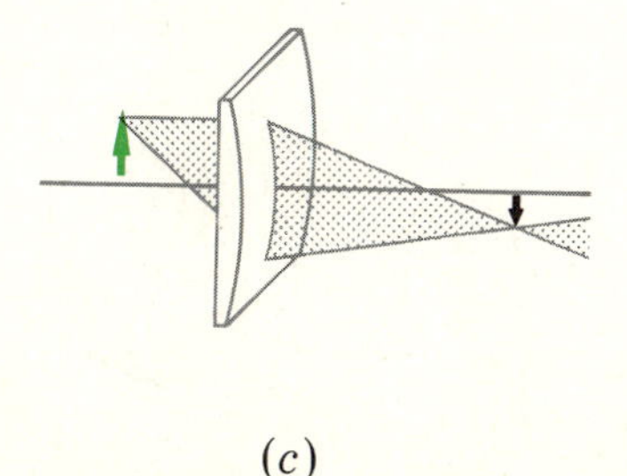

(*c*)

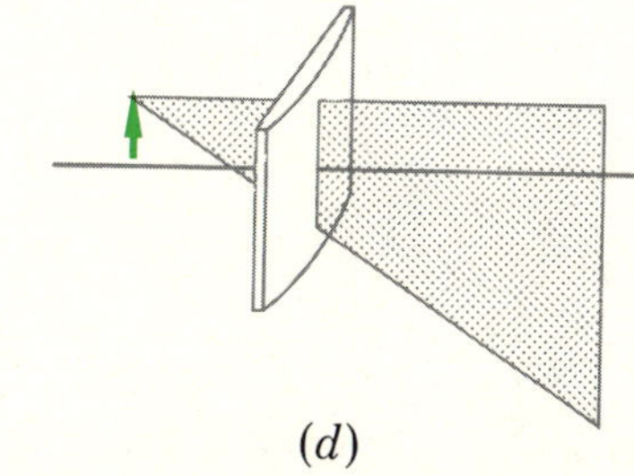

(*d*)

FIGURE 24.3

Cylindrical lenses form line images rather than point images. Rays in various planes are focused differently, as shown in (*c*) and (*d*).

the focal point. But a slightly misshapen eye lens often acts like a spherical lens with a cylindrical lens superimposed upon it. We can see the effect of the cylindrical lens by examining Fig. 24.3*b*, *c*, and *d*.

Notice in part *b* that the lens focuses parallel rays into a line image rather than a point image. Rays in a plane perpendicular to the cylinder axis, as in *c*, are focused, but rays in a plane parallel to the axis, as in *d*, are not focused at all. The result of this is to cause the eye lens to focus the lines in part *a* at different positions. When a vertical line is in proper focus, a horizontal line, for example, may not be in focus. Consequently, lines oriented at different angles appear different to a person who has this type of eye defect. To compensate for this, an eyeglass which exactly cancels the eye's cylindrical lens part must be used.

Illustration 24.1 A farsighted man is able to read the newspaper only when it is held at least 75 cm from his eyes. What focal length must the lenses of reading glasses have for him?

Reasoning We want him to be able to see print clearly when the reading material is 25 cm from his eyes, i.e., at the normal near point. Hence, a lens is needed which will give a *virtual* image at $i = -75$ cm when the actual object distance is $p = 25$ cm. Using the lens equation,

$$\frac{1}{25} + \frac{1}{-75} = \frac{1}{f}$$

from which

$$f = 37.5 \text{ cm}$$

Notice that a converging lens is needed.

Illustration 24.2 What must be the focal length of a corrective lens for a woman whose far point is 50 cm?

Reasoning The corrective lens must form an image of a distant object in such a way that the image will be virtual and 50 cm from the eye. Therefore, $i = -50$ cm, $p \to \infty$, and so the lens equation is

$$\frac{1}{\infty} + \frac{1}{-50} = \frac{1}{f}$$

from which

$$f = -50 \text{ cm}$$

Notice that the required lens is diverging.

24.2 THE SIMPLE CAMERA

A camera operates very much like the human eye. It uses a lens to produce an image of an object upon a film. The film serves the purpose of the retina in the eye. A schematic diagram of a typical simple camera is shown in Fig. 24.4. The image is inverted on the film, and its size I is related to the object size by the usual relation

$$\frac{I}{O} = \frac{i}{p}$$

Unlike the eye, the lens of a camera cannot be made with variable focal length. Hence, to achieve good focus on the film, the lens must be moved back and forth as the distance to the object changes. Cameras which do not have movable lenses usually have only a very small hole open in front of the lens; they operate much like a pinhole camera, which has no lens at all but merely uses a small pinhole to admit the light to the film. The operation of such a camera is treated as a question at the end of this chapter.

Expensive cameras possess very complex systems instead of a single lens. The complexity is necessary if a camera is to give very sharp images and fast shutter speeds. It is clear why the first condition is advantageous. The second, fast shutter speeds, allows one to take pictures of swiftly moving objects. Any moving object will blur the image somewhat. But the shorter the time the camera shutter is open, the less blurred the image will be. Since the shutter must be open long enough to allow sufficient light to hit the film, fast shutter speeds mean that the lens must be large so that a large amount of light enters the camera.

Lens Defects

As we saw when discussing the lens equation, only the central portion of a spherical lens can be used if a clear image is desired. This becomes even more important if a camera is to be used to take close-up pictures since then the lens must be very convex. It is only by making a complicated combination lens that the focusing errors inherent to a single lens can be eliminated. We say that such a lens has been corrected for **spherical aberration.**

Another lens defect causes images to have colored edges. This is called **chromatic aberration.** It results from the fact that the speed of light in glass varies with wavelength. As a result, the index of refraction of the glass is not the same for all colors. Blue light is focused more strongly by the lens than

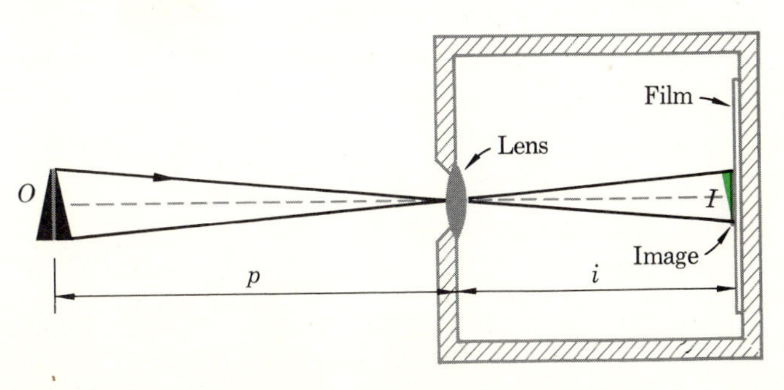

FIGURE 24.4

The simple box camera forms an inverted image on the film. In practice the lens hole would be much smaller than shown, and a shutter would cover the opening except when the picture was being taken.

red light is. This causes the colors in a beam of ordinary light to separate, and the image is therefore colored.

To correct for this defect, two or more types of glass must be layered together to form the lens. Expensive lenses consist of several individual lenses cemented together. A lens which has been partly corrected for chromatic aberration is called an **achromatic lens.** However, it is impossible to free a lens of this defect completely. Indeed, lenses used in expensive optical systems, such as a very good microscope, are very complex. They not only correct the system for spherical and chromatic aberration, but corrections are made for other lens defects as well. The design of precise optical instruments is very involved. You can read more about it in advanced texts concerned with geometrical optics.

24.3 LENSES IN CLOSE COMBINATION: DIOPTER UNITS

Perhaps you have had your eyes tested and noticed that the examiner often places several spectacle lenses, one in front of the other, in front of your eye at once. In order to make use of the observed best combination, he or she needs to know how to add the effects of thin lenses in close combination. The necessary simple formula can be derived readily, as we shall now show. We consider only the case where the focal lengths of the lenses are much longer than the lens separations.

Referring to Fig. 24.5, we find first the image formed by the first lens. Thus,

$$\frac{1}{p_1} + \frac{1}{i_1} = \frac{1}{f_1}$$

This image is then used as the object for the second lens. The object distance is $p_2 = -i_1$ if we ignore the small separation of the lenses. We use a negative sign because the object is on the wrong side of the lens and our sign conventions require a negative object distance in such cases. Writing, now, the lens equation for the second lens yields

$$\frac{1}{-i_1} + \frac{1}{i_2} = \frac{1}{f_2}$$

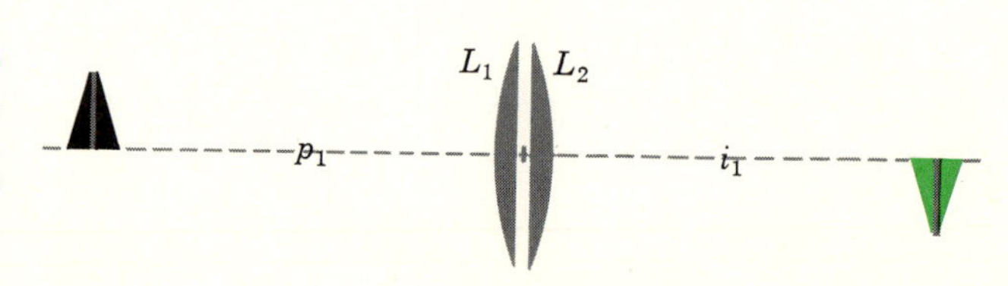

FIGURE 24.5
If the two lenses are very close together, their combined effect is to act as a single lens having a focal length $1/f = 1/f_1 + 1/f_2$.

If we add these last two equations, we have

$$\frac{1}{p_1} + \frac{1}{i_2} = \frac{1}{f_1} + \frac{1}{f_2} \qquad \text{or} \qquad \frac{1}{p_1} + \frac{1}{i_2} = \frac{1}{f}$$

where the combined focal length f of the two lenses is given by

$$\frac{1}{f} = \frac{1}{f_1} + \frac{1}{f_2} \qquad \text{close combination} \tag{24.1}$$

f for Lens Combinations

As we see, *two thin lenses in close combination behave like a single lens having a focal length given by Eq. (24.1).*

In order to save the eye examiner the trouble of adding reciprocals, a new unit using the reciprocal of the focal length is defined. *The* **power** *of a lens in* **diopters** *is the reciprocal of the focal length in meters.* For example, the power of a 20-cm focal-length diverging lens is $1/(-0.20)$, or -5 diopters.

DEFINITION

Illustration 24.3 Three lenses of focal lengths 20, -30, and 60 cm are placed in contact. Find the focal length of the combination.

Reasoning The powers of the lenses are 5, $-\frac{10}{3}$, and $\frac{10}{6}$ diopters. Their combined power is

$$\tfrac{30}{6} - \tfrac{20}{6} + \tfrac{10}{6} = \tfrac{20}{6} \text{ diopters}$$

Hence the combined focal length is $\frac{6}{20}$, or 0.30 m. It is interesting to notice that if the $\frac{10}{6}$-diopter lens had been negative, the three lenses would have substantially the same effect as a flat plate of glass.

24.4 THE MAGNIFYING GLASS (SIMPLE MAGNIFIER)

The normal person can see an object clearly only if it is 25 cm from the eye or farther. We show in Fig. 24.6 the image formed on the retina of the eye when the object is placed at that distance. If we could properly focus the image when the object was closer, the image on the retina would be much larger, as

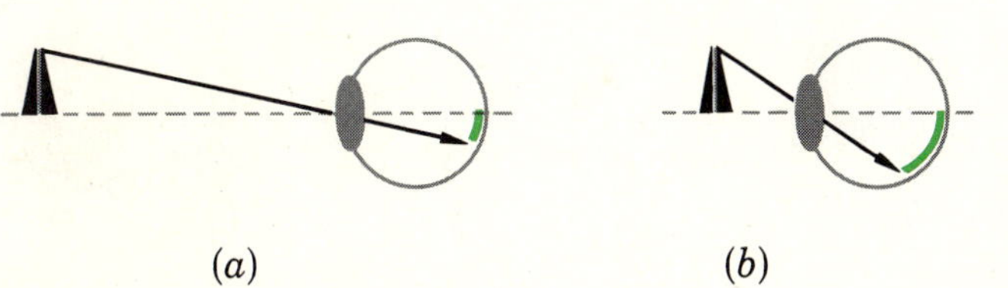

FIGURE 24.6
When an object is brought closer to the eye, the image on the retina is increased in size.

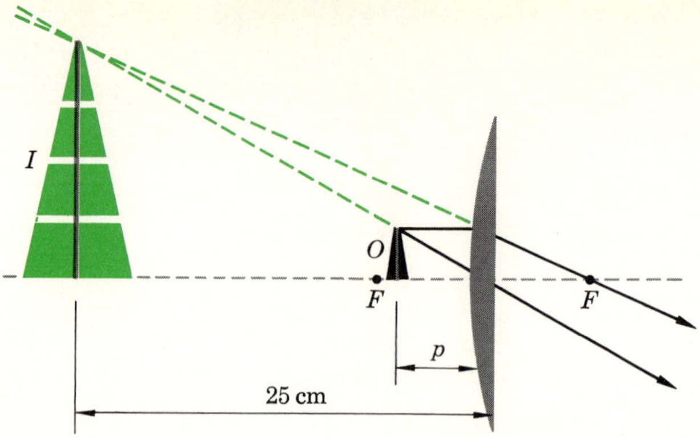

FIGURE 24.7
A magnifying glass effectively allows one to place the object being examined closer to the eye than would be possible with the eye alone.

indicated in part *b* of Fig. 24.6. It would then be possible to see fine detail on the object better, since it would appear much larger on the retina.

In order to see an object clearly, i.e., focus it on the retina, when it is closer than 25 cm, we need to use a converging lens to aid the eye lens. The effect of such a magnifying lens is shown in Fig. 24.7. We see that if we place the object inside the focal point of the lens, a virtual image of the object is formed much farther away from the lens. If the eye is placed just behind the lens, the image can be seen clearly if it is 25 cm or more away from the lens as shown.

Let us now see what effect this lens used in this way has upon the size of the image on the retina. In Fig. 24.8*a* we show the image formed on the retina when the object is 25 cm from the eye without the use of the magnifying lens. Part *b* of the same figure is redrawn from Fig. 24.7, with the eye shown in addition. The eye is actually looking at the virtual image formed by the magnifying glass. However, the size of the image on the retina is the same size as if the eye were focused on the actual object, which is now much closer than 25 cm. Hence, with the lens, the image on the retina is larger than without. This is purely the result of the fact that the lens has allowed us to view the object closer than at the normal closest distance, 25 cm.

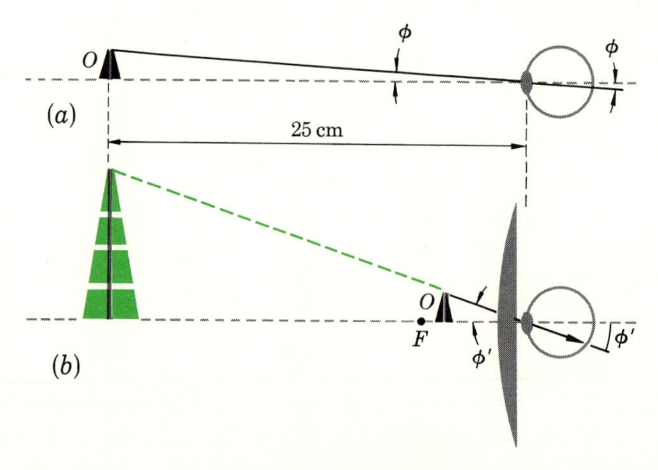

FIGURE 24.8
In both cases, the eye is focused on the triangle 25 cm from it. The actual object is much closer to the eye than 25 cm in (*b*), and so the angle subtended by it is much larger.

The **magnifying power** *M of an optical device is defined to be the ratio of the angle subtended by the image on the retina when the device is used to the angle subtended by the image on the retina when the object is viewed directly at 25 cm,* or, referring to Fig. 24.8,

DEFINITION
Magnifying Power

$$M = \frac{\phi'}{\phi}$$

The exact value of M will differ somewhat depending upon how the magnifying glass is held and how the eye is focused. A fair approximation in this case may be had by saying that the eye is just behind the lens and that the object is located close to the focal point of the lens. It turns out that for small angles such as we are usually concerned with, the angle is nearly equal to its tangent and its sine. Then from Fig. 24.8 we have

$$\phi = \frac{O}{25} \quad \text{and} \quad \frac{O}{f} \approx \phi'$$

which gives

$$M = \frac{\phi'}{\phi} = \frac{25}{f} \qquad f \text{ in cm}$$

A more exact treatment shows that $M = (25/f) + 1$, but for most purposes the approximate form given is adequate. (Why?)

M for a Magnifier

We therefore see that a lens of high magnifying power should have a short focal length—it should be highly converging. If f is 4 cm, the magnifying power would be 6.25 and the object would appear about six times larger with than without the magnifying glass.

24.5 THE MICROSCOPE

As seen in the previous section, the magnifying power of the simple magnifying glass is about $25/f$. Since the object must be placed within the focal length of the lens, it is not possible to make f extremely short. Hence, the power is limited by this as well as other practical considerations. The compound microscope allows larger magnification by the use of two converging lenses. A diagram of the microscope is shown in Fig. 24.9.

In this device the first lens, the **objective lens,** forms a real image I_1 of the object. Notice that the object is placed just outside the focal point of the objective. Since $I/O = i/p$, I_1 is a factor i_1/p_1 larger than the object. Since the object is usually placed fairly close to the focal point, $p_1 \approx f_1$. In addition, the image I_1 will be close to the second lens, as we shall soon see, and therefore i_1 is essentially equal to the length of the barrel of the microscope, about 18 cm in most cases. Therefore, I_1 is about $18/f_1$ times larger than the original object.

The second lens, the **eyepiece lens,** is used as a magnifying glass to look at the image formed by the objective. As such, I_1 must fall inside its rather small focal length. Moreover, for good viewing, the virtual image I_2 formed

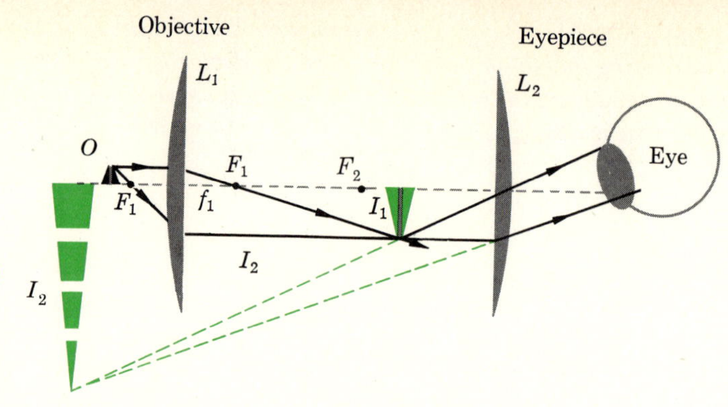

FIGURE 24.9

The eye sees the final image I_2. The microscope has greatly increased the angle subtended at the eye.

by the eyepiece must be about 25 cm from the eyepiece. As shown in the last section, the magnifying power of the eyepiece used as a magnifying glass is $25/f_2$. However, the objective lens had already magnified the object by a factor $18/f_1$, and so the angle subtended to the naked eye by I_1 was already $18/f_1$ times as large as the angle subtended by the original object. Therefore, the total magnifying power of the microscope is

M for a Simple Microscope

$$M = \left(\frac{18}{f_1}\right)\left(\frac{25}{f_2}\right)$$

where f_1 and f_2 are in centimeters.

As we see, both f_1 and f_2 should be made small for highest magnification. In order to achieve small f_1 and f_2, a good microscope uses quite complex combinations of lenses for the objective lens and for the eyepiece. Great care must be taken in the lens design, or various lens aberrations will so seriously distort and color the image as to make the instrument nearly worthless.

24.6 THE ASTRONOMICAL TELESCOPE

Two basic problems confront astronomers when they design a telescope to look at the moon or some distant planet. They would like the object to appear larger than when viewed with the naked eye. Also, they would like to increase the faint amount of light which reaches the naked eye directly from the planet. To overcome this latter difficulty, they use an objective, or first, lens which is very large so that it will collect a great amount of light from the star. One telescope, at the Yerkes Observatory, has an objective lens which is 40 in in diameter and has a 62-ft focal length. Clearly, such a large lens will gather much more light than the very small opening in the eye. A diagram of such a telescope is shown in Fig. 24.10.

As shown, the objective lens forms an image I_1 of the distant object. This image will be very close to F_1, the focal point of the objective lens, since the light from the distant object will be nearly parallel. The eyepiece lens acts, as usual, as a magnifying glass to look at the image cast by the first lens. Of

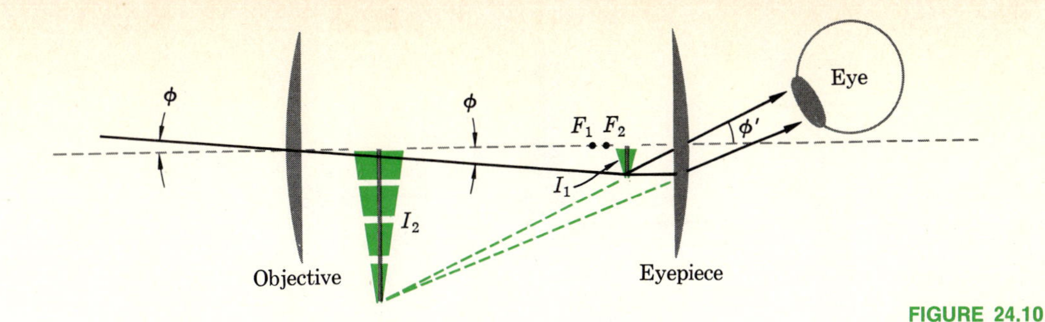

FIGURE 24.10
The two-lens telescope shown has considerably increased the angle subtended at the eye, but the final image is inverted.

course, the image formed by the eyepiece is virtual and about 25 cm from the eyepiece lens.

To find the magnifying power of this device, we notice that if the naked eye looks at the distant planet the angle ϕ subtended would be the same as the angle subtended at the eyepiece. (Why?) Hence, if ϕ is small so that it may be replaced by its tangent, we have

$$\phi \approx \frac{I_1}{f_1}$$

since the image is very close to the focal point. Also, since the focal point of the magnifying eyepiece is placed close to I_1, we have

$$\phi' \approx \frac{I_1}{f_2}$$

If we divide this latter equation by the former, we find the magnifying power of the telescope to be

M for a Telescope

$$M = \frac{\phi'}{\phi} = \frac{f_1}{f_2}$$

We see that the focal length of the objective should be long and that of the eyepiece should be small.

Since it is very difficult to make large, perfect lenses, many large astronomical telescopes use a concave mirror in place of an objective lens. One arrangement for such a telescope is shown in Fig. 24.11. The size of the

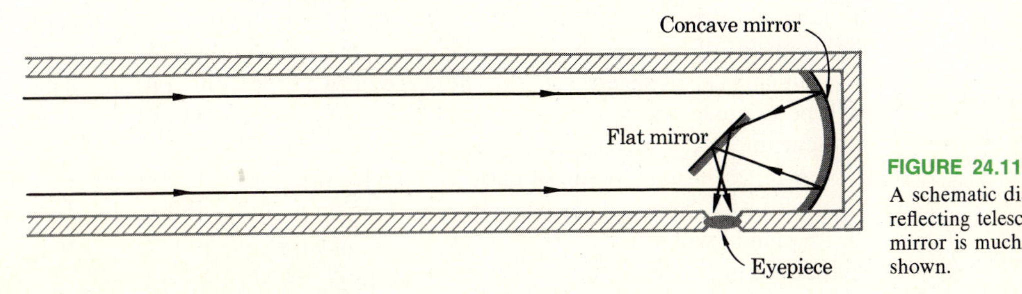

FIGURE 24.11
A schematic diagram of a reflecting telescope. The flat mirror is much smaller than shown.

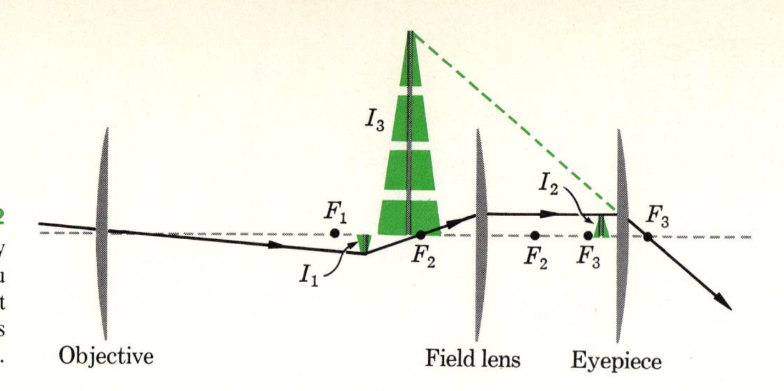

FIGURE 24.12
The terrestrial telescope. Only one ray is shown, but you should be able to verify that the images are at the positions indicated.

flat deflecting mirror is exaggerated. A reflecting telescope is not affected by chromatic aberration except for the eyepiece. Moreover, spherical aberration in the mirror is eliminated by using parabolic mirrors. Since it is considerably easier to produce a good large mirror than a large lens, it is clear why reflecting telescopes are sometimes preferred.

24.7 THE TERRESTRIAL TELESCOPE

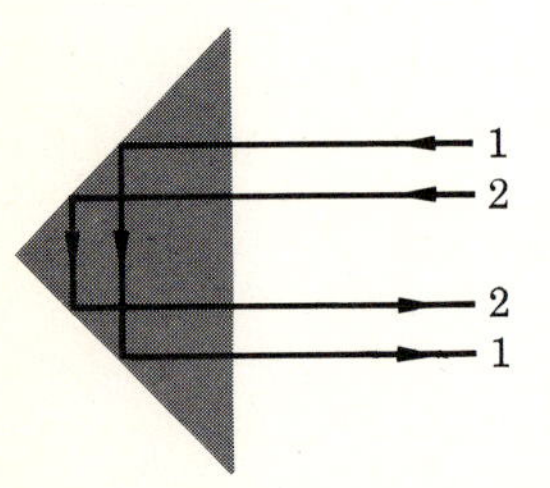

FIGURE 24.13
The inverted image in a binocular is turned right side up by the total internal reflection in a prism, as shown.

Although the astronomical telescope is satisfactory for star gazing, it is wholly unsuited to most work on the earth. The main difficulty is that the astronomical telescope inverts the objects, as may easily be seen by reference to Fig. 24.10. In order to see things right side up, the terrestrial (earth) telescope uses a third lens, as shown in Fig. 24.12. Notice that the field lens merely inverts I_1 to I_2 and that the final image seen by the eye, I_3, is erect.

This device is all contained in a rather long tube and is often referred to as a **spy glass.** It is inconvenient to use and has been mostly replaced by the prism binocular. In this latter device, the image is inverted by total internal reflection within two prisms. One prism turns the image right side up, as shown in Fig. 24.13, while another prism at right angles to it turns the beam around again and reverses the horizontal orientation of the image. Not only do the prisms return the image to its proper orientation, but they also greatly shorten the tube length between objective and eyepiece.

24.8 THE PRISM SPECTROSCOPE

In later chapters we shall speak frequently about the spectrum (or colors of light) given off by various atoms. These spectra are observed by means of instruments called **spectroscopes.** These are of two types, only one of which will be discussed now; the second will be treated in the next chapter. Our present interest is in the prism spectroscope.

FIGURE 24.14
The prism shown deviates the light beam through the angle D.

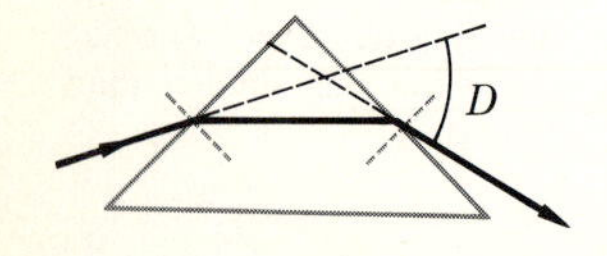

A prism, usually made of glass, is a device frequently used to separate light into its various colors. A diagram of a prism is shown in Fig. 24.14. A beam of light will usually be bent twice, once when it enters and once when it leaves the prism. We call the total angle through which the ray is bent the **angle of deviation.** It is shown as angle D in the figure.

(*Australian News & Information Bureau; Photograph by D. Moore.*)

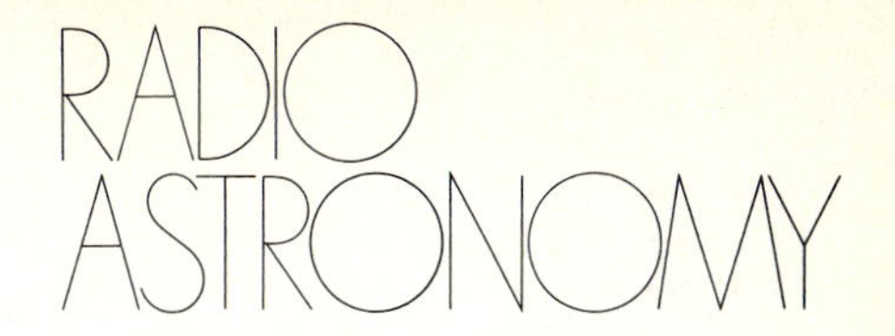

RADIO ASTRONOMY

Optical telescopes, until recently, were our only sensitive means for looking into the depths of space. They are limited in their sight to wavelengths of EM radiation which are visible to the eye or which can be photographed. This is, as we know, only a small portion of the total electromagnetic spectrum. As a consequence, optical telescopes are able to observe only those bodies in outer space which strongly emit light. These bodies must, of course, be white-hot in order to appear luminous.

We are led to wonder whether we are not missing much that exists in outer space since we with optical telescopes can see only the hot suns, the visibly glowing bodies. If we were able to "see" much cooler bodies in space, perhaps we would find new objects, such as suns which had cooled or suns which had not yet been born. To do this, we must have devices which can see longer wavelengths than visible light. In particular, telescopes which can see radio waves would be of tremendous value—hence the incentive for radio astronomy.

As we shall see in Chap. 25, in order to reflect EM waves, a surface should be smooth to within a wavelength of the radiation. The size of the reflector must be many wavelengths large. These restrictions require that a reflecting-type radio telescope use metal mirrors which are much larger than the wavelengths which it is to see. However, these mirrors need not be as smooth as those used for shorter wavelengths. In fact, wire-mesh surfaces can and have been used. The larger the mirror, of course, the more energy it can reflect and bring to focus. Therefore, a larger reflector is to be preferred.

A picture of a large radio telescope is shown in the figure. It is located in an isolated spot in Australia far from sources of extraneous earth-generated radio noise. This, as well as others elsewhere in the world, has found many heretofore unknown objects in the far reaches of space. The device illustrated makes use of wavelengths in the 10- to 20-cm range and is capable of detecting radio-wave sources which are 5 billion light-years away—10 times farther than is possible with present optical telescopes.

The 210-ft-diameter reflecting mirror, the parabolic metal grid, focuses the radio waves coming to it from space. Radio-wave detectors are used to monitor what the telescope sees. Provision is made for orienting the large reflector so that the telescope can be aimed at will toward any area of the heavens which is 30° above the horizon.

With the incident angle, the angles of the prism, and the refractive index of the glass known, it is possible, though difficult, to compute D by using

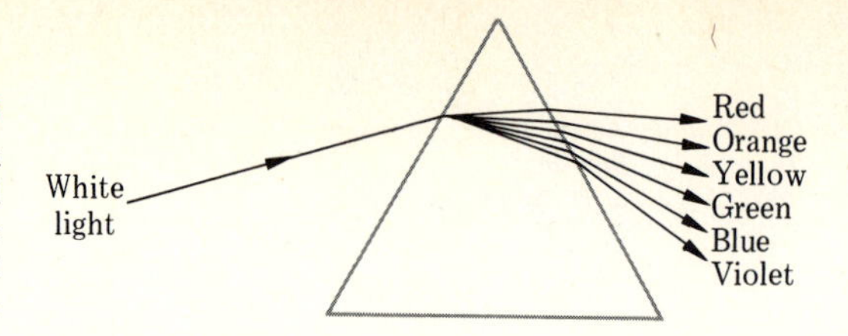

FIGURE 24.15
The angle of deviation by a prism is not the same for all wavelengths of light. Hence the prism disperses white light into its constituent colors.

Snell's law. However, it can be seen that the higher the index of refraction of the glass, the larger the deviation of the beam. This has important consequence, as we shall now see.

We mentioned earlier that the speed of light in most materials varies depending upon the wavelength of the light. This is equivalent to saying that the index of refraction of the material depends upon the color of the light. For most materials, the index of refraction for violet light is larger than for red light. Hence, violet light is bent more by a glass prism than red light is. Consequently, if a beam of white light enters a prism, as in Fig. 24.15, the light will be dispersed into its colors, as shown in the figure. The ability to disperse light varies from material to material. For high dispersion, the index of refraction must change markedly with wavelength.

In a prism spectroscope, a prism is used to disperse the wavelengths of light coming from the light source which is to be examined. In a typical device, such as the one shown in Fig. 24.16, the light from the source is used to illuminate a narrow slit. (Let us assume that a yellow sodium arc is shining on the slit. This gives off wavelengths of light very close to 5.89×10^{-5} cm, a characteristic yellow light.) Since the slit is at the focal point of the collimating lens, the light is parallel after leaving the lens. It is then refracted, as shown, by the prism. (Since we are assuming monochromatic light, i.e., single-color light, all the light will be bent the same.) The objective lens forms an image of the slit at its focal point. If a photographic plate is placed as shown, a single line, or image of the slit, will be photographed on the plate. Alternatively, the objective lens and photographic plate can be replaced by a telescope (since the light is parallel), and the yellow line, or image of the slit, can be seen visually.

If light from a mercury arc is used instead of sodium light, several lines

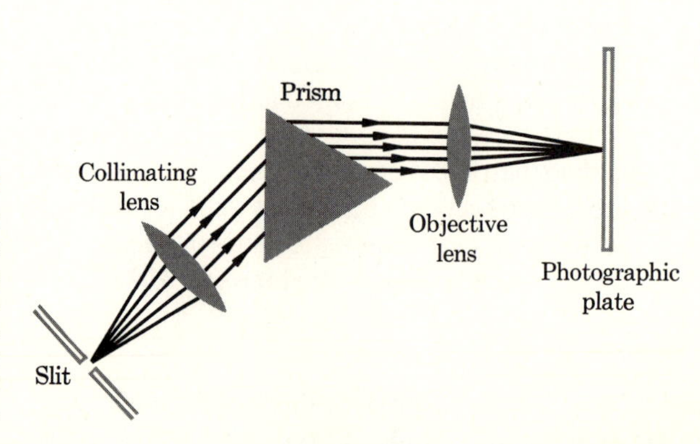

FIGURE 24.16
An image of the slit is obtained on the photographic plate of the prism spectrometer. If the light had contained more than one wavelength, multiple images would have been found in the photograph.

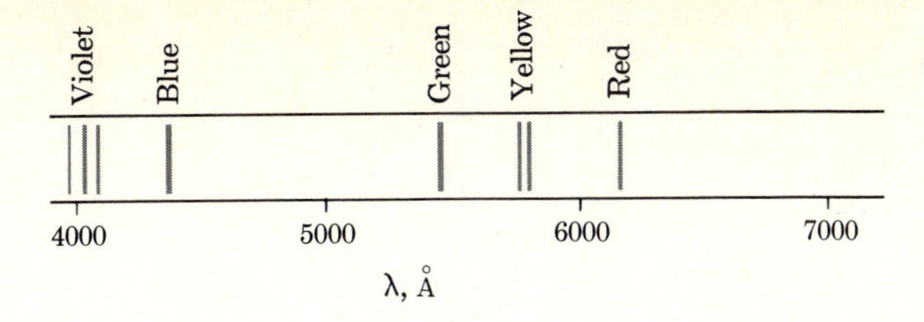

FIGURE 24.17

When a spectrometer is used to photograph a slit illuminated by a mercury arc, several images of the slit (or spectral lines) appear on the photograph, as shown here.

Spectral Lines

will be photographed on the plate. They will look as shown in Fig. 24.17. (This spectrum is also shown in color in the color plate insert.) Since each of the several discrete colors emitted by a mercury arc is deviated differently by the prism, several images of the slit, or lines, appear on the plate. These slit images are called **spectral lines.** Each line is the result of a certain color emitted by the light source. Every type of atom gives spectral lines which are characteristic of that type of atom. Not only does the emitted light when observed through a spectroscope allow us to tell what atoms are present, but also the lines or colors emitted tell us a good deal about the structure of the atom itself. We shall pursue this point further in a later chapter.

24.9 POLARIZED LIGHT

Many optical devices make use of the fact that light is a transverse vibration. As we shall see, this fact is of importance when light is transmitted through certain materials. It is also a factor when light is reflected. Although our previous discussion has not been concerned with this feature of light waves, it is fundamental to the behavior of light which we shall now discuss.

We saw in previous chapters that light is EM radiation. It consists of waves such as that shown in Fig. 24.18. The electric field vector is sinusoidal and perpendicular to the direction of propagation of the wave, as shown. If the wave is traveling along the X axis of the figure, the electric field vibrates up and down at a given point in space as the wave passes by. There is a magnetic field wave perpendicular to the page and in step with the electric field. We call a wave such as this a **plane-polarized wave.** It derives its name from the fact that the electric vector vibrates only in one plane, the plane of the page in this case.

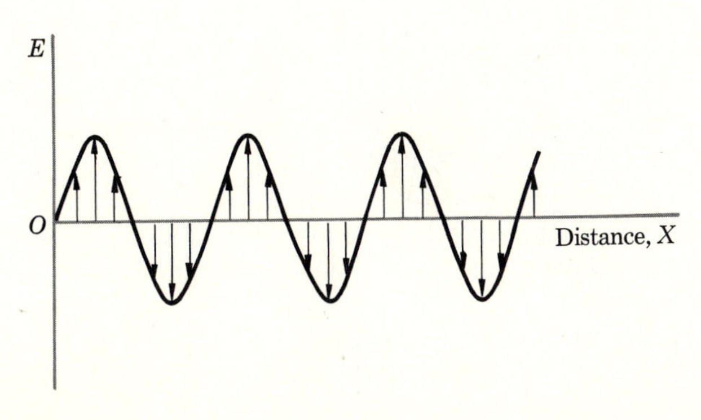

FIGURE 24.18

The electric vector vibrates in a single plane when a beam of light is plane-polarized.

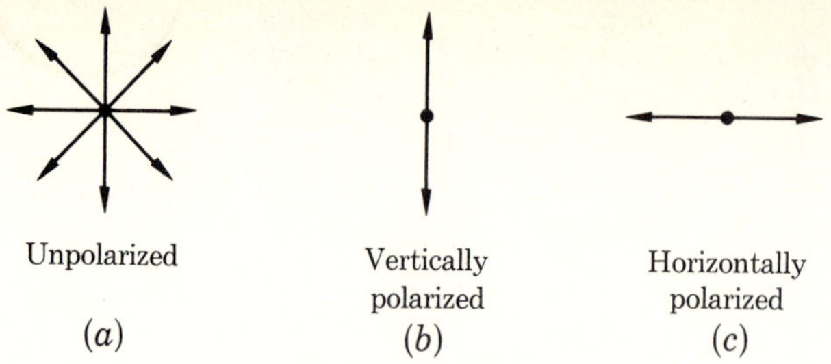

Unpolarized (a)

Vertically polarized (b)

Horizontally polarized (c)

FIGURE 24.19
If a pencillike beam of light is coming straight out of the page, the electric field vibration will be as shown for three types of beams.

Most light consists of many, many waves such as that shown in Fig. 24.18. If the direction of propagation is to the right, the electric vectors must all vibrate perpendicular to the X axis shown in Fig. 24.18. However, they need not all vibrate in the plane of the page, and actually most of them do not. Let us stand at the end of the X axis in Fig. 24.18 and look down it toward the point O, in other words, with the wave traveling straight toward us. The great multitude of waves coming toward us give rise to many individual electric vectors randomly oriented, as shown in Fig. 24.19*a*, where many more vectors than shown should actually be drawn. If the waves were plane-polarized vertically, i.e., in the plane of the page in Fig. 24.18, the approaching electric vectors would appear as shown in Fig. 24.19*b*. For a horizontally plane-polarized wave, the vectors would appear as in Fig. 24.19*c*.

Unpolarized light can be conveniently plane-polarized by use of a Polaroid sheet. This is a sheet of transparent plastic in which special needlelike crystals of iodoquinine sulfate have been embedded and oriented. The resulting sheet will allow light to pass through it only if the electric vector is vibrating in a specific direction. Hence, if unpolarized light is incident upon the sheet, the transmitted light will be plane-polarized and will consist of the sum of the electric vector components parallel to the permitted direction. (Before the invention of Polaroid in 1929 other methods were used, but because of its convenience and low cost, Polaroid has displaced them except in certain very exacting situations.)

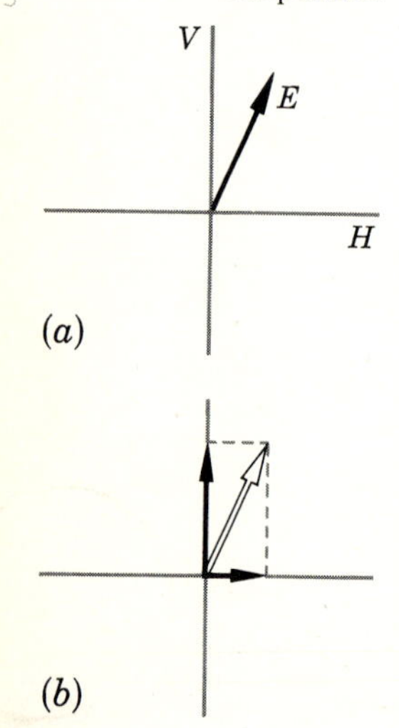

FIGURE 24.20
The electric field vector can be split into x and y components.

Any vector can be thought of as consisting of two perpendicular components. Hence, if the electric field is oriented as shown in Fig. 24.20*a*, it can be thought of as consisting of a vertical and horizontal component, as shown in part *b* of that figure. If we pass light vibrating at the angle shown through a Polaroid whose transmission direction is vertical, the vertical component of the vibration will pass through and the horizontal component will be stopped.

Consider what happens as unpolarized light is passed through two Polaroids as shown in Fig. 24.21. In part *a*, the polarizer (the first Polaroid) allows only the vertical vibrations to pass. These are also transmitted by the analyzer (the second Polaroid), since it, too, is vertical. However, in part *b*, the polarizer has been rotated through 90° and allows only horizontal vibrations to pass. These are completely stopped by the vertically oriented Polaroid. Therefore (almost) no light comes through the combination. We say that the polarizer and analyzer are **crossed** in this latter case.

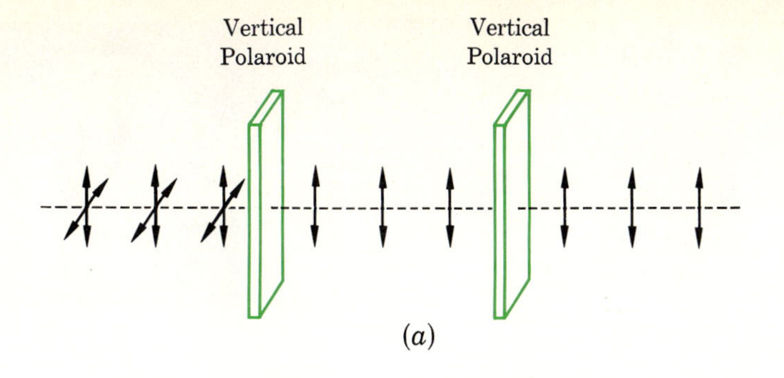

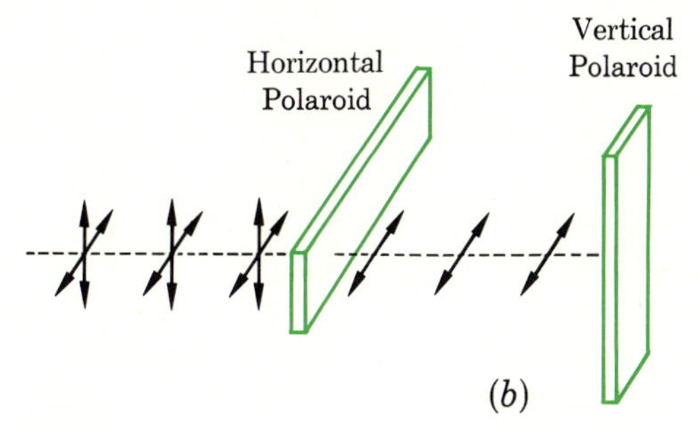

FIGURE 24.21

The unpolarized light is polarized by the first Polaroid, the polarizer. In (*b*) the second Polaroid, the analyzer, and the polarizer are crossed, and the beam is completely stopped by the analyzer.

Polarization of light is used in many technical and scientific applications. One use is in the determination of the concentration of optically active substances. **Optically active substances** *rotate the plane of polarization when polarized light passes through them.* For example, if a sugar solution is placed between the crossed polarizer and analyzer of Fig. 24.21*b*, it is found that the light is no longer stopped by the analyzer. However, by rotating the analyzer so that it is no longer oriented vertically, the light can be stopped. The sugar solution has rotated the original horizontal orientation of the direction of polarization somewhat toward the vertical. The amount of rotation of the plane of polarization has been found to be a direct measure of the concentration of the sugar in the solution. Hence we can measure the sugar concentration by this method. Many other substances are optically active. This activity can be used as a clue to the structure of the molecules composing the substance. Whether or not a molecule rotates the polarization plane depends upon the exact way in which the atoms are located relative to each other in the molecule. Some kinds of sugar exist in two forms; one rotates the polarization plane to the left, and the other rotates it to the right. Both types of molecule have the same chemical composition, but one is the isomer of the other; i.e., the structures differ only in spatial relationships on two or more chemically equivalent bonding sites within the molecule. There are many fascinating examples of this left- and right-handedness in nature. Some enzymes consume only one isomer of a molecule, upon which they thrive. Certain types of protein in particular are notable for their left- and right-handed structures in assuming a helical configuration which influences their physical behavior.

DEFINITION

Under a microscope the details of objects are often seen more clearly by

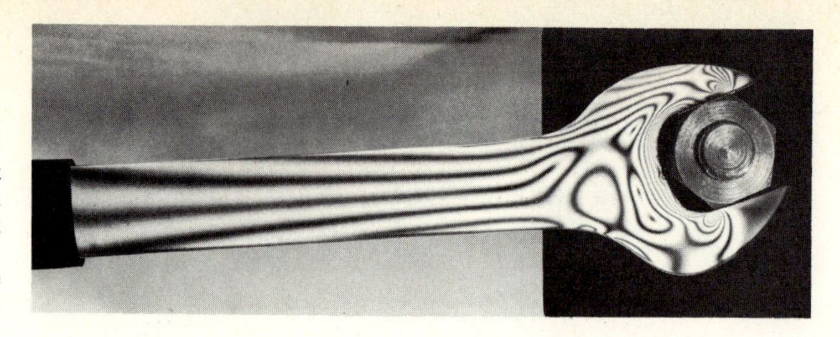

FIGURE 24.22
A strained transparent object viewed through crossed Polaroids shows alternate dark and bright bands. The stress is greatest where the bands are closest together.

examining them between crossed Polaroids. Portions of the object which appear the same in ordinary light may differ considerably in their ability to change the polarization of the transmitted light. Hence, these otherwise unobservable details can easily be seen. When a transparent object is under high stress, it often rotates the plane of polarization of the transmitted light. As a result, a nonuniformly stressed object observed between crossed Polaroids will show alternate dark and bright bands, as in Fig. 24.22. Where the bands are closest together, the stress is most uneven. By examining plastic models of strained objects like that in Fig. 24.22, it is possible to tell exactly how the stress is distributed. This is of great importance in the design of various parts for machines.

We seldom realize that the light coming from the blue sky is largely plane-polarized. Nor do we often remember that the glare of light from the surface of a lake or white concrete is considerably polarized. (This latter fact forms the basis for Polaroid sunglasses.) Since polarized light can be recognized only by use of an analyzer, our eyes do not tell us of its presence. The whole subject of polarized light is an interesting one and you may wish to pursue the subject further.*

*A readable booklet on the subject in the Momentum Book series is W. A. Shurcliff and S. S. Ballard, "Polarized Light," D. Van Nostrand Company, Inc., Princeton, N.J., 1965.

SUMMARY

In the human eye, the eye lens focuses an image on the retina. By changing the shape of the lens, the muscles of the eye can bring objects at various distances into focus. Nearsighted people are unable to focus well on distant objects; farsighted ones cannot focus nearby objects well. Astigmatism occurs when the eye lens system has a cylindrical distortion.

Rays passing through the outer portions of highly curved spherical lenses are not focused properly. This effect is called spherical aberration. Different colors are not focused exactly the same by a lens, and this effect gives rise to chromatic aberration.

The power of a lens in diopters is the reciprocal of the focal length in meters. For lenses in close combination, the power of the combination is equal to the sum of the powers of the individual lenses. As a result, $1/f = 1/f_1 + 1/f_2$.

The magnifying power of an optical device is the ratio of the angle subtended at the naked eye by the object to the angle subtended when viewed through the device. A magnifying glass allows one to view an object when it is closer than the near point of the eye. Its approximate magnifying power is $(25\text{ cm})/f$.

In a simple microscope, the objective lens forms a real image of the object. This image is then viewed by an eyepiece lens which acts as a magnifier. The magnifying power of such a microscope is $[(18\text{ cm})/f_0] \cdot [(25\text{ cm})/f_e]$. In a good microscope, each of the two lenses is actually a complex combination of lenses.

The prism spectroscope separates a beam of light into its constituent wavelengths (or colors). An image of the entrance slit is formed for each wavelength. Each image is called a spectral line and represents a distinct wavelength in the original light beam.

Ordinarily, light in a beam is unpolarized. Its electric vector vibrates in a complex way at all angles possible perpendicular to the direction of the beam. When unpolarized light passes through Polaroid, vibrations in only one direction are passed through. This transmitted beam, in which the electric vector always vibrates in the same direction, is called a polarized beam. Optically active substances have the ability to change the direction of polarization of a beam as the beam passes through them.

MINIMUM LEARNING GOALS

Upon completion of this chapter you should be able to do the following:

1. Sketch the important features of the eye and explain the function of each.

2. Explain what is meant by myopia, hyperopia, astigmatism, and lack of accommodation. Tell how each of these defects can be eased by use of lenses. Work problems such as Illustrations 24.1 and 24.2.

3. Sketch the construction of a simple camera. Explain how focusing is accomplished in a movable lens camera.

4. Give the meaning of spherical aberration and of chromatic aberration.

5. Define the diopter unit. Find the focal length of several given thin lenses in close combination.

6. Explain the operation of a magnifying glass. In relation to it, give the meaning of magnifying power. State the approximate magnifying power of a given converging lens.

7. Show how the two-lens microscope operates by sketching its optical system and drawing a ray diagram for it. Distinguish between the objective lens and the eyepiece lens.

8. Sketch the optical system for the astronomical telescope and locate the images it produces.

9. Explain what is meant by a prism spectroscope and show why it gives rise to line images. Describe how it separates colors and can be used to analyze a beam of light.

10. Distinguish between unpolarized and polarized light. Explain how polarized light can be produced. Define the following terms: Polaroid, crossed Polaroids, optically active substances.

IMPORTANT TERMS AND PHRASES

You should be able to define or explain each of the following:

Parts of the eye; iris, pupil, cornea, lens, retina
Myopia, hyperopia, astigmatism, lack of accommodation
Near point; far point
Spherical aberration; chromatic aberration
Power of a lens in diopters
Magnifying power
Objective; eyepiece
Prism spectroscope; spectral line
Polarized versus unpolarized light; plane of polarization
Crossed Polaroids; optically active substances

QUESTIONS AND GUESSTIMATES

1. Show that a real image formed of a man by a converging lens is inverted, but he and his image still have the same right hand. Show that exactly the reverse is true for an image formed by a plane mirror.

2. Clearer images are obtained in optical instruments when only a small portion of the lens is used. In the case of the pinhole camera, no lens is needed. To see how this is possible, draw a small bright object about 1 mm high at a distance of 10 cm from a 1-cm opening in a large opaque screen. Show how the bright spot cast by the object on a screen 5 cm behind the opening decreases in size as the opening is made smaller. Show that in the limit of a pinhole opening, two objects 1 cm apart and both 10 cm from the opening will give rise to well-defined images on the screen.

3. Show why a pinhole placed in front of a lens leads to a good image even when the image is not quite in focus. (See Question 2.)

4. Although all 35-mm cameras look about alike, their prices range from about $20 to several hundred dollars, even though they have virtually the same attachments. In what ways do these cameras differ?

5. A glass prism deviates a beam of blue light somewhat more than a beam of red light. Show by the means of wavefronts how this leads us to conclude that red light travels faster in glass.

6. Which of the following, as normally used, form real images: (*a*) eye, (*b*) camera, (*c*) microscope, (*d*) terrestrial telescope, (*e*) binocular, (*f*) projection lantern, (*g*) plane mirror, (*h*) concave shaving mirror, (*i*) searchlight mirror?

7. The wavelengths of light emitted by hot mercury vapor, for example, are called spectral lines. Explain clearly why they are called lines.

8. One can buy a cheap microscope for use by children. Invariably, the images seen in such a microscope have colored edges. Why is this so?

9. Suppose that the inside of a box camera is filled with water and that the lens is made stronger so that a focus will still occur at the location of the film. Will the pictures the camera takes be changed in any way? Repeat for a box with only a pinhole and no lens.

10. Why is the "speed" of a camera important? What design factors influence the speed? (Consider both the lens and shutter speeds.)

11. What happens to the light energy which is not transmitted by a Polaroid when unpolarized light is incident on it? Can you think of any drawback this might pose in using Polaroid?

12. How can one determine whether a beam of light is polarized? Whether it is composed of two beams, one polarized and the other not?

13. With a commercial camera with a 5-mm-diameter lens opening, i.e., aperture diameter, the proper exposure time for a scene is $\frac{1}{60}$ s. About what would be the exposure time for a pinhole camera with a 0.50-mm-diameter pinhole and the same type of film? (E)

14. You have available a long cylindrical cardboard mailing tube and two lenses with focal lengths 60 and 10 cm that can be fitted into the tube. Use these to design a toy telescope.

PROBLEMS

1. (*a*) When the eye-lens system is adjusted to view a distant object, what is its focal length? Assume the distance from eye lens to retina is 2.0 cm. (*b*) Repeat for the case where the object being viewed is at the near point, 25 cm from the eye.

2. An object which is 0.010 cm long is viewed by the naked eye when the object is at the near point, 25 cm. How long is the image on the retina? (Take the eye lens to retina distance as 1.0 cm.) Repeat if a 5.0-cm-focal-length magnifying glass is used and the object is placed close to its focal point.

3. A certain woman is able to see printing in a book clearly when the book is 60 cm away but not when it is closer. (*a*) Is she near- or farsighted? (*b*) What type and focal-length lens should she use to correct her sight?

4. A certain man is not able to see objects clearly unless they are closer than about 1.50 m. (*a*) Is he myopic or hyperopic? (*b*) What type and focal-length lens should he use to correct his sight?

5.* A little girl wears thick, magnifying-glass-type eyeglasses. Her older brother holds the eyeglasses in sunlight and obtains images of the sun. He finds each lens gives an image 30 cm from the lens. What are the girls' probable far point and near point without glasses?

6.* A teacher notices a child in class holds pages very close to the eye when reading. The usual position for this child is 11.0 cm. (*a*) Is the child near- or farsighted? (*b*) What kind and what focal-length lens should the child probably use in glasses?

7. An optometrist finds that the person being examined can see best when lenses with the following three focal lengths are stacked one in front of the other before the person's eye: 20, 50, and -100 cm. What focal length should be prescribed?

8. If a certain single-lens camera has a lens-to-film distance of 10.0 cm, and if it takes pictures which are 8 by 6 cm, how far from a painting which is 100 cm square must the camera be placed if the image of the painting is just to fit on the photograph?
9. When the camera from Prob. 8 is used to photograph a tree from a distance of 40 m, its image on the film turns out to be 2.0 cm high. How tall is the tree?
10. In a simple box camera the distance from the lens to the film is 10.0 cm. If the focal length of the lens is 9.50 cm, where must an object be placed to give the best image on the film?
11. Find (*a*) the power and (*b*) the focal length of the following lenses when placed in close combination: $f_1 = 20$ cm, $f_2 = -60$ cm, $f_3 = 40$ cm.
12. What focal-length lens is needed in close contact with a converging 50-cm lens to form a diverging lens of -100 cm focal length?
13. A certain magnifying glass gives an image of the sun at a distance of 3.0 cm from the center of the lens. What is the approximate magnifying power of the lens?
14. In a simple microscope, the objective has a 3-cm focal length, and the eyepiece has a 5-cm focal length. What is the magnifying power of the microscope?
15.* A boy makes a simple microscope by cementing a 5.0-cm-focal-length lens to one end of a 10-cm-long tube and a 3.0-cm-focal-length lens to the other. (*a*) If he uses the 3.0-cm lens as the eyepiece, about how far in front of the objective must he place the specimen he is looking at? (*b*) What will the approximate magnifying power of his microscope be?
16. What is the magnifying power of an astronomical telescope having a 1.50-diopter objective and a 20-diopter eyepiece lens?
17. A telescope at the Yerkes Observatory has an objective lens with a focal length of about 19 m. When observing the moon, how many kilometers on the moon corresponds to a 1.0-cm length on the image cast by the objective lens? (Distance to moon = 3.8×10^8 m)
18.** (*a*) By what factor is the light intensity increased in a telescope if the diameter of the objective lens is changed from 0.50 to 4.0 cm? (Assume that the other dimensions remain constant.) Suppose now, instead, that the focal length of the objective was tripled in going from the small to the large lens; (*b*) by what factor is the light intensity at a given point in the image changed?
19.** In Fig. 24.14 take the apex angle of the prism, i.e., the top angle, to be 60°. If $n = 1.50$ and the angle of incidence is 53°, find (*a*) the angle at which the beam leaves the prism and (*b*) the angle of deviation D.
20.* A certain piece of equipment contains a 48-cm-focal-length lens made of glass having $n = 1.50$. For purposes of cooling the lens, it is found expedient to immerse the portion of the apparatus containing the lens in a rectangular vessel of water. What focal-length lens made of the same type glass must be placed in contact with the original lens if the apparatus is to operate properly?

25 INTERFERENCE AND DIFFRACTION

In the last two chapters we discussed the behavior of lenses and mirrors, using the concept of light rays. We did not need to know whether the light consisted of particles or waves for the purposes of these discussions. This is not true of the topics to be treated in this chapter. We shall see that the wave nature of light gives rise to interference phenomena much like the interference effects we encountered in our study of wave motion and sound. The mere existence of these, as well as other effects to be discussed in this chapter, led to the final acceptance of the wave nature of light, as we shall see.

25.1 DIFFRACTION

Waves are capable of bending around corners, a phenomenon called **diffraction.** This fact is easily demonstrated visually by the use of water waves, as shown in Fig. 25.1. Notice how the waves pass through the hole in the barrier and then propagate into the whole region beyond. A similar situation is found for sound waves, which are also able to go around corners. Similarly, electromagnetic waves can spread out into a region beyond an obstacle. Later in this chapter we shall see that light displays this same property.

The underlying principle of diffraction phenomena is the fact that waves can interfere with each other. In order to prepare ourselves for a more meaningful discussion of diffraction, we shall first study how light waves can be made to interfere.

25.2 INTERFERENCE OF WAVES

You will recall that in our study of sound waves we showed that they could be made to interfere with each other. In particular, a single beam of sound waves could be split into halves by sending half of the sound down each of two pipes. If, later, the two waves were again combined, they would reinforce or cancel each other, depending upon the relative lengths of the two pipes.

For example, suppose that we have two sources of waves at points A and B in Fig. 25.2*a*. We assume these two sources to produce identical (or **coherent***) waves. In practice, we would probably use the wave from a single source and split it into two parts so that we could be sure the waves were identical in form. Clearly, if the waves A and B shown in Fig. 25.2*a* were joined together, they would reinforce each other, since they are in phase. The crests of wave A would add to the crests of wave B, producing large crests as shown for the "sum" wave in part *a*. *We call this situation, where two waves add so as to reinforce each other,* **constructive interference.**

Constructive Interference

If we move source B backward through half a wavelength, as shown in Fig. 25.2*b*, we see that the situation is quite different. Now, if we add the two waves, the crest of wave A will be added to a trough of wave B. As a result, waves A and B will cancel each other if they were originally of equal magnitude. The sum of these two waves, 180°, or $\lambda/2$, out of phase, is shown

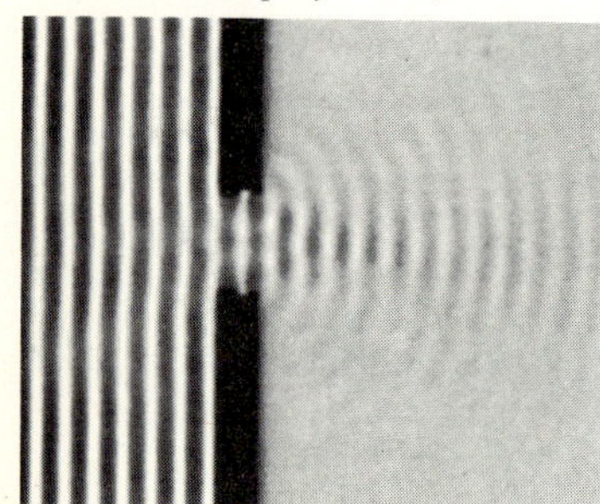
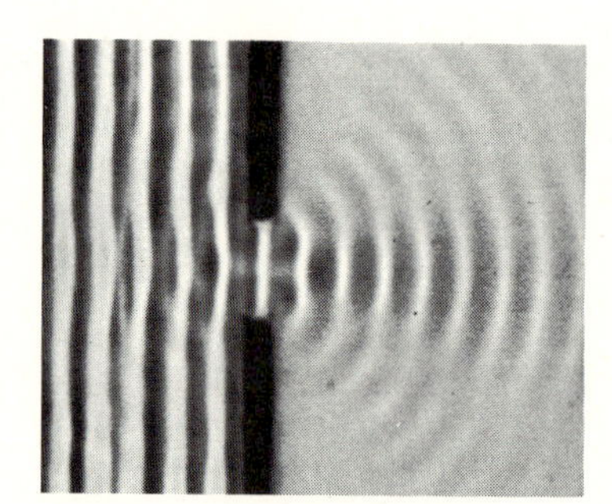
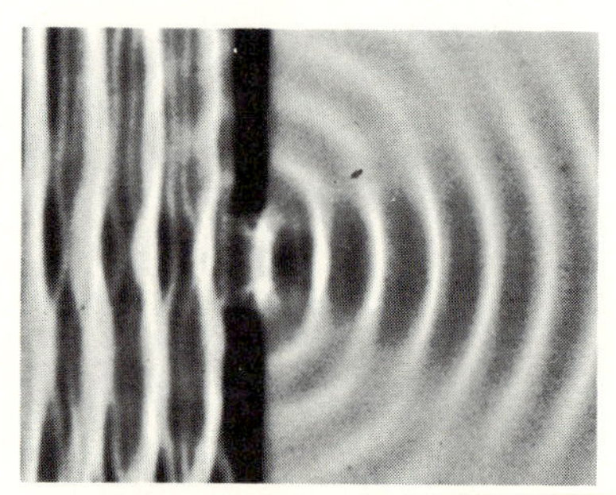

FIGURE 25.1
The water waves at the left are incident upon a hole in a barrier. Note that the waves spread out from the hole in the region at the right and that the waves spread to fill the whole region when the wavelength is comparable to the size of the hole. (*From "PSSC Physics," D.C. Heath and Company, 1965.*)

*See Sec. 27.13 for a further discussion of coherency.

in part b. As we see, their sum is zero, and we have the case of total **destructive interference.** *In total destructive interference, two waves exactly cancel each other when added together.*

Destructive Interference

If we hold back wave B through another half wavelength, it will be 360°, or one whole wavelength, behind, as shown in part c of Fig. 25.2. Here once again the two waves are exactly in step. If they are added together, they reinforce each other, as shown in part c of Fig. 25.2. From this type of reasoning, we see that waves which are 0°, 360°, (2)(360°), etc., out of phase tend to reinforce each other. Or, as commonly stated, if the waves have a phase (or path) difference 0, λ, 2λ, 3λ, etc., they reinforce.

Similarly, the situation shown in part b of the figure can occur if wave B is $\lambda + \frac{1}{2}\lambda$ behind wave A. *Cancellation occurs if the path difference is* $n\lambda + \frac{1}{2}\lambda$, where $n = 0, 1, 2, 3$, etc. *Destructive interference occurs if the path difference is* $\lambda/2$, $3\lambda/2$, $5\lambda/2$, *etc.*

A simple experiment to demonstrate the interference of water waves is illustrated schematically in Fig. 25.3. Plane water waves are incident from the left on the wall shown. Two slits in the wall, S_1 and S_2, allow the waves to pass through at these points. Hence, the slits act as new sources for waves in the region to the right of the wall. These sources generate coherent waves, since they are the result of the same incident wave. When a wave crest hits S_1, a circular wave crest is sent out from S_1 into the region at the right. Several of these circular wave crests emanating from S_1 and S_2 are shown in the figure in black.

Notice that the crests from the two sources are in phase and reinforce each other at the black dots indicated along lines OA, OB, OB', etc., in Fig.

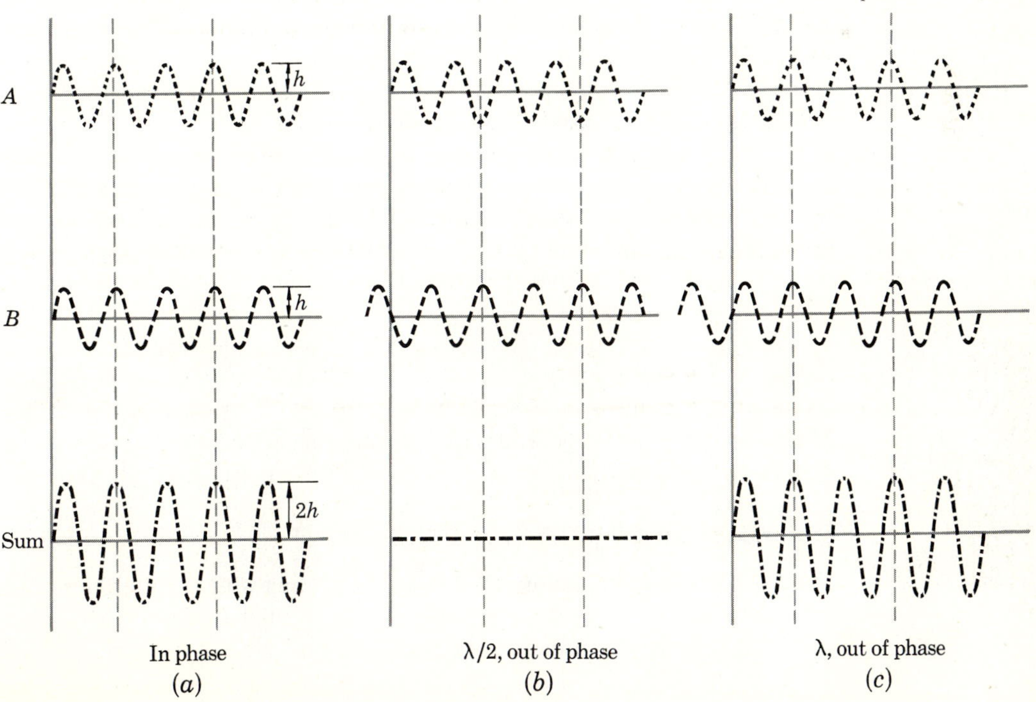

FIGURE 25.2
Identical waves can reinforce or cancel each other, depending upon their relative phase.

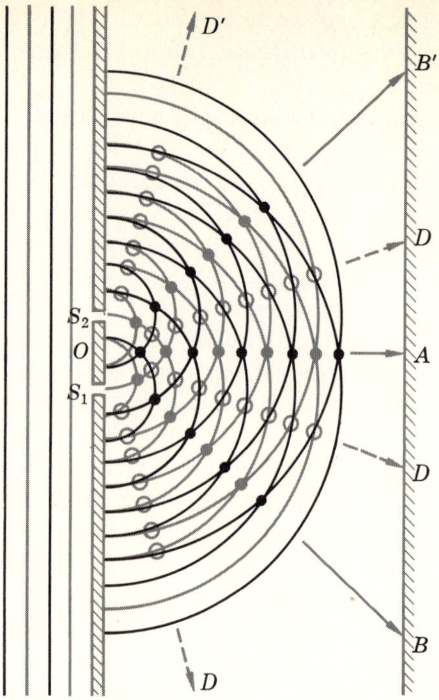

FIGURE 25.3
Plane waves are incident from the left upon the two slits S_1 and S_2. The crests of the waves are black, and the troughs are lighter. These slits send out new wave crests as shown. The waves reinforce along the lines of solid dots and cancel along the broken lines of outline dots.

25.3. Halfway between the crests, troughs exist. These also reinforce each other along these same lines at the points indicated by the lighter dots in Fig. 25.3. As we see, the waves from the two sources constructively interfere with each other along the lines OA, OB, OB', etc. Because of this, the waves hitting the wall at the right at points A, B, B', etc., have a large amplitude, and these points will correspond to antinodes.

If we examine Fig. 25.3 further, we see that at certain points the waves cancel each other. These points lie along the lines OD and OD'. At the positions of the open circles, a crest from one source is canceling a trough from the other. Since there is no displacement of the water surface along lines OD and OD', the points of intersection of these lines with the wall on the right are nodes. At the point labeled D, no wave motion exists. It is a simple matter to show that these points are always points of zero motion. (Can you prove this fact?)

To show this behavior still more clearly, consider the experiment shown in Fig. 25.4. Water waves are sent out from two coherent sources. These take the place of the slits S_1 and S_2. As we see, the two sets of waves give rise to alternate regions of large and small disturbance along the water surface.

FIGURE 25.4
The two sources send out coherent water waves much like the waves sent out from the slits in Fig. 25.3. (*From "PSSC Physics," 3d ed., D.C. Heath and Company, 1965.*)

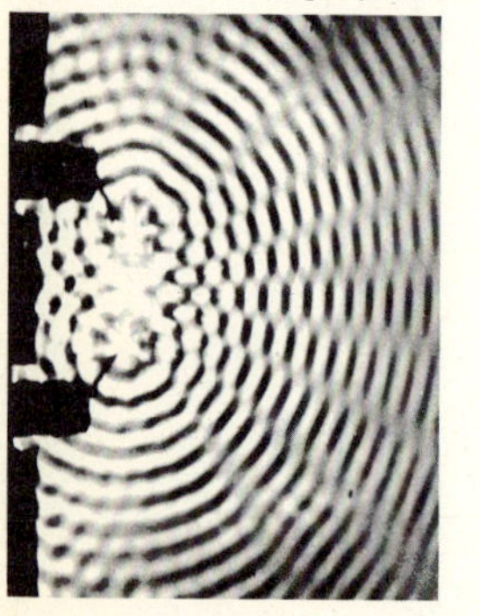

25.3 YOUNG'S DOUBLE-SLIT EXPERIMENT

The experiment described in the last section, dealing with the interference of coherent waves generated by two slits, is not peculiar to water waves. You will recall that the two prongs of a tuning fork can give rise to interference of sound waves. The explanation for this phenomenon is similar to the description of the interfering water waves, except that the waves are compressional sound waves rather than transverse water waves. It should be clear that any coherent set of waves, transverse or longitudinal, is capable of exhibiting interference phenomena.

As we have mentioned previously, Newton believed light to be corpuscular in nature. He pictured light to be a stream of particles shot out from light sources. Naturally, these particles must travel in straight lines, much as baseballs would do. Although Grimaldi had shown as early as 1660 that light can be diffracted, i.e., bent around objects, Newton was able to contrive an explanation of this fact in terms of his light corpuscles. His explanation was not very satisfying, but most people accepted his pronouncements concerning the nature of light. It was not until after 1803 that the wavelike nature of light became widely accepted.

In 1803 and 1807 Thomas Young published the results of his experiments demonstrating the phenomenon of interference of light. Using a narrow beam of sunlight passing through a hole in a window shutter, he allowed the light to fall upon two narrow, parallel slits in a piece of card paper, as illustrated in Fig. 25.5. To the right of the slits he placed a screen. He observed that alternate bright and dark spots appeared imaged on the screen as shown. This is very reminiscent of the nodes and antinodes observed when coherent waves pass through two slits.

Clearly, if light is wavelike in nature, the coherent waves generated at slits S_1 and S_2 in Fig. 25.5 will interfere with each other. At a point of brightness on the screen, the waves from the two sources must be in phase. Since the two sources are being activated by the same essentially plane incident wave, the sources will be generating waves which are in phase. The distances of S_1 and S_2 to point 0 on the screen are equal. As a result, the waves reaching 0 from S_1 and S_2 travel the same distance and therefore are in phase when they reach point 0. Hence they will reinforce each other, and 0 will be a light antinode, i.e., a bright spot.

In order to obtain the bright spot denoted as 1 on each side of the central bright spot (or maximum), the waves from S_1 and S_2 must once again be in phase. The situation at one of these points is shown in Fig. 25.6. (In practice the separation of the slits d would be only a fraction of a millimeter, while D would be a meter or so long. We have distorted the actual geometry so that the various distances can be seen more clearly.)

As shown in the figure, the light from S_1 has to travel a distance Δ farther than the light from S_2. From what has been said previously, it should be clear that the waves will reinforce when Δ is one wavelength long.

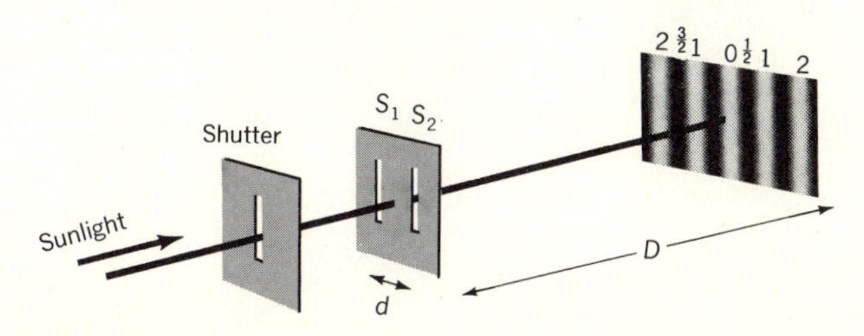

FIGURE 25.5
The two slits S_1 and S_2 act as sources for two coherent waves. In the case of light waves, the interference fringes are usually only a few millimeters apart. (Compare with Fig. 25.4 for water waves.)

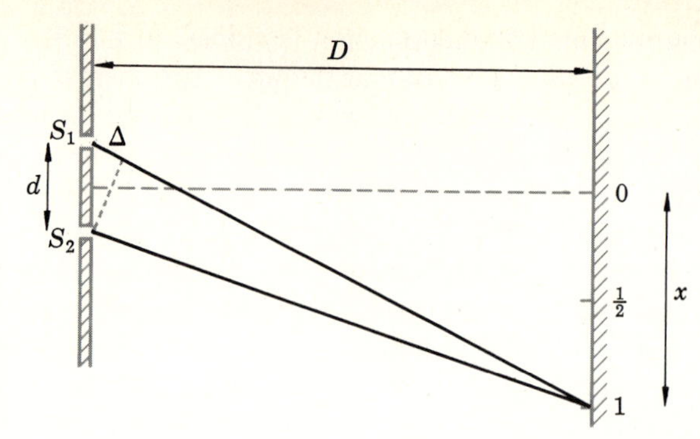

FIGURE 25.6

Where the difference in path length Δ is λ, the waves reinforce each other and brightness is observed.

If Δ is one wavelength long, the light waves from S_1 reaching 1 will be held back one full wavelength in comparison with the light from S_2. As shown in Fig. 25.2*c*, the waves will therefore reinforce each other. Point 1 will be a bright spot. Moreover, at point $\frac{1}{2}$, the wave from S_1 will travel a distance $\frac{1}{2}\lambda$ further than the light from S_2. At this point the waves will cancel, producing a dark spot. Similarly, at point $\frac{3}{2}$ in Fig. 25.5, Δ will be $\frac{3}{2}\lambda$, and the waves will cancel each other here, too. However, at point 2, $\Delta = 2\lambda$, and reinforcement (brightness) will result.

Let us now derive a relationship between the geometry of the slit, the nature of the light, and the position of the bright spots. To do this, we refer to Fig. 25.7. Remembering that d is much, much smaller than D, we see that the two shaded triangles are nearly similar. Moreover, since x is small in comparison with D, the hypotenuse of the large triangle is nearly equal to D. Hence we have

$$\frac{\Delta}{d} \approx \frac{x}{D} \tag{25.1}$$

This is the equation basic to Young's double-slit experiment.

If the distance x in Fig. 25.7 is to be the distance from the central bright spot to a side bright spot, Δ must be λ, or 2λ, or 3λ, etc. In general, for the nth bright spot counting from the zero bright spot at the center

$$\Delta = n\lambda$$

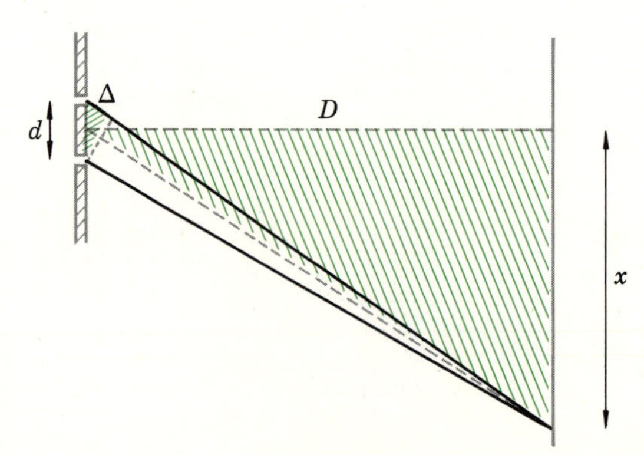

FIGURE 25.7

Since distance D is in practice very much larger than d and x, the fact that the two shaded triangles are very close to being similar triangles yields $\Delta/d = x/D$.

However, if x is the distance to a dark spot, Δ must be $\lambda/2$, $3\lambda/2$, $5\lambda/2$, etc., for the first, second, third, etc., dark spots from the center.

Illustration 25.1 In a particular double-slit experiment, yellow light from a sodium arc is used in place of sunlight. The geometry is such that $D = 100$ cm, $x = 0.50$ cm, and $d = 0.023$ cm, where x is the distance from the central bright spot to the second-order maximum, i.e., spot 2 of Fig. 25.5.* Find the wavelength of the sodium light.

Reasoning Using Eq. (25.1) and remembering that $\Delta = 2\lambda$ for the second-order bright spot, we have

$$\frac{2\lambda}{d} = \frac{x}{D}$$

or

$$\lambda = \frac{(\frac{1}{2})(0.023\text{ cm})(0.50\text{ cm})}{100\text{ cm}} = 5.8 \times 10^{-5}\text{ cm}$$

This is close to the accepted value for sodium yellow light, namely 5890 Å. Notice how extremely short the wavelengths of light are. Their lengths were first measured by Young using this method.

25.4 INTERFERENCE PATTERNS

As we saw in the last section, two slits can give rise to alternate bright and dark interference bands, or fringes, on a screen. A typical pattern for a Young's double-slit experiment using monochromatic (single-color or single-wavelength) light is shown in Fig. 25.8. Such a fringe system can be recorded by replacing the screen of Fig. 25.7 by a photographic plate. Notice that the fringes are equally spaced and the zero-order maximum, the central fringe, is the most intense.

The intensity of the light in the interference pattern is plotted in the lower portion of Fig. 25.8. We can use a photograph like that in Fig. 25.8 to compute the wavelength of the monochromatic light. Use is made of Eq. (25.1). The distance x is the distance from the center of the zero-order fringe to the center of the fringe under consideration. From Eq. (25.1) we find

$$x = \frac{D}{d} n\lambda$$

where n is the order number of the bright fringe being considered, i.e., the number above the fringe in Fig. 25.8. We see at once that the fringes will be

*The order number of a bright spot (zero, first, second, etc.,) is equal to the number of wavelengths' difference in the path lengths from the two adjacent slits to the bright spot in question. Since the path lengths from the two slits to the central bright spot are identical, the central bright spot or fringe is said to be the zero-order fringe.

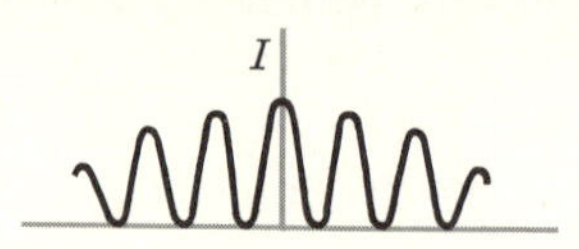

FIGURE 25.8

Interference fringes produced by a double-slit system. (*After Jenkins and White.*)

most widely spaced if d, the distance between slits, is small. Of course, the larger the distance D from the slits to the screen, the larger the fringe spacing, or separation.

It can be seen from the expression for x that the fringe spacing depends also upon the wavelength of the light. If one illuminates the slits by the yellow light from a sodium arc, the light is composed almost entirely of two wavelengths, 5890 and 5896 Å. These wavelengths are so close together that their fringe positions nearly coincide. Hence, Fig. 25.8 would be appropriate for this case.

However, light from a bluish mercury lamp consists of several colors, as seen in the color plate insert. It is principally constituted of a yellow wavelength (5790 Å), a greenish-yellow (5461 Å), a blue (4358 Å), and a violet (4047 Å). If this light is shone on a double-slit system, four sets of fringes will be set up, one for each color. The spacing of the fringes will be determined by the relation

$$x = \frac{D}{d} n\lambda$$

To show the effect of this complication, we shall consider the fringe patterns resulting from the two extreme wavelengths, 5790 and 4047 Å. The intensity pattern for the two sets of fringes is shown in Fig. 25.9, where the solid curve is for the yellow fringes and the dashed curve is for the violet fringes.

It should be clear from Fig. 25.9 that for light containing more than one

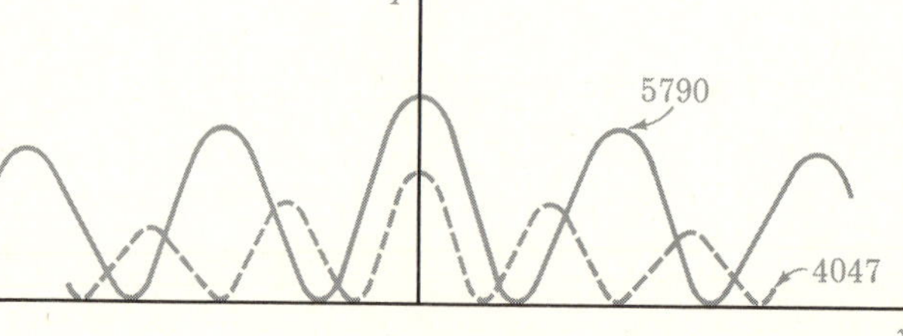

FIGURE 25.9

When two wavelengths of light are shone through a double slit at the same time, two individual patterns occur, one for each color.

wavelength the interference fringes for different colors will not fall on top of each other. This was noticed by Young, since he used sunlight for his experiments. As can be seen on the color plate insert, sunlight is composed of all colors. Hence the fringes he observed were highly colored. By noting the positions of the colors in the interference pattern he was able to show that red has the longest wavelength and violet the shortest. Of course, his wavelength measurements were not very accurate, because these highly colored fringes are broad and diffuse. Even with monochromatic light the fringes are quite broad, as shown in Fig. 25.8. For greater accuracy in measuring wavelengths one uses a diffraction grating consisting of thousands of slits, as we shall see in a later section.

25.5 MICHELSON'S INTERFEROMETER

Omit

Interference between beams of coherent light forms the basis for one of the most precise methods available for measuring lengths. The Michelson interferometer is shown schematically and simplified in Fig. 25.10. A beam of light from the monochromatic source is split into two parts by the semitransparent mirror P. About half the beam is reflected up to mirror M_1 and back, as shown. The other half is transmitted to mirror M_2 and reflected back to P, where it is reflected down, as shown. (Of course, other reflected beams occur too, but we are interested only in the two shown.)

If beam 1 travels exactly as far as beam 2, they will be in phase when they enter the eye of the observer and brightness will be seen. Suppose, now, that mirror M_2 is moved to the right through a distance of $\lambda/4$. Since beam 2 must travel this added distance twice (going down and coming back), beam 2 will now be traveling a distance $\lambda/2$ farther than beam 1. When they are joined together at the eye, they will completely cancel each other, since they are $\lambda/2$ out of phase. The eye will see only darkness.

Moreover, if mirror M_2 is moved another equal distance $\lambda/4$ to the right, beam 2 will then be traveling a distance λ longer than that traveled by beam 1. As a result, the waves will reinforce each other, and brightness will be seen

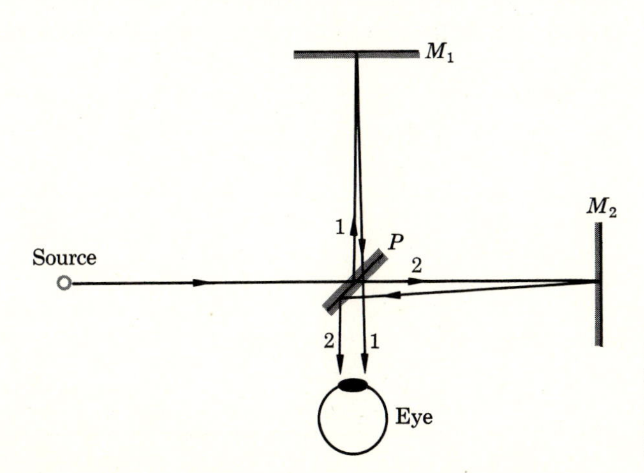

FIGURE 25.10
In the Michelson interferometer, the two halves of the beam travel different paths and may therefore interfere with each other.

once again. Clearly, as M_2 is moved slowly to the right, alternate brightness and darkness will be seen by the eye of the observer. Each time that the mirror is moved through $\lambda/2$, the bright fringe will give way to darkness and then the brightness will return. If the observer counts 1000 bright fringes appearing to the eye as the mirror M_2 is slowly moved to the right, the observer will know that the mirror has been moved a distance of $1000(\lambda/2)$.

We see, then, that one can easily measure the distance moved by mirror M_2 to within an accuracy of $\lambda/2$. Since λ might be blue light of wavelength 4×10^{-5} cm, it is possible to measure movements as small as 0.00002 cm. Actually, by special techniques, this device can be used to measure lengths to an accuracy of nearly one-hundredth this value. Although we have considerably oversimplified the pattern seen by the eye of the observer, this instrument is basically quite simple. Not only is it of value in the precise measurement of lengths, but it can also be used for the measurement of refractive indices of gases.

25.6 INTERFERENCE FROM THIN FILMS

The beautiful colors reflected from a thin film of oil on a water puddle and the colorful appearance of a soap bubble in a bright light are both the result of interference of light caused by a thin film. We shall now investigate how these interference fringes arise and see how they can be used to measure small distances.

Suppose that one has two very flat glass plates and forms a wedge of air between them, as shown in Fig. 25.11. Actually the angle between the plates would be smaller than shown. If the plates are illuminated from above by monochromatic light, an observer looking at the plates from above will see alternate dark and bright fringes, as shown by the letters D and B in the figure.

These fringes are the result of the interference of beams reflected from the upper and lower sides of the air wedge, as shown by the beams a and b of Fig. 25.11. When brightness is seen at B, rays a and b are either in phase or a whole number of wavelengths out of phase. In going from fringe B to B', we have held back ray b through one whole wavelength. We did this by

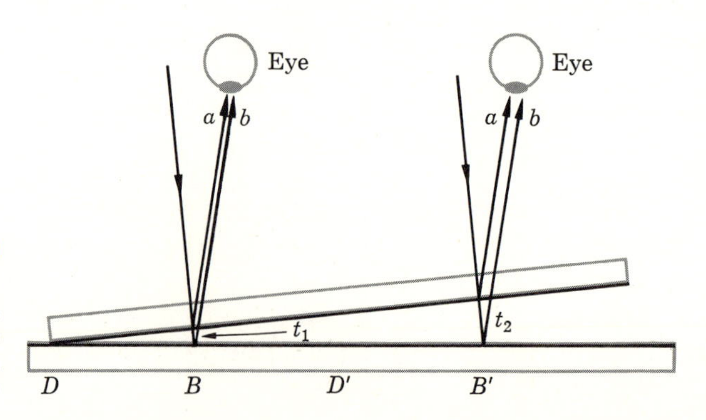

FIGURE 25.11

The two rays reflected by the two sides of the air wedge can interfere with each other. The diagram is only schematic. (In what respect is it not correct?)

making ray b at B' travel a distance $2(t_2 - t_1)$ farther than did ray b at B. (We neglect the small angle which the rays make with the vertical.) Hence we see that from one bright fringe to the next in an air wedge, the thickness has changed by an amount $\lambda/2$. (The distance is $\lambda/2$ and not λ because ray b travels the extra distance twice, once down and once up.) Similarly the dark fringes occur at $\lambda/2$ differences in thickness.

If multicolored or white light is used in this type of experiment, the different colors will reinforce at different places. As a result, the fringes are highly colored, an effect one often observes for soap and oil films in sunlight. (Can you show that the blue fringes will be more closely spaced than the red?)

If the wedge of Fig. 25.11 had been filled with liquid instead of vacuum (or air to a good approximation), ray b would have been delayed even longer relative to ray a. Since the speed of light in a liquid of index of refraction n is c/n, it takes longer for a beam to travel a given distance in liquid compared with vacuum. In the time taken for a beam to travel a distance d in the liquid, the beam would have moved a distance nd in vacuum. For this reason, *when light travels a distance d in a material of refractive index n, we say that the* **equivalent optical-path length** *in the material is nd.* As we see, beam b in Fig. 25.11 will appear to have traveled a distance (n)(twice the thickness) further than beam a.

Optical-Path Length

Previously we discussed the case of the air-filled wedge. For that case $n = 1.00$, and so the optical-path-length difference between beams b and a was simply (2)(thickness). However, for a liquid with refractive index n filling the wedge, the optical-path difference will be (n)(2)(thickness). As a result, if t_1 and t_2 are at adjacent bright spots, we have that

$$2(t_2 - t_1)n = \lambda$$

In other words, the change in thickness between bright fringes is $\frac{1}{2}\lambda/n$ instead of $\frac{1}{2}\lambda$.

Notice also that the point of contact of the glass plates in Fig. 25.11 is dark. Since ray b travels essentially the same distance as ray a at this point, we would ordinarily expect it to be bright. Early investigators tried in vain to polish the surfaces in such a way that brightness would exist at this point. However, the better the contact between the plates, the more perfectly dark this region becomes. We now know that *a beam reflected by an optically dense material, i.e., high value of n, undergoes a 180° phase change in the process.* However, this will not concern us, since we are primarily interested in differences of thickness such as are shown in the illustration below.

Illustration 25.2 When a plane convex lens is placed upon a flat glass plate, as shown in Fig. 25.12a, a phenomenon known as **Newton's rings** is demonstrated. If the system is illuminated and viewed from above, a series of interference rings are observed, as shown in part b. If light of 5890 Å is used, what is the thickness of the air gap at the position of the tenth dark ring? Repeat if the gap is filled with water $n = 1.33$.

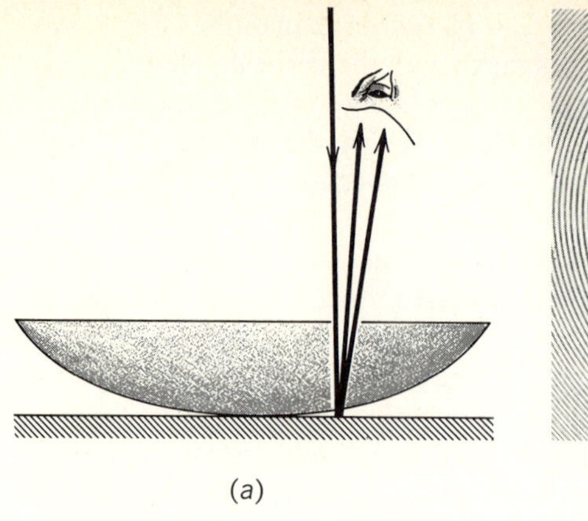

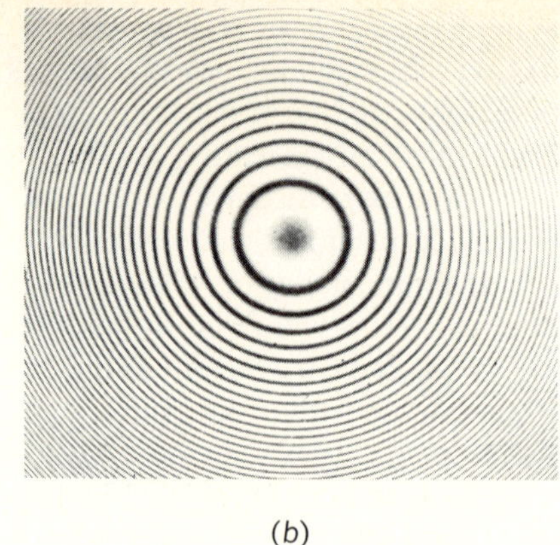

FIGURE 25.12
The interference fringes in (*b*), Newton's rings, are formed by the interference of the light rays reflected by the air wedge shown in (*a*). (Not to scale.) [(*b*) *Bausch & Lomb Optical Co.*]

Reasoning The center spot is dark, and the gap width there is zero. Going out to the first dark ring, the ray reflected from the lower surface must have been held back through a distance λ. Hence the thickness of the gap there is $\lambda/2$. Similarly, going from the first to the second ring, the wedge must increase once again by a thickness $\lambda/2$, and the wedge thickness is $2(\lambda/2)$. Clearly, at the tenth dark ring the wedge thickness will be $10(\lambda/2)$. We therefore find the gap to have a thickness

$$(10)\left(\frac{5.890 \times 10^{-5}\ \text{cm}}{2}\right) = 2.945 \times 10^{-4}\ \text{cm}$$

at the site of the tenth dark ring. In practice, the smoothness of the circles or other interference fringes for an air wedge can be used to determine how perfectly flat the surfaces are.

For the water-filled wedge, the equivalent optical-path length is (1.33)(2)(thickness). As a result, the gap thickness will be 1/1.33 times the value found above. It is therefore 2.21×10^{-4} cm.

25.7 THE DIFFRACTION GRATING

In order to measure the wavelength of light accurately with a practical instrument, a diffraction grating is often used. A device constructed for this purpose is called a **grating spectrometer.** *The grating* itself *is basically a large number of parallel evenly spaced slits in an opaque screen.* A common grating might have 10,000 slits in a distance of 1 cm. Hence the separation d between the centers of adjacent slits would be 0.0001 cm. We shall see that the operation of this device is not too different from Young's double slits.

A schematic diagram of a common grating spectrometer (or spectrograph) is shown in Fig. 25.13. The light source *Sc* illuminates the slit S_1 at the end of a tube C called a **collimator** which is used to make the rays of light parallel. A lens L_1 is placed with S_1 at its focus. Hence the light coming from the slit is made parallel, i.e., is collimated, by the lens L_1. This parallel beam

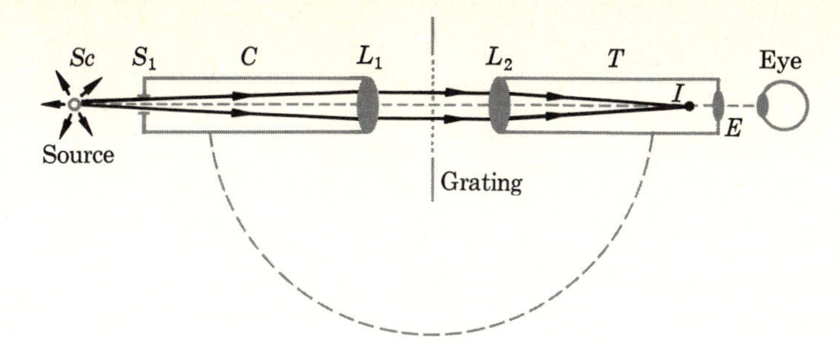

FIGURE 25.13

A schematic diagram of a grating spectrometer.

of light then passes through the grating and enters a telescope T. Lens L_2 forms an image of the slit at I, and this image is observed through the eyepiece E. Provision is made for the telescope to be rotated through accurately known angles, as shown in Fig. 25.14.

Of course, when θ is zero, one sees a clear image of the collimator slit. This direct-through image is called by various names, the **central maximum,** the **zeroth-order maximum,** and the **central image.** It has the same color as the light source. If white light is being used, the central image is also white. When the cross hairs of the telescope are coincident with this image, θ in Fig. 25.14 is zero.

We would expect in a multiple-slit device such as this to observe interference effects somewhat like those seen with a double slit. This is actually the case. However, with a grating having thousands of slits the interference pattern observed is much sharper than that found for a double slit. If the yellow light from a sodium arc is used to illuminate the slit of the collimator, one observes the light-intensity pattern shown in Fig. 25.15 as a function of the angle θ.

For a good instrument, the images at 0, 1, and 2 would be much sharper than shown. This figure should be compared with Fig. 25.8 showing a similar diagram for the double slit. *The interference pattern of a grating is much wider in angle and considerably sharper than the double-slit pattern.* We call the images seen through the telescope at equal angles on both sides of the central maximum the **high-order maxima.** For example, the images indicated by 1 in Fig. 25.15 are called the **first-order maxima,** or spectrum. Similarly, images 2 are called the **second-order maxima.** Depending upon the grating, only one or perhaps several orders may be seen before the 90° position is reached. We

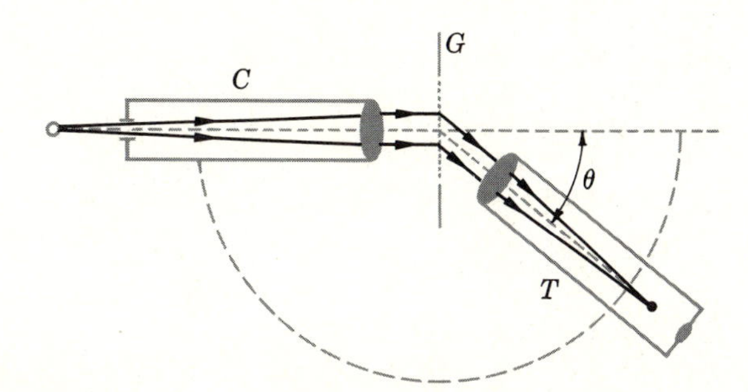

FIGURE 25.14

When the telescope is rotated on the arc of a circle, as shown, an image of the slit is formed by interference at an angle θ to the straight-through beam.

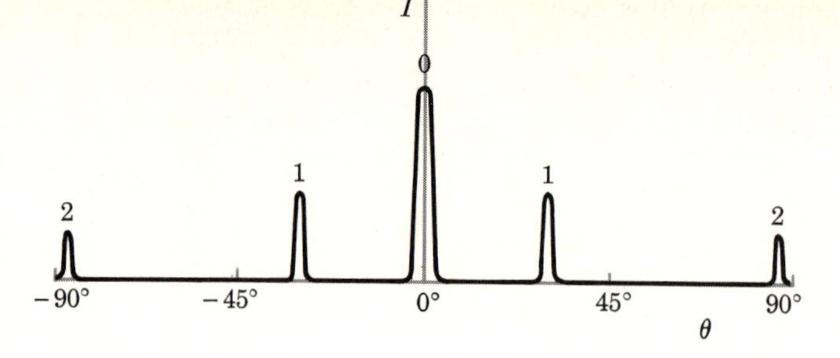

FIGURE 25.15
The diffraction grating gives a central image and symmetrical side images. For a good grating the lines would be much sharper than shown.

shall now derive the relation between the positions of the orders and the wavelength of the light.

Referring to Fig. 25.16, we consider four of the thousands of slits in the grating. Clearly, the direct, or straight-through, rays shown in part *a* are in phase and will reinforce each other when brought together within the telescope. However, the rays shown coming from the slits in part *b* will travel different distances to the telescope. As shown, ray *f* travels a distance Δ farther than *g*. Ray *g* travels a distance Δ greater than *h*, and so on. If Δ is exactly a whole number of wavelengths long, the light from all the slits will reinforce. The first-order maximum occurs for $\Delta = \lambda$; the second-order occurs for $\Delta = 2\lambda$; and, in general, the *n*th-order maximum (if possible) occurs for $\Delta = n\lambda$.

We see at once from the little triangle in part *b* of Fig. 25.16 that

$$\sin\theta = \frac{\Delta}{d}$$

Since $\Delta = n\lambda$ at the *n*th-order maximum, *we have the following equation for the angular position θ_n of the nth maximum:*

The Grating Equation

$$\sin\theta_n = \frac{n\lambda}{d} \tag{25.2}$$

This is called the **grating equation.**

Since one can measure with high precision the angle θ_n at which the

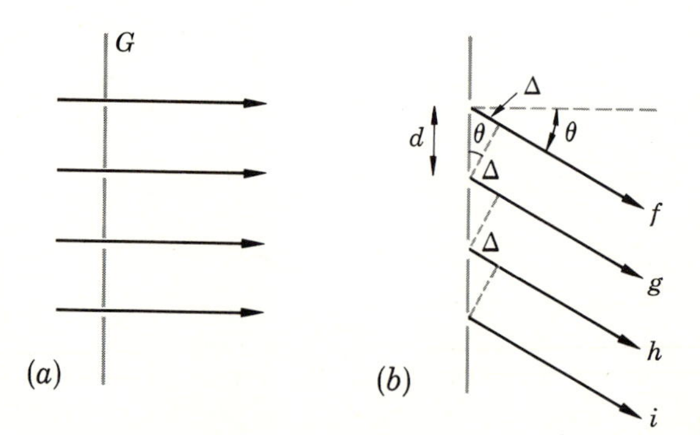

FIGURE 25.16
When Δ in (*b*) is a whole number of wavelengths long, the rays all reinforce each other. It is at these angles that the grating forms maxima.

nth-order maximum occurs, it is necessary to know only the grating spacing d in order to determine λ accurately. For example, if one uses sodium light in even a simple spectrometer, it is not difficult to see that the sodium light gives *two* slit images (or lines) at each order position. These lines are very close together and have wavelengths of 5890 and 5896 Å. The mere fact that one is able to see these two lines as distinct images provides some measure of the potential accuracy of such a device.

Types of Spectra

If mercury light is used in a grating spectrometer, several different colored lines are seen in each order. As pointed out previously, the mercury light consists of a yellow line (5790 Å), a greenish-yellow line (5461 Å), a blue line (4358 Å), and a violet line (4047 Å). Other fainter lines are also visible. (*These lines in the spectrum of a light source, called* **spectral lines,** *are actually images of the spectrograph slit.* These images appear as bright lines in a photograph, and hence the name *line* is used.) In later chapters we shall see that other types of light sources also exist. While *the light from a mercury arc consists of several easily observed discrete lines* (*called a* **bright-line spectrum**), *the light from an incandescent bulb contains all the colors and no sharp lines are observed. The incandescent source gives off a* **continuous spectrum,** *since a continuous band of color is seen when it is used in a spectrometer.* These facts are shown clearly in the color plate insert.

Illustration 25.3 A particular grating has 10,000 lines per centimeter. At what angles will the 5890-Å line appear?

Reasoning The grating space d is $\frac{1}{10,000}$ cm. Using the grating equation, we have

$$\sin \theta_1 = \frac{5.89 \times 10^{-5}}{10^{-4}} = 0.589$$

From the tables of sines, $\theta_1 = 36°$, and so the first-order images will be found at this angle on each side of the central maximum. For the second order we have

$$\sin \theta = (2)\left(\frac{5.89 \times 10^{-5}}{10^{-4}}\right) = 1.178$$

Because it is impossible for the sine of an angle to be greater than unity, this and higher-order images will not exist.

25.8 DIFFRACTION BY A SINGLE SLIT

From what has been said in the earlier portions of this chapter, it should come as no surprise to learn that a beam of light does not provide a sharp image of a slit through which it passes. For example, if a beam of light is sent

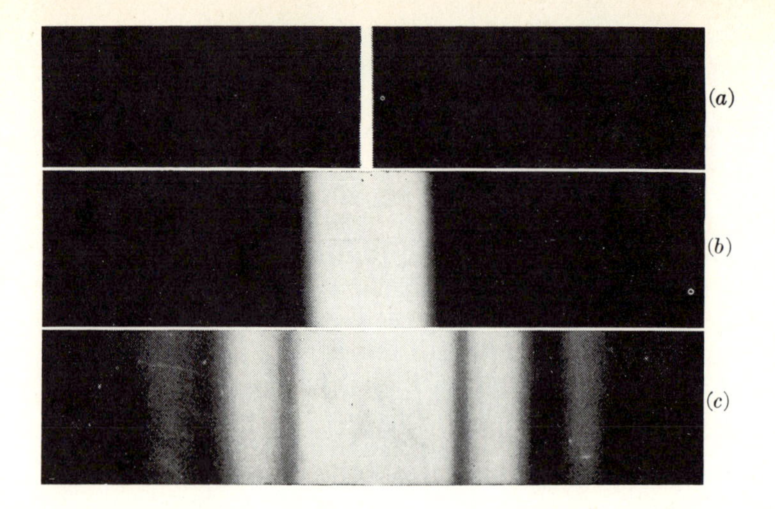

FIGURE 25.17

A photograph of a single-slit diffraction pattern for the arrangement shown in (*d*). The slit width is shown in (*a*). (*b*) and (*c*) are photographs of the slit using different exposure times. (*After Jenkins and White.*)

through a single slit, as shown in Fig. 25.17*d*, the resultant images on a photographic plate are as shown in parts *b* and *c* of the figure. We see that the central bright spot is considerably wider than the slit. Moreover, bright bands occur on each side of the central image and must result from some sort of interference effect. Let us now see what is involved in this situation.

As shown in Fig. 25.18*a*, the light rays which go straight through the single slit will all be in phase with each other. For this reason, the straight-through position is bright and gives rise to the central bright spot shown in Fig. 25.17. However, at an angle θ to the straight-through beam, rays from various parts of the slit will travel different distances to the film or photographic plate. The most important situations are shown in Fig. 25.18*b*, *c*, and *d*.

In part *b*, ray B from the middle of the slit is half a wavelength behind ray A. As a result, these two rays cancel each other. But that is not all, because we see that rays leaving the slit from positions just above A and B will also cancel since they too will have a path difference of $\lambda/2$. In fact, each ray leaving the lower half of the slit has a corresponding ray leaving the upper half which will cancel it. Hence, at this angle θ, no light will reach the film from the slit, and therefore one will observe darkness. As seen from the figure, this situation occurs when $\sin\theta = \lambda/b$, where b is the slit width. Notice that if the slit width b is equal to the wavelength of the light, the dark spot will occur at $\theta = 90°$. In other words, *if the slit width is decreased until it is as small as* λ, *the image of the slit will spread to become infinitely wide.*

If b is considerably larger than λ, as pictured in Fig. 25.18, a side bright

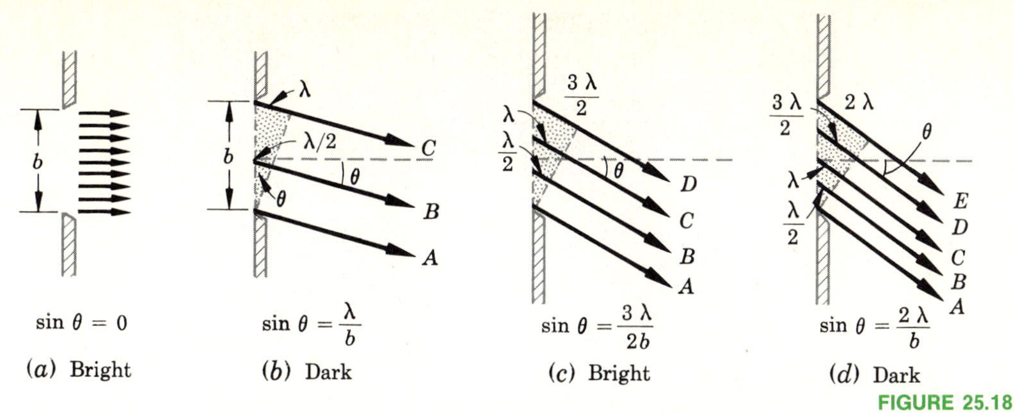

FIGURE 25.18

In analyzing the single-slit pattern qualitatively, we section the slit into portions whose rays differ by $\lambda/2$ in the path length. Why?

fringe will occur for the angle θ shown in part *c*. In this case the rays from the bottom third of the slit cancel those from the center third while the top third is left uncanceled. Darkness is again achieved at the larger angle shown in part *d*. Here the slit can be thought of as being divided in fourths. The bottom one-fourth slit is canceled by the portion just above it. Similarly the two upper sections also cancel. Hence darkness is observed at this angle. These, then, are the interpretations of the various features of the single-slit diffraction pattern seen in Fig. 25.17.

25.9 DIFFRACTION AND THE LIMITS OF RESOLUTION

Ordinarily, we think of light as traveling in straight lines. When light is shone through the two holes left by missing boards in a fence, we expect to see two separate beams of light coming through the holes, as shown in Fig. 25.19*a*. Two bright spots *A* and *B* should appear on a wall opposite the fence. This is actually what we observe. However, if the slits are made very narrow and brought very close together, as in a double slit, the situation is quite different. No longer do two distinct images appear. Instead we see the interference pattern illustrated in part *b* of Fig. 25.19.

Another example showing that light does not always travel in straight lines was pointed out in the last section, where diffraction by a single slit was discussed. There are many other examples of this type of behavior. A very

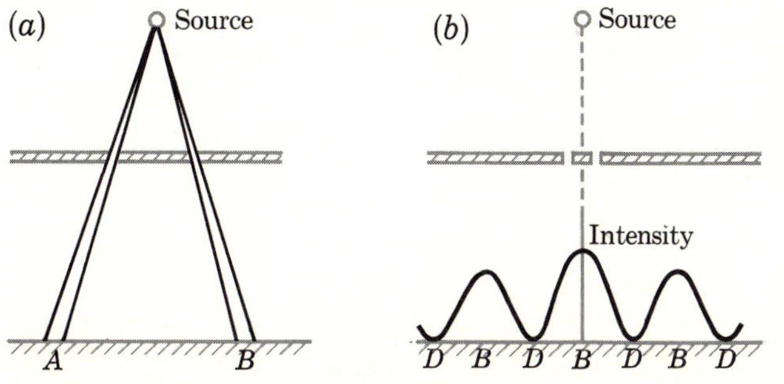

FIGURE 25.19

If the separation of the two slits and their size is large compared with the wavelength of light, images of the slits are formed at *A* and *B*, as shown in (*a*). When the slit width and separation are comparable with the wavelength of light, an interference pattern like that plotted on the diagram in (*b*) is found. (Not to scale.)

FIGURE 25.20
Shadow of a star-shaped washer. Diffraction bands are seen inside the hole and around the outer edge. (*Courtesy Bausch & Lomb Optical Co.*)

complicated, but typical, example is shown in Fig. 25.20. In all cases where a shadow is being formed, a similar situation exists. *For large objects, the shadow is relatively sharp. As the object becomes very small, interference or diffraction patterns are noticed near the edge of the shadow. For an object comparable in size with the wavelength of the radiation being used, the shadow can no longer be seen, because the interference pattern has become nearly infinitely large.*

Limit on Visible Detail

We must conclude, then, *that it is impossible to obtain images of objects with detail comparable in size with the wavelength of radiation being used.* It is for this reason that even the best microscopes cannot discern details comparable with the wavelength of light or smaller. Although a precise statement of the ability of optical devices to resolve the details of very small objects is beyond the scope of this book, *a rough rule of thumb is that detail smaller than a few wavelengths of light cannot be seen.*

Some very interesting effects result from diffraction, and we shall mention one of the most amazing here. If a shadow is cast by a penny in a narrow beam of light, tiny diffraction maxima and minima, i.e., bright and dark regions, occur near the edges of the shadow. But if one examines the center of the circular shadow carefully, one will see a very small bright spot. Actually, interference of the light passing around the penny causes a bright spot in the center of the shadow. This and many other similar effects are discussed in texts dealing with physical optics.

25.10 DIFFRACTION OF X-RAYS BY CRYSTALS

In Sec. 25.3 we found the distance from the central maximum to the first-order maximum in a Young's double-slit experiment to be given by

$$x = \frac{\lambda}{d} D$$

Thus, if x is to be appreciable in size, the ratio λ/d cannot be too small. Hence we see that the slit separation should be made as small as possible if interference effects are to be observed.

Similarly, in the case of interference fringes caused by a thin film, the film should not be too thick or the fringes will be so close together that they cannot be seen easily. It will be recalled that the distance between fringes is determined by the distance one must move along the film to change its thickness by $\lambda/2$. In addition the exact angle at which one views the film is of importance for thick films. To obtain well-separated fringes, the film thickness cannot be too much larger than the wavelength of the light used.

These facts are of importance if we wish to observe interference with x-rays. As we stated in Chap. 22, x-rays are EM waves of very short wavelength, close to 10^{-8} cm. If we wished to do a Young's double-slit experiment, but with x-rays, we should have slits separated by about 10×10^{-8} cm, or 10 Å. Since atoms have diameters of 2 to 5 Å, clearly such a slit system is impossible to obtain. Moreover, except in very special instances, it is out of the question to produce ordinary films separated by such small distances.

However, it is possible to obtain x-ray interference effects. We make use of a crystal such as that shown in cross section in Fig. 25.21*a*. The atoms in the crystal (perhaps it is rock salt) are uniformly spaced in planes. They are a distance d apart. If a beam of x-rays is incident on the crystal as shown in part *b* of Fig. 25.21, beam *a*, reflected from the top layer of atoms, will not travel as far as beam *b*. If the excess distance traveled by beam *b*, $2d \sin \theta$, is a whole number of wavelengths, the beams will reinforce. When that happens, all the layers will reflect beams which reinforce each other. Hence, strong reflection will occur when

Bragg Equation

$$n\lambda = 2d \sin \theta \qquad (25.3)$$

where n is an integer. This is called the **Bragg equation** after W. H. Bragg and his son W. L. Bragg, who first made extensive use of it in 1913.

Notice that Eq. (25.3) is similar to the grating equation (25.2). However, it differs by a factor of 2. In addition, the angle θ is defined differently in the two cases.

Equation (25.3) has been basic to many fundamental measurements. For example, consider what happens in Bragg reflection from a rock-salt crystal. Since the spacing of the atom layers in rock salt can be found from the

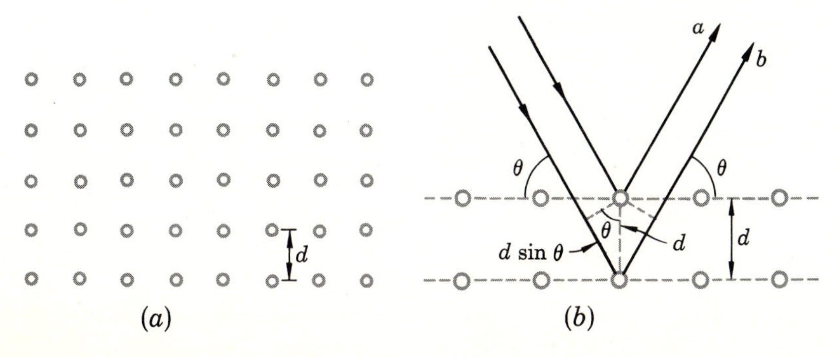

FIGURE 25.21
The atoms in crystals lie in evenly spaced planes. A simple example is shown in (*a*). When the atomic planes reflect x-rays, as shown in (*b*), the reflected rays give rise to interference effects.

density of salt and the mass of its atoms, the distance d is known in that case. Since both n and θ can be measured, the wavelength of the x-rays can be found by using Eq. (25.3). This is one way in which we have learned that the wavelengths of x-rays are in the range near 1 Å. Of course, if λ is known, the distance d can be measured. This is the basis for the field of x-ray crystallography, in which the structure of crystals is measured by using x-rays.

FIGURE 25.22

Many possible parallel-layer systems of atoms can exist in a crystal. Three such layer systems are shown in the figure.

Actual x-ray techniques for crystal-structure determination are quite involved. Notice that many possible layers of planes exist in the crystal. A few are shown in Fig. 25.22. When one considers the crystal in three dimensions, the situation is even more complex. Depending upon how the crystal is used, the photograph of the diffraction pattern may be a series of spots, a set of circular rings, parallel lines, and so on. Two types of photographs obtained are shown in Fig. 25.23. The photograph in part a was taken by sending a beam of x-rays through a single crystal, while the other photograph was obtained in a similar way by using a polycrystalline material. Interpretation of such photographs is a rather involved science in itself. However, most of our knowledge concerning the structure of crystals has been obtained from the analysis of photographs such as these.

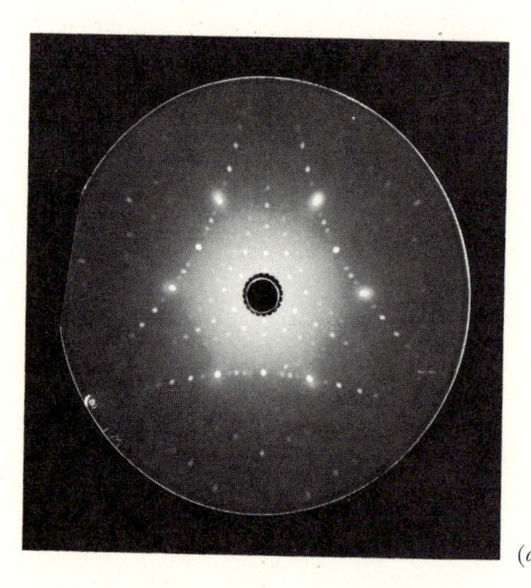

(*a*)

(*b*)

FIGURE 25.23

When a beam of x-rays is shone on a single crystal, a Laue diffraction picture is obtained. (*b*) If a crystalline powder is used instead of a single crystal, a Debye-Scherrer diffraction pattern is obtained.

SUMMARY

Waves can bend around obstacles into regions shadowed by the obstacle. This phenomenon is called diffraction. It is a result of the interference property of waves.

Two or more identical waves will give rise to interference effects when brought together. Waves which are in phase or $n\lambda$ out of phase (where n is an integer) will interfere constructively. Two waves which are 180°, or $\frac{1}{2}\lambda$, out of phase interfere destructively. Destructive interference also occurs if the phase difference is $\frac{1}{2}\lambda + n\lambda$, where n is an integer.

Identical waves, coherent waves, can be obtained by splitting a single wave disturbance into two parts. In the Young's double-slit experiment, a single wave incident on two slits gives rise to two coherent waves. These then interfere with each other to produce a series of bright and dark fringes. The positions of the fringes depend upon wavelength and can therefore be used to measure λ.

Two coherent waves can be obtained by partial reflections of a single beam from two different surfaces. In the Michelson interferometer, two portions of a beam produced in this way follow different paths and are then brought back together. By changing the length of one path, extremely accurate distance measurements can be made. Reflections from the two surfaces of a thin film or gap can also produce two interfering beams. Since the interference effects depend upon wavelength, the fringes produced by white light in such phenomena are highly colored.

The equivalent optical-path length for a distance d in a material of index of refraction n is nd. A beam will travel a distance nd in vacuum during the time it takes to travel a distance n in the material.

A diffraction grating consists of thousands of parallel, closely spaced slits. The interference pattern formed by it is much sharper than the double-slit pattern. Use is made of diffraction gratings in spectrometers for analyzing light. In such a spectrometer, images of the slit (called spectral lines) are formed at angles θ_n to the original beam given by $n\lambda = d \sin \theta_n$. The wavelength λ incident on a grating with slit separation d gives images at angles θ_n, where n is the order number, an integer.

Light from hot gases contains waves of only certain wavelengths. Each wavelength gives rise to a line, or image, in the spectrometer. This series of lines is called a bright-line spectrum. Hot, incandescent solids give off a continuous range of wavelengths which results in a continuous band of color in the spectrometer. This is called a continuous spectrum.

Diffraction places a limit on our ability to discern detail in optical devices. As a rule of thumb, it is impossible to see detail smaller than the wavelength of radiation being used.

Interference of x-rays can be carried out by using crystals. The atomic planes within the crystal act as reflecting surfaces. Use is made of the interference patterns so obtained in studying the structure of crystals.

MINIMUM LEARNING GOALS

Upon completion of this chapter you should be able to do the following:

1. Describe a water-wave experiment which illustrates the phenomenon of diffraction.

2. Show the phase relation of two identical waves if they are to interfere constructively or destructively.

3. Explain how two coherent beams are obtained in Young's experiment. Using a diagram, show why these two beams can interfere destructively and constructively at various points. From a consideration of the diagram, justify the relation $n\lambda/x = D/d$ for the bright fringe positions.

4. Use a double-slit interference pattern to determine λ if sufficient data are given.

5. Outline the construction and operation of the Michelson interferometer.

6. Explain how interference can be produced by a thin film or wedge. Tell why the fringes formed in white light are colored. Compute the thickness difference between two adjacent bright or dark fringes in an air wedge.

7. Tell what a diffraction grating is and show how it is used in a grating spectrometer. Explain why brightness is observed at angles for which $n\lambda = d \sin \theta_n$. Relate n to the order of an image.

8. Explain what is meant by a spectral line. Distinguish between a bright-line spectrum and a continuous spectrum.

9. Describe what happens to a beam of light transmitted through a slit as the slit is made very narrow. Pay particular attention to what happens when the slit width approaches λ. Explain the importance of this effect in our ability to observe details of objects.

10. When given the Bragg relation, explain the parameters in it. From a consideration of reflection from crystal planes, show how the relation arises. Explain why x-rays must be used, rather than visible light, to obtain interference effects from crystal planes.

IMPORTANT TERMS AND PHRASES

You should be able to define or explain each of the following:

Diffraction
Constructive interference; destructive interference
In phase; out of phase; 180° or $\frac{1}{2}\lambda$ out of phase
Young's double-slit experiment; $\Delta = n\lambda$ for maxima
$\Delta/d = x/D$
First-, second-, zeroth-order fringes
Michelson interferometer
Equivalent optical-path length nd
Newton's rings
Diffraction grating; grating spectrometer; $n\lambda = d \sin \theta_n$
Spectral line; line spectrum; continuous spectrum
Detail smaller than λ cannot be seen
Bragg equation

QUESTIONS AND GUESSTIMATES

1. The two loudspeakers shown in Fig. P25.1 are connected to the same oscillator. They therefore send out identical sound waves. Under what conditions would you be able to notice an interference effect as you walk along line AB? What if the loudspeakers are replaced by light bulbs?

2. Why is it impossible to obtain interference fringes in a double-slit experiment if the slit separation is less than the wavelength of light being used?

3. Devise a Young's double-slit experiment for sound using a single loudspeaker as a wave source.

4. Why does a glass or metal surface which has a thin oil film on it often reflect a rainbow of color when white light is reflected from it?

5. Very thin films are placed on the surface of coated lenses of expensive cameras to reduce reflection. Suppose a film of transparent ($n = 1.3$) material having an equivalent optical-path length of 1250 Å is coated onto a lens. Why will this reduce reflection from the lens? What other feature of the coating is important if it is to be most effective? Why does the reflected light from coated lenses appear colored?

6. Very thin films are sometimes deposited on glass plates. The thickness of the film can be controlled by observing the change in color of white light reflected from the surface as the film's thickness is increased. Explain.

7. Thin oil films on water separate white light into its

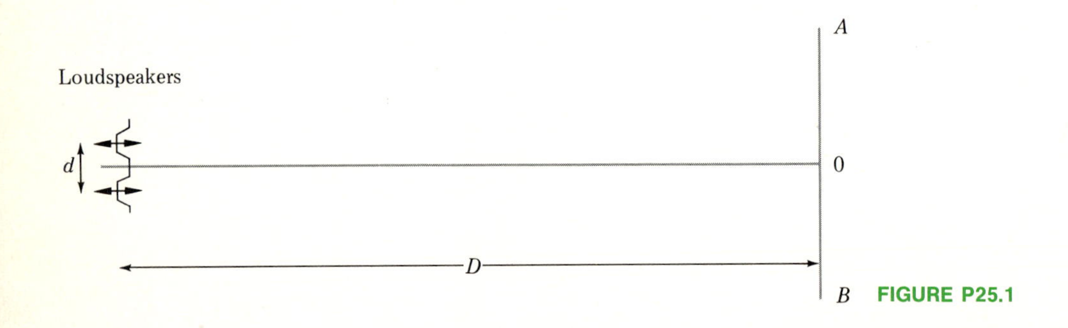

FIGURE P25.1

colors. Is there any similarity between this method of separating colors and the action of a prism? Explain.

8. Explain the following statement: the difference in thickness between two bright fringes in a thin-film interference pattern is zero or $\frac{1}{2}\lambda/n$, where λ is the wavelength of light used and n is the index of refraction of the film.

9. Should a microscope have any better resolving power when blue rather than red light is used? Explain.

10. Suppose that you are given a diffraction grating whose characteristics are unknown. How can it be used to determine the wavelength of an unknown spectral line?

11. Using two pieces of flat glass (microscope slides are ideal), press them together in various ways and estimate how close together the surfaces are from observation of the interfering reflected light. (You can see the interference pattern easily in any lighted room *provided* you get the plates close enough together.)

12. How could one use a Michelson interferometer to measure the index of refraction of air?

13. When viewed in reflected light, the coated lens of a camera appears violet with a reddish hue (magenta). The coating is magnesium fluoride, which has an index of refraction of 1.25. Estimate the thickness of the coating. (E)

PROBLEMS

1. Suppose the two sound sources in Fig. P25.1 send out identical in-phase waves with $\lambda = 50$ cm. Maxima and minima of sound are heard as one walks along the line AB. What is the path difference from the two sources at: (*a*) point 0; (*b*) the first maximum away from 0; (*c*) the third maximum; (*d*) the third minimum?

2. In Fig. P25.1, as stated in the previous problem, the sound sources send out in-phase waves with $\lambda = 50$ cm. If $d = 5.0$ m and $D = 25$ m, how far along AB from 0 is (*a*) the first-order maximum and (*b*) the first-order minimum?

3. In a double-slit experiment the slit separation is 0.20 cm, and the slit-to-screen distance is 100 cm. Calling the position of the central bright fringe zero, locate the positions of the first three maxima on both sides of the central maximum. The wavelength is 5000 Å.

4. For the double-slit experiment described in Prob. 3, find the positions of the first three minima.

5. What slit separation in a double-slit experiment would give a second-order maximum 1.00 cm from the central bright spot? The screen-to-slit distance is 2.0 m, and $\lambda = 500$ nm.

6. When sodium yellow light ($\lambda = 589$ nm) is used in a double-slit experiment, the first-order maximum is 0.030 cm from the central maximum. When the light is replaced by a source of an unknown wavelength, the second maximum occurs at 0.040 cm. (*a*) What is the wavelength of this latter light? (*b*) What region of the spectrum is it in?

7. Two parallel glass plates are originally in contact and viewed from directly above with 500-nm light (green) reflected nearly perpendicularly by the surfaces. As the plates are slowly separated, darkness is observed at certain separations. What are the first four of these values? *Hint:* Darkness is observed when the plate separation is zero.

8. Referring to Fig. P25.2, the left-hand edge is in contact. If the light used is blue light with $\lambda = 4400$ Å, about how far apart are the plates at the last dark band on the right? Assume the space between the plates is air-filled.

9.* When blue light of 400-nm wavelength is reflected from an air wedge formed by two flat plates of glass, the bright fringes are found to be 0.50 cm apart. How thick

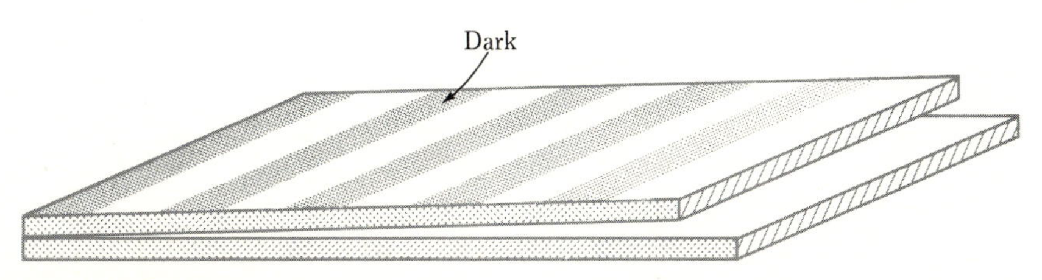

FIGURE P25.2

is the air wedge at a distance of 4.0 cm from the line of contact of the plates? Assume that the wedge is viewed at normal incidence.

10.* Repeat Prob. 9 if the wedge is filled with water rather than air.

11. To determine the pitch of a high-precision screw, the screw is used to move one mirror in a Michelson interferometer. It is found that 1 rev of the screw results in 2023 fringes ($\lambda = 5460$ Å, green) passing the field of view. How far does one turn of the screw move the mirror?

12.** One leg of a Michelson interferometer has in it an evacuated glass tube 2.0 cm long. A gas is slowly let into the tube, and the number of times the field of view changes from bright to dark back to bright is counted to be 210 (that is, 210 fringes pass). If yellow light with $\lambda = 5790$ Å is being used, what is the refractive index of the gas?

13.* A very thin wedge of plastic shows interference fringes when illuminated perpendicularly with white light. Two adjacent blue fringes ($\lambda = 4500$ Å) are separated by 0.40 cm. If the index of refraction of the plastic is 1.48, (*a*) what is the difference in thickness of the wedge between the positions of these two fringes? Two answers are possible. (*b*) How could you determine which is correct?

14.* Expensive cameras have nonreflecting coated lenses. Typically, the lens surface is covered with a thin layer of magnesium fluoride ($n = 1.25$). How thick must this layer be for destructive interference to occur between the light reflected from the two surfaces of the layer when 5000-Å light is used? [Why, qualitatively, would a layer of Lucite ($n = 1.48$) be unsuitable for this purpose? *Hint:* Compare the amount of light reflected by the Lucite and by the glass.]

15. A certain diffraction grating has 5000 lines per centimeter. At what angle does the second-order spectrum of the sodium yellow line occur ($\lambda = 5890$ Å)?

16. A certain grating has 4000 lines per centimeter. What is the angular separation between the blue (4358-Å) and green (5461-Å) mercury lines in (*a*) the first-order spectrum and (*b*) the second-order spectrum?

17. For a particular grating it is found that the second-order mercury blue line (4358 Å) lies exactly at 30°. At what angle will the first-order yellow line (5790 Å) be found?

18.* Modern grating spectrometers often use reflection gratings. These are mirrors on the surface of which there are a series of reflecting lines (equivalent to the slits of the transmission grating). This situation is shown in Fig. P25.3, where the distance between the centers of the reflecting lines is d. Find the grating equation for this device, i.e., the angles θ at which interference maxima occur.

19.** (*a*) Is it possible to design a grating so that the first-order 6000-Å red line will lie on top of the second-order 4000-Å violet line? (*b*) If so, how? (*c*) If not, could it be done for any other combination of orders? (*d*) If so, how?

20.** A diffraction grating (10,000 lines per centimeter) is used in a large tank of water. At what angles (in the water) will the blue mercury line (4358 Å) appear?

21.** Steel sheds often have a corrugated metal surface with corrugation repeating every 10 cm or so. Under appropriate conditions this type of wall can act as a reflecting diffraction grating for sound waves (see Prob. 18). What wavelength waves at normal incidence will give rise to a first-order maximum at an angle of 30° to the normal?

22.* A known wavelength of light (600 nm) falls on two slits (separation unknown), together with an unknown wavelength. It is found that the fourth-order maximum of the known wavelength falls at the same position on a screen as the fifth-order maximum of the unknown. What is the wavelength of the unknown?

23.* A double-slit system immersed in water is illuminated by 600-nm light. An interference pattern is formed on a screen at 2.0 m distance in the same tank of water. What is the distance from the central maximum to the first-order maximum on the screen if the separation of the slits is 0.040 cm?

24. A beam of x-rays is reflected from a crystal of NaCl by using the crystal planes, which are separated by a distance of 2.820 Å. The angle of incidence for strong reflection is 40°. (*a*) What are the possible wavelengths of the x-rays? (*b*) How could one determine which of these alternatives was correct?

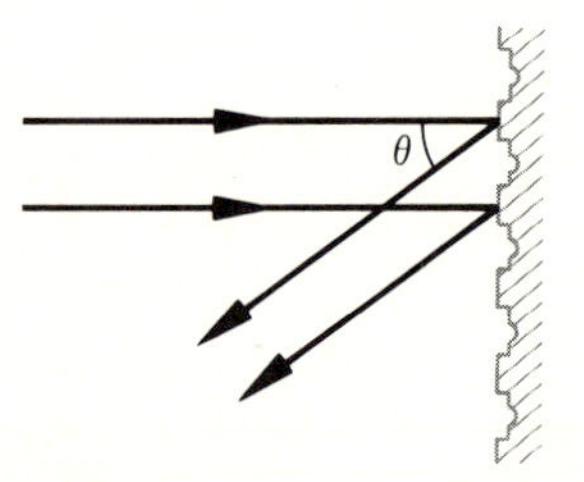

FIGURE P25.3

THE BIRTH OF MODERN PHYSICS

26

A clear distinction has been made in previous chapters between waves and particles. We shall see in the present chapter that under certain circumstances particles behave like waves, and waves behave like particles. The discovery of these facts opened new vistas for physicists. In this chapter, the development of these new concepts will be discussed. We shall also present the basic ideas of the theory of relativity. These concepts will be applied to the atom and its nucleus in later chapters.

26.1 PLANCK'S DISCOVERY

By the year 1900 many scientists felt that most of the great discoveries in physics had already been made. To be sure, a few vexing problems still remained to be solved, but it appeared that nearly all the fundamental physical laws of nature had been found. As we shall see in this chapter, such a view was completely incorrect. Vast areas of nature's physical behavior were still unknown at that time.

The first notable inkling that all was not serene in the field of physics was provided in an unexpected way in 1900. Max Planck (1858–1947), along with others, had been trying to interpret the radiation given off by hot, nonreflecting objects—so-called **blackbodies.** Careful measurements of the intensity of light (as well as infrared and ultraviolet radiation) given off by red-hot objects indicated that the intensity varies with wavelength as shown in Fig. 26.1. As we see, only a small fraction of the emitted radiation has wavelengths in the visible range. Most is in the infrared (or heat) wavelength range. Furthermore, the curves show that as the temperature is increased, the radiation maximum shifts from the infrared to the visible. This agrees with our experience that a white-hot body is hotter than a red-hot one.

In order to interpret these curves we are led to ask what sort of transmitting antenna could be sending out EM radiation from the hot object. Since the wavelengths involved are very short, the frequency of the vibrating charges must be very large. For example, at a wavelength of 1000 nm we have

$$\text{Frequency} = \nu = \frac{c}{\lambda}$$

or

$$\nu = \frac{3 \times 10^8 \text{ m/s}}{10^{-6} \text{ m}} = 3 \times 10^{14} \text{ Hz}$$

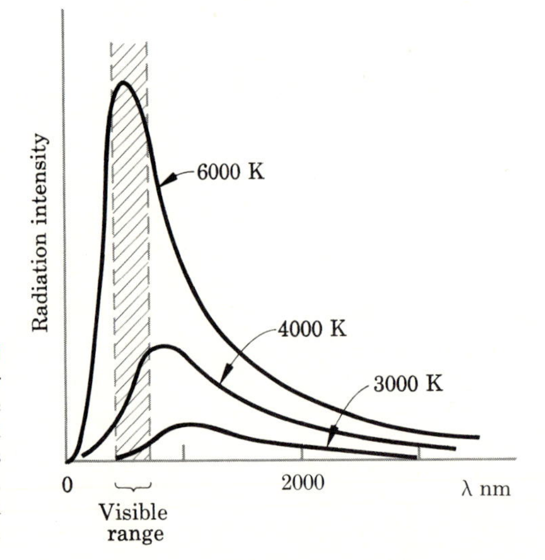

FIGURE 26.1
Blackbody radiation. For comparative purposes, the temperatures correspond as follows: 6000 K (sun's surface), 4000 K (carbon arc), 3000 K (very hot tungsten lamp).

where we have used the symbol ν (Greek nu) for frequency in place of f. Notice how very high this frequency is. Only in atomic-size antennas can charges be oscillated this fast. As a result, we expect that the EM radiation is being emitted by the vibrating charges within atoms and molecules composing the hot object.

There are many models we could postulate for these atomic or molecular vibrators. For example, if the object was composed of diatomic polar molecules, the vibrating molecule would be represented as shown in Fig. 26.2. The two atoms are held together by a springlike force, and since the molecule is polar, its two atoms carry equal and opposite charges. As the atoms vibrate back and forth, they act like vibrating charges on an antenna and therefore emit EM radiation of frequency ν_0, where ν_0 is the natural frequency of vibration of the molecular spring system. At least, this is the way Planck and his contemporaries reasoned.

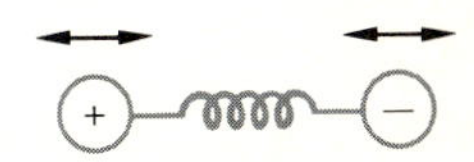

FIGURE 26.2
As the dipolar molecule vibrated back and forth, it was thought to send out EM waves.

It turns out, however, that all theories of radiation based upon this model failed to describe the radiation from hot objects accurately. The theories were capable of duplicating the curves of Fig. 26.1 at long wavelengths but gave completely incorrect predictions at small λ. It was Max Planck who discovered how the theory must be modified if agreement with experiment is to be achieved. His modification is easily understood but difficult to justify. In fact, his only justification for it was that it gives the correct answer. Let us now see what he had to assume to get agreement between theory and experiment.

As we know, the amplitude of vibration of a system such as that shown in Fig. 26.2 depends upon the energy of the system. This is true for all oscillators, whether they are masses on springs, a pendulum, etc. In all cases, the more energy the system has, the bigger the amplitude of the vibration is. Although the frequency of vibration is always ν_0, the natural vibration frequency, the amplitude increases as the energy increases. *Planck found that he had to assume that* a vibrating system could not vibrate with just any energy; he found instead that

Planck's Assumption

> A vibrator of natural frequency ν_0 can vibrate only with the energies $h\nu_0$, $2h\nu_0$, $3h\nu_0$, . . . and with no others.*

The quantity h is simply a constant (it is not the amplitude of vibration!) which has to have the value

$$h = 6.626 \times 10^{-34}\ \mathrm{J \cdot s}$$

if agreement with experiment is to be achieved. It is now called **Planck's constant.**

This is a truly astonishing assumption. It means that a vibrator can oscillate only with certain amplitudes and with no others. For example, since the total energy of a pendulum is mgH, where H is the height to which the bob swings, Planck says that mgH can only be $h\nu_0$, $2h\nu_0$, etc., but nothing in

*Zero energy is not allowed since this leads to conflict with the uncertainty principle mentioned later.

between. To see what this means, let us consider a pendulum which has a natural frequency $\nu_0 = 1$ cps and a bob of mass 100 g. Then the heights to which this pendulum could swing would be

$$H_1 = \frac{h\nu_0}{mg} = \frac{(6.6 \times 10^{-34}\ \text{J} \cdot \text{s})(1\ \text{s}^{-1})}{(0.10\ \text{kg})(9.8\ \text{m/s}^2)} = 6.7 \times 10^{-34}\ \text{m}$$

and

$$H_2 = 2H_1 = 13 \times 10^{-34}\ \text{m}$$

while

$$H_3 = 3H_1 = 20 \times 10^{-34}\ \text{m}$$

etc. No intermediate maximum vibration heights are possible.

Notice that the difference between successive heights of vibration as predicted by Planck is only about 10^{-34} m. This is far too small a difference to measure. As a result, we can never tell whether Planck is correct by observing the vibration of a pendulum. The gaps between the allowed energies are too small to be measured. This turns out to be true for all common vibrating systems. Hence, *we can neither prove nor disprove Planck's assumption using laboratory-type vibrating systems.*

Planck was therefore faced with a rather disturbing situation. He could obtain a suitable theory for the radiation from a hot object provided he was willing to make the assumption outlined above. The experimental test of it for other vibrating systems appeared to be impossible. Therefore, at the time, it was viewed both by Planck and his contemporaries as a rather curious result but one of doubtful validity. We shall see, however, that it appears to be correct and of extreme importance.

26.2 EINSTEIN'S USE OF PLANCK'S CONCEPT

Five years after Planck's discovery, another natural phenomenon was shown to involve Planck's constant h. This discovery was made by Albert Einstein.* As we shall see below, he was able to explain in a detailed way an experiment first performed by Heinrich Hertz. In so doing, he postulated that light had corpuscular as well as wave properties. This postulate, later verified, has become an integral part of the new physics.

Photoelectric Effect

It was discovered in 1887 by Hertz (who also produced and detected the first radio waves) *that light could dislodge electrons from a metal plate.* We now know this to be a general phenomenon—*short-wavelength EM energy incident upon a solid can cause electrons to be emitted by the solid. This is called the* **photoelectric effect.**

FIGURE 26.3
When light strikes the plate P of the photocell, electrons are emitted from it.

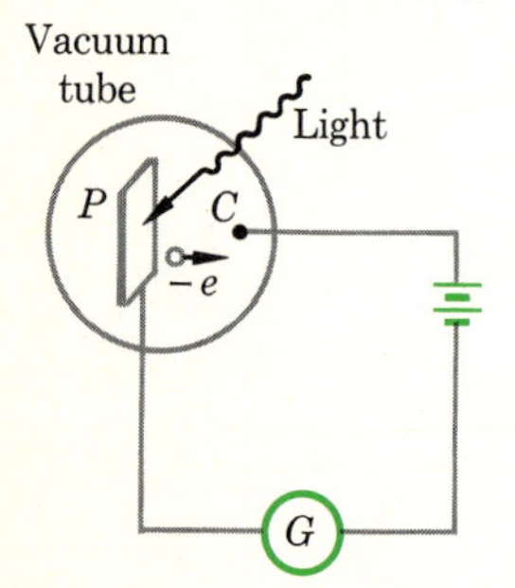

An experiment for observing the photoelectric effect is shown in Fig. 26.3. A metal plate P is sealed in a glass vacuum tube, together with a second small wire C, the collector. These elements are connected in a battery and galvanometer circuit, as shown. When the tube is covered so that no light enters it, no current flows through the galvanometer, since the portion of the circuit from P to C inside the bulb lacks a connection. The vacuum space between P and C has essentially infinite resistance.

* In the same year, 1905, Einstein proposed his theory of relativity and also gave the first adequate theory of Brownian motion. He was 26 years old at the time and was employed as a patent clerk.

If short-wavelength light is incident upon the plate *P*, as shown, it is found that the galvanometer needle deflects. The direction of flow of the current shows that electrons are leaving the plate *P* and traveling through the vacuum tube to the collector *C*. One's first thought is that the light heats up the plate and that when it becomes hot, thermionic emission occurs as in an ordinary diode. This is not the case, however. Careful experiments have shown that no matter how feeble the light and no matter how massive the metal plate, *a stream of electrons is emitted from the plate as soon as the light reaches it. No heating is required.*

It is further observed that *the number of electrons emitted from the plate is proportional to the intensity of the light.* If the battery in the circuit is large enough to attract all the emitted electrons to the collector, the current in the galvanometer is directly proportional to the light intensity. (It is for this reason that a photoelectric cell, such as this, is used to measure light intensity.)

A more startling feature is shown in Fig. 26.4. When the current through the galvanometer is plotted against the wavelength of the incident light, with the intensity held constant, *no electrons are emitted by light of wavelength larger than* λ_0. *This wavelength is called the* **photoelectric-threshold wavelength.** No matter how intense the light, if its wavelength is even just slightly longer than λ_0, no electrons are emitted. No matter how weak the light, if the light has wavelength shorter than λ_0, electrons are emitted essentially as soon as the light is turned on. *The particular value of* λ_0, *the critical wavelength for electron emission, depends upon the material from which the plate is made.*

Another interesting observation results from reversing the battery in the circuit of Fig. 26.3. Now the photoelectrons will be repelled from the collector. When the reverse voltage across the tube is V, an energy of Ve joules will be required for an electron to travel (now uphill) from the plate to the collector, where e is the electron charge. An electron will reach the collector only if its KE just after it has been thrown out of the plate is great enough so that $\frac{1}{2}mv^2$ is equal to or greater than Ve. That is, the KE of the electron when it is at the plate must be large enough to carry it uphill to the reversed-polarity collector.

When this experiment is done, *it is found that there is a stopping potential*

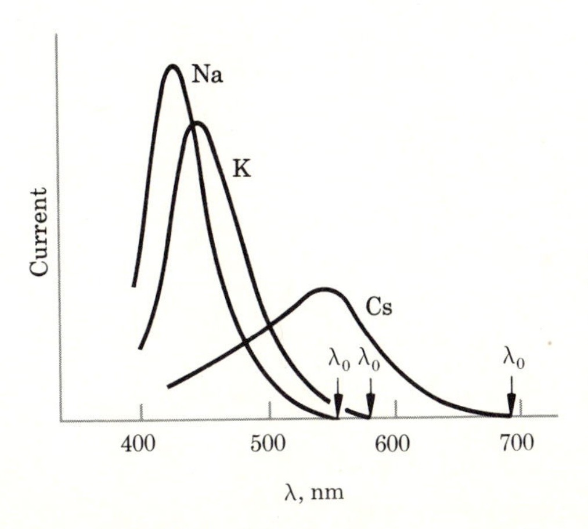

FIGURE 26.4
The current in the circuit of Fig. 26.3 varies with wavelength as shown. Data for three different metals (sodium, potassium, and cesium) are shown. What is the meaning of the λ_0 value indicated in each case?

V_0 that is just large enough to stop any electrons from reaching the collector. It is related to the wavelength of the incident light by the equation

$$V_0 e = \frac{A}{\lambda} - B \tag{26.1}$$

where A and B are numerical constants. We know from what has been said before that $V_0 e$ equals the KE with which the fastest electron is emitted from the plate. Hence we could write

$$V_0 e = (\text{KE})_{\text{max}}$$

or, after substituting this in Eq. (26.1) and rearranging,

$$\frac{A}{\lambda} = B + (\text{KE})_{\text{max}} \tag{26.2}$$

Many attempts have been made to explain all these observations in terms of the wave nature of light. None has been successful. Two basic difficulties are encountered by any wave interpretation:

1. How can one conceive of waves giving rise to a threshold wavelength? Light with λ just slightly less than λ_0 does not differ appreciably from light with λ just slightly greater than λ_0. Yet wavelengths slightly shorter than λ_0 cause electrons to be emitted, while those just slightly longer do not.
2. How can even the weakest possible beam of light cause electrons to be emitted as soon as the light is turned on? The light energy seems to localize on one electron instantaneously and causes the electron to break free from the solid.

Thus it appeared that a new approach was needed to explain the photoelectric effect. This bold, imaginative step was taken by Einstein.

In order to resolve this dilemma, Einstein seized upon Planck's ideas of quantized oscillator energies. Planck, it will be recalled, postulated that an oscillator with natural frequency ν_0 could take on only certain discrete energies, namely, $h\nu_0$, $2h\nu_0$, . . . , where $h = 6.626 \times 10^{-34}$ J · s. We say that the oscillator's energies are **quantized** and that the possible energies differ by an **energy quantum** $h\nu_0$. As we have seen, the EM radiation (including light) emitted by a hot object was considered to be emitted by atomic and molecular oscillators composing the object.

Einstein reasoned that if these atomic oscillators were to emit radiation in the way Planck visualized, the energy must be emitted in little bursts or packets. For example, since EM radiation carries energy, an oscillator emitting light must be sending out energy. However, since an oscillator can have only certain discrete energies, it cannot throw out energy continuously. It must throw out the energy in bursts of magnitude $h\nu_0$ because this is the spacing between the allowed energies of the oscillator.

The Photon Concept

To be specific, suppose an oscillator has an energy $37h\nu_0$. If it loses energy by sending out radiation, its energy can change to $36h\nu_0$ but not to anything in between since the oscillator's energies are quantized. But in so doing, the oscillator must have thrown out a pulse of light or other radiation, the energy of the pulse being $h\nu_0$. *We call such a pulse of EM energy a* **light quantum** *or a* **photon.** Hence we see that there is some justification for thinking that a beam of light consists of a series of energy packets or photons; each photon has an energy $h\nu_0$.

Einstein therefore postulated the following character for light:

A beam of light with wavelength λ (and frequency $\nu = c/\lambda$) consists of a stream of photons. Each photon carries an energy $h\nu$.

We shall see later how the photon energy is related to the structure of atoms and molecules. Let us now apply Einstein's model for a light beam to the photoelectric effect.

If light does consist of little particles of energy, these quanta, or photons, will collide with individual electrons as the light beam strikes a substance. When the energy of the photon is larger than the energy needed to tear an electron loose from the substance, electrons are emitted the instant the light is turned on. When the energy of a light quantum, or photon, is less than that value, no electrons are emitted no matter how intense the light. (The chance of *two* photons hitting the same electron simultaneously is practically zero.) We see at once that the energy needed to tear an electron out of the plate is exactly equal to the energy of a light quantum having the threshold wavelength. Hence, the work function ϕ, which is the work needed to tear an electron loose from a solid, as discussed in Chap. 22, is given by

$$\phi = \frac{hc}{\lambda_0} = h\nu_0$$

In the event that the light quantum has more energy than this, i.e., if λ is smaller than λ_0, not only can an electron be knocked out of the plate, but it will have KE to spare. That is to say, the energy hc/λ of the photon (or light quantum) is lost partly in doing work ϕ, in tearing the electron loose, and the remainder of the energy appears as the KE of the electron. We may therefore write

$$\frac{hc}{\lambda} = \phi + \tfrac{1}{2}mv^2 \qquad (26.3)$$

The Photoelectric Equation

Notice that Eq. (26.3) is the same as Eq. (26.2). The experimental relation between the wavelength and the energy of the fastest electron therefore follows at once if light is considered to consist of energy quanta, or photons. Moreover, even the constants predicted by *Eq. (26.3), the* **photoelectric equation,** are found to be those given by experiment. We also see that wavelengths of light greater than λ_0 will not have enough energy to tear the electrons loose from the material, and so no photoelectrons will be emitted by them. As we see, all the various details of the photoelectric effect make good sense if we are willing to accept that light waves of wavelength λ act as energy quanta, or photons, of energy $hc/\lambda = h\nu$.

Illustration 26.1 When light of wavelength 5×10^{-5} cm is incident upon a particular surface, the stopping potential is 0.6 V. What is the value of the work function for this material?

Reasoning We make direct use of the photoelectric equation (26.3). Using the SI units yields

$$\frac{(6.62 \times 10^{-34}\ \text{J} \cdot \text{s})(3 \times 10^{8}\ \text{m/s})}{5 \times 10^{-7}\ \text{m}} = \phi + (0.6\ \text{V})(1.6 \times 10^{-19}\ \text{C})$$

where $\frac{1}{2}mv^2$ has been replaced by V_0e. Solving for ϕ yields

$$\phi = 3 \times 10^{-19}\ \text{J} = 1.9\ \text{eV}$$

The conversion factor between joules and electronvolts is the charge on the electron, as you recall. Most metals have a work function several times larger than this value. However, various oxides and more complex compounds have work functions in this range.

26.3 THE COMPTON EFFECT

Since light and x-rays are both EM waves, the photon concept should apply to x-rays as well. Direct evidence for the x-ray photon was first provided by A. H. Compton in 1923. He noticed that when a monochromatic, i.e., single-wavelength, beam of x-rays was shone upon a graphite block, two kinds of x-rays were scattered from the block. Most of the x-rays have the same wavelength as the original waves. These can be pictured as arising in the following way: the oscillating electric field in the incident beam causes the charges in the atoms to oscillate with the same frequency as the wave itself. These oscillating charges act as antennas, radiating waves of the same frequency and wavelength. Hence the scattered x-rays are reradiated waves from the oscillating atomic charges.

FIGURE 26.5 In the Compton effect, a photon collides with an electron. Both energy and momentum are conserved in the collision.

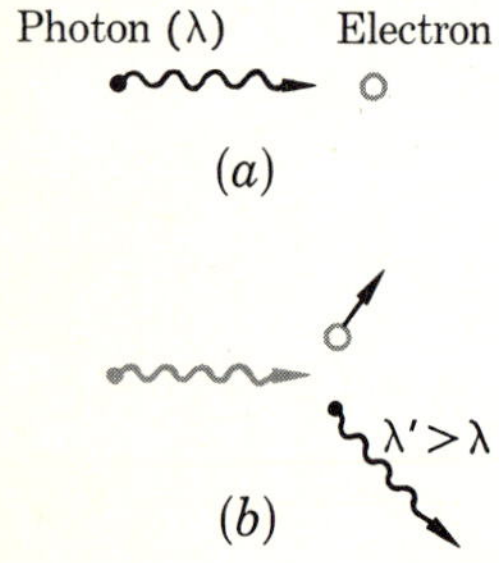

Photon (λ) Electron

(*a*)

$\lambda' > \lambda$

(*b*)

In addition to this relatively intense beam of scattered x-rays, there is another type of scattered x-ray which has a slightly longer wavelength. The exact wavelength of these anomalous x-rays varies in a precise and relatively simple way, depending upon the angle at which they are scattered. No explanation for their existence appeared possible by using simply the wave picture of x-rays.

A simple explanation for this phenomenon was simultaneously and independently presented by Compton and P. Debye. They considered that the x-ray beam consists of photons, each of energy $h\nu$, and that the photon collides with the electron much as two balls would collide, as shown in Fig. 26.5. It gives up some of its energy to the electron and bounces off as shown in part *b* of the figure. Since its energy is less after the collision, its wavelength must be longer.

They then treated the problem mathematically exactly as one would treat the problem of the elastic collision of two balls; i.e., both energy and momentum must be conserved. Thus,

$$\text{Energy before collision} = \text{energy after collision}$$

and

$$\text{Momentum before collision} = \text{momentum after collision}$$

These equations, then, could be solved directly for the loss in energy of the photon as a function of the angle at which it was scattered. The result could be used to compute the wavelength of the scattered photon, by using the fact that a photon's energy is hc/λ. When this was done, the computed wavelengths of the scattered photons were found to coincide exactly with the measured values. Here again was a striking confirmation of the particle properties of EM waves.

26.4 POSTULATES OF RELATIVITY

In the same year (1905) that Albert Einstein presented his theory of the photoelectric effect, he outlined his famous theory of relativity. This theory is a magnificent example of momentous deductions from a clear analysis of experimental fact. *Einstein recognized the following two statements as being experimental facts:*

The Two Postulates

1. The speed of light in vacuum is always measured to be the same ($c = 2.998 \times 10^8$ m/s) no matter how fast the light source or observer may be moving. (Of course, accurate measurements are assumed.)
2. Absolute speeds cannot be measured. Only speeds relative to some other object can be determined.

These are the two basic postulates of Einstein's theory of relativity.

It is not possible to prove these postulates directly. They are the consensus of all the experimental facts known. We consider it possible, though unlikely, that some experiment will sometime be found to disprove one of them. But they are supported by many unsuccessful attempts to disprove them. Moreover, as we shall see, they lead to astounding conclusions which have been well verified by experiment.

The second postulate needs some explanation, perhaps. It is easy to measure the relative speeds of objects. A car's speedometer tells us at once how fast the car is moving relative to the roadway. But this is not an absolute speed. The earth is moving due to its rotation on its axis and also due to its motion around the sun. Since we know both these speeds, if required, we could find the car's motion relative to the sun.

But the sun itself is moving in our galaxy, the Milky Way. And the galaxy center is in motion relative to the more distant stars. There appears to be no way to define a definite, absolute, speed of an object since everything appears to be moving. We can only state how fast one object is moving relative to another.

There is another way to state the second postulate which gives us an inkling of its fundamental importance. This alternate statement is usually made in terms of reference frames. *A* **reference frame** *is any coordinate system relative to which measurements are taken.* For example, the position of a sofa, table and chairs can be described relative to the walls of a room. The room is then the reference system or frame used. Or, perhaps a fly is sitting on a window in a moving car. We can describe the fly's position in the car using the walls of the car as a reference frame. Alternatively, we can describe the position of a spaceship relative to the positions of the distant stars. A coordinate system based on these stars is then the reference frame.

DEFINITION

The second postulate can be stated in terms of reference frames in the following way:

> **2** The basic laws of nature are the same in all reference frames moving with constant velocity relative to each other.*

Inertial Reference Frames

Often this statement is shortened by using the term inertial reference frame. *An* **inertial reference frame** *is a coordinate system in which the law of inertia applies:* a body at rest remains at rest unless an unbalanced force on it causes it to accelerate. The other laws of nature also apply in such a system. *To a very good approximation, all reference systems moving with constant velocity relative to the distant stars are inertial frames.*

> **2** The basic laws of nature are the same in all inertial reference frames.

You can understand the relation between these two alternate ways of stating the second postulate by considering the following. When we say that only relative speeds can be measured, a lack of bias in reference frames is being assumed. For example, a spaceship may be heading for the moon at a speed of 10^5 km/day relative to the moon. But it is also true that the moon is heading toward the ship at a speed of 10^5 km/day relative to the ship. The fact that one is moving relative to the other is easily ascertained. But both statements are equivalent, and neither object can be said to be at rest.

Suppose, though, that some law of nature depended on the speed of the reference frame. The people in the spaceship could use the law to obtain an indication of their speed. People on the moon could do likewise. The two measured speeds would be different. As a result, they would be capable of measuring more than just their relative speed. In fact, the law could be used to set up an absolute ranking of speeds. But this would contradict the second postulate. We therefore conclude that all of nature's laws must be the same in all inertial reference frames.

26.5 c IS A LIMITING SPEED

If we believe Einstein's two postulates, we can prove by logic alone that

> No material object can be accelerated to speeds in excess of the speed of light in vacuum.

---

* For example, Newton's second law is expressed as $F = ma$ in any of these frames. However, as we shall see, m (or F or a) may not have the same value in each.

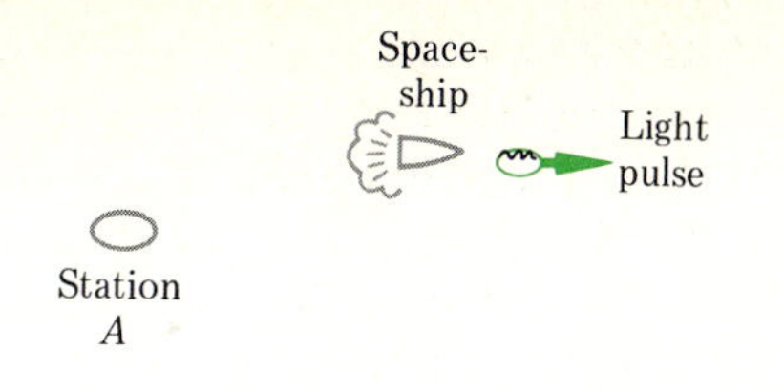

FIGURE 26.6
What is the maximum speed with which the ship can pass between the two space stations?

The validity of this statement is easily demonstrated in the following simple way. We prove it by the technique called **reductio ad absurdum,** in which we disprove a proposition (in this case, that an object can travel faster than c) by showing that the proposition leads to a known false result (in this case, that an observer will measure a value different from c for the speed of light).

Suppose we have two stations in space, shown as A and B in Fig. 26.6. The inertial observers at A and B have instructed the spaceship operator to follow a straight-line path between A and B. The ship is to travel at its top constant speed and is to send a light pulse from the front of the ship toward B as it passes A. Of course, A and B working in partnership can determine the speed of the spaceship by timing its flight from A to B. Let us now make the false assumption that they find the speed of the ship to be $2c$.

The spaceship sent out a pulse of light as it passed A, and since the laws of nature must apply to all three inertial observers (A, B, and the person in the ship), the light pulse must behave in a normal way to each of them. In particular, the light pulse must precede the ship and must reach B before the ship does. Therefore A and B working together would find that the light pulse is moving faster than the ship. But they measure the ship as moving with speed $2c$, and so they find that the speed of the light pulse is greater than $2c$. But this is a completely impossible result, since it contradicts the known fact that all observers will obtain c for the speed of light. We therefore conclude that our original assumption was false; the spaceship could not have been moving between A and B with a speed of $2c$.

This experiment will always lead to this contradiction as long as we insist that the speed of the ship exceed c. We therefore conclude that the spaceship cannot exceed the measured speed of light c. Indeed, one can enlarge this line of reasoning to include all material objects and signals which carry energy. As a result we can state:

A Limiting Speed Exists

> Nothing which carries energy can be accelerated to the speed of light c.

As we proceed, we shall see that this result of Einstein's theory has repeatedly been tested carefully and has been found correct in every test.

26.6 SIMULTANEITY

We shall see in this section that the basic postulates of relativity force us to conclude that events which are simultaneous in one inertial reference frame may not be simultaneous in another. To show this simply, we again resort to a thought experiment. The progress of a light pulse as noted by two inertial observers will form the basis for our experiment.

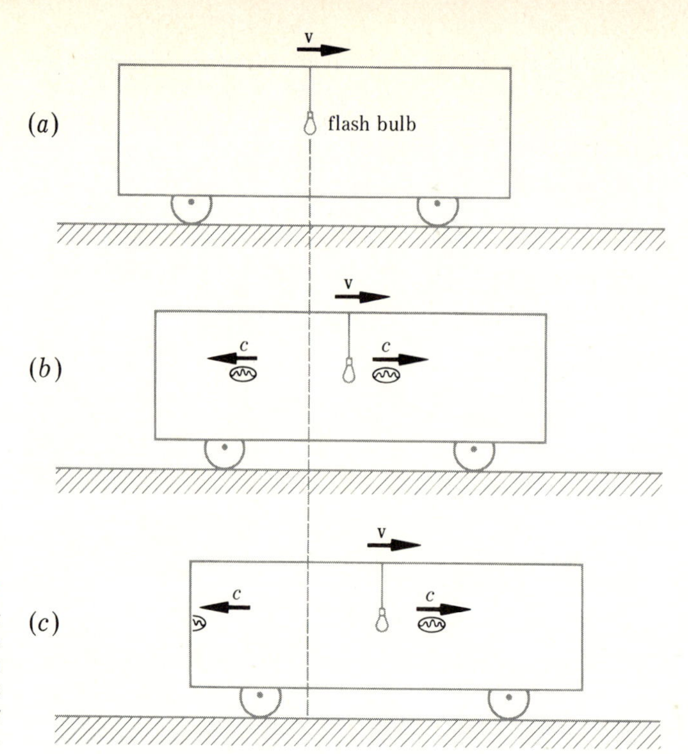

FIGURE 26.7
Unlike the inertial observer in the moving frame, the observer stationary on earth does not see the light pulses strike the ends of the car simultaneously.

As shown in Fig. 26.7*a*, suppose a boxcar is traveling to the right at a very high constant velocity. At the exact center of the car is a high-speed flashbulb which has reflectors so it will send out light pulses to the right and left when it explodes. The boxcar is fitted out with photocells at each end so a man in the boxcar can detect when the light pulses strike the ends of the car. By some ingenious device, a woman at rest on the earth is also able to measure the progress of the two pulses. Notice that both observers are in inertial reference frames (one is the moving boxcar, the other is the earth), and so they must both see the light pulses behave "normally" in their reference frames. Of course, "normal" for the woman on earth is that the light pulses travel with speed c to the right and left from the flashbulb. "Normal" for the man in the car is that the two light pulses strike the detectors at opposite ends of his car simultaneously.

Consider first the man in the car. To him, the experiment is very simple. The flashbulb is at rest relative to him in the center of his car. When the bulb explodes, two pulses travel the equal distances to the two ends of the car in equal times. (Remember, for him the experiment must be the same whether or not the car is moving since he cannot tell.) Hence *the light pulses hit the two ends of the car simultaneously.*

Now let us consider how the woman stationary on the earth sees the experiment. Her measurements show the experiment to proceed "normally" (for her), and so the situation progresses as shown in parts *b* and *c* of Fig. 26.7. Notice that the pulses travel equal distances in equal time to the right and left. But since the boxcar is moving to the right, the distance to the left end is shortened. As a result, the observer stationary on the earth measures the pulse on the left to strike the end of the boxcar before the other pulse strikes the opposite end. According to her, *the light pulses do not hit the two ends of the car simultaneously.*

We must therefore conclude that time is not a simple quantity because:

Events which are simultaneous in one inertial system may not be simultaneous in another.

Further considerations show that this situation exists only if the two events occur at different locations. In the present case, one event took place at one end of the car and the other was at the opposite end.

26.7 MOVING CLOCKS RUN TOO SLOW

As you might suspect from the results of the previous section, time is not a simple quantity. Einstein pointed this out when he showed that a clock ticks out time differently for an observer who holds the clock and for one moving past it. We shall demonstrate this effect in a thought experiment using a very special clock. But it was proved to be true in general by Einstein.

Consider the clock held by the woman in Fig. 26.8. It consists of a pulse of light reflecting back and forth between two mirrors in a cylindrical vacuum tube. Each time the light pulse strikes the lower mirror, it clicks out a unit of time which we shall call a "click." If the tube is 1.5 m long, the woman can compute easily that

$$1 \text{ click} = \frac{2d}{c} = \frac{3.0 \text{ m}}{3.0 \times 10^8 \text{ m/s}} = 10^{-8} \text{ s}$$

Suppose several copies of this clock are made, and one is being used by a man in a spaceship. The woman with her identical clock looks out the window of her laboratory (which is in another spaceship) and sees the man shoot past her with speed v. She is pleased to see that he is using a clock

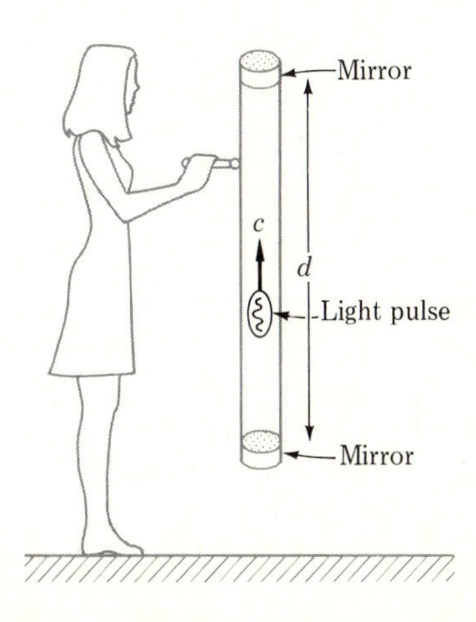

FIGURE 26.8

The light clock registers one click each time the light pulse is reflected from the lower mirror.

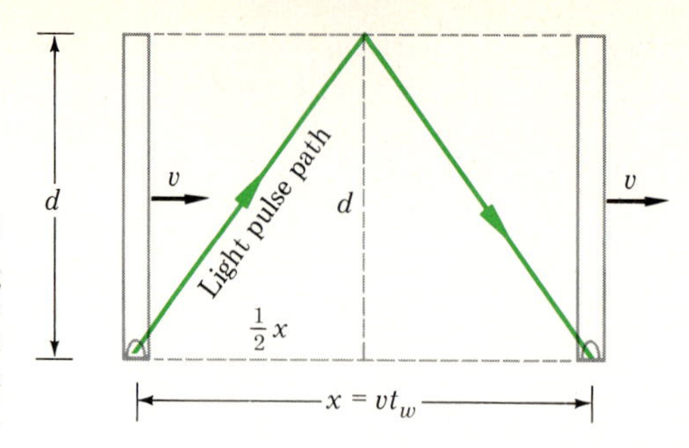

FIGURE 26.9

The light pulse in the moving clock must travel a distance larger than $2d$ during one click interval. The light-pulse path length is $2\sqrt{d^2 + (\frac{1}{2}vt_w)^2}$.

similar to hers and contacts the man by radio. He tells her the clock is functioning well and is ticking out time as usual, one click each $2d/c$ seconds.

After thinking about it a bit, the woman discovers there is something very peculiar about this. She concludes that the man's clock must be ticking out time more slowly than hers. We can understand her reasoning as follows.

Since the man's clock is operating properly for him, she knows it must be operating as shown in Fig. 26.9. We see there the clock in its positions at two consecutive clicks. Notice that the woman knows that the light pulse moves along the path indicated. Although the man sees the pulse to move straight up and down in the clock, the woman knows that the pulse moves to the right as well, because of the movement of the clock to the right.* The woman computes the time between clicks on the man's clock as follows.

According to the woman, the pulse moves a distance given by the colored line in the figure. From the pythagorean theorem and the dimensions given in the figure we see that the

$$\text{Pulse path length} = 2\sqrt{d^2 + (\tfrac{1}{2}x)^2}$$

But the woman knows that the man's clock is traveling with speed v past her. Further, according to her clock a time t_w will be taken to move from the one position to the other. Therefore, she knows that $x = vt_w$. As a result, according to the woman, the

$$\text{Pulse path length} = 2\sqrt{d^2 + (\tfrac{1}{2}vt_w)^2}$$

Further, she knows that a light pulse always travels through vacuum with speed c. According to her, then, the time taken for the change in position shown in the figure should be

$$t_w = \frac{\text{pulse path length}}{c} = \frac{2\sqrt{d^2 + (\frac{1}{2}vt_w)^2}}{c}$$

We can solve for t_w in this equation and find (after squaring both sides,

* You might ask: Who is right? They both are, as we shall soon see. Both are describing the behavior correctly as measured in their own reference frames.

rearranging, and taking square roots)

$$t_w = \frac{2d/c}{\sqrt{1 - (v/c)^2}}$$

But we recognize $2d/c$ to be the time the man insists it takes for his clock to make one click. We therefore have the following result:

$$\begin{pmatrix}\text{Time interval on}\\ \text{stationary clock}\end{pmatrix} = \left(\frac{1}{\sqrt{1 - (v/c)^2}}\right)\begin{pmatrix}\text{time interval}\\ \text{on moving clock}\end{pmatrix}$$

For example, suppose the man is moving at a speed of $0.75c$ past the woman. Then $\sqrt{1 - (v/c)^2}$ has a value 0.66, and the inverse of this is 1.51. Under these conditions the woman's clock will tick out 1.51 clicks during the time she knows the man's clock takes to tick out one click. As we see, the moving clock ticks out time more slowly than the stationary clock.

A clock moving with speed v ticks out a time of $\sqrt{1 - (v/c)^2}$ seconds during a time of 1 s on a stationary clock.

After arriving at this unexpected result, the woman contacts the man by radio and informs him that she has discovered that moving clocks tick out time too slowly. Before she can give him the details, he states that he has been thinking along the same lines. He has discovered that her clock, which was moving past him with speed v, was ticking out time too slowly. Then they both recall that only relative motion has meaning. Neither clock is special.

Time Dilation

Any clock moving relative to an observer will appear to tick out time too slowly compared to a clock stationary with respect to the observer.

We call this effect **time dilation** since time is stretched out, so to speak, for moving clocks.

This astonishing result applies to all timing mechanisms, no matter how complex. If the man had been using the growth rate of a fungus as a clock, the woman would have found the fungus growth rate to be slowed by its motion. Even aging of the human body will be slowed by motion at high speed, as we shall see in one of the following illustrations.

But there is one point we should always remember. A good clock always behaves normally to a person at rest relative to the clock. Other observers moving past the clock may claim it ticks out time too slowly. In spite of this, the clock still ticks out time properly as viewed by an observer stationary relative to it.

Illustration 26.2 One striking example of time dilation is obtained by measuring how long unstable particles "live." For example, a particle called the pion lives on the average only about 1.8×10^{-8} s when at rest in the laboratory. It then changes to another form. How long would such a particle live when shooting through the laboratory at a speed of $0.95c$?

Reasoning In the second case, the pion is moving with a speed $0.95c$ relative to the observers in the laboratory. Experiments should show that the internal clock of the pion, which controls how long it lives, should be slowed because of its motion. A time of 1.8×10^{-8} s read by the moving clock should be as follows when timed by the laboratory clock:

$$\text{Life according to lab clock} = \frac{1.8 \times 10^{-8}\ \text{s}}{\sqrt{1 - (0.95)^2}}$$

which turns out to be 5.76×10^{-8} s. As we see, the moving pion should live about three times longer than a stationary one. This experiment and variations of it have been carried out. The results found by experiment agree with the computed results.

Illustration 26.3 The star closest to our solar system is Alpha Centauri, which is 4.3×10^{16} m away. Since light moves with a speed of 3×10^{8} m/s, it would take a pulse of light 1.43×10^{8} s, or 4.5 yr, to reach there from the earth. (We say that the distance to the star is 4.5 light-years.) How long would it take according to earth clocks for a spaceship to make the round trip if its speed is $0.9990c$? According to clocks on the spaceship, how long would it take?

Reasoning To a good approximation, we can take the spaceship speed to be c for this computation, and so the round trip would require 9.0 yr according to earth clocks.

The spaceship clocks will appear to run too slow by the relativistic factor

$$\sqrt{1 - (0.999)^2} \approx 0.045$$

Therefore the spaceship clocks will read the 9.0 yr as (0.045)(9.0), or about 0.4 yr. As a result, the journey would seem to take only about 5 months according to the crew of the spaceship—far more tolerable than the 9.0 yr which people on earth would record.

The Twin Paradox

Incidentally, the twin of one of the crew who was left behind on the earth would age 9.0 yr during the time of the voyage. The twin in the spaceship, however, would only age 5 months. This phenomenon, the so-called **twin paradox,** has been discussed at length by scientists. They generally agree that this result is valid and that the two twins actually will age differently.

Illustration 26.4 Graph the relativistic factor $\sqrt{1 - (v/c)^2}$ as a function of v. Explain why we do not ordinarily observe relativistic time dilation.

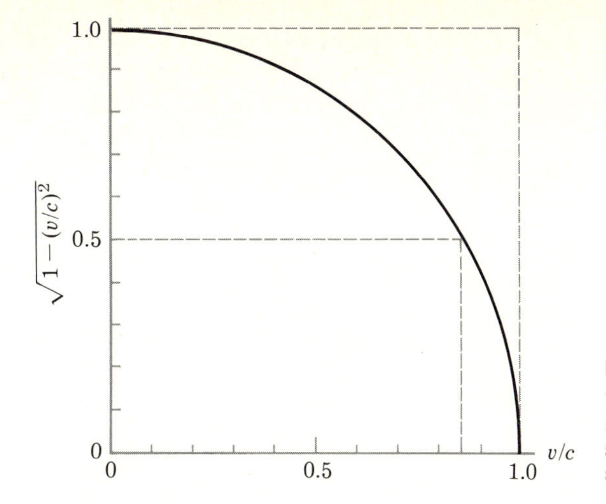

FIGURE 26.10
The relativistic factor differs appreciably from unity only at speeds which approach the speed of light.

Reasoning The appropriate graph is shown in Fig. 26.10. Notice that the relativistic factor departs from unity only at extremely high speeds. For most purposes, speeds below $0.10c$ are too small to show appreciable relativistic effects. In everyday life, our clocks never come anywhere close to such high speeds.

26.8 RELATIVISTIC LENGTH CONTRACTION

It turns out that the time-dilation effect implies a peculiar effect involving measured lengths. To see what this effect is, consider once again the man and woman of the previous section. Let us say that the woman is on the earth while the man is traveling with speed v along a straight line from earth to the nearest star, Alpha Centauri. Astronomers based on earth tell us the star is a distance $d = 4.3 \times 10^{16}$ m away from the earth. Therefore, the woman knows that the time taken for the man to travel from earth to star (as measured on her clock) is d/v.

But how long will this trip take according to the man's clock? The woman knows that his clock will tick out a time of only $\sqrt{1-(v/c)^2}$ of her value. Therefore, the time measured by the man for this trip is $(d/v)\sqrt{1-(v/c)^2}$ instead of d/v. We shall return to this fact soon.

Before that, though, we must point out that the man and woman both agree on their relative speeds. This must be true according to Einstein's second postulate. Therefore, both agree that the speed of relative motion is v. The woman says the man is moving relative to the earth and star, both stationary in her reference frame. But the man's reference frame is at rest at the position of the man; it is the earth-star system which is moving relative to him with speed v. Let us now combine this fact with the previous result.

The man knows he is moving relative to the earth-star system with speed v. As we have seen, his clock tells him it takes a time $(d/v)\sqrt{1-(v/c)^2}$ for the earth-to-star distance to pass by him. He can, from these facts, compute the distance from earth to star. It is

$$\text{Earth-star distance for man} = (\text{speed})(\text{time})$$
$$= (v)\left[\frac{d}{v}\sqrt{1-\left(\frac{v}{c}\right)^2}\right] = d\sqrt{1-\left(\frac{v}{c}\right)^2}$$

This is the distance from the earth to Alpha Centauri as computed by the man. The man comes up with a smaller distance than the distance d measured by the woman; it is smaller by the relativistic factor $\sqrt{1-(v/c)^2}$.

Einstein found this to be a general result. We can summarize it as follows:

Length Contraction

Objects moving with speed v past an observer are measured by the observer to have contracted along the line of motion. They contract by a factor $\sqrt{1-(v/c)^2}$.

Notice that the contraction occurs only along the line of motion. No such contraction is observed perpendicular to the direction of the motion.

Illustration 26.5 A man traveling at high speed in a spaceship holds a meterstick in his hand. What does he notice about the length of the stick as he rotates it from a position parallel to the line of motion to a position perpendicular?

Reasoning He notices no change in the stick's length. The length-contraction effect concerns objects moving at high speed relative to the observer. The meterstick is at rest relative to the man.

26.9 THE RELATIVISTIC MASS-ENERGY RELATION

The postulates of relativity tell us that no object can be accelerated to speeds in excess of the speed of light. It can be seen at once that this conflicts with prerelativity ideas. For example, consider an object of mass m being accelerated by a constant force F. We would usually state that the acceleration $a = F/m$. As a result, the object's final velocity after a time t is simply

$$v = v_0 + at = v_0 + \frac{F}{m}t$$

This relation must be wrong. It predicts that the velocity can increase without limit. As long as the force keeps acting, the velocity will keep on increasing. This contradicts the fact that the object's speed cannot exceed c. Something is obviously wrong with our nonrelativistic ideas for objects moving at very high speeds.

A consideration of the postulates of relativity leads to the source of the difficulty. It turns out that the mass of an object varies with the speed of the object. *At rest, the object has a mass m_0, its* **rest mass.** *But, at high speeds, the object's mass is larger.*

Relativistic Mass

An object of rest mass m_0 has an apparent mass $m_0/\sqrt{1-(v/c)^2}$ when moving with speed v past the observer.

This variation of mass with speed is shown in Fig. 26.11.

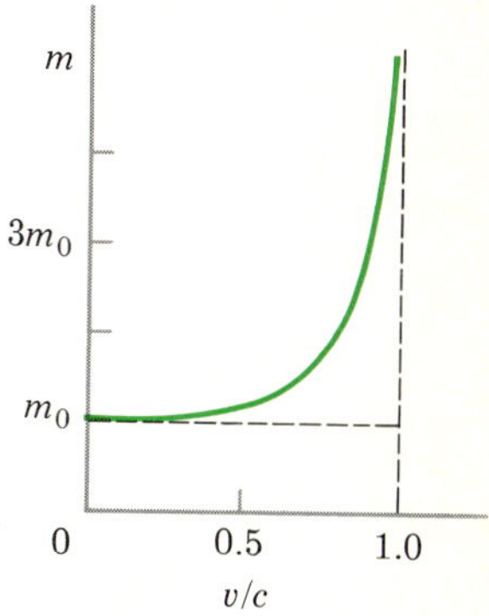

FIGURE 26.11
$m \to \infty$ as $v/c \to 1$.

Notice that the usual relativistic factor, $\sqrt{1-(v/c)^2}$, is important. The variation of mass is extremely small until speeds near c are reached. As v approaches c, the mass becomes increasingly large. If $v = c$, the relativistic factor will be zero and $m/m_0 \to \infty$. *The mass of an object becomes infinite as the object's speed approaches the speed of light.* An infinite mass would require an infinite force to accelerate it. Since infinite forces are impossible to obtain, it is apparent that an object cannot be accelerated to the speed of light. Of course, we already knew this. But we see now a reason for this restriction on speed.

The force acting to accelerate an object gives energy to the object. At low speeds we know that the work done by the applied force equals the increase in KE of the object provided changes in PE and friction work are negligible. This is still true at speeds close to c. But now the KE of an object is no longer given by $\frac{1}{2}m_0v^2$. Nor is it, as one might guess, $\frac{1}{2}mv^2$. Instead it is found that the kinetic energy of an object is given by

Relativistic KE

$$\text{KE} = (m - m_0)c^2 \tag{26.4}$$

It turns out that Eq. (26.4) reduces to $\text{KE} = \frac{1}{2}m_0v^2$ when $v \ll c$. We see, then, that the correct expression for KE is given by Eq. (26.4). The expression $\text{KE} = \frac{1}{2}m_0v^2$ is very accurate for speeds low enough to ensure that $m \approx m_0$. But when $v \to c$, the value $(m - m_0)c^2$ must be used for KE.

It is possible to show that a relation similar to Eq. (26.4) applies to all types of energy. Einstein showed that *for any change in energy of an object, there is a corresponding change in mass.* The result is that *for a change in energy ΔE, the object's mass changes by an amount Δm, given by*

Relativistic Mass-Energy Relation

$$\Delta E = \Delta m\, c^2 \tag{26.5}$$

(This is often written as $E = mc^2$.) As an example, if you increase the PE of an object, its mass increases in accordance with Eq. (26.5). In fact, this equation predicts that mass can be created by providing energy (we shall see examples of this later); or, more spectacularly, mass can be destroyed to provide energy. In either case, the change in mass Δm is equivalent to a change of energy in the amount $\Delta m\, c^2$. Perhaps you already know that the nuclear energy of a reactor or nuclear bomb results from the fact that an amount of mass Δm is destroyed to produce an energy $\Delta m\, c^2$.

Illustration 26.6 The available chemical energy in a 100-g apple is about 100 kcal (the nutritionists leave off the kilo- and call them calories). We learned in our study of heat that a calorie is 4.184 J of energy, and so an apple contains about 420 J of available energy. Compare this with the energy one could obtain by changing all the mass to energy.

Reasoning According to the mass-energy relation

$$\text{Energy} = \Delta m\, c^2$$

In this case $\Delta m = 0.10$ kg, and $c = 3 \times 10^8$ m/s, giving

$$\text{Energy} = 9 \times 10^{15}\ \text{J}$$

We see from this that when we eat an apple we obtain only about a fraction (10^{-13}) of its total energy.

Illustration 26.7 The light given off by a TV tube comes from electrons shooting down the tube and hitting a fluorescent screen at its end. Their speeds are of the order of one-third the speed of light. What is the apparent mass of such a high-speed electron ($m_0 = 9.1 \times 10^{-31}$ kg)?

Reasoning The electrons are moving with a speed $v = c/3$ relative to the person watching the TV set. As a result,

$$m = \frac{m_0}{\sqrt{1 - (\frac{1}{3})^2}} = \frac{m_0}{\sqrt{0.89}} = 1.06m_0$$

and so

$$m = 9.6 \times 10^{-31}\ \text{kg}$$

Even at this very high speed the electron mass has increased by only 6 percent.

26.10 MOMENTUM OF THE PHOTON

Let us now return to a discussion of the light quantum, the photon. Since a photon has energy, we would expect it to have momentum and mass as well. We can see, however, that the rest mass of a photon must be zero. Since it travels with speed c in vacuum, one has

$$m = \frac{m_0}{\sqrt{1 - (v/c)^2}} = \frac{m_0}{\sqrt{1 - 1}} = \frac{m_0}{0}$$

If m_0 were anything but zero, the photon would have infinite mass. But since $E = mc^2$, infinite mass implies infinite photon energy, and we know this to be untrue. Therefore, we must conclude that *the rest mass of the photon is zero.*

To find the momentum of the photon, we make use of the two relations for the photon energy.

Photon's Rest Mass

$$\text{Photon energy} = (m - m_0)c^2 = mc^2 \qquad \text{and} \qquad \text{Photon energy} = h\nu$$

If we equate these two expressions, we can solve for the momentum of the photon mc. We then have for *the photon's momentum p*

Momentum of Photon

$$p = \frac{h\nu}{c} = \frac{h}{\lambda} \tag{26.6}$$

This is the value used by Compton and Debye in their theories of the Compton effect.

We therefore have a method for assigning both energy and momentum to the photons of electromagnetic waves. Although these waves appear to possess a particlelike nature under certain circumstances, the *photons owe all their mass to their kinetic energy. They have no rest mass.* Any attempt to slow them down either will be unsuccessful or will cause them to give up their energy in a different form.

A striking example of what happens to a photon when it interacts in such a way as to lose its identity is found in the phenomenon of **pair production.** By means of experiments to be described in Chap. 28, it is sometimes observed that a very-high-energy photon disappears with the creation of two electrons. This is pictured in Fig. 26.12.

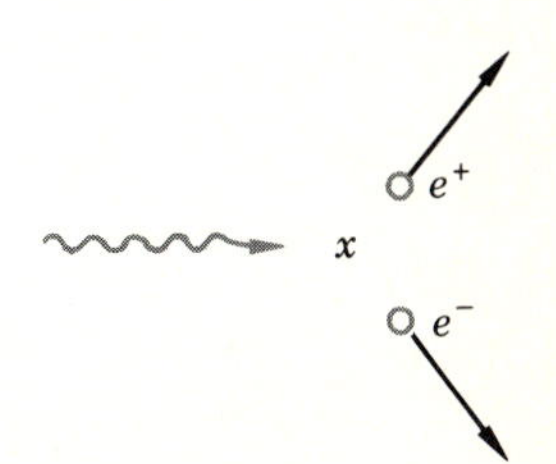

FIGURE 26.12
The photon transforms into a positron and electron when it reaches point x. This is called pair production.

A photon traveling through matter is found to disintegrate at point x into a negative electron and a positron (a positive electron). Notice that no net charge is created in this process, so that charge for the whole system is conserved. Energy and momentum must be conserved as well, of course. This condition can be satisfied only if the disintegration occurs near a nucleus. Although charge, momentum, and energy are conserved in pair production, rest mass is actually created in the process, since the original photon had zero rest mass, while the final products have a rest mass equal to twice the mass of an electron.

26.11 PARTICLE WAVES

As we have seen, EM waves, under certain conditions, act as though they were composed of particles. This wave-particle duality may seem rather strange. However, there is no known case where contradictions arise because of it. In any given experiment one is always able to say which type of behavior will exist. The general behavior can be summarized by the statement that wavelike behavior predominates during propagation of the radiation, while particlelike behavior predominates during interaction with matter.

A somewhat analogous situation can exist among people. No matter how well we think we know a man, we are unable to state a priori what his actions will be when he is confronted by a new situation. If we are surprised by his actions, we rationalize our surprise by saying that we did not know him as well as we thought we did. Similarly with light waves: In 1900 we believed that we knew them quite well. We were therefore surprised to find that they behaved in unexpected ways in new situations, as in the Compton and photoelectric experiments.

Once the particle nature of light waves had been discovered, it was perhaps reasonable that someone should ask whether particles might sometimes act like waves. This question was first treated by Louis de Broglie in 1923. (He was at that time 31 years old, and this work constituted his doctoral thesis.) He reasoned that since the momentum of a photon was given by h/λ, it might be possible that the momentum of a material particle, an electron, for example, would be equal to a similar quantity. In that case, λ would be the wavelength of an electron—or at least of some wavelike property associated with the electron. He therefore wrote $mv = h/\lambda$, from which

de Broglie Wavelength

$$\text{Wavelength associated with a particle} = \frac{h}{mv} \tag{26.7}$$

This is called the **de Broglie wavelength** *of a particle.*

We shall see in the next chapter that he had available a method for checking this assumption. In particular, Niels Bohr had postulated a model for the hydrogen atom in which the electron was assumed to behave in a very definite fashion. This model had proved exceptionally successful in explaining the properties of the atom, but Bohr was unable to offer any explanation of why the electron behaved in the way he had to assume it did. De Broglie was able to show that if the electron had wave properties with wavelength given by Eq. (26.7), Bohr's postulate followed at once from this wavelike property. We shall return to this problem after a discussion of the Bohr atom in the next chapter.

A direct proof of the wavelike character of material particles was obtained by Davisson and Germer in 1927. They were investigating the scattering of electrons from metal crystals. They fired a beam of electrons at a metal crystal (nickel, in their experiment) and observed the number of electrons scattered at various angles after the beam collided with the metal. They found that under certain circumstances the electron beam was scattered very selectively, many electrons coming off at certain angles and very few at others. They at first had no interpretation for these results and reported them as unexplained.

When it was suggested to Davisson and Germer that these results might be wave-interference effects resulting from the electron-wave character postulated by de Broglie, additional measurements were begun to check this point. It was soon confirmed by various investigators that electrons are reflected from crystals in the same fashion as x-rays are reflected. We saw in Chap. 25 how the angles at which x-rays will be strongly reflected from a crystal can be computed from the crystal-lattice spacing and the wavelength of the x-rays. By using de Broglie's expression for the wavelength to be associated with an electron, the angles of strong reflection of electrons were predicted, and excellent agreement was found with the experimental results. Hence we have a reasonably direct proof of the fact that electrons sometimes behave as waves and that their wavelike character can be predicted from Eq. (26.7).

Illustration 26.8 An electron is accelerated through a potential difference of 182 V. How large is its associated wavelength?

Reasoning We first find the speed of the electron from the usual relation

$$Vq = \tfrac{1}{2}mv^2$$

(It will be allowable to use nonrelativistic methods here, since v will be much smaller than c.) Substituting, one finds $v = 8 \times 10^6$ m/s. The associated wavelength is

$$\lambda = \frac{h}{mv} = \frac{6.6 \times 10^{-34}\ \mathrm{J \cdot s}}{(9.1 \times 10^{-31}\ \mathrm{kg})(8 \times 10^6\ \mathrm{m/s})}$$
$$= 0.91 \times 10^{-10}\ \mathrm{m} = 0.91\ \text{Å}$$

This is of the same order of magnitude as x-ray wavelengths. It is therefore clear why electrons will show diffraction effects similar to x-rays.

Illustration 26.9 A 50-g baseball rolls along a table with a speed of 20 cm/s. How large is its associated wavelength?

Reasoning Applying Eq. (26.7) and using SI units, we find $\lambda = 6.6 \times 10^{-32}$ m. In order to perform an interference experiment with such a ball, we would have to pass it through slits about this far apart. Since this length is many orders of magnitude smaller than atomic dimensions, it is obviously impossible to carry out such an interference experiment with the ball.

Similar calculations for any laboratory-sized object will convince one that the wavelengths associated with the object are far too small to allow observation of the wave nature of the object. It is only when m is very small, for electrons and atoms, that the wavelength becomes large enough for these effects to be measurable.

26.12 THE UNCERTAINTY PRINCIPLE

Since the time of the discovery of the wavelike nature of the electron, many experiments have been carried out to see whether other particles also exhibit this behavior. As we saw in Illustration 26.9, it is impossible to devise an experiment to observe the wave nature of large objects. However, particles of atomic size can be investigated relatively easily for wavelike effects. No exception to de Broglie's wavelength equation has ever been found. In fact, it is now commonplace to use electrons and neutrons as well as x-rays in diffraction experiments designed to investigate crystal structure.

The wave nature of all particles leads to a great philosophical principle. Prior to this discovery, philosophers had often argued about whether the fate of the universe was completely determined. *Could we, in principle at least, determine the position, speed, and energy of all the particles in the universe and then predict the course of all future events? It appears that the wave nature of all particles requires us to give a negative answer to this question. This fact is embodied in the* **Heisenberg uncertainty principle,** which we shall now examine.

Let us consider how we would locate the position, speed, and KE of an object or a particle. In order to locate the particle, we must either touch it with another particle or look at it in a beam of light. Let us make the light beam as weak as possible so that its momentum will not disturb the object at which we are looking. To that end, we shall look at the object by using a single photon. Or if we choose, we shall touch the object with a single, extremely small particle. We shall call the photon, or particle, which we use to investigate the object the **probe particle.**

To minimize the disturbance which the probe particle will cause, we shall use as low an energy as we possibly can. There is, however, a lower limit on this energy, because the wavelength of the probe particle must be smaller than the object we are looking at. Otherwise, as we saw in Chap. 25, interference and diffraction effects cause the waves associated with the probe to cast extremely blurred images of the object. In particular, we saw in Chap. 25 that the finest detail we can see using waves (either light or particle waves) is detail of the same size as the wavelength. Hence, the position of the object at which we are looking may be in error by an amount

$$\Delta x \approx \lambda$$

Moreover, the momentum of the probe particle (whether photon or material) is given by $p = h/\lambda$. When it touches the object we are looking at, some of this momentum will be transferred to the object and the momentum of the object may be altered because of this disturbance. The uncertainty in the momentum of the object Δp will therefore be given by

$$\Delta p \approx \frac{h}{\lambda}$$

If we multiply the expressions for Δp and Δx together, we find

$$\Delta p \times \Delta x \approx h$$

In other words, when we use the most precise experiment imaginable to locate the position of an object and measure its momentum simultaneously, the product of the errors in these two measurements will be approximately as large as Planck's constant h. This appears to be a perfectly general relation, and it is one form of Heisenberg's uncertainty principle.

A second form of the uncertainty principle can be obtained through similar reasoning. As the probe particle passes by the object we are looking at, the position of the object will be uncertain to within a distance of about λ, as we have just seen. Of course, λ is the wavelength associated with the probe particle. If the speed of the probe particle is v, the time taken for the particle to pass through this distance of uncertainty is λ/v. Therefore, the exact time when the object is at a particular position will be uncertain by an amount

$$\Delta t \approx \frac{\lambda}{v}$$

In addition, the energy of the probe particle will be partly lost to the object under observation when the two come into contact. As a result, the uncertainty in the energy of the object will be of the order of the probe particle's energy. Therefore,

$$\Delta E \approx \frac{hv}{\lambda}$$

Multiplying these two expressions together yields

$$\Delta E \times \Delta t \approx h$$

This is the alternate form of the uncertainty principle.*

Uncertainty Principle

Heisenberg uncertainty principle: *In a simultaneous measurement of coordinate x and momentum p of a particle*

$$\Delta x \, \Delta p \geq \frac{h}{2\pi} \tag{26.8}$$

where Δx and Δp are the errors in x and p. Similarly, if the energy E of a particle at time t is measured, then the errors ΔE and Δt are such that

$$\Delta E \, \Delta t \geq \frac{h}{2\pi} \tag{26.9}$$

It is impossible, then, even in principle, to know everything about an object. There will always be uncertainty about its exact energy at a given time and its exact momentum at a given place. This is one of the fundamental results inherent in the concepts of light quanta and particle waves. Clearly, a new formalism is needed to describe atomic particles and light quanta in situations where these effects are important. The methods of **quantum** or **wave mechanics** have been devised in order to handle these phenomena.

* More exact calculation shows that h should be replaced by $h/2\pi$ in each case.

26.13 QUANTUM MECHANICS

Let us examine a simple experiment to see exactly where the classical methods of mechanics do or do not apply. For this purpose, consider a modified double-slit experiment, as shown in Fig. 26.13. A uniform beam of light is incident on the two slits, as indicated, and the light passing through the slits strikes the screen behind them. As we learned in Chap. 25, if the slit widths and slit separation d are much larger than the wavelength of the light, clear shadows are cast. Two bright spots would be noticed on the screen, and these would be rather clear-cut images of the slits. This is shown in Fig. 26.13*a*.

Similarly, if a parallel *beam of electrons* is incident on two slits, they will behave as shown in Fig. 26.13*a* provided that the wavelength associated with the electrons is much smaller than the slit separation. Hence, the situation shown in Fig. 26.13*a* is exactly what one would predict from classical particle mechanics. Either a beam of baseballs or a beam of electrons would equally well pass through the holes and hit the screen within a well-defined region. Thus, classical newtonian mechanics is valid when the particle wavelength is much smaller than the geometrical dimensions involved in the experiment.

If we consider, however, the behavior of a light beam when the slit separation is comparable with the wavelength of the light, we observe a wide interference pattern on the screen. As shown in Fig. 26.13*b*, images of the slits are no longer observed. Similarly with the electron beam: if the wavelength associated with the particles is comparable to the slit separation, the electron beam spreads and hits the screen in an interference pattern, as illustrated in part *b* of Fig. 26.13. The intensity of the interference pattern in the case of the light waves is analogous to the number of electrons hitting the screen in the case of the particles. No particles strike where the intensity of light with identical λ is zero, and a maximum number of particles strike the screen where the light intensity would be maximum. This behavior is completely different from what newtonian mechanics would predict. Hence, classical mechanics is not applicable in this situation. *Classical mechanics becomes*

Validity of Classical Mechanics

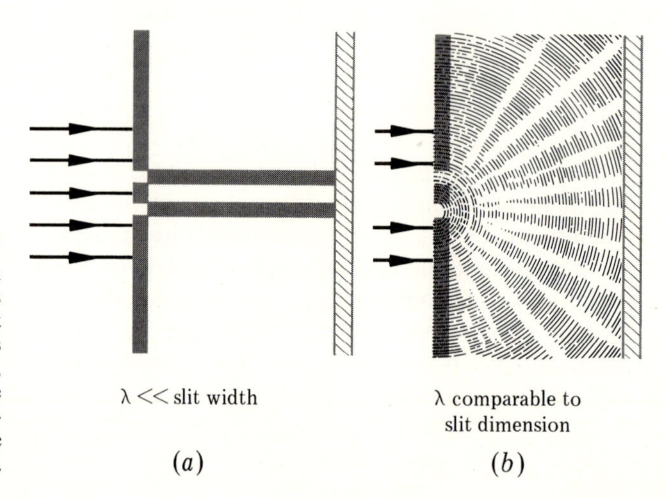

FIGURE 26.13

(*a*) When the wavelength associated with a particle is much smaller than the slit width, clear images of the slits are formed. (*b*) However, when λ is comparable to the slit width, typical wave-interference phenomena are observed.

invalid when the particle wavelength becomes comparable with the geometrical dimensions involved in the experiment.

It would appear, however, that the behavior of the light beam is always adequately described in terms of wave phenomena, at least in this experiment. Since the particle behavior can be described in terms of the associated wave in both these cases, while newtonian mechanics can describe only the case shown in part *a*, it is apparent that the wave viewpoint is more generally applicable.

Soon after de Broglie suggested the wave nature of particles, Erwin Schrödinger developed an equation to describe the behavior of particles in terms of their wave nature. **Schrödinger's equation** forms the basis of quantum mechanics. The equation is a differential equation, and it is an extension of a similar equation used to decribe the behavior of electromagnetic waves. With slight modification, it is now believed that Schrödinger's equation will predict the observed behavior of particles under all conditions. We should perhaps therefore discard all the mechanics we have learned in this text and start once again from this new basic equation. However, that would be foolhardy. Schrödinger's equation is exceptionally difficult to use to obtain practical answers except in the simplest problems. *Physicists therefore retain newtonian and classical concepts and use them to solve most problems. Only when particle speeds become near that of light do they worry about relativistic effects. Only when geometrical dimensions are comparable with particle wavelengths do they worry about quantum-mechanical effects.* In this case, in spite of the difficulties involved, quantum mechanics must be used in order to obtain reliable answers. We shall see in the next chapter that the internal behavior of atoms is one such case.

26.14 THE ELECTRON MICROSCOPE

As we have seen, electrons and other atomic-size particles possess a wavelike character. For ordinary beams of electrons, the associated wavelength is in the very short x-ray range, as calculated in Illustration 26.8. If one could use a beam of electrons rather than light in a microscope, the microscope should be able to "see" particles of very small size since the limiting size is comparable to the wavelength of the waves used. Whereas light microscopes are capable of seeing detail larger than about 1000 nm, a microscope using a beam of electrons should be able to see detail as small as a few tens of nanometers.

The basic principle of an electron microscope is nearly identical to that of a light microscope. You will recall that in a light microscope an objective lens forms an image of the object and then this image is examined using an eyepiece, basically a magnifying glass. This situation is reviewed in part *a* of Fig. 26.14. A similar situation exists in the electron microscope, as shown in Fig. 26.14*b*. However, the second lens focuses the electron beam onto a fluorescent screen so that an image (much like that on a TV screen) is formed. The operator then photographs or observes this image on the screen. Of course, the electron-beam path must be in vacuum so that the electrons

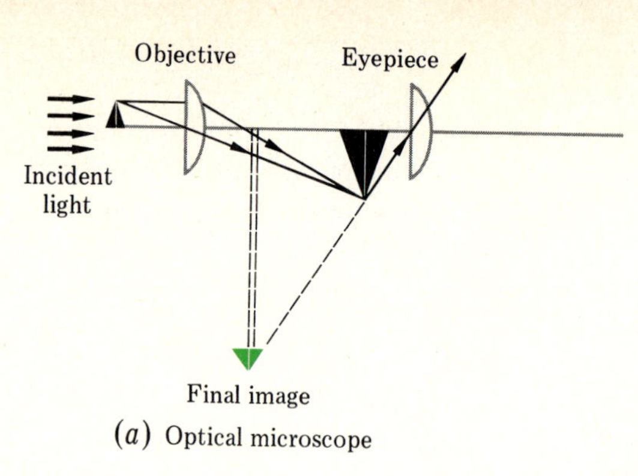

(*a*) Optical microscope

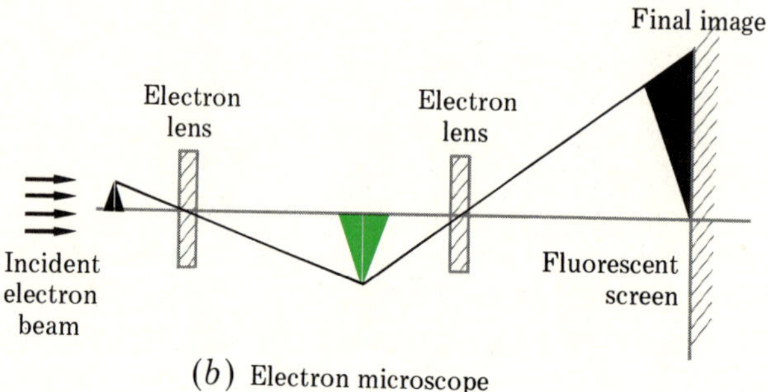

(*b*) Electron microscope

FIGURE 26.14
In the electron microscope the final image is real, while in the optical microscope it is usually virtual.

will not undergo collisions. A typical microscope and a photograph of an image produced by it are shown in Fig. 26.15. Notice that a light microscope would be incapable of showing any of the detail given there since the wavelength of light is larger than the objects being viewed.

The electron lenses in Fig. 26.14 can be of two types, electrostatic or magnetic. Basically a lens is a device to focus the beam, electrons (charged particles) in this case. These two types of lenses are shown schematically in Fig. 26.16. You should examine each diagram to see exactly how the focusing action is achieved. Unfortunately it is difficult to correct these devices for lens defects, and so at present the theoretical limits of resolution for the electron microscope have not yet been achieved. Detail smaller than a few angstroms is still not readily observed although in principle the electron microscope should be capable of doing so.

(*a*)

(*b*)

FIGURE 26.15

(*a*) In this electron microscope the electron beam is accelerated through $\sim$75,000 V. It is capable of a resolution of 10^{-7} cm. The stacked cylinders shown in the photograph contain magnetic lenses. The source of electrons is at the top, and the final, magnified image is cast upon a fluorescent screen at the bottom. Photographic plates can be inserted in this plane to obtain a photographic record. Focusing is accomplished by varying the current in the magnetic lenses. (*Photograph from I. M. Freeman, "Physics: Principles and Insights," McGraw-Hill, 1968.*) (*b*) Each of the molecules of poliomyelitis virus photographed here has a diameter of about 200 Å. (*Photograph by R. C. Williams, Virus Laboratory, University of California, Berkeley.*)

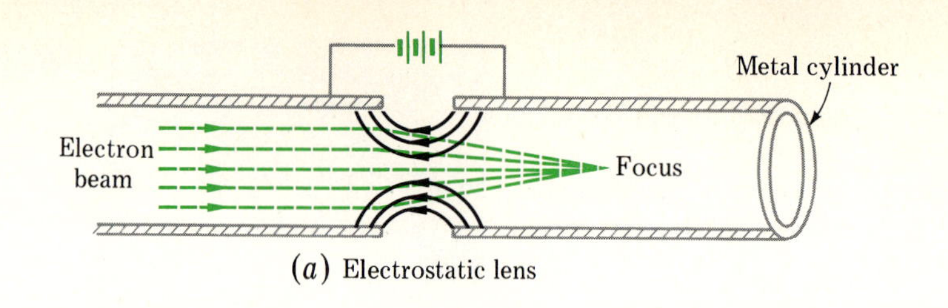

(a) Electrostatic lens

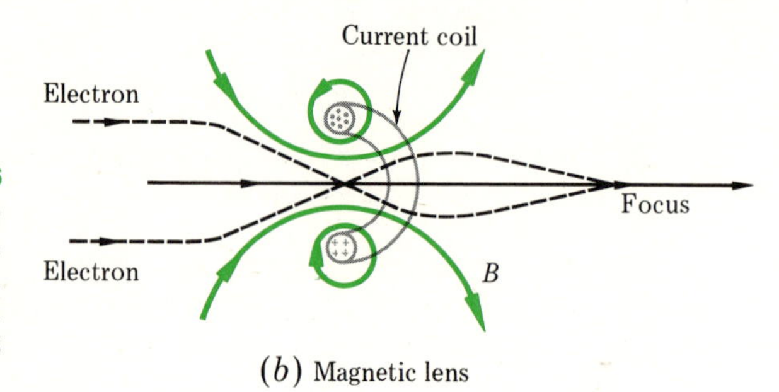

(b) Magnetic lens

FIGURE 26.16

In the magnetic lens, the B field of the coil causes the electrons to follow the path which loops around the coil axis in a spiral of decreasing radius.

SUMMARY

Planck found that an oscillator with natural frequency ν_0 can only have energies $nh\nu_0$, where n is an integer. For visible-size oscillators, these energies are so close together that they appear continuous.

Einstein showed that a beam of EM radiation is equivalent to a stream of photons. Each photon has an energy hc/λ. In the photoelectric effect, a photon collides with an electron in a solid. If the photon has energy in excess of the work-function energy, it may eject the electron from the solid. In the Compton effect, a photon scattered by collision with a nearly free electron loses part of its energy. These scattered photons have larger λ than the incident photons.

Relativity is based upon two apparent facts: the speed of light in vacuum is always measured to be c; only relative speeds can be measured. The second of these is equivalent to stating that the laws of nature are the same in all inertial reference frames.

Among the important conclusions arrived at from these facts are the following: no material object can be accelerated to speeds in excess of c; events which are simultaneous in one inertial frame may not be simultaneous in another; a moving clock appears to be slowed by a factor of $\sqrt{1-(v/c)^2}$; dimensions of objects are shortened along the line of motion by a factor $\sqrt{1-(v/c)^2}$; an object's mass varies with speed according to the relation $m/m_0 = 1/\sqrt{1-(v/c)^2}$; energy and mass are related through $E = mc^2$; the kinetic energy of an object is $(m - m_0)c^2$, and this reduces to $\frac{1}{2}m_0v^2$ at low speeds.

A photon has zero rest mass, but its momentum is given by h/λ. De Broglie applied this relation to material objects and found that an object moving with speed v has a wavelength $\lambda = h/mv$ associated with it. For atomic particles, the wavelength becomes large enough to create easily measured interference and diffraction effects.

The wave nature of particles gives rise to the law of nature summarized by the Heisenberg uncertainty principle. The principle is represented by the relations $\Delta x\, \Delta p \geq h/2\pi$ and $\Delta E\, \Delta t \geq h/2\pi$.

Quantum mechanics takes account of the wave nature of matter. It is therefore applicable even when classical, newtonian mechanics is not. Classical mechanics is no longer valid when the particle wavelength becomes comparable with the geometrical dimensions involved in an experiment.

MINIMUM LEARNING GOALS

Upon completion of this chapter, you should be able to do the following:

1. Compute the allowed energies (according to Planck) for an oscillator with known natural frequency provided Planck's constant is given. Explain why the energy of a pendulum appears to be continuous.

2. Sketch a graph of radiation intensity versus λ for a hot object. Show how the graph changes with temperature.

3. Describe the photoelectric effect and point out what is meant by the photoelectric threshold. State the energy of a photon in terms of its wavelength. Explain how the photon concept applies to the photoelectric effect. Compute the threshold wavelength from a knowledge of the work function.

4. Describe the Compton effect and explain how it can be interpreted in terms of photon scattering.

5. State the two basic postulates of relativity. Explain the meaning of inertial reference frame and express the second postulate in terms of laws of nature as applied to such a frame.

6. State the conclusions relativity gives rise to in respect to the following: maximum speed of objects; simultaneous events; time dilation; length contraction; mass variation with speed; KE; mass-energy conversion. Compute the answers to simple problems involving these conclusions.

7. State the rest mass and momentum of the photon.

8. Give the de Broglie wavelength of a particle of known mass moving with a known speed. Explain why the wave properties of electrons are easily noticed while those of a baseball are not noticeable.

9. State both forms of the Heisenberg uncertainty principle in both words and symbols.

10. Explain under what conditions newtonian classical mechanics must be replaced by quantum mechanics. Reasoning from the interference effects observed for light, show why newtonian mechanics breaks down under these conditions.

IMPORTANT TERMS AND PHRASES

You should be able to define or explain each of the following:

Light quantum; photon
Photoelectric effect; threshold wavelength
Work function
Compton effect
Postulates of relativity
Reference frame; inertial frame
Time dilation
Relativistic factor
Length contraction (relativistic)
Rest mass; $m = m_0/\sqrt{1-(v/c)^2}$
$\Delta E = \Delta mc^2$
Photon momentum $= h/\lambda$
de Broglie wavelength $= h/mv$
Davisson-Germer experiment
Uncertainty principle
Electron microscope

QUESTIONS AND GUESSTIMATES

1. Discuss how our world would be affected if nature suddenly changed in such a way that Planck's constant became 10^{32} times larger than it is. Consider the situation from two different aspects: (*a*) quantization of energy of oscillators; (*b*) the uncertainty principle.

2. How does the photon picture of light explain the following features of the photoelectric effect: (*a*) critical wavelength; (*b*) stopping potential is inversely proportional to wavelength?

3. How can one measure the work function of a metal? Planck's constant?

4. Make a list of experiments in which light behaves like a wave and a list of experiments in which its quantum character is important. Is there any experiment in your list which can be explained from both standpoints?

5. Suppose an astronaut has perfect pitch so that he can recognize at once that a particular tuning fork gives off a sound of middle C when struck. What would he hear if he listened to the tuning fork inside his spaceship while traveling through space at a speed of $0.9c$?

6. Before leaving the earth, a vibrating tuning fork in a spaceship is used to mark time intervals by printing out a number on a chart each time it passes its centerpoint. A similar tuning-fork system with an identical period is kept on the earth as the rocket ship goes out into space, travels around for a few years at nearly the speed of light, and returns to earth. How will the two charts for the two tuning forks compare when the ship returns to earth?

7. Most human beings live less than 100 yr. Since the maximum velocity one can acquire relative to the earth is c, the speed of light, it is impossible for a person on earth to travel farther than 100 light-years into space before he becomes 100 years old. Does this necessarily mean that no person from earth will ever be able to travel farther from earth than 100 light-years? (A light-year is the distance light travels in a time of 1 yr, or 9.46×10^{15} m.)

8. Suppose the speed of light were only 20 m/s and all the relativistic results applied using this speed for c. Discuss how our lives would be changed.

9. A spaceship moving past the earth at speed v_{earth} shoots a projectile straight ahead with a speed u_{ship} relative to the ship as viewed by a person within the ship. At ordinary speeds, the speed of the projectile relative to the earth is $v_{\text{earth}} + u_{\text{ship}}$. Show that this cannot be correct at relativistic speeds. (The correct relation at high speeds is

$$\frac{v_{\text{earth}} + u_{\text{ship}}}{1 + v_{\text{earth}}u_{\text{ship}}/c^2}$$

You can find how it is derived by referring to a more advanced text.)

10. It should be clear from this chapter that the statement "matter can neither be created nor destroyed" is false. What can one say instead?

11. When light is shone on a reflecting surface in vacuum, a pressure is exerted on the surface by the light. Explain. Would the pressure be different if the surface was black so that it absorbed the light?

12. If all the mass energy of a fuel could be utilized, about how many kilograms of the fuel would be needed to furnish the energy required by a city of 300,000 people for 1 yr? (E)

13. If we believe the uncertainty relations, why must we refuse to believe that all molecular motion ceases at absolute zero of temperature?

14. Estimate the power change for a local radio-station antenna system as it changes from one quantized oscillation energy state to an adjacent state. What energy photons does the station radiate? What wavelength photons? What frequency? (E)

15. Ultraviolet light causes sunburn while visible light does not. Explain why. Some people insist they sunburn easiest when their skin is wet. Do you see any reason for this?

PROBLEMS

1. The hydrogen chloride molecule acts in many ways like two balls joined by a spring as shown in Fig. P26.1. It has a back-and-forth vibration (stretching and compressing the spring) with a natural frequency of 8.5×10^{13} Hz. What is the energy gap between allowed energies for this oscillator? Express your answer in both joules and electronvolts.

2. The helium-neon laser emits red light with $\lambda = 6.328 \times 10^{-7}$ m. What is the energy of a photon in such a beam? Express your answer in both joules and electronvolts.

3. It is convenient to remember that photons with 1 eV energy have a wavelength of 1240 nm. (*a*) Prove this fact. (*b*) What portion of the spectrum is this? (*c*) What wavelength corresponds to a 3.0-eV photon?

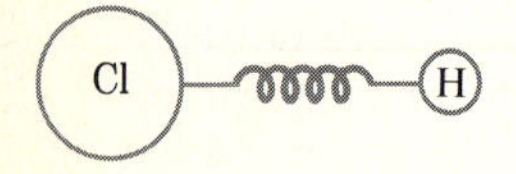

FIGURE P26.1

4. Find the work function for a material which has a threshold wavelength of 3.3×10^{-7} m. Express your answer in electronvolts.

5. (*a*) Find the threshold wavelength for a material (gold) which has a work function of 4.8 eV. (*b*) What portion of the spectrum is this in?

6. (*a*) Find the energy of the EM-wave quanta sent out by a radio station operating at 1 Mc. Express the answer in electronvolts. (*b*) What would be the speed of an electron which has this amount of kinetic energy?

7. The average thermal translational KE of a particle is $\frac{3}{2}kT$. (*a*) What photon wavelength is equivalent to this average thermal energy at 27°C? (*b*) What type of radiation is this?

8. Light with wavelength 400 nm is shone upon a surface having a work function of 2.0 eV. Find the speed of the fastest photoelectron emitted from the surface.

9. The stopping potential measured when 400-nm light is shone on a certain substance is 0.30 V. What is the work function of the substance?

10.* The energy of the carbon-carbon bond in organic molecules is about 80 kcal/mol. If all the energy of a photon could be utilized in breaking this bond, what wavelength photons could just accomplish it? *Hint:* 80 kcal/mol equals $80{,}000/(6 \times 10^{23})$ cal per bond.

11. An energy of 13.6 eV is needed to tear loose the electron from a hydrogen atom, i.e., to ionize the atom. If this is to be done by striking the atom with a photon, what is the longest-wavelength photon which can accomplish it? (Assume all the photon energy to be effective.)

12. According to astronomers on earth, the distance to the star Arcturus is about 40 light-years. How long, in years, would it take for a spaceship to reach there if its speed is $0.990c$ relative to the earth according to (*a*) earth clocks and (*b*) spaceship clocks?

13. A neutron alone and at rest lives about 1000 s before it changes into something else. Will a neutron shot out from the sun at a speed $0.995c$ be able to reach the planet Saturn before it changes form? (Saturn is about 1.5×10^{12} m from the sun.)

14. ** The K^+ meson lives about 10^{-8} s before changing form. How fast must such a meson be moving through the laboratory if it is to live long enough to travel the length of the laboratory, a distance of about 15 m?

15.* In modern nuclear accelerators, particles are sometimes accelerated to energies of billions of electron-volts. (*a*) What is the mass of a 2×10^9-eV proton? (*b*) How fast is it moving? ($m_0 = 1.67 \times 10^{-27}$ kg.)

16. To melt 1 g of ice at 0°C requires an energy of 80 cal. By how much does the mass of the ice increase because of the energy added to melt it?

17. Chemists sometimes say "the mass of the reactants equals the mass of the products" in a chemical reaction. When 2 g of hydrogen is burned with 16 g of oxygen to form 18 g of water, the reaction gives off about 60,000 cal of heat energy. How much mass is lost in the process?

18. At low speeds, if a man moving with speed v relative to the earth shoots a projectile out along his line of motion with speed u relative to himself, the speed of the projectile relative to the earth will be simply $v + u$. This cannot be correct at speeds near c since speeds in excess of c would be predicted. (For example, if $v = 0.7c$ and $u = 0.7c$, the speed relative to the earth would be $1.4c$, an impossibility.) Einstein showed the speed to be given by the formula

$$\frac{v + u}{1 + vu/c^2}$$

If a spaceship has a speed $0.7c$ past the earth and shoots a projectile out from itself in its line of motion with speed $0.9c$, what is the speed of the projectile relative to the earth?

19. Repeat Prob. 18, replacing the projectile by a light pulse.

20. What is the wavelength associated with an electron which has fallen through a potential difference of 6 V?

21. Atoms are of the order of 1 Å in radius. How fast must an electron be moving if its associated wavelength is to be smaller than 1 Å?

22.** Nuclei are of the order of 4×10^{-15} m in diameter. In order for a neutron to be within the nucleus, its inherent uncertainty in position should be less than 4×10^{-15} m. This means the neutron should have an associated wavelength less than this. Neglecting relativistic effects, find the speed with which a neutron must be moving for its associated wavelength to be 1×10^{-15} m. (Neutron mass $= 1.67 \times 10^{-27}$ kg.)

23.** A photon with wavelength λ strikes an electron (originally at rest) head on, and both rebound elastically. Assuming that all the motion occurs along the same straight line, (*a*) show that the velocity v of the electron after impact and the final wavelength of the photon are given by

$$v = \frac{1.46 \times 10^{-3}}{\lambda} \text{ m/s} \qquad \text{provided that } v \ll c$$

and

$$\frac{1}{\lambda'} = \frac{1}{\lambda} - \frac{mv^2}{2hc}$$

(*b*) Evaluate both v and λ' if $\lambda = 1.00$ nm.

27 ATOMIC STRUCTURE AND THE EMISSION OF LIGHT

We saw in the last chapter that the years following 1900 were years of rapid advances in physics. In this chapter we shall see how these developments were extended to explain the structure of atoms and their interaction with EM waves. Before doing so, however, we shall first discuss an important experiment which contributed greatly to our understanding of atomic structure.

27.1 THE NUCLEAR ATOM

At the beginning of this century, the internal structure of atoms was still a mystery. It was recognized that each atom has associated with it a number of electrons Z equal to the atomic number of the element in question. Since each atom is electrically neutral, the charge $-Ze$ carried by the electrons in an atom must be balanced by a positive charge $+Ze$. Moreover, the Z electrons provide only a small fraction of the total mass of an atom. Hence, in addition to its electrons, the atom must have within it a mass nearly equal to the atom's mass and a positive charge Ze.

The arrangement of the electrons, positive charge, and mass within atoms was first learned from the definitive experiments of Ernest Rutherford and his associates in 1911. We illustrate the basic idea behind these experiments in Fig. 27.1. As we see there, a beam of α particles (alpha particles) is incident upon a thin film of gold atoms. The α particles are shot out from the radioactive element radium. These particles were known to carry a charge $+2e$, and they have a mass about equal to that of the next to smallest atom, helium. Since a gold atom has a mass about 50 times larger than an α particle, one might first picture the situation to be that shown in Fig. 27.1*a*.

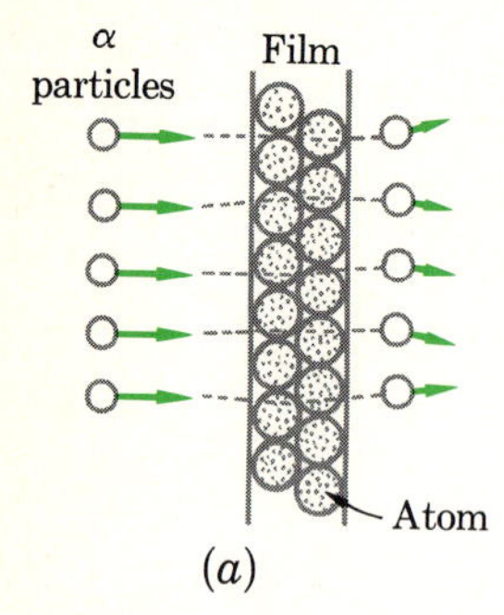

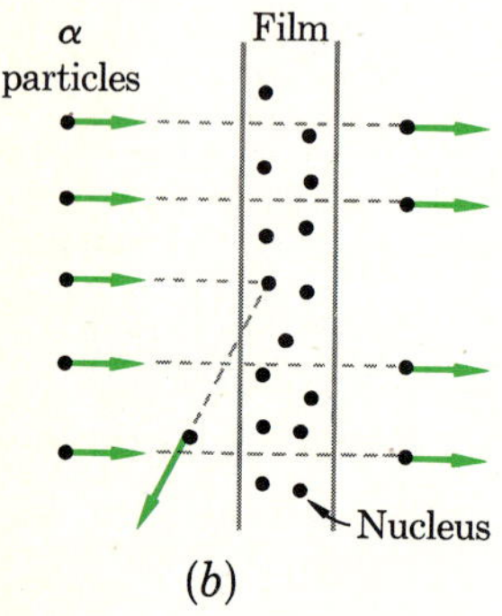

FIGURE 27.1
(*a*) According to Thomson's model of the atom, the incident α particles should be deflected only slightly. (*b*) Rutherford actually observed this result; the size of the nucleus is exaggerated, and the electrons are not shown.

Shown there is the so-called **Thomson model** of the atom (after its proposer, J. J. Thomson). It assumes the atom to consist of a more or less uniform spherical mass having the atom's positive and negative charges distributed throughout it. Since the target film used was of the order of 100 atoms thick, the Thomson atom model would predict the α particle to be slowed and perhaps even stopped by the film. The experimental results, however, turned out to be quite different.

In the first place, it was found that only a small percentage of the α particles were affected in any way by the presence of the film. It was as though the film were full of holes and most of the α particles, having gone through the holes, were slowed down hardly at all. The second unexpected result had to do with what happened to the few particles which apparently did hit something. Some of these were deflected through very large angles as though they had struck a very massive object. For example, as shown in Fig. 27.1*b*, some of the particles were deflected nearly straight backward.

The Nuclear Atom

In order to explain these results, Rutherford postulated the so-called **nuclear atom.** His measurements showed that the atom must contain a very tiny massive core which carries all the atom's positive charge and nearly all its mass. The diameter of this positively charged ball (or **nucleus**) at the center of the atom had to be of the order of 10^{-14} m or less. However, the diameter of atoms as computed from the density of crystals was known to be of order 10^{-10} m.* Therefore, if the nucleus is pictured to be a positively charged ball about 1 mm in diameter, the atom drawn to this same scale would have a radius of about 10 m. It is impossible, therefore, for us to draw a scaled diagram of the nuclear atom since, if the atom is to fit on the page, the nucleus will be too small to be seen.

Atomic Number and Number of Electrons

The atoms of the periodic table were known from their chemical

*The number of atoms per unit volume can be obtained by dividing the mass per unit volume, i.e., the density, by the mass of an atom. The volume taken up by each atom can then be found by taking the reciprocal of the number of atoms per unit volume.

behavior to contain a number of electrons equal to the atomic number for that element. For example, hydrogen is the first element in the table, and since its atomic number is 1, it must possess one electron. Similarly, gold, being the seventy-ninth element in the table, must possess 79 electrons. Rutherford postulated that since the atom must be electrically neutral, the nucleus must carry a charge of $+79e$. The immense region outside the nucleus contains the 79 electrons of the atom, as shown schematically in Fig. 27.2. Since the electrons are also very small, the major portion of the atom behaves like empty space and does not deflect the bombarding particles. Since Rutherford's time, his nuclear picture of the atom has been fully confirmed. We shall make use of this model to explain other facets of atomic behavior.

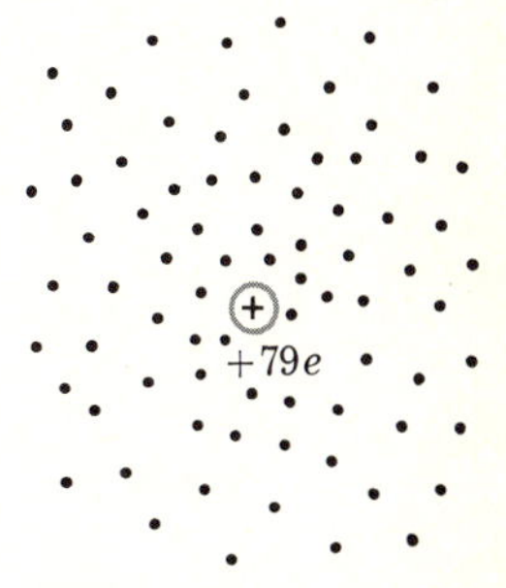

FIGURE 27.2

The Rutherford atom model pictured the positive nucleus at the center of a spherical region containing the electrons. Both the electrons and nucleus should be much much smaller than shown.

A schematic picture of the carbon atom which is consistent with Rutherford's result is shown in Fig. 27.3. The size of the nucleus and the electrons would be smaller than pinpricks if the radius of the atom were as large as shown in the figure. This picture is not acceptable as it stands, however, because it is not stable mechanically. The positive nucleus will attract the six negative electrons, and they will fall into the nucleus and be neutralized. In fact, it can be proved that electrostatic forces by themselves are not capable of holding a stationary group of charges in static equilibrium. Hence, there is no possible way in which stationary electrons and a nucleus could be arranged in space so that they would remain fixed in place.

One might speculate that perhaps the electrons are circling about the nucleus much as the planets circle about the sun. The inverse-square gravitational force supplies the centripetal force needed to hold a planet in its orbit. A similar situation could apply to the atom, with the inverse-square Coulomb attraction replacing the gravitational force. Even though this idea had much to recommend it, there was a nearly overwhelming reason for not believing it.

When an electron rotates about a positive nucleus, it creates, in effect, an oscillating pair of charges. Maxwell's equations predict that an oscillating charge system will radiate EM waves in the manner we have discussed for an antenna. For an atom, the waves would be of very high frequency, probably light waves. But if the atom oscillator radiated energy, it would be bound to run down. The electrons would spiral into the nucleus, and the atom would collapse. Since this does not happen, the whole idea appears to be wrong.

On the other hand, if the atom did radiate energy, this would be a convenient explanation for the fact that atoms sometimes give off light. Unfortunately, though, the frequency would change as the electrons fell in toward the nucleus, contradicting the fact that a given atom radiates only very well-defined wavelengths. As you can see, the situation was far from satisfactory in 1913, when people were still groping for an answer to this problem.

FIGURE 27.3

A possible but incorrect picture of the carbon atom.

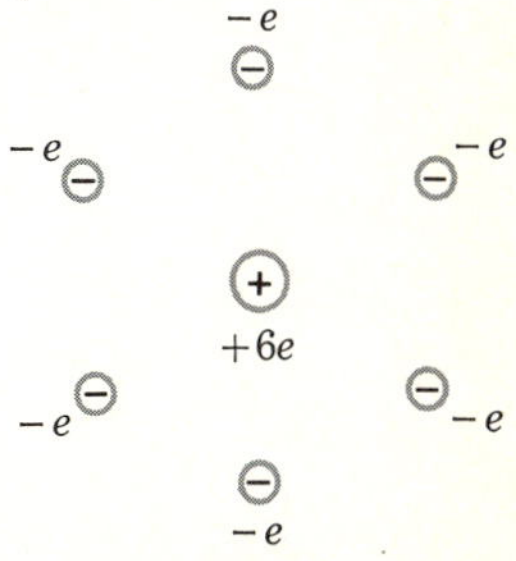

27.2 THE SPECTRUM OF HYDROGEN

In spite of the fact that the mechanics of the atom was not well understood in 1913, there were available detailed and precise data concerning the light given off by atoms. Grating and prism spectrometers had been widely used for years, and the field of optical spectra (light radiation by atoms and molecules) had been a field of diligent experimental research. In principle, researchers found any gaseous mass of atoms could be made to emit light by sending a spark, or discharge, through it by means of two high-voltage probes. For materials which are normally solid, a gaseous mass could be created by first vaporizing them in a hot arc. The wavelengths of light given off by these hot gases, i.e., their spectrum, could be investigated by use of a spectrometer, as discussed in Sec. 25.7.

Since hydrogen is the simplest of all atoms, having only one electron, it was natural that great interest should be manifested in its spectrum. As it turns out, this atom does in fact have a simple spectrum. When measured, it was found to consist of the series of spectral lines shown in Fig. 27.4. (Recall from Sec. 25.7 that a spectral line is actually an image of the slit of the spectrometer. Each wavelength gives a separate image.) The lines in the near ultraviolet were visible only in photographs, of course.

Series Limit

Notice that the lines at shorter wavelengths are closer and closer together. However, no lines of wavelength shorter than $\lambda = 364.6$ nm occur (in this region), and this shortest wavelength of the series is called the **series limit.** According to the theory we shall present shortly, there should be an infinite number of lines in this series. About 40 have actually been resolved. The remainder are too close together to be seen distinctly.

Since these spectral lines seem to have a definite sort of order, it is perhaps natural to try to fit their wavelengths to an empirical formula. This was first done by Balmer in about 1885, and this series is now known as the Balmer series. He found that the wavelengths of the lines could be expressed by the following remarkably simple formula,

Balmer Series Formula

$$\frac{1}{\lambda} = R\left(\frac{1}{2^2} - \frac{1}{n^2}\right) \tag{27.1}$$

where R is a constant of value 1.0974×10^7 m^{-1} and $n = 3, 4, 5$, and so on. Of course, 2^2 is simply 4. If n is set equal to 3, λ turns out to be 656 nm according to the formula. This is also the first line in the Balmer series, shown in Fig. 27.4. For $n = 4$, λ is given as 486.2 nm, and so on. The integers from 3 to infinity when placed in Eq. (27.1) yield the lines of the Balmer series. When n is set equal to infinity, the formula yields the series limit, 364.6 nm. The empirical constant R is called the **Rydberg constant** in honor of the man

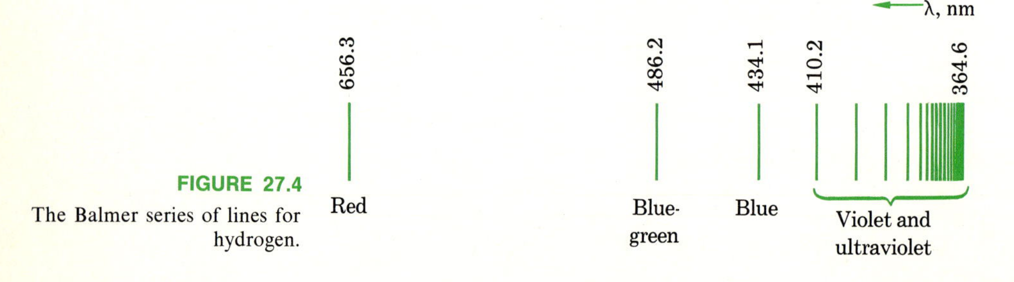

FIGURE 27.4
The Balmer series of lines for hydrogen.

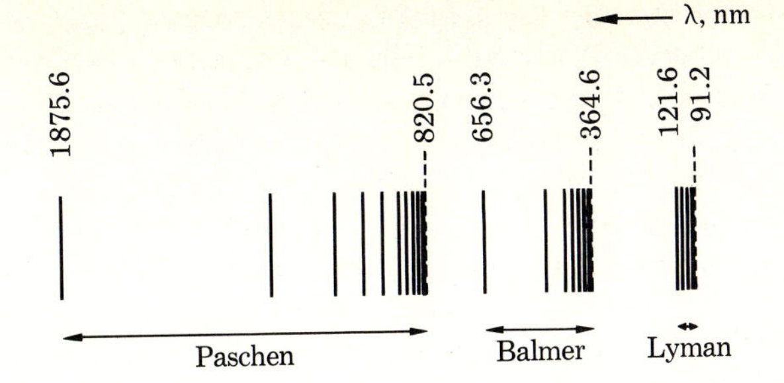

FIGURE 27.5
The three shortest spectral series of lines given off by hydrogen atoms.

who accurately determined its value. It is customary to denote the lines in this series as the H_α line, H_β line, H_γ line, etc.

Later, it was found that hydrogen atoms emit wavelengths other than those found in the Balmer series. The Lyman series occurs in the far ultraviolet, while the Paschen series is in the infrared. These series are illustrated in Fig. 27.5. Other series still farther in the infrared have also been found. Amazingly enough, these series also follow formulas very much like Balmer's. It is found that

$$\begin{aligned}
&\text{Lyman:} \quad &\frac{1}{\lambda} = R\left(\frac{1}{1^2} - \frac{1}{n^2}\right) \qquad &n = 2, 3, \ldots \\
&\text{Balmer:} \quad &\frac{1}{\lambda} = R\left(\frac{1}{2^2} - \frac{1}{n^2}\right) \qquad &n = 3, 4, \ldots \\
&\text{Paschen:} \quad &\frac{1}{\lambda} = R\left(\frac{1}{3^2} - \frac{1}{n^2}\right) \qquad &n = 4, 5, \ldots
\end{aligned} \tag{27.2}$$

and so on, with $R = 1.0974 \times 10^7\ \text{m}^{-1}$.

It is truly more than mere coincidence that such simple formulas should apply to such a complicated phenomenon as light emission. Clearly some great simplicity in atomic behavior must be responsible for this remarkable set of relations. They should therefore furnish a simple test of atomic theories. As we shall see, they furnished Niels Bohr with the clues he needed to provide us with the first workable picture of the atom.

27.3 THE BOHR ATOM

In 1913, when he was a 28-year-old research student at Cambridge, Niels Bohr presented a novel picture of the hydrogen atom. He built it upon the concept of an electron revolving in a circular orbit about the nucleus as center. The centripetal force needed to hold the electron in orbit was furnished by the electrical attraction between the nucleus and the electron. The concept is shown in Fig. 27.6, where the charge on the nucleus is taken to be Ze. In the case of hydrogen the atomic number Z is unity.

FIGURE 27.6
In the Bohr atom, the electron is assumed to travel in a circle about the nucleus. The centripetal force is furnished by the attraction to the nucleus.

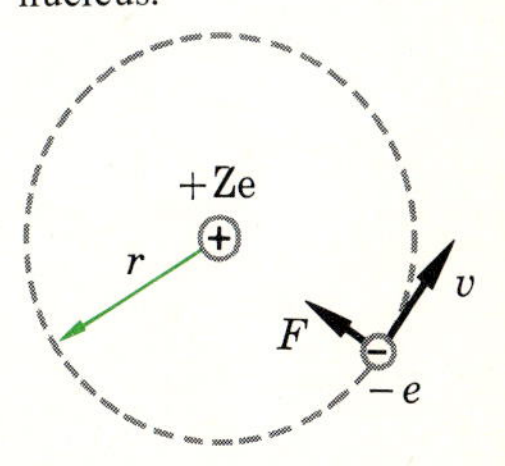

Coulomb's law tells us that the electrical force on the electron is $k(Ze^2/r^2)$ toward the center, where k is $9 \times 10^9\ \text{N} \cdot \text{m}^2/\text{C}^2$. This force furnishes the centripetal force mv^2/r, and hence we can write

$$\frac{kZe^2}{r^2} = \frac{mv^2}{r}$$

Solving for mv^2 yields

$$mv^2 = \frac{kZe^2}{r} \tag{27.3}$$

These ideas were nothing new, of course. As stated earlier in this chapter, an oscillating electron should radiate energy, and hence the atom would run down and collapse. It was for this reason that this general picture had not been widely accepted. To circumvent this difficulty, Bohr took the attitude that perhaps the atom does not obey ordinary electric laws. He tentatively assumed that the oscillating electron does not radiate energy, even though this is contradictory to the behavior observed for the oscillating electrons in antennas and in other nonatomic systems.

Stable Orbits Assumed

Let us suppose, with Bohr, that there are certain special orbits in which the electron is stable, i.e., does not radiate energy. These orbits would be somewhat as shown in Fig. 27.7. If the electron is in orbit 2, it might, under certain circumstances, fall to orbit 1. Actually, since the electron is attracted by the nucleus, it will fall "downhill" and will lose PE as it falls from orbit 2 to orbit 1. *If the total energy of the electron in orbit n is E_n and in orbit p is E_p, the electron will loose an energy $E_n - E_p$ when it falls from orbit n to orbit p.*

Photon Emission Assumed

Since energy can be neither created nor destroyed, the energy lost $E_n - E_p$ must go somewhere. Bohr postulated it to be radiated as a single quantum of light energy, $h\nu$. He therefore wrote for the energy of the emitted photon,

$$h\nu = \frac{hc}{\lambda} = E_n - E_p \tag{27.4}$$

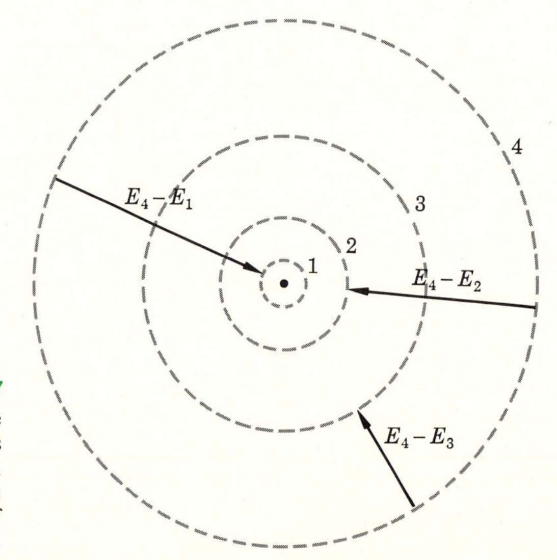

FIGURE 27.7
Bohr assumed that there are certain stable circular orbits about the atom nucleus. An electron in orbit 4 can leave the orbit by means of any of the three transitions indicated.

Let us now see what conclusions we can draw from Eqs. (27.3) and (27.4).

To find E_n, we notice that the electron has both KE and PE. Its KE is $\frac{1}{2}mv^2$, or, from Eq. (27.3), we have

$$\text{KE} = \frac{Ze^2k}{2r}$$

where $k = 9 \times 10^9\ \text{N} \cdot \text{m}^2/\text{C}^2$. The PE of the electron is negative, since we define its PE to be zero at infinity and it falls to a lower PE as it comes closer to the nucleus which is attracting it. We have

$$\text{PE} = \frac{-Ze^2k}{r}$$

Adding these two contributions together gives us the total energy, namely,

$$E = \frac{-Ze^2k}{2r} \tag{27.5}$$

Notice that the energy becomes less, i.e., more negative, as r becomes smaller. This is the result of the fact that PE is lost by the electron as it falls toward the nucleus.

Substitution in Eq. (27.4) yields

$$\frac{hc}{\lambda} = \frac{Ze^2k}{2}\left(\frac{1}{r_p} - \frac{1}{r_n}\right)$$

or

$$\frac{1}{\lambda} = \frac{Ze^2k}{2hc}\left(\frac{1}{r_p} - \frac{1}{r_n}\right) \tag{27.6}$$

Equation (27.6) is very similar to the experimental relation, Eq. (27.2). Both equations equate the reciprocal of λ to a difference of two terms. Since the radii of the orbits r_p and r_n are not known, it should be possible to choose them so that Eq. (27.6) would be exactly the same as Eq. (27.2). In view of Planck's and Einstein's success in picturing energy to be quantized, i.e., come in small packets, or quanta, Bohr looked for some other quantity which, when quantized, would provide the proper values for the orbit radii. He found that the proper choice was to say that the angular momentum of the electron mvr must be an integer multiple of $h/2\pi$, where h is Planck's constant. He then wrote

Angular Momentum Quantized

$$mvr_n = \frac{nh}{2\pi} \tag{27.7}$$

with n equal to any integer and where r_n is the radius of the nth orbit.

Solving for v in Eq. (27.7) and substituting for v in Eq. (27.3), we find r_n to be given as

$$\frac{1}{r_n} = \left(\frac{4\pi^2Ze^2mk}{h^2}\right)\left(\frac{1}{n^2}\right) \tag{27.8}$$

Bohr's Balmer Type Formula

Substituting this (and a similarly derived expression for r_p) in Eq. (27.6) yields

$$\frac{1}{\lambda} = \left(\frac{2\pi^2 Z^2 e^4 m k^2}{h^3 c}\right)\left(\frac{1}{p^2} - \frac{1}{n^2}\right) \tag{27.9}$$

According to this, when an electron falls from orbit n to orbit p, light is emitted and its wavelength is given by Eq. (27.9). We notice at once that Eq. (27.9) is exactly the same in form as Eq. (27.2), the experimental relation. Furthermore, when the constant in Eq. (27.9) is evaluated, it proves to be exactly equal to the Rydberg constant R of Eq. (27.2).

Bohr's derivation of Eq. (27.9), an equation which predicts almost exactly the known wavelengths emitted by hydrogen atoms, was a remarkable success. For the first time physicists were able to derive a quantitative picture of light emission by atoms. In deriving it, *Bohr had made two unorthodox assumptions, in addition to using Planck's and Einstein's ideas concerning the energy of a quantum of light. He assumed that certain stable electron orbits exist in which the electron circles the nucleus without emitting energy. Moreover, he postulated that these orbits are such that the angular momentum of the electron is equal to an integer multiple of* $h/2\pi$.

If pressed for a justification of these assumptions, Bohr could support them only with the argument that they seemed to work. The reason for their validity was not known. However, while the assumptions seem to contradict our experience with ordinary objects, an atom is not an ordinary object. We have no real reason not to believe that electrons in atoms behave as Bohr postulated. The fact that his postulates lead to valid results forces us to conclude that they may perhaps be correct. As we shall see later in this chapter, the modern trend is to accept Bohr's postulates in a general sort of way but to replace his detailed picture of electron orbits with more complicated spatial patterns.

27.4 EMISSION OF LIGHT BY BOHR'S ATOM

As we saw in the last section, according to Bohr the hydrogen electron can revolve in any one of a large number of stable orbits about the nucleus. The radii of these orbits are given by Eq. (27.8). If we evaluate the constant in that relation, we find that the stable (or allowed) electronic orbits in hydrogen are as follows:

$$\begin{aligned} n = 1: &\quad r_1 = 0.53 \times 10^{-10}\ \text{m} \\ n = 2: &\quad r_2 = (4)(0.53 \times 10^{-10})\ \text{m} \\ n = 3: &\quad r_3 = (9)(0.53 \times 10^{-10})\ \text{m} \end{aligned}$$

and so on. These orbits are pictured in Fig. 27.7. Notice that the radii of the orbits are in the ratio $1 : 2^2 : 3^2 : 4^2$, etc.

We learned in mechanics and thermodynamics that a system will continue to lose energy until further energy loss is impossible. This is why a pendulum eventually stops swinging and why a ball falls to the earth. Similarly, the electron in a hydrogen atom will eventually lose as much energy as it can. As we have seen, the electron loses energy as it falls closer to the nucleus. However, it can fall no lower than orbit 1, since there is no orbit smaller than this to which it can fall. An atom in which the electron has fallen to its lowest energy level, orbit 1, is said to be **unexcited.** The atom is stable in this energy state and does not emit light. Its total energy is E_1.

Work is required to raise (or excite) the electron from this lowest energy state (orbit) to one of the higher orbits or energy levels. The electron cannot by itself move to orbit 2 or any of the higher orbits. However, if the atom is bombarded by other particles or atoms, a collision with the electron may give it enough energy so that it is lifted to an orbit of higher energy.

Suppose, for example, that the electron has been knocked out to orbit 4, as shown in Fig. 27.7. Although the electron is relatively stable in this state, it will fall to a lower energy state (or orbit) within a very small fraction of a second. When it does so, it loses energy and a quantum of light, i.e., a photon, is emitted. For example, the electron could fall directly from orbit 4 to orbit 2. According to Eq. (27.9) the wavelength of the light emitted as a result of this transition is

$$\frac{1}{\lambda} = R\left(\frac{1}{2^2} - \frac{1}{4^2}\right)$$

where the constant has been replaced by its measured value, the Rydberg constant. Notice that this wavelength, according to Eq. (27.2), is the second line of the Balmer series.

Similarly, if the electron had fallen directly to orbit 1, the wavelength would have been

$$\frac{1}{\lambda} = R\left(\frac{1}{1^2} - \frac{1}{4^2}\right)$$

which is the third line in the Lyman series. Notice that the electron has fallen through a greater energy difference in this case, and so the light emitted will be of higher energy and shorter wavelength. This fits in very well, since we know that the Lyman series is in the far-ultraviolet portion of the spectrum, while the Balmer series is in the visible portion of the spectrum.

Of course, the electron might also fall from state 4 to state 3. This would be a line in the Paschen series. We see that if an electron falls from an outer orbit to the first orbit, a line in the Lyman series is emitted. If it falls to energy state 2, a wavelength in the Balmer series is given off. For a transition down to level 3, a line in the Paschen series results. This is shown in Fig. 27.8.

Origin of Series Limit

The series-limit line is emitted when the electron falls from outside the atom to the lowest energy state for that series. When the electron falls from orbit 3 to 2, the first line in the Balmer series is emitted. From orbit 4 to 2, the second line of the Balmer series results. Actually, the energy difference between the levels decreases rapidly as we go to higher and higher orbits. Hence, nearly as much energy is emitted when the electron falls from orbit 10

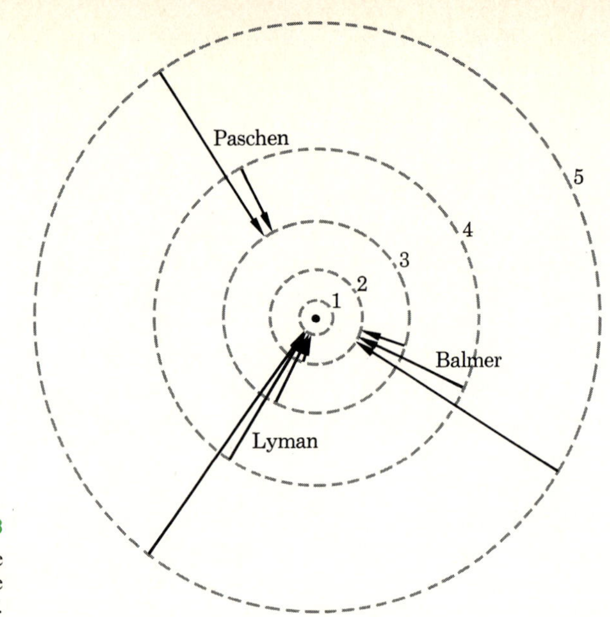

FIGURE 27.8
In a single spectral series the electron always falls to the same inner orbit.

to 2 as when it falls from orbit 100 to 2. This means that the lines in the Balmer series become very closely spaced as we go to the wavelengths emitted by transitions from the outermost orbits to orbit 2. Of course the most energy is emitted if the electron falls from outside the atom ($n \rightarrow \infty$) to orbit 2. This results in the emission of the series-limit wavelength.

27.5 ENERGY-LEVEL DIAGRAMS

It is often convenient to focus one's attention on the energies of the electron (and the atom) without bothering about the geometry of orbits, etc. In fact, usually one is interested only in the energies since these alone influence the wavelengths of light emitted by the atom. We therefore make use of an **energy-level diagram** for an atom (and for many other systems as well). Let us now see what such a diagram is for the hydrogen atom.

As we have seen in Eq. (27.5), the total energy of the hydrogen atom when the electron is in the nth orbit is

$$-E_n = \frac{Ze^2k}{2r_n} \tag{27.5}$$

But $1/r_n$ can be found from Eq. (27.8). When it is substituted in Eq. (27.5), one finds

$$E_n = -\left(\frac{2\pi^2k^2e^4Z^2m}{h^2}\right)\left(\frac{1}{n^2}\right) \tag{27.10}$$

Upon evaluating the constant (recalling that $Z = 1$ since the charge on the hydrogen nucleus is $+1e$) one finds the energy in electronvolts to be

$$E_n = -\frac{13.6}{n^2} \quad \text{eV} \tag{27.11}$$

Bohr Atom Energies

From this we see that the energy of the atom is zero when $n \to \infty$. This you will recall is the result of the way we defined our zero for potential energy; the PE is zero when the nucleus and electron are infinitely far apart. When $n \to \infty$, $r_n \to \infty$. Further, when $n = 1$ and the electron is in the first orbit, $E_1 = -13.6$ eV. We call this the **ground state** of the atom. Similarly, when $n = 2$, the electron is in the second orbit and $E_2 = -3.4$ eV, and so on. The energy-level diagram used to illustrate these energies for the hydrogen atom is shown in Fig. 27.9*a*. It is simply a series of horizontal lines so drawn as to show the energies which the atom can assume.

Part *b* of the figure shows how the various spectral lines emitted by hydrogen can be represented on such a diagram. On this type of diagram the length of the transition arrow is a direct measure of the energy involved. Hence the Lyman series arrows are longer than those of the Balmer series, telling us at once that the Lyman series wavelengths will be the shorter. It is also easily seen from this diagram that the spectral lines in a series corresponding to transitions from the higher n values will lie very close together since these energy levels have nearly the same energies.

Illustration 27.1 Singly ionized helium is a helium atom which has lost one of its two electrons. Draw the energy-level diagram for this ion.

Reasoning The singly ionized helium atom will be much like a hydrogen

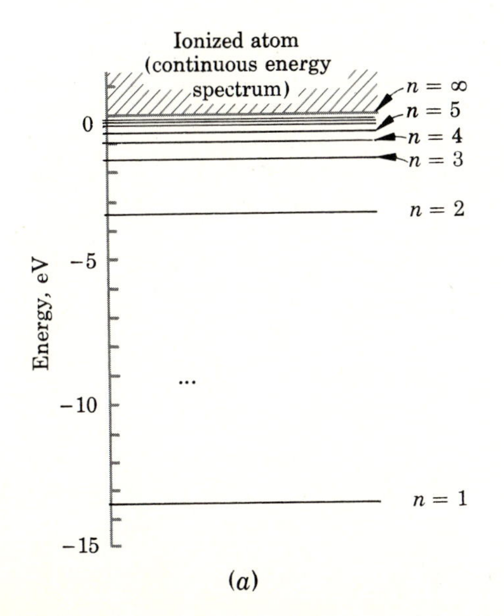

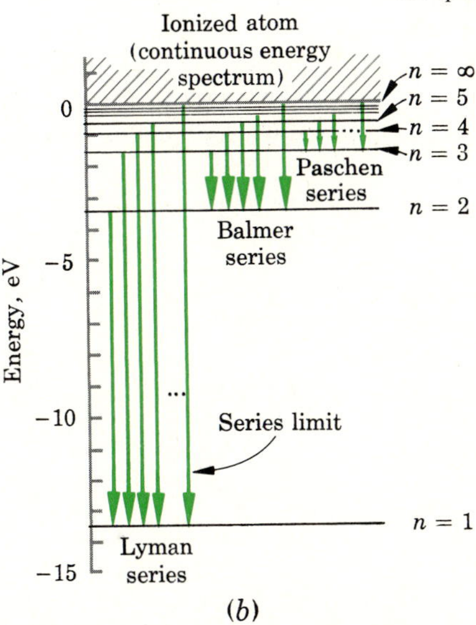

FIGURE 27.9
(*a*) The energy-level diagram for hydrogen. (*b*) Origin of the spectral series.

atom except that the charge on the nucleus is $+2e$, and so $Z = 2$. From Eqs. (27.10) and (27.11) we have, after setting $Z = 2$,

$$E_n = -\frac{54.4}{n^2} \quad \text{eV}$$

This then yields $E_1 = -54.4$ eV, $E_2 = -13.6$ eV, $E_3 = -6.04$ eV, etc. These values are shown in the energy diagram of Fig. 27.10.

Illustration 27.2 Find the wavelength of the first line of the series equivalent to the Lyman series in singly ionized helium.

Reasoning The Lyman series corresponds to the electron falling from the higher levels to the $n = 1$ level. In the present case we are interested in the first line of the series, and so we need the $n = 2$ to $n = 1$ transition. From the previous illustration (or Fig. 27.10) one has $E_1 = -54.4$ eV and $E_2 = -13.6$ eV. Therefore the energy of the emitted photon will be $54.4 - 13.6 = 40.8$ eV. This is equivalent to 65×10^{-19} J. Since the photon energy is $h\nu$ or $h(c/\lambda)$, we have

$$\lambda \approx \frac{hc}{65 \times 10^{-19}\ \text{J}} = 30.0 \text{ nm}$$

We could have done the computation much more easily if we had known that 1 eV corresponds to $\lambda = 1240$ nm. Then, since λ varies inversely as

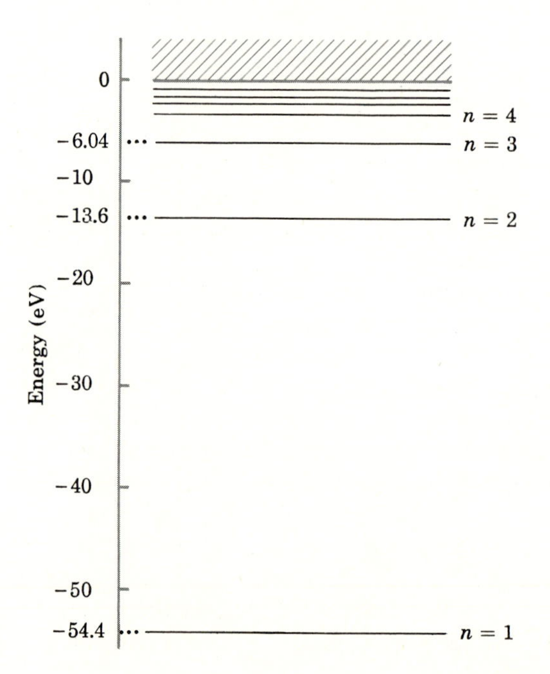

FIGURE 27.10
The energy-level diagram for singly ionized helium.

energy, the wavelength can be found by proportion:

$$\frac{\lambda}{1240 \text{ nm}} = \frac{1 \text{ eV}}{40.8 \text{ eV}}$$

from which

$$\lambda \approx 30.0 \text{ nm}$$

This conversion, that 1 eV corresponds to 1240 nm, often proves convenient in computations such as this.

27.6 ABSORPTION OF LIGHT BY BOHR'S ATOM

Unexcited hydrogen atoms do not give off light, because the electron is in the lowest energy level at orbit 1, that is, the atom is in the ground state. As noted previously, *the atom will emit light only if it has somehow been excited into one of its higher energy levels. Then the electron can fall back toward the nucleus, emitting light in the process.*

There are several ways in which the electrons can be excited to higher energy levels. At very high temperatures the atoms will have so much KE that upon collision with each other the electrons will become excited to a higher level. Very hot gases give off light because the atoms are excited by this means.

Alternatively, if the atoms of a gas are placed in a discharge tube, the electrons can be excited to higher levels by collisions with fast-moving ions or electrons. As illustrated in Fig. 27.11, the high voltage between the two electrodes accelerates the small number of ions and free electrons which have been formed from neutral atoms by stray cosmic or nuclear radiation shooting through the gas. If the voltage is high enough, the accelerated charges will collide with other atoms and ionize them, i.e., tear electrons loose from them. A large number of high-speed charged particles will soon be created, and an electrical discharge occurs. When the electrons recombine with the ions, light is given off. Other levels of excitation occur, as well as ionization. These, too, give off light when the electron falls to lower energy levels. The emission of light by atoms in a gas-discharge tube is the source of light in neon signs and similar devices.

FIGURE 27.11

The high voltage across the tube causes the electrons and ions to accelerate. If the voltage is high enough, these moving charges will ionize other atoms by collision.

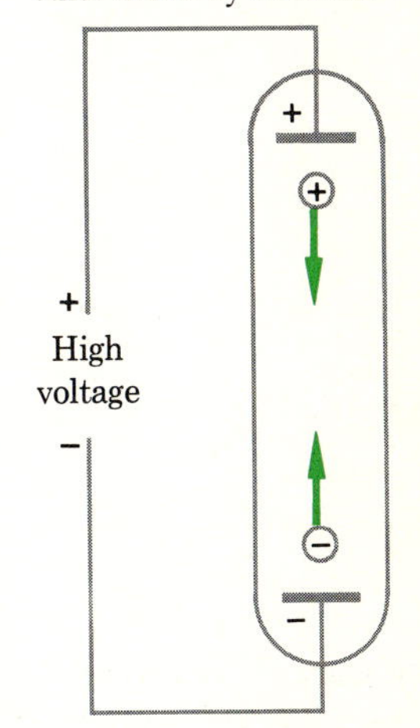

Atoms can also be excited by shining light on them. We can think of a beam of light as consisting of photons, with energy $h\nu$. If one of the photons collides with an atom, it might possibly give its energy to the electron and excite it to a higher orbit. Except in rare instances, namely, in the Compton effect, a photon must lose all its energy or it loses none whatever. Hence, a photon will give energy to the electron in an atom only if the photon energy $h\nu$ is just equal to the difference in energy between the original level and some higher level.

Because of this fact, atoms of hydrogen (as well as other gases) are excited only by special wavelengths of light. Since most of the electrons will be in the lowest energy level or state, the ground state of the atom, the electron will require at least an energy $E_2 - E_1$ to be excited. Exactly this amount of energy is required to lift the electron from level 1 to level 2 (see Fig. 27.9). Photon energies less than this will not be able to raise the electron to level 2, and hence they cannot be absorbed.

If the photon energy is slightly greater than $E_2 - E_1$, it cannot be absorbed, for the electron would have to be raised to a nonexistent energy level, i.e., a level different from those which Bohr found to be allowed. The next higher photon energy capable of exciting the atom is $E_3 - E_1$. This photon would raise the electron to level 3. Similarly photon energies of $E_4 - E_1$, $E_5 - E_1$, and so on, are capable of exciting the electron. These photons (or the equivalent wavelengths of light) will be strongly absorbed by hydrogen atoms. No other photons will be absorbed.

A photon of energy $E_\infty - E_1$ will tear the electron completely loose from the atom. In other words, the atom will be ionized by this wavelength of radiation. We call this difference in energy the **ionization energy** of the atom. Bombarding photons, electrons, or any other particles must have an energy $E_\infty - E_1$ or larger to ionize the atom. From Fig. 27.9 we see that this energy is 13.6 eV for hydrogen. If the atom is given more energy than this, the electron will be thrown loose and will have energy to spare. This energy appears as KE of the freed electron. Such energy is not quantized and is represented by the continuum for $E > 0$ in Fig. 27.9.

There is a very precise relation between the wavelengths of light which an atom will absorb and those which it emits. For example, unexcited hydrogen atoms cannot absorb photons of energy less than $E_2 - E_1$. However, a photon of exactly this energy will be strongly absorbed, since it is just correct for raising the electron to level 2. In fact, this transition gives rise to the longest-wavelength line in the Lyman series, as we have seen. Clearly, the wavelength of the first line of the Lyman series is not only emitted by an excited atom but may also be absorbed by an unexcited atom. Similarly, all the other wavelengths we have shown to be strongly absorbed are also members of the Lyman series. As we would expect from this discussion, excited atoms emit the wavelengths of light that are highly absorbed by the unexcited atoms.

Illustration 27.3 What is the longest wavelength of light capable of ionizing a hydrogen atom.

Reasoning The desired wavelength has enough energy to tear the electron loose from the atom, i.e., to raise it to the $n \to \infty$ energy level. This same energy is emitted when the electron falls from orbit ∞ to orbit 1. Clearly, the wavelength is given by Eq. (27.9) with $p = 1$, $n = \infty$, and the constant $R = 1.097 \times 10^5\ \text{cm}^{-1}$,

$$\frac{1}{\lambda} = (1.097 \times 10^7)\left(\frac{1}{1} - \frac{1}{\infty}\right)\ \text{m}^{-1}$$

or

$$\lambda = 0.912 \times 10^{-7}\ \text{m} = 91.2\ \text{nm}$$

This, of course, is also the series limit for the Lyman series.

This answer could have been obtained more quickly by noting from Fig. 27.9 or Eq. (27.11) that $E_\infty - E_1 = 13.6$ eV. Then using the proportion discussed in Illustration 27.2,

$$\frac{\lambda}{1240 \text{ nm}} = \frac{1 \text{ eV}}{13.6 \text{ eV}}$$

from which $\lambda = 91.2$ nm.

27.7 DE BROGLIE'S INTERPRETATION OF BOHR'S ORBITS

As we have seen, Bohr's theory for the hydrogen atom was very successful in its ability to describe the behavior of the atom. Moreover, since singly ionized helium and doubly ionized lithium (for which $Z = 3$) are also compatible with Bohr's model, it is no surprise to find that they too correspond very well to the predictions of the model. In addition, we shall see that Bohr's model serves to correlate observations concerning more complex atoms. Despite these successes, the Bohr model is seriously deficient in several ways. One of its most serious weaknesses has to do with Bohr's assumption concerning orbits. Let us now see how de Broglie's concept of the wave nature of particles leads to a more satisfactory picture of atoms.

You will recall that a particle of mass m and speed v is characterized by its de Broglie wavelength, $\lambda = h/mv$. It seemed to de Broglie that the puzzling quantization phenomena found by Planck and later by Bohr had an analogy in our common experience. In the vibration of strings, tubes, etc., the vibrations are quantized in the sense that resonance occurs only at certain special frequencies. Invariably these frequencies are those for which the wave involved "fits" properly on the string, in the tube, or whatever. With this fact in mind, *de Broglie thought of Bohr's orbits as being the resonance paths for the electron wave.*

Bohr's Orbits as Electron-Wave Resonances

In order for a wave to resonate on a circular path, the path must be exactly one wavelength long, or two wavelengths long, or three, etc. Only then will the wave crests appear always at the same points on the path so that they will always resonate crest upon crest and trough upon trough. This situation is shown in Fig. 27.12. For that case the path length is four wavelengths long. In general, if the path is n wavelengths long, we have

$$2\pi r_n = n\lambda$$

or since $\lambda = h/mv$,

$$2\pi r_n = n\frac{h}{mv}$$

But this can be written as

$$mvr_n = n\frac{h}{2\pi}$$

and *this is exactly Bohr's assumption,* Eq. (27.7)!

FIGURE 27.12

If the orbit length $2\pi r$ is an integral number of wavelengths, the wave will reinforce itself when it returns to the starting point A. In the case shown, $2\pi r = 4\lambda$.

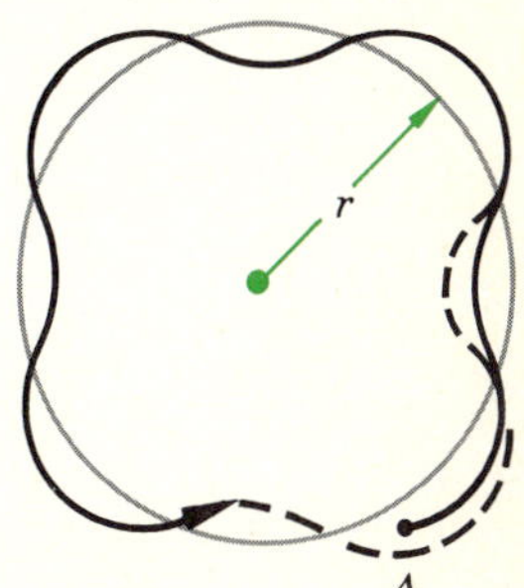

We find, therefore, that *the Bohr orbits can be considered the resonances of the electron's wave circling the nucleus.* As such, the orbits are no longer the result of an ad hoc assumption but are traceable to the more fundamental fact that particles have wave properties. Although the Bohr theory is made more plausible by this explanation of the orbits, we shall see in the next section that Bohr's theory is grossly oversimplified.

27.8 RESONANCES OF WAVES IN ATOMS

Before continuing with our discussion of the role of de Broglie waves in an atom, let us examine the meaning of the de Broglie wave itself. For many years there was a great deal of discussion as to what the wave properties of particles really told us. It is now generally accepted that the *de Broglie waves are related to material particles in basically the same way that electromagnetic waves are related to photons.*

For example, when a beam of light is sent through two slits, an interference pattern is observed upon a screen beyond the slits. In fact, this type of phenomenon is noticed for all types of waves, as we have seen. In past chapters we have been content to compute the positions of the maxima and minima in the interference patterns. However, photons strike all positions where any light is observed; the number of photons striking a given point in the pattern is proportional to the intensity of the light at that point. Hence, *the light-intensity pattern can be thought of as a diagram showing the proportions of the photons in the incoming beam which hit the screen at various positions.*

To compute the light intensity for the complete pattern one must solve a rather complex mathematical problem by use of the **wave equation.** The solution of this equation results in finding a quantity called the **wave function,** which is represented by the symbol ψ (a Greek letter, pronounced "psigh"). It is a quantity which when squared gives the intensity of light one should find at any point in space. But since the light intensity is proportional to the number of photons, the wave function also tells us how many photons will be found at each point in space. In summary, when one solves the wave equation for light waves passing through a double slit or some similar device, one finds the wave function ψ for the light waves. This wave function gives the light intensity everywhere in the interference pattern. In turn, this tells us how many photons exist in each portion of the pattern.

A similar situation applies to de Broglie waves. The wave equation which governs their behavior is the Schrödinger equation, already mentioned in Sec. 26.7. Like the wave equation for light, Schrödinger's equation can be solved to yield a wave function ψ. When squared, this wave function gives the intensity of the de Broglie waves. As before, *the intensity tells us the distribution of particles in space.* Whereas the light intensity and its wave function describe the motion of photons, the Schrödinger equation wave function describes the motion of electrons (or whatever other particle is considered).

To see what this means in an actual situation, let us consider the hydrogen atom. At best, Bohr's theory can only be telling us where the maxima in the interference pattern will be. As in the case of the double-slit or any other interference experiment, this is only a small part of the story since photons (for light) or electrons (for de Broglie waves) also exist elsewhere. *When Schrödinger's equation is solved for the hydrogen atom* (not a very simple chore), *the intensity pattern is found to depend upon just how the electron wave is resonating within the atom. To describe these resonances one makes use of sets of integers called* **quantum numbers.**

Principal Quantum Number

The so-called principal quantum number n is identical to the n used in Bohr's theory. It tells us the energy of the atom. In both the Bohr theory and the wave theory of the atom, the energy is

$$E_n = -\frac{13.6Z^2}{n^2} \qquad \text{eV}$$

When $n = 1$, Bohr assumed the electron to be in its first orbit, a circle. However, the wave solution gives us the whole intensity pattern and tells us that the electron is likely to be found in many different positions but is most likely to be found in those regions where the wave function ψ is large. This intensity pattern corresponding to $n = 1$ for the hydrogen atom is shown in Fig. 27.13*a*. We show there the chance (or probability) that the electron will be found at given radii from the nucleus. Bohr's theory, of course, says the electron will be found only at the radius of the first Bohr orbit, 0.53 Å.

As we expected, *the wave theory shows that the Bohr's orbit radius is the radius at which the electron is most likely to be found.* It is, so to speak, the location of the maximum in the intensity pattern. *However, the electron can be found at many other places in the atom as well.* (Recall that in a light-interference pattern, the photons can be found elsewhere than at the maxima.) So we see that Bohr's picture of an electron circling the nucleus in orbit is not realistic. In fact, the intensity pattern found from the wave theory is not even a ring about the nucleus. Instead, it constitutes a sphere centered on the nucleus. An attempt is made to show this in Fig. 27.14*a*. To get the

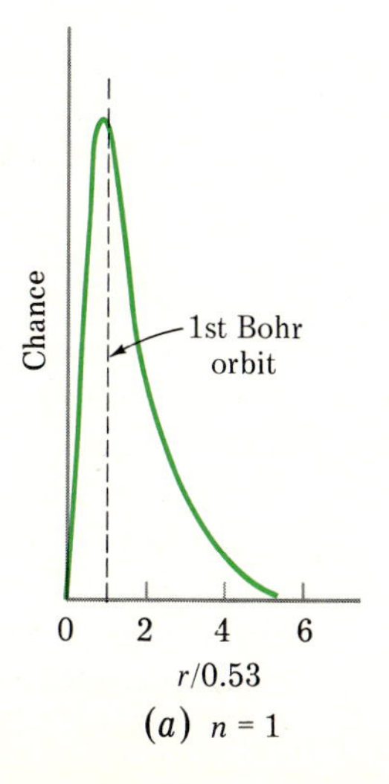

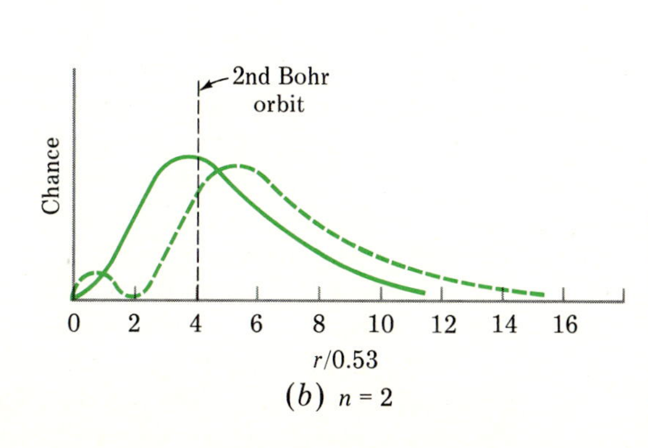

FIGURE 27.13

(*a*) The wave theory predicts the probabilities shown that the electron will be found at various radii (r is measured in angstroms). (*b*) Two different resonance forms for $n = 2$.

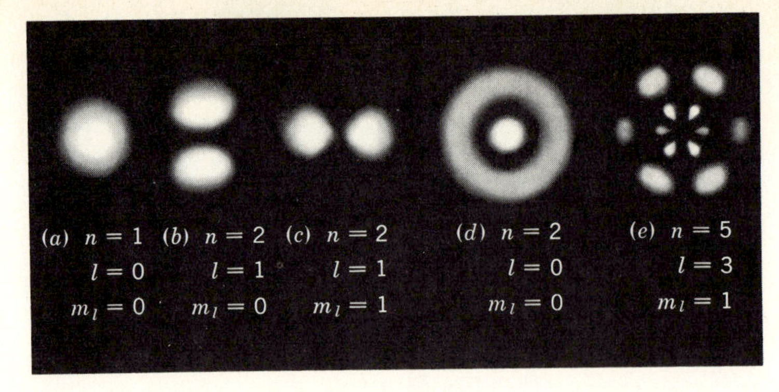

FIGURE 27.14

To obtain the electron distribution in space, the figures shown must be rotated around a central vertical axis. All the distributions are constant on any circle with the axis as center.

three-dimensional picture of the physical situation, the pattern should be rotated about a vertical axis through the center; the intensity of the pattern in space gives the chance of finding the electron there.

Let us now go on to consider the resonance pattern within the atom when $n = 2$. You will recall that Bohr's theory predicts the electron to be in an orbit of radius 4×0.53 Å in this case. The wave-theory prediction is shown in Fig. 27.13*b* by the solid curve. Notice that, as before, the electron is most likely to be found at the Bohr orbit but can also be found elsewhere. This is not unexpected. However, as in the $n = 1$ case, the actual intensity pattern in space is not shaped like a circular orbit. Instead, it consists of a dumbbell shape as shown in Fig. 27.14*b*. Recall that the space diagram is obtained by rotating the diagram about a vertical axis. As you can see, the electron is restricted to two lobes on the axis.

However, this is not the only way in which the electron's de Broglie wave can resonate for the energy with $n = 2$. Two other resonance forms are also found; all three belong to the same group since their principal quantum number n and hence their energies are the same. The intensity patterns for these other two resonances are shown in Fig. 27.14*c* and *d*. Together they give rise to the electron distribution shown by the broken curve in Fig. 27.13*b*.

When we go to $n = 3$, again there turn out to be several resonances possible with this energy. As we saw, for $n = 1$ there was one resonance form. For $n = 2$, there are three resonance forms. If $n = 3$, one finds six different resonance patterns. Some of them become quite complicated; a typical one is shown in Fig. 27.14*e*.

We see from all of this that *the Bohr theory is indeed a gross oversimplification for the electron behavior in the hydrogen atom. In particular, Bohr's concept of fixed orbits is untenable. However, the energy levels of the atom are predicted correctly by the Bohr theory* and the principal quantum number n introduced by Bohr has great importance. Although we should always keep the limitations of the Bohr model in mind, it offers a framework for a systematic description of atoms and we shall make frequent reference to it.

Illustration 27.4 Estimate the position uncertainty for the electron in the hydrogen atom when it is in the $n = 1$ energy state.

Reasoning To do this, let us use the uncertainty principle in the form

$$\Delta p\,\Delta x \geq \frac{h}{2\pi}$$

We know the particle energy in the $n = 1$ state is -13.6 eV, or about -2.2×10^{-18} J. Moreover,

$$\text{KE} = \tfrac{1}{2}mv^2 = \frac{p^2}{2m}$$

But we found in the development of Eq. (27.5) that the KE of the Bohr electron was numerically equal to the total energy of the atom. Therefore $\text{KE} = 2.2 \times 10^{-18}$ J. Equating this to $p^2/2m$ and solving for p, we find

$$p = \sqrt{(2)(9 \times 10^{-31})(2.2 \times 10^{-18})} = 2.0 \times 10^{-24}\ \text{kg} \cdot \text{m}$$

If we assign the maximum value to the momentum uncertainty Δp, we shall obtain the least possible value for Δx. Let us therefore equate Δp to p itself. Then from the uncertainty relation we find

$$\Delta x \geq 0.53 \times 10^{-10}\ \text{m} = 0.53\ \text{Å}$$

Since the radius of the atom is about 0.5 Å, we see that there is no possibility at all that we can pinpoint the position of the electron in the atom. The wave theory of the atom reflects this fact by showing us that the electron can be found throughout the atom.

27.9 QUANTUM NUMBERS AND THE PAULI EXCLUSION PRINCIPLE

As we have seen in the previous sections, the hydrogen atom and its electron can exist in certain discrete energy levels characterized by an integer n. In particular, we have seen that the energy levels are given by the relation

$$E_n = \frac{-13.6Z^2}{n^2} \qquad \text{eV}$$

The integer n ranges from 1 to ∞ as the atom assumes its various allowed energies. Although we arrived at this result by use of Bohr's model, the wave picture, based on the solution of the Schrödinger equation, also leads to this same result. Hence it is seen that n is a fundamental parameter needed to describe the state of a hydrogen atom. As mentioned earlier, it is called the principal quantum number. Notice that it characterizes the energy level in which the electron is to be found. Bohr pictured each value of n to be associated with a particular orbit for the electron, but this proves untenable, as pointed out in the last section. Nevertheless, it is common usage to say that

Atom Shells

each value of n corresponds to a particular **shell** (rather than orbit) about the nucleus. For example, when the atom is in the $n = 3$ energy level, it is customary to say that the electron is in the $n = 3$ shell.

We also saw in the last section that more than one wave resonance form is possible for the same energy, i.e., for the same value of the principal quantum number n. The wave theory shows that two other quantum numbers must be specified in order to designate a particular wave resonance within the atom. One of these, the **orbital quantum number,** is related to the angular momentum of the Bohr electron in its resonance orbit. It is represented by the letter l and can assume integer values from 0 to $n - 1$. For example, when $n = 1$, the possible values for l are limited to a single value, namely $l = 0$. When $n = 2$, it is apparent that l can take on the values 0 and 1 since $n - 1 = 1$ in this case.

Orbital Quantum Number

The third quantum number, the so-called **magnetic quantum number,** can assume the values $0, \pm 1, \pm 2, \ldots, \pm l$. It is represented by m_l. When $n = 4$, for example, the largest possible value for l is 3 and hence m_l can take on the values $-3, -2, -1, 0, +1, 2, 3$. This means that when the atom is in the $n = 4$ energy level, there are seven different resonance forms possible which have $l = 3$. In addition, there are five resonance forms with $l = 2$, three resonance forms with $l = 1$, and one resonance form with $l = 0$. Hence the atom can exist in $7 + 5 + 3 + 1 = 16$ different resonances each having the same energy, namely, the energy of the $n = 4$ level.

Magnetic Quantum Number

Finally, a quantum condition exists for the electron itself. As we have mentioned before, the electron acts like a small magnet because of its spin about an axis through its center. This magnet can take up only two orientations relative to an external magnetic field in which the atom may find itself. It can either align parallel or antiparallel to the field line direction. We characterize this by assigning a **spin quantum number** designated $m_s = \pm\frac{1}{2}$; the two signs represent the aligned and antialigned positions.

Spin Quantum Number

We see therefore that four quantum numbers are needed to describe the state of an electron in an atom:

Principal:	$n = 1, 2, \ldots$
Orbital:	$l = 0, 1, \ldots, n - 1$
Magnetic:	$m_l = 0, \pm 1, \ldots, \pm l$
Spin:	$m_s = \pm\frac{1}{2}$

For a given set of n, l, and m_l values, there is a very definite wave resonance pattern for the electron wave within the atom. To specify the electron completely, however, we must know whether the electron spin is aligned with ($m_s = \frac{1}{2}$) or opposite to ($m_s = -\frac{1}{2}$) the magnetic field. We call each combination of the quantum numbers n, l, m_l, and m_s an electronic **state** of the atom. We shall now see that an extremely important law of nature applies to the behavior of electrons in the available states.

The importance of designating these states as we have done was first appreciated fully by Pauli in 1925. His discovery can be stated very simply and is known as the **Pauli exclusion principle:**

No two electrons in an atom can have the same four quantum numbers; i.e., no two electrons can exist in the same state.

Pauli Exclusion Principle

This principle is basic to an understanding of the electronic structure of atoms, as we shall see in the next section.

27.10 THE PERIODIC TABLE

Until now we have been primarily concerned with an atom which has only one electron. This might be hydrogen, singly ionized helium, doubly ionized lithium, and so on. We are now in a position to discuss how the additional electrons are arranged in the multielectron atoms found in nature and listed in the periodic table. To do this, we shall once again make use of the concept of electron shells about the nucleus; each value of n has associated with it a shell. Moreover, we shall assume that the same resonances found for the single-electron atom can be carried over qualitatively to more complex atoms. That is to say, we shall make use of electronic states specified by the n, l, m_l, and m_s quantum numbers described in the previous section.

The question we must now answer is: How do the electrons arrange themselves in the various atomic states when more than one electron exists in an atom? For example, there are six electrons in each carbon atom. In which energy levels and electronic states are they to be found? This question can be answered by making use of the following three rules, which we have already discussed:

A neutral atom has a number of electrons equal to its atomic number Z.

In an unexcited atom, the electrons are in the lowest possible energy states.

No two electrons in an atom can have the same four quantum numbers (the exclusion principle).

Let us now use these rules to determine the electronic structure of the unexcited atoms in the periodic table.

Hydrogen **($Z = 1$)** Its single electron will be in the $n = 1$ level. This is the lowest possible energy level, and no violation of the exclusion principle occurs.

Helium **($Z = 2$)** Its two electrons can both exist in the $n = 1$ level since they can have the following, nonidentical quantum numbers:

ELECTRON	n	l	m_l	m_s
1	1	0	0	$\frac{1}{2}$
2	1	0	0	$-\frac{1}{2}$

However, since these are the only combinations of quantum numbers possible for $n = 1$, a third electron cannot enter this shell. The shell is filled.

Lithium ($Z = 3$) This atom has three electrons, and so the third must go into the $n = 2$ shell. We have

ELECTRON	n	l	m_l	m_s
1	1	0	0	$\frac{1}{2}$
2	1	0	0	$-\frac{1}{2}$
3	2	0	0	$\frac{1}{2}$

Since this third electron is in the second energy level, it is much more easily removed from the atom than the first two are. Hence lithium loses one electron in chemical reactions and is univalent.

Obviously there are quite a few possible combinations for the quantum numbers when $n = 2$. If you count them, you will find there are eight as follows:

n	l	m_l	m_s
2	0	0	$\pm\frac{1}{2}$
2	1	0	$\pm\frac{1}{2}$
2	1	+1	$\pm\frac{1}{2}$
2	1	−1	$\pm\frac{1}{2}$

Therefore eight electrons can exist in the $n = 2$ shell. This means that the shell will not become closed until element $Z = 10$, neon, is reached. You probably know that this is an unreactive gas, unreactive since it has a closed shell. The next element, $Z = 11$, is sodium. This is univalent since its extra electron is alone out in the $n = 3$ shell and is rather easily removed.

As one proceeds to the very high-Z elements in the table, the concept of shells becomes less useful. The trouble arises primarily because the separation between energy levels is relatively small at high n values. In these cases the repulsions between the various electrons in the atom contribute energies large enough to sometimes cancel out the influence of energy differences between shells. Despite this complication, the shell approach still proves useful for qualitative considerations.

27.11 PRODUCTION OF X-RAYS

We know that atoms emit light or other EM radiation when an electron falls from a high to a lower energy level. For hydrogen and one-electron ions, the spectra are quite simple, since their energy levels are given by the relation

$$E_n = -\frac{13.6Z^2}{n^2} \qquad \text{eV}$$

In other atoms, the innermost electrons are largely influenced by the nearby nucleus, and so this relation applies rather well for them. One should notice in particular that the charge of the nucleus is very influential in determining these inner levels. For example, in the case of $Z = 100$, the energies involved are 10,000 times larger than for hydrogen. Consequently, *photons emitted when an electron falls from the $n = 2$ shell to the $n = 1$ shell in a high-Z element will have energies in the range of 100,000 eV. This corresponds to a wavelength of 1240 nm/100,000 or about 0.012 nm, a wavelength in the x-ray region.*

Although the atoms of heavy elements are capable of emitting x-rays, they do not do so unless properly excited. In order for an x-ray photon to be emitted, an electron must fall from an outer shell of the atom to a vacancy in an inner shell. Since unexcited atoms have no vacancies in these shells, x-rays are not emitted by them. The atom must first be excited in such a way that an electron in the $n = 1$ or $n = 2$ shell is thrown out of the shell, thereby providing a vacancy into which an outer electron can fall. This is usually accomplished with an x-ray tube like the one shown in Fig. 27.15*a*.

Characteristic X-Rays

As shown there, electrons emitted from the hot filament are accelerated through potential differences of the order of 10^5 V. When these high-energy electrons strike the high-Z atoms in the target, electrons are knocked out of the inner shells of the atoms. As other electrons fall into the vacancies, x-ray photons are emitted. *The x-rays so generated have wavelengths characteristic of the energy differences between the various shells within the atom.* That is, the emitted photons carry an energy equal to the difference in energies between the two shells which act as starting point and end point for the electron which falls into the vacancy. *X-rays emitted by this process are referred to as* **characteristic x-rays.**

DEFINITION

DEFINITION

Bremsstrahlung

Another type of x-ray emitted from a target when it is bombarded by electrons is referred to as **bremsstrahlung,** *from the German "braking radiation." As the name implies, these x-rays are emitted by the bombarding electrons as they are suddenly slowed upon impact with the target.* We know that any accelerating charge emits electromagnetic radiation (a charge oscillating on an antenna, for example). Hence these impacting electrons also emit radiation as they are strongly decelerated by the target. Since the rate of deceleration is so large, the emitted radiation is correspondingly of short wavelength and so the bremsstrahlung is in the x-ray region. However, unlike the characteristic x-rays, the bremsstrahlung has a continuous range of wavelengths. This reflects the fact that the deceleration process can occur in a nearly infinite number of different ways, so that the energy released varies widely from one impact to another.

In Fig. 27.16 is shown a graph of the radiation emitted from a molybdenum target when bombarded by 35,000-eV electrons. The two sharp peaks are the characteristic x-rays emitted as electrons fall to the $n = 1$ shell of this atom from its $n = 2$ and $n = 3$ shells. The shorter wavelength, of course, corresponds to the higher-energy transition, i.e., the $n = 3$ to $n = 1$ transition. Bremsstrahlung is the cause of the lower-intensity radiation spread over all wavelengths longer than λ_m. Since the energy of the electrons in the impacting beam was 35,000 eV, the emitted photons cannot have energies larger than this value. Using our conversion based upon 1240 nm being equivalent to 1 eV, we find that 35,000 eV corresponds to (1240 nm) /35,000 $\approx$ 0.035 nm. As we see from Fig. 27.16, the highest-energy bremsstrahlung does indeed have this wavelength.

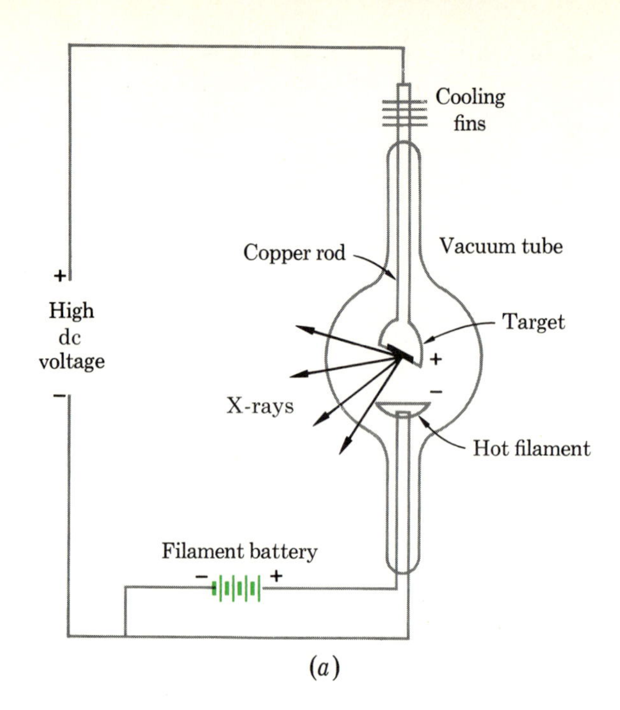

(*a*)

(*b*)

FIGURE 27.15

(*a*) Electrons emitted by the hot filament bombard the target. The target then emits x-rays. (*b*) An x-ray system for routine use. This system was used to obtain the photograph shown on page 80. [(*b*) *Hewlett-Packard Co.*]

Illustration 27.5 From the data in Fig. 27.16, find the energy difference between the $n = 1$ and $n = 2$ levels in molybdenum.

Reasoning As we saw in our discussion of Fig. 27.16, the long-wavelength peak in that figure 0.70 Å results from the $n = 2$ to $n = 1$ transition. Therefore the photon of wavelength 0.07 nm carries away the energy lost by an electron as it falls from the $n = 2$ to the $n = 1$ shell. Since 1240 nm corresponds to 1 eV, 0.070 nm corresponds to an energy of 1240/0.070, or about 18,000 eV. Therefore the energy difference between these two shells in molybdenum atoms must be about 18,000 eV.

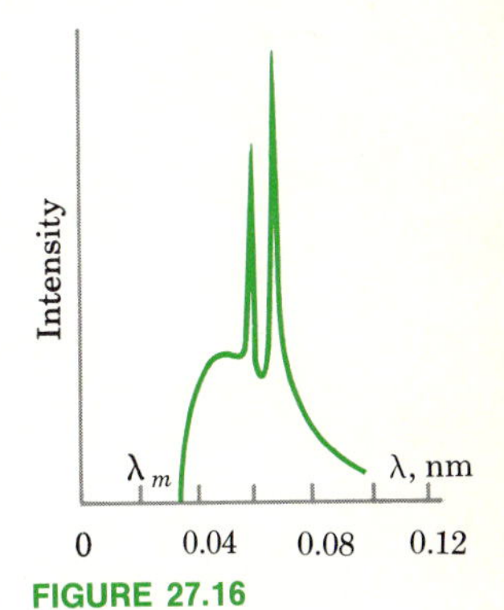

FIGURE 27.16

X-rays emitted from a molybdenum target when bombarded by 35,000-eV electrons.

27.12 LINE, BAND, AND CONTINUOUS SPECTRA

As we saw in the previous section, the energy difference between the $n = 1$ and $n = 2$ shells is tens of thousands of electronvolts for atoms which contain many electrons. Since the energy differences between these inner shells is so large, the small effect of electron-electron interaction is often negligible. However, as we proceed outward in the atom to the higher-n shells, the energy differences between shells become comparable to the interaction energies between the electrons in the same atom. As a result, each Bohr energy level is split into a multitude of discrete levels, which vastly complicates the energy-level diagram for the atom. This effect causes the outer-electron energy levels to become almost hopelessly complex for atoms such as iron and gold. Since the energy-level diagrams of such atoms lack the basic simplicity of the hydrogen diagram, the spectrum of emitted wavelengths from these atoms is also far less simple.

For most of these high-Z atoms, a tremendous number of spectral lines is observed in the radiation they emit. No obvious regularities such as the Balmer series are discernible in the spectra. An example of this complexity is shown in Fig. 27.17, where we give a very small portion of the spectrum emitted by iron atoms vaporized in a hot electric arc. This is typical of the spectra of high-atomic-number atoms. In spite of this complexity, the line spectra of the elements are of great practical utility. Since each element gives off its own distinctive spectral lines, its presence is easily detected by spectroscopic means. For example, a material can be analyzed for its atomic composition by vaporizing it in an arc and measuring the spectral lines given off by the excited atoms in its vapor. This technique, emission spectroscopy,

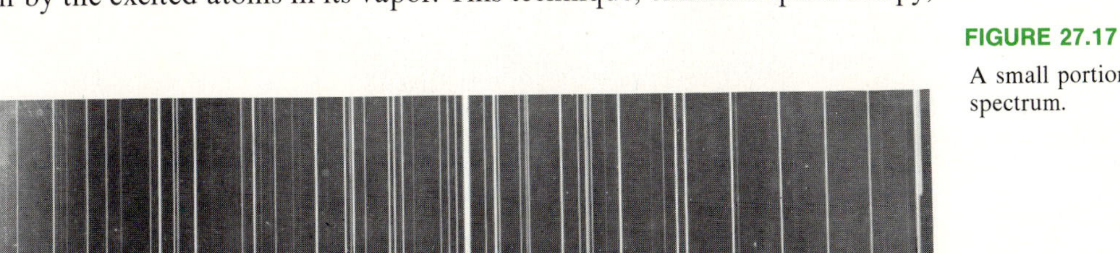

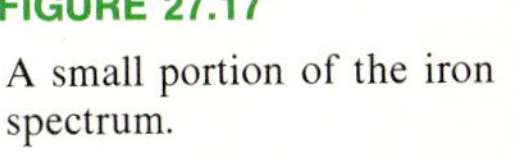

FIGURE 27.17

A small portion of the iron spectrum.

Line Spectra

is used routinely to determine the quantitative as well as qualitative composition of inorganic materials. Notice that in this technique the atoms are vaporized, and their spectrum consists of discrete spectral lines, a so-called **line spectrum.** See the color plate insert for examples of the line spectra.

Before proceeding to the spectrum emitted (and absorbed) by atoms in molecules and solids, let us briefly summarize the requirements for a line spectrum. It originates in the case of widely separated atoms, a situation in which it is possible to speak of distinct electronic shells about the nucleus. When an electron falls from one shell to another, a very definite amount of energy is lost. Hence, a distinct wavelength of light is emitted, the same for all like atoms, and the light appears as a line in the spectrometer. *The spectrum of isolated atoms is therefore a line spectrum.* Once we begin to pack atoms together in a liquid or solid, however, this picture is no longer valid.

Consider the two atoms and their shells illustrated in Fig. 27.18. The electrons in the third shell are seriously affected by the neighbor atom, and we can no longer say that the electron energy in this shell is the same as it is in a single, isolated atom. For simplicity, however, we shall assume that the inner shells are not too seriously overlapped by the neighbor atom and that their energies are not changed much.

Transitions to lower shells give rise to short-wavelength emission, usually ultraviolet or x-ray. Visible and infrared radiation is the result of transitions in the outer shells. It follows, therefore, that the emitted visible light is seriously dependent upon the closeness of the atoms, while the ultraviolet and x-ray emission is not much affected by this factor. Exact calculations confirm this reasoning. As the separated atoms are brought closer, the possible energies of the electrons in outer shells become smeared out. Hence, *electron transitions between these outer shells in a solid or liquid no longer give rise to the emission of sharp wavelength lines.*

Continuous Spectrum

When a solid or liquid is heated enough to give off light, the light is emitted by the outer atomic electrons as they fall from one outer shell to another. Since these shells and their energies are diffuse, a whole range of wavelengths is emitted. The spectrum no longer contains definite, visible spectral lines but consists, instead, of a broad, continuous blur of color, with no individual lines discernible. *This type of spectrum is called a* **continuous**

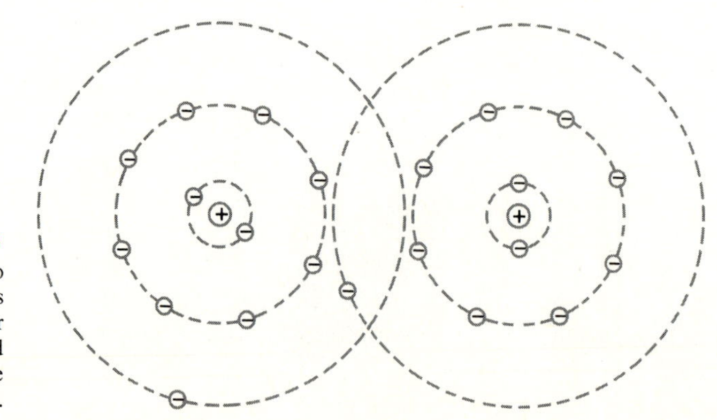

FIGURE 27.18

In a solid, the atoms tend to overlap. Hence the energies characteristic of the outer orbits are poorly defined, and the spectral lines become diffuse.

spectrum. *It is emitted by incandescent objects such as molten metals or the hot filament of a light bulb.* An example of this type of spectrum can be seen on the color plate insert. Since no individual spectral lines can be seen in a continuous spectrum, it is not useful for analysis of the molecular composition of the solid.

Band Spectra

A third type of spectrum, a **band spectrum,** *is emitted by molecules.* This case is intermediate between those of a solid and free atoms. To a first approximation, the atomic emission spectrum is not much changed by inclusion of the atom in a molecule. Only the energy levels characteristic of the valence electrons are modified seriously. However, inclusion of the atoms within a molecule gives rise to a whole new group of energy levels. These are levels characteristic of vibrations involving bonds holding the atoms together in the molecule.

You will recall that Planck found that any system which vibrates with natural frequency ν_0 possesses a series of energy levels spaced an energy $h\nu_0$ apart. To a good approximation, two adjacent atoms within a molecule act like two masses held together by a spring. They have a natural frequency of vibration ν_0 relative to each other and therefore give rise to a series of energy levels $h\nu_0$ apart. In fact, each different chemical bond has its own characteristic vibration frequency and gives rise to a distinctive series of energy levels. Typical examples are frequencies of 3.24×10^{12} Hz for the C—N bond, 3.4×10^{12} Hz for the C—C bond, and 6.5×10^{12} Hz for the C=C bond.

As a vibrating molecule falls from one vibrational energy level to another, a photon with energy $h\nu_0$ is emitted. This energy is 0.014 eV for a C—C bond and gives rise to a photon with $\lambda = 88{,}000$ nm, a wavelength in the infrared. The emitted photons are characteristic of the molecular bond and can be used to identify bonds within molecules. This technique, called **infrared spectroscopy,** is used routinely for the analysis and identification of organic compounds.

27.13 COHERENCY

As we have seen, light and other EM radiations are emitted by atoms and molecules as they fall from an excited state to a lower-energy state. In each transition, a photon is emitted. The frequency of the wave in the energy pulse is given by energy $= h\nu$, but the wave in the pulse extends only a short distance in space. We have tried to represent the situation schematically in Fig. 27.19*a*.

As we show there, a monochromatic light beam consists of a stream of light quanta. Each one is of a finite length and contains a limited number of wavelengths in its length (more than shown, however). These pulses all have nearly the same wavelengths since they result from a single atomic transition. However, each is emitted by a different atom. As a result, there is no definite relation either in time or position between the pulses. They were emitted at random by the atoms and so their waves have random phase relations with respect to each other. *We call a beam such as this, where the photons have random phases (and/or) wavelength, an* **incoherent beam** *or* **pulse.**

DEFINITION

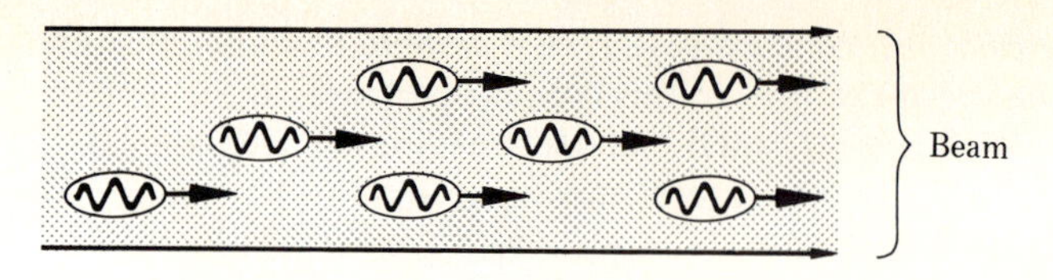

(a) Incoherent photons.

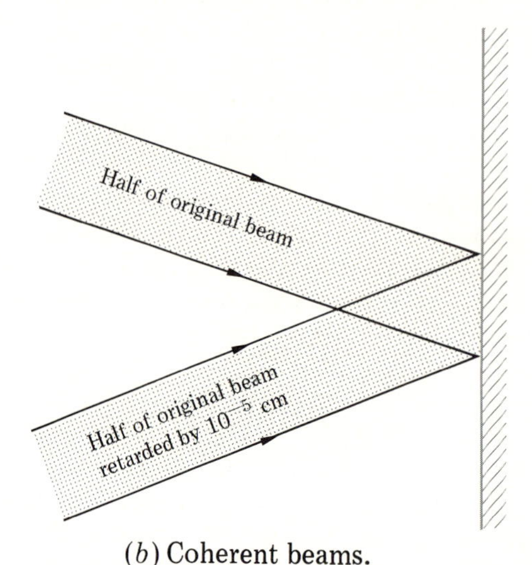

(b) Coherent beams.

FIGURE 27.19

(a) Schematic representation of an ordinary light beam (incoherent photons). (b) By use of a double slit or reflection from a thin film, it can be split into two coherent beams.

Incoherent Beam Pulses

Consider what happens as such a beam strikes a screen or the retina of one's eye. Two or more photons may strike the same spot at once. If their waves are in phase, they will add their amplitudes and reenforce. But, if they are 180° out of phase, they will cancel each other. Since the pulses are at random phase relative to each other, both reenforcement and cancellation (and everything in between) will occur as time goes on and the photons strike the screen. If we were able to record the light intensity and show the details of its variation to a time as short as 10^{-10} s, we would record a highly variable light intensity on the screen. This would be a record of the cancellations and reenforcements of the photons as they strike the screen.

But most light measuring devices, including the eye, record only an average intensity over a period of time much longer than 10^{-10} s. As a result an incoherent beam of photons appears bright but not nearly as bright as it could be if all the photons reenforced each other. And even though cancellation within the beam is taking place, it occurs for such a short time that darkness is not perceived. Even so, it is possible to produce destructive interference with such a beam, as we shall now see.

Suppose the beam of Fig. 27.19a is split into two parts by means of a double slit or reflection from the two surfaces of a thin film. Let us further suppose that each photon consists of a wave pulse which is 30 cm long,* i.e.,

* For example, suppose it takes an atom 10^{-9} s to fall from one state to another. During this time, a pulse is being emitted. The leading end of the pulse will travel a distance $= ct = 0.30$ m in this time, and so the pulse will be 30 cm long.

the pulses in the figure are 30 cm long. If the beams are now rejoined (as shown in Fig. 27.19*b*) after one has traveled only a short distance farther than the other, the two halves of each photon will be rejoined, slightly out of phase. In effect, each photon has been sliced in half lengthwise. The amplitude is smaller in each but otherwise the photons are unchanged. One half has been held back relative to the other; then the two halves are rejoined. As long as the path-length difference is small in comparison with the length of a wave pulse, the half beams will show obvious interference effects when they are rejoined. If the path difference is $n\lambda$ (where n is an integer), constructive interference and brightness will exist at the screen. If the path difference is $n\lambda + \frac{1}{2}\lambda$, destructive interference will occur and darkness will be observed on the screen. This is, of course, exactly what happens in the interference experiments described in Chap. 25.

The two beams shown in Fig. 27.19*b* give rise to visible interference effects at the screen. This is a result of the fact that the wave pulses hitting the screen from the two beams are identical and have a fixed phase relation to each other. We say that the two beams are coherent. **Coherent beams** *consist of wave pulses corresponding identically in the two beams and which maintain a fixed phase relation to each other.*

DEFINITION

Coherent Beams

It is interesting to note that the beams lose their coherency if one beam is retarded too far. Each wave pulse is limited in length. If the path-length difference for the two beams is too large, the pulses will no longer be joined on top of each other when the beams are rejoined. Instead, unrelated photons will be brought together, and only random phase differences will exist. Long-term cancellation and reenforcement will no longer be observed.

As we see, *the individual wave pulses in an ordinary beam of light are not coherent; the wave in one pulse has phase relations that differ widely from those in the waves in the other pulses.* The light we observe from such a beam is the average result of random, very fast interference effects. But if the beam is split into two parts, the two parts can be given a definite phase relation to each other. If the separated beams are rejoined in such a way that the two beams are coherent, pronounced interference effects are visible. Although reenforcement can occur between the two parts of the beam, the beam's intensity is still limited by the fact that its individual wave pulses are not coherent. In the next section we shall learn about a light source which gives out coherent wave pulses.

27.14 THE LASER

Let us now turn our attention to a remarkable type of device which makes use of the fact that, under very special circumstances, atoms can be made to emit light waves which are all in phase with each other. In almost all light sources the atoms act independently; the emission of a photon by one atom is not coordinated with the emission by other atoms. As a result, the light beam consists of a complex mixture of electromagnetic waves from the various atoms. Of course these waves are not all in phase with each other, and so they sometimes cancel and sometimes add. We saw in the last section that this causes the light beam to be much less intense than it would be if all the atoms emitted their waves in phase. A very intense beam would result if all the atoms could be persuaded to emit their waves together in phase. One light source comes close to achieving this; it is called a **laser.**

There are many types of lasers available, but they all operate on the principle from which they get their name, *l*ight *a*mplification by *s*timulated *e*mission of *r*adiation. We shall describe a helium-neon gas laser. This gives a continuous light beam. The basic outline of the helium-neon gas laser is shown in Fig. 27.20 together with an illustration of the narrow pencillike quality of the beam it gives off.

The heart of the laser is a glass tube containing helium and neon gas at relatively low pressure. At the ends of the tube are extremely flat glass plates accurately parallel to each other. Each plate is coated to act like a mirror; but one end plate is coated lightly enough for about 1 percent of the light striking it from inside to leak through the mirror and leave the tube.

Not just any two gases will work in a laser. The two chosen in this case have a very special relation to each other. Each has its own function in the tube. One type of atom, helium, acts as an energy source for the other, neon in this case. In turn, the helium atoms must be given energy. This is done by means of a very-high-frequency electrical discharge in the tube. Basically, the

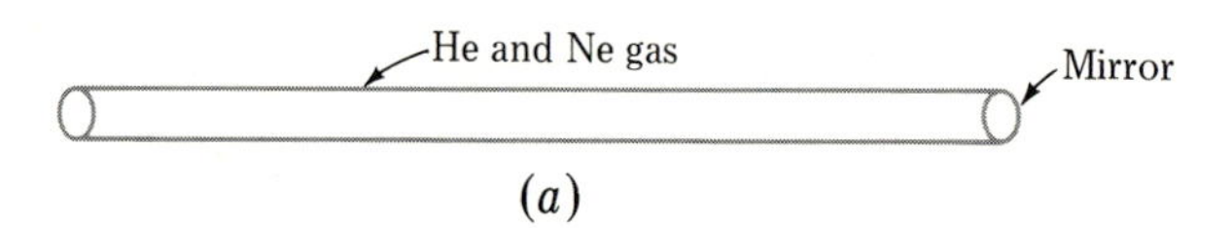

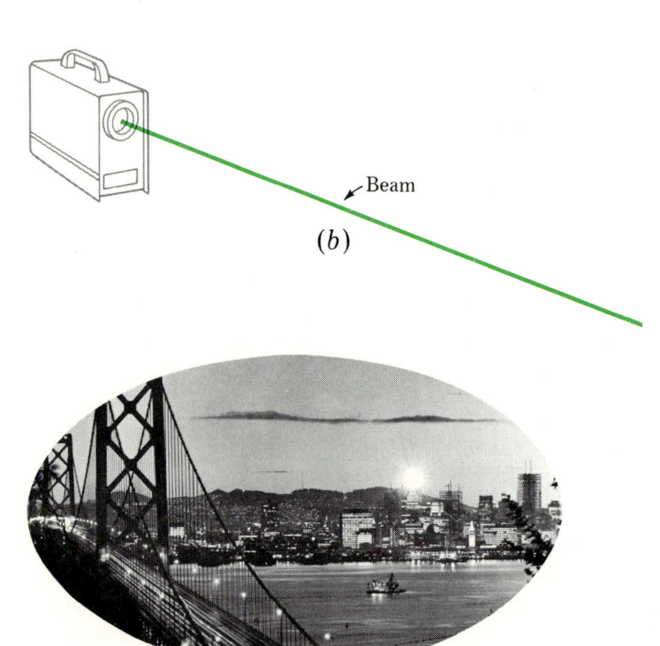

FIGURE 27.20

Even though the Electro Optics Associates laser (*b*) has an output of only 0.0005 W, its narrow, pencillike beam shines brightly in the distance (*c*). For comparison, notice the less bright light coming from the high-intensity lamps on the San Francisco-Oakland Bay Bridge. (*Electro Optics Associates, Palo Alto, Calif.*)

helium in the tube is acting like the gas in any gas-discharge tube, such as a mercury-vapor lamp. The high voltage on the tube produces a gas discharge, and many energetic ions and electrons are made to shoot around inside the tube. The excited helium atoms then give off the usual helium spectrum as the electrons fall back to the $n = 1$ level. However, the light from this effect is only a very small fraction of the total light emitted from the tube.

Of course the electrons in the helium atoms are thrown up to all the possible energy levels and states of the atom by means of collisions inside the tube. It turns out that one state which has an energy 20.6 eV higher than the $n = 1$ state is what is called a **metastable state.** In such a state, the electron resists falling to the lower states, and so the atom exists in this state for an abnormally long time. However, the energy-level structure of a neon atom is just right so that, upon collision between the neon and helium atoms, the excited helium atom gives up its energy to the neon atom. Actually, the excited neon atom has an energy slightly greater than that of the helium atom, and this excess energy is furnished by the KE it acquires upon collision.

What has happened is this. The helium atoms are excited in the gas discharge. Some find themselves in a metastable state and wander around until they collide with a neon atom. During the collision, the neon atom takes the energy from the helium atom and is then in an excited state. This state turns out to be a rather stable state in the neon atom, and so the electron in it does not fall to a lower state immediately. As time goes on, the helium atoms excite a large number of neon atoms and they are then simply waiting to radiate their energies. When they do, they fall to a level 1.96 eV lower and, in the process, emit electromagnetic radiation with wavelength 6328 Å, in the red. This in itself is no advantage since we would then simply recover the energy the helium atoms alone would have radiated. The advantage occurs because we can now control the neon atoms and make them emit their waves in phase.

In spite of the fact that the neon atoms are reluctant to fall to the lower energy state, some eventually do so. As a result, a pulse of EM radiation with $\lambda = 6328$ Å is released in the tube. On its way through the tube, it passes other excited neon atoms and subjects them to an oscillating EM field of frequency identical to the frequency of the light which the excited atom should emit. Such an oscillating field is very effective in causing the excited neon atom to fall to its lower state. It does this in such a way that the photon it sends out has its wave exactly in phase with the radiation which *stimulated* it to emit the photon.

As these waves are reflected back and forth between the ends of the tube, they are vastly augmented by identical waves which they stimulate other neon atoms to emit. Consequently *there exists in the tube a very strong coherent set of wave pulses* with wavelength 6328 Å. A small portion of this radiation leaves the tube through the slightly leaky mirror at one end, and this is the laser beam. *Unlike in an ordinary source, the neon atoms have all added their radiation together so that the waves are all cooperating. The wave pulses all reenforce each other and so the resultant beam is very intense.*

In addition, the fact that the beam reflects back and forth many times between the parallel mirrors causes the light rays to come straight out from the end of the tube. Any which diverge from the axis of the tube are lost out the sides during the many trips back and forth through the tube. The fact that the beam is a fine pencil of rays is also of great importance. *Unlike light from a bulb, the energy does not spread out in space. Instead, it flows out into space through a thin cylinder and maintains its strength over very long distances.* For example, the laser shown in Fig. 27.20 has a power output of only 0.0005 W, but the light energy is confined to such a narrow pencil that the intercepted beam far outshines the high-power bulbs in the foreground. Laser beams sent to the moon and reflected back to the earth are currently used to make measurements on the moon.

In recent years very powerful lasers with outputs of many watts have found widespread use in many situations. You no doubt are familiar with their use in eye surgery, tumor destruction, and similar medical applications. Surgeons now make use of them routinely. The more powerful lasers are currently used for welding metals by means of the intense heat generated as the beam is absorbed in matter. Many other uses could be cited, and the list grows almost daily. For example, the communication of information over large distances by means of laser beams is nearly reality at this time. In research, the laser has already become a useful tool. It is interesting to realize that it would have been impossible to conceive this very useful light source without the vast amount of basic knowledge of atoms, their energy levels, and behavior amassed by many scientists over nearly a half century. This is a typical example of how increasing knowledge of nature leads us to better ways of utilizing its laws.

SUMMARY

The atom is of the order of 10^{-10} m in diameter. It consists of a positively charged nucleus of diameter about 10^{-15} m. If the nuclear charge is $+Ze$, there are Z electrons in that portion of the atom outside the nucleus.

Excited hydrogen atoms emit light. The wavelengths of the light constitute several series of spectral lines. These series can be represented by the relation $1/\lambda = R(1/p^2 - 1/n^2)$, where $n > p$ and both are integers. When $p = 1$, the Lyman series in the ultraviolet is found. For $p = 2$, the Balmer series results; this series is partly in the visible region. When $p = 3$, the Paschen series in the infrared is found.

Bohr was able to explain the hydrogen atom spectrum by making the following assumptions: the electron revolves around the nucleus in certain stable orbits in which no energy is radiated; the radii of these orbits is given by a quantum condition on the angular momentum, $mvr_n = nh/2\pi$; the atom loses energy as the electron falls from an outer orbit to an inner one, and this energy is emitted as a photon with energy $h\nu$. He was able to show that the energy levels of the hydrogen atom are given by $E_n = 13.6/n^2$ eV.

Atoms can be excited to higher energy states by collisions at high temperature or in an electric discharge. Incident photons can also excite an atom. But the photon must have an energy exactly equal to that needed to lift the atom to the higher level. Hydrogen atoms absorb the Lyman series of wavelengths.

De Broglie postulated that an electron has an associated wavelength $\lambda = h/mv$. The Bohr orbits can be interpreted as the resonance paths for the electron waves. This interpretation justifies Bohr's quantum condition for angular momentum.

Schrödinger's equation formulates the wave nature of particles in a quantitative way. Solution of it tells us how electrons are distributed within atoms. It shows that Bohr's circular orbits are only qualitatively correct. The Schrödinger solution shows that four quantum numbers are of importance: n, l, m_l, and m_s. They can take on the following values: $n = 1, 2, 3, \ldots$; $l = 0, 1, \ldots, n - 1$; $m_l = 0, \pm 1, \pm 2, \ldots, \pm l$; $m_s = \pm \frac{1}{2}$.

The structure of multielectron atoms is strongly influenced by the Pauli exclusion principle. It states that no two electrons in an atom can have the same set of quantum numbers. Because of this, only two electrons can exist in the $n = 1$ shell, only eight in the $n = 2$ shell, and so on. The shell structure of an atom greatly influences its chemical properties.

X-rays are produced when high-energy electrons bombard solids. Characteristic x-rays are emitted by high-Z elements when electronic transitions occur in the innermost shells. Gaseous atoms emit line spectra whereas hot, incandescent solids and liquids emit a continuous spectrum.

Two waves are coherent if they are of identical form and maintain the same relative phase. Coherent waves give rise to interference effects which persist for extended lengths of time. Ordinary light consists of noncoherent wave pulses. Laser light-wave pulses are not only coherent but also in phase with each other. As a result, laser beams can be made very intense. They can also be made to have nearly zero divergence of the rays.

MINIMUM LEARNING GOALS

Upon completion of this chapter, you should be able to do the following:

1. Describe Rutherford's experiment and explain how it leads to the concept of the nuclear atom.

2. Give the approximate diameter of an atom.

3. Sketch the lines of the Balmer series and write down the Balmer formula. Compute the wavelength of a given line of the Balmer series provided the Rydberg constant is given. Repeat for both the Lyman series and the Paschen series. State which series are partly in the visible region.

4. Describe Bohr's model for the hydrogen atom. In your description, give his postulates in regard to stable orbits and emission of light. Explain on the basis of the model how the various spectral series arise.

5. Draw the energy-level diagram for a simple one-electron atom when the formulas for its energies are given. Show on the energy-level diagram for hydrogen how the various spectral series arise.

6. Explain why hydrogen atoms normally absorb the wavelengths of the Lyman series but not those of the Balmer series.

7. Describe how the de Broglie wave concept can be used to explain Bohr's choice of stable orbits.

8. Explain the meaning of an electron-distribution diagram such as those shown in Fig. 27.14.

9. State the Pauli exclusion principle and show why it predicts lithium to be univalent.

10. Describe how x-rays are produced in an x-ray tube. Distinguish between characteristic x-rays and bremsstrahlung. Compute the shortest wavelength x-rays emitted by a target impacted by electrons of a given high energy.

11. Distinguish between a line and a continuous spectrum. Point out which kinds of light sources emit each.

12. Explain the difference between coherent and noncoherent waves.

13. State the important features of a laser beam in regard to coherency, phase, and pencillike quality. Point out how these features lead to specialized uses for lasers.

IMPORTANT TERMS AND PHRASES

You should be able to define or explain each of the following:

Nuclear atom

Spectral series; series limit

Lyman, Balmer, and Paschen series

Bohr atom and its postulates

$$\frac{1}{\lambda} = R\left(\frac{1}{p^2} - \frac{1}{n^2}\right)$$

Energy-level diagram

Atomic shells
Characteristic x-rays; bremsstrahlung
Line spectrum; continuous spectrum
Coherent waves and beams
Laser

Ionization energy; ground state
Pauli exclusion principle

QUESTIONS AND GUESSTIMATES

1. Why doesn't the hydrogen gas prepared by students in the laboratory glow and give off light?
2. Given a glass tube containing two electrodes sealed through its two ends. The gas inside is either hydrogen or helium. How can you tell which it is without breaking the tube? If the gas is at high pressure, what difficulty might you have?
3. Hydrogen gas at room temperature absorbs light of wavelength equal to the lines in the Lyman series but does not absorb the wavelengths of the Balmer series. Why not?
4. When white light is passed through a vessel containing hydrogen gas, it is found that wavelengths of the Balmer series as well as the Lyman series are absorbed. We conclude from this that the gas is very hot. Why can we draw this conclusion? (This is actually the basis for one method of measuring the temperature of a hot gas.)
5. The spectral lines emitted from an arc formed in hydrogen gas at very low pressure are much sharper than those generated in a high-pressure discharge tube. Explain.
6. Explain clearly why x-ray emission lines in the range of 1 Å are not observed from an x-ray tube using a low-atomic-number metal as the target in the tube.
7. Why do hot solids give off a continuous spectrum, while hot gases give off a line spectrum?
8. A steel company suspects that one of its competitors is adding a fraction of a percent of a rare-earth element to its (the competitor's) product. How can the element quickly be identified and its concentration determined?
9. It is suspected that a certain lot of benzene contains a trace of acetone as impurity. What practical methods exist for testing this suspicion?
10. In the helium atom, the two electrons are in the same shell but avoid each other well enough so that their interaction is of only secondary importance. Estimate the ionization energy (in electronvolts) for helium, i.e., the energy required to tear one electron loose. Also, estimate the energy needed to tear the second electron loose. Which of these two values is most reliable? (E)
11. The ionization energies for lithium, sodium, and potassium are 5.4, 5.1, and 4.3 eV, respectively, while those for helium, neon, and argon are 24.6, 21.6, and 15.8 eV, respectively. Explain in terms of atomic structure why these values are to be expected.
12. Estimate how much energy a photon must have if it is to be capable of expelling an electron from the innermost shell of a gold atom. (E)
13. The diameter of a nucleus is about 10^{-15} m. Estimate the least momentum a proton must have if it is to be a part of the nucleus. (E)

PROBLEMS

1.* Rutherford and his coworkers shot α particles ($q = +2e$) at gold atoms ($Z = 79$). Some of the particles had a KE of 4.8 MeV. (*a*) What is the PE (in terms of r) of an α particle at a distance r from the gold nucleus? (*b*) How close can Rutherford's α particles come to the center of the gold nucleus? Assume that the gold nucleus remains essentially stationary and neglect the effect of the (relatively distant) atomic electrons.
2.** The density of gold is 19.3 g/cm^3, and its atomic mass is 197. (*a*) What is the mass of a gold atom? (*b*) How many gold atoms are there in a 1-cm^2 area of gold film which is 0.010 cm thick? (*c*) The diameter of a gold nucleus is about 10^{-14} m. Assuming no overlap, how much area of the 1.0-cm^2 total area do the gold nuclei cover? (*d*) If he used a film of this thickness, about what fraction of the α particles would Rutherford have observed to be strongly deflected?
3. Compute the wavelength of (*a*) the fourth line in the Lyman series and (*b*) the fifth line of the Balmer series. (Use $R = 1.097 \times 10^7$ m^{-1}.)

4. Confirm the series limit given for the Paschen series in Fig. 27.5 by computing it from the Paschen series formula. Use $R = 1.097 \times 10^7\ m^{-1}$.

5. Call $\Delta\lambda$ the wavelength difference between the seventh line of the Balmer series and the series limit. Find the ratio of $\Delta\lambda$ to the series-limit wavelength. Use $R = 1.097 \times 10^7\ m^{-1}$.

6. (*a*) What is the formula for the energy levels of doubly ionized lithium written in the form $E_n = -\text{const}/n^2$ eV? (*b*) How much energy (in electronvolts) is needed to remove the last electron from doubly ionized lithium? (*c*) What wavelength photon would just be capable of knocking this electron free?

7. Suppose an electron to be rotating in a circular path of radius 0.10 nm about the hydrogen nucleus. (*a*) What speed must the electron have if the Coulomb force is to furnish the centripetal force? (*b*) What is the frequency of the electron in the orbit? (*c*) On the basis of classical theory, to what wavelength of radiation should this give rise?

8. Suppose that the Bohr theory can be applied to the innermost electron in a gold atom ($Z = 79$) by neglecting the presence of all the other electrons. (This is really not too bad an approximation.) (*a*) Show that the energy needed to remove this electron from the atom is $(13.6)(79)^2$ eV. (*b*) What is the radius of the first Bohr orbit for this atom?

9. An atomic subshell consists of those electrons in an atom which have the same n and l but different m_l and m_s. How many electrons exist in the $n = 3$, $l = 2$ subshell of gold?

10. How many electrons are needed to fill (*a*) the $n = 4$ shell, and (*b*) the $n = 3$, $l = 1$ subshell? (See Prob. 9 for definition of subshell.)

11. Modern color TV sets often have electron beams accelerated through nearly 20,000 V. What are the shortest-wavelength x-rays generated by a 20,000-V beam as it hits the end of the TV tube? Some early TV sets were not properly shielded and leaked appreciable amounts of x-rays outside the set.

12.** A beam of 0.50-Å-wavelength γ rays shines on a gas of hydrogen atoms. It expels electrons (photoelectrons) from the hydrogen atoms. (*a*) What is the energy of the expelled photoelectrons? (*b*) What is their speed?

13.* An ordinary x-ray tube might easily operate at 30,000 V with a current of 10 mA bombarding the plate. (*a*) How much heat is produced in the plate per second by this bombardment? (*b*) If the specific heat capacity of the metal of the 200-g plate is 0.10 cal/g, what temperature rise will occur in 1 min if no heat is lost?

14.* Some very-high-energy x-ray tubes circulate water through the inside of the metal plate in order to cool it. Suppose that a certain tube is operating at 30 mA and 200,000 V. How much would the cooling water be heated if it circulated through the plate at 1 liter/min?

15.* A room-temperature gas of hydrogen atoms is bombarded by a beam of electrons which has been accelerated through a potential difference of 12.9 V. What wavelengths of light will the gas emit as a result of the bombardment?

16. The series limit for a certain x-ray series in copper is about 1.3 Å. What is the energy required to expel an electron from the atom when the electron is in the lowest energy level of this series?

17. (*a*) From the data of Fig. 27.16, determine the energy difference between the $n = 2$ and $n = 3$ levels in molybdenum. (*b*) If you wished to construct the energy-level diagram for this atom, what further data would be needed?

18. An atom may take a time of 10^{-9} s to emit a photon. Suppose the atom emits a wave pulse during this time with the front end of the pulse emitted 10^{-9} s earlier than the rear end. If the wavelength of the wave is 500 nm, how many wavelengths long is the photon?

19.** Two different laser beams will be coherent if the lasers emit exactly the same wavelength. Even if the wavelengths are slightly different, the two beams will show an interference effect. When joined together, they will give a resultant beam which fluctuates in time back and forth from brightness to darkness. This is similar to the phenomenon of beats discussed in Chap. 15 for sound waves. If one beam has a wavelength of exactly 600 nm, what must the wavelength of the other beam be to produce maximum brightness once each second? [You may want to use the fact that, for $x \ll 1$, $1/(1 \pm x) \approx 1 \mp x$.]

THE ATOMIC NUCLEUS

28

As we saw in the previous chapter, Rutherford's experiments showed that the atomic nucleus is a very small entity at the center of the atom. Within it is concentrated all the positive charge of the atom as well as most of its mass. In this chapter we shall examine the details of the nucleus. Furthermore, we shall become acquainted with some of the properties of nuclei such as radioactivity, fission, and fusion. Various applications of nuclear physics will also be given.

28.1 STRUCTURE OF THE NUCLEUS

In the last chapter we saw how Rutherford was able to infer the structure of the atom from the results of scattering experiments. Since that time, the nuclear model of the atom has been confirmed in numerous ways. *We therefore picture the atom as follows. At its center is a tiny nucleus; it is here that all of the positive charge of the atom is located, together with more than 99.9 percent of its mass. The atom itself has a radius of the order of 10^5 times larger than that of the nucleus. It is in this outer region that the electrons of the atom are found.* Their behavior was discussed in the last chapter. We now proceed to a discussion of the atomic nucleus.

The nucleus is composed of neutrons (n) *and protons* (p). As we have learned previously, *a proton is simply the nucleus of the hydrogen atom. It has a charge of* $+e = 1.602 \times 10^{-19}$ *C and a mass of* 1.673×10^{-27} *kg. The neutron has no charge and its mass is* 1.675×10^{-27} *kg.* We shall make frequent use of a more tractable mass unit called the **atomic mass unit** (u), the exact definition of which will be given later. *The proton's mass is 1.007 u,* and *the mass of a neutron is 1.009 u.* For many purposes we shall round off the masses of these particles to unity.

Since the charge on a nucleus is Ze for an element of atomic number Z, we see that *there are Z protons in the nucleus.* These Z protons furnish Z units of mass (in atomic mass units) to the nucleus. However, the mass M of a nucleus exceeds Z for all elements except hydrogen, and we attribute this extra mass to the neutrons in the nucleus. *We refer to the neutrons and protons in a nucleus as* **nucleons.** *The total number of nucleons within the nucleus is called the* **atomic mass number** *of the element in question and is represented by the symbol A.* Since the masses of the proton and neutron are close to 1 u, the value of A will be close to the mass of the nucleus measured in atomic mass units.

Mass Number A

The size of the nucleus can be measured by shooting high-energy particles at it. Such measurements show that *the nucleus can be approximated roughly as a sphere with radius R given by*

Nuclear Radius

$$R = (1.2 \times 10^{-15})(A^{1/3}) \qquad \text{m}$$

It is reasonable that the nuclear radius should vary as $A^{1/3}$ since the mass of the nucleus is proportional to the atomic mass of the element. If we assume the nuclei of the various elements to have the same density ρ, then

$$A \propto \text{mass} = (\tfrac{4}{3}\pi R^3)\rho$$

and so it follows that

$$R \propto A^{1/3}$$

The fact that R does vary in this way, at least for the larger nuclei, indicates that the nuclei of the elements have about the same density.

Returning now to the simplest of all atoms, hydrogen, we note that its mass number is unity. Its nucleus is simply a proton. For convenience, we shall usually express atomic masses in atomic units and charges in multiples of the charge quantum, $+e$. Hence the hydrogen nucleus has unit mass and unit charge.

Following hydrogen in the periodic table,* the next heavier element is helium. Helium has two electrons outside its nucleus and is a chemically inert gas. Since the nuclear charge must exactly balance the electronic charge, helium nuclei must carry a charge of +2. The mass spectrometer (discussed in the next section) shows the mass of helium nuclei to be very close to 4 u. Helium nuclei must contain two protons to account for the +2 charge on each nucleus. The other two units of nuclear mass are the result of the two neutrons composing the remainder of the nucleus. Hence we see that two units of positive charge and two of the four units of mass in the helium nucleus are contributed by the two protons, while the remaining two units of mass are the result of the two neutrons.

As we proceed through the periodic table, we shall see that all nuclei can be thought of as composed of protons and neutrons. The following terminology is used as a shorthand method for describing a nucleus. *The nucleus of an element X, for example, is described by the symbolism*

Symbol for Nucleus

$$^{\text{mass}}_{\text{charge}}\text{X} \quad \text{or} \quad ^{A}_{Z}\text{X}$$

That is, one would symbolize hydrogen as $^{1}_{1}\text{H}$ and helium as $^{4}_{2}\text{He}$. The charge number is the same as the atomic number Z, since you will recall from chemistry that the atoms of the periodic table are listed in terms of the number of electrons in the atom.

As one of the higher elements in the periodic table, we select uranium as an example. All uranium atoms have a nuclear charge of 92, and uranium is therefore listed as element 92 in the periodic table. (Note that the nature of the element is determined by its chemical reactivity. This in turn is determined by the electron structure of its atom. Hence the chemist is concerned only with the charge on the nucleus, because this determines how many electrons the atom will have available for chemical reactions.) As we shall see in the next section, not all nuclei of the same element have the same mass. However, let us consider uranium 238, the uranium nucleus having a mass of 238 u. We would designate this nucleus by the symbol $^{238}_{92}\text{U}$.

It should be clear from the symbol that this nucleus contains 92 protons. They furnish a charge of +92 and 92 of the total 238 mass units. The remaining $238 - 92 = 146$ units of mass must be furnished by neutrons. We therefore see that the nucleus $^{238}_{92}\text{U}$ contains 92 protons and 146 neutrons.

28.2 ISOTOPES

Early investigators thought that nuclear masses were essentially integer multiples of the hydrogen mass, but, as time passed, more accurate measurements of the atomic masses indicated that this was not true. We should

*See Appendixes 3 and 4 for the periodic table and a table of isotopes.

understand that the atomic mass (nearly the same as the nuclear mass, since the electrons have such small mass) was at first determined by chemists. For example, if 1 g of hydrogen combined with 35 g of chlorine to form 36 g of hydrogen chloride, HCl, the chemist reasoned that the chlorine atom was 35 times as massive as the hydrogen atom. The atomic masses in the periodic table have been determined in much this way. (Actually, the chemists for many years chose oxygen to have a mass of exactly 16 u and based the masses of all the other elements upon this. On that basis, the hydrogen mass was found to be 1.0080 u instead of 1.0000. It has now been agreed to base all masses upon another standard, ^{12}C. *In this, the* **unified system,** *the mass of a* ^{12}C *atom including electrons is by definition taken exactly as 12 u.* For many purposes, the difference between these two systems is negligible.)

Atomic Mass Unit

However, as measurements became more precise, it became obvious that the atomic masses were not integer multiples of the hydrogen mass. For example, the mass of chlorine is 35.5 u, which is far from integral. Although the lower-atomic-number elements have masses which are reasonably close to integers, even boron, element 5, has an atomic mass of 10.8 u, which is certainly not a whole number of proton and neutron masses. This state of affairs caused a considerable amount of difficulty in the interpretation of nuclear structure. It was not until the invention of the mass spectrograph that the difficulty was resolved.

Mass Spectrograph

The mass spectrograph is constructed as shown in Fig. 28.1. Atoms are ionized in the ion source. When a positive ion strays out through slit S_1, it is attracted to the negative plate and slit S_2. If it is a singly charged ion, its charge will be $+e$ and upon reaching S_2 its kinetic energy will be Ve, where V is the potential through which the ion falls. We therefore have

$$\tfrac{1}{2}mv^2 = Ve \tag{28.1}$$

When the ion passes through slit 2, it enters a magnetic field perpendicular to the paper. The ion will be deflected in a circular path, as shown. As discussed in Chap. 19, the radius of the path can be found from the fact that the centripetal force is furnished by the magnetic force on the ion,

$$Bev = \frac{mv^2}{r} \tag{28.2}$$

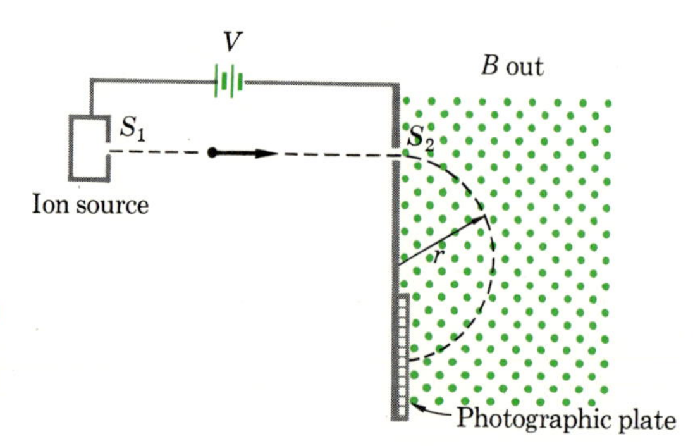

FIGURE 28.1

In the mass spectrograph, positive ions are deflected by a magnetic field.

Substitution from (28.1) in (28.2) yields the following expression for the radius of the circular path:

$$r = \sqrt{\frac{2Vm}{B^2 e}} \tag{28.3}$$

Since the ion will be detected on the photographic plate, as shown in Fig. 28.1, r can be measured. Of course V, B, and e would be known, and so the mass of the ion can be found. Most modern spectrographs use an electronic detector rather than a photographic plate.

When the mass spectrograph was put to use, an interesting discovery was made. Many high-purity elements were found to act like several species appearing under the same chemical identity. For example, pure chlorine was found to consist of two types of ions. Two separate beams of ions were found in the mass spectrometer, as illustrated schematically in Fig. 28.2. From the radii, upon using Eq. (28.3), the masses of the two types could be computed. In addition, from the brightness of the image on the photographic plate, the relative abundance of the two types of ions could be determined.

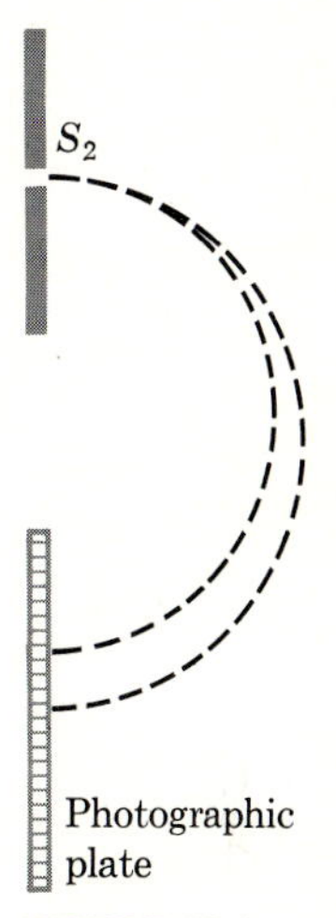

FIGURE 28.2
The two isotopes of chlorine describe different paths in the mass spectrograph, as shown here (with some exaggeration).

	TYPE 1	TYPE 2
Mass, u	35.0	37.0
Relative abundance	0.754	0.246

If we take the average mass of these two species (by multiplying the masses by the relative abundance of each and adding), we obtain an average mass of 35.5 u. This is exactly the mass determined by the chemists. We conclude that the atomic masses listed in the periodic table are the average masses of these individual species.

We see at once that the masses of the individual species determined by the mass spectrometer are essentially integers. This proves to be the case for all the elements examined in this way. Since we know that all types of chlorine have a nuclear charge of 17 (because the atomic number of chlorine is 17), the chlorine nucleus must contain 17 protons. It is therefore concluded that the nucleus of species 1 given above contains $35 - 17 = 18$ neutrons, while species 2 contains 20 neutrons. *We call nuclei of the same element which contain different numbers of neutrons* **isotopes** *of the element.* In this terminology, chlorine found in nature contains two isotopes designated $^{35}_{17}Cl$ and $^{37}_{17}Cl$.

DEFINITION

All isotopes of an element behave similarly in chemical reactions. This follows from the fact that all the isotopes have the same nuclear charge and thus the same number of electrons in their atoms. The atomic mass of each isotope is very close to an integer, since its nucleus is made up of neutrons and protons, each of which has essentially unit mass. Most of the elements occurring in nature are mixtures of isotopes, and it is these mixtures which are used in ordinary chemical reactions. The chemically determined atomic masses in the periodic table are nonintegers because they are the average of the isotopic masses found in nature. In Appendix 4 you will find an abbreviated list of the isotopes found in nature.

28.3 MASS DEFECT AND BINDING ENERGY

With the advent of extremely sensitive mass spectrographs, the masses of nuclei were determined to high precision. The very exact results thus obtained provided a striking confirmation of Einstein's concept of mass-energy interchange. Indeed, these measurements showed the possibility for the utilization of nuclear energy, an energy source previously not even suspected. We can appreciate the principle involved by examining a specific situation.

To take a simple case, let us analyze data for the mass of the helium nucleus. Its total mass should consist of two neutron masses plus two proton masses.

$$\begin{aligned} 2n &= 2.01734 \text{ u} \\ 2p &= \underline{2.01456} \\ \text{Computed He mass} &= 4.03190 \\ \text{Measured He mass} &= \underline{4.00150} \\ \text{Mass difference} &= 0.03040 \text{ u} \end{aligned}$$

Clearly, the computed and measured masses do not agree. The discrepancy, 0.030 u, is far larger than one can attribute to experimental error. Where did this extra mass go when the neutrons and protons joined to form the nucleus? This is the question we shall now investigate.

The helium discrepancy is not the only one. Similar discrepancies occur for other nuclei as well, as illustrated in Fig. 28.3, where the mass change per nucleon $\Delta M/A = (M - M_{np})/A$ is plotted against atomic number. By mass change per nucleon we mean the true nuclear mass M minus the combined mass of the protons and neutrons M_{np} all divided by the total number of neutrons and protons, A. Of course ΔM is zero for hydrogen, and it lies off the top of the graph in Fig. 28.3.

We know of only one way in which mass can be lost in a situation like this. In Chap. 26 we saw that mass can be converted to energy and that the

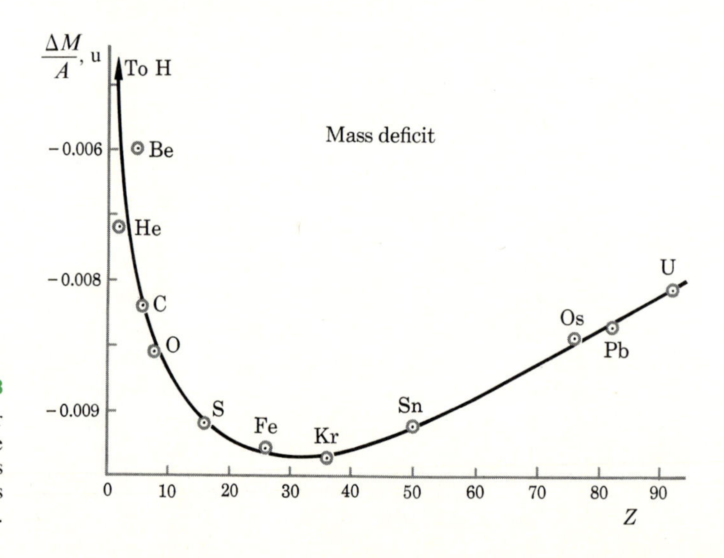

FIGURE 28.3

The mass discrepancy per nucleon for nuclei near the center of the periodic table is larger than that for elements at either end of the table.

relation between them is

$$\text{Energy} = \Delta m\, c^2 \tag{28.4}$$

where Δm is the lost mass and c is the speed of light, 3×10^8 m/s. Hence it would appear that when two protons and two neutrons are joined together to form a helium nucleus, energy must be given off. The amount of energy given off is related to the mass loss by Eq. (28.4).

As we see from Fig. 28.3, all the nuclei have less mass than the neutrons and protons which were joined together to form the nuclei. Mass was lost and energy was given off as the nuclei were formed. This also means that energy is required to tear the nucleons apart from each other since mass would need to be created in the process. *The energy needed to tear a nucleus apart into its constituent isolated nucleons is called the* **binding energy** *of the nucleus.*

DEFINITION

We can compute the binding energy per nucleon directly from the data of Fig. 28.3. For example, the mass loss per nucleon in forming a beryllium (Be) nucleus is 0.007 u. This is equivalent (from energy = $\Delta m\, c^2$) to an energy loss of 6.5 MeV. Hence the binding energy per nucleon of the Be nucleus is 6.5 MeV. That much energy multiplied by the number of nucleons in the Be nucleus must be furnished to tear this nucleus completely apart. In Fig. 28.4 we show a graph of the binding energy per nucleon for typical elements. Notice that *a large binding energy per nucleon implies a very stable nucleus.* We shall return to this topic when we discuss nuclear power.

Binding Energy Related to Stability

Illustration 28.1 How much energy is given off when one helium nucleus is formed from neutrons and protons?

Reasoning We saw in the last section that the mass lost in this process is 0.030 u. Since the proton has a mass of 1.008 u, or 1.67×10^{-27} kg, we have the conversion factor 1 u = 1.66×10^{-27} kg. Hence, the mass lost in forming

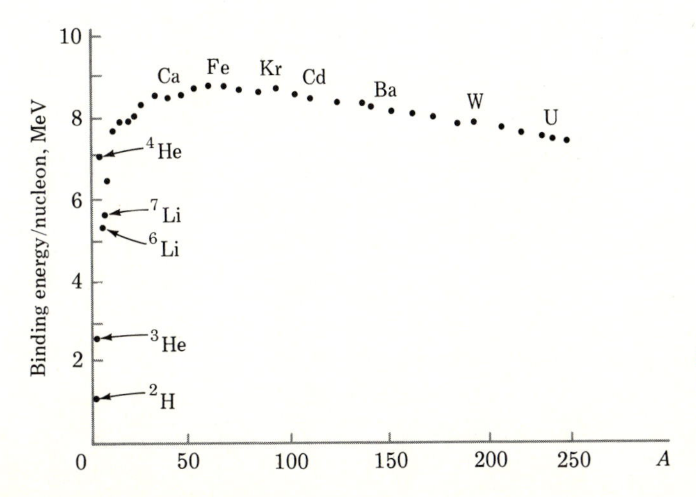

FIGURE 28.4
Binding energy per nucleon for representative elements. Note the very high stability of ^{4}He.

one helium nucleus is 5.0×10^{-29} kg. Using Eq. (28.4), we have

$$\begin{aligned} \text{Energy} &= (5.0 \times 10^{-29}\text{ kg})(9 \times 10^{16}\text{ m}^2/\text{s}^2) \\ &= 4.5 \times 10^{-12}\text{ J} = 2.82 \times 10^{7}\text{ eV} = 28.2\text{ MeV} \end{aligned}$$

It is convenient to notice that since a mass loss of 0.030 u gives rise to 28.2 MeV of energy,

$$1\text{ u} \rightarrow 931\text{ MeV} = 1.49 \times 10^{-10}\text{ J}$$

We also notice that since 4 g of helium contains 6×10^{23} nuclei, the energy liberated in the formation of 4 g of helium is

$$(6 \times 10^{23})(4.5 \times 10^{-12}\text{ J}) = 2.7 \times 10^{12}\text{ J}$$

This is a tremendous amount of energy. For comparison, the work done in lifting a 100-kg man a distance of 100 m is only about 10^4 J. The amount of energy liberated in forming 4 g of helium nuclei from protons and neutrons is large enough to lift 200 million 100-kg people through a distance of over 100 m.

28.4 RADIOACTIVITY

Certain naturally occurring elements are not stable but slowly decompose by throwing away a portion of their nucleus. We say that they are **radioactive.** The first discovery of a radioactive element was made in 1896 by Henri Becquerel when he found that uranium atoms ($Z = 92$) give off radiation which fogs photographic film and plates. Two years later, Marie and Pierre Curie succeeded in chemically isolating two new radioactive elements, polonium ($Z = 84$) and radium ($Z = 88$). It has since been found that all the elements having Z greater than 83 are radioactive to some extent.

There are many man-made nuclei which are radioactive. Nuclear reactors, in particular, produce copious quantities of nuclei which do not occur in measurable quantity on earth elsewhere. These nuclei are highly unstable and undergo rapid radioactive change. They may have existed early in the history of the earth but have long since disintegrated.

No matter what their source or age, all pure radioactive substances decompose (or decay) in accordance with the same mathematical law. The law is stated in terms of N, the number of nuclei still undecayed, and ΔN, the number of nuclei which decay in a small time interval Δt. *Experiment shows that*

The Decay Law

$$\Delta N = \lambda N\,\Delta t \tag{28.5}$$

where λ, *a constant for a given substance, is called the decay constant.* Since Eq. (28.5) can be rewritten as $\lambda = (\Delta N/N)/\Delta t$, the decay constant has a simple meaning: *the* **decay constant** λ *is the fraction of nuclei which decay in unit time.*

DEFINITION

Since λ is constant for a given substance, we see that the fraction which decays in unit time is always the same. For example, λ for radium is 0.000428 yr^{-1}. This tells us that in any sample of radium, no matter what its past history may have been, a fraction 0.000428 of it will decay in the following year. *In Fig. 28.5 we show a graph relating the number of still undecayed nuclei to time. It is an exponential decay curve.*

DEFINITION

Radioactive substances are often described in terms of their half-life. The **half-life** *of a substance is the time taken for half of the material to decay.* See Fig. 28.5, where this time is indicated. *It can be shown mathematically that the half-life is related to the decay constant in the following way:*

Half-Life

$$\text{Half-life} = \frac{0.693}{\text{decay constant}} \tag{28.6}$$

In the case of radium, $\lambda = 0.000428\ \text{yr}^{-1}$. Substitution of this value in Eq. (28.6) gives the half-life of radium to be 1620 yr. We know from this how much of a sealed sample of radium will still remain after 1620 yr; only half of the original quantity will remain as radium. The other half will have decayed to other nuclei. *No matter when you start timing, after one half-life has passed, only half of the original material will remain unchanged.*

Radioactive materials which are found in nature or which are made artificially show a wide range of half-lives. The half-life of ^{238}U (uranium 238) is about 4.5×10^9 yr. This is about the same as the age of the earth. We conclude that this isotope still exists on earth because only half of the original amount has decayed. Radium, on the other hand, only has a half-life of 1620 yr. It long ago should have decayed to an unmeasurably small quantity. However, as we shall see, new radium is being produced constantly in nature. Other isotopes, made in nuclear reactors, have half-lives of only a small fraction of a second. If they existed on the earth at its beginning, they have decayed to negligible amounts by now.

Illustration 28.2 Iodine 131 is a radioactive isotope made in nuclear reactors for use in medicine. It becomes concentrated in the thyroid gland when taken into the body. There, it acts as a radiation source in the treatment of hyperthyroidism. Its half-life is 8 days. Suppose a hospital orders 20 mg of ^{131}I and stores it for 48 days. How much of the original ^{131}I will still be present?

Reasoning Each 8 days, the iodine decays by half. We can therefore make the following table:

TIME, days	IODINE PRESENT, mg	TIME, days	IODINE PRESENT, mg
0	20	32	1.25
8	10	40	0.625
16	5	48	0.312
24	2.5		

After 48 days only 0.312 mg of the original 20 mg will remain.

FIGURE 28.5

A radioactive element decays exponentially. The half-life is the time taken for half of the nuclei to decay.

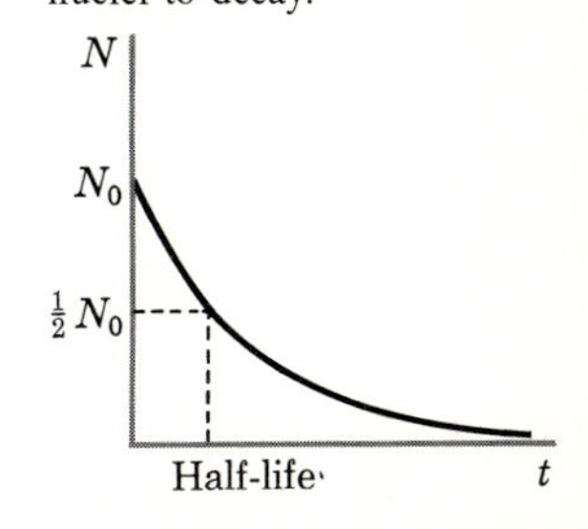

The Curie Unit

Illustration 28.3 *The amount of material in which 3.70 $\times$ 10^{10} disintegrations occur in 1 s is said to be a* **curie** *(Ci) of the material,* but 10^{-6} Ci, a microcurie (μCi), is more commonly used. What fraction of a curie is present in 1 g of radium? The half-life of radium is 1620 yr, or 5.1×10^{10} s.

Reasoning The number of atoms decaying in time Δt is

$$\Delta N = \lambda N \, \Delta t$$

Since there are Avogadro's number of atoms in one atomic mass of radium, 226 kg, the number of atoms N in 1 kg is

$$N = \frac{6 \times 10^{26} \text{ atoms}}{226 \text{ kg}} = 2.66 \times 10^{24} \text{ atoms/kg}$$

We still need to find λ. This can be done using Eq. (28.6). We have

$$\lambda(5.1 \times 10^{10} \text{ s}) = 0.693 \qquad \text{or} \qquad \lambda = 1.36 \times 10^{-11} \text{ s}^{-1}$$

Placing this in the first expression for ΔN yields

$$\begin{aligned} \Delta N &= (1.36 \times 10^{-11} \text{ s}^{-1})(2.66 \times 10^{21} \text{ atoms/g}) \, \Delta t \\ &= 3.6 \times 10^{10} \, \Delta t \qquad \text{atoms/g} \end{aligned}$$

when Δt is expressed in seconds. The number of disintegrations per second in 1 g of radium is therefore 3.6×10^{10}. Since 1 Ci of material is defined to have 3.7×10^{10} disintegrations per second, we see that 1 g of radium is equivalent to 3.6/3.7, or 0.97 Ci. Actually, the curie was originally thought to be exactly the number of disintegrations per second from 1 g of radium. However, more precise measurements showed that the curie as defined was slightly larger than this experimental value.

28.5 DECAY PRODUCTS AND RADIOACTIVE SERIES

γ Ray

When a nucleus emits a particle, it is obviously changed in some way. Radioactive nuclei found in nature emit α particles, β particles, and γ rays. The γ ray is nothing more than an x-ray and is pure energy of magnitude $h\nu$ or $h(c/\lambda)$. The nucleus loses no charge and no appreciable mass when it emits a γ ray. Hence, the nucleus is changed, but not greatly, by such an emission.

In the emission of a γ ray, the nucleus has merely adjusted its internal structure in such a way as to lose some energy. This is similar to the emission of light by an atom. In an atom, the energy is lost when the electron falls to a lower energy state. We are not quite sure what really happens in a nucleus when it throws out a γ ray. The nucleus has fallen to a lower energy state. However, we do not know enough yet about the structure of the nucleus to describe this type of transition nearly as precisely as in the atomic case. In any event, if nucleus ${}^A_Z\mathrm{X}$ loses a γ ray, the reaction is usually written as

$${}^A_Z\mathrm{X} \rightarrow \gamma + {}^A_Z\mathrm{X}$$

The final charge Z and mass A of the nucleus is the same as before the γ ray was emitted.

α Particle

When a nucleus emits an α particle, on the other hand, the mass of the nucleus decreases by 4 and its charge decreases by 2. This is the result of the fact that the emitted α particle is actually a helium-atom nucleus, ${}^4_2\mathrm{He}$. For this type of emission we would write

$${}^A_Z\mathrm{X} \rightarrow \alpha + {}^{A-4}_{Z-2}\mathrm{Y}$$

To take an actual example, radium emits an α particle, and so

$${}^{226}_{88}\mathrm{Ra} \rightarrow \alpha + {}^{222}_{86}\mathrm{Rn}$$

where the product nucleus is the inert gas radon.

Sometimes it is convenient to write mass and charge numbers for the particles:

$${}^{226}_{88}\mathrm{Ra} \rightarrow {}^4_2\alpha + {}^{222}_{86}\mathrm{Rn}$$

Notice, in such an equation, that the sum of the mass numbers must be the same on each side. The same is true for the charges.

β Particle

β particles are simply electrons. When a nucleus emits a β particle, we have a conceptual difficulty since no electrons are in the nucleus. We can imagine, however, that the neutron is a proton plus an electron. (This is purely an imaginary device which has utility but probably no basic validity.) When the β particle is emitted, the original neutron in the nucleus is replaced by a proton. Hence the emission of a β particle increases the net charge of the nucleus by 1 but changes the mass hardly at all (since the electron mass is so small). In a typical case

$${}^A_Z\mathrm{X} \rightarrow {}^{0}_{-1}\beta + {}_{Z+1}{}^{A}\mathrm{Y}$$

To illustrate this type of reaction we shall discuss one of the less abundant isotopes of lead, ${}^{214}_{82}\mathrm{Pb}$. It has a half-life of 26.8 min and decays by β emission. Thus

$${}^{214}_{82}\mathrm{Pb} \rightarrow {}^{0}_{-1}\beta + {}^{214}_{83}\mathrm{Bi}$$

The bismuth isotope formed in this reaction also decays by β emission. Both these isotopes are but steps in a series of decays.

A Decay Series

The uranium series is a dramatic example of the type of decay in which one nucleus decays to another, that other to still another, and so on. Uranium 238 is slightly radioactive, having a half-life of 4.5×10^9 yr, and decays according to the scheme shown in Fig. 28.6. You should follow this scheme

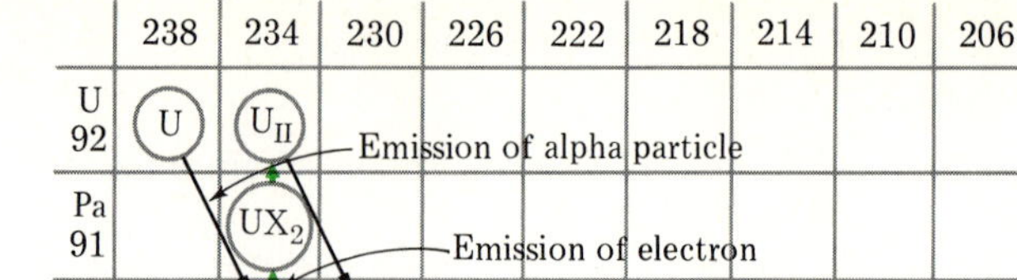

FIGURE 28.6

A typical radioactive series. It is called the uranium series since the parent nucleus is uranium.

step by step so that you understand its meaning. Notice that the names radium A, and so on, are used in place of the usual atomic names given at the left side of the figure. This is sometimes, but not usually, the way in which the members of this series are written.

In Fig. 28.6 the final nucleus of the series is lead. Two other series such as this exist in nature. They all start with a long-half-life nucleus. Perhaps others once existed and have long since decayed to imperceptible traces. In any event, the last member of the series is a stable nucleus. For each series the terminal nucleus is lead, but it is a different isotope in each case.

28.6 NUCLEAR REACTIONS AND TRANSMUTATIONS

When we wrote the radioactive decay equation

$$^{226}_{88}\text{Ra} \rightarrow \alpha + {}^{222}_{86}\text{Rn}$$

in the previous section, we were writing a **nuclear reaction.** Several factors determine whether a postulated nuclear reaction will actually occur. This particular reaction does occur, of course. We know that the total energy, rest energy included, of the original reactants (in this case, just $^{226}_{88}Ra$) is equal to the final energy of the products. For this to be true, the rest mass of the original radium nucleus must be greater than or equal to the rest masses of the two products, the radon nucleus and the α particle. If this were not true, we would be creating mass in the reaction. Since we are assuming that no external source of energy is acting upon the system, we would be creating mass from nothing and this is obviously impossible.

The fact that the total energy before reaction (including the equivalent energy of the rest masses) must equal the total energy after reaction is a useful tool in the study of nuclear reactions. For example, in one of the very first induced nuclear reactions, performed by Rutherford in 1918, he shot α particles at nitrogen nuclei and observed the reaction

$$^{14}_{7}N + ^{4}_{2}\alpha \rightarrow ^{17}_{8}O + ^{1}_{1}H$$

In other words, the α particle entered the ^{14}N nucleus, which then disintegrated by ejecting a proton. The original nitrogen nucleus was **transmuted** into oxygen. We now ask whether even very slow α particles could cause this reaction.

To learn more about this reaction we notice that the masses of the reactants are

$$\begin{aligned} \text{Mass of } ^{14}N &= 14.0031 \text{ u} \\ \text{Mass of } ^{4}He &= \underline{4.0026} \\ \text{Total mass before reaction} &= 18.0057 \text{ u} \end{aligned}$$

In the same way, we examine the masses after reaction,

$$\begin{aligned} \text{Mass of } ^{17}O &= 16.9991 \text{ u} \\ \text{Mass of } ^{1}H &= \underline{1.0078} \\ \text{Total mass after reaction} &= 18.0069 \text{ u} \end{aligned}$$

We see at once that the products have more mass than the original reactants, the difference being 0.0012 u. This mass could be created only if additional energy had been added to the reaction. Since 1.0 u is equivalent to 931 MeV, as shown in Illustration 28.1, we see that the increase in mass in this reaction required an external energy of $(\frac{931}{1})(0.0012) = 1.1$ MeV. The incident α particle must have had at least this amount of KE in order to make the reaction occur. Actually, since momentum must also be conserved in such a reaction, the end products will not be standing still. As a result, the particle must have more than 1.1 MeV of KE if the reaction is to be feasible.

Computations like this tell us a great deal about the feasibility of proposed nuclear reactions. Of course, a large variety of reactions are possible. In fact, the field of nuclear reactions in physics is as involved as the subject of organic reactions in chemistry.

28.7 THE NUCLEAR FORCE AND NUCLEAR STABILITY

Thus far in our study of physics we have learned about two fundamental forces. One of these, *the gravitational force,* is of great importance when massive objects are concerned. This force *is always attractive, but it is quite weak.* Two objects on a table attract each other, but the force is so small that it can be measured only with very delicate equipment.

The second fundamental force we have encountered is the electric force. It can be either attractive or repulsive depending upon the charges involved. *This force is quite large* and is easily demonstrated with electrostatic charges. For example, two charged balls exert easily measured forces on each other. At one time it was thought that magnetic forces were different from electric forces. However, by use of the theory of relativity, these two forces can be shown to be basically the same. It is the electric force which holds atoms, as well as liquids and solids, together. Most of the phenomena we encounter in everyday life are intimately related to this force.

There is yet another force which is probably not familiar to you. *The third force is the one which holds the nucleons together in the nucleus. Unlike the other two forces, it is not an inverse-square-law force.* You will recall that both the gravitational and electrical forces vary as $1/r^2$. They are long-range forces in the sense that they are finite and nonzero even at very large values of r.

The nuclear force on the other hand *is a very-short-range force. Two nucleons exert negligible force on each other until they are brought closer than about* 5×10^{-15} *m from each other. They attract each other strongly as they are brought closer than this. The nuclear attraction force is so large at small separation that it completely overpowers electrical repulsive forces.* Two protons, although repelling each other with their like charges, are held together strongly by the nuclear force. To a first approximation at least, the nuclear attractive force is the same between proton and proton, neutron and neutron, and between proton and neutron.

In spite of its strength, the nuclear force is unable to hold many protons together without a sufficient number of neutrons to "dilute" the positive charge. The electrostatic repulsion force is long-range, and each proton "feels" the effect of every other proton in the nucleus. But the short-range nuclear attractive force is felt only by nearest neighbor particles in the nucleus. For this reason, among others, only certain combinations of protons and neutrons form stable nuclei. In Fig. 28.7 we show the stable nuclei which are known.

Notice the axes in Fig. 28.7. If the nuclei had equal numbers of protons and neutrons, they would lie on the $N = Z$ line. As we see, however, the larger nuclei require more neutrons to dilute their charge in order to remain stable. Also shown in the figure are the modes of decay of unstable nuclei which deviate from those indicated. You should be able to show that the modes of decay shown tend to bring the unstable nuclei to a stable form. (In this regard, one should note that a positron has a charge of $+e$ and the mass of an electron.)

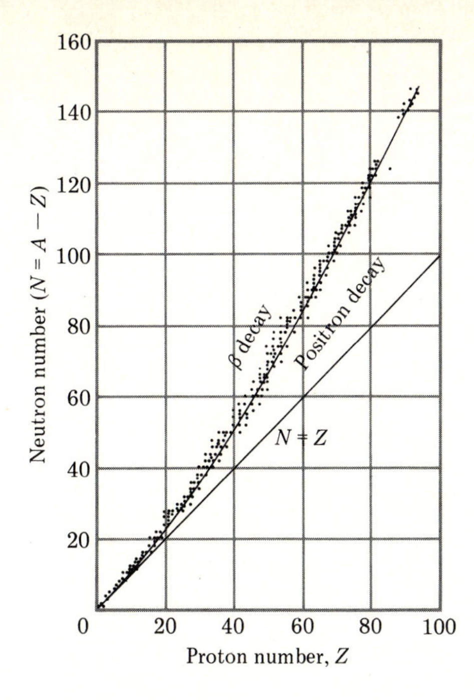

FIGURE 28.7

The nuclei shown as dots in this chart are stable or have half-lives in excess of 1000 years.

In addition to these considerations, we learned in Sec. 28.3 that the mass defect of nuclei provides us with another criterion for nuclear stability. As we saw in Fig. 28.4, the nuclei near the center of the list of elements have the highest binding energy per nucleon. These nuclei are intrinsically harder to tear apart than those near the two ends of the list of elements. As we shall see in that which follows, this has extreme practical implications and consequences.

28.8 NUCLEAR FISSION

Nuclear power sources are one of the leading options available in our search for energy. Let us now investigate this practical application of nuclear physics. If we refer back to Fig. 28.4, we see that *the nuclei at the two ends of the list of elements have more energy per nucleon than those near the center of the list. Suppose a high-Z nucleus such as uranium split in half.* Two equal-mass particles, each with charge $Z/2$, would be formed. Because of the mass-defect effect, these particles will have a mass much larger than the stable nucleus of the element of atomic number $Z/2$. In the end, to form two stable nuclei with atomic number $Z/2$, the unstable original halves must radiate energy equivalent to the mass defect. As we see, for this process to occur, a large amount of *energy would have to be liberated.* This liberated energy could then be utilized. Unfortunately, *this splitting of high-Z nuclei into intermediate-sized nuclei, a process we call* **fission,** *seldom occurs spontaneously.* Even though the products of the fission reaction have less energy than the original material, the reaction requires external influences if it is to proceed.

We have a similar situation in chemistry. Heat is given off when wood is burned in oxygen. Obviously, then, the reactants have more energy than the products, and hence the reaction should be possible. However, wood does not combine spontaneously with oxygen at room temperature—at least not to any great extent. The wood-oxygen chemical reaction must first be started by some external means, a hot flame from a match, for example. Once it has started, the reaction produces enough heat to keep itself going provided that the geometrical arrangement of the piece of wood is such that the heat generated does not escape too easily.

As it turns out, the nuclei of most of the heavy elements cannot easily be split. Striking them with very-high-energy particles does sometimes cause splitting. Even though more energy is given off when the fission reaction occurs than was needed to start it, this energy is not easily utilized to keep the reaction going. *Hence, even though all heavy nuclei are potential sources of energy, it is impractical to obtain the energy given off by the fission reaction in nearly all cases.*

There are, however, a few instances in which the fission reaction proves to be of practical importance. The first of these was discovered by Hahn and Strassmann in 1939. With the aid of others, *they found that an isotope of uranium, ^{235}U, would capture (or attach itself to) a slowly moving neutron and would then undergo spontaneous fission.* Here was a fission reaction which did not require high-energy bombardment. Moreover, when each ^{235}U nucleus splits, the reaction products contain about three neutrons,* and of course a large amount of energy is given off. Each ^{235}U nucleus gives off about 200 MeV of energy when it splits.

The Chain Reaction

We see, then, that *we can visualize a* **chain reaction** *in ^{235}U initiated by a single slow neutron.* This is shown diagrammatically in Fig. 28.8. Only the first two steps in the reaction are shown, but if all the neutrons are successfully captured by ^{235}U nuclei, the original neutron has already generated 3^2 neutrons. In an ideal situation after n steps in the chain reaction have occurred, 3^n neutrons will be available. If each step of the reaction takes 0.01 s, at the end of 1 s the total number of neutrons would be 3^{100}, or about 10^{48}. Since 235 g of ^{235}U contains only 6×10^{23} atoms, you can see that such a reaction could occur with explosive violence.

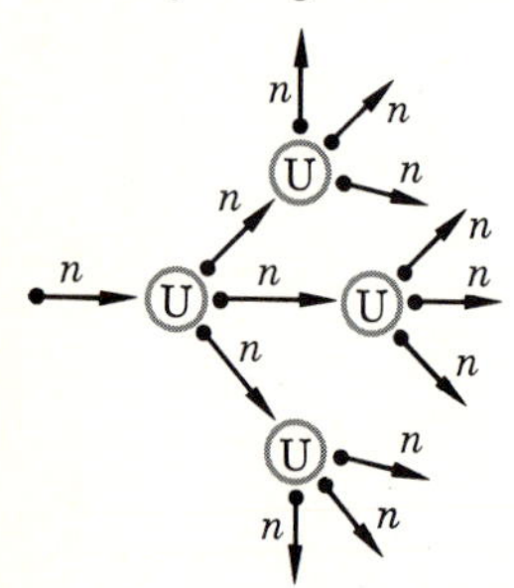

FIGURE 28.8
A chain reaction can be initiated by a single neutron.

If the reaction is to proceed smoothly and be self-sustaining, a **critical mass** of ^{235}U is needed, as is easily seen by reference to Fig. 28.9. When fission occurs in the thin film shown in part *a*, the neutrons produced easily escape into the air and so the reaction does not build up. In part *b*, the situation is somewhat more favorable, but here, too, the reaction stops after two steps because of the loss of neutrons to the surroundings. In part *c* we have a sphere of ^{235}U. If the sphere is large enough, most of the neutrons produced will be captured before they can get away. *The critical mass is that amount of ^{235}U in which, on the average, exactly one neutron from each*

*This isotope of uranium constitutes only 0.7 percent of the mixture of uranium isotopes occurring in nature.

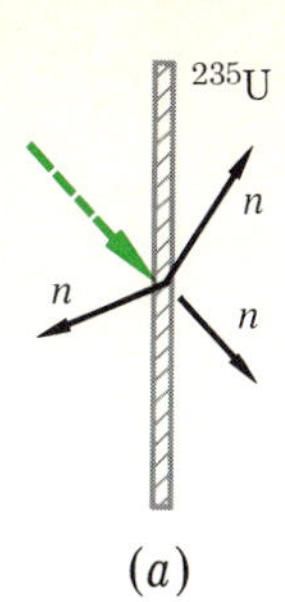

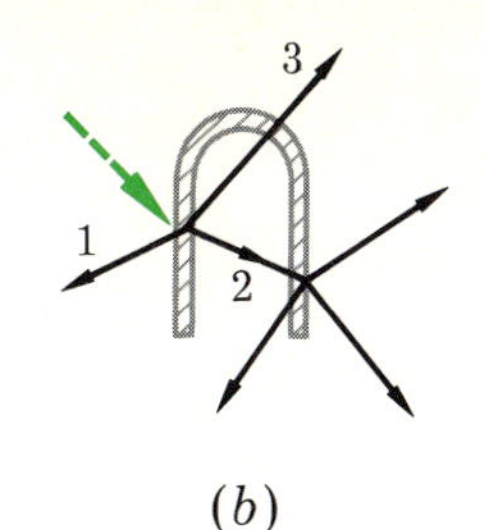

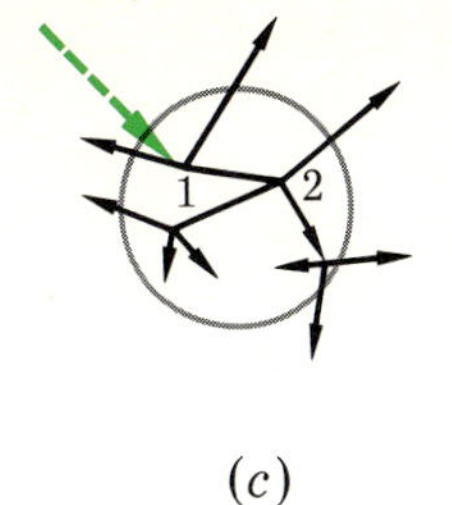

FIGURE 28.9

The efficiency of neutron utilization depends upon the size and shape of the piece of ^{235}U.

reaction causes a further reaction. In this case the reaction will proceed smoothly at its initial slow rate.

If a much larger than critical mass is involved, the reaction will build up at a fast rate and an explosion will occur. This is desirable, of course, if one is making a nuclear weapon. However, in the nuclear reactor, one wishes the reaction to proceed smoothly so that a steady but nonexplosive source of energy results. In practice, the number of reacting neutrons in a reactor is controlled by the use of neutron-absorbing rods. For example, cadmium rods readily absorb neutrons, thereby removing them from the reaction. Hence, if such rods are put into the reactor, the nuclear reaction will slow down. The reaction rate is readily adjusted by positioning rods of this sort in the reactor.

28.9 NUCLEAR REACTORS

The reactor in a nuclear power station serves the same purpose as the furnace in a steam generator. It acts as an intense source of heat, and that heat is used to generate steam. The steam in turn is used to drive the turbines of the electric-generator system. A schematic diagram of a typical reactor is shown in Fig. 28.10.

The heart of the reactor consists of the fissionable material, the fuel, sealed in cylindrical tubes. Originally, uranium 235 was the principal reactor fuel. Now, however, other fissionable materials are also in use as fuel rods. These rods are immersed in a material such as water, carbon, a hydrocarbon,

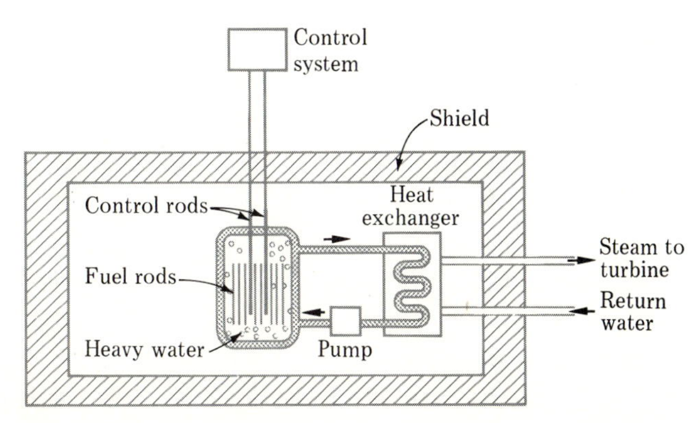

FIGURE 28.10

A schematic diagram of a reactor.

or some similar low-atomic-mass material. This material, the moderator, slows the fission produced neutrons and reflects them back into the fissionable material. (Water made from the isotope ^{2}H rather than ^{1}H is often used since it is less apt to remove neutrons from the reaction.) In the design shown, the moderator also acts as the heat-exchange fluid to carry heat away from the fuel rods.

When a nucleus undergoes fission within the fuel rod, highly unstable intermediate Z nuclei are formed. These undergo extensive radioactive decay and eject high-energy particles in the process. As these particles are slowed, their energy is changed to heat, thereby heating the reactor system. This heat is then carried away to a heat exchanger by the heavy water.

In the heat exchanger, the heat is transferred to ordinary water in a steam-boiler system. Steam is generated and this steam is then used to power electric turbines. As we see, the steam does not come in direct contact with the reactor core. For this reason, its level of radioactivity is low. But the fluid which circulates through the core is bombarded by radiation from the fission products. Like all other portions of the core material, it is often highly radioactive.

When the material in the fuel rods has been used for many months, its original fissionable material is much depleted. The rods are then removed and replaced by new ones. Unfortunately we still have no really satisfactory disposal method for the waste material in the old rods. This material consists of highly radioactive, fairly long-lived fission products. It takes centuries for the radioactivity to decay to harmless levels. Disposal of this waste is one of the major drawbacks of nuclear reactors.

However, reactors also can provide us with radioactive materials for medical, industrial, and other uses. Many of the radiation sources presently used by hospitals, industry, and research laboratories are made by placing suitable materials within the core of the reactor. In addition, research reactors exist in many parts of the world. The intense radiation in their cores can be "piped" outside the reactor to act as powerful beams of radiation. As we see, the fission process has vast potential as well as hazards for mankind.

28.10 THE FUSION REACTION

The fission reaction makes use of the fact that high-Z nuclei possess more energy per nucleon than intermediate-Z nuclei. Let us now consider another energy-releasing process which involves nuclei with small atomic numbers. As we saw in our discussion of the mass defect, *energy is released as protons and neutrons are fused together to form helium nuclei. A similar process is possible when any light nuclei are fused together to form nuclei with intermediate Z. Figure 28.4 tells us that energy must be released in such a fusion process.*

Unfortunately, *nuclear fusion is more difficult than fission.* The electrostatic charge on the nucleus aids the fission process since like charges repel. But *when we try to fuse two protons, for example, the Coulomb repulsion of the two nuclei impedes the fusion process. Only with great difficulty can we bring two light nuclei close enough together for the nuclear attraction force to hold them together.* Even though the overall fusion reaction gives off more energy than was put in, tremendous energies are needed to get the reaction started.

At the present time, the only practical use of the fusion reaction on earth is in the fusion-type nuclear bomb. In such a bomb, the fusion reaction is ignited by a fission-type bomb. The temperatures within the fission bomb are high enough to give the low-Z nuclei enough energy to fuse. Work is progressing throughout the world to find a practical means of producing and controlling the fusion reaction. If this quest proves successful, this reaction may well replace the fission reaction for producing nuclear power. However, this source of power is not expected to become available for many years to come, if ever. In spite of this, the fusion reaction does indirectly provide us with most of our present power since, as we shall see in the next chapter, it is the source of the sun's radiation.

28.11 RADIATION: EFFECTS AND DETECTION

As we make use of nuclear power and other sources of radiation, the effects of radiation on the human body and on materials we use becomes important. A good deal can be learned about high-energy radiations by studying the more or less typical properties of α particles, β particles, neutrons, and γ rays. For example, the interaction of protons with atoms is closely akin to the behavior of α particles, and similarities exist within the other groups mentioned.

An α particle is relatively large (4 u) and carries a double charge ($+2e$). When such a particle is shot at atoms, we would expect it to collide rather frequently. In fact it is found that α particles ionize air very rapidly. As they travel through the air, they occasionally strike an atom and tear an electron loose from it. (The word "strike" is used rather inexactly here. Even if the positive α particle passes close to an atom, the electrical attraction between it and the negative electron can cause ionization.)

By using methods to be discussed later in this section, the path of charged particles through air can actually be seen. In fact, even the individual ions thay produce can be counted. It is found that even quite energetic α particles cannot travel far through air before stopping. For example, the range of the 7.7-MeV α particles emitted from radium C is only about 7 cm in air. The range is, of course, much smaller in denser materials. In aluminum the same α particle would penetrate only about 0.004 cm. As we see, α particles are quite easily stopped.

We should point out that an α particle does not stop after one collision with an atom. Measurements show that about 35 eV of energy is lost for each atom ionized in air. Hence a 7.7-MeV particle would create about 0.2 million ions before coming to rest. The number of ions created by the particle, or alternatively its range, can be used as a measure of the energy of the particle.

A proton has properties similar to the α particle. Since it is only one-fourth as large in mass and one-half as large in charge, the proton would be expected not to ionize air as much as an α particle would. This turns out to be true. A proton will travel about 5 to 10 times as far through matter as an α particle of the same energy. From this it is apparent that protons are less than one-fifth as effective as α particles in ionizing atoms.

The β particle is considerably different from the two particles thus far discussed. It has only about $\frac{1}{1830}$ as much mass as the proton. A bombarding β particle is capable of tearing an atomic electron loose, but in this case this is the result of a repulsive force. In spite of this difference in mechanism, there is really not too much difference between the overall ionization effect of a positive and a negative bombarding particle provided that the bombarding particles have the same mass.

A great difference between the action of β and α particles is apparent, however, because of their mass differences. An α particle striking an electron is much like a 10,000-kg truck striking a 2-kg toy truck. The α particle continues on almost as though it had hit nothing. On the other hand, an electron or β particle colliding with another electron in an atom is like two equal-sized objects colliding. The β particle is deflected considerably when it undergoes a near-head-on collision. Depending upon how the particles collide, a large share of the β-particle energy may be lost in just one collision. However, head-on collisions are rather rare events, and so particle deflections are usually not too large. The β particle will undergo relatively few ionizing collisions, because its mass and charge are small.

The range of a β particle in air is considerably larger than that of an α particle with the same energy. As a rough order-of-magnitude estimate, a β particle will penetrate matter hundreds of times farther than an α particle of similar energy. Although a piece of paper will stop many α particles, it is not uncommon for a β particle to pass through absorbers much thicker than this. Compared with neutrons and γ rays, though, β particles are relatively easily stopped.

Neutrons have no charge and a mass very close to that of a proton. As a result, there is no ordinary electrostatic repulsion or attraction between these particles and the various portions of the atom. Consequently, a neutron will undergo collision only rarely. A direct hit on an electron or nucleus must occur before any disturbance to its travel is noticed. A β or α particle can ionize an atom by a near miss, since the electrostatic forces between the charges act strongly upon the atom even though a true collision does not occur. This is impossible for a neutron, however. Hence *the neutron is a highly penetrating, very slightly ionizing particle.*

Until now we have discussed the behavior of particles only. γ rays are quite different in character from any of these, since they are electromagnetic radiation, having neither charge nor rest mass. Nevertheless, we have already discussed instances where light and x-rays interact with matter, namely, in the photoelectric and Compton effects. *Since γ rays are merely short-wavelength x-rays, we would expect them to show similar characteristics.* This is indeed true.

A γ ray, being a photon, will ordinarily lose all its energy in one event, except in the case of Compton scattering. When a beam of γ rays, or photons, passes through a gas, many of them are stopped when they strike atomic electrons and eject them from the atom. This is simply the photoelectric effect, of course. As we know from our experience with x-rays, however, short-wavelength or high-energy electromagnetic radiation is extremely penetrating. The x-rays which penetrate and pass through our body when an x-ray photograph is taken are, in comparison with γ rays, relatively low-energy radiation, usually less than 0.10 MeV.

The γ rays emitted by nuclei are often several million electronvolts in energy. Since the penetrating ability of γ rays is generally greater for shorter wavelengths, it is clear that γ *rays are highly penetrating.* Very-short-wavelength, i.e., high-energy, γ rays are capable of penetrating several inches of concrete. Since the photoelectric effect accounts for a large fraction of the loss of γ rays from a beam at not too high energies, materials containing a large number of electrons per unit volume should be the best γ-ray absorbers. It is for this reason that the very heavy elements, lead in particular, are used as shielding against γ rays and x-rays.

We see from this discussion that most of these types of radiation cause ionization of atoms in their path. This fact serves as the basis for most methods of detecting them.

Photographic Emulsions When an ionizing particle or photon passes through the emulsion layer on a photographic plate or film, the ions formed in it act as loci for the formation of silver specks when the emulsion is developed. As a result, the ions along the particle's path leave a photographic record of the particle's passage.

Cloud and Bubble Chambers When a fog is about to form in a supersaturated vapor, or when a bubble is about to form in a superheated liquid, the droplets and bubbles form preferentially upon ions. Hence the ions along the paths of a particle can be seen as the droplets or bubbles form. Typical patterns observed in a cloud chamber are shown in Fig. 28.11. The patterns seen in a bubble chamber are much the same except that the path lengths are much shortened because of difference in density between liquid and gas. Often a magnetic field is superimposed on the chamber, causing the paths of charged particles to be curved and thereby yielding additional information about the particles.

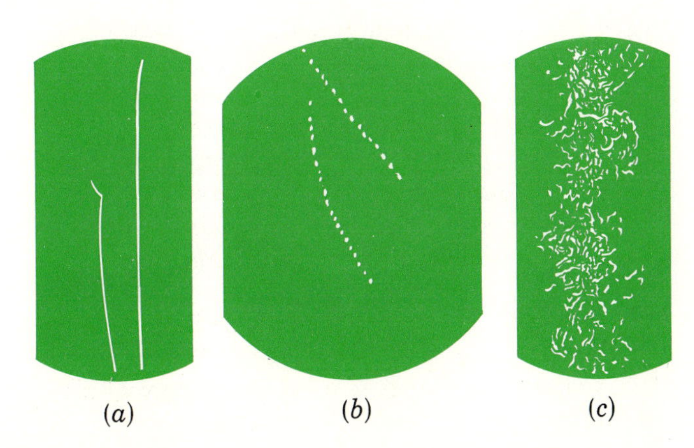

FIGURE 28.11

(*a*) The tracks of two α particles. (*b*) Two β-particle tracks. (*c*) A γ-ray beam leaves no track, but its path is shown by the β-particle tracks left by the electrons ejected by the γ rays.

Geiger Counter This portable device is widely used for detection of radiation. Its basic construction is shown in Fig. 28.12. A high potential difference exists between a metal tube and a wire on its axis. The tube is filled with a special gas mixture at reduced pressure. When a particle enters the counter through the thin entrance window, it forms ions in the tube. The current which flows because of these ions alone is extremely small, but the high voltage causes the ions to initiate a discharge within the tube and so, momentarily, the tube allows a current to pass. Due to reasons we cannot pursue here, the current stops in a small fraction of a second. Hence, each time a particle passes through the counting tube, a pulse of current passes through the circuit. Any recording device (a loudspeaker, meter, etc.) connected across terminals *ab* will register each time such a pulse occurs, and so the particles passing through the counter tube can be counted.

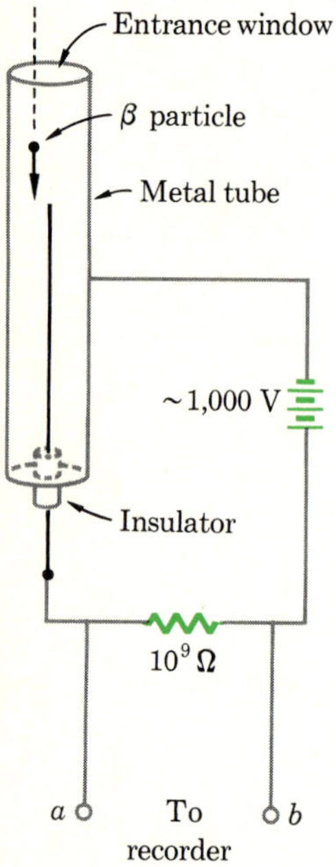

FIGURE 28.12
The Geiger counter.

Scintillation Counter When a fast-moving particle strikes a fluorescent material, a pulse of light is given off. This fact is basic to picture formation in a TV tube and is the principle upon which the scintillation counter is based. As shown in Fig. 28.13, a phototube is used to detect the light pulse, but since the light pulse from a single particle is so small, a special phototube called a **photomultiplier** tube is used. Its principle of operation is shown in the figure. This device, like the Geiger counter, is used widely for radiation detection.

Solid-State Detectors These devices are basically solid-state semiconductor diodes. As you recall, no current will pass through such a device in the reverse direction. However, if a fast-moving particle passes through the junction region between the *n*- and *p*-type semiconductor, a current pulse passes through the diode. This pulse can be used to count the particles in the same way as for the previous two devices.

28.12 RADIATION DOSE

As we saw in the previous section, high-energy radiation is capable of ionizing atoms, and it can tear molecules apart. The resultant damage is of

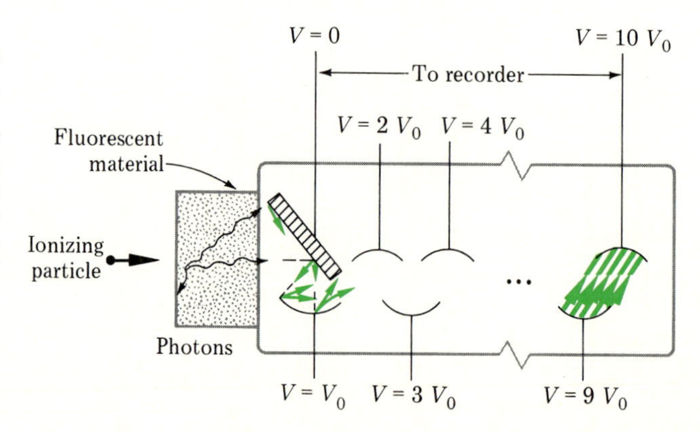

FIGURE 28.13
In the scintillation counter, light from the scintillator causes photoemission of electrons in the phototube. In the first stage, electrons accelerate through the potential V_0, and each liberates several secondary electrons. This multiplicative process continues and results in a comparatively large output current pulse, much larger than indicated in the figure.

importance and we shall discuss it in due course. First let us learn some of the units used to describe the effects of radiation.

In many applications of radiation, the beneficial (or deleterious) effects of the radiation are roughly proportional to the amount of radiation energy absorbed. Therefore, a unit is needed to measure the energy absorbed from a radiation beam by a material. *The unit of absorbed radiation energy is called the* **rad** (*rd*). It is defined as follows: *When 1 g of material exposed to the beam absorbs 10^{-5} J of radiation energy, the absorbed dose is* **1 rd.** *Notice that the rad is a measure of absorbed energy per unit mass.*

DEFINITION

Unit of Absorbed Radiation

Because of the way the rad is defined, the same beam will result in different doses for different materials. A beam passing through human flesh might well be less absorbed than when passing through bone. As a result, a single beam passing through a person would cause a higher dose to the bone through which it passed than to the flesh.

Unfortunately, the rad is not a very good unit for measuring the effect of radiation on people. The difficulty lies in the fact that different types of radiation cause different types of damage to human tissue. For example, a dose of 1 rd from an electron beam causes only about one-tenth as much damage as an equal dose from a beam of neutrons or protons. Although the rad is a convenient unit for comparing effects of the same type of beam, it does not lend itself well when different types of radiation are compared. For that reason, we make use of another unit.

The biological effect of absorbed radiation is measured in terms of a unit called the rem (rad equivalent-man). It is a comparative unit in the sense that it measures the effect by comparing it to the effect caused by a 1-MeV x-ray beam. When one rad of x-radiation is absorbed, a certain amount of biological damage occurs. *We define the* **rem** *in such a way that a dose of 1 rem causes biological damage equivalent to a dose of 1 rd of 1-MeV x-rays.*

The REM Unit

DEFINITION

To place the rem unit in a more useful relation to the rad, *a quantity called the* **quality factor** (QF) *of the radiation is introduced. It consists of a ratio* arrived at in the following way. The radiation in question is used to irradiate the biological material so as to cause an absorbed dose of 1 rd. This same type of material is then irradiated by a 1-MeV x-ray beam until an equal amount of biological damage is done. The absorbed dose in rads for this process is noted. We then define the quality factor as the ratio of these two doses:

$$\text{QF} = \frac{\text{equivalent dose of x-rays}}{\text{1 rd of radiation in question}}$$

The Quality Factor or RBE

As a typical example, an x-ray dose of 20 rd is required to cause the same damage as a 1-rd dose of fast α particles. The QF for fast α particles is therefore 20. (The QF is often called the RBE, relative biological effectiveness.) Typical approximate values of QF are given in Table 28.1. (By reference to the ideas of the previous section, can you explain qualitatively why α particles should be much more damaging than electrons?)

TABLE 28.1
APPROXIMATE VALUES OF QF (or RBE)

RADIATION	APPROXIMATE QF
X-rays, electrons	1.0
Fast neutrons and protons	10
Effect on eye	30
α Particles	10–20
Slow neutrons	4–5

From the definitions of the rem and QF, we can write the following relation:

$$\text{Dose in rem} = (\text{QF})(\text{dose in rad})$$

Notice that the units of QF are rems per rad.

Illustration 28.4 How large a dose of fast neutrons is equivalent to a 50-mrd dose of slow neutrons?

Reasoning Let us first find the number of rems in a 50-mrd dose of slow neutrons. We have

$$\text{Dose in rems} = (\text{QF})(\text{dose in rads})$$

with QF $\approx$ 4.5. Therefore, for slow neutrons,

$$\text{Dose in rems} = (4.5\ \text{rems/rd})(50 \times 10^{-3}\ \text{rd}) = 0.225\ \text{rem}$$

It is now possible to apply this same equation to the fast neutrons by use of QF = 10 and dose in rems = 0.225 rem. Then, for fast neutrons,

$$0.225\ \text{rem} = (10\ \text{rems/rd})(\text{dose in rads})$$

which gives 0.0225 rd as the required dose of fast neutrons.

DEFINITION

There is yet another unit often used to describe radiation. It is called the **roentgen** (R) and is used primarily for x-rays; *1 R is defined to be the amount of radiation which will produce 2.1 × 10*^{9} *ion pairs in 1 cm*^{3} *of air under standard conditions.*

28.13 RADIATION DAMAGE

Since radiation can tear apart molecules, it is capable of damaging materials. One of the most common types of radiation damage is due to the ultraviolet rays in sunlight. These lead to sunburn and tanning of the skin. The high-energy photons disrupt skin molecules upon impact and cause these easily observed effects. In this case, the damage is usually of little importance. Most of the sun's ultraviolet rays are absorbed by the ozone in the upper atmosphere so normal exposure to the sun's rays need not be avoided. However, in recent years we have become aware that a serious hazard could arise if we deplete the ozone layer with man-made chemicals. There is danger then that the increased ultraviolet radiation reaching us could increase the incidence of skin cancer.

Background Radiation

We are continuously exposed to other radiation in addition to sunlight. Nearly all materials contain a slight amount of radioactive substances. As a result, your body is unavoidably exposed to a low level of background radiation. Typically, each person experiences a background radiation dose of about 0.15 rem each year. Let us now examine the effects of different levels of radiation dose upon the body.

Radiation Damage to Human Beings

High levels of radiation covering the whole body disrupt the blood cells so seriously that life cannot be maintained. For whole-body doses in excess of 500 rem, death is likely to occur. Even a whole-body dose of 100 rem can cause radiation sickness of a very serious, although nonfatal, nature. Blood abnormalities occur for doses in the range of 30 rem and above. At still lower whole-body doses, the overall effects on the body are less apparent but nevertheless can cause serious consequences.

Even very low radiation doses reaching the reproductive regions of the body are potentially dangerous. The giant molecules in our bodies which carry reproductive information can be disrupted by a single radiation impact. If enough of these molecules are damaged, defective reproduction information will be furnished to a fetus as it develops. As a result, birth abnormalities will occur. Even though there is some evidence that a low level of reproduction abnormalities may be beneficial to mankind, most birth defects are not desirable. For this reason, *no one of child-bearing age should be exposed to unnecessary radiation of the reproductive organs.* Of course a properly given arm x-ray, for example, presents no such danger.

In addition to causing birth abnormalities, low levels of radiation present two other hazards. First, there appears to be a delayed cancer effect. Although cancer may not appear at once, low levels of radiation may cause cancer to develop many years later. Second, a child is particularly vulnerable to radiation. Because the child is growing rapidly, any cell mutations caused by radiation could have serious consequences. For this reason, most doctors are reluctant to prescribe x-ray scans for children unless absolutely necessary.

There is no "safe" limit of body exposure to radiation. It can only be said that radiation should be kept to the least value possible within reason. For example, since we are all subjected to a background radiation of about 0.15 rem/yr, there is no reason to disrupt our lives to avoid radiation doses less than this. Even though a person who lives in the mountains may experience an annual background dose 0.05 rem higher than at sea level, the difference is not large enough to warrant moving. In the last analysis, one must often make a compromise between radiation safety and other considerations. Despite that fact, maximum occupational radiation doses are of value and have been specified. As a rough rule, the maximum yearly dose, except for the eyes and reproductive organs, is about 15 rems.

28.14 BASIC PARTICLES

Thus far in this text we have presented well-understood laws of nature and their consequences. Although from time to time we have pointed out areas of uncertainty, these have been mostly peripheral. Now, however, we move to a topic fraught with uncertainty and conjecture—the basic particles which exist in nature.

About a half century ago, the situation appeared simple. We knew of the electron, proton, neutron, and photon. These were considered to be the basic particles of the universe. But as the years have gone by, many new particles have been discovered. Powerful accelerating machines have thrown particle against particle with tremendous momentum. The net result has been the production (or creation) of many types of new particles ranging in rest mass from zero to several atomic mass units. Their half-lives vary from less than 10^{-16} s to infinity. We are almost certain that other new particles will be discovered in the future.

At present we have no satisfactory proved interpretation for these particles. Theories for their interrelation exist, but discoveries are occurring so rapidly that no single theory has remained acceptable, unchanged or unextended, for long. You can obtain a good idea of the present situation by referring to a recent readable article on the subject.* By the time you read it, it, too, will probably be out of date.

In spite of this chaos, there are some results which we know to be important. They involve fundamental facts applicable to all particles. A final—or at least moderately acceptable—theory of these particles will make use of them. Let us see what some of these basic facts are.

1 All particles obey the conservation laws of energy, linear momentum, and charge. When particles decay into other particles, or when new particles are formed in bombarding reactions, these laws still apply.

2 All particles obey the law of conservation of angular momentum; i.e., spin is conserved. Moreover, spin is quantized. A particle can have an angular momentum (pictured as being due to its spin on its axis) of $s(h/2\pi)$, where $s = 0, \frac{1}{2}, 1, \frac{3}{2}, \ldots$ and h is Planck's constant. We call s the spin of the particle, and each particle always has the same value for s. The electron, proton, and neutron all have $s = \frac{1}{2}$. They are called spin-$\frac{1}{2}$ particles. The photon has $s = 1$. Mesons have zero spin, and so on. In any process, the angular momentum of the reactants must equal that of the products.

3 Baryon number is conserved. Spin-$\frac{1}{2}$ particles with masses equal to or larger than the proton are called **baryons.** Each baryon is assigned a number, either $+1$ or -1, in a way described in the next paragraph. In any process, the baryon number at the start must be the same as at the end.

4 Each particle has an antiparticle. For every particle there is another particle which is "identical but opposite" to it. Not only does an electron exist, but so does an antielectron, which we call the positron. The positron has the same mass as the electron, but it has opposite charge $+e$. When an

---

*S. L. Glashow, Quarks with Color and Flavor, *Sci. Am.*, October 1975, p. 38.

electron meets a positron, they annihilate each other. The proton also has an antiparticle, the antiproton. For the neutron there is an antineutron, and so on. In assigning baryon numbers, particles are $+1$ and antiparticles are -1.

5 All spin-$\frac{1}{2}$ particles obey Fermi-Dirac statistics. By this we mean that no two of these particles can have the exact same set of quantum numbers in any one system. For electrons in the atom, this reduces to the Pauli exclusion principle discussed previously. However, the exclusion principle also applies to spin-$\frac{1}{2}$ particles in the nucleus and in other particle aggregates.

In addition to these facts, there are others less well established. Other conserved quantities have been proposed; strangeness and iso spin are two of them. Despite their difficulties in solving the problem of the fundamental particles, physicists still remain in good humor. Perhaps as an unconscious mechanism to keep spirits high, they describe properties of particles in such whimsical terms as "strangeness," "charm," "color," and so on. Nevertheless, theirs is research of the highest caliber and most intense effort. The study of the basic particles probes the deepest depths of our universe. We expect that a way will soon be found to bring order out of the present state of chaos.

SUMMARY

The nucleus of an atom with atomic number Z and atomic mass number A contains Z protons and $A - Z$ neutrons. Its radius is given approximately by $R = 1.2 \times 10^{-15} A^{1/3}$ m.

On the atomic-mass scale, the carbon 12 atom has a mass of exactly 12 u. The proton and neutron masses are close to 1.0 u.

A nucleus of element X with mass number A and atomic number Z is represented by ^A_ZX. Two nuclei having the same Z but differing in A are isotopes of the same element. They act the same in chemical reactions but differ in the number of neutrons in their nuclei. The masses of isotopes (in atomic mass units) are close to integers. The atomic masses given in the periodic table are averages of the isotopes existing in nature.

All nuclei have less mass than the combined masses of their individual protons and neutrons. The mass Δm lost in assembling the nucleus is related to the binding energy (BE) of the nucleus through $\text{BE} = \Delta m\, c^2$. In effect, this lost mass must be created when the nucleus is separated into its parts.

Nuclei of intermediate Z elements have the highest binding energy per nucleon. As a result, energy is given off in the fission and fusion reactions. Nuclear reactors make use of the fission reaction to generate energy.

All elements with $Z > 83$ are radioactive. Radioactive elements follow an exponential decay law. The time taken for half of the nuclei to decay is called the half-life. The half-life $= 0.693/\lambda$, where λ is the decay constant. During a short time interval Δt the number of nuclei which decay, ΔN, is related to the number of nuclei available for decay N by $\Delta N = \lambda N\, \Delta t$.

Radioactive materials formed in nature emit α and β particles and γ rays; α particles are helium nuclei; β particles are electrons; γ rays are the same as x-rays. In order of decreasing ionizing ability and increasing penetration ability, these are α, β, and γ.

The nuclear force holds the nucleons together. Unlike the electric and gravitational forces, it is a short-range force. It acts only over distances of less than about 10^{-14} m. At such small distances, however, it is much larger than the Coulomb force.

Absorbed radiation dose is measured in terms of a unit called the rad. One rad is equivalent to an absorbed dose of 10^{-5} J in each gram of material. The dose in rems = (QF) (dose in rads), where the quality factor (QF) measures the relative biological effectiveness of the radiation in question.

MINIMUM LEARNING GOALS

Upon completion of this chapter you should be able to do the following:

1. Give an order-of-magnitude value for the radius of a nucleus.
2. Distinguish between atomic number Z and atomic mass number A. Relate them to the number of protons and neutrons in a nucleus.
3. Interpret symbols like ${}^{6}_{3}\text{Li}$.
4. Explain what is meant by an isotope.
5. Define the atomic mass unit and give the approximate masses of the proton and neutron in this unit.
6. Sketch a graph of the mass defect per nucleon as a function of Z.
7. Sketch a graph of the binding energy per nucleon versus Z and explain why it is related to the graph in 6.
8. Sketch a graph of N versus t for a radioactive substance. Show the half-life on the graph. Compute the fraction of material not decayed after a length of time equal to a given integer number of half-lives.
9. Define each quantity in the equation $\Delta N = \lambda N \Delta t$ and be able to use the equation in simple situations. Give the relation between λ and the half-life.
10. Define α particle, β particle, γ ray, positron.
11. Write the nuclear reaction equation for a given nucleus which emits one of the following: α particle, β particle, γ ray.
12. Prepare a diagram such as Fig. 28.6 for a series in which the starting nucleus and the emitted particles (α, β, and γ) are given.
13. Compare the nuclear force to the gravitational and Coulomb forces in regard to strength, range, and dependence on r.
14. Explain why the nuclear fission reaction should release energy by reference to the mass-defect and binding-energy graphs. State what is meant by a fission chain reaction and relate this to why ${}^{235}\text{U}$, but not ${}^{238}\text{U}$, is usable in a nuclear bomb.
15. Sketch a schematic diagram of a nuclear power reactor showing fuel rods, moderator, control rods, heat exchanger, and output to turbine and explain the function of each.
16. Explain why the nuclear fusion reaction should release energy by reference to the mass-defect and binding-energy graphs. State why fusion is much more difficult to achieve in a laboratory reactor than fission.
17. Compare the range and ionization effect of α, β, and γ radiation when passing through matter.
18. Define each of the following: rad, rem, quality factor (or relative biological effectiveness). Give the equation relating the three. Use the equation in simple problems such as Illustration 28.4.
19. Describe the following: Geiger counter, scintillation counter, cloud chamber, bubble chamber.
20. Explain why radiation can be harmful to people. In your explanation, point out which regions of the body and which type of person should be particularly well shielded from radiation.
21. State five conservation laws obeyed by all basic particles in all reactions.

IMPORTANT TERMS AND PHRASES

You should be able to define or explain each of the following:

Atomic number Z; atomic mass number A
Isotope
Atomic mass unit u
Mass defect; binding energy
Radioactive; half-life
Decay constant; $\Delta N = \lambda N \Delta t$
Curie unit
α particle; β particle; γ ray; positron
Nuclear force
Fission; chain reaction
Nuclear reactor
Fusion reaction
Rad, rem, and QF (or RBE)
Radiation is particularly dangerous for those of child-bearing age and to children

QUESTIONS AND GUESSTIMATES

1. How many neutrons are there in the nucleus of $^{39}_{19}K$? How many protons? How many electrons does this atom have?

2. Why do chemists consider different isotopes to be the same element even though their nuclei are not the same?

3. Would the optical spectra of ^{35}Cl and ^{37}Cl atoms differ in any major way? Explain.

4. Even though an α particle is attracted rather than repelled by an electron, it is still capable of knocking an electron loose from the atom without a direct collision. Explain.

5. Why is a given thickness of lead a better absorber of 1-MeV α, β, and γ rays than the same thickness of water? Why is water a better shielding barrier against neutrons than lead?

6. The artificially produced isotope ^{102}Ag has a half-life of 73 min. It decays in two alternative ways. Part of the nuclei emit a positive electron, a positron. The rest capture one of the electrons in the first Bohr orbit and take it into the nucleus. In what way are the two equivalent? What would one notice in the laboratory for each process?

7. Tritium is the ^{3}H isotope of hydrogen. Its atomic mass is 3.016; the atomic mass of ^{1}H is 1.0078. Using the fact that the neutron mass is 1.00867, what do you predict about the stability of tritium? Repeat for ^{2}H, deuterium, which has a mass of 2.0141.

8. Before an x-ray survey is made of the gastrointestinal tract, the patient must drink a solution of a barium compound. Why?

9. When an x-ray photograph is taken of a person's arm, the bones are clearly shown on the photograph. Why do they show up in this way? After all, the arm is no thicker where the bone is than elsewhere.

10. A small amount of radium is sealed in an evacuated glass tube. When it is later broken open, the tube is found to contain some gas. The mass spectrometer shows the gas to consist of a very-low-molecular-weight species and a very-high-atomic-weight species. What are they? Estimate their relative abundance.

11. In atomic mass units the mass of an electron is 0.000549, of a proton 1.00728, and of a neutron 1.00867. Is it feasible for a proton and electron both at rest to combine and form a neutron? Explain. Would it matter if they were close together or far apart when they were at rest?

12. Why is the fusion reaction so much more difficult to initiate than the fission reaction?

13. Is there any possibility of causing the fusion of two small nuclei without making them collide with extremely high energy? Defend your answer.

14. A nuclear reactor produces energy chiefly in the form of heat. Explain how this heat is generated as a result of nuclear fission.

15. The fission and fusion reactions appear to create energy. How can we reconcile this fact with the statement that energy can neither be created nor destroyed?

16. It is possible for a man working with x-rays to burn his hand so seriously that he must have it amputated, and yet the man may suffer no other consequences. However, an x-ray overexposure so slight as to cause no observable damage to his body could cause one of his subsequent offspring to be seriously deformed. Explain why.

17. Most radiologists feel that women beyond child-bearing age can safely be exposed to much more x-radiation than young women. How can they justify such an opinion?

18. Low-energy (soft) x-rays are used to treat skin cancer, while high-energy (hard) x-rays are used to treat cancer deep within the body. Why are soft x-rays not used for this latter purpose, even though they can penetrate deeply enough to kill the cancerous region?

PROBLEMS

1. (*a*) Estimate the radius of a proton and (*b*) determine its average density in grams per cubic centimeter.

2. Compute the approximate radius of the ^{235}U nucleus and its density in grams per cubic centimeter.

3. In a mass spectrograph like that shown in Fig. 28.1, it is found that r for ^{12}C is 10.00 cm. How large would r be for ^{16}O? Assume identical charges and accelerating potentials.

4. Potassium found in nature contains essentially only two isotopes. One has an atomic mass of 38.975 and constitutes 93.4 percent of the whole. The other 6.6 percent has a mass of 40.974 u. Compute from these data the atomic mass which the chemists list in the periodic table.

5. Two isotopes exist in commercial copper. One constitutes 70 percent of the whole and has an atomic mass of 62.96 u. The chemists assign a mass of 63.58 u to copper. What is the mass of the second isotope?

6. The α particle emitted by radium has an energy of 4.79 MeV. Compute the speed of the particle.

7. A radioactive series in addition to that shown in Fig. 28.6, the thorium series, starts with ${}^{232}_{90}\text{Th}$ and emits in succession one α, two β, four α, one β, one α, and one β particles. What is the final product of the series?

8. Approximately what fraction of a sample of radium will be left after 16,000 yr?

9. A tiny ampoule of radon gas contains 3.0×10^{12} atoms of radon. The half-life of radon is 3.8 days. (*a*) How many disintegrations occur in the ampoule each second? (*b*) What is the activity of this sample in curies?

10. Watches with numerals visible in the dark often have radioactive material in the paint used for the numerals. A student estimates from measurements using a Geiger counter that 1000 disintegrations occur each second on the watch face. How many curies of radioactivity exist on the watch if the student's figures are correct?

11.** Cobalt 60 has a half-life of 5.3 yr. What is the mass of 1 Ci of this material? (Cobalt 60 emits γ rays and is widely used in medical and industrial applications where very penetrating x-rays are required.)

12. The mass of the most abundant lithium nucleus is 7.014 u. Using 1.0073 and 1.0087 u for the masses of the proton and neutron, respectively, compute the binding energy of lithium in electronvolts.

13. The mass of carbon 12 atoms is defined to be exactly 12.000 u. Subtracting the mass of its six electrons shows that the nucleus of ${}^{12}\text{C}$ has a mass of 11.997 u. Using the data from Prob. 12, compute the least possible energy required to separate this nucleus into its constituent neutrons and protons.

14. Consider the nuclear reaction

$$\underset{13.9993}{{}^{14}_{7}\text{N}} + \underset{4.0016}{{}^{4}_{2}\alpha} \rightarrow \underset{16.9948}{{}^{17}_{8}\text{O}} + \underset{1.0073}{{}^{1}_{1}\text{H}}$$

where the masses *of the nuclei* in atomic mass units are also given. What is the least energy the α particle must have if the reaction is to be possible?

15. When a nucleus throws out a γ ray of energy 2.0 MeV, by how much does the nuclear mass change?

16.** A neutron with speed 10^6 m/s hits a stationary deuterium atom (heavy hydrogen, ${}^{2}_{1}\text{H}$) head on in a perfectly elastic collision. (*a*) Find its speed after collision. (*b*) Repeat if the deuterium atom is replaced by an oxygen atom, ${}^{16}_{8}\text{C}$. Notice that light nuclei are most effective in slowing neutrons.

17. The background radiation at Dallas, Texas, can be divided into two types. It furnishes a yearly dose of about 30 mrd of γ radiation and about 1.5 mrd of particles having a QF of about 20. How many rems will a resident of Dallas experience each year from this cause?

18. Some authorities believe that the maximum allowed dose to a person's hands should be 70 rems/yr. Express this as dose in rads if the radiation being used is (*a*) x-rays; (*b*) γ rays; (*c*) fast electrons: (*d*) fast neutrons.

19.** Suppose 1 kg of deuterium (heavy hydrogen, ${}^{2}\text{H}$) is combined to form 1 kg of helium according to the reaction

$$\underset{2.0141}{{}^{2}_{1}\text{H}} + \underset{2.0141}{{}^{2}_{1}\text{H}} \rightarrow \underset{4.0026}{{}^{4}_{2}\text{He}}$$

where the atomic masses are given. (*a*) How much energy is liberated? (*b*) If the confined helium has a specific heat capacity of 0.75 cal/(g)(°C), by how much does its temperature increase as this energy is added to it?

20.** Neutrons are frequently detected by allowing them to be captured by a boron nucleus. This reaction is

$$\underset{1.0087}{{}^{1}_{0}n} + \underset{10.0129}{{}^{10}_{5}\text{B}} \rightarrow \underset{7.0160}{{}^{7}_{3}\text{Li}} + \underset{4.0026}{{}^{4}_{2}\text{He}}$$

where the mass of the atoms is given without subtracting the electron masses. (Since there are the same number of electrons on each side of the equation, this will be of no consequence. Why?) Energy given off in the reaction appears as KE of the products. The resulting swiftly moving helium ions (α particles) are then counted, using ordinary techniques. Compute the speed of the α particle, assuming the original reactants to be essentially at rest. *Hint:* Compute the energy liberated in the reaction, and then write down the laws of conservation of energy and momentum.

29 PHYSICS OF THE UNIVERSE

The past several chapters have been concerned with the smallest bits of matter found in nature—atoms, nuclei, and the basic particles. Now we turn our attention to the other extreme, the universe itself. In this chapter we shall examine its major features and some theories of its origin.

29.1 THE PRIMORDIAL FIREBALL

Scientists discover how nature behaves; they correlate these discoveries to bring order to a multitude of seemingly disconnected facts; they construct theories to explain the order found; they test the theories by predicting the results of experiments still not performed. Experimental results are at the heart of all science. Without experiment, science is no longer science but is guesswork, philosophy, or theology.

Astrophysics is greatly hampered by the lack of our ability to perform the really crucial experiments. The most important act of all, the formation of the universe, was performed several billions of years ago and is still going on. We cannot duplicate this process. All our knowledge of the origin of the universe must be gleaned from this one "experiment" over which we have no control. Many of the pertinent data should have been noted billions of years before mankind existed. Other data will not become available until billions of years in the future. We have no alternative but to work with the few data we can acquire by the limited means available. For that reason, most of what is said in this chapter must be viewed with caution; many of our interpretations may later be proved incorrect.

The Big-Bang Theory

Until recently there were several competing theories for the earliest history of the universe. At present only one is widely accepted, that formulated in its current form in 1948 by G. Gamow. It is called the **big-bang theory.** In the beginning, it pictures the whole universe to have been confined to a ball with a diameter perhaps about 10 times that of the present sun. One may object that this is impossible: all the material of the universe could not be packed into such a small region. However, if one recalls that the atom is nearly all empty space, any idea based upon the bulkiness of matter as we know it is unreliable.

For example, in Probs. 1 and 2 of the last chapter we learned that the density within a nucleus is of the order of 2×10^{14} g/cm^3. Since the present estimate of the average density of the universe lies in the range between 10^{-29} and 10^{-31} g/cm^3 while its radius is thought to be of the order of 10^{10} light-years,* the total mass of the universe is probably within a few orders of magnitude of 10^{54} g. Packing all this mass into a volume 10 times the diameter of the sun would give a density of about 10^{18} g/cm^3. This is close enough to the density within nuclei (which themselves may be largely empty space) for the original concept certainly not to be ruled out by the enormous density it would involve. At such high densities, matter as we know it could not exist. Certainly there would have been no atoms or molecules or even nuclei at that stage. The energy would have been so great that the effective temperature in the fireball would have been at least 10^{12} K. At these very high energies the nuclei of all atoms would be torn apart; their binding energies correspond to only a small fraction of the thermal energy at this temperature.

We therefore picture the universe to have been an extremely hot fireball at the outset (it is only a hypothesis of course). It was a cauldron of energy and

---

*It will be recalled that one light-year is the distance light travels in a year, namely, 9.46×10^{15} m.

charge. If particles existed, they would have been the very-high-energy particles which we can produce in the laboratory only with huge accelerators. In a sense, at least, we can think of the fireball as an extremely hot gas of highly energetic particles and photons. *Like any gas, the fireball then began to expand.*

As it did, work was performed against gravitational forces. The huge pressure within the fireball did work during the expansion against the tremendous gravitational forces holding the fireball together. As a result, the material of the fireball lost thermal KE as the gravitational PE increased. *Consequently the temperature of the fireball dropped rapidly as the expansion continued.* Nevertheless, the expansion process accelerated the outgoing matter to speeds near the speed of light. The universe appears to be expanding even now with a speed close to c.

It is interesting to speculate on the eventual fate of our expanding universe. Although the expansion is still being slowed by gravitational forces, we do not know enough about the universe to say whether it is expanding with energies large enough to provide the escape velocity for the matter within it. If it has enough energy, the universe will continue to expand forever. But if not, the expansion will eventually stop. Gravitational forces will begin to draw the universe back again into the primordial fireball from which it came. If this second alternative applies, eventually the fireball will be recreated and the whole process will be repeated. In fact, the fireball we have visualized may simply be the latest of a series of pulsations of a pulsating universe. It would appear that we can never learn the history of the universe during previous pulsations even if they existed. Can you say why?

29.2 GALAXIES AND STARS

As the original fireball expanded into space, its temperature soon reduced enough to permit particles to be formed. A typical reaction in which this occurs is the process called **pair production,** in which a photon is changed into an electron and positron. (This is the reverse of the antiparticle annihilation reaction.) There are other reactions of this general type where, in effect, energy is changed into rest mass and equal but opposite charges are created. The end result, as the fireball cooled, was to form a gas composed primarily of protons, neutrons, electrons, and other high-energy particles. The temperature was still too high to permit formation of hydrogen atoms, and, of course, nuclei for larger atoms were still unformed.

We can estimate the temperature of the fireball when hydrogen atoms began to form. The ionization energy of hydrogen is 13.6 eV. Clearly, thermal energy kT could not be much larger than this if hydrogen atoms were to survive. Since 300 K is equivalent to $\frac{1}{40}$ eV, 13.6 eV would correspond to a temperature of about 160,000 K. As its temperature fell below this value, the fireball became a cloud of hot hydrogen gas together with neutrons and other basic particles. As we shall see later, the cloud of gas filling the universe has become so cold its temperature is now of the order of 3 K.

So far we have discussed the expanding cloud as a smooth, homogeneous entity. In all probability, the gas was not completely uniformly distributed throughout space. Certain regions must have had higher densities than others as a result of random factors, e.g., thermal motion. Although we have no direct proof of the hypothesis, it seems reasonable that regions of unusually high density acted as focal points for what one might best describe as a gravitational condensation. Over huge regions of space, unbalanced gravitational forces were produced and made the matter start moving toward regions of higher density. Of course, superimposed upon this was the continued radial motion of the material of the cloud out from the origin of the fireball. *These huge, noncompact regions into which matter began to collect are presently huge* **galaxies,** *aggregate systems containing many stars.*

Formation of Galaxies

As the cloud began to form regions of higher density, the masses had to obey the usual laws of motion. One, the law of conservation of angular momentum, had a profound effect upon the behavior of the condensing cloud. Any small net angular motion ω_0 of the mass within the huge region originally occupied by the galaxy must be multiplied by the moment of inertia of that region I_0 to obtain the original angular momentum $I_0\omega_0$. However, as aggregation into a galaxy took place, the radius r of gyration for the galaxy's mass became much smaller, thereby decreasing I since $I \propto r^2$. Since the angular momentum must be conserved, $I\omega = I_0\omega_0$; and as I decreases, ω must increase. Therefore, *as condensation took place, the material began to spin about the center of the condensation.*

Spinning of the condensing cloud mass would be fastest for regions with the largest rotation at the outset. We can therefore expect some galaxies to be spinning very little while others have relatively large angular speeds ("large" in this context still means millions of years per rotation). As with any spinning system (a stone on a string or a twirling pizza, for example), the spinning causes the system to flatten into a disk with the axis of rotation perpendicular to its plane. For the fastest-spinning galaxies, the spiral galaxies, this effect is very pronounced (Fig. 29.1). *Our own solar system is part of a somewhat less spiraling galaxy, the Milky Way, and is located about two-thirds of the way out from the galaxy center.* (There are about 100 billion stars in our galaxy, the diameter of which is of the order of 100,000 light-years. It takes our sun about 200 million years to rotate once around the galactic center with its approximately 150 mi/s orbital speed.)

While a galaxy as a whole is in the process of formation, localized regions in it are condensing much more rapidly into centers of mass, which eventually become the stars in the galaxy. It is easy to see how they become intensely hot. As matter is pulled toward the center by gravitational forces, PE is converted to KE. This means that a tremendous amount of energy is carried to the condensation center by the aggregating mass. Hence the temperature of the aggregate can become extremely high and in fact exceeds the temperatures needed for complete ionization of all the hydrogen atoms in the aggregate. This huge mass, perhaps many times larger than our sun, is now a white hot, very dense "soup," called a **plasma,** composed of protons, neutrons, electrons, and other basic particles. This is the birth of a star.

FIGURE 29.1

The spiral galaxy M81. (*Lick Observatory photograph.*)

29.3 STELLAR EVOLUTION

Pulled radially inward by gravitational forces, the star will continue to contract until its internal pressure balances the gravitational pressure. As in any gas, the internal pressure increases with rising temperature. For small aggregates of matter, the gravitational forces are small, and so equilibrium is reached at rather low temperatures. For the large stars of interest, however, contraction continues until the temperature in the star reaches a few million degrees. At this high temperature the average thermal energy of a proton is about 1000 eV. In spite of this rather low average energy, enough high-energy protons exist for a fusion reaction to become possible.

Fusion Reactions in Stars

At the earliest stages, the fusion reaction of importance is

$$^1_1\text{H} + {}^1_1\text{H} \rightarrow {}^2_1\text{H} + \text{positron} + \text{neutrino}$$

where the neutrino is a zero-charge particle which has no rest mass. The deuteron, ${}_1^2H$, reacts again with a proton:

$$ {}_1^2H + {}_1^1H \rightarrow {}_2^3He + \gamma $$

and this isotope of helium reacts as follows:

$$ {}_2^3He + {}_2^3He \rightarrow {}_2^4He + 2\ {}_1^1H $$

In other words, protons are fused together in this reaction to form helium; six protons react to form a helium nucleus and two protons. As with any fusion reaction of this type, large amounts of energy are released. The star is now capable of supplying energy to itself without further gravitational collapse.

As soon as the fusion reaction has become powerful enough to increase the thermal pressure within the star to a point where it balances the gravitational pressure, the star stabilizes. The temperature at the very center of the sun, the hottest point, is estimated to be close to 15 million kelvins. Although the proton reaction in the sun has been proceeding for about 4.5 billion years, apparently enough protons remain for the reaction to continue steadily for about that long in the future. In stars more massive than our own sun, the interior temperature is higher. (Why?) At these higher temperatures, other nuclear fusion reactions are possible.

Red Giant Stars

Eventually the proton reaction uses up most of the available protons, and the reaction slows down. As a result, the thermal pressure decreases and the unbalanced portion of the gravitational pressure causes the star to begin to contract again. When this happens, the outer (formerly cooler) regions of the star are heated enough for the protons there to begin reacting through the fusion process previously outlined. The resulting thermal pressure in this portion of the star causes the outer layers to expand. Hence the outer portion of the star enlarges while simultaneously cooling. At this stage the star changes in appearance from a white-hot star to a considerably larger, redder star, called a **red giant.**

Although the exterior surface of the star has cooled during this transition, the interior has heated as a result of the inner contraction. The core is largely ${}^4He$, a resultant product of the burned-out proton-fusion reaction. Only after the star has reached a temperature of about 100 million kelvins does the helium begin to fuse. At that stage begins the reaction

$$ {}_2^4He + {}_2^4He \rightarrow {}_4^8Be $$

The Triple Alpha Reaction

The beryllium then combines with helium:

$$ {}_4^8Be + {}_2^4He \rightarrow {}_6^{12}C $$

This reaction is called the **triple alpha** reaction because the stripped helium atom is an α particle.

Later stages of development are uncertain, but laboratory experiments using large accelerators show that reactions between ${}^{12}C$ and the other high-energy particles in the star can lead to the formation of larger nuclei. In

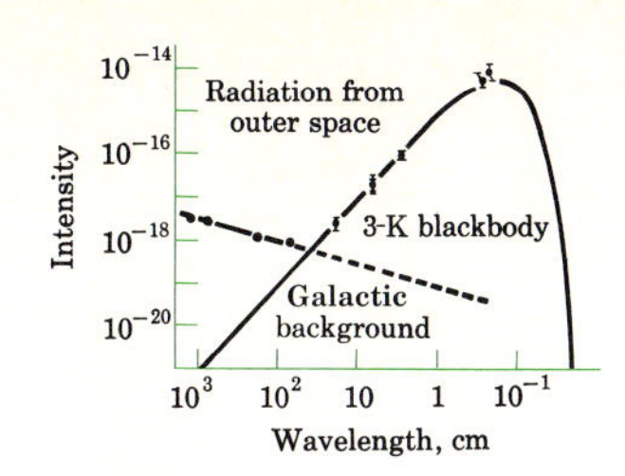

TEMPERATURE OF SPACE

We can consider the whole universe to be like the interior of a huge oven; the oven's temperature is the average temperature of the universe. As we saw in Chap. 26, the radiation emitted from a blackbody has a wavelength dependence characteristic of the temperature of the blackbody. Knowing the wavelength distribution of the electromagnetic radiation striking us from outer space, we can determine the temperature of the oven about us, the universe. The rather sketchy experimental data presently available indicate that the average temperature of the universe is close to 3 K. This is seen in the figure, where the points represent the data and the solid curve is predicted for a 3-K blackbody by Planck's radiation law. The maximum in the intensity curve occurs at a wavelength near 1 cm; this is in the short radar or microwave range. Measurements in this range are extremely difficult for two reasons: (1) the instrumentation needed is near the limit of that available for electronic detection; (2) the earth's atmosphere absorbs radiation in this short-wavelength range so strongly that it is almost impossible to obtain reliable data on earth. When an observatory is established on the moon or in space, this difficulty will be eliminated. We see, however, that the presently available data strongly suggest that the average temperature of the universe is close to 3 K.

---

fact, it seems likely that the stable elements of the periodic table are formed at this stage of star life. Since our own sun and the planets contain reasonably comparable amounts of the heavy elements, a clue exists as to the evolutionary stage of our own solar system.

White-Dwarf Stars

Eventually the fusion reactions in the interior of a red giant must die out as the available fuel becomes exhausted. At that stage the gravitational forces are no longer balanced by sufficient thermal pressure, and the star will contract. As it does so, it heats up further, because of the conversion of potential to thermal energy. After a few tens of million years, the star will have shrunk to a very dense, white-hot body, called a **white dwarf.**

We have good theoretical reasons to believe that a white dwarf cannot be stable if its mass is greater than about 1.2 times the mass of our sun. Although definitive evidence is lacking for this hypothesis because the masses of only a few white dwarfs have been measured, the existing data do not contradict it. However, since many red giants have masses much in excess of this limiting mass, they must somehow lose mass as they contract to the white-dwarf stage. Exactly how this is accomplished is not known.

One possibility consistent with observation is that, through processes observed as novas and supernovas, the collapsing red giant undergoes explosions as it approaches the white-dwarf stage. These explosions send out into space great masses of gas composed of nuclei. As a result, the galaxy acquires a cloud composed mainly of hydrogen and helium but also containing the nuclei of the heavier elements. This cloud could then undergo gravitational condensation, and new stars would be formed. Consequently the whole process of star evolution outlined above could be repeated. If one believes this hypothesis, our own sun appears to be a star of this type since it contains the nuclei of heavy elements. Eventually (perhaps in 4 or 5 billion years), our sun should become a red giant; still later, it should contract to become a white dwarf. Of course, during the red-giant stage, the earth would become too hot for human habitation.

29.4 THE EXPANDING UNIVERSE

According to the big-bang theory, the universe is expanding. If so, we should notice a Doppler effect in the light of receding stars. Let us compute the relation between the wavelength of, say, a line in the hydrogen spectrum from a source on earth λ_0 as compared to the wavelength as observed for a source on a star. Suppose the star to be receding from the earth with speed v.

Consider a crest of the wave emitted by the star source. It will travel toward earth a distance ct in a time t, where c is the velocity of light. However, the stellar source will emit a wave crest once every τ_0 seconds as measured by a timer on the star. According to relativity theory, this time τ_0 will actually be read as $\tau_0/\sqrt{1-(v/c)^2}$ by an earth clock. As a result, the distance the first crest will move during the time between its emission and the emission of the next crest will be (according to an earth observer)

$$\text{Distance} = ct = \frac{c\tau_0}{\sqrt{1-(v/c)^2}}$$

During this time, the stellar source will have moved a distance vt, and so the second crest will actually be a distance $ct + vt$ behind the first. This, then, is the wavelength one would observe on the earth, namely,

$$\lambda = ct + vt \qquad \text{or} \qquad \lambda = \frac{(c+v)\tau_0}{\sqrt{1-(v/c)^2}}$$

But since the stellar source is itself in an inertial system, the usual laws must apply there. Hence, $\tau_0 = \lambda_0/c$, and so *the wavelength λ of the star's line as observed on earth is related to the wavelength λ_0 of the line from a stationary source by the equation**

$$\lambda = \lambda_0 \frac{1+v/c}{\sqrt{1-(v/c)^2}}$$

* If you compare this with the Doppler effect equation for sound obtained in Chap. 15, you will see that they are different. At low v/c values, however, where relativistic effects can be ignored, they do coincide.

The Red Shift

We see that λ *is larger than* λ_0. *In other words, the wavelengths of the spectral lines from a receding star will appear lengthened, i.e., shifted from blue toward the red. This is often referred to as the* **red shift.**

If we examine the light reaching us from the stars, we do indeed find the wavelengths emitted by the atoms to be shifted to the red. We interpret this to mean that all stars are receding from us. By use of the red-shift equation, it is possible to compute the recession speeds of the stars. It is found that the more distant a star is from the earth, the faster it is receding from us. For the most distant stars, v is quite close to c, the speed of light. For them, lines in the blue are shifted into the infrared.

A rather simple experimental relation is found between the recession speed v of a star and its distance s from earth. It is

$$s = 3.6 \times 10^{17} v \qquad \text{m}$$

Let us now interpret this relation in terms of the big-bang theory. We shall see that it allows us to compute the age of the universe.

According to the big-bang theory, the earth is part of this now quite cold expanding cloud. It is a property of such an expanding system that everything within it is separating from everything else, as illustrated by the diagram of Fig. 29.2. If two objects are moving along the same line but at different speeds v_1 and v_2, they will be separating at a speed $v_2 - v_1$. Assuming they started from the same point at time $t = 0$, that is, when the fireball first exploded, their present separations should be given by

$$s = (v_2 - v_1)t$$

But this is identical in form to the experimental relation found from the red-shift data, namely,

$$s = (v)(3.6 \times 10^{17})$$

From our definition of $v_2 - v_1$, it is the same as v. Hence one has

$$t = 3.6 \times 10^{7}\ \text{s} = 1.1 \times 10^{10}\ \text{yr}$$

Age of the Universe

as the time from the first explosion of the fireball until now. This gives an age of 11 billion years for the universe. Although this age has been arrived at

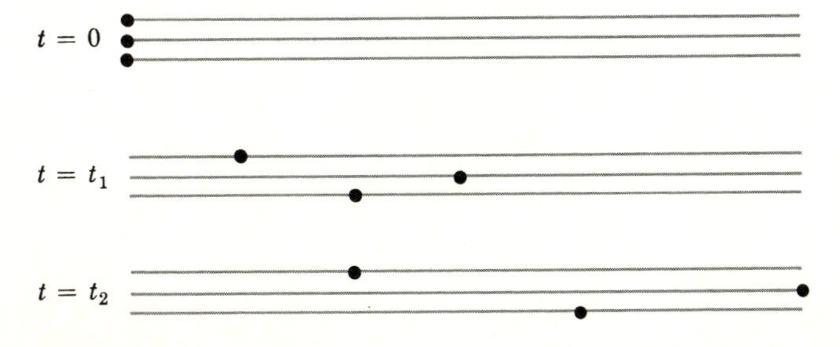

FIGURE 29.2

Starting from the same place at the same time, the spots recede from each other because their speeds are different.

from rather tentative assumptions, other data tend to support it at least approximately. One such supporting evidence is given in the next section.

29.5 AGE DETERMINATION FROM RADIOACTIVITY

In principle it is a simple matter to determine the age of a rock or other substance if it contains radioactive material together with its decay products. For example, at various places on the earth one can find rocks which still contain measurable amounts of ^{238}U. This element has a half-life of 4.5 billion years and decays to the end product ^{206}Pb, one of the less abundant isotopes of lead. Intermixed with the original ^{238}U is the final decay product ^{206}Pb.

Age of the Earth

It so happens that on earth we find about as much ^{238}U in such rocks as we find ^{206}Pb. We must conclude, therefore, that half the uranium has decayed to lead in the time since the rock was formed. Noting that half the uranium has decayed, we conclude that the rock is about one ^{238}U half-life old, namely 4.5 billion years.* This then gives us the time since the earth cooled sufficiently to allow rocks to solidify near its surface. How much older than this the earth is can be estimated only if it is known how the earth was formed. This we do not know for certain. The other natural radioactive elements which can be used to estimate the age of the earth's rocks all agree within reason with the result just given.

Radiocarbon Dating

Radioactive dating methods can also be used to date much younger substances. Carbon-containing substances can be placed within a certain time scale provided the substance came originally from plant or animal life. This method, called **radiocarbon dating,** is based upon the use of the rare isotope ^{14}C.

Cosmic radiation striking the upper portion of the earth's atmosphere leads to the formation of ^{14}C through the reaction

$$^{14}_{7}N + ^{1}_{0}n \rightarrow ^{14}_{6}C + ^{1}_{1}H$$

This unstable isotope of carbon has a half-life of 5730 yr. After formation, it circulates through the atmosphere, sometimes as carbon dioxide, CO_2. Plants take up CO_2 and incorporate it in their leaves and other structures. As a result, all plant life has some ^{14}C in it. Before atomic-bomb testing began, the ratio of ^{14}C to ^{12}C in plant life was 1.5×10^{-12}. Hence all plant life is now slightly radioactive.

If one has an old piece of wood and with very delicate measuring techniques finds it to be only half as radioactive from ^{14}C as presently grown objects, one assumes that the wood has been dead for a long enough time for half its ^{14}C to have decayed. Since this takes one half-life, 5730 yr, one concludes that to be the age of the wood. Of course we have used easy numbers for computation purposes, but the method is not limited to them. An artifact which is very old cannot be dated this way since its radioactivity

* More accurately, about 3.9×10^9 yr.

has decreased to too low a level. At the other end of the time scale, the object must be at least old enough for its radioactivity to have decreased measurably.

This gives only a glimpse of dating methods based upon radioactivity. Many other systems exist, and often checks can be made of one method against the other. As far as the earth and moon are concerned, we can now say with some certainty that the ages of the oldest rocks on both are about the same, 3.9 billion years.

29.6 ORIGIN OF THE EARTH

Although the age of the oldest rocks on the earth is now fairly well agreed to be about 3.9 billion years, the circumstances which led to the creation of the solar system and earth are largely the subject of speculation. As we have seen, there is reason to believe that our sun is the product of a second-stage condensation; it appears to have condensed from the remnants of the explosion associated with the formation of one or more white dwarfs. If we accept that hypothesis, several alternate theories for the formation of the planets can be suggested. The most likely one appears to be the following.

As the cloud began to contract to form the sun, conservation of angular momentum required the cloud to rotate more and more swiftly. In order to preserve the angular momentum of the solar system as it is now, if the planets had not formed, the angular speed of the sun would need to be much larger than it is at present. Apparently, during the contraction process, it proved more advantageous from a kinetic and energetic standpoint for the rotating cloud to separate into rather distinct parts while only the centermost part contracted to form the sun. The outer portions underwent local condensation to form the planets. This picture predicts that all the planets would be rotating in the same way about the sun since they must preserve the original angular momentum of the cloud. Such is indeed the case.

Reasoning as we did previously in the formation of the stars, large thermal energies are generated by the contraction of the cloud to form the sun and the planets. The more massive the object, the greater the gravitational forces and therefore the higher the temperature of the object formed. In the case of the sun, the temperature rose high enough to ignite the fusion reaction. But since the planets are much less massive objects, their final temperatures were much lower than that of the sun, far too low to ignite the fusion reaction. We are not certain whether the initial temperature of the earth was high enough to cause it to be a molten mass.

The Molten Earth

There is much evidence to show that the earth was molten at some stage in its development, but this could easily have occurred later. In fact, much of the core of the earth is liquid even now. The very high temperatures generated there are thought to be the result of energy given off in the radioactive decay of the small amount of unstable nuclei still found on the earth. When the earth was formed, the radioactivity was considerably higher than now. It is believed that the earth was maintained in a molten state for a time by the energy released in radioactive decay.

At this early time the earth consisted of all of the elements, but lighter ones predominated since they were formed in larger quantities in the fireball. Since atoms of many light elements had speeds larger than the escape speed at this high temperature, much of the nonreactive helium boiled away, and only a small fraction of the original hydrogen failed to escape. Many of the other elements which have low boiling points and which do not react readily to form high-melting-point compounds were also lost from the earth in this time. The major portion of the earth left behind is iron, silicon, oxygen, magnesium, and aluminum. In view of the fact that oxygen is a rather light gas, it may seem surprising that it remains in large quantities. The reason for this is not far to seek; oxygen combines with silicon to form rather heavy silicate ions and molecules, which were retained.

As time went on, the earth cooled by radiation of heat out into space. The crust of the earth solidified. In this process of cooling, material rose to the surface. This process was instrumental in apportioning the materials inside the earth. The molten iron and silicon moved toward the center while the lighter silicates and similar compounds moved to the surface.

The situation within the earth today is known to us (albeit sketchily) from several sources. The reflection and penetration into the earth of waves sent out from sites of earthquakes and man-made explosions show that the surface of the earth has a thin crust of the order of 20 miles thick, as indicated in Fig. 29.3. Below that is a thick solid region called the **mantle.** Still lower is the molten core, with perhaps a rather small solid inner core. Let us talk about each in turn.

The composition of the crust is varied but is characterized by the fact that about 45 percent is oxygen in combination. Most of the rocks on earth are primarily compounds of oxygen and silicon as in silicates, although compounds of oxygen with aluminum, iron, calcium, etc., are also found.

Extending about 2000 mi inward below the crust is the mantle, which is now also solid. Like the crust, it is still warmed from the heat that radioactive materials have produced in the earth. (Only the surface of the earth is greatly heated by the sun.) We believe the mantle to be composed mainly of oxygen, magnesium, and silicon in the compound magnesium silicate. Although the

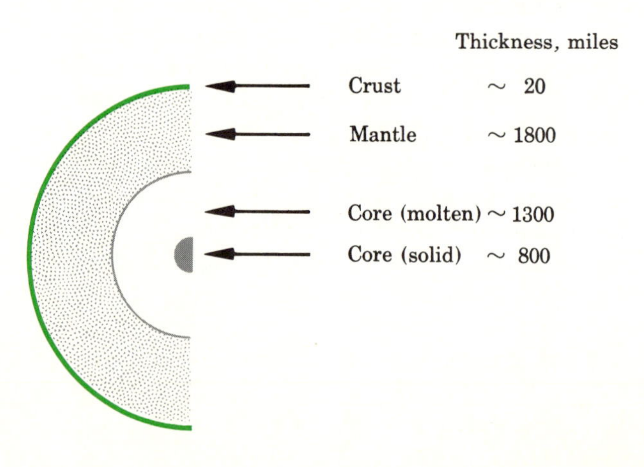

FIGURE 29.3
Tentative picture of the structure of the earth.

temperature in the lower mantle is red hot, the material is under great pressure and is solid.

Beneath the mantle and extending to nearly the center of the earth is the molten core. It is primarily iron (we believe) but probably contains appreciable quantities of nickel and silicon as well. Contemporary theory supposes that electrical currents flow in the molten metallic core, and these currents are thought to be the source of the earth's magnetic field, but this process is still conjectural.

Our knowledge of the earth is still woefully incomplete. The moon and the planets are even less well known. At this writing, and despite the recent landings, we still have no accepted theory for the origin of the moon. And the structure and history of the planets is an almost total mystery. In view of our lack of understanding of the earth itself, it is not likely that the full character of the moon will be known in the near future. Clearly, a great deal remains to be learned about our own solar system.

29.7 THE END OF THE BEGINNING

In the previous sections we obtained a glimpse of the wide areas of astro- and geophysics. These, together with many others, such as biophysics and particle physics, are only a few of the topics we would like to be able to discuss in detail. But a book must have an end. Even though the field of physics and its importance to you are much too large to be contained within the covers of a single book, you have at least begun to understand their scope in this volume.

Looking forward to the years ahead, we know that the physics of today is only the beginning of the physics which will be known a century from now. We expect that, in years to come, physics, biology, medicine, chemistry, and psychology will more and more merge into one vast science. As we learn more about molecules and how they combine to form cells, we shall be in a better position to learn the basis of life itself.

Yet this is but one avenue along which the physics of the future will advance. Our knowledge of the universe is still in its infancy. We are only beginning to understand the mechanism by which it was created. The behavior of distant galaxies, the history and future of the stars are but a few of the questions which intrigue the astrophysicists of our day. We cannot even begin to guess what course their investigations will follow in the years ahead.

And while some are investigating the immense reaches of the universe, many others will be answering questions about the smallest entities of which we know. Our understanding of the nucleus is still incomplete. We have only begun to learn of structure *within* the "primary" particles. Many people think we shall never know what (if anything) goes on inside a neutron, for example. The scientists of 1900 could hardly be expected to predict the discoveries so important to present-day physics—the photon, relativity, the wave character of matter. We, too, can predict only that the days of exciting discoveries have not yet ended.

You, of course, will be intimately concerned with these developments, since you will be living in a world greatly influenced by them. It is to be hoped that you will use the knowledge gained from this book as a footing upon which to build in the future and that you will relate the discoveries of science to the needs of society. Many people speak of science as a monster which will eventually destroy all of society. This is a definite possibility which we must all face, scientist and nonscientist alike.

As human beings, we cannot hide our heads in the sand and hope that scientific discoveries will cease or peacefully melt away. Science will continue to advance. Whether these advances will be for the improvement or destruction of mankind is dependent upon how we, as members of society, react to them. To make proper use of our scientific discoveries will require the active cooperation of scientist, politician, philosopher, theologian—all citizens. We must understand each other if we are to help each other in this monumental task. Even those of you who do not proceed further in physics will now be better prepared to meet this challenge. We hope that, with the additional knowledge and understanding you have obtained from this book, you will be better able to take your rightful place in the universe of your future.

SUMMARY

According to the big-bang theory of the universe, our universe existed as a highly dense, extremely hot ball about 11 billion years ago. The ball expanded rapidly and cooled. It is still expanding. Whether it will continue to expand forever or will eventually contract back to its original form is not known.

As the universe expanded and cooled, local fluctuations in mass gave rise to regions of condensing matter which formed galaxies. Within the galaxies, hot suns (or stars) were formed as the material pulled into a condensation point by huge gravitational forces generated heat upon collision. The temperatures so generated were often high enough to ignite the nuclear fusion reaction, and this then furnished additional energy to the star. As the star contracted and went through several stages of development, the higher-atomic-number elements found in the universe were formed and ejected.

The age of ancient rocks on the earth can be determined from their radioactivity. Such measurements tell us that the earth's crust solidified about 3.9 billion years ago. At one stage, the earth was molten. Part of the necessary heat was the result of the gravitational forces which formed it, and part was due to energy furnished by radioactive decay. Even now the center of the earth is molten, and the heat is furnished by radioactive substances.

MINIMUM LEARNING GOALS

Upon completion of this chapter you should be able to do the following:

1. Give an outline of the big-bang theory including the following: original situation; expansion and cooling; galaxy formation; rotation; star formation; energy source of stars; production of He and higher-Z elements; time span since the big bang.

2. Describe the process by which the material of stars is heated to the point where the fusion reaction can ignite. State at least one major way that our sun and the stars maintain their high temperatures.

3. Explain how the red shift implies that the universe is expanding.

4. Explain how the age of a uranium-bearing rock can be determined from an analysis of its constitution. Give the basic idea of radiocarbon dating of organic materials.

5. Sketch a cross section of the earth and show its major layers. Give a rough estimate of the thickness of the crust. State which parts are molten.

IMPORTANT TERMS AND PHRASES

You should be able to define or explain each of the following:

Big-bang theory
Age of the universe
Galaxy
Sun and stars
Red giant; white dwarf
Temperature of space at present
Red shift
Crust, mantle, core of earth

QUESTIONS AND QUESSTIMATES

1. What factors must be known in order to compute whether our universe will pulsate or expand forever?
2. Why can't one determine whether the universe has pulsated several times before?
3. Our present knowledge indicates that the primordial fireball took about an hour to cool to the range 10^{10} K. At that time, how large was the average thermal energy kT, in electronvolts? Compare this to the binding energy of a ^{4}He nucleus. What can you say about nuclear stability at that time?
4. One can estimate the temperature of the surface of the sun by measuring the wavelength dependence of the light given off by it. Explain how such an estimate can be made.
5. We know that charge is conserved. Does this mean that there is the same number of positive and negative charges in the universe now as there were in the original fireball?
6. When we look into space, we find that all the galaxies are flying away from us. Does this mean that we are at the location of the original fireball?
7. Since there is no sign of life in our solar system other than that on earth, the nearest neighbors we could have would live on a planet of the nearest star, Alpha Centauri, which is 4×10^{16} m away. About how long would it take one of our present spaceships to reach there? A spaceship with speed $0.99c$? (E)
8. When it was formed about 5 billion years ago, why was the earth much more radioactive than it is now? Would you expect the radioactivity types today to be the same as then?
9. Ordinarily we assume that the heating of a radioactive substance will not alter its half-life. Estimate how hot the substance must be in order for this assumption to be wrong. (You may wish to use the fact that at 300 K thermal energy is about $\frac{1}{40}$ eV.) (E)
10. If the temperature of outer space is actually about $-270°$C, why don't astronauts freeze to death during space journeys? On the moon, they actually require refrigeration. Why?
11. Why doesn't the moon have any atmosphere?
12. The temperature inside deep mines is uniform throughout the year. What does this tell us about the source of the heat in the mine?
13. If the original temperature of the fireball was 10^{12} K and its size was 10^{10} m, estimate the mass the universe must have if it is to pulsate. Is yours an upper or lower limiting estimate? (E)
14. About how old is a piece of wood if its ^{14}C content is about one-tenth that in trees today? What factors could make your estimate be in error?
15. The earth receives energy from the sun at an average rate of about 2 cal/(m^2)(min). Estimate how much hydrogen is being converted to helium in the sun each day.

APPENDIX 1 CONVERSION OF UNITS

It often happens that we wish to convert a quantity expressed in one set of units to another set of units. Typically, we might want to know how many kilometers is equivalent to 20 mi. For conversions like this, we make use of conversion factors. Let us see what a conversion factor is.

We know, for example, that

$$100 \text{ cm} = 1 \text{ m}$$

Division by 100 cm gives

$$1 = \frac{1 \text{ m}}{100 \text{ cm}} = 0.010 \text{ m/cm}$$

This is a conversion factor between meters and centimeters. Notice that the conversion factor is unity. When we multiply any quantity by a conversion factor, it is the same as multiplying by unity. It does not change the value of the quantity in question. We shall now give a few examples of the use of conversion factors.

1 How many meters are there in 30 mi? The appropriate conversion factor is obtained from 1.60 km = 1 mi to be 1.60 km/mi. Since this is simply unity, we can multiply or divide by it without changing the value of a quantity. Therefore

$$30 \text{ mi} = (30 \cancel{\text{mi}})\left(1.60 \frac{\text{km}}{\cancel{\text{mi}}}\right) = 48 \text{ km}$$

2 How many hours are there in 200,000 s? We know that 1 h = 60 min and 60 s = 1 min. Therefore

$$200{,}000 \text{ s} = (200{,}000 \cancel{\text{s}})\left(\frac{1 \text{ min}}{60 \cancel{\text{s}}}\right) = \frac{20{,}000}{6} \text{ min}$$

$$= \left(\frac{20{,}000}{6} \cancel{\text{min}}\right)\left(\frac{1 \text{ h}}{60 \cancel{\text{min}}}\right) = 55.6 \text{ h}$$

Typical conversion factors are given on the inside cover of this book.

APPENDIX 2 POWERS OF TEN

It is often inconvenient to write numbers like 1,420,000,000 and 0.00031. These same numbers can be written as 1.42×10^9 and 3.1×10^{-4}, as we shall now show.

Consider the identities

$$\begin{aligned} 1 &= 10^0 \\ 10 &= 10^1 \\ 100 &= (10)(10) = 10^2 \\ 1000 &= (10)(10)(10) = 10^3 \\ 1{,}000{,}000{,}000 &= 10^9 \end{aligned}$$

If we wish to write the number 4,561,000,000, we have at once that this number is equivalent to multiplying 4.561 nine times by 10. (Each time we multiply by 10 we move the decimal point one place to the right.) Hence,

$$4{,}561{,}000{,}000 = 4.561 \times 10^9$$

In general, if we move the decimal place on a number q places to the left, we must multiply the number by 10^q if it is to remain unchanged. For the example just given, $q = 9$.

The procedure for writing numbers less than unity is similar. We make use of the fact that

$$\begin{aligned} 0.1 &= \frac{1}{10} = 10^{-1} = 1 \times 10^{-1} \\ 0.01 &= \frac{1}{100} = \frac{1}{10^2} = 10^{-2} = 1 \times 10^{-2} \\ 0.001 &= \frac{1}{1000} = \frac{1}{10^3} = 10^{-3} = 1 \times 10^{-3} \\ 0.000{,}000{,}01 &= \frac{1}{10^8} = 10^{-8} = 1 \times 10^{-8} \end{aligned}$$

Clearly, when we move the decimal point q places to the right, we must multiply by 10^{-q}. In the case originally treated, $0.00031 = 3.1 \times 10^{-4}$, q was 4, since the decimal point was moved four places.

APPENDIX 3 PERIODIC TABLE OF THE ELEMENTS

The values listed are based on $^{12}_{6}C = 12$ u exactly. For artificially produced elements, the approximate atomic weight of the most stable isotope is given in brackets.

GROUP		I	II	III	IV	V	VI	VII	VIII	0
PERIOD	SERIES									
1	1	1 H 1.00797								2 He 4.003
2	2	3 Li 6.939	4 Be 9.012	5 B 10.81	6 C 12.011	7 N 14.007	8 O 15.9994	9 F 19.00		10 Ne 20.183
3	3	11 Na 22.990	12 Mg 24.31	13 Al 26.98	14 Si 28.09	15 P 30.974	16 S 32.064	17 Cl 35.453		18 Ar 39.948
4	4	19 K 39.102	20 Ca 40.08	21 Sc 44.96	22 Ti 47.90	23 V 50.94	24 Cr 52.00	25 Mn 54.94	26 Fe 27 Co 28 Ni 55.85 58.93 58.71	
	5	29 Cu 63.54	30 Zn 65.37	31 Ga 69.72	32 Ge 72.59	33 As 74.92	34 Se 78.96	35 Br 70.909		36 Kr 83.80
5	6	37 Rb 85.47	38 Sr 87.62	39 Y 88.905	40 Zr 91.22	41 Nb 92.91	42 Mo 95.94	43 Tc [98]	44 Ru 45 Rh 46 Pd 101.1 102.905 106.4	
	7	47 Ag 107.870	48 Cd 112.40	49 In 114.82	50 Sn 118.69	51 Sb 121.75	52 Te 127.60	53 I 126.90		54 Xe 131.30
6	8	55 Cs 132.905	56 Ba 137.34	57–71 Lanthanide series*	72 Hf 178.49	73 Ta 180.95	74 W 183.85	75 Re 186.2	76 Os 77 Ir 78 Pt 190.2 192.2 195.09	
	9	79 Au 196.97	80 Hg 200.59	81 Tl 204.37	82 Pb 207.19	83 Bi 208.98	84 Po [210]	85 At [210]		86 Rn [222]
7	10	87 Fr [223]	88 Ra [226]	89–103 Actinide series†						

Lanthanide series	57 La 138.91	58 Ce 140.12	59 Pr 140.91	60 Nd 144.24	61 Pm [147]	62 Sm 150.35	63 Eu 152.0	64 Gd 157.25	65 Tb 158.92	66 Dy 162.50	67 Ho 164.93	68 Er 167.26	69 Tm 168.93	70 Yb 173.04	71 Lu 174.97
†*Actinide series*	89 Ac [227]	90 Th 232.04	91 Pa [231]	92 U 238.03	93 Np [237]	94 Pu [242]	95 Am [243]	96 Cm [247]	97 Bk [247]	98 Cf [251]	99 E [254]	100 Fm [253]	101 Md [256]	102 No [254]	103 Lw [257]

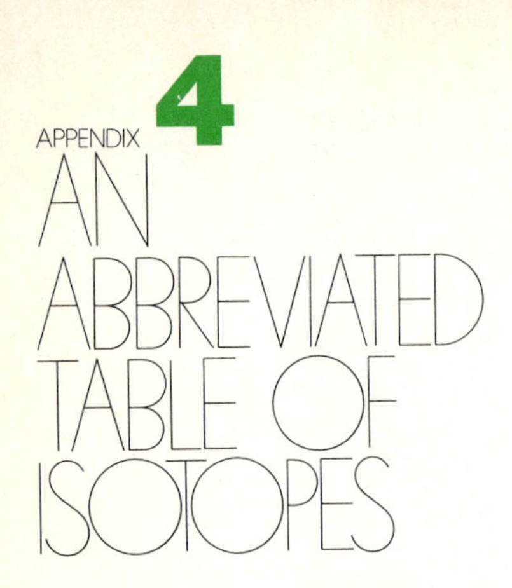

APPENDIX 4 AN ABBREVIATED TABLE OF ISOTOPES

The values listed are based on ${}^{12}_{6}C$ = 12 u exactly. Electron masses are included.

ATOMIC NUMBER Z	SYMBOL	AVERAGE ATOMIC MASS	ELEMENT	MASS NUMBER A	RELATIVE ABUNDANCE, %	MASS OF ISOTOPE
1	H	1.00797	Hydrogen	1	99.985	1.007825
				2	0.015	2.014102
2	He	4.0026	Helium	3	0.00015	3.016030
				4	100—	4.002604
3	Li	6.939	Lithium	6	7.52	6.015126
				7	92.48	7.016005
4	Be	9.0122	Beryllium	9	100—	9.012186
5	B	10.811	Boron	10	18.7	10.012939
				11	81.3	11.009305
6	C	12.01115	Carbon	12	98.892	12.0000000
				13	1.108	13.003354
7	N	14.0067	Nitrogen	14	99.635	14.003074
				15	0.365	15.000108
8	O	15.9994	Oxygen	16	99.759	15.994915
				17	0.037	16.999133
				18	0.204	17.999160
9	F	18.9984	Fluorine	19	100	18.998405
10	Ne	20.183	Neon	20	90.92	19.992440
				22	8.82	21.991384
11	Na	22.9898	Sodium	23	100—	22.989773
12	Mg	24.312	Magnesium	24	78.60	23.985045
13	Al	26.9815	Aluminum	27	100	26.981535
14	Si	28.086	Silicon	28	92.27	27.976927
				30	3.05	29.973761
15	P	30.9738	Phosphorus	31	100	30.973763
16	S	32.064	Sulfur	32	95.018	31.972074
17	Cl	35.453	Chlorine	35	75.4	34.968854
				37	24.6	36.965896
18	Ar	39.948	Argon	40	996	39.962384
19	K	39.102	Potassium	39	93.08	38.963714
20	Ca	40.08	Calcium	40	96.97	39.962589
21	Sc	44.956	Scandium	45	100	44.955919
22	Ti	47.90	Titanium	48	73.45	47.947948
23	V	50.942	Vanadium	51	99.76	50.943978
24	Cr	51.996	Chromium	52	83.76	51.940514
25	Mn	54.9380	Manganese	55	100	54.938054
26	Fe	55.847	Iron	56	91.68	55.934932
27	Co	58.9332	Cobalt	59	100	58.93319
28	Ni	58.71	Nickel	58	67.7	57.93534
				60	26.23	59.93032

ATOMIC NUMBER Z	SYMBOL	AVERAGE ATOMIC MASS	ELEMENT	MASS NUMBER A	RELATIVE ABUNDANCE, %	MASS OF ISOTOPE
29	Cu	63.54	Copper	63	69.1	62.92959
30	Zn	65.37	Zinc	64	48.89	63.92914
31	Ga	69.72	Gallium	69	60.2	68.92568
32	Ge	72.59	Germanium	74	36.74	73.92115
33	As	74.9216	Arsenic	75	100	74.92158
34	Se	78.96	Selenium	80	49.82	79.91651
35	Br	79.909	Bromine	79	50.52	78.91835
36	Kr	83.30	Krypton	84	56.90	83.91150
37	Rb	85.47	Rubidium	85	72.15	84.91171
38	Sr	87.62	Strontium	88	82.56	87.90561
39	Y	88.905	Yttrium	89	100	88.90543
40	Zr	91.22	Zirconium	90	51.46	89.90432
41	Nb	92.906	Niobium	93	100	92.90602
42	Mo	95.94	Molybdenum	98	23.75	97.90551
43	Tc	*	Technetium	98		97.90730
44	Ru	101.07	Ruthenium	102	31.3	101.90372
45	Rh	102.905	Rhodium	103	100	102.90480
46	Pd	106.4	Palladium	106	27.2	105.90320
47	Ag	107.870	Silver	107	51.35	106.90497
48	Cd	112.40	Cadmium	114	28.8	113.90357
49	In	114.82	Indium	115	95.7	114.90407
50	Sn	118.69	Tin	120	32.97	119.90213
51	Sb	121.75	Antimony	121	57.25	120.90375
52	Te	127.60	Tellurium	130	34.49	129.90670
53	I	126.9044	Iodine	127	100	126.90435
54	Xe	131.30	Xenon	132	26.89	131.90416
55	Cs	132.905	Cesium	133	100	132.90509
56	Ba	137.34	Barium	138	71.66	137.90501
57	La	138.91	Lanthanum	139	99.911	138.90606
58	Ce	140.12	Cerium	140	88.48	139.90528
59	Pr	140.907	Praseodymium	141	100	140.90739
60	Nd	144.24	Neodymium	144	23.85	143.90998
61	Pm	*	Promethium	145		144.91231
62	Sm	150.35	Samarium	152	26.63	151.91949
63	Eu	151.96	Europium	153	52.23	152.92086
64	Gd	157.25	Gadolinium	158	24.87	157.92410
65	Tb	158.924	Terbium	159	100	158.92495
66	Dy	162.50	Dysprosium	164	28.18	163.92883
67	Ho	164.930	Holmium	165	100	164.93030
68	Er	167.26	Erbium	166	33.41	165.93040
69	Tm	168.934	Thulium	169	100	168.93435
70	Yb	173.04	Ytterbium	174	31.84	173.93902
71	Lu	174.97	Lutetium	175	97.40	174.94089
72	Hf	178.49	Hafnium	180	35.44	179.94681
73	Ta	180.948	Tantalum	181	100	180.94798
74	W	183.85	Tungsten	184	30.6	183.95099
75	Re	186.2	Rhenium	187	62.93	186.95596
76	Os	190.2	Osmium	192	41.0	191.96141
77	Ir	192.2	Iridium	193	61.5	192.96328
78	Pt	195.09	Platinum	195	33.7	194.96482
79	Au	196.967	Gold	197	100	196.96655
80	Hg	200.59	Mercury	202	29.80	201.97063
81	Tl	204.37	Thallium	205	70.50	204.97446
82	Pb	207.19	Lead	208	52.3	207.97664
83	Bi	208.980	Bismuth	209	100	208.98042
84	Po	[210]	Polonium	210		209.98287
85	At	*	Astatine	211		210.98750
86	Rn	*	Radon	211		210.99060
87	Fr	*	Francium	221		221.01418

ATOMIC NUMBER Z	SYMBOL	AVERAGE ATOMIC MASS	ELEMENT	MASS NUMBER A	RELATIVE ABUNDANCE, %	MASS OF ISOTOPE
88	Ra	[226]	Radium	226		226.02536
89	Ac	*	Actinium	225		225.02314
90	Th	[232.038]	Thorium	232	100	232.03821
91	Pa	[231]	Protactinium	231		231.03594
92	U	[238.03]	Uranium	233		233.03950
				235	0.715	235.04393
				238	99.28	238.05076
93	Np	*	Neptunium	239		239.05294
94	Pu	*	Plutonium	239		239.05216
95	Am	*	Americium	243		243.06138
96	Cm	*	Curium	245		245.06534
97	Bk	*	Berkelium	248		248.070305
98	Cf	*	Californium	249		249.07470
99	Es	*	Einsteinium	254		254.08811
100	Fm	*	Fermium	252		252.08265
101	Md	*	Mendelevium	255		255.09057
102	No	*	Nobelium	254		254
103	Lw	*	Lawrencium	257		257

*The atomic masses of unstable elements are not listed unless the isotope given constitutes the major isotope.

APPENDIX 5
TRIGONOMETRIC FUNCTIONS

ANGLE deg	SINE	CO-SINE	TAN-GENT
0°	0.000	1.000	0.000
1°	.018	1.000	.018
2°	.035	0.999	.035
3°	.052	.999	.052
4°	.070	.998	.070
5°	.087	.996	.088
6°	.105	.995	.105
7°	.122	.993	.123
8°	.139	.990	.141
9°	.156	.988	.158
10°	.174	.985	.176
11°	.191	.982	.194
12°	.208	.978	.213
13°	.225	.974	.231
14°	.242	.970	.249
15°	.259	.966	.268
16°	.276	.961	.287
17°	.292	.956	.306
18°	.309	.951	.325
19°	.326	.946	.344
20°	.342	.940	.364
21°	.358	.934	.384
22°	.375	.927	.404
23°	.391	.921	.425
24°	.407	.914	.445
25°	.423	.906	.466
26°	.438	.899	.488
27°	.454	.891	.510
28°	.470	.883	.532
29°	.485	.875	.554
30°	.500	.866	.577

ANGLE deg	SINE	CO-SINE	TAN-GENT
31°	.515	.857	.601
32°	.530	.848	.625
33°	.545	.839	.649
34°	.559	.829	.675
35°	.574	.819	.700
36°	.588	.809	.727
37°	.602	.799	.754
38°	.616	.788	.781
39°	.629	.777	.810
40°	.643	.766	.839
41°	.658	.755	.869
42°	.669	.743	.900
43°	.682	.731	.933
44°	.695	.719	.966
45°	.707	.707	1.000
46°	.719	.695	1.036
47°	.731	.682	1.072
48°	.743	.669	1.111
49°	.755	.656	1.150
50°	.766	.643	1.192
51°	.777	.629	1.235
52°	.788	.616	1.280
53°	.799	.602	1.327
54°	.809	.588	1.376
55°	.819	.574	1.428
56°	.829	.559	1.483
57°	.839	.545	1.540
58°	.848	.530	1.600
59°	.857	.515	1.664
60°	.866	.500	1.732

ANGLE deg	SINE	CO-SINE	TAN-GENT
61°	.875	.485	1.804
62°	.883	.470	1.881
63°	.891	.454	1.963
64°	.899	.438	2.050
65°	.906	.423	2.145
66°	.914	.407	2.246
67°	.921	.391	2.356
68°	.927	.375	2.475
69°	.934	.358	2.605
70°	.940	.342	2.757
71°	.946	.326	2.904
72°	.951	.309	3.078
73°	.956	.292	3.271
74°	.961	.276	3.487
75°	.966	.259	3.732
76°	.970	.242	4.011
77°	.974	.225	4.331
78°	.978	.208	4.705
79°	.982	.191	5.145
80°	.985	.174	5.671
81°	.988	.156	6.314
82°	.990	.139	7.115
83°	.993	.122	8.144
84°	.995	.105	9.514
85°	.996	.087	11.43
86°	.998	.070	14.30
87°	.999	.052	19.08
88°	.999	.035	28.64
89°	1.000	.018	57.29
90°	1.000	.000	∞

Appendix 6 Answers

Chapter 1

1. 10 blocks; 53° north of east
3. 71 km west; 71 km south
5. 14.2 cm at 257°
7. 953 mi north; 550 mi west
9. 3.38 N at 331°
11. 13.1 lb; 10.5 lb
13. 5.0 lb at 90°
15. 5.8 cm/s at 31° east of south
17. 50 lb; 100 lb
19. 432 lb
21. W_2; 3
23. 28 N; 80 N
25. 88 N; 38 N, 94 N, 32 N

Chapter 2

1. 91.4 mi/h = 147 km/h
3. 0.49 cm/yr; 49 cm/century upward
5. 1 cm/s; 0.86 cm/s; −0.40 cm/s; −1.0 cm/s; zero
7. 80 m
9. 1.4 m/s^2; 280 m
11. 3.2×10^5 m/s^2; 4.7×10^{-4} s
13. $L/200$ s^{-2}; 156 s
15. 100 ft; 2.5 s
17. 198 m/s
19. 2.76 s; 20 m/s
21. 8.89 s; 133 m
23. 2.67 m/s^2
25. 0.495 s; 4.85 m/s

Chapter 3

1. 140 N
3. 50 lb
5. 1000 N = 225 lb
7. 3.6×10^4 lb
9. 210 lb; 110 lb
11. 0.98 ft
13. 6.8 ft/s^2
15. 3.82 m/s^2; 9.6 N; 0.90 m/s^2; 14.2 N
17. 47 N; 1.43 s
19. 4.9 m/s^2; 14.7 N
21. 1.73 m/s^2; no
23. $F/2m$; $(F - 2f)/2m$
25. 1.84 m/s^2; 3.68 m/s^2; 0.735 N; 2.39 N

Chapter 4

1. 1.86×10^{-40} N; $F_G/W = 1.14 \times 10^{-14}$
3. 1.34
5. 501 lb
7. 79 m
9. 411 lb
11. 2920 N
13. 78 m
15. 33 m; $6.2W$
17. (*a*) 64 m; (*b*) 3.70 s
21. 2.71 m/s^2; 50 N
23. 0.977 m/s^2; 17.6 N; 105 N

Chapter 5

1. 88 J
3. 2×10^4 ft · lb
5. (*a*) 3.23×10^8 J; (*b*) 181 hp
7. 2250 lb
9. (*a*) 8.2×10^4 ft · lb; (*b*) $\frac{1}{41}$
11. 0.163 W

13. 835 W
15. (*a*) Yes; (*b*) 20.4 m
17. 4.43, 2.42, and 3.13 m/s
21. 7.2 m/s
23. 6.8 m/s
25. (*a*) 20; (*b*) 15; (*c*) 0.75
27. (*a*) 8.33; (*b*) 10; (*c*) 83%
29. 0.59 hp
31. (*a*) 265 N; (*b*) 0.027 N

Chapter 6

1. $m\sqrt{2gh}$
3. (*a*) 1600 lb; (*b*) 200 ft
5. 7200 N
7. (*a*) 0.143 m/s; (*b*) 0.048 N
9. 673 m/s
11. 0.30 m/s
13. Zero
15. h; h
17. $-\frac{1}{2}v_0$; $\frac{1}{2}v_0$
19. (*b*) Two masses fly off
21. Four times atmospheric

Chapter 7

1. (*a*) 25°, 0.069 rev, 0.436 rad; (*b*) 464°, 1.29 rev, 8.1 rad; (*c*) 263°, 0.73 rev, 4.59 rad
3. (*a*) 0.80 rev/s^2; (*b*) 160 rev
5. 1.13 rev/s
7. (*a*) 463 m/s; (*b*) zero
9. 565 in.
11. 90 rev
13. 9000 lb
15. 0.57
17. 2.42 m/s
19. 9.5×10^{-8} N
21. 4.74 N
25. 0.55°

Chapter 8

1. (*a*) L, $0.77L$, $0.33L$, $0.29L$; (*b*) $-FL$, $0.77FL$, $0.33FL$, $0.29FL$
3. 70 lb
5. (*a*) 1170 N; (*b*) 752 N, 100 N
7. (*a*) $2.77F$; (*b*) $-0.45F$, $3.47F$
9. 1450 N; 1000 N; 500 N; 2410 N
11. (*a*) 4.25 kg · m^2; (*b*) 7.06 kg · m^2; (*c*) 0.92 m, 1.19 m
13. 1×10^{-3} kg · m^2
15. 0.80 N · m
17. (*a*) 0.125 lb · ft; (*b*) 20 rad
19. (*a*) 7200 slug · ft^2; (*b*) 15.99 lb
21. (*a*) 3.03 m/s; (*b*) 9.65 rev/s
25. 0.388 rev/s

Chapter 9

1. 7.92 g/cm^3
3. 0.070 m^3
5. 9.1×10^{-4} m
7. 1.39×10^{-4}
11. (*a*) 6.1×10^{-7}; (*b*) 10^7 N/m^2
13. 6.08×10^3 N = 1370 lb
15. (*a*) 10.34 m; (*b*) 10.34 m
17. 0.87 g/cm^3
19. 0.95 g/cm^3
21. 0.75 g/cm^3
23. 21.6 g
25. 0.895
29. (*a*) 4.87×10^{-6} N; (*b*) 1.334×10^{-5} N; (*c*) 4.28×10^{-3} m/s
31. (*a*) 56.7 ft/s; (*b*) 3.9×10^{-3} ft^3 = 6.8 in^3
33. 3.81 N

7. 177°C
9. 0.449 cm^3
11. (*a*) 1.30×10^{-25} kg; (*b*) 6.8×10^{21}
13. (*a*) 5.5×10^{-11} kg · mol; (*b*) 3.3×10^{10}
15. 1.36×10^6 N/m^2 = 13.4 atm
17. 22.4×10^{-3} m^3
19. (*a*) 4.1×10^{-22} atm; (*b*) 6.2×10^{-23} J; (*c*) 273 m/s
21. (*a*) 8.15×10^4 N/m^2, 1.92×10^4 N/m^2, 0.066×10^4 N/m^2; (*b*) 1.01×10^5 N/m^2
23. 42,300 kg/kg · mol

Chapter 10

1. (*a*) 25°C, 298 K; (*b*) −35°C, 238 K
3. −38°F; 675°F
5. (*a*) 131°C; (*b*) 10.5 atm

Chapter 11

1. (*a*) 24.8 cal; (*b*) 0.0984 Btu
3. (*a*) 48.4 Btu; (*b*) 12,200 cal
5. 65°C
7. 78 g
9. 8.0 g
11. 46.9 g
13. 8400, 8400, 8400, 8300, 8500, 8400, and 8400 J/(kg · mol)(K)
15. 24.3°C
17. 7.86 g
19. 0.55 cal/g
21. 0°C
25. 0.075%
27. 395°C
29. 1.23 cm^3
31. 5820 lb/in^2
33. 15.1 cal
35. 2.01×10^5 cal
39. (*a*) 13.50 g/m^3; (*b*) 79%
41. 21%

Chapter 12

1. (*a*) 5020 J; (*b*) larger
3. (*a*) 45 J; (*b*) zero
5. (*a*) 606°C, 1192°C, 215°C; (*b*) yes, no, no, yes
7. 6960 J
9. (*a*) 1.00×10^6 N/m^2; (*b*) 460°C
11. (*a*) 9.2 hp; (*b*) 5.2 g
13. 0.58
15. (*a*) no; (*b*) 2; (*c*) 4; (*d*) 7
17. (*a*) 9.8×10^{-4}; (*b*) 9.8×10^{-3}
19. (*a*) Zero; (*b*) 1.38×10^{-23} ln 5 or 2.22×10^{-23} J/K

Chapter 13

1. (*a*) 0.333 Hz; (*b*) 3.0 s; (*c*) 7.0 cm
3. (*a*) 1.93×10^{-4} m; (*b*) 36 Hz
5. 2.19×10^{-3} J
7. (*a*) 0.090 m/s^2; (*b*) 0.052 m/s
9. (*a*) 3.0 Hz; (*b*) 15 cm; (*c*) 2.83 m/s; (*d*) zero
11. (*a*) 53 m/s^2; (*b*) zero; (*c*) 53 cos θ m/s^2
13. (*a*) 9.8 m/s^2; (*b*) 1.76 Hz
15. (*a*) 1.28 Hz; (*b*) 2.54 ft/s
17. (*a*) 1.80 Hz; (*b*) 1.89 ft/s
19. 0.389 m
23. 0.0316 m

Chapter 14

1. 1.80 s
3. (*a*) 30 m; (*b*) 600 m/s
5. 8.0×10^{-3} N
7. (*a*) 0.53 m; (*b*) 267 m/s
9. (*a*) Lowest frequency; (*b*) 4
11. 20, 40, 80, and 100 Hz
13. 5000 Hz
15. 1250, 3750, and 6250 Hz
17. 0.32, 0.64, and 0.95 Hz

Chapter 15

1. 2040 m
3. 4.4×10^{-10} m^2/N
5. 2.9%
7. 73 dB
9. (*a*) 10^{-6} W/m^2; (*b*) 1.33
11. (*a*) ±34, ±68, and ±102 cm; (*b*) ±17, ±51, and ±85 cm
13. 113, 340, and 567 Hz
15. 1700 Hz
17. 0.065 and $n \times 0.065$ Hz
19. 31 m/s
21. (*a*) 2202 and 2198 Hz
23. 2988 Hz

Chapter 16

1. 1200 N
3. (*a*) 8.2×10^{-8} N; (*b*) 2.2×10^{6} m/s
5. 4.79 N at −45°
7. (*a*) -9×10^{5} N/C; (*b*) 3.11×10^{5} N/C
9. 6.4×10^{6} N/C down
11. (*a*) 9.6×10^{-18} J; (*b*) -9.6×10^{-18} J
13. (*a*) 2.0 cm; (*b*) 4.8×10^{-17} N
15. (*a*) 2.4×10^{5} V; (*b*) zero
17. (*a*) 2 A, 5 A, 27 V; (*b*) 50 μC -1.80×10^{6} V
19. (*a*) 1.225×10^{5} V/m; (*b*) 3.2×10^{-19} C
21. (*a*) 25 cm from 3 μC; (*b*) unstable

Chapter 17

1. (*a*) 2.05×10^{6} m/s; (*b*) 12 eV
3. (*a*) 1500 eV; (*b*) 3500 eV; (*c*) 3000 and 2000 eV
5. (*a*) 2400 C; (*b*) 1.5×10^{22}
7. 240 Ω
9. (*a*) 0.0161 Ω; (*b*) 0.48 V
11. 24 Ω
13. 130 Ω
15. 1.8×10^{-7} C
17. 4.73×10^{-8} F
19. (*a*) 36 nC; (*b*) 13.3 V; (*c*) 24×10^{-9} J; (*d*) 24×10^{-9} J

Chapter 18

1. (*a*) 6 Ω; (*b*) 2.0 A
3. (*a*) 12 Ω; (*b*) 8 Ω
5. 1.33 μF
7. 3.33, 0.50, and −3.83 A
9. 0, 0 A; 42 μC
11. (*a*) 2.15 Ω; (*b*) 29 V
13. (*a*) 1.14 A; (*b*) 6.3 V
15. 15.3 A
17. (*a*) 2 A, 5 A, 27 V; (*b*) 50 μC
19. 2, 14, 3, and 9 A
21. (*a*) 12 V; (*b*) 24 μC; 48 μC
23. $\frac{5}{3}$ Ω

Chapter 19

1. (*a*) 0.024 N; (*b*) into earth
3. (*a*) 0.0083 N; (*b*) west
5. (*a*) 1.2×10^{-6} T; (*b*) north
7. 1250 A
9. (*a*) 0.0314 T; (*b*) 1.57 T; (*c*) 2.2×10^{-5} Wb
11. (*a*) 1×10^{-5} T; (*b*) 2×10^{-4} N, repulsion; (*c*) 2×10^{-4} N
13. 11.2×10^{-4} T
15. Along the line $y = -2x$
17. 0.167 m
19. Circle perpendicular to B; $r = 1.14$ m
21. Helix with $r = 0.52$ cm and pitch = 5.7 cm
23. (*a*) 6.3×10^{14}; (*b*) 1×10^{-9} T; (*c*) clockwise
25. 3.5×10^{6} m/s

7. $0.188 \text{ A} \cdot \text{m}^2$
9. (*a*) 1.91×10^{-4} N · m; (*b*) north
11. 5.0 V
13. 1.50 T
15. (*a*) 2.0 V; (*b*) 0.10 H
17. (*a*) 0.0036 V
19. (*a*) 110 V; (*b*) 24 A

Chapter 20

1. 0.050 Ω shunt
3. 1950 Ω in series
5.

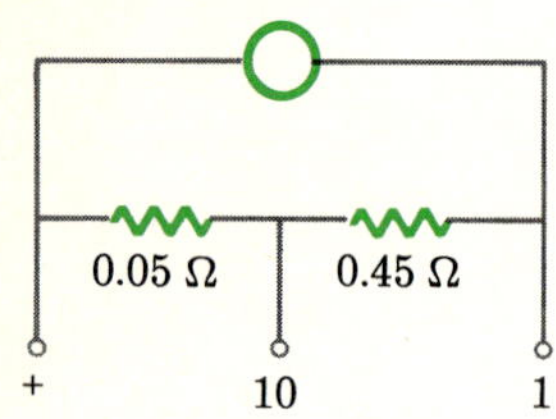

Chapter 21

1. (*a*) 8.0 s; (*b*) 8.0 s; (*c*) 1.6×10^{-5} C; (*d*) 2.0×10^{-6} A
3. 2.83 A
5. 250 W
7. 0.45 A
9. (*a*) 377 V; (*b*) 3.77×10^6 V
11. (*a*) 0.036 A; (*b*) 3.6 A; (*c*) zero
13. 2.36 A
15. (*a*) 70 μF; (*b*) 7.0 H
17. (*a*) 0.66 H or 0.74 H; (*b*) 47 mH
19. (*a*) 0.10 A; (*b*) 0.10 W; (*c*) 0.10
21. (*a*) 180 W; (*b*) 0.092 H
23. 41.7 on 120 V; no
25. 100

Chapter 22

1. (*a*) 5.2×10^{-20} J; 0.32 eV
3. (*a*) 3.5×10^8 N/m²; (*b*) 3500 times larger
5. 0.20 or 20%
7. (*a*) 0.80 Hz; (*b*) 3.75×10^8 m
9. 14.7 s
11. Of the order of 1×10^{-6} V
13. (*a*) 8.3×10^{-8} m; (*b*) long x-rays

Chapter 23

1. Right
3. 2.03×10^8 m/s
5. 3.9×10^8 m
7. (*a*) 24°; (*b*) 0°
9. 37°; 29.3°
11. (*a*) 1.414 at least
13. (*a*) −8.33 cm; (*b*) 0.50 cm; (*c*) virtual, erect; (*d*) −6.67 cm, 1.00 cm virtual, erect; (*e*) −3.33 cm, 2.0 cm, virtual, erect
15. (*a*) 200 cm; (*b*) no; (*c*) same
17. (*a*) 30 cm; (*b*) 10 cm
19. (*a*) 25 cm, 0.75 cm, real, inverted; (*b*) 40 cm, 3.0 cm, real, inverted; (*c*) −20 cm, 6.0 cm, virtual, erect
21. (*a*) $p = 2f$; (*b*) real
23. (*a*) 75 cm past second lens; (*b*) 0.50
25. (*a*) 11.2 cm past first lens; (*b*) 0.75
27. Left of lens: 4, 4, and 8 cm

Chapter 24

1. (*a*) 2.0 cm; (*b*) 1.85 cm
3. (*a*) Farsighted; (*b*) converging, 43 cm
5. Infinity; 150 cm
7. 16.7 cm
9. 8.0 m
11. (*a*) 5.83 diopters; (*b*) 17.1 cm
13. 8.3
15. (*a*) About 17 cm; (*b*) about 3.5
17. 200 km
19. (*a*) 45°; (*b*) 38°

Chapter 25

1. (*a*) Zero; (*b*) 50 cm; (*c*) 150 cm; (*d*) 125 cm
3. ±0.025, ±0.050, and ±0.075 cm
5. 0.020 cm
7. 250 , 500, 750, and 1000 nm
9. 1.6×10^{-6} m
11. 5.52×10^{-4} m
13. (*a*) Zero or 152 nm
15. 36.1°
17. 19.4°
19. (*a*) No; (*c*) yes; (*d*) ratio of orders must be 3:2
21. 5.0 cm
23. 0.226 cm

Chapter 26

1. 5.6×10^{-20} J = 0.35 eV
3. (*a*) Infrared; (*b*) 4130 Å
5. (*a*) 2590 Å; (*b*) ultraviolet
7. (*a*) 3.2×10^{-5} m; (*b*) infrared
9. 2.8 eV
11. 91.2 nm
13. Will live 10,013 s; yes
15. (*a*) 5.2×10^{-27} kg; (*b*) $0.948c$
17. 2.8×10^{-12} kg
19. c
21. 7.3×10^{6} m/s
23. (*a*) 1.46×10^{6} m/s; (*b*) 1.005 nm; 1.46×10^{6} m/s

Chapter 27

1. (*a*) $(3.6 \times 10^{-26})/r$ J; (*b*) 4.7×10^{-14} m
3. (*a*) 9.50×10^{-8} m; (*b*) 3.97×10^{-8} m
5. 0.052
7. (*a*) 1.59×10^{6} m/s; (*b*) 2.53×10^{15} Hz; (*c*) 1.18×10^{-7} m
9. 10
11. 0.062 nm
13. (*a*) 72 cal/s; (*b*) 215°C
15. 1879, 656, 486, 122, 103, and 97 nm
17. (*a*) 3000 eV
19. $600 \pm 1.200 \times 10^{-12}$ nm

Chapter 28

1. (*a*) 1.2×10^{-15} m; (*b*) 2.3×10^{14} g/cm^3
3. 11.55 cm
5. 65.03 u
7. $^{208}_{92}$Pb
9. (*a*) 6.3×10^{6} s^{-1}; (*b*) 170 μCi
11. 8.9×10^{-4} g
13. 92 MeV
15. 2.15×10^{-3} u
17. 60 mrems
19. (*a*) 5.7×10^{14} J; (*b*) 1.82×10^{11}°C

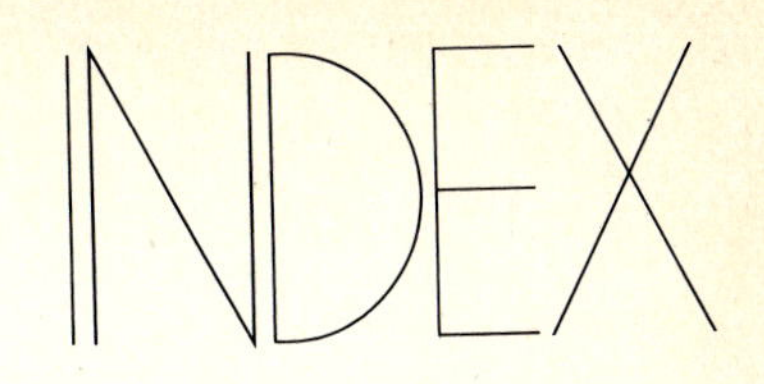

INDEX

12.00